THE PHYSIOLOGY OF PLANTS UNDER STRESS

THE PHYSIOLOGY OF PLANTS UNDER STRESS

Soil and Biotic Factors

David M. Orcutt
Erik T. Nilsen
Virginia Polytechnic Institute and State University

With additional contributions by

Thomas Ting Lei
Cynthia Lipp
Shawn Semones

JOHN WILEY & SONS, INC.

New York • Chichester • Weinheim • Brisbane • Singapore • Toronto

A NOTE TO THE READER
This book has been electronically reproduced from digital information stored at John Wiley & Sons, Inc. We are pleased that the use of this new technology will enable us to keep works of enduring scholarly value in print as long as there is a reasonable demand for them. The content of this book is identical to previous printings.

This book is printed on acid-free paper. ♾

Published simultaneously in Canada.

To order books or for customer service please, call 1(800)-CALL-WILEY (225-5945).

The Library of Congress already has cataloged as follows:

Library of Congress Cataloging in Publication Data:

Nilsen, Erik T.
The physiology of plants under stress / Erik T. Nilsen, David M. Orcutt.
p. cm.
Includes bibliographical references (v. 1, p.).
Contents: v. 1. Environmental factors.
ISBN 0-471-03152-6 (cloth : alk. paper)
1. Plant physiology. 2. Plants, Effect of stress on. I. Orcutt, David M. II. Title.
QK711.2.N55 1996
581.2—dc20 96-5443
v. 1 0-471-03152-6
v. 2. ISBN: 0-471-17008-9

Printed in the United States of America

10 9 8 7 6 5 4 3 2

With Love
for Pam and Andrea
—DMO

With Love
for Karen and Joel
—ETN

CONTENTS

PREFACE

In order to cover the complexities of plant-environment interactions, we have separated processes based on whether they occur above or below ground, and whether they are dominated by abiotic or biotic factors. This separation of topics has not been done to reflect the natural world but for pedagogical reasons. We are acutely aware that all of these environmental attributes interact among themselves, resulting in a complex field of interacting processes. The influence of soil and biotic factors on plant physiology and ecology is an important part of that complex plant response to environmental stress. The evolution of physiological, molecular, morphological, and anatomical traits in response to stress is one of the primary processes of microevolution in plants. Therefore, some of the major forces that have sculpted the structure and function of plants are soil and biotic stressors. The applied significance of research on plant response to stress is exemplified by simply observing the enormous impact of herbivory on agricultural productivity and the amount of funds that are expended annually to control herbivores. Major efforts to breed for traits that confer tolerance of nutrient and salinity stress, as well as protection against pathogens and herbivores, occur each year throughout the world. An understanding of the mechanisms that regulate form and function, and the significance of those processes to plant physiology, ecology, and agriculture, must include knowledge of plant stress physiology. This is the reason why we have written these books and have taught plant stress physiology for the past 18 years.

The purpose of this book is to provide the following:

1. A textbook that for plant physiology, plant pathology, and plant ecophysiology classes covers plant stress physiology as it responds to the soil and biotic environments.
2. A reference resource for students and scholars in a diversity of botanical fields other than stress physiology. We hope that the information contained in this volume will be a valuable resource that can be used when scholars are interpreting the significance of their research to broader issues in plant sciences.
3. A starting point for graduate students who aspire to careers in plant physiology, plant pathology, and plant ecology.

The first version of this book was written by two scientists who were active in agricultural research (M. G. Hale and D. M. Orcutt). Their work was one of the first textbooks that focused on the physiology of plants under stress. However, as time passed, the amount of material in the field expanded exponentially, and new concepts were discovered. It became clear that a new, revised version of the book was required. The second version of this text was conceived over the last 18 years, during which the authors have taught plant stress physiology to students at Virginia Tech. During those years, the new authors came to realize that it is very important to link the fields of agricultural plant physiology and physiological ecology in order to maximize our understanding of plant response to stress. We were not able to identify other books that brought together these two fields for our plant stress physiology class. Thus, our motivation arose for producing this new version. This version is a completely revised text with a large expansion of material. Two volumes became necessary, and this volume is a companion to *Physiology of Plants Under Stress: Abiotic Factors*. We redesigned the whole text from the ground up, based on our experiences in the classroom and in research. Emphasis is placed on physiological response to environmental stress and on both ecological and agricultural interpretation of those responses. The significance of evolutionary processes in the development of plant stress responses has been expanded in this version. We believe that *The Physiology of Plants Under Stress* is a strong and effective presentation of plant stress physiology, particularly in relation to joining concepts and expertise from physiological ecology and agricultural plant physiology.

THEMES AND PHILOSOPHIES

The major theme of this text is an evaluation of the various mechanisms used by plants to cope with environmental stress. Our definition of environmental is used in the broad sense, including both biotic and abiotic processes experienced by plants. In order to discuss these concepts, a framework for organizing topics had to be developed. We subdivided the field into all the major stressors, including abiotic factors above ground, abiotic factors in soil, and biotic ones. It is not possible to cover all aspects of this field in a comprehensive manner and still produce a reasonably sized volume. Therefore, we divided out treatment into two books. *Abiotic Factors*, the first book, published in 1996, was designed for use by classes in ecological physiology and plant response to the environment. The second book, *Soil and Biotic Factors*, is designed for use by plant pathology, plant ecophysiology, and plant ecology classes. Taken together, the two books offer a comprehensive coverage of the field that serves as a resource for scholars in all aspects of botanical research and teaching. In each chapter, we develop the basic physical, chemical, and biologically important aspects of the environmental component under consideration. Then detrimental influences of the particular environmental factor on plant metabolism are considered, followed by an analysis of the mechanisms that have evolved in plants to cope with these detrimental effects. Throughout this volume, we have emphasized both the interactive influences of long-term evolutionary processes on metabolism and the short-term responses of metabolism to environmental stress. Another general theme of this text is to bring out the multiple levels of adaptation (behavioral, morphological, anatomical, physiological, and biochemical) that occur in plants in response to any environmental factor.

Our philosophy on literature was to cite some of the important foundation articles in the particular topic and to present some of the current examples. It is impossible to present all possible research in each topic. For example, the number of articles found by a key word search of the citation index using "salinity" is second only to that for photosynthesis. Thus, there are many excellent pieces of work that we were not able to incorporate into the text. We hope our peers understand that if their work was not cited, that does not mean we think it is of lesser importance than the articles we have cited.

We have tried to carefully link figures and tables to text material. Figures and tables were judiciously selected to provide the best possible illustrations for the text. In many cases, we have generated our own material, but in others, we have reproduced figures or tables from other sources. We have been very careful to correctly cite all sources for the tables and figures and to obtain permission to use any of those that were reproduced. We have also been careful to cite those sources that gave us ideas about the design of our own figures.

ORGANIZATION OF THE TEXT

This text of this book is organized into three main sections. Before entering into our discussion of the stressors, we present some introductory material about the field and some preliminary information that we feel is a critical foundation for the material in later chapters. In the first part (Chapters 2–5), we present information on soil stressors and their influences on plant physiology. We provide one chapter on the processes by which plants interact with the soil environment. This is done to set the broad stage for the discussion of soil-root-plant interaction. We then include two chapters related to responses of plants to elemental chemistry of the soil environment. Plant response to mycorrhizae is included to represent a soil-based biotic interaction.

The second part of the text (Chapters 6–10) covers plant responses to biotic factors of the environment. We have divided this material on the basis of the organisms that are interacting with plants. The chapter on pathogens as a stress on plants includes viruses, bacteria, and pathogenic fungi. The influence of herbivores is discussed with an emphasis on phytophagous insects. The interaction of plants with each other is discussed in chapters on allelopathy, competition, and vascular plant parasites.

The third part (Chapters 11 and 12) concerns stresses that have been induced by human populations. These anthropogenic stresses are divided into soil-related problems (pesticides, heavy metals), and atmospheric stressors (air pollutants and elevated CO_2 levels). We include these topics because they are important in shaping the performance of plants and can induce microevolution of stress adaptation traits in plants.

IN-TEXT LEARNING AIDS

Two main mechanisms of study assistance are built into the text. The Study-Review Outline at the end of each chapter is meant to provide the student and scholar with the main topical information for the chapter. This outline summarizes and rephrases main concepts to reinforce learning and assist in studying the material. Also, Self-Study Questions are provided. Students should try to answer these questions on their own and then refer to the text to see if they have grasped the concepts presented in the text. A mimeographed set of answers to these questions will be available from the author at a nominal cost. (Contact E. T. Nilsen, Biology Department, Virginia Tech, Blacksburg, VA 24061.) Supplementary readings are also listed at the end of each chapter to provide alternate sources for material concerned with the topics covered in the chapter. The Glossary, at the end of the book, covers many of the specialized terms utilized in the text.

Although the material included in this volume has been reviewed by peers and students, the real test of any text is how effective it is in helping students and scholars learn. We hope this text will be useful for classes on topics such as plant physiology, plant pathology, plant stress physiology, agronomy, and physiological ecology. Also, we hope this book will serve as a useful reference source for scientists in a wide variety of applied and basic disciplines in the plant sciences. We welcome comments and criticisms from professors and students who utilize this text. These comments are valuable for any future revisions.

ACKNOWLEDGMENTS

Special contributions to this work have been made by:

Thomas Ting Lei (Chapter 8)
Cotton Research Unit, CSIRO, Locked bag 59, Narrabri, NSW 2390, Australia

Cynthia Lipp (Chapter 9)
Dynamac Inc., ESEPA, 200SW 35th Street, Corvallis, OR 97333, USA.

Shawn Semones (Chapter 4)
Biology Department, Virginia Polytechnic Institute and State University, Blacksburg, VA 24061, USA.

David Orcutt expresses his gratitude to the many students and colleagues who have influenced the thinking and development of this book. Particular thanks are extended to Dr. Maynard G. Hale, a mentor and friend, who initiated the original concept of this book.

It is impossible to measure the contributions made to such an endeavor by one's family, but suffice it to say that without the support, understanding, and love of my wife, Pam, and daughter, Andrea, the culmination of this work would not have been possible.

Erik Nilsen thanks all those students and colleagues who have helped his conception of plant stress physiology to gel. The many spirited and thoughtful discussions the author has enjoyed with students and colleagues provided the impetus for preparing this book.

Needless to say, this book could not have become a reality without the thoughtfulness, tolerance, understanding, and love of my wife, Karen, and son, Joel.

Please send your comments directly to us at either of the following addresses.

Dr. David M. Orcutt
Department of Plant Pathology, Physiology, and Weed Science
Virginia Polytechnic Institute and State University
Blacksburg, VA 24061

Dr. Erik T. Nilsen
Biology Department
Virginia Polytechnic Institute and State University
Blacksburg, VA 24061

PART I
Soil Processes and Plant Stress Physiology

1 Introduction and General Concepts

OUTLINE

OBJECTIVES

1. Introduce the extent of stress conditions in natural and agricultural environments.
2. Discuss the motivations, both applied and theoretical, for studying the physiology of plants under stress.
3. Outline some general concepts in plant stress physiology.
4. Discuss the significance of functional-type theory to plant stress physiology.
5. Outline some general concepts of plant response to biotic factors.

I. INTRODUCTION

There are not any locations in the world where plants are isolated from environmental stress. Moreover, the planet is replete with locations of relatively extreme environmental stress. Plants grow where air temperatures dip to −35°C in Antarctica or soar to over 55°C in Death Valley of the United States, where herbivores consume 90% of plant tissues every year, and where pollutants are profuse or the environment is pristine. Evolutionary processes have sculpted plants with elaborate combinations of morphological, anatomical, physiological, and behavioral traits that can optimize their fitness under these diverse environmental conditions. However, biophysical and biochemical limitations to plant structure and function constrain the number of possible suites of traits plants can have to minimize the detrimental effects of particular environments. Furthermore, there are positive and negative interactions between specific adaptive syndromes (particularly compatible trait combinations) and growth rate, competitive capacity, defense against herbivores, or reproductive potential. For example, engaging a particular defensive strategy against herbivores can have far-reaching costs for plant fitness. Thus, long-term processes (on an evolutionary scale) have brought together particularly favorable combinations of traits due to selective pressure by environmental conditions (both abiotic and biotic). The number and nature of these trait combinations are constrained by physiochemical limitations of life and negative interactions between adaptation to climate and adaptation to biotic factors. The short-term performance of species in response to a particular environmental change is reflective of the longer-term legacy of evolutionary processes.

Many of the disciplines included under the umbrella "plant sciences" are affected by the relationship between environment and plant physiology. This relationship is the theoretical basis for *physiological ecology,* as recently defined by Lambers et al. (1998):

> Plant ecophysiology is an experimental science that seeks to describe the physiological mechanisms that underlie ecological observations. In other words, ecophysiologists, or physiological ecologists, address ecological questions about the controls over the growth, reproduction, survival, abundance, and geographical distribution of plants as these processes are affected by the interactions between plants with their physical, chemical, and biotic environment.

Traditionally, plant ecologists have considered the combination of physiological responses of a plant population to a variety of physical or chemical factors to be the *fundamental niche* (Figure 1.1). These multidimensional surfaces, which plot physiological responses to environmental characteristics, represent the physiological framework defined by genetic attributes of individuals in the population. In nature or in agricultural systems, no species or population can realize all of its fundamental capabilities. This is because biotic factors of the environment such as competitors, herbivores, or parasites limit species acquisition of resources. Furthermore, not all combinations of resource availability are present in the environment, restricting the

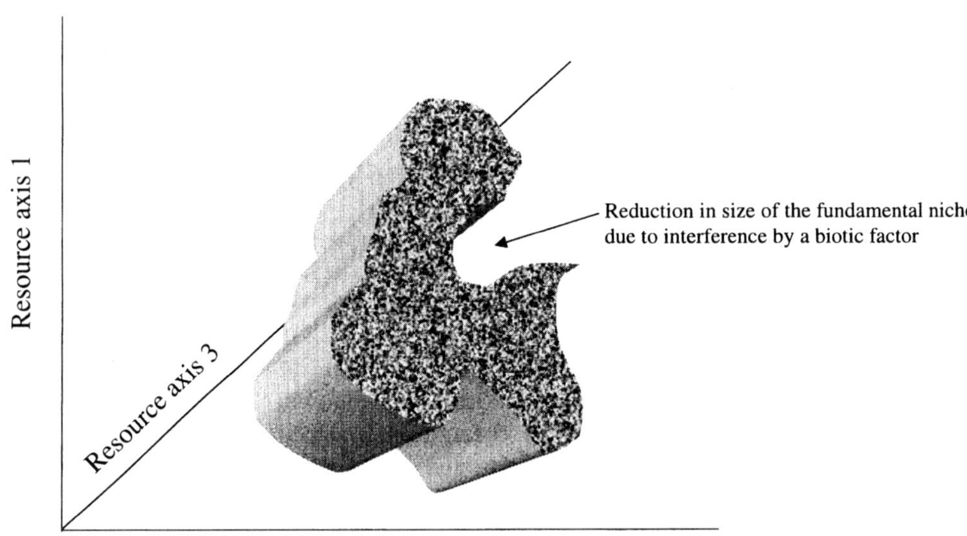

Figure 1.1. A theoretical representation of a realized niche of any species in relation to three dimensions. The realized niche occupies less space than the fundamental niche due the effects of a biotic stress.

population from inhabiting all possible resource availability combinations. Thus, all species have *realized niches* that are smaller than their fundamental capabilities.

The plant–environment interaction is also important for a wide diversity of subjects within the science of plant physiology. In fact, plant response to changes in the environment is the foundation of this science. A common definition of *plant physiology* is found in Noggle and Fritz (1976): "Plant physiology is the study of natural phenomena in living plants. It is the science concerned with processes and functions, the response of plants to changes in the environment, and the growth and development which result from the responses." Therefore, plant physiologists study physiological pathways or metabolic processes that are dependent upon the structure and function of the compounds involved, the molecular regulation of genes that determine the synthesis of those compounds, and the physiochemical constraints on the performance of the system. Each species must have the same primary physiological processes to maintain a functional metabolism (photosynthesis, respiration, nitrogen metabolism, etc.), but the specific nature of the compounds involved and their regulation can be different among species within the bounds of physical and chemical constraints.

The definitions of these two sciences overlap on the basis of plant response to the environment. However, the historical trends in the development of these sciences have followed two different paths (Figure 1.2). Plant ecophysiology began with studies of whole-plant responses to environmental conditions. There was a tendency to perform research in relatively extreme environments in order to discover whole-plant responses to intense stresses. In recent years, the science has moved toward a close association with population biology and ecosystem processes. Many studies in plant ecophysiol-

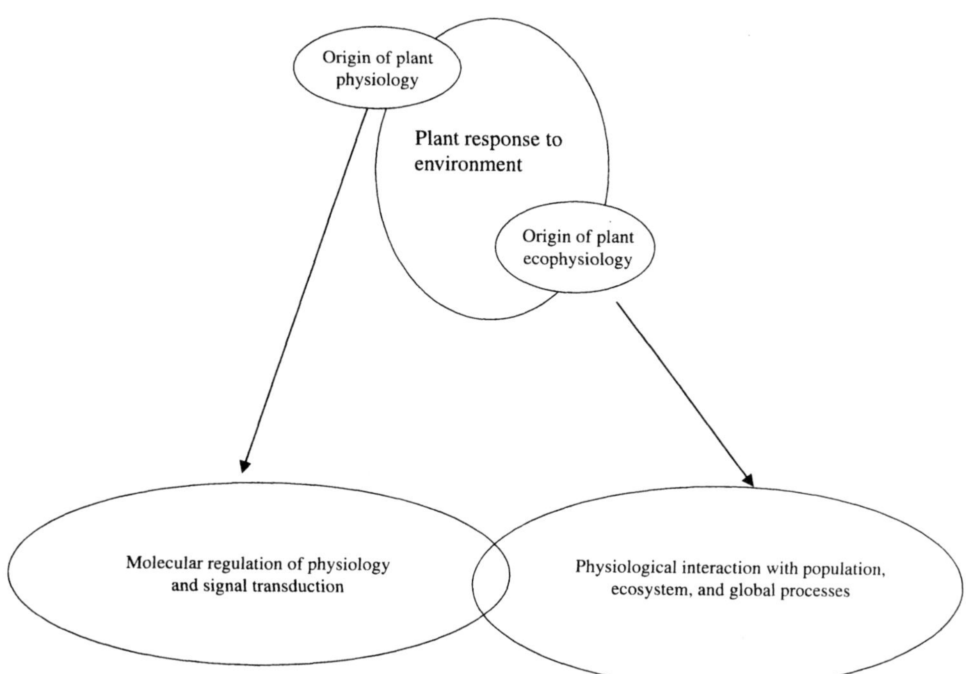

Figure 1.2. This figure represents the evolution of the science of plant physiology and plant ecophysiology (sometimes called physiological ecology) in relation to plant stress physiology. Notice that the overlap between the sciences has greatly reduced over time.

ogy now concern how whole-plant responses to changing environmental conditions (particularly CO_2 rise, or other global change phenomena) influence population regulation and ecosystem processes. The science of plant physiology is older than ecophysiology, but also has its roots in understanding the response of whole-plant processes to environmental conditions. Charles Darwin's research on coleoptile bending toward light is good example. In later years, plant physiology has moved toward a strong emphasis at the molecular level. Many studies now concern the molecular regulation of physiological pathways and the signal transduction process that serve as the interface between the environment and the genome. This divergence of the two fields has led to a reduced interaction between them such that some research on molecular regulation of growth and development is exclusively done at the subcellular level and is difficult to interpret at the whole-plant level. Moreover, some research at the whole-plant interface with ecosystem processes could be greatly enhanced by an understanding of the genetics and molecular processes that regulate the whole-plant responses. The interaction of these two fields is one goal of this book, and should be a fertile area of research in the future.

Plants growing in *natural systems* have very different stress factors in comparison with plants growing in *agricultural systems.* In natural systems, a complex interplay between biotic factors and abiotic stressors has resulted in suites of physiological and morphological traits by adaptation, acclimation, and speciation. In agricultural systems, the domestication process has limited genetic diversity and promoted specific trait combinations that maximize yield under relatively optimum abiotic growth conditions. Combinations of traits that may increase success in natural environments may not result in high yield because of the costs involved in protecting the plant from biotic factors and resource heterogeneity. Accordingly, combinations of traits that maximize yield in agricultural fields may not be appropriate for coping with locations or times when abiotic resources are limited or biotic factors are strong. Therefore, those scientists studying plant response to agricultural environments may come to different conclusions than those studying plant response to natural environments even though both groups of researchers may be studying the same phenomena.

II. MOTIVATIONS FOR STUDYING PLANT STRESS PHYSIOLOGY

Motivations for studying plant stress physiology relate to the multiple significances of stress physiology to basic and applied science (Figure 1.3). In the most basic sense, the study of plant stress physiology can reveal mechanisms by which the physiology of plants changes to optimize fitness in response to the complex biotic and abiotic environment. Once these mechanisms are understood, the information can be applied toward understanding the processes that regulate the distribution of species across the landscape. The study of plant physiology is intimately related to plant stress physiology because many of the environmental factors that influence plant stress physiology are also the cues that regulate plant development. Moreover, signal transduction pathways (a major interest of plant physiologists) are stimulated by a change in environment such as an increase in light duration, herbivore damage, virus infection, etc., and these changes are an integral part of plant stress physiology. One focus of *population biology* is to understand how the fitness of particular genotypes changes through time, and what is its relationship to genetic and phenotypic diversity. The fitness of individual genotypes is manifest as their reproductive success in a given environment. Physiological adaptation to specific stresses of a given environment will

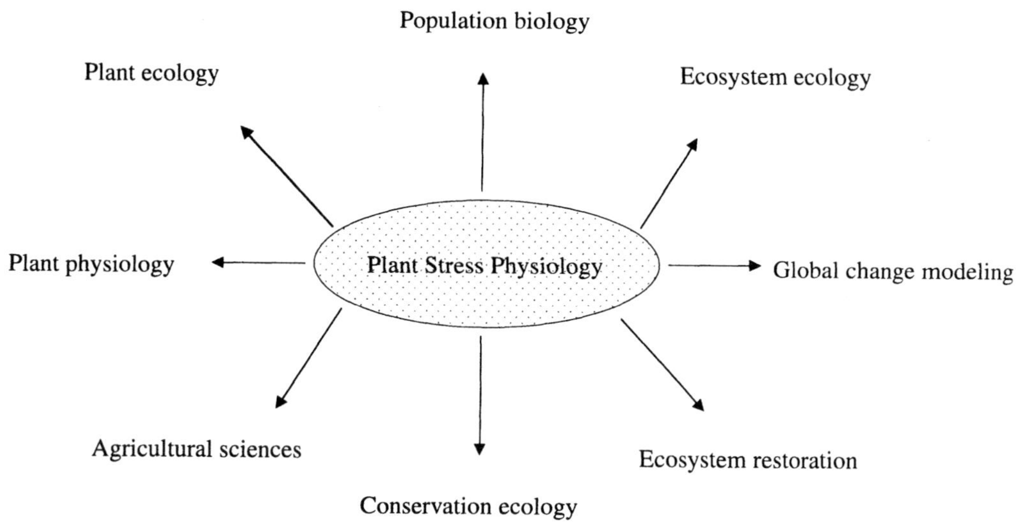

Figure 1.3. Plant stress physiology has application to a diversity of different sciences. The relative importance of plant stress physiology to these sciences may be different, and there are other sciences not mentioned in this representation.

be a major moderator of plant fitness. Community ecology is directly concerned with the assemblages of species and how those assemblages change on a spatial and temporal scale. Assemblages of plant species (communities) have inherent interactions among species through competition, allelopathy, herbivory, and other biotic stresses. The physiological processes that plants use to defend themselves against these negative interactions is a major subject in plant stress physiology. An important aspect of ecosystem science is *biogeochemistry* and the factors that regulate those processes. Biogeochemical cycles concern the transformations, flux rates, and pool sizes of elements (such as calcium) as those elements reside in and move through biotic and abiotic components of ecosystems. Plants are intimately involved with biogeochemical cycles of important elements such as nitrogen, carbon, phosphorus, and others. Thus, the way plants respond to variations in biotic processes and abiotic resources of an ecosystem will influence biogeochemical cycles. Some important issues in *global ecology* are the rise in CO_2, increasing ecosystem fragmentation, increased atmospheric concentrations of pollutants such as ozone, and an increase in ultraviolet radiation. The way that the physiology of plants respond to these anthropogenic stresses is a critical body of information needed for the development of models that are designed to predict future consequences of these global changes. Therefore, the significance of plant stress physiology extends to all scales of plant sciences from molecular regulation of plant development to global modeling of ecosystem processes.

Motivations for studying plant stress physiology also include applied sciences. The process of discovering stress induced physiological changes can result in new technology for evaluating other physiological processes. For example, the enzyme used in polymerase chain reactions (a technique used to multiply DNA by cycling between high and low temperature conditions) was discovered by determining the mechanism by which a bacterium could tolerate high heat conditions. In agriculture, it is increasingly important for scientists to improve crop tolerance of environmental stress. In general, crops are limited to about 25% of their potential yield by the impacts of environmental stress (Boyer 1982). Also, as towns and cities grow, they cover some of the best arable land. Urban sprawl and poor management of agricultural land have slowly degraded the quality (environmental conditions) of land used in crop production in developed countries, and in underdeveloped countries arable land is already of low quality. In order to improve the ability of crops to cope with environmental stress, understanding the mechanisms by which natural plants respond to the same stresses is critical. Scientists can discover new genes that enhance stress tolerance by studying the physiology of plants under extreme environmental conditions. Once stress tolerance mechanisms are known, locating genes in native species that provide specific metabolic mechanisms for increased yield under stressful conditions, is an important first step to genetic engineering. Breeding programs cannot be designed in an educated manner unless the physiological traits that are required for coping with the environmental stress in question are known. The ability to engineer the genetics of crops in order to enhance the crop's ability to cope with biotic stress factors has important applied and economic implications. Increasing internal plant defenses to herbivores can have profound effects on crop production. Enhancing plant resistance to pesticides or herbicides can strongly influence agricultural practices and serve as a large financial resource for chemical companies. Therefore, the study of plant stress physiology is important for the development of technology, new crop varieties, field management practices, and industrial economics of agriculture.

Land-use practices around the world have often resulted in degraded ecosystems that will not rapidly return to their original state. For example, restoration of surface mining sites is a long-term endeavor due to the lack of soil development and the abundance of heavy metals. For this reason, research on restoring ecosystems is a major component of conservation and environmental sciences (engineering, policy, etc.). Commonly, disturbed ecosystems have many stressors that impact the fitness of plants trying to reestablish a foothold on the site. In fact, these disturbed sites have many of the same environmental attributes of extreme habitats in natural landscapes. Restoration of degraded locations can be assisted by judiciously incorporating species that have high stress tolerance. Therefore, knowledge of what physiological processes are required to cope with extreme environments will enable a more effective protocol for restoring damaged ecosystems.

III. GENERAL CONCEPTS IN PLANT STRESS PHYSIOLOGY

The definition of *physiological stress* can be simply stated as a set of conditions that cause an aberrant change in physiological processes eventually resulting in injury. Under some conditions this definition can be difficult to use because physiological processes change in response to environmental characteristics without necessarily having a detrimental effect on plant performance. For example, if plants are placed in the sun and water is withheld, leaves may wilt. Wilting could be detrimental to the plant in the short term because carbon gain is inhibited. But, wilting may be critical for survival in the long term because leaf temperature may be kept low enough to avoid permanent damage (less light energy is absorbed by a wilted leaf). Therefore, wilting is both a detrimental and an advantageous response to low water availability and high light. This gray distinction between adaptive mechanism and detrimental affect on physiology prompted the employment of the additional term *strain* (Levitt 1980, 1982).

Differences between the terms *stress* and *strain* can be explained by applying the engineering meaning of these terms to their use for biological systems. The biophysical definition of *stress* is the applied force divided by the area of the force (Niklas 1994), or pressure. The term *stress* in a plant physiological sense is therefore reflective of the amount of environmental "pressure" for change placed on an organism's physiology. *Strain* is defined as the proportional change in a substance as a consequence of stress (Niklas 1994). Strain can be characterized as physiological change that occurs in response to environmental factors (Levitt 1982).

Other authors define stress as the change in physiology that occurs when species are exposed to extraordinarily unfavorable conditions that need not represent a threat to life but will induce an alarm response (Larcher 1980). *Alarm responses* are defensive and or adaptive responses to the stimulus. For example, the induction of the transcription of a defensive compound after some herbivory damage could be classified as an alarm response. In this text, we take a broader view, in part because the distinction between stress and strain has not become a foundation for research in plant stress physiology, and because of the broad diversity of scales involved in plant stress physiology. Here we define a *stressor* as an aspect of the biotic or abiotic environment that can induce an alteration in plant physiology. Stress occurs when the stressor induces enough physiological change to result in reduced growth, reduced yield, physiological acclimation, species adaptation, or a combination of these processes.

There is a diversity of types of stressors that may impact plant physiological processes. In general, these stressors can be classified into four groups (Table 1.1). Physical attributes of the environment can impact physiological patterns in a positive and negative manner. For instance, an increase in temperature can cause an increases in enzyme activity within a certain range, above which any further increase in temperature will cause a strong inhibition of enzyme activity. Many chemical attributes of the environment that are important resources at low concentration (salinity, copper, iron, etc.) can act as stressors at relatively high concentration. Biotic stressors are those concerned with the mechanism of interaction between populations. Some of these (pathogens, herbivory) are of particular interest to agricultural systems. Human habitation and culture alters the environment and produces other stressors (anthropogenic stressors). Increases in toxic gases in the atmosphere and pesticides or heavy metals in the soil are some of these factors. However, all plants will be exposed to a number of these potential stressors at one time, and interactions between them are highly likely. In this book, we will be primarily concerned with stressors in the soil environment, biotic stressors, and anthropogenic stressors.

Physiological response to stressors can be divided into two possibilities. In one case, *tolerance,* plants have mechanisms which maintain high metabolic activity (similar to that in the absence of stress) under mild stress and reduced activity under severe stress (Osmond et al. 1987). In contrast, mechanisms of *avoidance* involve a reduction of metabolic activity, resulting in a dormant state, upon exposure to extreme stress. In agricultural systems, breeding and other domestication technologies have maximized stress avoidance mechanisms in plants. In natural systems, the diversity

TABLE 1.1 Some of the Stressors That Plants Experience Organized into Basic Classes

Physical	Chemical	Biotic	Anthropogenic
Drought	Salinity	Competition	Air pollution
Temperature	Nutrient limitation	Herbivory	Pesticides
Radiation	Soil pH	Pathogens	Heavy metals
Flooding	Soil organics	Allelopathy	Fires
Wind	Atmospheric gases	Mycorrhizae	Alien invasions

of potential habitats has resulted in a development of both tolerance and avoidance mechanisms. Commonly, a plant species may have several tolerance or avoidance mechanisms or a combination of both. Species that have evolved in the same habitat may not have the same set of mechanisms to cope with the local combination of stressors even if they have a similar outward appearance (Nilsen et al. 1984). Responses to stressors also may be induced on a temporal scale. This is particularly true for defenses against herbivores and pathogens.

Another important concept in plant stress physiology is scale, both spatial and temporal. Over the evolutionary history of plants, their structure and physiology, within phylogenetic lines, have changed in response to a multitude of environmental attributes including the diversity of biotic and abiotic stressors. The legacy of those evolutionary influences has shaped the nature of current-day responses. When studying plant physiological response to stressors, the evolutionary time scale and the significance of phylogeny should not be ignored. Many responses to stressors also are manifest at the population level in natural systems. *Adaptation,* or the change in genotype frequency of a population in response to a stressor, involves processes of population genetics. Moreover, agriculture has utilized *selective breeding* (artificial selection of genotypes with high fitness) to accelerate, and influence the direction of, adaptation in crops. Understanding the variation among genotypes in a population in response to a stressor is critical to understanding the potential for population success in a changing environment. Plant stress physiology also includes processes that occur at the level of the individual. For example, changes in allocation of resources to specific organs within plants can be highly critical to coping with environmental stress. Of course, studies at the scale of cells over seconds of time (physiological, biochemical) are more familiar scales for the student of plant stress physiology. However, to understand the reasons why specific physiological or biochemical changes occur in response to a stressor, one cannot avoid considering the implications of evolutionary, phylogenetic, and population-level processes.

The influence of a particular stressor on the physiological processes of a particular plant species is not always equal. At different times during their life cycles, plants will be differentially sensitive to particular stressors. As seedlings emerge near the soil surface, they may be sensitive to the quantity of water in the upper soil layers. However, following growth to maturity, any sensitivity to surface soil moisture may be completely absent. Leaves may be vulnerable to herbivores when young, but after maturity leaves are much more resistant. Herbivory that occurs late in the development of a crop will have little to no impact on yield, but the same level of herbivory occurring early in crop development may severely reduce yield. Therefore, there are *windows of sensitivity* during the life cycle of plants when they are particularly sensitive to the influences of stressors. Sensitivity may also vary with seasonal patterns of physiology. In the winter, leaves on evergreen species of temperate environments may have a very low freezing point and a high tolerance of freezing, yet those same leaves may have a high freezing point in the summer and be very sensitive to frost damage. Thus, the developmental status of plants, the age of the organs in question, and the impact of seasonal environmental patterns will all influence the sensitivity of plant physiology to stressors.

The *intensity of stress* (pressure to change exerted by a stressor) is not easily quantified. Stress could occur at a low level, creating conditions that are marginally nonoptimal with little effect expected. However, if this mild stress continues for a long time, becoming a *chronic* stress, the physiology of plants is likely to be altered. In contrast, conditions could become difficult quickly, resulting in an *acute* condition.

This shock pattern of stress is likely to induce temporary changes in a short time frame. A good example is heat-shock proteins that appear shortly after a rapid rise in temperature and disappear shortly thereafter. Toxicologists, particularly in the realm of pollution studies, have developed the concept of dose, which has important application to stress physiology. The *dose* is defined as the magnitude of perturbation times the length of time the stress is applied. Stress may cause equally dramatic changes when it is applied for short duration and high intensity, or when it is applied for long duration at low intensity. Alternatively, plant responses to chronic stress and acute stress may be very different even though the dose is the same. Both possibilities can occur in natural systems.

Heterogeneity of stress is an important consideration in plant stress physiology. In some cases, particularly in controlled experiments, stress is constant and can be defined by its negative effect on plant performance. However, over time the physiological characteristics of the plant change in response to the stress. Consequently, the stress gradually decreases (even though the intensity of the stressor has remained unchanged) because the plant's physiology has acclimated to a new level of performance. Also, the intensity of stress in any natural or agricultural environment is variable both spatially and temporally. At any one time, each leaf on a plant will be receiving different intensities and qualities of herbivory. Specific roots in the soil may be flooded while other roots remain dry. Just because of the size of plants and their modular nature, there will be potential variation in stress among different parts of the plant at the same time. One of the most difficult challenges in plant stress physiology is to integrate all of these spatially explicit stress factors. Furthermore, stress factors are not constant in time. Herbivore populations rise and fall on a seasonal and annual cycle. The frequency of freeze–thaw events in the winter will vary from year to year. Specific locations of sulfur dioxide plumes will change on a minute-to-minute basis. Nutrients in soils are extremely patchy from one capillary pore to the next. The spatial and temporal heterogeneity of resources and stressors can have a large impact on the patterns of physiological response.

No plant resides in a location where only one factor dominates the way the environment affects plant development, fitness, and survival. Plants respond to a multitude of abiotic and biotic factors that vary spatially and temporally. Some combinations of these factors will have interactive effects on plant physiological processes. Synergistic effects may be particularly deleterious. For example, increased temperature may be exacerbated by low water availability because plants cannot transpire enough water to cool their surfaces. High temperature can be intensified by high irradiance because plant surfaces absorb radiation, which causes the temperature of leaves in the sun to rise further than that of leaves in the shade. Antagonisms among stressors may also occur. For example, periods of low water availability can protect plants against the deleterious effects of freezing temperature. Plant adaptation to low nutrient availability can produce leaf structure and function that protects against herbivory.

In plant stress physiology, an important distinction must be made between ultimate (adaptation) and proximal (acclimation) responses. *Adaptation* occurs by various evolutionary mechanisms at the genetic level in populations over many generations. Microevolutionary processes change gene frequencies of a population over time. In a stressful environment, it is logical to assume that specific genotypes with appropriate gene combinations (those that confer the ability to survive and reproduce) are dominant in the population. Populations that have adapted through evolutionary processes acting at the genetic level to a particular climatic regime are by no way static systems. In contrast, plants have an incredible ability to adjust physiological and structural attributes on the scale of seconds or seasons within a single genotype (*acclimation*). On a long-term scale, acclimation is enhanced in plants because of the modular nature of growth. Plant parts can be abscised and regrown with a new morphology, chemistry, or anatomy, specific organs can be enhanced by increasing their numbers or size, and biomass can be preferentially accumulated in organs that have specific functional attributes. On a short-term basis (seconds or minutes), protein populations can ebb and wane, growth regulators can be released or activated, and transcription or translation can be up- or down-regulated.

Acclimation is a phenotypic response to different combinations of environmental characters. *Phenotypic plasticity* is an index of the amount of acclimation that is possible in a particular character suite, such as photosynthesis, to changing environmental characteristics within one genotype. It has been argued that populations with high genetic diversity (many genotypes) are characterized by low phenotypic plasticity. However, there is no reason to believe that a trade-off between phenotypic plasticity and genetic heterogeneity exists (Schlicting and Pigliucci 1998). Rather, phenotypic plasticity seems to be genetically regulated. This means that some genotypes, or specific genetically regulated characters, have greater phenotypic plasticity than others. For example, genotypes of American cranberry from bogs with very low genotypic diversity do not have a greater phenotypic plasticity than those from bogs where the American cranberry population has a high genetic diversity (Stewart and Nilsen 1995a, 1995b).

There are countless examples of morphological changes that occur in plants when they are moved to, or grown in, different climatic regimes. In fact, it is necessary to grow plants that have different physiological or morphological attributes (potentially induced by environmental differences) in a *common garden* under one climatic regime to determine whether the character difference is genetically regulated (Figure 1.4). Another experiment that is used to determine the ability of a character to acclimate is the *reciprocal transplant.* In this case, individuals from a population that have different morphology or physiology are reciprocally transplanted into the area where the other population is located. If the morphological or physiological traits of each individual genotype revert to that of the other when grown in the other's habitat, acclimation has occurred. The potential for phenotypic plasticity can be evaluated with a *reaction norm experiment* (Figure 1.4). In this experiment each genotype is grown in a number of different environmental conditions. A comparison of the morphological or physiological characteristics among plants grown in these different climatic regimes indicates the magnitude of phenotypic plasticity. These experiments are useful for determining the importance of genotypic or phenotypic plasticity on plant response to stress.

IV. FUNCTIONAL TYPES AND PLANT STRESS PHYSIOLOGY

Ecologists and plant stress physiologists have been interested in grouping plants on the basis of function for many years. This interest has a long history, but the modern theory on grouping plants into functional types dates back to the use of the term *guilds* in reference to plants (Root 1967). Guilds of plants are those species that use the same resources in a similar way, or, in other words, species that overlap significantly in their functional niches. Friedel et al. (1988) defined these groups as those species that respond similarly to the same perturbation. In a recent general treatment of functional types, Gitay and Nobel (1997) suggested (from a broad review of the

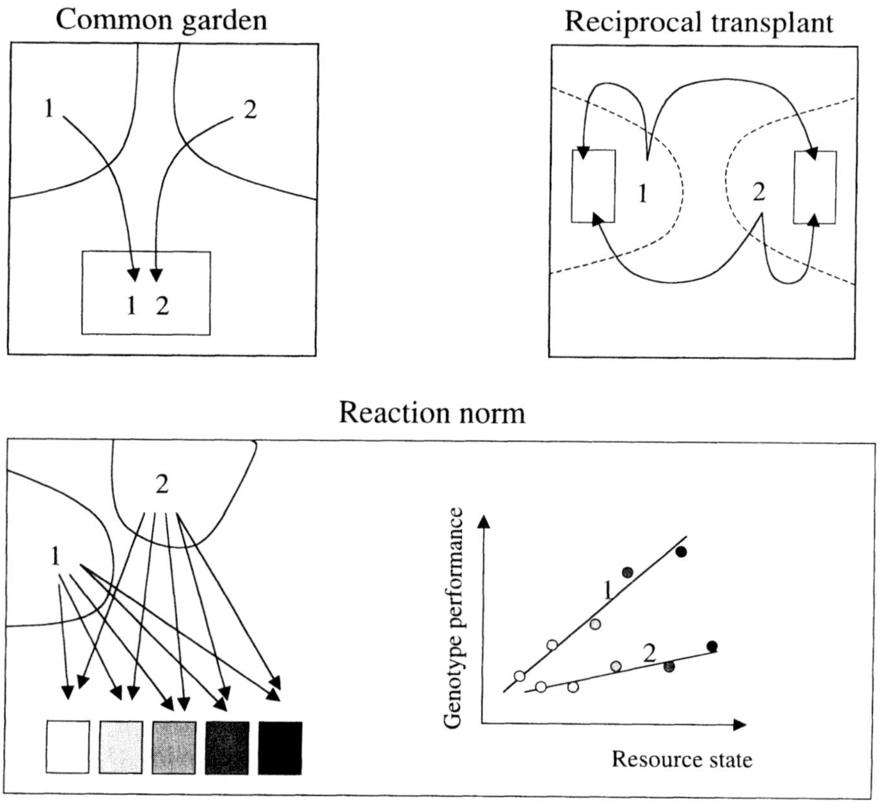

Figure 1.4. Three types of experiments that are used primarily to determine whether differences in physiological, morphological, anatomical, or behavioral traits in different environments are due to genetic causes. The reaction norm experiment is used to determine if phenotypic plasticity of plants from different environments is different. Reaction norm experiments are also used to determine what proportion of total variation in a trait expressed by a population is due to genetic or environmental causes.

literature) that functional types be defined as follows: "Thus, a functional group is a non-phylogenetic classification leading to a grouping of organisms that respond in a similar way to a syndrome of environmental factors." This idea has been of interest to ecologists because of their emphasis on using nonphylogenetic classifications of organisms when considering the function of ecosystems. For example, an increment of nitrogen in an ecosystem could be the result of nitrogen fixation. The functional type "nitrogen fixers" includes many unrelated organisms such as legumes, non-legume species harboring actinomycete bacteria such as alder species, and cyanobacteria. Furthermore, in order to understand global processes and how human habitation has affected them, models must be developed to integrate our concepts and provide predictive power. It is not feasible to produce models for every ecosystem or every species. Therefore, it is important to be able to group species on the basis of their functional response to the environment in order to make global modeling attainable.

The theory of functional types also has direct application to understanding plant stress physiology. When responses of one plant in a particular environment are determined, we would like to know whether the same pattern occurs in other species in similar environments at different locations. *Convergent evolution* is a process that would result in species that have similar functional responses but different phylogenetic histories. One of the best examples of convergent evolution in plant stress physiology is the succulent CAM (crassulacean acid metabolism) species of the New World (mostly cacti) and the Old World (mostly euphorbs) deserts. Grouping species into stress response functional groups has led to a separation between response groups and functional groups. *Response groups* are collections of species that respond to an environmental perturbation in the same way, for example, a group of species able to induce the synthesis of defensive compounds in response to herbivore damage, or a group of species able to reduce transpiration in response to water stress. Response groups are amenable to the purposes of ecosystem and community ecology. In contrast, a *functional group* is a collection of species that respond to a perturbation with the same mechanism. Examples are a group of species that induce alkaloid synthesis upon herbivore damage, or a group of species that increase abscisic acid (ABA) transport during water stress in order to decrease transpiration by causing stomata to close. Functional groups are more interesting for plant stress physiology than response groups because the former deals with mechanisms that regulate responses and the latter concerns the response only.

Grouping plants into functional types can be used only in a broad sense. Upon close examination of a collection of plants that seem to express similar response syndromes, one often observes significant differences in mechanisms. For example, a collection of phreatophytes in a desert wash would be expected to follow a similar mechanistic response to the environment because they utilize the same water resources in the same manner (very deep root systems in a relatively constant water supply). However, upon close examination of the water relations parameters of these species, many significant differences are found in the mechanisms by which the species cope with the desert climate (Nilsen et al. 1984). Therefore, functional groups are valuable tools for discussing broad-brush mechanistic types for application to a large-scale interpretation. However, when one considers specific physiological responses to environmental stress, phylogenetic history and constraints become a more dominant factor.

V. GENERAL COMMENTS ON BIOTIC STRESSORS

One of the most complex aspects of the environment is the web of biotic interactions that influence the structure and function of plants. Biotic factors range from those crucial to plant fitness (pollination vectors), to those completely detrimental to plant fitness (fungal pathogens), and all possibilities between these extremes. Moreover, the interactions between biotic factors are exceedingly complex. Is a negative interaction between two species in the field due to allelopathy or to competition? Are some herbivores also pollinators, and what determines which interaction is more important for the fitness of the plants? Many questions such as these can be posed about the diversity of species interactions in natural and agricultural situations. Interactions among species have been classified into categories based on whether the result is positive or negative for either of the interacting parties (Table 1.2). This classification is useful for population and community ecology, and sets the stage for plant stress physiology, but is not useful for plant stress physiology research per se because no mechanistic interactions are included in the definition of the classification.

Biotic stressors share some similarities with abiotic stressors. The consequence to plants from biotic stress can be positive and negative depending on the intensity of

TABLE 1.2 A Common Classification Scheme for Biotic Interactions That Occur Between Plants and Other Species in Natural Communities

	Fitness[a]			
Interaction	Of Species A When in Contact with Species B	Of Species B When in Contact with Species A	Of Species A When Not in Contact with Species B	Of Species B When Not in Contact with Species A
Neutralism	0	0	0	0
Competition	−	−	0	0
Protocooperation	+	+	0	0
Unnamed	+	+	0	−
Mutualism	+	+	−	−
Commensalism	+	0	−	0
Amensalism	0 or +	−	0	0
Parasitism/predation	+	−	−	0
Unnamed	+	0	0	0

[a]Species A is considered the target plant species under consideration, and species B is another species, which could be plant, animal, fungus, bacterium, or virus. The fitness of species when in contact with each other and not in contact with the other species is determined at a series of different densities of individuals. + = positive result, − = negative result, 0 = neutral result.

Source: Patterned after Burkholder 1952.

the stress. For example, low-level herbivory can improve plant fitness by stimulating compensatory growth, but intense herbivory can kill plants. This is similar to the way that temperature increase at moderate temperature causes an increase in growth, but the same temperature increase at high temperature can be lethal. Biotic factors interact with each other just as abiotic factors do. For example, herbivore damage can create tissue damage that promotes pathogenic attack. In contrast, infection by a pathogen can cause the induction of metabolic pathways leading to the synthesis of molecules that serve to limit the growth of pathogens as well as inhibit herbivore populations. Heterogeneity is a prominent characteristic of both abiotic and biotic stressors. Herbivore populations rise and fall with time, pathogenic fungi come and go in specific locations over time, and certain microsites are optimum for bacterial growth. Similarly to abiotic stressors, many biotic stressors stimulate the same signal transduction pathway. Several abiotic stressors (water, high temperature, chilling) induce a response through the ABA-regulated signaling mechanism. Likewise, several biotic stressors (herbivory, fungal pathogens, bacteria) induce a response in plants through the systemin signaling pathway. Abiotic stress causes a change in plants by acclimation which then changes the future stress pattern for the same abiotic factor. Limited water availability induces a reduction in turgor pressure, but over time turgor pressure returns without a change in water availability. In a similar manner, herbivore damage induces a defensive response that changes the future interaction between herbivore and plant.

Although we have noted several similarities between the interaction of abiotic or biotic factors and plants, there are also differences. Biotic stressors are not independent factors as are abiotic characteristics. For example, herbivore populations are regulated both by the plants they feed upon and the predators that feed upon them. Abiotic factors that influence the herbivore population structure also will have a secondary effect on the plant population. This contrasts with abiotic factors that are usually independent of the plants they are influencing. Biotic factors can be very species-specific or very general. For example, some herbivores feed exclusively on one species of plant, while others are generalists feeding on many plant species. Abiotic factors can selectively affect some species over others (acting as a species filter), but abiotic factors have not evolved specific attributes to make them more effective with one plant species than with other plant species. Generalist herbivores or pathogens can switch from one plant species to another when the former plant species becomes less suitable. In contrast, abiotic stressors cause a specific pattern for all species in the region and do not switch between affecting one species to affecting another species. Although both abiotic and biotic factors influence natural populations and agricultural crops, biotic factors are more influential in agriculture than are abiotic factors. In natural populations, the diversity of species and genotypes reduces the severity of biotic interactions. Furthermore, many biotic factors are present at one time, so any one of these factors is not likely to have a large effect. This relationship changes when alien biotic stress factors are introduced into natural ecosystems. The introduced biotic factor can rapidly dominate the community and cause major changes. Such was the case with chestnut blight, hemlock woolly adelgid, and gypsy moth infestations in the northeastern temperate forest of the United States. However, in agricultural situations (particularly crops), species diversity and genetic diversity is very low. The low genetic diversity of plants causes a low species diversity of herbivores and pathogens, but the populations of those biotic factors can rise quickly and cause significant problems. Moreover, artificial selection for high yield and productivity has reduced the capacity of crops to defend against pathogens and herbivores. This is because resources allocated to defend against pathogens and herbivores cannot be allocated to yield. Thus, artificial selection for increased yield in agricultural crops has resulted in greater vulnerability to biotic stress.

VI. SUMMARY

The study of plant stress physiology is demanding and exciting. The multitude of different stressors, their spatial and temporal patterns, their variation in intensity and dose, and their potential interactions yield an abundance of scientific questions. Since all plants, whether in agricultural or natural systems, are experiencing this complex of interacting stressors, the significance of discovering mechanisms of adaptation is great. Results from research on the physiology of plants under stress have important ramifications for other fields such as agriculture, theoretical biology, toxicology, ecology, and developmental biology. Some of the interesting aspects of plant stress physiology are scale, sensitivity, dose, reciprocal interaction, and heterogeneity. Generalities about the interaction between plant and biotic or abiotic stressors are both similar and dissimilar. In this book, we hope to pass on this excitement about the field of stress physiology in plants, and help students and scholars gain an appreciation for the significance of plant stress physiology to a diversity of other disciplines in plant sciences.

STUDY-REVIEW OUTLINE

Introduction

1. All plants experience stressful conditions at some time during their existence.
2. Plant stress physiology is a major component of many other fields of plant sciences.

3. Both plant physiology and plant ecophysiology began with the study of plant response to the environment.
4. *Plant physiology* has evolved toward a molecular approach and a focus on signal transduction pathways.
5. *Plant ecophysiology* has evolved toward the population and ecosystem scale, particularly as it relates to global climate change.
6. Plant species do not attain their fundamental capabilities in natural or agricultural systems. Instead they realize a portion of that fundamental capacity because of interference by other species or other stressors in their environment.
7. In agricultural systems, crops are limited to approximately 25% of their potential due to environmental stress.
8. Environmental stress has helped to shape the physiology, anatomy, and behavior of plant species by the processes of adaptation, acclimation, and speciation.
9. Any study of the immediate responses of plants to an environmental stressor must also consider the evolutionary history of the species.
10. Responses to stress occur as suites of traits, the nature of which is constrained by physical and chemical limitations to life.
11. Due to the extent of artificial selection of crops, patterns of response to environment discovered in crops may be different than those of plants in natural habitats.

Motivations for Studying Plant Stress Physiology

1. The basic mechanisms that plants have to cope with environmental perturbation will be elucidated through research on plant stress physiology.
2. Changes in plant development (and physiological processes) are regulated by signal transduction pathways, which ultimately respond to environmental changes.
3. Information from studies of plant response to stress at the physiological level form the basis for an integrated view of plant function at the whole-plant level.
4. Plant response to environmental parameters is used to develop predictive models that can forecast the consequences of climate change on plant performance.
5. The discovery of stress induced physiological processes can lead to new technology in agriculture and genetic engineering.
6. Population-level changes in genotype frequency are in part determined by the variation among genotype responses to environmental stresses.
7. Processes that occur at the ecosystem level (biogeochemical cycles) are intimately involved with plant response to the environment.
8. Understanding plant response to stress can lead to the development of bioassay technology for toxicology testing.
9. When genes regulating tolerance of stress conditions are discovered in wild plants, they may be able to be transferred into crop species in order to enhance crop productivity.
10. Breeding programs aimed at selecting for stress tolerant genotypes cannot be effective without a basic understanding of the physiological-mechanisms of stress tolerance.
11. The restoration of damaged ecosystems requires knowledge of plant stress physiological mechanisms and the plants that harbor those mechanisms.

General Concepts in Plant Stress Physiology

Terminology

1. The term *stress* is difficult to define. According to the definition we use in this book, stress occurs when environmental changes lead to reduced growth, reduced yield, physiological acclimation, or species adaptation.
2. The term *strain* has been used in the past to represent the magnitude of physiological change that occurs in plants in response to environmental stress.
3. A *stressor* is a characteristic (or combination of characteristics) of the environment that causes stress in plants.

Types of Stressors

1. Stressors can be divided into four basic classes: physical, chemical, biotic, and anthropogenic.
2. *Tolerance* occurs when a plant under stress is able to maintain high metabolic activity similar to that attained under optimum natural conditions.
3. *Avoidance* occurs when plants have a severe reduction in metabolism during stress, and often enter into a dormant state.
4. Tolerance and avoidance mechanisms are not mutually exclusive; both can occur in the same plant at the same time.

Scale

1. The study of plant stress physiology must include a wide range of temporal and spatial scales.
2. Information about a plant's evolutionary phylogeny is needed to understand why a particular stress tolerance or avoidance mechanism exists.
3. Adaptation at the population level, or acclimation at the level of the individual, may also depend upon processes at the molecular level.

Sensitivity

1. The greatest response to a stressor occurs the first time a plant experiences a certain level of stress. Subsequent impacts by the same stressor at the same intensity will not have as large an impact on physiological processes.
2. At specific developmental stages, plants are either more or less sensitive to particular stressors. The sensitive stages of development are called *windows of sensitivity.*
3. The same intensity of a stressor can cause differential response among organs on the same individual. For example, outer canopy leaves of a tree respond to light differently than do leaves within the canopy of the same tree.
4. Plant species can be sensitive to a certain level of stress by a stressor during one season and insensitive to the same stress level during another season.

Intensity

1. *Chronic stress* occurs when the "pressure" placed by a stressor on a physiological process is mild but is maintained for an extended period of time.
2. An *acute stress* occurs when intense pressure is placed on a physiological process by sudden exposure to extreme conditions.
3. Plants can respond in a different way to acute or chronic stress.
4. The acute and chronic stress definitions constitute the endpoints for a continuum of different intensity and duration of stress.
5. *Dose*, originally defined by toxicologists, is the magnitude (analogous to the concentration of toxin) times the length of exposure. This concept accounts for the influence of both intensity and duration on physiological performance.

Heterogeneity

1. Stresses are not homogeneous in the environment. There is distinct spatial and temporal heterogeneity in resource availability or toxin abundance in the environment.
2. Temporal and spatial variation in biotic stresses is one factor responsible for the evolution of inducible defenses.

Interaction

1. All plant species are exposed to several stressors at the same time.
2. Stresses can interact with each other in an antagonistic, a synergistic, or a neutral manner.

3. One antagonistic interaction occurs when water-stressed plants gain protection against freezing stress.
4. One synergistic interaction occurs when high light intensity heats leaves and water stress interrupts transpiration causing the leaf temperature to rise further.

Ultimate/Proximal

1. Ultimate causes of plant response to environmental factors are due to long-term *adaptation* and phylogenetic history.
2. Proximal causes of plant response to environment are due to the ability of a genotype to *acclimate* (change) physiological processes without genetic changes.
3. The *phenotypic plasticity* of a particular genotype is a measure of its capability to change physiological, morphological, or anatomical traits in response to environmental change.
4. High genetic diversity does not necessarily result in low phenotypic plasticity.

Experimental Technique

1. If differences between physiological traits of different genotypes remain when they are grown in a common garden, then those differences are most likely genetically determined.
2. A reciprocal transplant experiment is an excellent way to determine genetic and environmental regulation of adaptive traits.
3. Reaction norm experiments are used to evaluate differences in phenotypic plasticity among genotypes.

Functional Types and Plant Stress Physiology

1. Plants can be grouped into *response types* based upon similar response patterns to environmental perturbations.
2. Plants can be grouped into *functional types* if the mechanisms they use to cope with particular environmental perturbations are the same.
3. Response types are best suited to broad-brush considerations of plant response to environment, and are most useful for community or ecosystem ecology.
4. Functional type groupings are most useful for the study of plant stress physiology and plant ecophysiology.
5. Care must be taken when assuming that groups of species using the same resources and having the same overall morphology fall into the same functional group.

General Comments on Biotic Stressors

1. There is a web of biotic factors that influence plant physiological pathways and plant fitness.
2. Classification of biotic stressors is based upon the positive and negative interactions that occur for each party to the interaction.
3. Both positive and negative interactions are possible for many biotic environmental factors.
4. The influence of one biotic stressor can lead to either greater vulnerability or greater tolerance to other biotic stressors.
5. Heterogeneity is an important aspect of biotic stressors because the latter is variable in space and time and plant response is also variable seasonally, developmentally, and among tissues.
6. The signal transduction pathways responsible for defense against several different biotic stressors can be the same. For example, the jasmonic acid pathway is the signaling mechanism that induces defenses against herbivores, fungal pathogens, and bacterial infections.
7. Biotic stressors are regulated both by the plant attributes and by higher trophic levels such as predators.
8. Environmental conditions can modify the intensity and persistence of a biotic stressor.
9. Biotic stressors can be specific (influencing only one species), or general (able to switch from one species to another). This is in contrast with the way abiotic stressors influence plants in a natural environment.

10. Biotic stressors are more important to agricultural systems than they are to natural systems. This is due to the uniform genetic makeup of agricultural populations compared with natural populations. Moreover, selecting for high yield in agricultural crops often reduces plant defenses against biotic stressors.

SELF-STUDY QUESTIONS

1. How can the information gained from studies of plant stress physiology be useful for population ecology, ecosystem science, global ecology, and agriculture?
2. What factors regulate the sensitivity of plants to a stressor?
3. Differentiate between the terms *stress, strain*, and *stressor.*
4. What are some positive and negative interactions between stresses in relation to plant physiology?
5. Describe the major factors that can regulate the intensity of stress.
6. Why can stress decrease with time when plants are exposed to a chronic stressor?
7. What is the difference between response groups and functional groups?
8. How are biotic stressors classified?
9. What are some similarities between abiotic and biotic stressors?
10. Why are abiotic and biotic stressors different?

SUPPLEMENTARY READING

Bazzaz, F. A. 1996. Plants in Changing Environments: Linking Physiological, Population, and Community Ecology. Cambridge University Press, New York.

Dirzo, R., and J. Sarukhan. 1984. Perspectives on Plant Population Ecology. Sinauer Associates, Sunderland, MA.

Hall, D. O., J. M. O. Scurlock, H. R. Bohlhar-Nordenkampf, R. C. Leegood, and S. P. Long. 1993. Photosynthesis and Productivity in a Changing Environment: A Field and Laboratory Manual. Chapman & Hall, New York.

Levitt, J. 1980. Responses of Plants to Environmental Stress, 2 vols. Academic Press, New York.

Osmond, C. B., O. Bjorkman, and D. J. Anderson. 1980. Physiological Processes in Plant ecology: Toward a Synthesis with *Atriplex.* (Ecological Studies, Vol. 36). Springer-Verlag, Berlin.

Pianka, E. R. 1994. Evolutionary Ecology. HarperCollins, New York.

Schlichting, C. D., and M. Pigliucci. 1998. Phenotypic Evolution: A Reaction Norm Perspective. Sinauer Associates, Sunderland, MA.

Smith, T. M., H. H. Shugart, and F. I. Woodward. 1997. Plant Functional Types: Their Relevance to Ecosystem Properties and Global Change. Cambridge University Press, New York.

2 Soil/Plant Relationships

OUTLINE

OBJECTIVES

1. Discuss in general terms the geological, physical, and biological factors involved in soil formation processes.
2. Describe how the physical and chemical characteristics of soils impact and affect soil ecosystems.
3. Indicate how soil composition and structure impacts soil biota.
4. Describe the types of flora and fauna that occur in various types of soils.
5. Discuss trophic levels of interaction between microflora and fauna.
6. Address the relationships between higher plant roots and soil microbiota.

I. INTRODUCTION

Soils are complex media that support the growth of a myriad of plants, animals, and microorganisms. These edaphic, biological, physical, and chemical interactions provide an unlimited number of microhabitats and ecosystems that are in a constant state of change. Miles (1985) pointed out the dynamic nature of soil/plant relationships in suggesting that "an ecosystem has no absolute reality, other than as a concept" and that "any ecosystem must be defined arbitrarily in an *ad hoc* manner." When addressing soil/plant relationships, it is impossible to discuss the subject without some consideration of climate, geologic factors, the processes involved in soil formation, soil biology, and the relationship of these processes to natural and agricultural plant and animal communities. All of these factors interact, and each affects the other (Figure 2.1). This chapter will attempt to give the reader a basic appreciation of the many soil/plant factors that influence plant growth in natural and agricultural settings. Geologic/climatic, soil, and biological factors will be the focus of consideration.

II. GEOLOGIC AND CLIMATIC FACTORS

A. Parent Material and Soil Genesis

Soils have been defined as the superficial, unconsolidated part of the Earth's crust that supports plant life and is comprised of a heterogeneous mixture of inorganic and organic (living and dead) material derived from various forms of geologic parent material and biological organisms, respectively. The types of parent material from which soils are derived include igneous, metamorphic, and sedimentary rock formations. The rate of formation of soils depends on the type of parent material, the topographic relief, climatic factors (amount of rainfall, temperature, irradiation, and air pollutants), geologic events (volcanic activity, earthquakes, and glacial activity), and biotic and anthropogenic factors.

B. Biological, Chemical, and Climatic Factors Affecting Soil Formation

The most obvious impact of topographic relief is the rate of runoff of water from rainfall and snowmelt, which impacts the rate of erosion of parent material. The

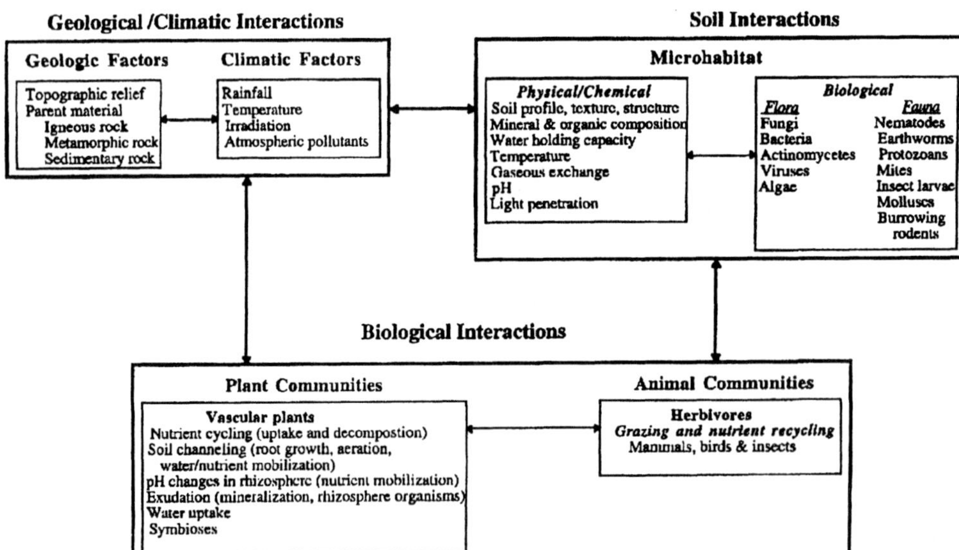

Figure 2.1. Geologic, climatic, soil, and biological interactions.

amount of rainfall can affect soil-forming processes by affecting erosion and abrasive processes particularly in high-relief areas. Acidic rain may impact soils by decreasing soil pH, and high rainfall may result in anaerobic conditions in the soil leading to the production of microbial by-products that lower the soil pH. This, in turn, can frequently reduce the diversity of the microflora and -fauna that are important in the degradation of organic soil constituents and can lead to a buildup of organic material in soils. In areas of low rainfall, soils are usually characterized by alkaline pH, exhibit low microbial floral and faunal activity, are sparsely inhabited with macro-flora and-fauna and, hence, exhibit low organic soil components. Extremes in pH characteristically lead to low recycling rates of nutrients and may reflect low productivity of such soils as a result of nutrient deficiencies and/or toxicities. Acidic and alkaline conditions may also directly impact soil-forming processes through chemical reactions with parent material or soil particles.

Temperature impacts soil formation directly through freezing-thawing and expansion processes as well as by affecting biological activity in and on soils. The amount and quality of irradiation can affect temperature and influence biological activity in and on the soil surface. Although light may penetrate soils for only a few millimeters, it is nonetheless important to many photoautotrophic bacteria and algae that contribute to the soil-forming process. Biological degradation of organic materials and the mineralization and nutrient recycling processes are integral parts of the soil-forming process. Geologic events such as volcanic activity, earthquakes, and glacial activity can have a variety of effects relative to the deposition, erosion, and exposure of geologic materials, which ultimately affect the type and composition of soils.

III. PHYSICAL, CHEMICAL, AND BIOTIC INTERACTIONS IN SOILS

A. Mineral and Organic Components of Soils

The interactions that comprise the microhabitat of soils are quite complex. Soils are dependent upon their physicochemical nature and on the biological components, all of which interact. Soil types and textures are based on the size, amount, and types of mineral and organic components. The mineral components of soils include clay (<0.002 mm), silt (0.002–0.02 mm), and sand (0.02–2.0 mm). Sand and silt particles are mineralogically similar (quartz, SiO_2), differing only in size. Silts and sands are further subdivided into silicates, micas, and feldspars. Clays generally consist of clay minerals such as kaolinite, illite, and montmorillonite. Clays are comprised of crystalline tetrahedral and octahedral layers of Si and Al, respectively, in sheetlike structures in either 1:1 or 2:1 arrangements (Barber 1995). The octahedral layer can also substitute Mg, Fe, Zn, and Li for Al. The 2:1 arrangement has an octahedral layer sandwiched between two tetrahedral layers, while the 1:1 arrangement has a single octahedral layer and a single tetrahedral layer (Figure 2.2). Clays contribute considerably to the cation exchange capacity (CEC) and electrostatic charge capacity of soils (Eijsackers 1994). This is important relative to nutrient uptake by plant roots and other organisms. Soils exhibiting differing percentages of minerals are classified into various categories, as indicated in Figure 2.3. Soil organic components consist of humic and nonhumic substances. *Humus* is an amorphous material comprised of highly oxidized aromatic and polymeric organic material derived from the degradation of lignin by microbial activity (Metting 1993). *Nonhumic* substances consist of carbohydrates, proteins, amino acids, fats, waxes, alkanes- and low-molecular-weight organic acids. Organic matter can comprise less than 0.25% in desert soils to more than 80% in peat

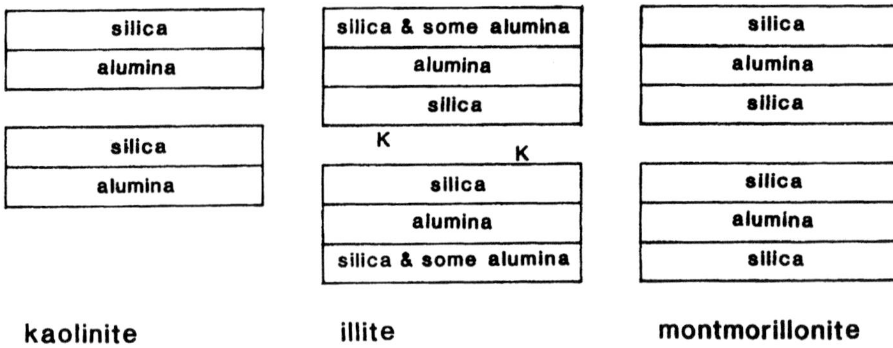

Figure 2.2. Structure and composition of clay types. (From Jeffrey 1987.)

soils. Temperate regions exhibit soil organic matter in the range of 0.4–10%, humid regions 3–4%, and semiarid regions 1–3% (Smith et al. 1993). High organic soils are generally associated with high-altitude/latitude cool environments susceptible to waterlogging or with warmer latitudes with poor drainage (Metting 1993). Mineral soils are described as seldom having more than 20–30% organic material, usually less than 5–20% and commonly 1% or less.

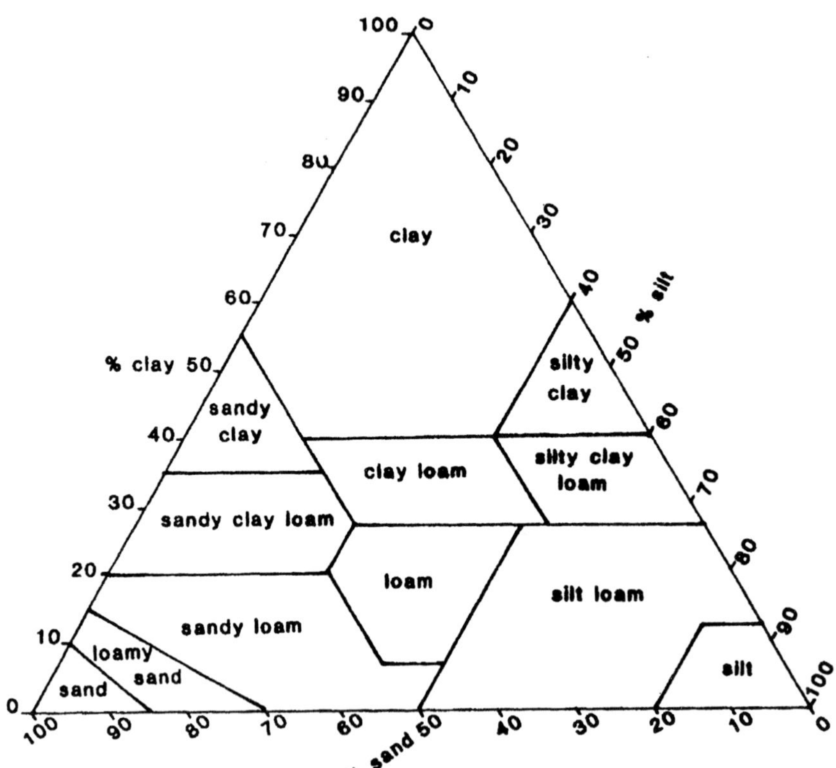

Figure 2.3. Textural triangle showing percentages of sand, silt, and clay in each textural class. (From Jeffrey 1987.)

B. Soil Structure, Profiles, and Water-Holding Capacity

Soil structure is related to the aggregate properties of soils, which in turn are related to the mineral and organic composition, soil biota, climate, parent material of origin, and anthropogenic activities. Soil aggregate properties are important since they determine pore space within and outside the aggregates that comprise the soil. Pore space is important relative to water movement, gaseous exchange, and nutrient availability, which influence survival of soil organisms and plants that grow in the soil. Secretion of root exudates and metabolites (polysaccharides) from organisms living in soils also impact the aggregation properties of soils. Table 2.1 outlines processes leading to aggregate formation as well as factors affecting aggregate stability.

Undisturbed soils take on characteristic profiles that support distinctive plant communities and are characterized by unique physical characteristics and quantifiable chemical compositions that form the basis of soil survey maps of interest to pedologists, agriculturists, and ecologists. Soil profiles are divided into horizons, which include the O horizon (organic horizon), A horizon (eluvial, or uppermost mineral layer), B horizon (illuvial), and C horizon (unconsolidated rock). Figure 2.4 illustrates how the various horizons are positioned and further subdivided. Eluvial horizons are zones of maximal leaching, and illuvial horizons are zones of maximal deposition of leached products from above. B horizons are the zones where Ca salts accumulate in arid and semiarid regions and where Fe and Al oxides and silicate clays form in tropical and temperate climate soils (Metting 1993). When the upper horizons are mixed, as would occur with agricultural practices, the mixed layers are referred to as the A_p horizon, or plow layer.

The ability of soils to retain and hold water depends on all the aforementioned soil characteristics. The water-holding capacity of soils is least for sandy soils and increases with decreasing soil particle and pore size as in a clay loam soil. The water-holding capacity of a soil is related to the *matric potential* (Ψ_m), which is the force restraining water in the soil and is related to the affinity of water for soil particles

TABLE 2.1 Soil Aggregate Formation and Stabilization

Forces Leading to Aggregate Formation
Expansion of water in freeze–thaw cycles
Shrinkage and dilation of clays in wetting–drying cycles
Formation of fecal pellets and casts by soil fauna
Components of Aggregate Stability
Capillarity of water–hydrogen bonding between water molecules and soil particles
Divalent cations. Aggregates between clay particles are held by Ca^{2+} bridges between adjacent platelets. Disaggregation can be induced if flooding with monovalent cations occurs, e.g., sodium.
Colloidal oxides and hydroxides of iron and manganese
Humic colloids
Exuded polysaccharides of plant or microbial origin
Microbial filaments
Mucus produced by soil fauna
Fibrous organic matter

Source: From Jeffrey 1987.

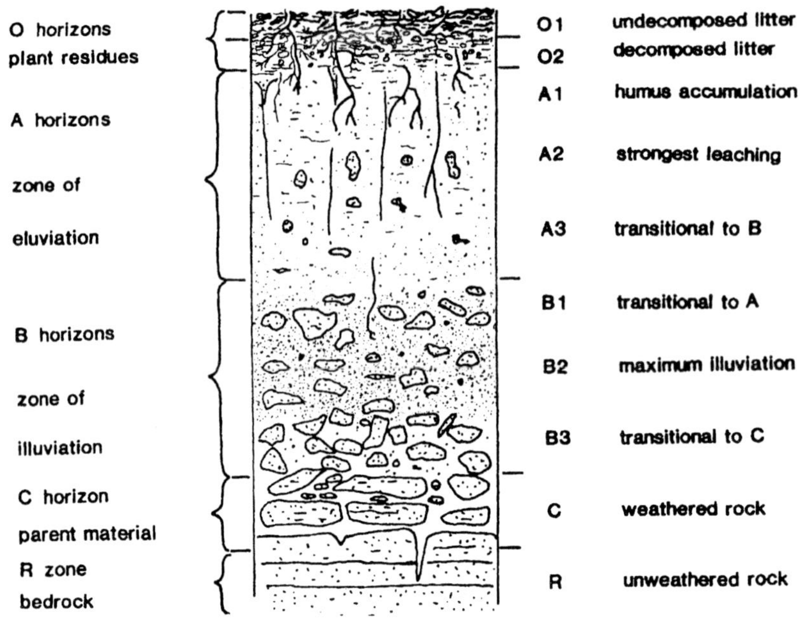

Figure 2.4. Soil profile and horizon development. (From Jeffrey 1987.)

both organic and inorganic. Figure 2.5 illustrates the relationships between soil water content, Ψ_m, and pore size. Thus, the availability of water, and the capacity of vascular plants and of soil micro- and macroflora and -fauna to extract water and nutrients from the soil, depend to a large extent on soil characteristics.

C. Soil Temperature and Heat Transfer

Soil temperature will vary with the amount of radiative input and the physico/chemical characteristics of the soil. Figure 2.6 summarizes the physical aspects of how radiative energy is exchanged between the soil surface and the atmosphere. Heat energy can be lost from soils by reflection, reradiation, convective transfer, and evaporative cooling. The degree and mechanism by which heat energy is absorbed by soils depend on several factors, including the physical and chemical characteristics of the soil, the topographic relief, the soil water content, the amount of irradiation received, the amount of and type of vegetative coverage, and whether snow or ice covers the surface of the soil. The conductive properties of soils will have an impact on the rate of heating and cooling of various regions of the soil profile. Areas deep in the soil profile may act as storage reservoirs for heat and exchange heat energy to the upper horizons. The zone of greatest biological activity occurs in the rooting zone and represents a region susceptible to considerable fluctuation in temperature. Regions deeper in the soil profile may also exhibit seasonal and diurnal fluctuations in temperature, but may lag behind upper horizons relative to the magnitude and temporal aspects of temperature change. Diurnal and seasonal fluctuations in soil temperature impact on the physicochemical nature of soils as well as the soil biota. Highly conductive soils such as sandy soils show the greatest diurnal variations in soil temperature.

Table 2.2 shows some of the conductive properties of soil components. It is quite apparent that sandy soils are better conductors of heat than soils high in organic matter,

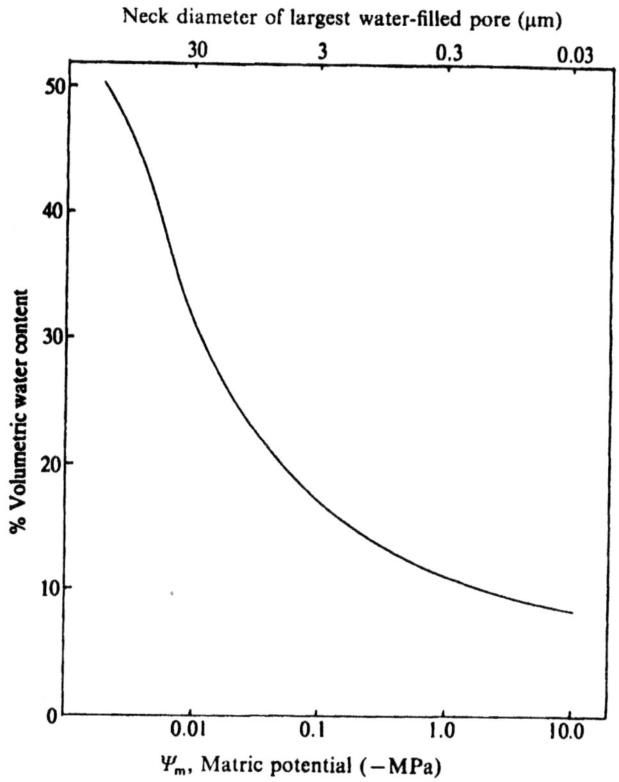

Figure 2.5. Relationship between soil pore size, soil matric potential, and soil water content. (From Killham 1994. Reprinted with permission of Cambridge University Press.)

and that soils with high moisture levels are going to be better conductors of heat than soils in which the pore structure is filled with air. Although snow is not as good a conductor of heat energy as ice, snow insulates the soil and prevents the loss of trapped heat energy but at the same time reflects considerable amounts of incident radiation. Soil temperature also affects the diffusion rates of gases, nutrients, and water in soil pores. This, in turn, can have a direct effect on the growth of plant roots and other soil biota that are involved in mineralization and recycling processes. Changes in soil temperature can change the viscosity of soil water, which can impact gaseous exchange, soil water pH, soil redox potentials, nutrient mobilization in the soil, and nutrient uptake and availability. Soil freezing and thawing can impact soil structure by affecting aggregate properties as well as by causing movement of soil components as a result of ice expansion. Freezing of soil water also reduces available water for plant roots and other soil biota and can result in the concentration of soil nutrients, which may have detrimental effects on soil organisms.

D. Soil Porosity, Water and Nutrient Mobility, and Gas Exchange

Soil porosity affects gaseous exchange between the soil and the atmosphere. The amount of water in the soil also affects the diffusion of gases through the soil since the diffusion of gases through water is about 10^4 times slower than through air (Jeffrey 1987). Soil porosity is related to a number of factors that are dependent upon soil structure and composition as well as on the activity of soil biota (earthworm activity,

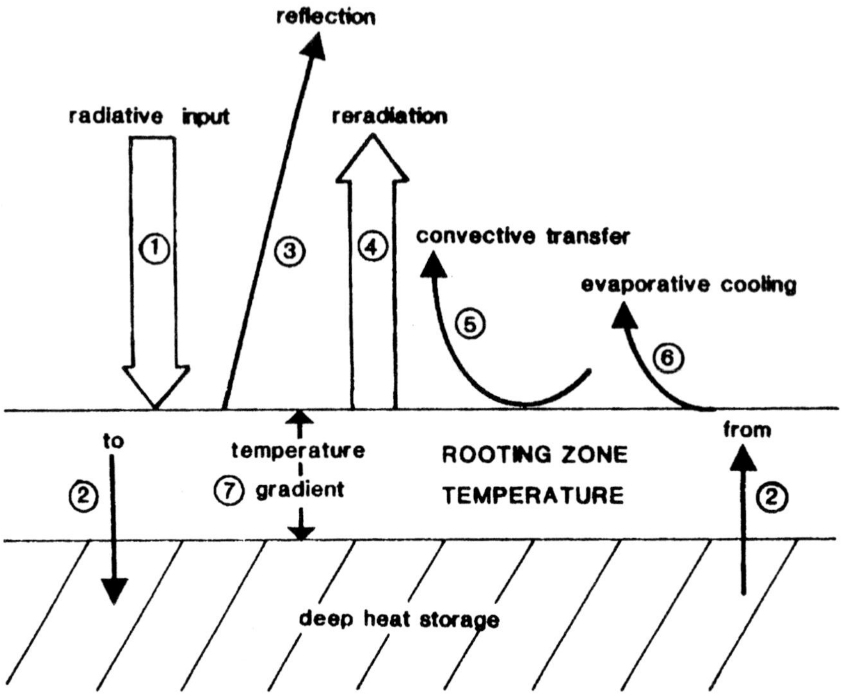

Figure 2.6. Energy exchanges that contribute to soil temperature. (From Jeffrey 1987.)

decaying root structures, burrowing organisms, and soil aggregate formation as influenced by exudation of organic polymers). The exchange of gases within soil aggregates is slower than in areas surrounding them, thus creating microhabitats potentially low in O_2 and high in CO_2 as a result of microbial respiration. Although a number of gases are associated with soils under differing conditions (O_2, CO_2, N_2, N_2O, NH_3, SO_2, H_2S, CH_4, CO, and C_2H_4), the three major gaseous constituents of soils are O_2, CO_2, and N_2. The importance of soil gaseous exchange relates to the effects on organisms growing in the soil. Plant roots and micro-flora and -fauna utilize O_2 and produce CO_2 as a by-product of respiration. In a well-aerated soil, O_2 concentrations are around 20% and CO_2 levels about 1%. Soils under anaerobic conditions—such as would occur when waterlogged, compacted, or with a high clay component—are

TABLE 2.2 Thermal Conductivities of Soil Constituents at 10°C

Mineral	Thermal Conductivity (W m^{-1} K^{-1}
Quartz	8.8
Other minerals (average)	2.9
Organic matter	0.25
Water	0.57
Air	0.025
Ice (0°C)	2.2
Snow	0.15–1.7

Source: From Jeffrey 1987 as derived from data in Hillel 1980.

depleted of O_2, and levels of CO_2 can reach as high as 10% (Killham 1994). It is under these conditions that other gases, such as those mentioned above, are produced by anaerobic bacteria and can be detrimental to the growth of vascular plants and other soil organisms. Under anaerobic conditions, organisms utilize a variety of electron acceptors resulting in the production of energy and different reductants as by-products. This occurs as a consequence of the change in the oxidation-reduction potential of the soil. The redox potential (E_h), measured in volts or millivolts, is defined as the tendency of a substance to gain or lose electrons under a given set of conditions and depends largely on the number of free electrons. Several factors can influence the E_h of a soil including pH, temperature, pressure, and soil oxygen (Metting 1993). As soil E_h declines, a sequence of soil electron acceptors is utilized by various soil microorganisms resulting in the production of metabolic energy and reduced products (Table 2.3). The production of many of these products may limit growth through direct toxicity or produce deficiencies of essential elements.

Nitrogen gas obtained from the atmosphere is important in the nitrogen fixation process, which occurs in bacteria that form symbiotic relationships with roots of legumes (*Rhizobium*) or in free-living soil bacteria (*Clostridium*), actinomycetes (*Frankia*), and cyanobacteria (*Nostoc*). As with O_2 and CO_2, low gas exchange under anaerobic conditions can result in reduced fixation of nitrogen as a result of reduced diffusion of N_2 into the soil. In addition, soil NO_3 can be lost as gaseous N_2 into the atmosphere as a result of facultative anaerobic bacteria such as *Pseudomonas* and *Alcaligenes*.

Increases in atmospheric gases such as CO_2, CO, SO_2, NO_2, and O_3 are causing increasing concern relative to their direct and/or indirect effects on soils and soil organisms. Increases in atmospheric CO_2 may result in increased fixation of carbon especially by C_3 plants. This may in turn result in increased root growth and exudation resulting in greater primary productivity in soils, which could support greater populations of soil flora and fauna. However, the quality of the organic material may change relative to the potential reduction in nitrogen content of the plant material. This may alter the N:C balance in soil/terrestrial systems, which could lead to

TABLE 2.3 Sequential Electron Acceptors Used by Soil Organisms and Associated Redox Potentials in Anaerobic Environments

Terminal Electron Acceptor and Ultimate Reduced Product	Environmental Process	Redox Potential at pH 7 (mV)	Soil Biota Involved
$O_2 + e^- \rightarrow H_2O$	Aerobic respiration	+820	Plant roots, aerobic, microbes, animals
$NO_3^- + e^- \rightarrow N_2$	Denitrification	+420	*Pseudomonas*,
$Mn^{4+} + e^- \rightarrow Mn^{3+}$	Manganese reduction	+410	*Bacillus* etc.
Organic matter + $e^- \rightarrow$ organic acids	Fermentation	+400	*Clostridium* etc.
$Fe^{3+} + e^- \rightarrow Fe^{2+}$	Iron reduction	−180	*Pseudomonas*
$NO_3^- + e^- \rightarrow NH_4^+$	Dissimilatory nitrate reduction	−200	*Achromobacter*
$SO_4^{2-} + e^- \rightarrow H_2S$	Sulphate reduction	−220	*Desulfovibrio*
$CO_2 + e^- \rightarrow CH_4$	Methanogenesis	−240	*Methanobacterium*

Source: From Killham 1994. Reprinted with permission of Cambridge University Press.

changes in ecosystems above and below ground (Diaz 1995). Little information is available concerning the effects of O_3 on soil organisms or soilborne pathogens of plants. This probably reflects few identifiable point sources for O_3 (Smith 1990). Laboratory studies, involving different species of fungi, indicate that relatively high rates of O_3 are required to affect pathogen growth, with stage of fungal development and species differences being important in responses observed. Relationships between ozone exposure and susceptibility to infection by the soilborne root pathogen *Heterobasidion annosum* did not show any degree of correlation with oxidant damage in Jeffrey or Ponderosa pine. However, stump inoculation did correlate with O_3 exposure suggesting predisposition of the organism to oxidant damage (James et al. 1980). Ozone-fumigated Ponderosa pine seedlings were more susceptible to *H. annosum* than were nonfumigated seedlings (Miller 1977).

It has been estimated that 6×10^{14} g of CO are discharged into the atmosphere annually (Seiler 1974). This gas results from incomplete combustion processes, and soils act as an important sink for CO removal (Ingersoll et al. 1974). Forest soils appear to be the most important soil-vegetation type relative to removal, as indicated in Table 2.4. The specific components of the soil microflora involved in the removal of CO are not known, but several species of fungi have been shown to remove CO, and all microbes with hemoproteins and cytochrome oxidase apparently can fix this gas. Bacteria belonging to the group carboxydobacteria oxidize CO as an energy source and are thought to be related to the genus *Pseudomonas*. Depletion of the world's tropical rain forests could have a tremendous impact on the removal of atmospheric CO.

E. Physical and Chemical Factors Affecting Soil pH

Soil pH is determined by a number of interactive factors, including the type of parent material from which the soil was derived, environmental conditions, and the amount and type of biota supported by the specific soil. pH is the measure of the hydrogen ion concentration of a soil. In soils, hydrogen ions can exist in soil solution or in a number of different states of adsorption on solid particles such as clays and organic materials (Allaway 1957). Thus, total soil acidity can be divided into two parts: *active acidity,* which refers to the hydrogen ion concentration in the soil solution, and a larger component referred to as *potential acidity,* which takes into account the potential exchangeable hydrogen ions associated with clay–organic colloids comprising the soil. The hydrogen ion exchange capacity of the latter will vary with type of clay and the amount of organic material comprising the colloid component of the soil. Generally, fine-textured soils exhibit greater total soil acidity than do sandy soils or soils with low clay and organic matter content. Soil colloids exhibit a negative charge and thus attract positively charged cations. Thus, any positively charged cation, has the potential to be adsorbed to the surface of soil colloids. (H^+, Mg^{2+}, Ca^{2+}, K^+, Al^{3+}.) The replacement of one type of cation for another is called *cation exchange,* and the total amount of cations occupying a given amount of soil at any one time is called the *cation exchange capacity* (CEC).

Figure 2.7 illustrates how cation exchange reactions can occur as a result of the application of lime, $Ca(OH)_2$, to an acid soil to raise the pH. On the acid soil side of the diagram, one can see that H^+ ions are part of the soil solution (active acidity) and are adsorbed to the soil colloids as are other cations. The amount and types of cations binding to the soil colloids depend on the type of clay and amount of organic material that comprise the colloids of the soil. Generally, the finer the soil texture, the greater

TABLE 2.4 Potential Carbon Monoxide Uptake Rates of the Soils of the World

Soil-Vegetation Type	Area (10^6 mi^2)	Average CO Uptake ($\times$ 10 tons yr^{-1} mi^{-2})	Total CO Uptake ($\times$ 10^6 tons yr^{-1})
Tropical rain forest	6.55	805.6	5277.5
Tropical grassland	3.45	886.9	3062.5
Tropical deciduous forest	1.78	1,105.9	1969.6
Montane forest	4.66	175.7	817.8
Taiga forest	4.71	127.5	600.9
Desert	10.86	52.2	567.0
Mixed and broadleaf forest	1.32	254.8	410.0
Agricultural	4.60	86.0	395.4
Pasture	1.84	179.3	329.7
Temperate grassland	3.10	86.5	268.4
Steppe	3.45	76.7	264.5
Southern pine forest	0.29	505.8	145.2
Tundra	1.15	123.0	141.3
Covered by ice, water, roads, structures	9.20	0	0
Total	56.96		14,250.0

Source: From Smith 1990, data after Ingersoll et al. 1974.

the adsorbing surface area and hence the greater the CEC. As lime is applied to the soil, calcium replaces H^+ and Al^{3+} ions, forming a by-product of $Al(OH)_3$ and H_2O. Aluminum then persists in the soil as insoluble oxides or hydroxides and is not available to plants. As lime is added to the soil, the clay–organic colloid binding sites, previously occupied by Al^{3+} and H^+ ions, are saturated with Ca^{2+}, and the H^+ ions in the soil solution also are replaced by Ca^{2+}. Hydroxyl ion accumulation results in increases in pH, and when a balance between H^+ and OH^- ions is reached, the pH

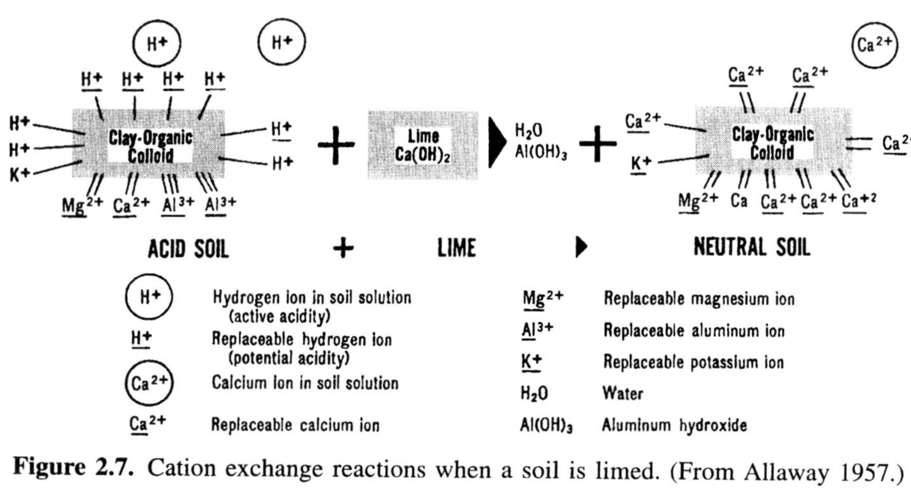

Figure 2.7. Cation exchange reactions when a soil is limed. (From Allaway 1957.)

of the soil becomes neutral. Ion exchange is not necessarily 1 for 1 but depends on the valence state of the ion. For example, a single Ca^{2+} ion will replace two H^+ ions, and three Ca^{2+} are required to exchange for two Al^{3+}.

The intent of the preceding very basic and overly simplified soil chemistry lesson is to provide the reader with sufficient background to better understand why some soils are acidic while others are basic in nature. Acid soils usually occur in areas of high rainfall, temperate forested areas, or areas prone to frequent flooding. Typically high rainfall causes leaching of basic cations such as Ca^{2+} and Mg^{2+} to lower layers or into the groundwater of the soil and are replaced by H^+ ions from the downward percolating water causing a reduction in pH. Soil biota may also contribute to this process since under anaerobic conditions, which would occur in flooded soils, fermentative processes would contribute organic acids to the acidification process.

Another factor that can contribute to acid soils is air pollution. In areas where high levels of sulfur dioxide and nitrous oxides are generated (regions where coal is burned and automobiles are concentrated), acidic rainfall is prevalent. These gases are converted into sulfuric and nitric acid, respectively, in the atmosphere in the presence of water and precipitated as acid rain. Although acid rain has not conclusively been show to cause direct injury to a number of plants, it may have indirect consequences relative to plant nutrition and soil ecology. For instance, increasing soil acidity may increase the availability of potentially toxic elements such as Al or other elements. In addition, if the sulfate ion is not metabolized by soil microbes or adsorbed to iron or aluminum oxides, it could be combined with Ca or other ions and leached beyond the root zone, resulting in nutrient deficiencies (Smith 1990). Both acids can be converted to utilizable sources of nitrogen and sulfur by soil organisms. Sulfur dioxide pollution has also been shown to affect soilborne root pathogens in forest ecosystems. Figure 2.8 illustrates the impact of SO_2 on the frequency of occurrence of several wood-rotting fungi on forest trees in Polish forests (Grzywacz and Wazny 1973). It is clear from these data that different fungi exhibit different sensitivities to SO_2 pollution, and that the closer the trees and pathogens are to the point source of pollution, the greater the reduction in occurrence. Whether this response has a direct effect on the fungus/host or is a soil pH effect is not clear. Host predisposition, fungal/host susceptibility have been shown to be important variables with respect to the effects of SO_2 on foliar pathogens of plants.

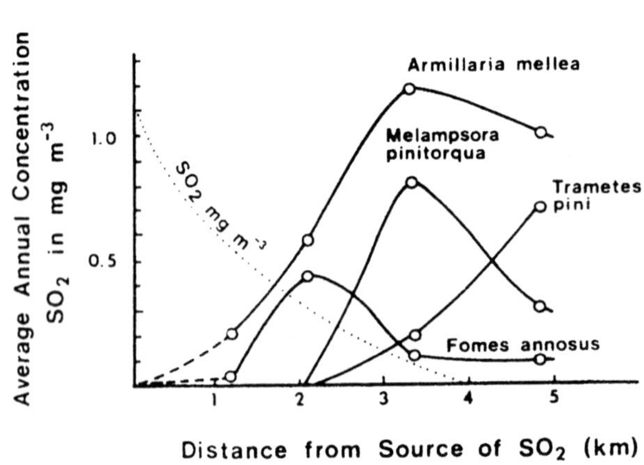

Figure 2.8. Average percentage of trees infected by various fungal pathogens in Polish forests at varying distances from sulfur dioxide point sources. (From Grzywacz and Wazny 1973.)

Soils characterized by high pH (alkali soils) occur in arid areas of very low rainfall. In these soils, basic mineral elements accumulate in the upper horizons because of limited rainfall and reduced leaching into the lower horizons. Also, movement of basic elements from the groundwater up into the upper horizons of the soil can occur by capillarity when rainfall is limiting. Consequently, fluctuations of basic elements up and down in the soil profile can occur, resulting in different levels of cations and fluctuating pH in different horizons at any give time. Many such soils are derived from limestone and marl parent material, and 60–70% of the soil formed from such material can be comprised of calcium and magnesium carbonate. The following reaction occurs when calcium carbonate is in the presence of water:

$$CaCO_3 + H_2O \Leftrightarrow Ca^{2+} + HCO_3^- + OH^-$$

In such a soil, H^+ ions are displaced from soil colloids and solution by Ca^{2+}, and an accumulation of OH^- ions occurs resulting in an increase in pH.

F. Impact of Light on Soils

Light penetration, apart from having a warming effect on soils, can influence photosynthesis of soil algae and photoautotrophic bacteria. Although such organisms contribute only a small portion of the organic material in soils, under certain conditions, algal blooms near the surface of soils may play a critical role in soil aggregate formation (Killham 1994). Light penetrates only a few millimeters into the soil, but the depth of penetration depends on a number of factors, including the aggregate properties of the soil, soil porosity, ground cover, tunneling activity of soil fauna, and degree of compaction. The quality of light penetrating the soil may be as important as its quantity. For example, seed germination is inhibited by far-red light in a number of photosensitive seeds. Soil filters out red wavelengths of light and allows only far-red wavelengths of light to penetrate. This is why some seeds in soils do not germinate until the soil is overturned, as when a field or garden is plowed. Other organisms in the soil may be affected by light quality and quantity, such as insects whose lifecycles have different stages.

IV. SOIL FLORA

A. Bacteria, Fungi, Actinomycetes, and Algae

The foregoing has considered the physicochemical aspects of the soil, while the following section will focus on the biological components and how they interact with the former. The biology of soils consists of a floral and faunal component. Table 2.5 summarizes the various biological components of soils relative to percent biomass and their relative decomposer contributions. It is clear from this table that the greatest biological components in soils are plant roots, bacteria, and fungi. Although contributions by each may vary with soil type and the physicochemical nature of the soil. Also, bacteria and fungi are the most important contributors to biomass decomposition. Figure 2.9 illustrates how various soil biota interact relative to organic cycling processes. With the exception of those soil organisms that obtain energy directly from sunlight (algae, cyanobacteria, photoautotrophic bacteria) and some chemotrophic bac-

TABLE 2.5 Decomposers and Other Soil Organisms with an Approximate for Biomass

% Soil Biomass	Organism Group	% Decomposer Biomass
66.0	Roots[a]	—
17.1	Bacteria	55.5
13.1	Fungi	38.9
1.4	Actinomycetes	4.2
0.1	Algae[a]	—
0.7	Protozoa	1.9
0.2	Nematodes	0.3
0.7	Earthworms and enchytraeids	2.0
0.3	Collembola, Acarina, Myriopoda, Opiliones, Hymenoptera, Diplopoda	1.0
0.3	Diptera, Crustacea	0.8
0.1	Vertebrates	0.4

[a] Primary producers

Source: From Jeffrey 1987.

teria, most bacteria and fungi rely on soil organic material to provide energy via heterotrophic processes.

Bacteria occupy a number of soil niches including inter- and intra-aggregate pore spaces of the soil. Depending on water, oxygen and nutrient availability different species may exist in various microhabitats. The most common soil-inhabiting bacteria belong to the genera *Pseudomonas, Bacillus,* and *Arthrobacter*. Bacteria are important in the recycling processes of the soil and require simpler forms of carbon for energy such as sugars and amino acids. Thus, large populations of soil bacteria, as well as other organisms, are generally associated in the rhizosphere, since plant roots exude organic products that can be used by bacteria as energy sources. Some soil bacteria are also root pathogens of plants, and under proper conditions may cause plant disease.

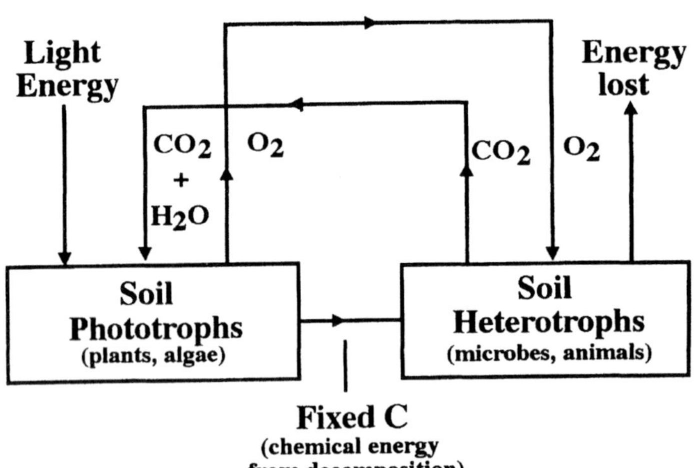

Figure 2.9. Organic cycling processes between soil phototrophs and heterotrophs. (From Killham 1994. Reprinted with permission of Cambridge University Press.)

Some examples of bacteria that are considered soilborne pathogens are species of *Erwinia, Pseudomonas,* and *Agrobacterium*. Other soil bacteria may be beneficial to plants and include the symbiotic and free-living nitrogen fixing bacteria that convert atmospheric nitrogen to a form that is available to plants. Cyanobacteria are also important nitrogen-fixing soil organisms and are capable of producing their own photosynthetic energy required in the fixation process.

Soil fungi are another important group of decomposers and are particularly prevalent in forest soil habitats, most are characterized by acid soil pH. They are usually filamentous in morphology (except for the unicellular yeast species), heterotrophic, and capable of degrading more complex organic molecules such as cellulose. Since fungi are larger than bacteria, they primarily occupy the intra-aggregate pore space of the soil. Bacteria occupy both the inter- and intra-aggregate pore space. Generally, depending on soil and environmental conditions, soil fungi are more prevalent than bacteria on a soil mass basis. Like bacteria, fungal species can be beneficial or harmful to plants. Some important soilborne fungal pathogens of plants include species belonging to the genera *Pythium* and *Phytophthora*. Fungi represent the most abundant class of microorganisms that cause disease in plants. Some fungi form symbiotic associations with plant roots known as *mycorrhizal associations*. These are of two types, endomycorrhizal and ectomycorrhizal. In *ectomycorrhizal associations,* the fungal hyphae do not enter the root but form an external mantle of hyphae around the root. *Endomycorrhizae,* on the other hand, penetrate the cellular spaces of the root and enter individual root cells. As a result of these nonpathogenic relationships, the vascular plant provides carbohydrates and possibly other organic substrates such as vitamins that help sustain the fungus, while the fungus increases the absorbing abilities of the root system of the plant by aiding in the acquisition of nutrients and possibly water.

The actinomycetes have characteristics in common with both bacteria and fungi. Like some fungi, they have mycelia and exhibit branching, but their size and cell wall compositions resemble some bacteria. They are mostly free-living saprophytes and most abundant in soils characteristic of semiarid to arid environments, where soil pH is high and water content low. One species, *Streptomyces scabies,* is an important pathogen of potatoes (common potato scab), while *S. griseus* is used for the commercial production of the antibiotic streptomycin. As a result of the antibiotic activity of *Streptomyces* species, biological fungicides are being developed and marketed (Mycostop) to control fungal plant diseases such as carnation wilt and potentially other fungal diseases of wheat and cotton (Glass 1993). Species of *Frankia,* another actinomycete genus, symbiotically fix nitrogen in alder (*Alnus*), wax myrtles (*Myrica*), and mountain lilacs (*Ceanothus*) (Raven et al. 1992) and, thus, like some species of bacteria and fungi, contribute to the nitrogen economy in vascular plants.

Soil algae are represented by a range of algae classes including Cyanophyceae, Chlorophyceae, Bacillariophyceae, and Xanthophyceae. Although many algae can synthesize most organic components for growth utilizing sunlight and inorganic constituents derived from soils (i.e., they are *photoautotrophic*), some are unable to produce all the required organic materials even though they are photosynthetic (*photoauxotrophic*). Some algae (Cyanophyceae and Bacillariophyceae) exhibit facultative *photoheterotrophy,* while others (Xanthophyceae and Chlorophyceae) obtain energy as facultative *chemoheterotrophs*. Most soil algae inhabit the soil surface or the first few millimeters down to 4–6 cm depending on the environmental conditions and the type of energy requirements needed by the individual species. Algae are important to soils in that they affect soil structure by secreting polysaccharides that affect soil aggregation properties. As symbionts with fungi, they form lichens, which inhabit rocks and soil surfaces and are involved with primary soil formation processes as a result of acid deposition. Algal surface crusts help prevent erosion, retain soil moisture, and provide a substrate for seed germination. They are an important food source for nematodes, protozoans, snails, insects, and earthworms that also live in the soil. They contribute to the nitrogen, carbon, and phosphorus economy of soils either directly or indirectly though increasing nutrient solubilities (Kumar 1985).

V. SOIL FAUNA

A. General Types and Classification

Soil fauna can be divided into micro-, meso-, and macrofauna. *Microfauna* are organisms that range in body size from 20 μm to 200 μm and include primarily protozoans, but in the upper size range includes mites, nematodes, rotifers, tardigrades (microscopic arthropods), and copepods. *Mesofauna* include organisms ranging in size from 200 μm to 1 cm and include mites, springtails, larger nematodes, rotifers, tardigrades, spiders, pseudoscorpions, harvestmen, pot-worms, insect larvae, and small millipedes and isopods. *Macrofauna* are those soil-dwelling organisms in excess of 1 cm and include earthworms, molluscs, arachnids, and soil-dwelling vertebrates (Wallwork 1970). Figure 2.10 illustrates the diversity of the soil flora and fauna relative to size and feeding type. Soil fauna have also been classified on the basis of feeding activity, which includes carnivores (predators and animal parasites), phytophages (plant-eating animals), saprophages (those feeding on dead organic matter), and microphytic feeders (those that feed on bacteria, fungi, algae, lichens, and spores). Soil microhabitat also provides a means of classification based on habitat preference. For example, some soil fauna occupy water-filled spaces and surface films in and around soil particles (protozoans, rotifers, tardigrades, copepods, crustaceans, and some nematodes), while others prefer the drier environment of soil air spaces (meso- and macrofauna). Locomotion throughout the soil system also varies with different soil fauna. Some rely on the presence of soil moisture as a medium through which flagella, cilia, or amoeboid movement can propel them through the soil, while others travel through preexisting channels or tunnels formed from decomposing plant roots or burrowing animals, the latter of which represent animals that can *eat* or *excavate* their way through the soil. The following will describe some of the general characteristics of some of the more common soil fauna.

Protozoa can be further subdivided into ciliates, flagellates, and amoebae. Most of these organisms live in regions of the soil where bacteria and algae proliferate since protozoa rely on these organisms, in part, as a food source. Most are heterotrophic organisms in that they depend on preexisting organic matter as a food source. However, some flagellates, such as *Euglena*, are photoautotrophic. Protozoa are usually motile in that they utilize specialized structures (cilia, flagella, or pseudopodia) to move around in the soil water in search of food. Thus, the availability of water in the soil is critical to the survival of these organisms, although many exhibit specialized stages in their life cycles that allow them to survive drought or low food source availability. Meso- and macrofauna include organisms such as oligochaetes (earthworms and enchytraeid worms), nematodes, arthropods, and molluscs. These organisms all have important functions relative to recycling, mineralization, and

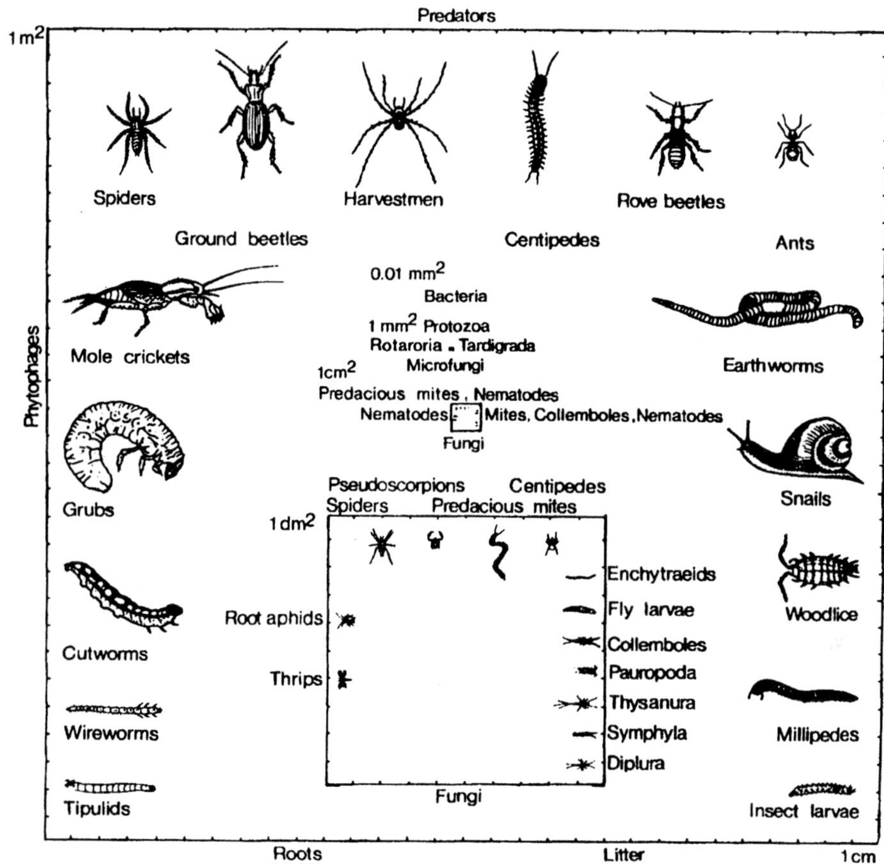

Figure 2.10. Soil faunal types, size and feeding types. (From Eijsackers 1980.)

soil-forming processes. For example, earthworms are very important in the decomposition of soil organic matter, increasing soil aeration, providing channels for water movement, enhancing the mineralization process (particularly nitrogen levels) and aiding in the development of soil structure. Some nematodes are important plant pathogens and can be responsible for considerable economic losses to agricultural plants. They feed on plant tissues, soil bacteria, algae, protozoans, fungi, and even other nematodes. Arthropods include such animals as millipedes, which are *saprophagous* (feeding on dead organic matter); centipedes, which are carnivorous; and mites, which are of both types. They contribute to the decomposition processes in soils, either directly through production of degradative enzymes or indirectly by degradation caused by gut-associated microbes. Molluscs include snails and slugs. These organisms feed on a variety of organic materials including living plant tissues, fungi, detritus, carrion, and some are predators of earthworms (Wallwork 1970). Snails exhibit high cellulase activity, thus contributing significantly to the recycling of organic matter as well as being important agricultural pests. The mucoid material produced by these organisms allows them to move across and through the soil more readily and may improve the soil aggregation characteristics.

VI. PLANT ROOT AND SOIL FLORA/FAUNA RELATIONS

A. Trophic Interactions of Plant Roots and Soil Microbiota

Figure 2.11 illustrates the complexity of interactions that can occur among plant roots and soil flora and fauna of a short-grass prairie soil environment. Plant roots and associated mycorrhizae are estimated to provide 40–85% of the net primary productivity in forest, shrub, and grassland ecosystems (Fogel 1985). Table 2.6 illustrates the contribution that roots make to the net primary productivity in various types of ecosystems. Net primary productivity is defined as the rate of storage of organic matter in plant tissues in excess of respiratory utilization during a given period (Odum 1971). This stored organic matter then becomes the source of all energy for soil flora and fauna, which constitutes a complex interactive food web where organisms feed on each other or on dead and decaying organic matter. Plant roots provide an important source of carbon for soil organisms through the production of exudate/secretion/lysate fractions containing sugars, amino acids, and organic acids (Bowen and Rovira 1991). Carbon losses to root exudation have been estimated to be as high as 20% of the total carbon fixed by the plant. These are important carbon and nitrogen sources for bacteria and fungi which generally proliferate in the root *rhizosphere* (that portion of the soil influenced by the root) or at the *rhizoplane* (the root-soil interface). This may either be beneficial or detrimental to the plant depending on the types of bacteria and fungi growing in the vicinity of the roots. Symbiotic relationships in which root nodule associations are formed may occur between some bacteria and legumes. The bacterium provides nitrogen for the plant, and the plant provides carbon for the bacterium. Fungal associations called *mycorrhizae* (see Chapter 4) appear to be associated with a wide variety of different plant species and have been shown to benefit the plant by improving P and N availability and by enhancing uptake of water and to provide a

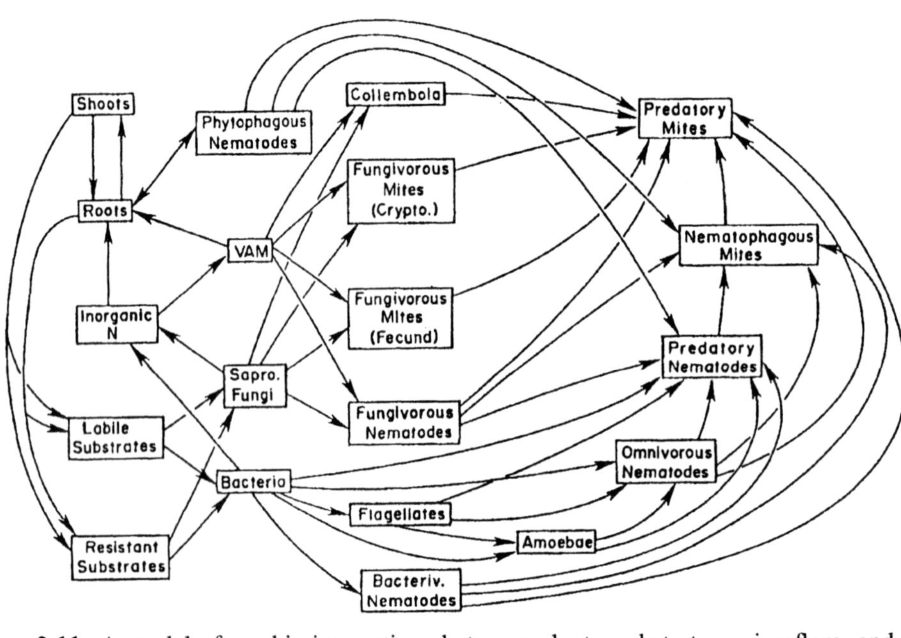

Figure 2.11. A model of trophic interactions between plants, substrates, microflora, and fauna in a short-grass prairie. (From Coleman 1985.)

TABLE 2.6 Net Primary Production of Roots in Different Ecosystems (% of Total)

Ecosystem	Age (years)	Contribution	Reference
Coniferous forest			
Douglas fir	55	73	3
Pacific silver fir	23	60	4
	180	71	4
Scots pine	14	60	1
Deciduous forest			
Yellow poplar	?	40	5
Shrub			
Atriplex steppe	?	65	2
Ceratoides steppe	?	66	2
Herbaceous			
Tall-grass prairie	?	50	6
Short-grass prairie	?	85	7

Source: From Fogel 1985.

network of nutrient transport between plants interconnected via mycorrhizal hyphae (Read et al. 1985). Mycorrhizal associations may help explain why some plants are able to survive otherwise very stressful environments such as extremely shaded conditions. The production of root exudates not only can attract beneficial organisms but can also support and attract the growth of bacterial and fungal pathogens. These organisms, in the presence of a host that is susceptible, injured, or predisposed to some other environmental stressor, may result in the development of plant disease.

Food web expansion continues to occur as a result of micro-and mesofauna feeding on the bacteria and fungi. Competition among soil organisms for nutrients results in continuously changing microhabitats in the rhizosphere of the soil. Animal–microbial interactions are common in soil and include such relationships as bacteria residing in the digestive tracts of insects and earthworms that facilitate the digestion of organic material consumed as the organism burrows or eats its way through the soil. Microbial dispersal may also occur as a result of passage of fecal material containing normal and ingested floral components through the digestive tract of the organism. Such relationships are important in the mineralization process where higher forms of nitrogen and carbon are decomposed to forms that are utilizable by other soil-inhabiting organisms as well as by plant roots. A good example of this was illustrated in a study in which soil microcosms were used to show the interactions of wheat roots with bacteria and protozoans. The findings of the study suggested that wheat roots produced carbon, which was used by soil bacteria as an energy source to mineralize organically bound soil nitrogen. The bacteria were then consumed by protozoans, which utilized the bacteria as an energy source and excreted nitrogen in the form of NH_4, which then could be used by the plant roots resulting in increased growth (Clarholm 1985). These relationships are illustrated sequentially in Figure 2.12(A–E). Other types of animal–microbe interactions include the grazing of nematodes or mites on mycorrhizae and root nodules, which could impact the symbiosis and nutrient benefits of the association. Research with Collembola (springtails) suggests that these insects have a preference for certain mycorrhizal fungi associated with pine roots. Such preferences could impact mycorrhizal species composition as well as the types of fungal decomposers that inhabit forest ecosystems (Shaw 1985). Selective removal of nematophagous mites, using miticides, has been shown to cause an increase in populations of

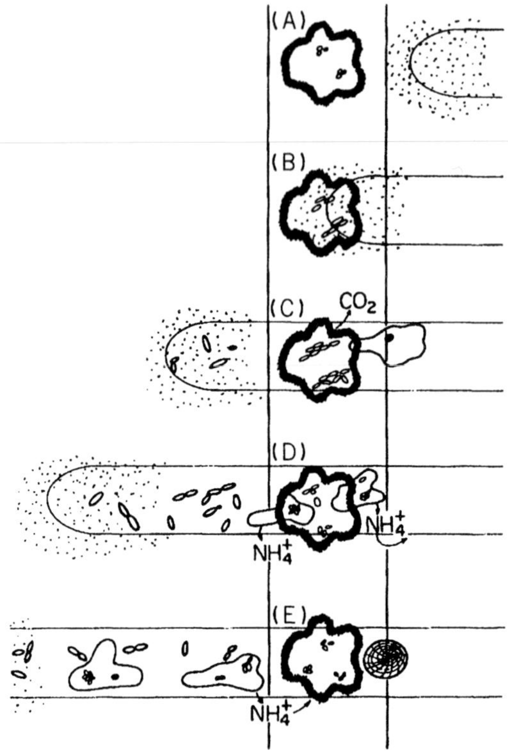

Figure 2.12. Model of interactions around the root initiated by root-derived carbon to release soil organic N for plant uptake. (From Clarholm 1985.)

nematodes that feed on bacteria, resulting in population reduction and reduced rates of soil litter decomposition. If the nematode were a pathogen, it could have implications relative to disease establishment in a crop or natural ecosystem. Reduced numbers of bacterial decomposers could have important consequences relative to the mineralization of organic material. Either effect could alter plant growth and productivity.

Microbial interactions in soils are particularly important relative to the recycling of elements essential to plants, such as nitrogen and sulfur. In the case of denitrification, where soil NO_3 is converted to N_2 under anaerobic conditions, three different species of bacteria are involved in sequential reductions carried out by a specific organism, each providing the substrate for the next until gaseous nitrogen is formed and lost to the atmosphere. Likewise, in the process of nitrification, NH_4, produced from ammonification, is used by one type of bacterium (*Nitrosomonas*) as an energy source, resulting in the production of NO_2^- which is then utilized by yet another bacterium (*Nitrobacter*) to form NO_3, which can then be taken up by plant roots. Similarly, several bacteria are involved in substrate/product oxidation-reduction reactions in the sulfur cycle.

Microorganisms also produce a number of chemical substances (toxins, antibiotics, hormones) that can impact the types and interactions of organisms occupying the rhizosphere (Figure 2.13). These substances may be important relative to the ability of microbes to compete for often limited growth resources. Such chemicals can impact not only microbes in the soil but the soil fauna and plant roots also. It is well estab-

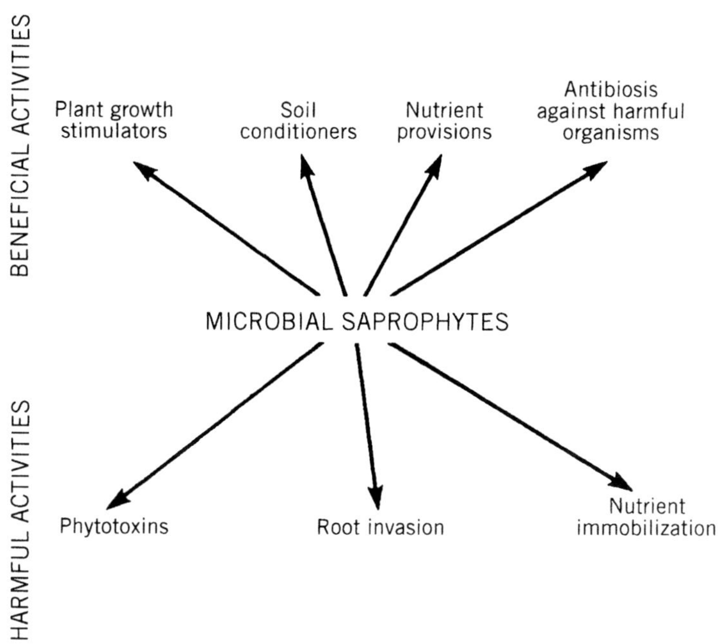

Figure 2.13. Effects of microbial saprophytes on plants. (From Lynch 1985.)

lished that bacteria and fungi are capable of producing plant hormones such as auxins, gibberellins, cytokinins, and ethylene (Nilsen and Orcutt 1996). In most cases, these substances are known to increase membrane permeability and may actually increase root exudation, which in turn could benefit the growth of soil organisms.

Ultimately, all plant material is consumed and its components recycled into the biosphere and atmosphere. Animal communities above and below ground have a considerable influence on these processes and both the animal and plant communities that currently exist on the Earth are a result of eons of natural selection processes. The topic of herbivory and its impact on plants and recycling in the biosphere will be the subject of another chapter.

VII. SUMMARY

From the foregoing, it should be obvious to the reader that relationships between plant roots and the soil are quite complex and are as equally important as interactions between the aboveground environment and aerial plant parts. However, because of the obvious difficulty in studying belowground plant parts, this aspect of plant physiology has experienced less progress in the understanding of physiological processes compared with what is known about aboveground plant structures. The physical and chemical characteristics of soils will in part dictate the plant communities that grow in specific localities. However, soil inputs from pollutants and effects from the aboveground environment influence soil properties and chemistry. The abundant flora and fauna that comprise soils will not only be affected by these factors but will also contribute to these complex interactions. The relationships of plant roots with soil biota are complex, but each depends on the other for growth and the recycling of nutrients. Imbalances in these structured ecosystems can result in the reorganization of plant and animal communities to the extent that entire ecosystems will change composition with the possible extinction and/or migration of some members. Anticipated increases in global temperatures and increased air pollutants (gases, metals, pesticides, etc.) are predicted to have, and indeed are having, significant effects on soil biota which in turn can affect the survival and growth of natural and agricultural plant communities. The ability of scientists to study and predict how anthropogenic pollution impacts biological systems requires many years of carefully planned and executed research. The speed and magnitude at which changes in our environment are occurring relative to the time it takes scientists to observe responses of organisms to pollutants should be some cause for concern.

STUDY-REVIEW OUTLINE

Introduction

1. Soil is a heterogeneous mixture of inorganic and organic material derived from various forms of geologic parent material and biological organisms.
2. Rates of soil formation depend on type of parent material, topography, climate, geologic events, biotic factors, and human activities.

Geologic and Climatic Factors

1. Soil types and textures are based on the size, amount, and type of mineral and organic components.
2. Mineral components of soils are mineralogically similar, consisting of clay, silt, and sand.

Physical, Chemical, and Biotic Interactions in Soils

Mineral and Organic Components of Soils

1. Three types of clays comprise soils: kaolinite, illite, and montmorillonite. Each type has its own characteristic crystalline structure.
2. Clay content contributes to the cation exchange capacity of soils which is important relative to availability of cations to plant roots.
3. Organic soil components consist of *humic* and *nonhumic* substances. The former consists of organic polymers derived from lignin degradation, while the latter consists of lower-molecular-weight carbohydrates, proteins, fats, waxes, and so on.
4. Organic soil components vary with temperature and rainfall, which relate to soil microbial activity.

Soil Structure, Profiles, and Water-Holding Capacity

1. Soil aggregate properties are important relative to soil structure and are related to mineral content, organic composition, soil biota, climate, and parent material.
2. Soil aggregate properties affect soil porosity, which in turn affects gaseous exchange, nutrient mobilization, and water availability in the soil environment.
3. Undisturbed soils are characterized by profiles comprised of O, A, B, and C horizons.
4. The eluvial horizon is the upper most mineral layer and the zone of maximal leaching, while the illuvial horizon is the zone of maximal deposition of leached products.
5. The water-holding capacity of soils is related to the *matric potential,* which is a measure of the affinity of water for soil particles, both organic and inorganic.
6. The water-holding capacity can influence the availability of water to plants and soil biota.

Soil Temperature and Heat Transfer

1. Soil temperature is important relative to root growth and the growth of soil organisms. Heat energy can be lost from soils by reflection, reradiation, convection, and evaporative cooling.
2. Soil heat absorption depends on the conductive properties of soils, topography, soil water content, amount of irradiation, vegetative cover, and ice or snow cover.
3. Changes in soil temperature can affect diffusion of gases, nutrients, and water and cause shifts in pH and redox potentials, all of which can impact the growth of plant roots and soil organisms.

Soil Porosity, Water and Nutrient Mobility, and Gas Exchange

1. The three major gases in soils are O_2, CO_2, and N_2, all of which are biologically important. Other gases such as N_2O, SO_2, H_2S, CH_4, CO, and C_2H_4 can be produced by microbes under anaerobic conditions and can be detrimental to growth of plants and other soil organisms.
2. Soil redox potentials affect the types of substrates available to soil microbes. As soil redox potentials decline, a sequence of electron donors is made available to different organisms that can utilize them as an energy source.
3. Atmospheric N_2 is important relative to the fixation of nitrogen in a form that is utilizable by plants. Free-living and symbiotic bacteria, fungi, and actinomycetes are important in this process.
4. Air pollutants such as CO_2, CO, SO_2, NO_2, and O_3 not only impact organisms above ground but may impact soil biota directly or indirectly.
5. Increases in CO_2 levels could result in increased production of root exudates particularly in C_3 plants. Such changes could alter the C/N balance in the soil/terrestrial systems and alter belowground ecosystems.
6. Little information is available regarding O_3 effects on soil organisms or soilborne pathogens of plants and what data are available are inconclusive.
7. Forest soils are major sinks for removal of CO pollutants from the atmosphere. Evidence suggests that several soil microbes can fix this gas and use it as an energy source.

Physical and Chemical Factors Affecting Soil pH

1. Soil acidity can be divided into two components: *active* and *potential acidity*. The former refers to the hydrogen ion concentration in soil solution. The latter refers to the potential hydrogen ion concentration associated with bound ions on soil particles.
2. Soil colloids exhibit negative charge characteristics and thus attract cations. The total amount of cations occupying a given amount of soil is referred to as the *cation exchange capacity* (CEC).
3. Finer-textured soils having high clay content usually exhibit a high CEC.
4. Acid soils generally occur in areas of high rainfall, temperate forest areas, or regions prone to frequent flooding.
5. Areas prone to high levels of air pollutants such as NO_2 or SO_2 may exhibit acid soils since such pollutants can be precipitated as acid rainfall.
6. Alkaline soils occur in regions of low rainfall and high temperature. Basic mineral elements accumulate in the upper horizons as a result of reduced leaching in contrast to acid soils, where such minerals are leached and replaced by hydrogen ions from abundant rainfall.

Impact of Light on Soils

1. Light not only has a warming effect on soils but also impacts seed and spore germination and possibly the life cycles of soil fauna as well.

Soil Flora

Bacteria, Fungi, Actinomycetes, and Algae

1. Generally, but not always, the largest contributors to soil biological mass are plant roots, fungi, and bacteria.

2. Bacteria and fungi are the most important contributors to biomass decomposition.
3. With the exception of photoautotrophs and chemotrophs, most soil organisms depend on organic material in the soil for energy.
4. Soilborne bacteria (free-living, plant pathogens, and symbiotic bacteria) are frequently associated with the root rhizosphere, which is usually high in energy-rich exudates.
5. Soil fungi are most abundant in forest ecosystems and can be both beneficial and harmful as in bacterial associations. *Mycorrhizal associations* with plant roots provide nutrients and water in a symbiotic relationship.
6. Fungi are important degraders of complex organic molecules such as cellulose, in contrast to bacteria, which utilize simpler forms of carbon as an energy source.
7. Fungi represent the class of organism that causes the greatest number of disease problems in plants.
8. Actinomycetes have characteristics common to both bacteria and fungi and are most prevalent in semiarid to arid environments.
9. Some species of actinomycetes can fix atmospheric nitrogen and form symbiotic associations with higher plants.
10. Most soil algae are *photoautotrophic,* while others are facultative *photoheterotrophs* or facultative *chemoheterotrophs.*
11. Algae can affect soil structure by secreting polysaccharides. Algae are symbionts of lichens, which are involved in soil formation processes, help retain soil moisture by forming surface crusts, and are important food sources for soil fauna.

Soil Fauna

General Types and Classification

1. Soil fauna can be classified according to size, feeding type, means of locomotion, and microhabitat preference.
2. Examples of *microfauna* include organisms such as protozoans, mites, nematodes, rotifers, some arthropods, and copepods.
3. Examples of *mesofauna* include larger mites, nematodes, rotifers, arthropods, spiders, insect larvae, and small millipedes.
4. *Macrofauna* include earthworms, molluscs, arachnids, and soil-dwelling vertebrates.
5. Microhabitat preference may range from the need for free water, to a preference for water bound to soil particles.
6. Methods of locomotion involve amoeboid movement, use of cilia or flagella, movement through preexisting tunnels, or active burrowing.

Plant Root and Soil Flora–Fauna Relations

Trophic Interactions of Plant Roots and Soil Microbiota

1. The diversity of soil flora and fauna creates complex food webs of organisms that are important in mineralization and the recycling of nutrients in soil ecosystems.
2. Plant roots and associated mycorrhizae constitute the primary producers of soil environments.
3. Exudates, secretions, and lysates provide necessary nutrients for the existence of other soil biota.
4. Complex interactions among higher plant roots and soil microflora and -fauna form intricate ecosystems in the soil that, if disturbed, could result in shifts in above- and belowground ecocommunities.

SELF-STUDY QUESTIONS

1. Discuss the geologic, climatic, and biotic factors that influence soil formation processes.
2. Describe the basic composition of soils, and indicate the types of mineral and organic material from which soils are derived.

3. Review the importance of soil porosity relative to factors that affect it and the way porosity influences the growth of soil biota.
4. Describe the various horizons that comprise a soil profile.
5. What is the water-holding capacity of a soil, and why is it important to plants and other soil organisms?
6. Outline the methods of heat exchange in soils, and review the factors that affect this process.
7. Name the various gases that can be produced in soils, and indicate the importance of gaseous exchange in the growth of plants and other soil organisms.
8. Describe how various air pollutants might impact belowground ecosystems.
9. Define soil acidity, and indicate how soil composition and climatic factors could affect pH.
10. What is CEC, and indicate why it is important to higher plants?
11. What factors affect whether a soil is acid or alkaline?
12. How does light affect soils and the organisms associated with them?
13. Characterize soil flora and fauna relative to classification, abundance, and function in soil ecosystems.
14. Describe what is meant by trophic food webs in soils, and indicate how the disturbance of such food webs might impact above- and belowground ecocommunities.

SUPPLEMENTARY READING

Barber, S. A. 1995. Soil Nutrient Bioavailability: A Mechanistic Approach. John Wiley & Sons, New York.

Barbour, M. G., J. H. Burk, and W. D. Pitts. 1987. Terrestrial Plant Ecology. Benjamin/Cummings Publishing Company, Reading, MA.

Burrows, C. J. 1990. Processes of Vegetation Change. Unwin Hyman, Boston, MA.

Clarholm, M. 1985. Interactions of bacteria, protozoa and plants leading to mineralization of soil nitrogen. Soil Biology and Biochemistry 17:181–187.

Coleman, D. C., E. P. Odum, and D. A. Crossley Jr. 1992. Soil biology, soil ecology, and global change. Biology and Fertility of Soils 14:104–111.

Curtis, P. S., E. G. O'Neill, J. A. Teeri, D. R. Zak, and K. S. Pregitzer (eds.). 1995. Belowground Responses to Rising Atmospheric CO_2: Implication for Plants, Soil Biota, and Ecosystem Processes. Kluwer Academic Publishers, Boston, MA.

Harley, J. L., and R. S. Russell (eds.). 1979. The Soil-Root Interface. Academic Press, New York City, NY.

Russell, R. S. 1977. Plant Root Systems: Their Function and Interaction with the Soil. McGraw-Hill Book Company, New York.

Tu, C. M., and J. T. Trevors. 1985. Invisible farmhands: soil microbes have a big job to do. Crops and Soils Magazine, February, pp. 12–15.

Volk, T. 1994. The soil's breath. Natural History 11:48–54.

Wright, R. J., V. C. Baligar, and R. P. Murrmann (eds.). 1991. Plant-Soil Interactions at Low pH. Kluwer Academic Publishers, Boston, MA.

3 Nutrient Deficiency Stress and Plant Growth and Development

Outline

Objectives

1. Discuss the role of soils in determining nutrient availability in plants.
2. Describe root tissues relative to the path of nutrient transport into vascular system and transport to the shoot.
3. Address the various mechanisms by which mineral nutrients are transported across cell membranes.
4. Examine whole-plant and cellular distribution of mineral nutrients.
5. Describe the physiological and biochemical functions of the essential elements.
6. Illustrate the functions of essential elements relative to affecting developmental processes in plants.
7. Discuss the role and functions of elements considered to be "beneficial" for some plants but not essential for normal growth processes in most plants.
8. Characterize the morphological and physiological characteristics of plants adapted to nutrient deficient conditions.

I. INTRODUCTION

In preparing this chapter, it became apparent that there is a revitalization of interest in the subject of mineral nutrition that probably reflects the growing concern regarding the widespread use of commercially produced fertilizers and the potential for soil and groundwater pollution. Related to this, and perhaps helping to drive this interest, is the development of the relatively new field of *phytoremediation,* which is the use of plants to remove or degrade pollutants. In addition, concern over the limited number of staple food crops grown in the world and the decline in arable lands on which to grow crops has sparked interest relative to identifying plants that are more efficient users of mineral elements and that can grow on marginal soils.

Normally, when considering plant mineral stress the topics of mineral deficiency and toxicity are generally the focal points. However, there are a number of other factors that influence the *acquisition* and effective *utilization* of mineral elements by plants. Thus a holistic approach that considers inherent physicochemical characteristics of soils, soil-root interactions, growth and the structure and function of roots, ion uptake and transport mechanisms, cellular and tissue distribution of mineral elements, physiological and biochemical functions of elements, and the role of nutrient elements in growth and developmental processes would seem to be essential relative to understanding how these factors interact to affect the acquisition and utilization of mineral elements. Such an approach will be taken in this chapter.

II. SOIL CHARACTERISTICS AND MINERAL STRESS

A. Soil Factors and Nutrient Acquisition

As indicated in Chapter 2, soils are very complex chemically, physically, and biologically. The ability of plants to acquire nutrients from soils, to a large extent, depends on these same characteristics. Soils that are high in organic matter and clay have a greater ion exchange capacity than soils that have high percentages of sand. In addition, the former types of soils have a greater water-holding capacity. Together, these two characteristics (ion exchange and water-holding capacity), along with the concentration of minerals that comprise soils, form the fundamental basis for availability and transfer of nutrients from soils to the roots of plants. Mineral elements occur attached to soil particles (solid phase) or dissolved in soil water (liquid phase) that occupies the pore structure of the soil. Soils high in clay content typically have negative charges associated with them. As a result, such soils have the capacity to attract and accumulate positively charged cations, while, negatively charged anions are repelled and become associated with the liquid phase. In order for plants to absorb nutrients from soils, they must first be in solution.

Figure 3.1 illustrates cation exchange occurring on a soil particle as a result of an influx of K ions and displacement of Ca into the soil solution. The concentrations of ions in solution depend to a large extent on the texture and composition of the soil, its ion exchange capacity, the amount of water in the soil, and soil pH. The last of these plays an important role in establishing the availability of plant nutrients. Figure 3.2 shows how pH and soil type influence the availability of several elements essential to plant growth. Note that in mineral soils, extremes in pH can either increase or decrease the availability of specific elements, thus potentially causing deficiency or toxicity problems for plants that may grow in such soils. For instance, alkaline pH limits Fe, Mn, and Zn, while an acid pH increases the availability of Al, Fe, and Mn. Extremes in pH can also cause the formation of insoluble salts, such as occurs with Ca and P at high pH and Al, Fe, and P at low pH. Soils high in organic matter tend to make ions available to plants over a broader pH range except for elements such as P, Mn, and B, which show low levels of availability between pH 7 and pH 8.

As plant roots absorb nutrients and water from the soil, the thermodynamic equilibrium between elements bound to soil particles and those in solution is disturbed creating an ionic and water gradient that favors removal of ions from the soil particles and replenishing the dissolved elements and water in the vicinity of the rhizosphere.

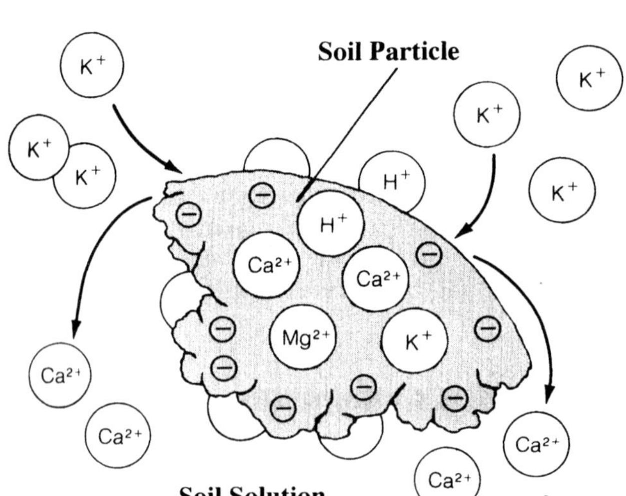

Figure 3.1. Ion exchange between K^+ and Ca^{2+} as a result of K^+ influx from the soil solution. (From Taiz and Zeiger 1991.)

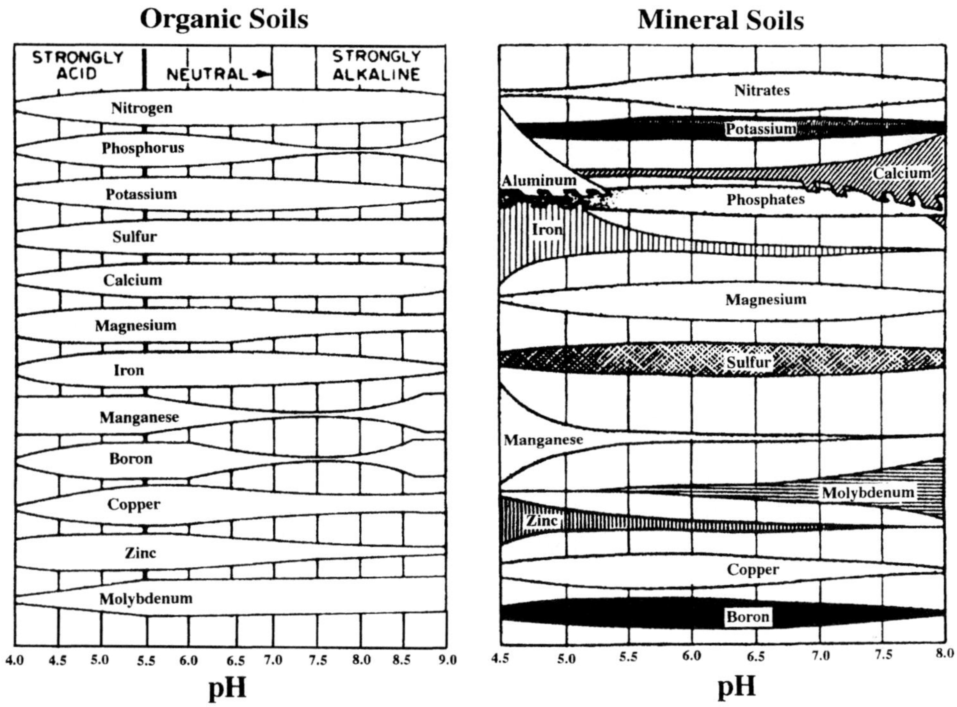

Figure 3.2. Influence of pH on nutrient availability in organic and mineral soils. Increased bar width signifies greater nutrient availability. Interlocking elements denote formation of insoluble precipitates. (From R. B. Clark, Plant response to mineral element toxicity and deficiency. Pp. 71–1442 in: M. N. Christiansen and C. F. Lews (eds.), *Breeding Plants for Less Favorable Environments.* Copyright © 1982 by John Wiley & Sons, Inc. Reprinted by permission of John Wiley & Sons, Inc.)

The driving force for this process is transpiration and the transport of nutrients into the plant.

B. Mass Flow, Diffusion, and Nutrient Uptake in Roots

There are three recognized mechanisms by which water and dissolved ions come in contact with roots: mass flow, diffusion, and root interception. *Mass flow* can be likened to the transport of a boat along a stream flowing downhill. Elements in solution follow the flow of water in the soil in a similar manner in that the plant's transpiration stream pulls water out of the soil into the plant, carrying with it dissolved ions. Once the ions reach the *rhizoplane,* or surface of the root, they are absorbed or, in some cases, accumulated at the root surface if in high enough concentrations, such as might occur for Ca or NaCl in alkaline or high saline soils.

Ions can also reach the rhizoplane of the root by *diffusion;* however, water must be available. Rates of diffusion vary with the diffusion coefficient of the element or ion, available soil water, soil temperature, soil porosity, and concentration of the ion in solution. Some elements and ions are more mobile in soils than others. As an example, NO_3 has a higher *diffusion coefficient* in soil (D_e $1.9 \times 10^{-10} - 10^{-11}$ m^2 s^{-1}) than does either K ($2 \times 10^{-11} - 10^{-12}$ m^2 s^{-1}) or H_2PO_4 ($0.9 \times 10^{-12} - 10^{-15}$

$m^2\ s^{-1}$). Diffusion is the main mechanism for the mobility of ions such as P and K in soils. Table 3.1 illustrates the respective diffusion minima and maxima of several cations and anions in soil.

Root interception is the third mechanism by which nutrients can reach roots. This is a function of roots coming in contact with areas of the soil solution as a result of growth. Root interception depends on the growth rate and habit of roots and, thus, is to some degree species dependent. Also, soil characteristics such as porosity and compaction can influence the ability of roots to explore different areas of the soil profile. Root interception is thought to supply only a small part of the total nutrient demand of plants (1% or less) and is, thus, thought to be a relatively insignificant factor in the over all nutrient supply to plants. Table 3.2 illustrates how mass flow, diffusion, and root interception contribute to the nutrient demands of a maize crop. It can be seen from this table that diffusion is the primary mechanism by which P and K are supplied to corn, while mass flow contributes to the major supply of N and Mg. Root interception has little influence on the supply of these nutrients except for Mg. However, mass flow provides an excess of this element, which suggests that supply exceeds demand for Mg and accumulation occurs at the rhizoplane. Table 3.3 outlines a number of other processes and factors that can influence the transfer of nutrients from the soil to the plant.

III. NUTRIENT ABSORPTION BY ROOTS

A. Root Structure and Ion Uptake

It is important to have an understanding of the morphological and anatomical structures of roots to fully appreciate how roots absorb nutrients into plant tissues and the various resistances encountered. Different plant species have varying requirements for nutrients in support of growth processes, and these differences could be reflected in the morphology and anatomy of roots and the changes that may occur in the various root tissues with stage of development and environmental conditions.

Figure 3.3 illustrates the various structures that comprise a plant root. Starting from the exterior of the root and progressing inward, the surface of the root often has mucilaginous material associated with areas near the root tip. Microorganisms are

TABLE 3.1 Range of Diffusion Coefficients ($cm^2\ s^{-1}$) for Various Ions in Soil

	Ion	Minimum	Maximum
Cations	Na^+	1×10^{-7}	1×10^{-5}
	K^+	2×10^{-7}	2×10^{-6}
	Rb^+	6×10^{-12}	7×10^{-6}
	NH_4^+	4×10^{-8}	1×10^{-6}
	Ca^{2+}	3×10^{-8}	3×10^{-7}
	Zn^{2+}	3×10^{-10}	2×10^{-7}
	Mn^{2+}	3×10^{-8}	2×10^{-7}
Anions	NO_3^-	5×10^{-7}	1×10^{-5}
	Cl^-	3×10^{-7}	1×10^{-5}
	$H_2PO_4^-$	1×10^{-14}	4×10^{-9}

Source: As summarized by Fitter and Hay 1987, from Barber 1974 and Fried and Broeshart 1967. Reprinted with permission of The University Press of Virginia and Academic Press.

TABLE 3.2 Relative Contribution of Root Interception, Mass Flow, and Diffusion in Supplying Nutrients to Corn (kg ha^{-1})

	Amount Needed for	Approximate Amounts Supplied By		
Nutrient	9500 kg ha^{-1} yield of grain	Root Interception	Mass Flow	Diffusion
Nitrogen	190	2	150	38
Phosphorus	40	1	2	37
Potassium	195	4	35	156
Calcium	40	60	150	0
Magnesium	45	15	100	0
Sulfur	22	1	65	0

Source: From *Soil Nutrient Bioavailability* by S. A. Barber. Copyright © 1984 John Wiley & Sons, Inc. Reprinted by permission of John Wiley & Sons, Inc.

usually abundant in these areas, which frequently represent regions of the root where exudates seep out from between cells or are actively excreted. The outermost layer of root cells is the *rhizodermis* (or epidermis). Occasionally, cells of the rhizodermis undergo extension growth to form *root hairs,* which are very important in extending the ability of roots to absorb water and dissolved nutrients. The rhizodermis and root hairs have a *cuticle,* which to some extent may restrict the flow of water and nutrients into cytoplasm of these cells. The cuticle does not appear to be uniformly continuous over the surface of the rhizodermis or the root hairs. This may be particularly true in regions associated with emerging lateral roots. Immediately beneath the rhizodermis is the *exodermis* (*hypodermis*), which around the cell walls exhibits varying degrees of *suberization* (deposits of suberin, a complex polymer of hydroxy acids and alcohols), which appears to vary with environmental conditions and stage of development.

TABLE 3.3 Processes and Factors Involved in Nutrient Transfer from Soil to Plant

Process	Factors
Root development	Root length Root distribution Root morphology, diameter, and hairs
Nutrient uptake	Concentration at root surface Kinetics of uptake
Transport from soil to root: Mass flow, diffusion	Transpiration Concentration of soil solution Concentration gradient Diffusion coefficient
Mobilization by roots: Desorption, dissolution, hydrolysis of organic compounds	Depletion of soil solution Root exudates (H^+, CHO_3^-, reducing agents, chelates, organic anions) Chemical soil composition pH of soil solution Enzymes (e.g., phosphatases)
Mobilization by associated organisms	Mycorrhizal infection Bacteria

Source: Reprinted from Jungk 1991, p. 456, by courtesy of Marcel Dekker, Inc.

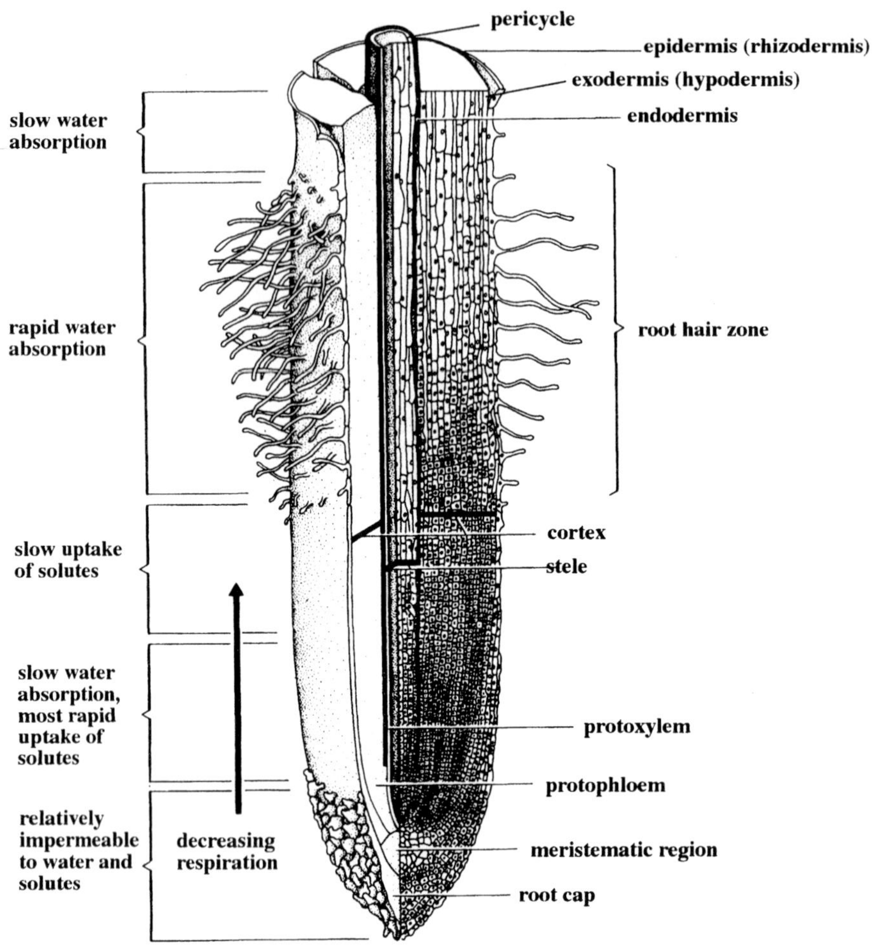

Figure 3.3. Illustration of a root showing various tissues and regions of nutrient and water absorption. (Modified from *Plant Physiology, 4th edition*, by Devlin/Witham. © 1983. Reprinted with permission of Global Rights Group, a division of Thomson Learning. Fax 800 730-2215.)

However, such suberization has potential for limiting water movement between the free space of the root tissues. The next layer of tissue encountered is the *cortex* and is several layers thick. Interestingly, this tissue may have limited capacity to absorb nutrients, although there is some debate as to whether ions are absorbed primarily in the outer layers of the rhizodermis and exodermis or in the inner layers of the cortex. The next layer is the *endodermis*. This layer of cells is responsible for root pressure in plants and in part for the selective absorption of ions into the inner tissues of the plant. The next layer of cells is the *pericyle,* which periodically becomes meristematic giving rise to lateral roots that have to break their way through the various tissue layers as they grow to the outside of the root. Areas of lateral root emergence have been hypothesized to provide relatively unrestricted flows of water and nutrients into the inner tissues of the root until the tissues respond to the injury, sealing off the accessible area. Finally, the last tissues interior to the pericycle are the xylem, phloem, cambium (species dependent), and associated parenchyma and support cells. The xylem is primarily responsible for the upward transport of water and dissolved mineral elements, while the phloem is rich in organics and responsible for movement of carbohydrates, nitrogen-containing compounds, hormones, and other molecules up and down the phloem depending on the need.

The endodermis and the concepts of apoplast (free space) and symplast are important in understanding the physical mechanism by which water and ions gain access to the xylem tissues of the root. *Apoplast* refers to the nonliving free space that comprises the root tissues. This involves areas that comprise the capillary space between the microfibrils of the cell walls, intercellular space, and the nonliving vessels and tracheids of the root xylem. The apoplast starts at the rhizoplane and extends between cells of the rhizodermis, exodermis, and cortex and is interrupted by the endodermis. At this point, the apoplast is discontinuous because of suberin layers (*Casparian strips*) that form around the circumference of each cell that comprises the endodermis. In effect, this forms a virtually impermeable seal between adjacent cells that prevents the passage of nutrients between the intercellular spaces of the endodermis. Thus, entry into the xylem has to be via the *symplast* system, which is the space on the cytoplasmic side of the plasmalemma (plasma membrane).

Unlike the apoplast, the symplast is continuous from the rhizodermis to the living cells of the early metaxylem. Continuity is maintained via *plasmodesmatal* connections between adjacent cells that comprise the root tissues. Theoretically, water or ions that transverse the membrane of the rhizodermis can reach the apoplastic xylem cells in the center of the root by passing from cell to cell via the plasmodesmatal connections. Plasmodesmata are extensions of the *endoplasmic reticulum* originating in one cell and extending to another. Figure 3.4 shows the structure of a plasmodes-

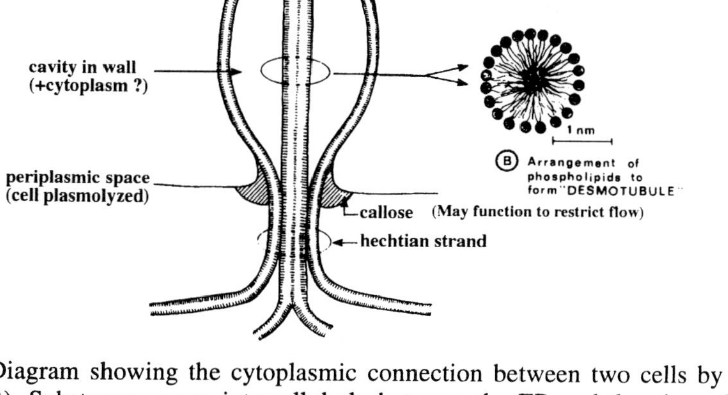

Figure 3.4. Diagram showing the cytoplasmic connection between two cells by endoplasmic reticulum (ER). Substances move intercellularly between the ER and the plasmalemma in the neck region A. Closure of the neck region (hechtian strand) by callose formation may regulate transfer between cells. (From Overall et al. 1982, with permission; reprinted from Clarkson 1991, p. 36, courtesy of Marcel Dekker, Inc.)

mata. Movement of ions in solution between cells occurs between the *desmotubule* and the plasmalemma. There is evidence that movement between cells can be restricted or controlled as a result of *callose* deposition at one end of the desmotubule. Callose can be induced under high Ca levels and may be involved with this process.

Seemingly, water and dissolved nutrients have free access to the tissues of the root through the apoplastic continuum, at least up to the endodermis. Obviously water and nutrients may enter the symplast anywhere along the apoplastic system, but the rate of uptake of ions from the apoplastic system may decline toward the endodermis as a result of declining O_2 tensions in the root tissues. As a result of differential absorption of ions into the symplast of the root, an osmotic and water potential gradient is maintained toward the center of the root. When ions reach the living early metaxylem cells of the stele, they in some manner either leak out or are transported from the symplast (early metaxylem cells) into the nonliving apoplast comprised of vessels and/or tracheids. It is not certain if there are specific carriers that transport ions into the xylem or whether leakage occurs as a consequence of low O_2 tensions or as a result of maturation of early metaxylem and subsequent release caused by death of the cells. The endodermis, thus, serves as a type of osmometer in which the low water potential difference on the xylem side of the endodermis results in the flow of water into the xylem from the cortex side of the endodermis (via the symplastic system). This creates a hydrostatic pressure in the xylem that results in root pressure, which in essence forces water and dissolved nutrients up into the xylem of the shoot. This pressure may be important in the transport of nutrients up into the shoot in the case of newly developing seedlings and small herbaceous plants. This topic will be addressed further in another section. Figure 3.5 illustrates how ions can enter the symplast from the apoplast and diffuse to the center of the root, where they may either leak out of the symplast or be actively transported into the xylem apoplast.

IV. ION TRANSPORT ACROSS MEMBRANES

A. Transport Proteins

Water and dissolved nutrients (ions and small molecules such as sugars and amino acids) can freely enter the apoplastic space of the root tissues. The cells that comprise the rhizodermis, exodermis, and cortex have negative charges associated with their cell walls as a consequence of their structure and composition. The cellulose and hemicellulose composition of the cell walls and association of proteins and lipids at the interface of the cell wall provide the negative charge that results in a type of *cation exchange capacity* (CEC) that differs among plant species. These charge characteristics attract positively charged ions and repel negatively charged ions. As a consequence, this mechanism provides a pool of ions in close proximity to the membrane that in some cases may be readily available for absorption. The types and amounts of ions that are adsorbed in this way depend on the composition of the cell wall, the types and concentrations of ions in the soil, and the demand for a particular ion during the growth of the plant. Generally, dicot roots have a higher CEC than do monocots, as indicated in Table 3.4. This capacity of roots to adsorb ions may have important ramifications relative to a plant's ability to remove and/or accumulate small amounts of nutrients that may be present in soils. Also, it may be a mechanism whereby high concentrations of ions, which may be otherwise toxic, can be sequestered and isolated from more sensitive tissues.

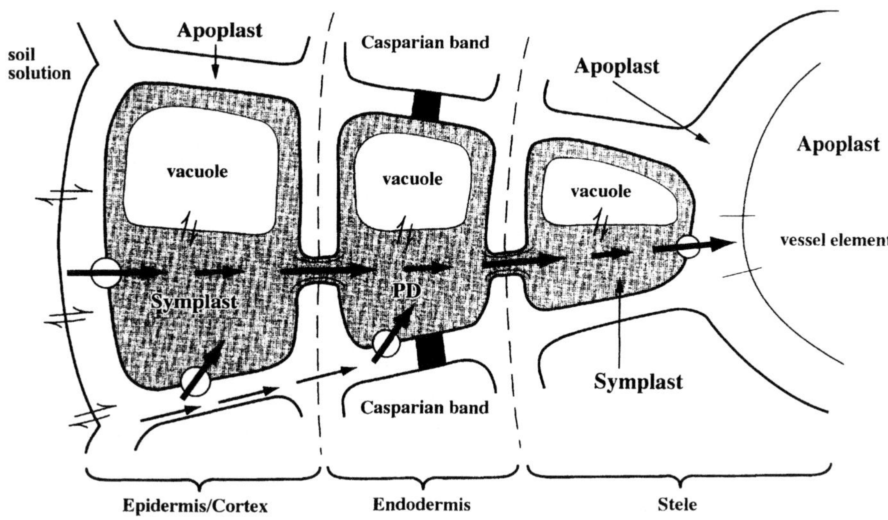

Figure 3.5. Pathway of ion absorption into plant roots. (Modified from *Introduction to Plant Physiology* by W. G. Hopkins. Copyright © 1995, John Wiley & Sons, Inc. All rights reserved. Reprinted by permission of John Wiley & Sons, Inc.)

Absorption, or the movement of ions across the membranes of root cells into the symplast, can occur passively or actively. *Passive transport* occurs along a chemical potential gradient in which the concentration of the solute molecule is higher in the external medium than in the root cells. Such gradients can be maintained by incorporation of the absorbed ion into an organic molecule or rapid movement of the ion out of the root into other tissue sinks. Passive transport may also occur along a electropotential gradient that may have been established by active transport mechanisms. *Active transport* requires the expenditure of metabolic energy and results in the accumulation of the solute in higher concentrations inside the cell compared to the soil or growth solution. Ion absorption is very specific and reflects the different types of *carriers, channels,* and *proton pumping mechanisms* associated with the plasmamembrane and the tonoplast. Table 3.5 shows how different ions are selectively accumulated or excluded from the roots of corn and bean as compared with the concentrations of the various elements in the growth medium. It is obvious that species differences are also prevalent relative to ion accumulation. As indicated in Table 3.5, one wonders if the differences in the ability of maize and bean to absorb K is related to differences in the CEC of the roots.

TABLE 3.4 Cation Exchange Capacity of Root Dry Matter of Different Plant Species

Plant Species	Cation Exchange Capacity (meq per 100 g dry wt)
Wheat	23
Maize	29
Bean	54
Tomato	62

Source: From Keller and Deuel 1957.

TABLE 3.5 Ion Concentration in Root Press Sap of Corn and Bean Compared with External Solution

	External Concentration (mM)			Concentration in the Root Press Sap (mM)	
		After 4 Days			
Ion	Initial	Maize	Bean	Maize	Bean
Potassium	2.00	0.14	0.67	160	84
Calcium	1.00	0.94	0.59	3	10
Sodium	0.32	0.51	0.58	0.6	6
Phosphate	0.25	0.06	0.09	6	12
Nitrate	2.00	0.13	0.07	38	35
Sulfate	0.67	0.61	0.81	14	6

Source: From Marschner 1995, with permission.

Membrane structure is paramount in understanding how ions are transported. The reader is referred to chap. 3 of Nilsen and Orcutt (1996) for a more complete review of membrane structure and function and how abiotic stress affects membrane fluidity. However, as a brief review, membranes are composed of phospholipids, sterols, glycolipids, sulfolipids, and proteins. Some proteins are associated with the surfaces of membranes while others may be incorporated and may penetrate the entire thickness of the membrane. Many proteins are involved with ion transport and function as ATPases involved with proton transport, whereby they establish pH and electropotential gradients, while other proteins may function as channels or carrier proteins that are driven by ion or electropotential gradients established by ATPases. Figure 3.6 gives the reader a better perspective of the types of ion transport proteins in membranes and where they are located. Both the plasmalemma and the tonoplast have ATP-driven proton pumps. In the vacuole, pyrophosphate-driven proton pumps are also thought to occur, and in the case of the plasmalemma a transmembrane redox mechanism may exist in which NAD(P)H (reduced nicotinamide adenine dinucleotide phosphate) is used as an electron donor, possibly to reduce Fe(III) at the apoplast side of the plasmalemma. In essence, these proton pumps function to maintain a physiological pH in the cytosol and to establish electropotential gradients (proton) across membranes that are used to drive ion transport proteins.

Generally, three types of transport proteins are recognized: antiports, symports, and uniports. *Antiports* transport different ions in opposite directions across a membrane, *symports* transport different ions in the same direction, and *uniports* transport a single ion in one direction. In the case of antiport and symport transport, the energy is provided by the proton gradient established via proton pump mechanisms, and the proton is the accompanying ion along with either another cation or an anion. In the case of uniports, the transport is in a downhill direction as a result of a low concentration of the ion in the cell and is essentially passive.

Other transport proteins that occur in membranes are referred to as channels and are thought to be involved with rapid transport of ions associated with *thigmonastic* (touch) and *nyctinastic* (sleep) movements, and stomatal opening and closing in plants. These proteins are envisioned as being *gated,* in that movement across them is controlled by environmental cues that open and/or close the gates, which allows for very rapid transport of ions such as K through the membrane channel (Figure 3.7). These gates can affect the influx or the efflux of ions across membranes. It is important to

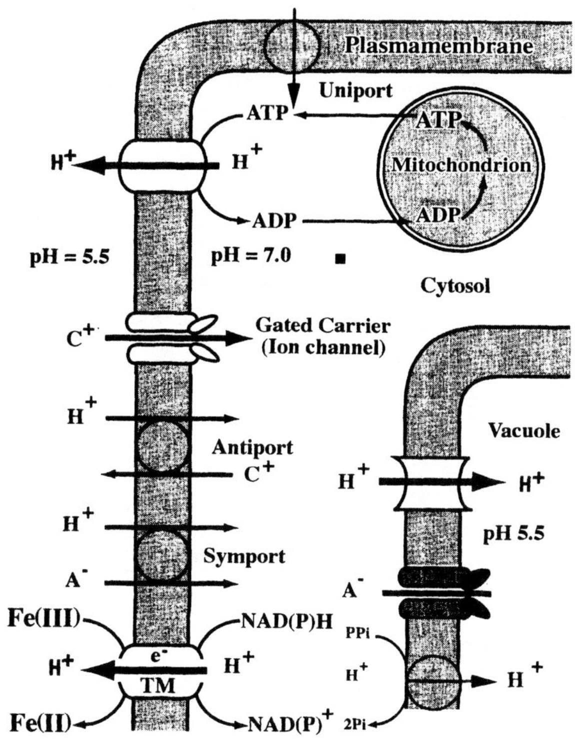

Figure 3.6. Diagram illustrating various types of ion transport proteins occurring in the plasmalemma and tonoplast. ATPase proton pumps move H^+ across the plasmalemma and tonoplast, creating proton gradients that provide the energy to drive antiport and symport carriers, which transport ions and uncharged molecules across membranes. Note that the energy required for this process (ATP) is generated by the mitochondrion and also the chloroplast. (Modified from *Introduction to Plant Physiology* by W. G. Hopkins. Copyright © 1995 John Wiley & Sons, Inc. All rights reserved. Reprinted by permission of John Wiley & Sons, Inc.)

note, at this point, that proton pumping mechanisms can be regulated by the concentrations of specific ions and the levels of ATP or PPi (inorganic pyrophosphate) that may be available to the cell. For example, the rate of ATPase activity in the tonoplast membrane is stimulated by Mg and Cl and is not affected (or is inhibited) by K and NO_3, while the activity of PPiase (inorganic pyrophosphatase) is affected in a reverse manner by these ions compared with that of Mg-ATPase with the exception that Mg is stimulatory. The overall activity of PPiase is considerably less than that of Mg-ATPase. Thus, deficient or excessive levels of these ions may affect membrane transport of other ions by affecting electropotential gradients generated by proton pumps.

V. CELLULAR AND WHOLE PLANT NUTRIENT DISTRIBUTION

A. Plant Homeostasis and Ionic Balances

The intracellular uptake and distribution of ions at the whole-plant level is affected by a number of interacting factors. These include biotic and abiotic environmental factors that interact with specific plant factors related to developmental, anatomical/

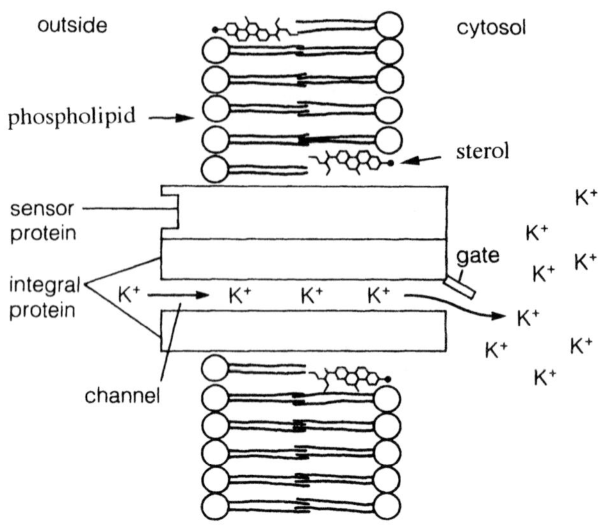

Figure 3.7. Model of a gated ion channel. Stimuli such as light, hormones, or possibly high Ca concentrations are sensed by the sensor protein and regulate the opening and closing of the channel gate. Ions diffuse down an electropotential gradient. Such channels are thought to control rapid transport of ions across membranes resulting in turgor changes that result in plant movements (e.g., Venus flytrap and sensitive plants). (Modified from *Plant Physiology, 4th edition,* by F. Salisbury and C. Ross. © 1992. Reprinted with permission of Global Rights Group, a division of Thomson Learning. Fax 800 730-2215.)

morphological, and biochemical differences that are under genetic control. Thus, a complete understanding of how deficient or toxic levels of ions affect plants requires some understanding of how plants regulate the distribution of mineral elements at the cellular and whole-plant level. Plants have adapted to specific locations in response to the prevalent environmental conditions in which they grow. Part of this selection process involves the nutrients available to the plant. In essence, plants have established a type of *homeostasis* with their environment. Since plants are sessile organisms, they have to constantly monitor the external environment and make internal physiological and biochemical adjustments to maintain a more or less homeostatic internal environment. The control of ionic balances at the cellular level and the distribution of essential elements throughout the plant has a central role in maintaining some sense of equilibrium in plants since mineral elements are required for structural, biochemical, and osmotic functions.

At the cellular level, there is evidence of homeostasis being established in terms of ionic balances in cellular compartments. Multiple compartments exist at the cellular level and include the cytoplasmic compartment established by the plasmalemma, the vacuole surrounded by the tonoplast as well as other membrane-bound organelles such as the nucleus, mitochondrion, chloroplast, endoplasmic reticulum, Golgi bodies, peroxisomes, and glyoxysomes. Each cellular compartment has its own unique metabolic role or roles and specific requirements for essential mineral elements that are paramount to its proper functioning. Such activity may be related to pH control, osmotic regulation, structural organization, and involvement in enzyme catalytic or cofactor functions.

B. Ion Compartmentation and pH Control

What seems to be one of the primary control mechanisms for maintaining a homeostatic environment in the cytoplasm (and hence the metabolic activity associated there) and in other compartments within the cytoplasm is the interaction between the plasmalemma and tonoplast to control ion influx and efflux. The vacuole seems to act as a kind of buffer zone against which cytoplasmic ion concentrations and pH are regulated. The maintenance of physiological pH in the cytoplasm is of great importance relative to cellular metabolism. Cytoplasmic pH is maintained at or around pH 7.3–7.6, and is kept in this range as a result of proton pumps in the plasmalemma and tonoplast that remove protons to the vacuole and/or the apoplast outside the plasmalemma. The result is the development of a proton (pH) gradient between the apoplast/cytoplasm (pH 5.5) and the vacuole/cytoplasm (pH 4.5–5.9). This is important in that the chemical potential gradient created as a result of proton accumulation drives the various ion transport carriers in the plasmalemma and the tonoplast. The activity of the proton pumps, as well as other ion carriers, appears to be regulated by the concentration of specific ions (referred to above), pH, ion ratios, temperature, oxygen concentrations, and, most likely, many other environmental influences.

C. Ionic Composition of Plant Cells and Tissues

The "generalized" plant cell cytoplasm is said to have a total inorganic concentration of 100–500 mOsm, a high K/Na ratio, free [H] and [Ca] of about 0.1 μM, and relatively low concentrations of NO_3, Cl, SO_4 and H_2PO_4. The composition of the vacuole reflects the multiple roles of this organelle relative to cell growth, storage of inorganic (Ca, Mg, NaCl, and NO_3) and organic nutrients (malic acid, sugars, and amino acids) and the repository of potentially toxic ions. Table 3.6 illustrates how ions are compartmentalized in cells and how toxic elements such as Na and Cl are isolated in the vacuole and kept at low relative ratios compared with K in the cytoplasm. Under conditions of limited K supply, other ions can accumulate in the vacuole to offset and balance the transport of stored K from the vacuole to the cytoplasm. Higher levels of Na, Ca, and H frequently accumulate in the vacuole as a result of low K.

Potassium is one of the most abundant elements in plant tissues and is very important for osmotic regulation and protein synthesis. Table 3.7 shows the distribution of K among various organelles associated with different plant species compared with the external medium in which they were grown. Potassium usually occurs in higher concentrations in the cytoplasm compared with the vacuole and is carefully regulated via reserves that can occur in the vacuole if K is readily available. Thus, it has been shown that plants grown under decreasing K concentrations appear to regulate K concentrations in the cytoplasm by removing K from the vacuole in an attempt to maintain a constant level of K in the cytoplasm. Also, it is interesting to note that K is maintained at high levels in the nucleus and chloroplast (Table 3.7).

Figure 3.8 illustrates how the levels of K are maintained at a fairly constant level, under two different K regimes in barley. Through the vegetative stage of development, K remains relatively constant and only increases as the reproductive stage ensues and senescence begins. Plant species may differ relative to the optimum K levels in the cytoplasm and have different critical levels below which control is lost. Concentrations of K in the cytoplasm are in the order of 100–200 mM, while those in the vacuole

TABLE 3.6 Cytoplasmic and Vacuolar Ion Concentration in Various Plant Species

Ion	$[ion]_c$	$[ion]_v$	Plant
K^+	92 m*M*	95 m*M*	Barley
	71–119 m*M*	9–12 m*M*	Low-salt barley
		150 m*M*	*Hippeastrum*
	100 m*M*	83 m*M*	Onion
Na^+	8 m*M*	32 m*M*	Barley
Ca^{2+}	10.2 m*M*	1.3 m*M*	Onion
		6 m*M*	*Hippeastrum*
Mg^{2+}	3.25 m*M*	3.3 m*M*	Onion
		30 m*M*	*Hippeastrum*
H^+	0.1 μM	10 μM	
Cl^-	10 m*M*	150 m*M*	Carrot
	30 m*M*	24 m*M*	Onion
NO_3^-		97 m*M*	Barley
		87 m*M*	Barley
SO_4^{2+}	0.7 m*M*	0.5 m*M*	Carrot
	0.3 m*M*	1.3 m*M*	*Memna*
Pl	25 m*M*		Barley
	20 m*M*	25 m*M*	

Source: Glass and Siddiqi 1984. *Advances in Plant Nutrition,* V.1, P. B. Tinker and André Läuchli (eds.).

are usually lower, vary more, and rarely fall below 10–20m*M*. Table 3.8 illustrates the tissue water concentrations of K in a range of plants grown in flowing nutrient solution containing 1 m*M* K. The data indicate that different species may have different optima at which they control K levels in the tissues (cytoplasm). Also, it appears that grasses may control K levels at higher concentrations than do dicots. Compartmentation of ions can occur not only intracellularly but intercellularly as well. It

TABLE 3.7 Distribution and Concentration of Potassium in Cellular Compartments in Various Plant Species

Plant	Units	Cytoplasm	Nucleus	Chloroplast	Vacuole	External, m*M*
Nitella }					40–80	
Chara }	m*M*	93–125	—	84–150	59	0.1
Hydrodictyon }	Mean	113		108		
Saccharomyces cerevisiae	mmol kg^{-1} fresh wt	192	195	—	92	
Hordeum vulgare	m*M*	100	—	—	20	0
(low-salt roots)	m*M*	71–119	—	—	9–12	0
Zea mays (mesophyll cell)	count rate	1355	1765	1620	500	6.0
Marine algae	m*M*	210–480	—	—	7–625	10–13

Source: Adapted with permission from Flowers and Läuchli 1983 and references therein.

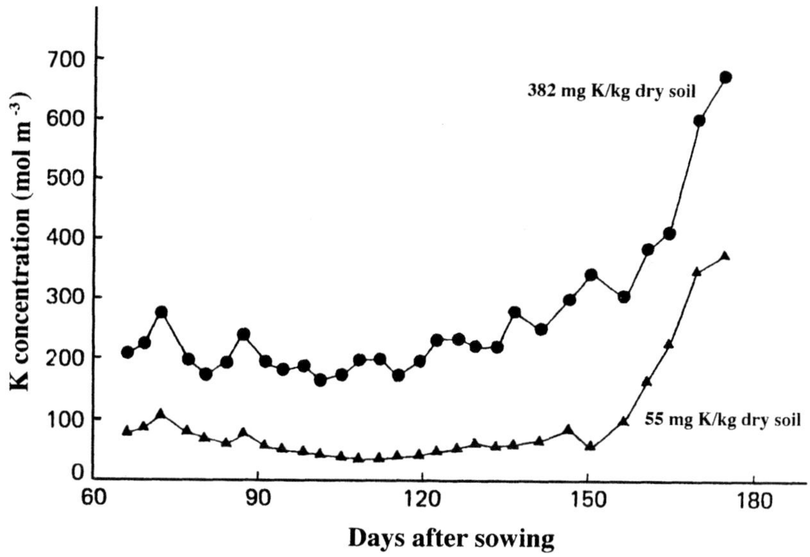

Figure 3.8. Changes in K concentrations (based on tissue water) in shoots of barley plants grown at two K levels. (Modified from Leigh and Storey 1991. Reprinted with the permission of Cambridge University Press.)

appears that, at least in the leaves of barley, specific ions can be selectively absorbed by different tissues. Table 3.9 shows that the mesophyll cells of barely selectively accumulated P while Cl and Ca are excluded. While in the adaxial and abaxial epidermal cells, Cl and Ca were localized in these cells. Further studies of this in salt-tolerant and susceptible varieties of barley revealed that the susceptible cultivar exhibited higher concentrations of Cl in the mesophyll cells than did the tolerant cultivar. Thus, how plants compartmentalize nutrients and ions at the cellular level has significant implications relative to understanding how plants may acclimate or adapt to differing excessive and low mineral environments.

TABLE 3.8 Average K Concentrations Maintained in Tissue Water of Shoots of Various Crops Harvested During Vegetative Growth

Crop	Average K Concentration (mol m^{-3})
Hordeum vulgare	184
Triticum aestivum	170
cv Brimstone	
cv Square Heads Master	178
Avena sativa	152
Solanum tuberosum	155
Vicia faba	118
Brassica napus	100

Source: Leigh and Storey 1991.

TABLE 3.9 Localization of P, Cl, and Ca Between Epidermis and Mesophyll Cells in Barley Leaves as Determined by X-ray Microanalysis

Cell Type	Number of Cells Analyzed	Percent of Cells with Detectable Level of Element		
		P	Cl	Ca
Mesophyll	97	71	0	0
Epidermis				
Adaxial	127	0	62	72
Abaxial	109	0	67	61

Source: Leigh and Storey 1991.

D. Ionic Signaling and Regulation of Ionic Balance

It is hypothesized that there are feedback and/or signaling mechanisms that may be concentration or ion specific that help to govern the balance of ions in the cytoplasm, vacuole, and apoplast of the stele. Figure 3.9 illustrates how such a feedback mechanism might function. In this model, the solid lines illustrate ion fluxes mediated by transporter proteins. The dashed lines illustrate feedback concentrations from within the cytoplasm, tonoplast, and stele. Obviously, such ion balances are going to be very much related to the transpiration and growth rates of the tissues involved and will change diurnally. As ions are transported into the stele, in response to demand for ions to support growth, ion levels in the cytoplasm are lowered, which could signal ion transporters in the plasmalemma to take up ions from the soil or remove ions

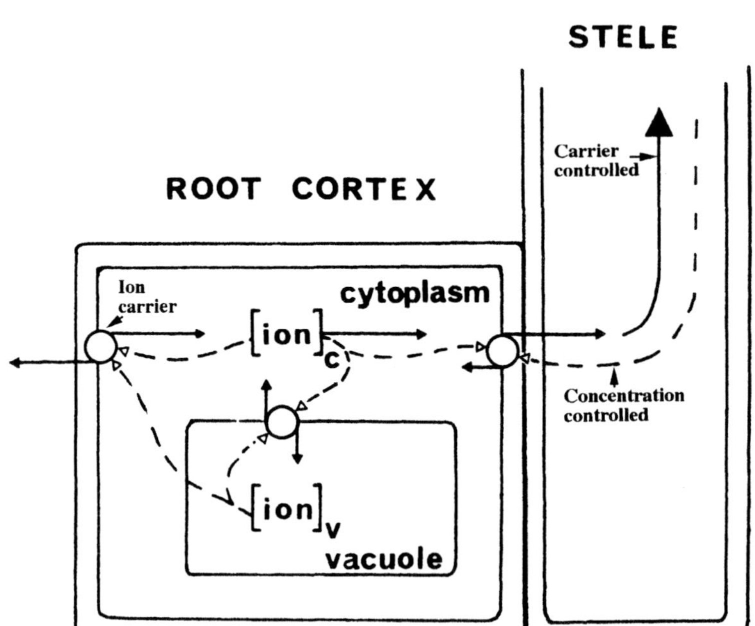

Figure 3.9. Model illustrating possible feedback mechanisms for the regulation of ion fluxes in roots. Solid lines = fluxes controlled by ion carriers in membranes, dashed lines = feedback regulation from ion concentrations in the cytoplasm and vacuole. (Modified from Advances in Plant Nutrition, V.1, by A. D. M. Glass and M. Y. Siddiqi. Copyright © 1984 by Praeger Publishers. Reproduced with permission of Greenwood Publishing Group, Inc., Westport, CT.)

from the vacuole. As demand is reduced for ions, feedback concentrations of ions cause other transporters to work in the opposite directions to cause an efflux of ions or to accumulate them in the vacuole.

VI. LONG-DISTANCE TRANSPORT OF MINERAL ELEMENTS

A. Xylem Transport

Transport of ions and nutrients in plants occurs via the xylem and phloem tissues of plants. Movement in the xylem (nonliving cells) is driven by hydrostatic pressures generated by root pressure and transpiration rates. The direction of flow of water in the xylem occurs along a water potential gradient that is higher in the soil and becomes progressively negative in the roots, stems, and leaves. This creates considerable negative tensions during the day in the xylem. Movement in the phloem (living cells) is bidirectional but usually occurs in a downward direction. The mechanism of phloem transport has been referred to has the *pressure flow* or *mass flow* hypothesis. In essence, a hydrostatic pressure is generated in the phloem as a result of loading sugars and other solutes into the phloem in the leaves and removing them in the cells of the root. As a result of more negative water potential in the phloem of the leaves, water moves into the phloem creating a positive pressure. In the roots, solutes are actively unloaded and used for respiration or stored in an osmotically inactive form such as starch. The removal of sugars from the phloem creates a less negative water potential in the phloem compared with the xylem and, thus, water moves into the xylem and is recirculated back to the shoot as a result of transpiration and root pressure. The net result is a type of circulatory system in plants where water and nutrients can be recirculated throughout the plant in response to the respective needs of roots and shoots.

Transport of ions out of the roots into the shoots occurs in the xylem, although some nutrients may get into the phloem and be transported via phloem tissues. As indicated above, movement in the xylem is driven by transpiration rates and root pressure. The latter may be particularly important in young seedlings and small herbaceous plants and during dark periods when transpiration demand is low. Also, all ions and molecules are not transported at the same rate in the transpiration stream. For example, uncharged molecules are usually transported faster than ions or charged molecules. Ions such as Na, Ca, and Si seem to be more readily transported in the transpiration stream than K, NO_3, or P.

Other factors that can influence ion transport include the growth rate of the plant and developing sinks (meristems, fruits, etc.). Directional movement of ions into these sinks may be under the control of phytohormones such as cytokinins and auxins. En route to the leaves and other shoot tissues, ions may be absorbed by living cells in the stem, which may be in close proximity to the nonliving xylem cells. Ions can be removed by specialized cells called *transfer cells* and in some cases may serve as a mechanism for removing toxic ions from the transpiration stream before they reach the leaf tissues, which may be metabolically sensitive to such ions. As indicated above for barley leaves, uptake may be selective for specific ions. Also, as in root tissues, cells of the stem may serve as cation exchangers by exchanging ions with the surfaces of cell walls, thus removing them from the xylem solution.

Movement of ions in the stem xylem depends on valence, xylem sap pH (ranges 5–7), conjugation of metals with soluble organic molecules, charge density of negative groups in the transport system (species specific, dicots vs. monocots) and relative complexing activity of the ions involved. Some plants may deal with high ion concentrations within leaf tissues by forming insoluble salts in the apoplast, developing salt glands, or unloading salts through formation of guttation water at the tips of leaves. In addition, the cellular components of the xylem (tracheids only, or vessels and tracheids) can affect the rate of ion movement by restricting the flow of nutrients in the xylem and thus potentially increasing uptake of ions along the stem path. The stem appears to play an important role in the storage of mineral nutrients when nutrients are not limiting. When they become limiting, the reserves in the stems are used to restore nutrient balances. Nodes and the root–shoot interface also represent areas within the stem where restricted flow occurs and frequently represent sites of accumulation of mineral nutrients. Such restricted flow probably reflects differences in the composition and relative sizes of the cells that comprise the vessels and tracheids in these areas.

Flow of xylem water and solutes also differs with types of tissues. For example, newly forming and expanding leaves are going to be the primary sites of nutrient delivery simply as a result of greater rates of mass flow driven by transpiration. Organs such as fruits, nonexpanded leaves, and buds will have relatively lower transpiration rates and thus will be the sites of lower concentrations of mineral elements. Since the vascular tissues often end near the margins of leaves, these areas represent potential regions where ions may accumulate in toxic concentrations causing marginal scorch symptoms. Conversely, tissues in which low transpiration rates may occur could develop deficiency symptoms such as "blossom end rot" of tomato fruits, which has been associated with Ca deficiency. Other fruit and vegetable disorders related to the deterioration of tissues in the centers of fruits and vegetables probably reflect the minimal transport of essential elements required for normal growth and development.

B. Phloem Transport

Phloem transport is essential for the movement of photosynthate and other essential nutrients to the roots. Phloem sap is comprised primarily of organic molecules, the majority of which is sucrose or other sugars depending on species (Table 3.10). Other important organics include amino acids, citric acid, malic acid, succinic acid, proteins, hormones, vitamins, and ATP. All elements that occur in the xylem are also found in the phloem. As in the xylem, the mobility of ions in the phloem differs, as indicated in Table 3.11. Generally, micronutrients are less mobile in the phloem than macronutrients, which may reflect a higher degree of reactivity of these elements compared with elements such as K. However, the mobility of micronutrients also seems to vary depending on stage of development. Higher levels of these ions seem to be transported when plants are in the reproductive state and may reflect stronger sinks during this period than at other times of development. Sink strengths may also reflect the level of hormonal activity as well. Usually, the solute concentrations of ions and organics in the phloem are several times that of the xylem, and the pH of the phloem (7.9–8.2) is more basic than that of the xylem (4.9–6.3). The composition of the phloem differs with plant age, the site in the plant from which the sample is taken, the initiation of reproduction, diurnal and seasonal cycles, and the availability of nutrients in the soil. Movement in the phloem can be bidirectional, and it changes in response to growth and sink development. Since the xylem and phloem in the vascular system are separated by only a few cells, interchange of components between the xylem and phloem can occur readily. Since the concentration of nutrients in the phloem is higher than the xylem, transfer energetically is downhill from phloem to xylem and uphill

TABLE 3.10 Xylem and Phloem Sap Composition of *Lupinus albus*

	Xylem Sap (Tracheal) (mg L^{-1})	Phloem Sap (Fruit Bleeding) (mg L^{-1})
Sucrose	ND[a]	154,000
Amino acids	700	13,000
Potassium	90	1,540
Sodium	60	120
Magnesium	27	85
Calcium	17	21
Iron	1.8	9.8
Manganese	0.6	1.4
Zinc	0.6	5.8
Copper	Tr[b]	0.4
Nitrate	10	ND[a]
pH	6.3	7.9

[a]ND = not detectable
[b]TR = trace
Source: From Pate 1975, with permission.

from xylem to phloem, and in the latter case metabolic energy is expended when phloem loading occurs. Such exchanges can occur throughout the root-shoot continuum and are very important relative to maintaining optimal ionic and nutrient balances in the plant.

One of the more interesting examples of nutrient exchange and circulation is the interaction between K and NO_3. Basically, nitrate is absorbed in the roots and can be either reduced in the roots to form amides or ureides, which are then transported to the leaves where amino acids can be produced, or NO_3 can be transported to the leaves and reduced to form amino acids, which are in turn circulated back to the roots via the phloem. Where NO_3 reduction occurs is species specific. In the absorption of NO_3 by the roots, K acts as a transport counterion to form KNO_3, which is the translocatable form of NO_3 in the xylem. In the shoot, NO_3 is reduced to form NH_3, releasing K again to serve as a counterion to balance the negative charges of organic acids produced in carbon fixation that are loaded into the phloem. Potassium salts of these acids are then translocated to the roots, where they may be incorporated into

TABLE 3.11 Phloem Mobility of Mineral Elements

Mobile	Intermediate	Immobile
Potassium	Iron	Lithium
Rubidium	Manganese	Calcium
Sodium	Zinc	Strontium
Magnesium	Copper	Barium
Phosphorus	Molybdenum	Boron
Sulfur		
Chlorine		

Source: From E. Epstein, *Mineral Nutrition of Plants: Principles and Perspective*. Copyright © 1972 by John Wiley & Sons, Inc. Reprinted by permission of John Wiley & Sons, Inc.

the respiratory cycle to produce energy for the roots. The K is then freed to be used again as a counterion for the transport of more NO_3 to the leaves. This process is illustrated in Figure 3.10. The ability of the vascular system to circulate and exchange ions and nutrients provides a mechanism whereby changes in the quality and/or quantity of nutrients and ions could serve as signaling mechanisms to alter root absorption and/or metabolic rates.

Related to this is the involvement of essential mineral elements in the control of phytohormone production in plants (Nilsen and Orcutt 1996, chap. 4). For example, cytokinins and abscisic acid are produced in the roots of plants and may act as signals between the roots and shoots via transport in the vascular system in controlling stomatal opening and closing in plants in response to water deficit stress. The concentrations of both of these phytohormones can be influenced in plants by the levels and source of nitrogen (NO_3 or NH_4) on which they are grown. Changes in the levels of K, P, and Ca have also been shown to influence the concentrations of cytokinins, while changes in Zn, K, and P influence the levels of abscisic acid in some plants. Not only can specific elements affect the concentrations of phytohormones in plants, but in actively growing tissues (sinks), high hormone concentrations may be involved as signals in the transport of elements and organics such as amino acids and sugars to these active sinks. Two classes of hormones—cytokinins and auxins—have been most frequently implicated in such responses, but so have abscisic acid, ethylene, and the gibberellins. Certainly, the interdependence of specific elements and hormones on nutrient balances in plants seems to be an important area for further inquiry.

The relative importance of xylem and phloem in providing nutrients for the growth of plant tissues appears to depend, in part, on tissue type and the mobility of the element in the xylem or phloem. The xylem provides most of the nitrogen in the form of NO_3 to the leaves, but nitrogen can be recirculated via the phloem in various forms as described above. The forms and concentrations circulated depend on the amounts

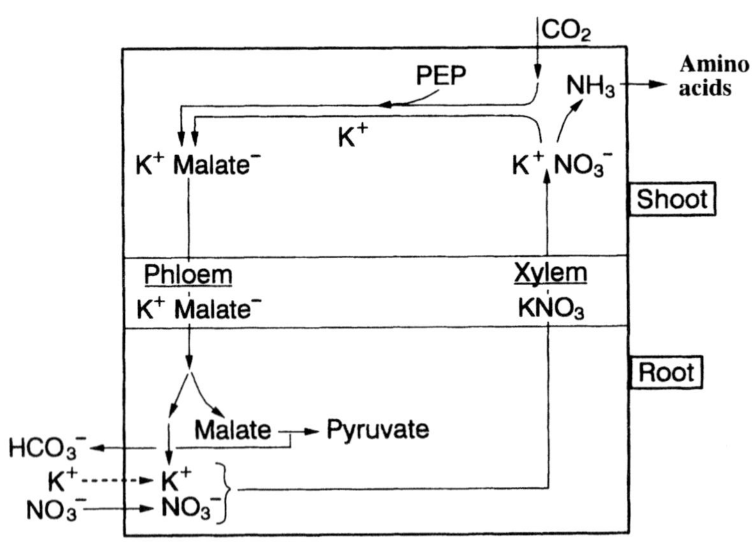

Figure 3.10. Model illustrating circulation and counterion function of K relative to the movement of malate and NO_3 in the xylem and phloem. (From Ben-Zioni et al. 1971; Kirkby and Knight 1977, © ASPP, reprinted with permission. Modified by Marschner 1995.)

and forms of nitrogen available to the plants. In barley fed NO_3, 79% of the nitrogen fed to the plants was recirculated to the roots as reduced nitrogen, 21% of that was incorporated by the roots, and the remainder was recirculated back to the leaves. In castor beans, the xylem provided the major source for Mg and Ca, while the phloem was the major source for K in the nutrition of the terminal bud. However, in young leaves both xylem and phloem provided K nutrition equally while the xylem provided most of the Mg and Ca. Thus, nutrition via xylem or phloem may differ with tissues and relative mobility of nutrients in the respective vascular compartments.

VII. INTERACTIONS OF MINERAL ELEMENTS

A. Antagonisms and Synergisms

Ions can interact in the soil and/or the plant causing deficiencies and/or toxicities that can modify plant growth. Ion interactions in the soil depend on types of ions, concentrations, soil type, and pH. Frequently, such interactions result in precipitation as a result of ionic or covalent bonding between ions and subsequent reduced availability to plants. In plants, interactions can result as a consequence of competition for cation exchange sites on or in the root, competition for carriers that transport ions across membranes, antagonisms relative to activation and deactivation of transporters, interactions of ions transported in the vascular system (pH related and formation of chelates), competition or substitution of ions in metabolic functions, and precipitation of ions as insoluble salts within tissues. Cations and anions differ in their affinity for ligands as a result of differences in their electron configuration. This reactivity is part of the basis for specificity of function of elements and may also influence interaction and ionic competition. Potassium, Na, Mg, Ca, and Mn exist mainly as free or reversibly bound ions, while Fe, Cu, Zn, and Mo occur as structural chelates or metalloproteins. Calcium and Mg have high affinities for carboxylate or phosphate-O-ligands: Na and K have high affinities for ether, alcohol, and carbonyl-O-ligands; and the transition metals have high affinities for N and S ligands. Anions such as Cl^- are weakly bound to hydrophobic sites, while ions such as $H_2PO_4^-$ may form moderately strong complexes through H bonding and condensation.

Some examples of how deficiencies and high levels of ions may influence the uptake of other ions to affect growth in plants include the following:

1. Deficiencies in Fe, Mo, Zn, or K can cause reduced N uptake and utilization.
2. Deficiencies in Fe or Zn reduce the uptake of K.
3. High levels of K, B, or Mn can reduce the uptake of Fe.
4. High Fe, Zn, or Cu concentrations can reduce the uptake of Mn.
5. High P or B reduce the uptake of Zn.

Examples of ions that can partially substitute for each other include: (1) Mn/Mg (some enzyme reactions) and (2) K/Na (osmotic relations). Examples of ions that are antagonistic include Ca/Na (membrane permeability), and Fe/Zn and Cu/Fe (enzyme function). The reader is referred to Shuman (1994) for additional information relative to ion interactions in plants.

VIII. PHYSIOLOGICAL AND BIOCHEMICAL FUNCTIONS OF ESSENTIAL ELEMENTS

A. Essential and Required Levels of Elements

Generally, 17 elements are considered essential for normal plant growth and development. These include the macronutrients (C, H, O, N, P, K, S, Ca, and Mg) and the micronutrients (Cu, Fe, Cl, B, Mn, Zn, Mo, and Ni). The distinction between macro- and micronutrients is based on the amounts of these nutrients needed to sustain normal growth and development. Generally, *micronutrients* are required at concentrations ranging from 0.1 μg–0.1 mg/g of dry weight of leaf tissue, while *macronutrients* range from 1–15 mg/g dry weight. Table 3.12 summarizes the macro- and micronutrients, their mobility characteristics within plants, critical levels, and available

TABLE 3.12 Critical Levels, Sufficient Range, Toxicity Level, Mobility, and Available Forms of Elements Known to Be Essential in Plants

Element	Macro or Micro[a]	Mobility[b]	Critical Level[c]	Sufficient Level[c]	Toxicity Level[d]	Available Form
			Oxides			
C	Mac	M	—	—	—	CO_2
H	Mac	M	—	—	—	H_2O
O	Mac	M	—	—	—	CO_2, H_2O
N	Mac	M	<2.0%	2.0–5.0%	NT	NO_3, NH_4^{+e}
S	Mac	I	<0.1%	0.1–0.3%	NT	SO_4^{2-}
P	Mac	M	<0.2%	0.2–0.5%	NT	$H_2PO_4^-$ HPO_4^{2-}
Mo	Mic	(M)	<0.1 ppm	0.1–0.5 ppm	>0.5 ppm	MoO_4^{2-}
B	Mic	(I)	<10 ppm	10–100 ppm	>100 ppm	BO_3^- $B_4O_7^{2-}$
			Metallic Ions			
K	Mac	M	<1.0%	1.0–5.0%	NT	K^+
Ca	Mac	I	<0.1%	0.1–1.0%	NT	Ca^{2+}
Mg	Mac	M	<0.1%	0.1–0.4%	NT	Mg^{2+}
Fe	Mic	I	<50 ppm	50–250 ppm	NT	Fe^{2+}, Fe^{3+}
Mn	Mic	(M)	10–20 ppm	20–300 ppm	>300 ppm	Mn^{2+}
Cu	Mic	I	3–5 ppm	5–20 ppm	>20 ppm	Cu^{1+}, Cu^{2+}
Zn	Mic	I	15–20 ppm	20–100 ppm	>400 ppm	Zn^{2+}
Ni	Mic	M	0.1–1 ppm	1–10 ppm	10–50 ppm	Ni^{2+}
			Halides			
Cl	Mic	(M)	<0.2%	0.2–2.0%	>2.0	Cl^-

[a] Mac = macronutrient, Mic = micronutrient.
[b] M = mobile, I = immobile, ( ) = uncertain relative to mobility.
[c] % = % of dry weight.
[d] NT = Nontoxic relative to direct effects.
[e] NH_4 is a hydride and not an oxide.

Source: Developed from information in Bennett 1993; and A. D. M. Glass, *Plant Nutrition: An Introduction to Current Concepts,* 1989, Sudbury, MA: Jones and Bartlett Publishers. www.jbpub.com. Reprinted with permission.

forms. Although these elements are considered to be essential for most photosynthetic plants, there are other elements that are essential and/or beneficial to a few plants relative to growth and development. Some of these elements include Na, Si, Va, and Se and will be addressed later in this chapter.

The elemental composition of plants will differ with plant part, stage of development, species, concentrations of elements in soils, and time of year. Usually, elements are concentrated in plant tissues at levels that are several magnitudes higher than those in the soil and may also reflect the nutrient composition of the soil environment. Table 3.13 illustrates how different species and classes of plants growing in a similar habitat differ relative to the accumulation of different ions. The fact that some ions accumulate in plant tissues that are not considered essential for the nutrition of plants (Cd, Pb, and Cr) illustrates that frequently the composition of the plant reflects the soil environment in which it grows. In some cases, however, plants such as *Armeria maritima* and *Minuartia verna* have been specifically associated with high levels of Cu and are useful as indicator plants for detecting high levels of elements such as Cu in soils. Table 3.14 illustrates how specific tissues in a single organism can vary with respect to the relative concentration of elements. It is interesting to note in this table that concentrations of Al and Fe are several-fold higher in the young roots compared with any other tissues of the pine analyzed. Since these elements can be toxic to photosynthetic tissues at high levels, this may reflect a mechanism whereby the plant can isolate these elements from more sensitive tissues. Figure 3.11 illustrates how concentrations of different elements change in the leaves of deciduous trees. The increase

TABLE 3.13 Trace Element Composition of Different Plant Species/Groups Growing in the Same Forest Ecosystem[a] Site

	Elements (ppm dry wt)								
Plants	B	Cd	Cu	Fe	Mn	Pb	Zn	Cr	Ni
Grass: *Agrostis alba*	3.4	0.6	4.2	80	740	1.2	59	1.0	1.8
Clover: *Trifolium pratense*	9.0	0.7	6.0	115	136	2.8	99	1.0	2.2
Plantain: *Plantago major*	7.0	1.9	9.8	135	100	2.4	97	1.4	3.0
Mosses:									
Polytrichum juniperinum	3.4	0.8	9.2	800	176	22.4	69	2.0	2.0
Entodon schreberi	3.2	0.7	10.3	425	180	13.0	77	2.8	1.6
Lichens:									
Parmelia physodes	2.4	0.4	5.0	1100	62	17.0	78	3.2	4.8
Lobaria pulmonaria	2.4	0.5	7.5	1450	66	28.0	74	3.2	2.4
Edible fungi:									
Cantharellus cibarius	4.0	1.0	24.5	49	19	1.2	150	0.4	2.2
Leccinum scabra	0.8	2.7	18.0	44	6	<0.1	125	0.4	1.8
Inedible fungi:									
Tylopilus felleus	3.5	1.6	35.0	50	14	0.4	180	0.4	1.2
Russula veternosa	6.4	1.0	32.0	28	18	1.0	175	0.4	1.0

[a] Pine and birch forest on light sandy soil near Warsaw, Poland.

Source: Reprinted with permission from Kabata-Pendias. 1992. *Trace Elements in Soils and Plants*, 2nd ed. Copyright Lewis Publishers, an imprint of CRC Press, Boca Raton, Florida.

TABLE 3.14 Trace Element Composition of Different Pine Tissues

Plant Tissues	Element[a]										
	Al	B	Co	Cr	Cu	Fe	Mn	Ni	Pb	Ti	V
Needles											
1 year old	400	18.0	0.9	4.8	4.2	150	430	6.0	0.2	15	0.6
Older	200	24.0	0.8	4.0	2.5	370	740	2.1	0.5	30	1.2
Branches	400	6.0	0.6	1.6	3.0	650	430	1.1	0.6	25	1.8
Knots	120	4.5	0.2	0.8	1.2	78	185	0.3	0.1	6	0.8
Bark	230	4.5	0.4	1.0	2.0	100	123	0.4	0.3	15	2.8
Wood	7	0.9	0.1	0.3	0.6	5	61	0.3	0.1	1	0.2
Roots											
1 mm diameter	1430	6.5	0.1	0.9	3.5	7171	134	1.1	0.3	46	0.6
5 mm diameter	82	3.2	0.7	0.6	1.2	46	50	0.4	0.1	6	0.5

[a] Samples from a pine forest on old alluvial sands in Ukraine, USSR.
Source: From Dolobovskaya 1975 as modified by Kabata-Pendias. 1992. *Trace Elements in Soils and Plants,* 2nd ed. Copyright Lewis Publishers, an imprint of CRC Press, Boca Raton, Florida.

in Ca and the reduction of N, P, and K probably reflect the greater mobility of the latter three elements from leaves late in the growing season and the lack of mobility of Ca.

Essential elements may be classified into the following general categories of function:

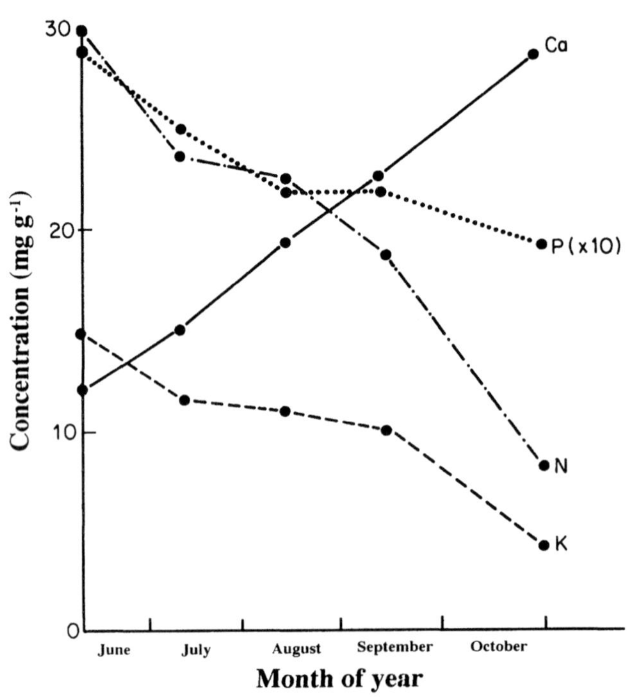

Figure 3.11. Mean seasonal changes in elemental composition of nine broad-leaved deciduous trees. (From Alway et al. 1934, as modified by Fitter and Hay 1987.)

1. structural (components of all organic cellular constituents)
2. osmotic (controlling cell turgor and water relations)
3. electroneutrality (maintaining optimum ionic and pH balance)
4. transport (assisting in mobilization of organic and mineral nutrients)
5. cofactors or enzyme constituents (catalyzing biochemical reactions)

B. Structural Functions

Carbon, H, O, N, S, and P are all supplied to plants as oxides (see Table 3.12). Carbon, H, and O comprise approximately 94% of the dry weight of plant tissues (44%, 6%, and 44%, respectively). Table 3.15 summarizes the percent mineral composition of the main classes of biochemical components of corn. By far the major components are C, H, and O. Virtually all organic compounds in plants have these three elements as components with the exception of carotenes and aliphatic hydrocarbons, which have only C and H. Figure 3.12 shows examples of representative compounds containing C, H, and O in the three major groups of organic molecules: carbohydrates, lipids, and proteins.

Nitrogen, S, and P together comprise about 3% of the total dry weight of plants (2%, 0.5%, and 0.4%, respectively). Nitrogen is available to plants from the soil in the form of NO_3, NH_4, urea, and in legumes and some grasses via microbial fixation of N_2 through symbiotic associations. After NO_3/NH_4 is absorbed by the roots, it ultimately is incorporated into low-molecular-weight amino acids, amides, peptides, amines, and ureides (Figure 3.13). However, NO_3 has to be reduced to NH_3 before this can occur. Figure 3.14 illustrates how NO_3 and NH_4 are assimilated in plants. Note that NO_3 has to be reduced to NH_3 before it can be introduced into carbon skeletons supplied by the Krebs cycle. This process requires an expenditure of energy, unlike the absorption of NH_4, which does not have to be reduced. The initial N-containing compounds produced are further used to synthesize higher-molecular-weight substances utilized in a myriad of biochemical and physiological processes. Representative examples of some important N-containing compounds in plants are

TABLE 3.15 Whole-Plant Analysis of Corn Illustrating Elemental Composition of Major Organic Constituents

Element	Carbohydrates	Fats	Protein	Nucleic Acid	Plant
O	51.4	11.3	24	30	44.4
C	42.1	76.5	52	39	43.6
H	6.5	12.2	7	3.5	6.2
N	—	—	16	17	1.5
S	—	—	0.1	—	1.2
K	—		—	—	0.9
P	—	—	—	10	0.2
Mg	—	—	—	—	0.2
Fe	—	—	—	—	0.1
Other	—	—	0.9	0.5	1.7
Totals	100	100	100	100	100

Source: From E. Epstein, *Mineral Nutrition of Plants: Principles and Perspective.* Copyright © 1972 by John Wiley & Sons, Inc. Reprinted by permission of John Wiley & Sons, Inc.

(A) Carbohydrates (C, H, O)

Galactose: a monosaccharide

$C_6H_{12}O_6$

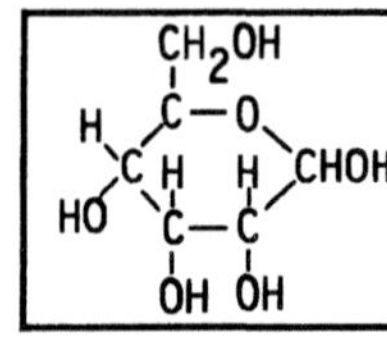

Sugars serve as energy sources and metabolic intermediates

(B) Proteins (C, H, O, N)

Glycine: an amino acid

$C_2H_5O_2N$ NH_2CH_2COOH

Amino acids are required for enzyme sytnthesis

(C) Lipids (C, H, O)

Eicosadienoate: an unsaturated fatty acid

$C_{20}H_{38}O_2$ $CH_3(CH_2)_7CH = CH - CH = CH(CH_2)_7COOH$

Fatty acids serve as storage products and membrane components

Figure 3.12. Representative classes of major organic compounds occurring in plants that contain C, H, and O.

given in Figure 3.15. As indicated in this figure, N is a component of organic molecules at numerous levels of biochemical function (i.e., nucleic acids, amino acids, enzymes, reducing power, hormones, and secondary plant products that are critical in support of metabolic processes such as photosynthesis and respiration).

Sulfur is primarily found as a component of the S-containing amino acids methionine and cysteine. These two amino acids are important in forming disulfide bonds, which are involved in the maintenance of stability and tertiary structure of proteins. Methionine is also a precursor of the phytohormone ethylene. As a component of cysteine, S is part of several Fe-S proteins that serve as electron donors to $NADP^+$ (as a component of ferredoxin in the light reactions of photosynthesis), nitrite reductase, sulfite reductase, N_2 reduction, and glutamine:2-oxoglutarate aminotransferase (GOGAT). In addition, S is a component of coenzyme A (involved in the conversion of pyruvate to acetyl CoA in respiration), biotin (required for fatty acid synthesis), thiamine pyrophosphate, glutathione (part of the ascorbic acid antioxidant pathway

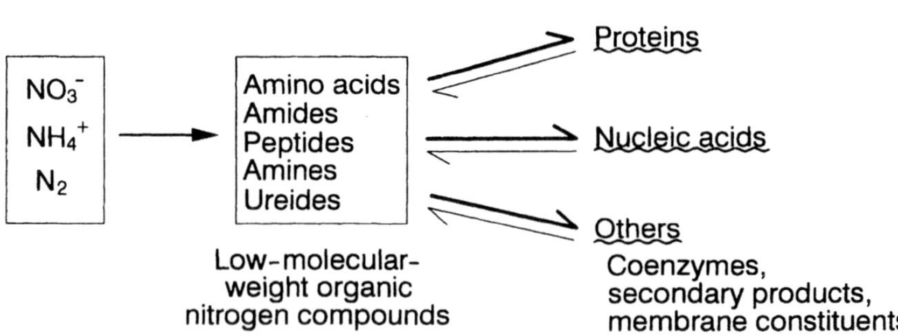

Figure 3.13. General categories of nitrogen-containing compounds found in plant tissues. (From Marschner 1995, with permission.)

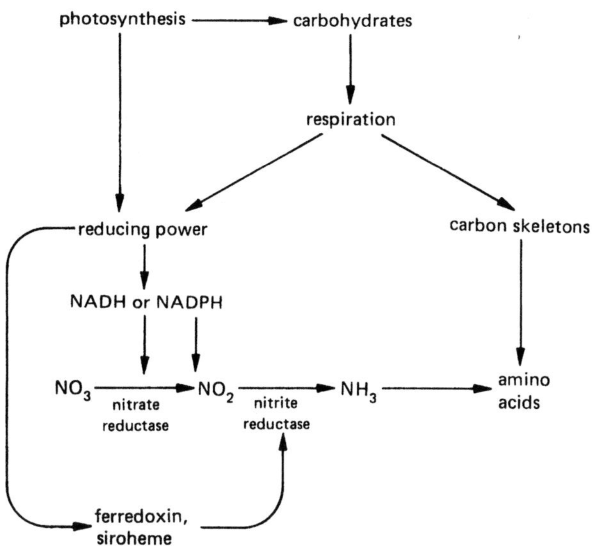

Figure 3.14. Generalized scheme for NO_3 reduction to NH_3 and subsequent incorporation into amino acids. (From *Plant Physiology, 4th edition*, by Devlin/Witham. © 1983. Reprinted with permission of Global Rights Group, a division of Thomson Learning. Fax 800 730-2215.)

that scavenges oxidative free radicals, prevents destructive oxidation reactions in plants, and is a precursor of phytochelatins that chelate toxic and low levels of metals to decrease or increase nutrient availability to plants), sulfolipids (prevalent in thylakoid membranes, regulators of ion transport across membranes, and correlated with salt tolerance in some plants) and secondary S compounds (alliins and glucosinolates, which give the characteristic odors to *Allium* species and plants in the family Brassicaceae). Figures 3.15c–e and 3.16 illustrate some of the aforementioned structural and chemical functions of S in plants.

Phosphorus is absorbed in plants as H_2PO_4 and remains in plants as free inorganic phosphorus (P_i) or as carboxyl or pyrophosphate esters. It forms ester linkages between nucleotides of nucleic acids, is important in energy transformation reactions involving AMP/ADP/ATP and $NAD(P)^+$, and is a component of phospholipids that comprise biomembranes (may aid as a divalent cation transporter in membranes). P_i regulates starch synthesis and carbohydrate transport in the chloroplast, is a component of numerous sugar intermediates of the Calvin-Benson cycle, and is concentrated in some seeds in the form of phytic acid (hexainositol phosphate), which may also bind with other ions such as Ca and Mg. Figure 3.17 illustrates important organic molecules and physiological and biochemical processes in which P plays a structural role. Table 3.16 illustrates the concentration of important P-containing compounds in the aquatic plant *Spirodela* and their synthesis and turnover rates. It can be seen that the amounts of RNA, phospholipid, and Pi are the highest in this plant, as are the turnover rates for RNA and DNA. Synthesis rates are higher for ATP.

C. Osmotic Functions

Potassium is a very important element relative to regulating turgor in plants. As indicated earlier, this element is precisely regulated relative to the concentrations in the vacuole versus the cytoplasm. The levels of K in the plant are important in regulating

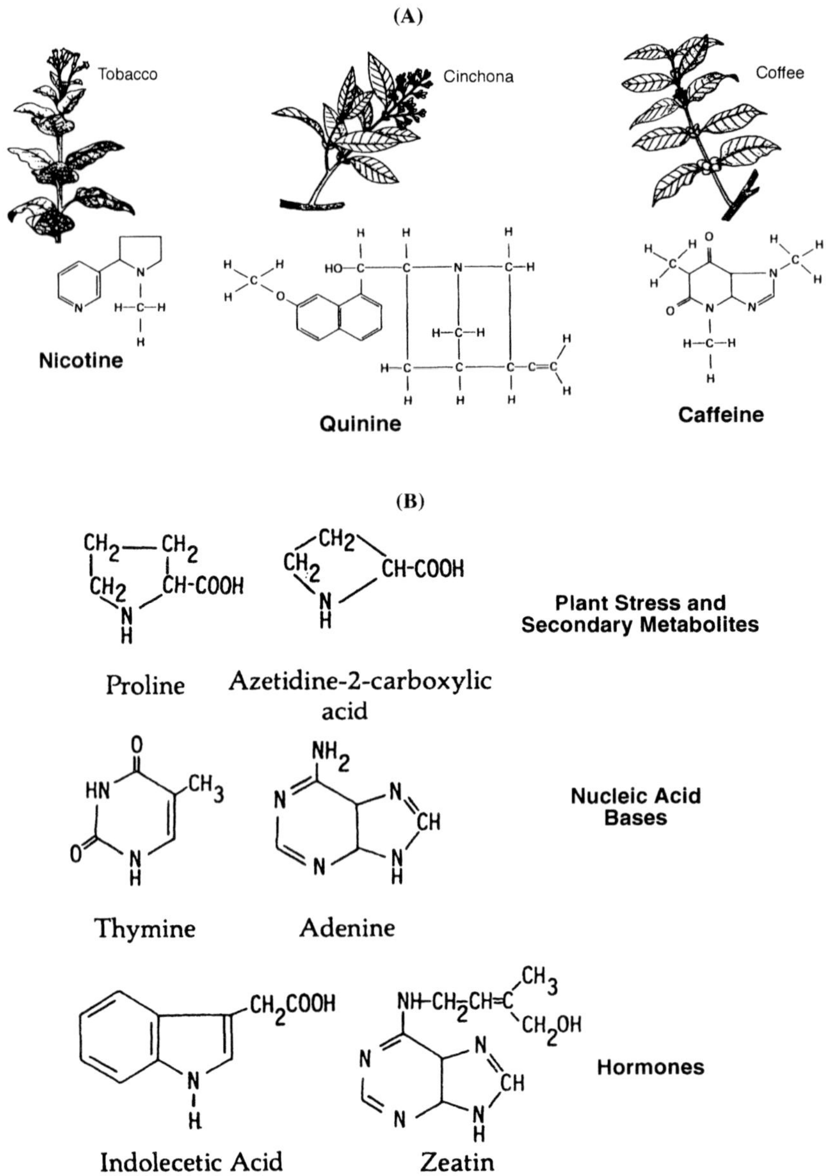

Figure 3.15. (A) Several common nitrogen containing alkaloids found in plants. (From Ehrlich and Raven 1967, with permission.) (B) Important nitrogen-containing compounds that occur in plants and are involved in plant stress and secondary metabolism, cell division and growth and development. (C) Biotin (vitamin H) is a N-S-containing molecule that is important in fatty acid synthesis. (From Ucko 1986, with permission.) (D) Coenzyme A (vitamin B_1) is a N-S containing compound required in the conversion of pyruvate to acetyl CoA. (From Ucko 1986, with permission.) (E) As with coenzyme A, thiamine pyrophosphate is an additional cofactor required in the synthesis of acetyl CoA from pyruvate. (From Mahler and Cordes 1966.)

(C)

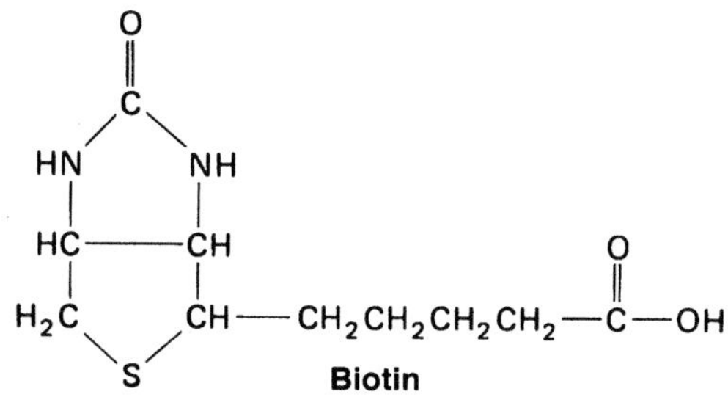

Biotin

(D)

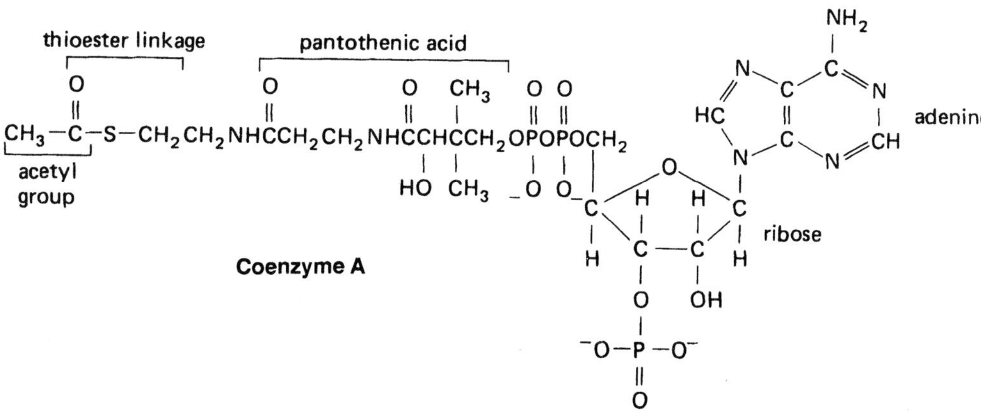

(E)

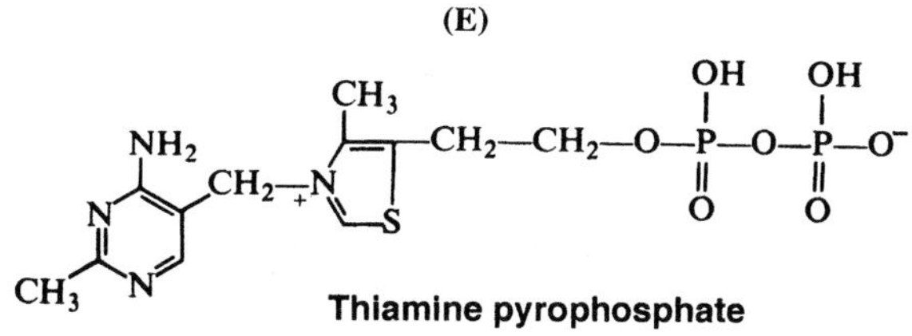

Thiamine pyrophosphate

Figure 3.15. (*Continued*)

changes in water potential between the guard cells and subsidiary cells in leaves and, thus, controlling stomatal opening and closing (Figure 3.18). In some plants, Cl may act in conjunction with K to control osmotic relations between guard cells and subsidiary cells. Potassium is also involved in regulating leaf movements in response to touch (*thigmotropism*), such as those observed in the sensitive plant. Calcium may also be involved with signal transduction in this response.

Soil salinity occurs as a result of high concentrations of NaCl, $NaCO_3$, $NaSO_4$, or salts of Mg and Ca. Plants adapted to these high-saline environments use high salinity to their advantage by storing abundant ions in the vacuole and using them to balance the water potential within the plant against that of the soil. Calcium compartmentation is metabolically essential, with high levels of Ca being stored in the vacuole compared with the cytoplasm. Plants adapted to alkaline soils, rich in $CaCO_3$, may store excess Ca in the vacuole to help in osmotic regulation and utilize what K is available for

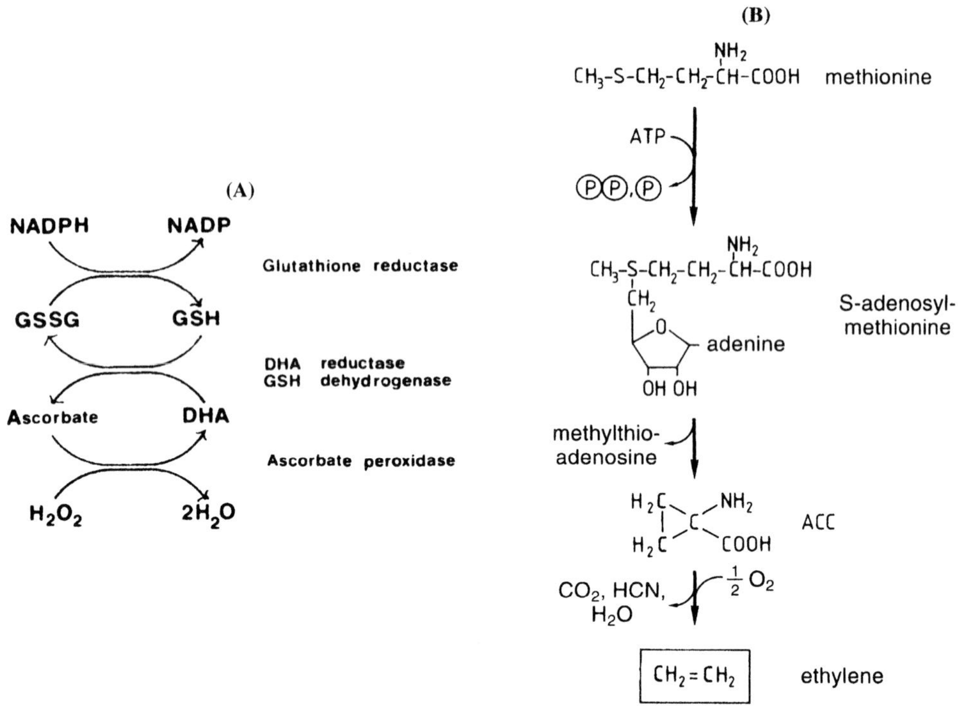

Figure 3.16. (A) Ascorbate-glutathione pathway showing the peroxidation of H_2O_2 as well as reduced (GSH) and oxidized (GSSH) glutathione-containing sulfur. H_2O_2 can be highly toxic to plants and is a by-product of the function of SOD, which is important in removing free-radical oxygen species. (From Smith et al. 1989.) (B) The sulfur-containing amino acid methionine is used as a precursor in the production of the plant hormone ethylene. (From Mohr and Schopfer 1995, with permission.)

other essential metabolic functions. Also, under NO_3-reducing conditions, high levels of organic acids produced have to be balanced either with counterions or converted to oxalic acid, which can be stored as osmotically inactive Ca oxalate in the vacuole. This, in essence, provides an effective way to control ionic and/or osmotic balance within plant cells. Since Ca is an integral component of cell membranes, it is important in the regulation of permeability. Calcium and Na compete for membrane sites, and soils where Na/Ca ratios are high show an increase in membrane permeability and leakage of ions and other important solutes from cells. Thus, Ca affects osmotic relations through the maintenance of membrane integrity and ion fluxes.

Although NO_3 is not typically viewed as an ion involved with osmotic adjustment, it is stored principally in the cell vacuoles of older leaves. As alluded to earlier, NO_3 is important in the circulation of K as ion pairs from roots to shoots and could potentially impact the osmotic relations directly or indirectly throughout the plant system. In addition, plants that are deficient in K tend to accumulate polyamines (e.g., putrescine, spermidine, spermine, cadaverine), which are polycationic nitrogen-containing molecules and are thought to replace the function of K with respect to adjusting osmotic balances. They may also be important in increasing the drought tolerance in plants, either by retaining water within the plant or by stabilizing the plant membranes under dehydration stress.

D. Ionic Balances

The maintenance of ionic balances within plant cells is important for several reasons. Metabolically, plants require specific ions for normal biochemical and physiological functions, and at the same time ions are needed to maintain osmotic relations required for growth and turgidity, the establishment of electropotential gradients to facilitate nutrient and ion transport, and the maintenance of physiological pH in various compartments of the cell. Plants appear to try and maintain a type of homeostasis within cells as well as throughout the whole plant with respect pH, cellular ion concentrations within compartments (i.e., vacuole, cytoplasm, mitochondrion, chloroplast), and perhaps tissue concentration. The ability of the plant to establish and maintain homeostasis depends on numerous metabolic and environmental factors including the types and availability of ions in the soil. The electropotential differences across membrane systems that separate compartments in plants is important in establishing pH differences and ion balances on respective sides of membrane systems. Electropotential gradients are important in the transport of ions between compartments. Figure 3.19

(A)

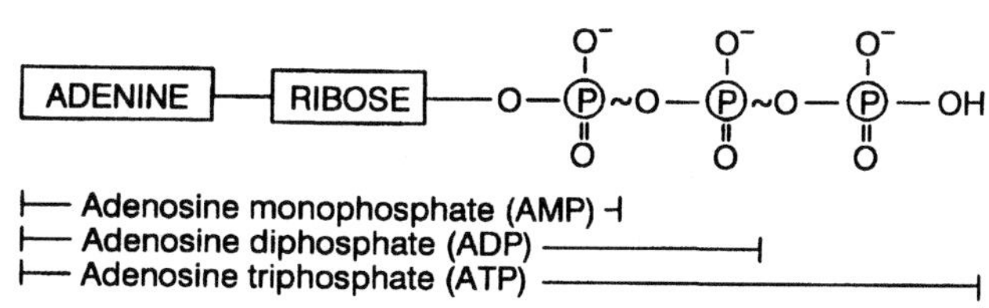

(B)

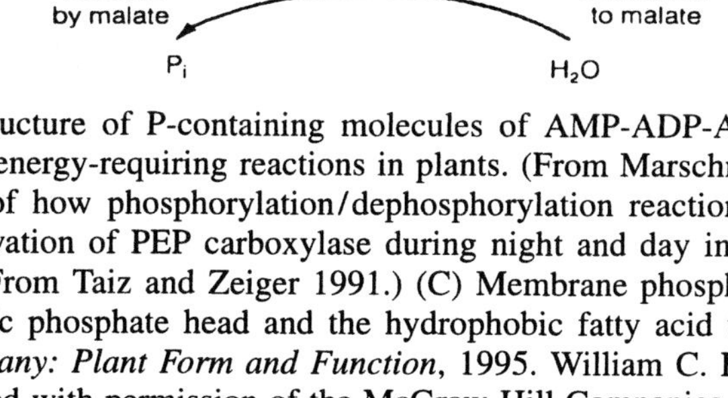

Figure 3.17. (A) Structure of P-containing molecules of AMP-ADP-ATP used in numerous phosphorylation and energy-requiring reactions in plants. (From Marschner 1995, with permission.) (B) Example of how phosphorylation/dephosphorylation reactions are involved in the activation and deactivation of PEP carboxylase during night and day in the crassulacean acid metabolism plants. (From Taiz and Zeiger 1991.) (C) Membrane phospholipid molecule illustrating the hydrophilic phosphate head and the hydrophobic fatty acid tails. (From R. Moore and W. D. Clark, *Botany: Plant Form and Function*, 1995. William C. Brown Publishers, Dubuque, IA. Reproduced with permission of the McGraw-Hill Companies.) (D) Phytic acid molecule showing esterification of hydroxyl groups of myoinositol with phosphate. Phytic acid concentrations are high in grains and seeds and function to store P, Fe, and Z. (E) Diagram illustrating the role of Pi in the regulation of starch and sucrose synthesis in the chloroplast and cytoplasm, respectively. High Pi inhibits starch formation and enhances transport of triose phosphates out of the chloroplast and used to produce sucrose. (From Walker 1980, with permission; as modified by Marschner 1995.)

(C)

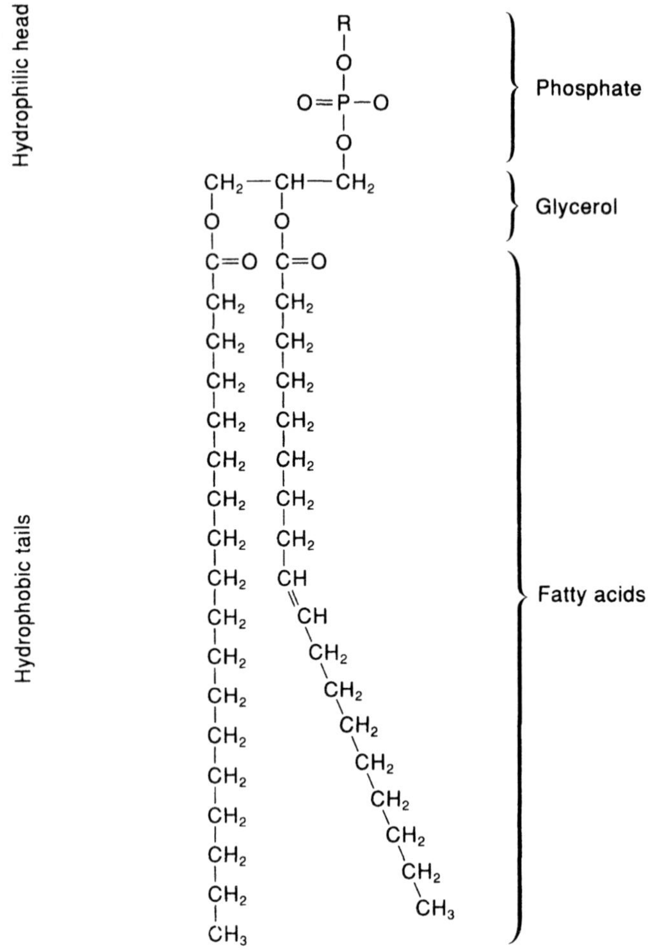

(D)

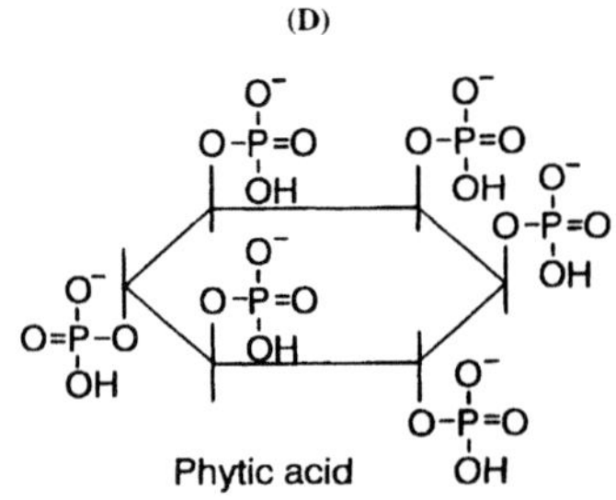

Figure 3.17. *(Continued)*

(E)

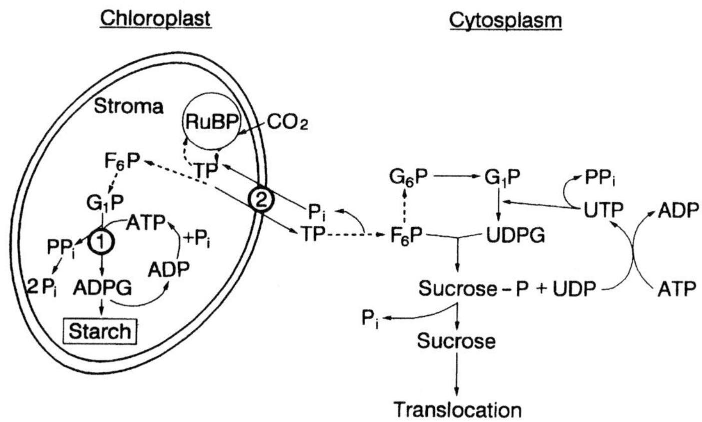

Figure 3.17. *(Continued)*

illustrates how the chloroplast, mitochondrion, vacuole, and cytoplasm are coupled relative to energy relations and ion transport among these four important compartments of the cell. Note that the chloroplast and mitochondrion provide the necessary ATP required for establishing proton gradients in the cytoplasm, the extracellular space, and the vacuole that are important in the absorption of ions into the cell and storage in the vacuole. Thus, there is a fine line between the need to establish and maintain pH, and the requirement for ions needed for metabolic functions. Many different ions are involved in this process, but some important ones include OH, H, K, Na, Cl, NO_3, H_2PO_4, SO_4, and also organic acids, such as malic and aspartic acid.

TABLE 3.16 Synthesis and Turnover Rates of P-Containing Compounds in *Spirodela*

Phosphorus Fraction	Amount (nmol g^{-1} fresh wt)	Turnover Time (min)	Synthesis Rate (nmol P g^{-1} min^{-1})
	From Tissue Analysis		
ATP + UTP (γP)	170	0.50	340
NTP + NDP (βP)	300	3.5	85
G6P	670	7	95
3PGA	330	6.5	50
Adenosine αP	200	ca. 160	1
Uridine αP	320	ca. 160	2
Phosphoryl choline	70	6	12
Phospholipid	2700	130	20
RNA (estimated)	4900	ca. 900	5
RNA (minimum rate)	4900	2800	2
DNA (mininum rate)	560	2800	0.2
	From Respiration Rate		
"ATP"	170	0.25	650
G6P	670	40	17
Pi	(3400)	5	650

Source: Modified with permission from Bieleski and Ferguson 1983.

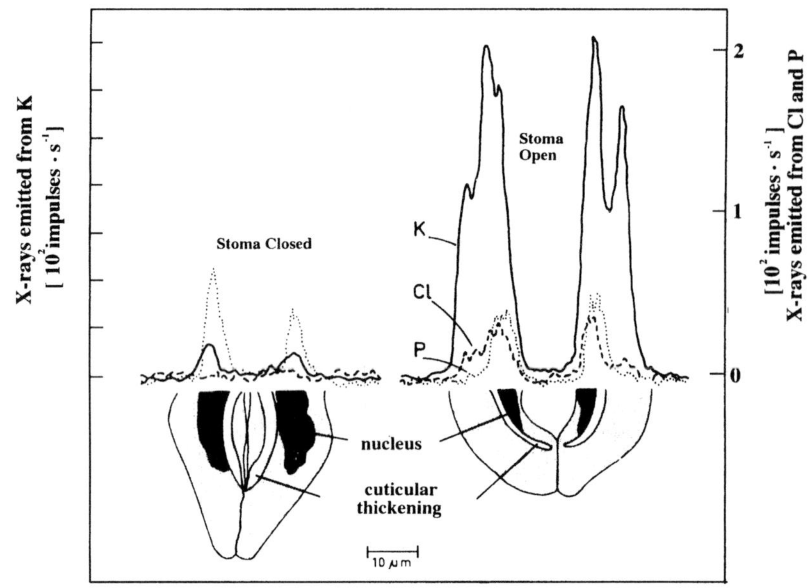

Figure 3.18. X-ray microanalysis of broad bean guard cells showing changes in concentration of K, Cl, and P in the open/turgid and closed/flaccid condition. (Modified from Humble and Raschke 1971, © ASPP. Reprinted with permission.)

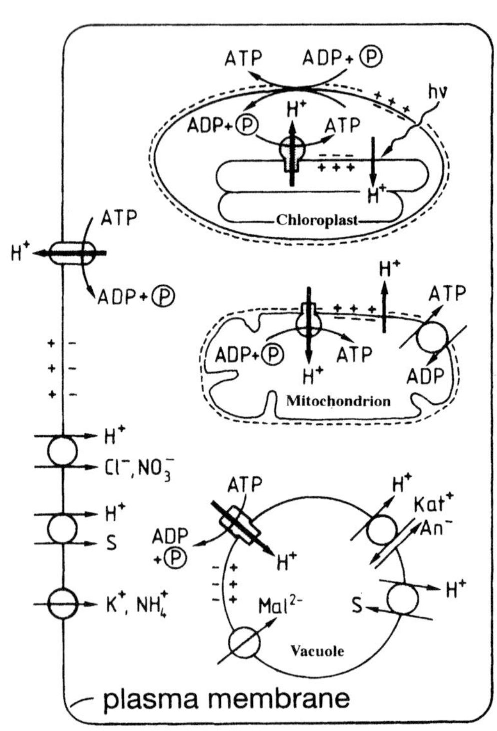

Figure 3.19. Plant cell illustrating the coupling of the chloroplast, mitochondrion, vacuole, and plasmalemma relative to ion transport. Note that ATP generated by the chloroplast and mitochondrion are essential in establishing electropotential proton gradients that provide the necessary energy to transport ions and other molecules such as sucrose (S). (Modified from Mohr and Schopfer 1995, with permission.)

Under limitations of specific ions, substitutions of ions can also occur, particularly with respect to ions principally involved in turgor relations in plants (e.g., substitution of Na for K when the latter is limiting). This suggests that plants are considerably plastic relative to the utilization of available nutrients in attempts to maintain an internal physiological balance.

E. Transport

The only ion that has been associated with a specific function of transport in plants is B. Boron forms complexes with hydroxyl radicals in compounds that have two OH groups in the *cis* configuration, as would occur in sugars, sugar alcohols, phosphate esters, and *o*-diphenols (Figure 3.20). Such associations were thought to affect the charge characteristics of the boron complex resulting in molecules that are more easily transported across membranes and through the vascular system. This is still a controversial issue that has not been completely resolved. A precise function of B is not known, but deficiencies in this element have been correlated with a broad spectrum of physiological and biochemical responses in plants including membrane permeability, phenol metabolism, cell wall synthesis, lignification, respiration, indoleacetic acid metabolism, carbohydrate metabolism, RNA metabolism, as well as a role in cell division associated with apical meristems. Figure 3.21 illustrates the complexity of these interactions. Boron does not appear to act as a cofactor or coenzyme in its function, nor does it appear to facilitate rates of enzyme reactions.

The only other roles that ions may have in transport processes have to do with chelation with organic compounds at the root surface or within the symplast/apoplast system, which could help facilitate the uptake and translocation of ions and/or organic compounds in a similar fashion suggested for B. An example of this was given earlier (Figure 3.10) in discussing the circulation a K/malate complex from the shoot to the roots via the phloem and the recirculation of K as KNO_3 via the xylem back to the shoot. In view of the ionic exchange capacities of vascular tissues, this may be very important for the transport of both mineral and organic nutrients in plants, and appears not to have been studied to a great extent.

F. Cofactors and Enzyme Constituents

All of the micronutrients (Fe, Mn, Cl, Zn, Cu, Mo, Ni) fall into the category of cofactors and enzyme constituents. These ions either function as part of a protein/enzyme structure or are required to activate an enzyme. Iron functions as a component of proteins categorized as hemoproteins (heme proteins) and iron-sulfur proteins. The heme proteins are important in oxidation-reduction reactions (cytochromes) that occur in the light reactions of photosynthesis and oxidative phosphorylation during respi-

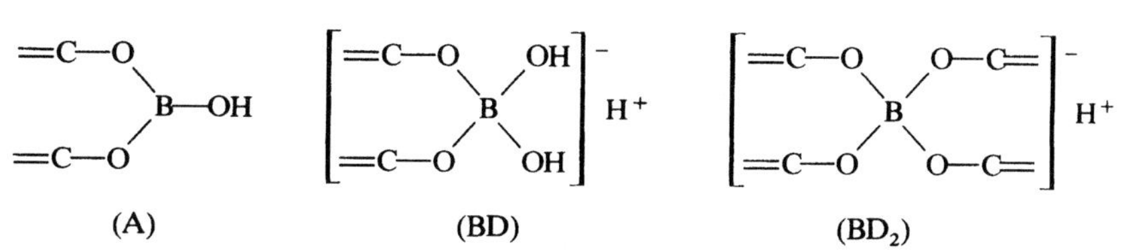

Figure 3.20. Types of chemical associations formed between borate and sugars or sugar alcohols. BD_2 occurs with high diol/borate ratios, while BD-type associations are formed with low diol/borate ratios. (From Dugger 1983, with permission.)

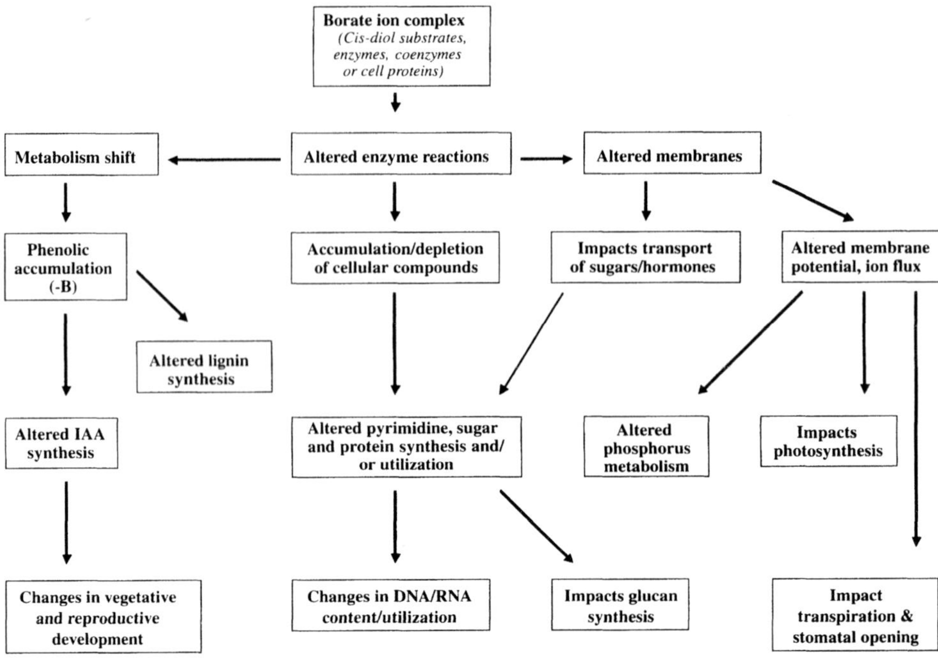

Figure 3.21. Cascading effect proposed as a result of B deficiency resulting in several key metabolic functions being altered. (From Dugger 1983, with permission.)

ration in the mitochondrion. Two important enzymes, sulfite and nitrite reductase, are considered to be heme proteins (also iron-S proteins) and are involved in the reduction of sulfate and nitrate in plants. Lehaemoglobin, another heme protein found in the nodules of legumes, is similar in structure and function to hemoglobin associated with mammals. This protein is responsible for the pink color frequently observed in the nodules of legumes and is thought to have the function of providing oxygen to support respiration in the bacteriods that exist in symbiosis with the host plant. Peroxidase and catalase enzymes are also haem proteins. An example of an important peroxidase enzyme in plants is IAA oxidase, which is involved with the destruction of indole-acetic acid (IAA) and probably represents an important mechanism in the control of the concentrations of this important hormone in plants. Catalase is ubiquitous in plants and important in the dismutation of H_2O_2 to water and O_2. This is important since H_2O_2 is a strong oxidant and reacts readily with double bonds in membrane lipids resulting in their destruction. Catalase has important functions in the dismutation of H_2O_2 produced in the conversion of glycolate to glycine and in reactions where superoxide dismutase (SOD) enzymes scavenge free-radical O_2 (at high light intensities). Ferrodxin is perhaps the most familiar iron-S protein, in that it is required for the reduction of NADP in the light reactions of photosynthesis and also functions as an electron transmitter in a number of other metabolic processes such as nitrite and sulfite reduction and N_2 reduction. Other enzymes that contain Fe include riboflavin (a possible light-absorbing pigment involved in phototropism) and superoxide dismutase. Iron is also required in the synthesis of haem coenzymes, chlorophyll, and ethylene. Figure 3.22 illustrates many of the functions of iron in plants.

Manganese serves as a cofactor for an estimated 35 different enzymes. Some of the general types of reactions these enzymes catalyze include hydrolytic, oxidation-reduction, and decarboxylation reactions. An important oxidation-reduction function is the role of Mn in the Hill reaction of photosynthesis. As a component of the oxygen evolving complex (OEC), Mn donates electrons to tyrosine of the D1 protein, which loses an electron to P680, which has a deficiency of electrons (oxidized) as a result of excitation by light. The electron deficit in Mn, associated with the OEC, results in the removal of electrons from water, resulting in hydrolysis and the formation of hydrogen and O_2. Manganese also functions to bind ATP to enzyme complexes such

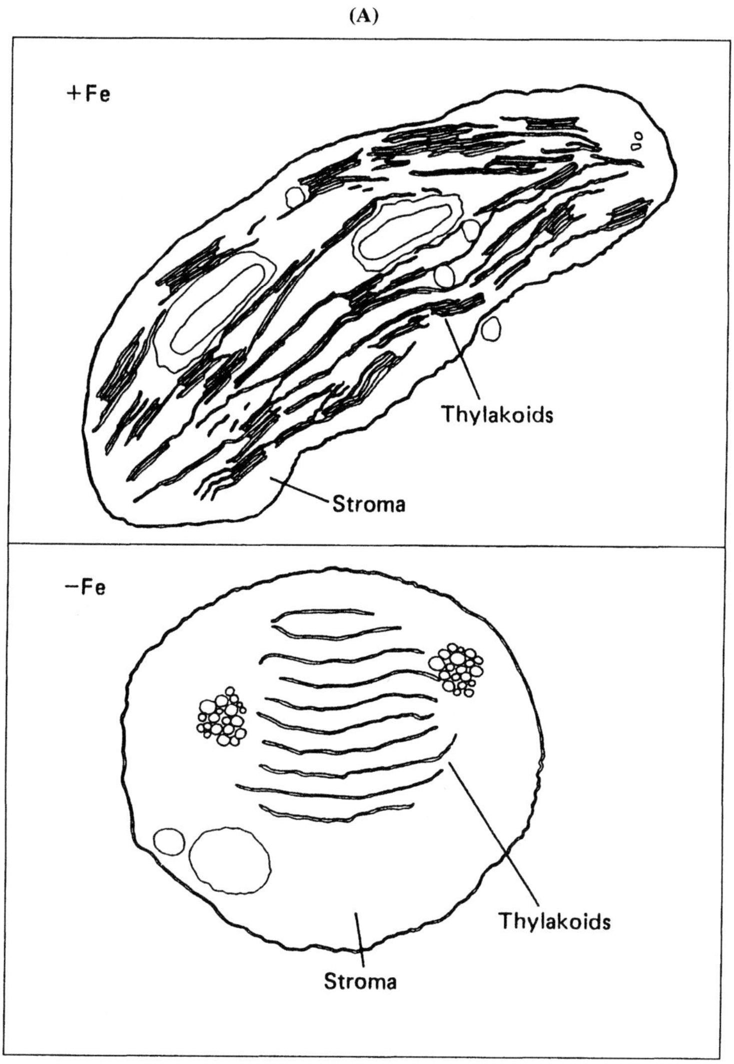

Figure 3.22. (A) Plant chloroplast showing the effects of Fe deficiency on thylakoid structure. (From Spiller and Terry 1980, © ASPP. Reprinted with permission. Modified by Terry and Rao 1991. Reprinted with the permission of Cambridge University Press.) (B) Nitrate reduction showing the role of Fe and Mo in electron transport. (From Ullrich 1983, as modified from Guerrero et al. 1981.) (C) Structure of ferrodoxin showing bonding between sulfur in cysteine and Fe. Ferrodoxin is important in redox reactions in the light reactions of photosynthesis and nitrate reduction. (From Blomstrom et al. 1964 and Tanka et al. 1964, as modified by Mahler and Cordes 1966.)

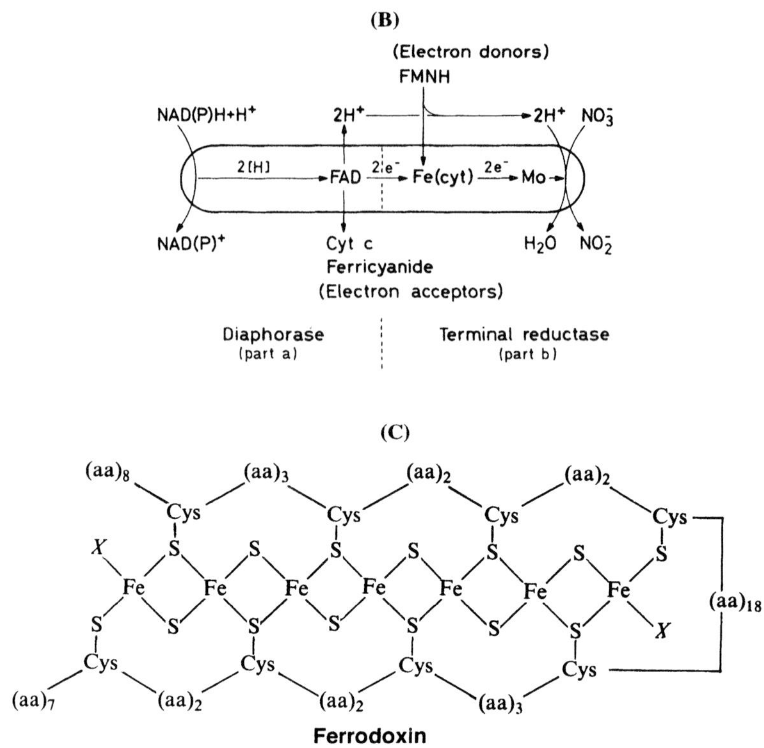

Figure 3.22. *(Continued)*

as phosphokinases and phosphotransferases. An example of an important kinase in plants is phosphofructokinase, which is important in regulating carbohydrate partitioning between sucrose synthesis and glycolysis. Manganese also has a catalytic role in several decarboxylation reactions in the Krebs cycle and activates enzymes involved in the shikimic acid pathway (aromatic amino acids, IAA, flavonoids, lignin), the terpenoid pathway (sterols, carotenoids, abscisic acid, gibberellins), and fatty acid synthesis.

Magnesium is a central atom in the structure of chlorophyll that binds the four tetrapyrrole rings of the chlorophyll molecule together via N atoms. Depending on environmental conditions, which can impact chlorophyll levels in plants, 6–25% of the total Mg in plants occurs in the chlorophyll molecule. There are numerous enzymes in which Mg either promotes or is required for activity. Some examples include glutathione synthase, PEPase, ATP synthesis, ribulose bisphosphate carboxylase, and RNA polymerase. In many reactions, Mg serves as a cofactor to bridge ATP to enzymes involved in phosphorylation reactions. It also activates dehydrogenases and enolases as well. Magnesium also stabilizes ribosomes during protein synthesis. Some enzymes may be activated or stimulated by either Mn or Mg, since they are similar in their physiochemical properties, while other enzyme reactions are very specific for one or the other. Figure 3.23 illustrates some of the roles of Mn and Mg in plants.

Chloride has an important function relative to the photolysis of water, which is somehow linked to Mn in this respect. It may function to stabilize the oxidized state of Mn associated with OEC, or it may be a functional component of the polypeptides involved with this process. Its actual function is not clear. Chloride also functions as

(A)

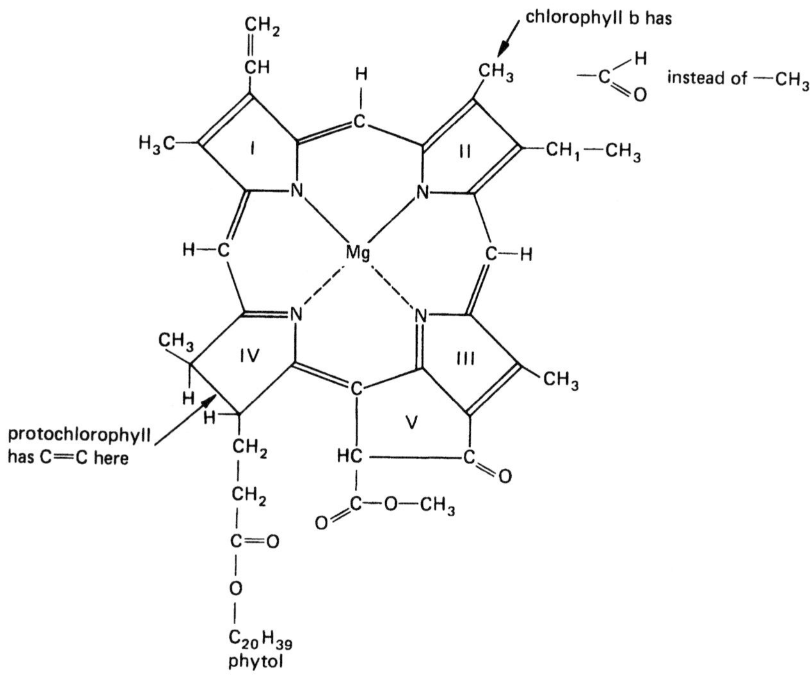

(B)

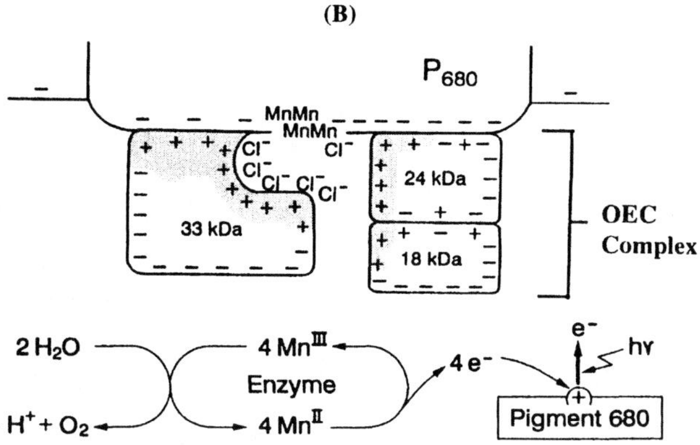

Figure 3.23. (A) Chemical structure of chlorophyll a and b. Note the Mg atom connecting the four N atoms of the tetrapyrrole rings. (From Bidwell 1979.) (B) Schematic and chemical representation of the role of Mn in the transfer of electrons from water during photolysis to electron-deficient reaction centers of photosystem II. Note that the function of Cl is not fully understood but may aid in the stabilization of the oxygen-evolving complex. (From Itoh and Uwano 1986, with permission; modified by Marschner 1995.) (C) Magnesium coupling of ATP to enzyme complex prior to the phosphorylation reaction. (From Mengel and Kirkby 1982, with permission.) (D) Two reactions in Krebs cycle in which enzymes are activated by either Mn or Mg. (E) Light-regulated exchange of H^+/Mg in the thylakoid and subsequent stimulation of rubisco and increased carbon fixation. Note the response is pH and Mg dependent. (From Sugiyama 1969, as modified by Marschner 1995; reprinted with permission.)

(C)

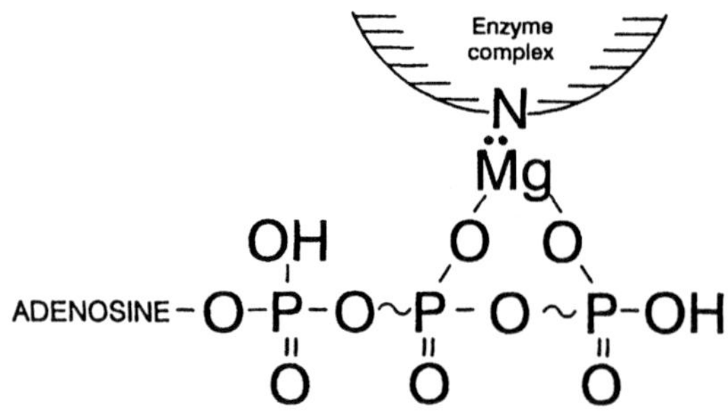

(D)

$$\text{Malate} + \text{NADP}^+ \xrightarrow[\text{Mn}^{2+},\ \text{Mg}^{2+}]{\text{Malic enzyme}} \text{Pyruvate} + \text{NADPH} + \text{H}^+\ \text{CO}_2$$

$$\text{Isocitrate} + \text{NAD(P)}^+ \xrightarrow[\text{Mn}^{2+},\ \text{Mg}^{2+}]{\text{Isocitrate dehydrogenase}} \text{2-Oxaloglutarate} + \text{NAD(P)H} + \text{H}^+\ \text{CO}_2$$

(E)

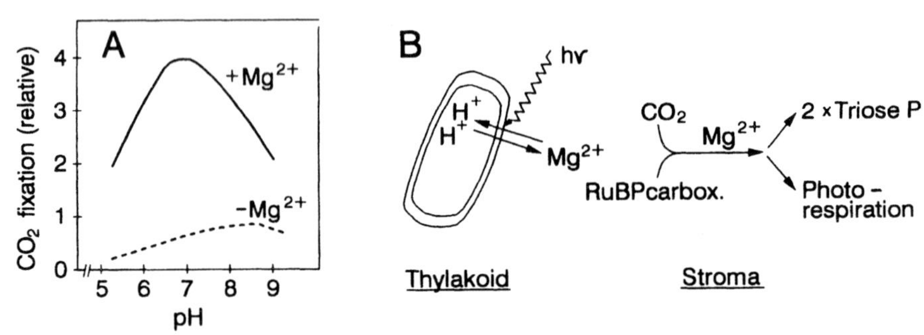

Figure 3.23. (*Continued*)

an osmoregulator in plants and, in particular, is important in the stomatal opening and closing in some plants. Related to its role as an osmoregulator, it specifically activates proton pumps in the tonoplast of plant cells. This may be important relative to the storage of high levels of this potentially toxic ion in plants. Chloride may also be important in the regulation of cell division in the growth of leaves and roots and has been shown to regulate some aspects of nitrogen metabolism such as the stimulation of asparagine synthase, which is involved in the synthesis of asparagine from glutamine. Chlorine has also been shown to be a component of 4-chloroindoleacetic acid, which occurs in plants and may have some hormonal function.

Zinc is a component of, or acts as a cofactor in, many enzyme reactions. For instance, it is involved as a component of carbonic anhydrase (CA), which has the function of hydrating CO_2 to form HCO_3, which is used as a substrate in carbon fixation reactions of C_3 and C_4 plants. Deficiencies in Zn can control the rates of activity of CA, which appears to be more critical for C_4 than for C_3 plants. Differences in the localization of this enzyme in C_3 and C_4 plants, and the high requirements for HCO_3 as a substrate for PEP carboxylase in C_4 plants, may be, in part, the reasons for the different responses in these plants. Also, Zn has been shown to be a component of alcohol dehydrogenase, DNA synthase, superoxide dismutase, alkaline phosphatase, phospholipase, carboxypeptidase, and RNA polymerase. Zinc is also thought to be important in maintaining the structural integrity of ribosomes and is some way involved in the synthesis, inhibition, and degradation of indoleacetic acid. Plants deficient in Zn accumulate starch and sugar, an effect thought to be related to the shifting of Calvin cycle intermediates away from dihydroxyacetone phosphate (DHAP) and phosphoglyceraldehyde (PGAL) and toward the production of starch and sugar. Since Zn appears to protect sulfhydryl groups from oxidation, it has the important function of protecting membranes comprised of sulfur-containing amino acids from oxidative enzymes, which also accumulate under Zn deficiency. Figure 3.24 illustrates some of the functions of Zn in plants.

Copper is a component of many enzymes in plants and as such generally functions in oxidation-reduction reactions. Plastocyanin is one example of a protein in which Cu is involved in redox reactions (Figure 3.25). This protein is important in the electron transport system of the light reactions of photosynthesis. More than half of the Cu found in plants is associated with plastocyanin in the chloroplast. Copper is also a component of the haem a/a_3 complex that comprises cytochrome oxidase in mitochondria and is thought to be involved in respiration processes involving low O_2 tensions in organisms. Copper, as a component of superoxide dismutase (SOD), functions directly in the detoxification of the superoxide radical that is produced by photosynthesis under high light intensities. This function of Cu contrasts with that of Zn in this enzyme, since the latter serves a structural role in SOD. Copper is also a component of such enzymes as ascorbate oxidase, which is involved with the oxidation of ascorbate to dehydroascorbate. This enzyme is widely distributed in plants and appears to be attached to leaf cell walls, but its function is widely unknown. It is tempting to speculate that it may be in some way connected to the ascorbate-glutathione redox chain associated with leaf chloroplasts that functions to remove the highly oxidative metabolite H_2O_2, which is produced as a result of SOD activity. Copper is a constituent of polyamine oxidases, which are involved in the metabolism and degradation of polyamines. These polycationic amines, thought by some to be plant growth regulators, increase under K deficiency and drought stress. These compounds may also function as membrane stabilizers and osmoregulating molecules in plants under stress. Phenol oxidases are enzymes that are involved in oxygenation reactions in plants in which Cu plays an important role. Polyphenol oxidases are important in lignin synthesis, the formation of protectants (phytoalexins) against plant pathogens and physical injury, and the control of auxin oxidase levels, a fact which has important ramifications relative to growth and development. The latter may be related to observations that low phenol oxidase activity as a result of Cu deficiency results in reduced flowering and delays maturity in plants. Copper may also play a role in DNA, carbohydrate, and protein synthesis, since all three decline during vegetative growth in plants. Copper may also be important in the synthesis and/or stability of chlorophyll, and there is some indication of the need for Cu in symbiotic N_2 fixation.

Only a few enzymes are known that have Mo as a component. These include nitrate reductase (Figure 3.22c), which is required in the reduction of NO_3 to NO_2; nitrogenase, which is involved in the fixation of nitrogen by nitrogen-fixing bacteria; and xanthine oxidase/dehydrogenase, which converts purines to ureides, which are important N-containing compounds in legumes. Thus, plants deficient in Mo are prone to nitrogen deficiency as well.

Nickel is usually not limiting to plant growth since it is required in very small amounts. However, in higher plants, Ni has been shown to be a component of the

(A)

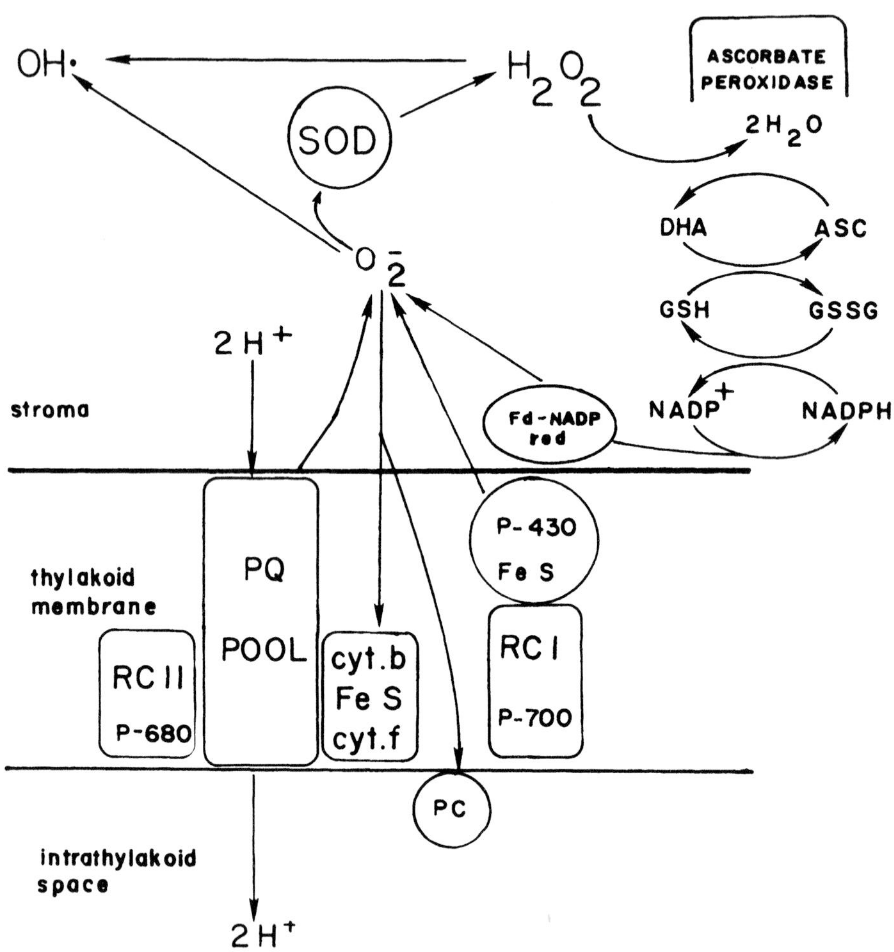

Figure 3.24. (A) Superoxide dismutase (SOD) converts free-radical oxygen species, produced by photosynthesis under high light intensity, to H_2O_2, and the latter is peroxidized in the ascorbate pathway to H_2O. Zinc, Fe, Cu, and Mn are components of different forms of SOD. (From Rabinowitch and Fridovich 1983 as modified by Salin 1988. Reproduced by permission of the American Society for Photobiology.) (B) Polypeptide showing the role of Zn in tertiary structure formation. (From Coleman 1992 and Valle and Falchuk 1993, as modified by Marschner 1995.) (C) The enzyme alcohol dehydrogenase contains Zn as a structural and catalytic component. This enzyme is produced under anaerobic conditions and may be important in detoxifying the highly toxic acetaldehyde accumulation under flooding conditions. (D) Synthesis of the phytohormone IAA from tryptophan and the proposed, but uncertain, requirement for Zn. (E) Carbonic anhydrase, a Zn-requiring enzyme, catalyzes the interconversion of various pH-dependent forms of inorganic carbon providing CO_2 for photosynthesis.

(B)

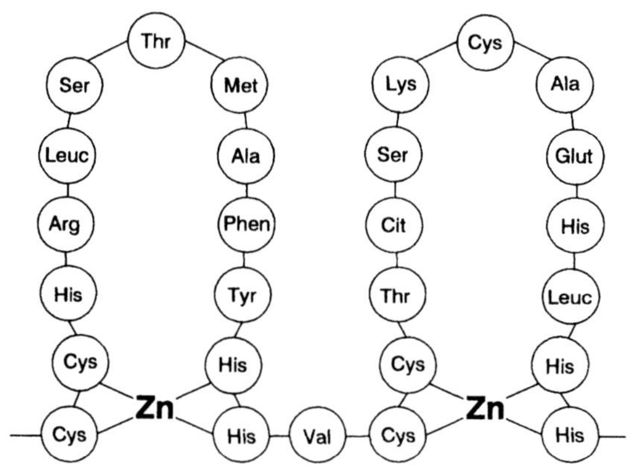

(C)

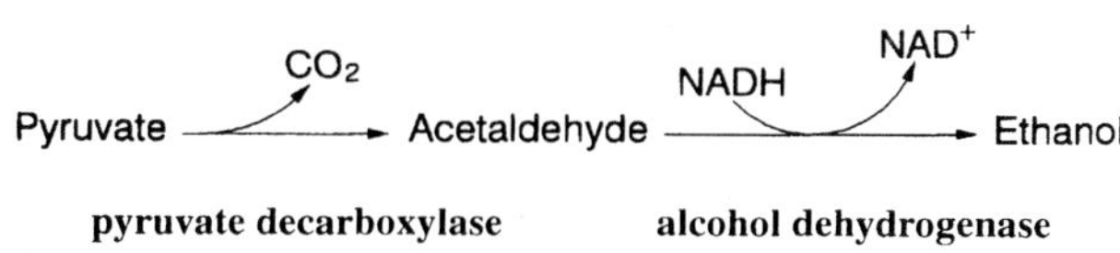

(D)

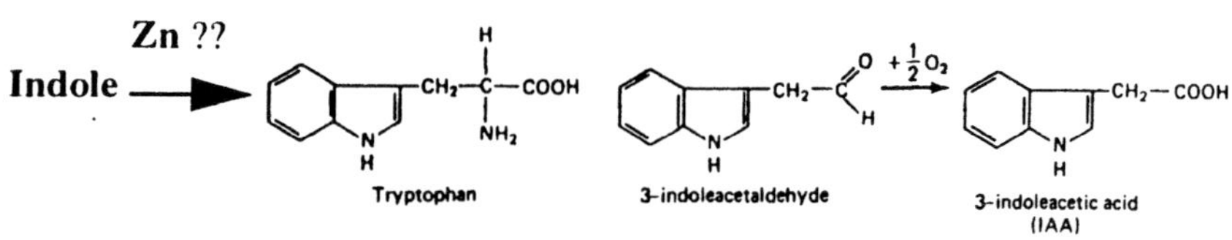

(E)

$$CO_2 + H_2O \overset{Zn}{\longleftrightarrow} HCO_3^- + H^+$$

carbonic anhydrase

Figure 3.24. (*Continued*)

enzyme urease, which is required in the hydrolysis of urea to water and NH_3. Urea is a normal metabolic end product derived from the metabolism of several nitrogen-containing compounds including arginine, agmatine, canavanine, allantoic acid, and γ-guanidinobutyrate. The accumulation of urea can be toxic to plants, thus Ni and urease are important relative to maintaining low levels of urea in plant tissues.

Calcium has a dual role in plants as an enzyme activator, and also has important structural functions in cell walls and in membrane stabilization and integrity (Figure 3.26). One of the important functions of Ca is the activation of calmodulin, an important protein involved in the regulation of nicotinamide adenine dinucleotide (NAD) kinase reactions, microsomal Ca transport, and microsomal Ca-ATPase reactions. Calcium appears to act as a second messenger via some primary messenger (i.e., light or hormones) to activate calmodulin. Regulation of Ca concentrations among compartments in the cell appears to be an important component of this process and specific

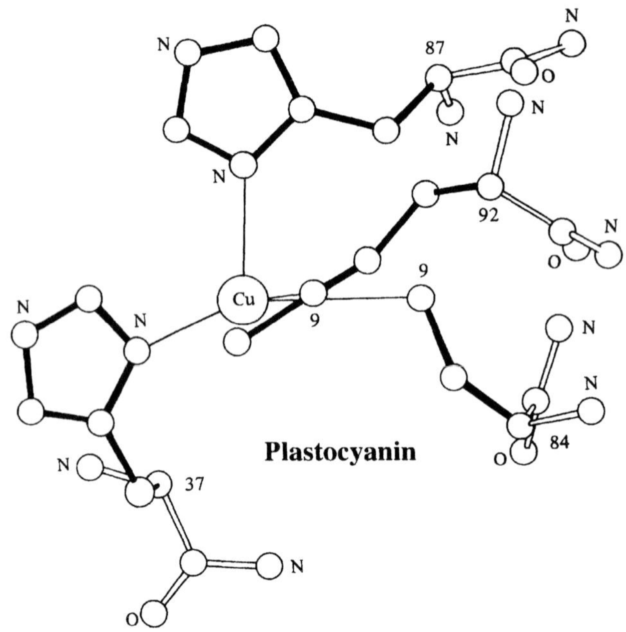

Figure 3.25. Proposed structure of plastocyanin showing Cu as an integral component. (From Coleman et al. 1978.)

membrane-bound Ca-ATPases and Ca channels are thought to be involved. The mitochondria, endoplasmic reticula, nucleus, and vacuole all appear to be organelles that are particularly high in Ca while the cytoplasm is maintained at very low levels. It appears that calmodulin is activated as a result of movement of Ca from storage compartments in response to some external stimulus. Graviperception in roots is thought to be under the control of calmodulin, which is thought to activate Ca and auxin pumps in specialized root cap cells called *statenchyma,* which in turn results in a disproportionate transport of auxin in the lower cells of the root resulting in differential growth responses. Calcium is also known to directly activate protein kinases, which are important in the phosphorylation of other enzymes such as ATPases, which are important in proton pumping. Figure 3.26a illustrates how specific signals may be involved in a cascade of events leading to a specific response involving calmodulin activation by Ca. Calcium is also a component of the enzyme α-amylase, which is important in the degradation of starch to simple sugars during the germination process in seeds. This is an example where hormones appear to regulate the movement (gibberellic acid increases, abscisic acid decreases) of Ca from storage compartments (endoplasmic reticulum) into the cytoplasm. High Ca in turn stimulates α-amylase activity.

Calcium is also a structural component of cell walls and membranes. As a component of pectate, it provides strength to the cell wall by cross-linking the pectic chains in the middle lamellae. Calcium is also an important element for regulating membrane permeability. It is thought that Ca forms linkages with the negative charges of some phospholipid head groups and, depending on the concentration of Ca, controls the permeability (fluidity) of cell membranes by the degree of condensation of membrane lipids (Nilsen and Orcutt 1996, chap. 3). Calcium is also important in cell growth in that it controls the alignment and orientation of cytoskeleton components

(A)

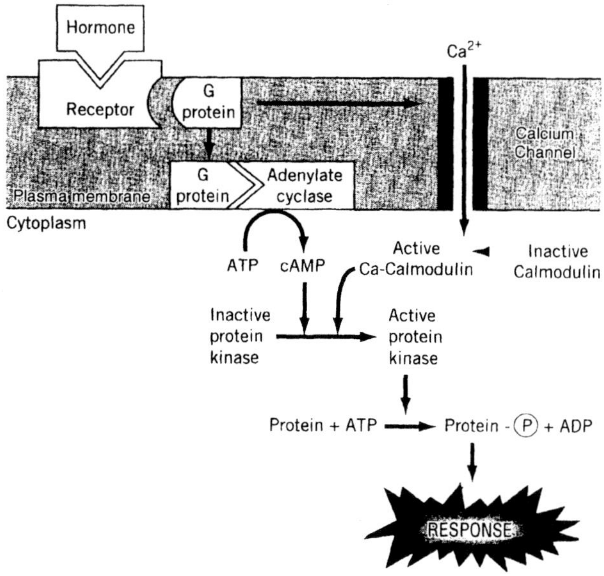

(B)

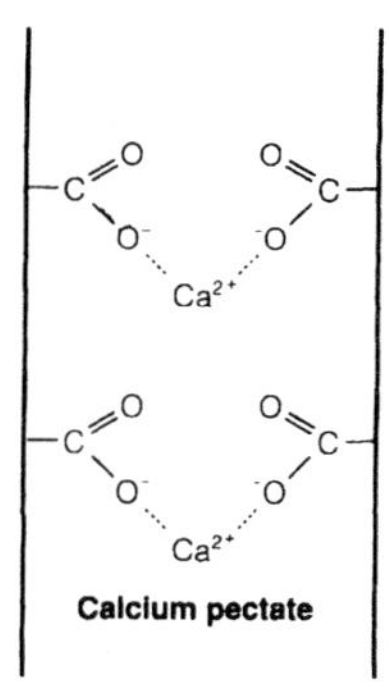

Figure 3.26. (A) Generalized model of Ca activation of calmodulin in regulating physiological response. (Modified from *Advances in Plant Nutrition,* V.1, by A. D. M. Glass and M. Y. Siddiqi. Copyright © 1984 by Praeger Publishers. Reproduced with permission of Greenwood Publishing Group, INC., Westport, CT.) (B) Role of Ca in binding pectin molecules in the middle lamella of a cell. (Modified from *Introduction to Plant Physiology* by W. G. Hopkins. Copyright © 1995 John Wiley & Sons, Inc. All rights reserved. Reprinted by permission of John Wiley & Sons, Inc.) (C) Response of *Fucus* zygote to light and the development of Ca flux leading to the establishment of polarity and cell division. (From Taiz and Zeiger 1991.) (D) Geotropic response in roots illustrating Ca-calmodulin-IAA interaction to effect differential growth response in lower root cells. (From Hasenstein and Evans 1988. © ASPP. Reprinted with permission. Modified by Taiz and Zeiger 1991.)

(C)

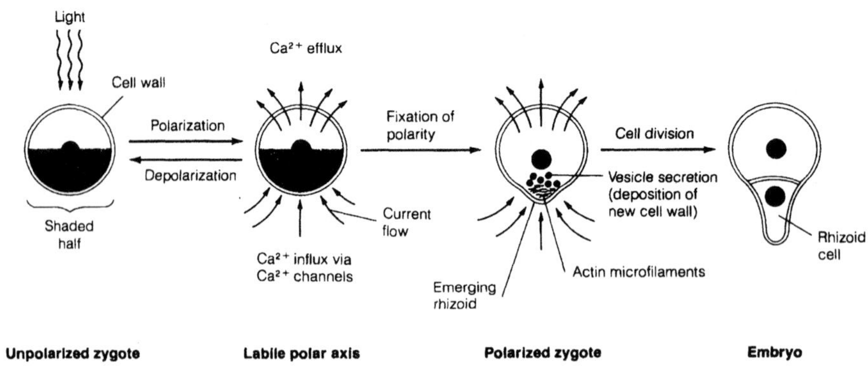

(D)

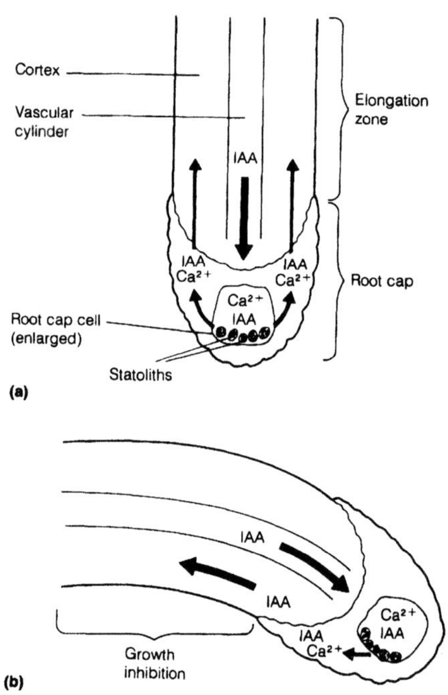

Figure 3.26. (*Continued*)

and, thus, controls the orientation of cell wall deposition. Calcium has also been shown to be involved in determining the polarization of cell growth in early embryonic stages, is required for muscilage production by root caps (which is important for facilitating root growth in soils), is needed for secretory vesicle production in the synthesis of cell walls and membranes, is important in callose production initiated in tissues sustaining physical or biotic injury, and is involved with auxin-induced acid growth resulting in loosening of the cell wall and stimulation of enzymes involved in new cell wall synthesis.

IX. FUNCTIONS OF NUTRITION IN PLANT GROWTH AND DEVELOPMENT

A. Mineral Nutrients and Cell Division and Expansion

As indicated previously (Nilsen and Orcutt 1996 chap. 4), plant nutrition (i.e., mineral nutrients), can have profound effects on phytohormone levels and function. This in turn can affect many aspects of growth, development, and differentiation processes in plants. Conversely, phytohormones such as cytokinins and auxins appear to have some influence on the directional movement of nutrients, both organic and inorganic, to sink areas in plants where high rates of growth are occurring (i.e., meristems, leaves, root tips, and fruit and seed development). Thus, there appears to be a rather complex interaction between plant hormones and the elemental nutrition of plants, and little is known about how these processes are regulated.

As with the effects of nutrient elements on hormone levels and functions in plants, mineral nutrition can affect growth and development processes indirectly through their various activities in plants as a result of regulation of numerous enzyme reactions, controlling pH, regulating turgor, and so on. If one considers the light reactions of photosynthesis alone (Figure 3.27), essentially all of the mineral elements we have discussed previously are involved in this process, and limiting any single element can impact the energy relations of the plant and thus growth and development. Sometimes the question is raised, can mineral elements affect growth and development directly, or is their effect secondary via phytohormone interactions or through limiting photosynthetic efficiency or other key metabolic processes? Obviously, the answer depends on the element in question.

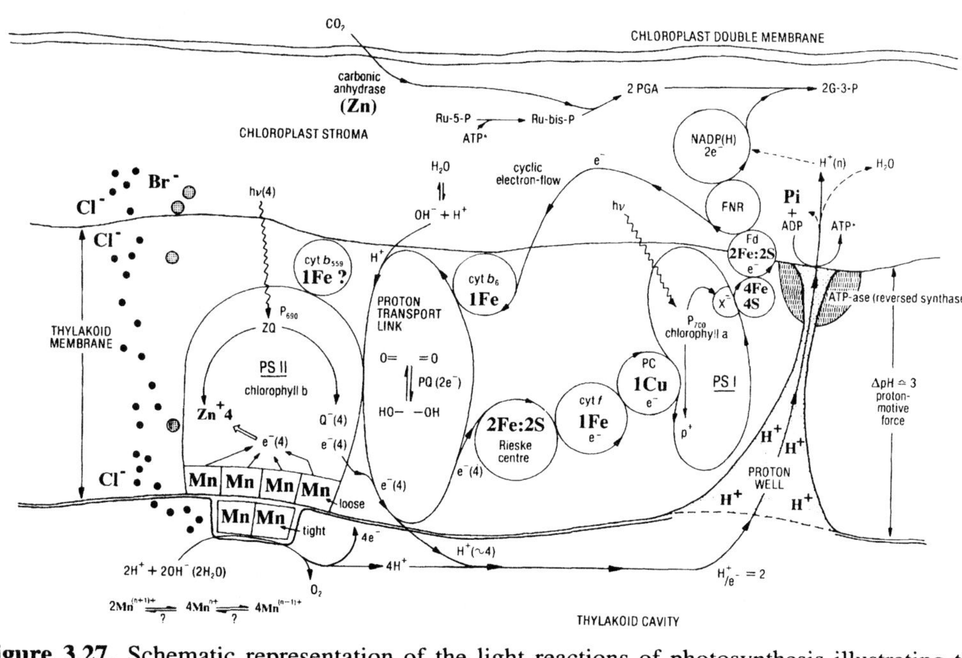

Figure 3.27. Schematic representation of the light reactions of photosynthesis illustrating the various functions of essential elements in this process. (From Bould et al. 1983.)

For example, deficiencies in elements such as Mn and Cl that are required in the photolysis of water in the light reactions of photosynthesis would suggest that limitations of these elements cause a primary effect on photosynthetic processes, thus limiting carbon fixation required for growth. In other instances, limitations in N reduced leaf area, growth of tillers, lateral branching, and increased leaf number. At the cellular level, growth reductions can occur as a result of reduced cell division and/or cell expansion. In the case of N deficiency, it would appear that both processes can be inhibited or only a single process affected, but what is interesting is that growth can be affected before photosynthetic rates decline. Whether cell division or expansion is affected may depend on when the stress is applied relative to development and cell cycle stage. Plants deficient in N also exhibit reduced leaf turgor and negative leaf water potentials. This has also been associated with reduced root hydraulic conductance. Thus, the reduced capacity to absorb water and maintain cell turgor can affect cell expansion.

Deficiencies in P have also been correlated with reduced leaf growth and may be related to reduced cell expansion since root hydraulic conductivity and cell turgor both decrease as a result of low P levels. Growth reductions also precede reductions in photosynthesis in P-deficient plants.

Calcium deficiency could affect both cell division and expansion. Calcium is thought to have a role in the polymerization of microtubules involved in the process of mitosis. Calcium-deficient plants show reduced activity of apical meristems and increased rosetting as a result of reduced cell division and expansion. Abnormal cell divisions can result in deformed appearances in leaves and lack of leaf expansion, which may be related to the correlation between Ca concentrations within cells and the plane of cell division. Cell expansion may be influenced by Ca levels as a result of Ca's function in regulating membrane permeability and stability as well as its role as a component of Ca pectate, which functions to bind cell walls together and may impact the extensibility of cell walls during cell expansion. Potassium deficiency can affect cell water potentials since it is one of the primary osmoregulating elements in plants. Thus, cellular growth by expansion can be affected by low levels of K as well as of other elements such as Na or Cl, which may be used in osmotic adjustment under certain conditions. Boron is also required for the production of nucleic acids, and deficiencies in this element can impact cell division in plants. Thus, several elements are directly involved in the growth of plants through their effects on cell division and expansion.

B. Nutrient Deficiencies and Root/Shoot Growth

Certain elements have also been shown to affect organ growth and development. For example, several nutrient deficiencies can affect root to shoot ratios, usually in favor of roots. This effect may simply favor reduced shoot growth compared to root growth, or it may actually enhance root growth. Whatever the effect on growth, it clearly is to the advantage of the plant to consolidate its resources toward root development in an effort to explore more soil area to obtain more of the limiting resource. Also, other aspects of development including numbers of leaves, tillers, lateral shoots, and flowers, and their rates of development are known to be influenced by mineral deficiencies. As an example, N deficiency can reduce the numbers of leaves produced in wheat, barley, and sugar beet, and low P and Cu can decrease leaf number in wheat. Table 3.17 shows how low N levels can cause a reduction in number and emergence rate of barley leaves. The response appears to be more pronounced in some plants compared with others, and in some instances no effect of N has been reported.

TABLE 3.17 Leaf Number and Emergence Rates of Two- and Six-Row Barley, and Time of Emergence of Flag Leaves

	Two-row		Six-row	
	High[a]	Low	High[a]	Low
Nitrogen treatment				
No. of leaves	8.6**	7.1	8.1**	6.9
Emergence rate (days per leaf)	4.9*	5.7	4.8**	5.9
Time of appearance of flag leaf (days)	42	41	39	41

[a] Significant difference between N treatments: $*p < 0.05$, $**p < 0.01$.
Source: Moorby and Besford 1983.

The impact of mineral availability on apical meristem development is relatively unexplored. Much of how the availability of elements impacts apical development may relate to how mobile a specific element is in the plant and the fact that the apical growing region in shoots is not highly vascularized. It appears that apical growing regions may try to establish and maintain a type of mineral homeostasis for high-demand nutrients such as N and P, as has been observed for Ca and K at the cellular level. Thus, under nutrient stress, senescence induction of older leaves is frequently encountered, which probably results in the transport of mobile elements from these leaves to newly developing tissues in the shoot apex to help to maintain the required balance of nutrients. Deficiencies of nutrients also can affect the rate at which organs such as leaves and lateral branches or tillers emerge and develop. As an example, as P levels increased from 25 mg P kg^{-1} soil to 300 mg P kg^{-1} soil, the rate of emergence of leaves on the main stems of barley increased (Figure 3.28).

C. Nutrient Deficiencies and Initiation of Lateral Shoots, Tillers, and Leaves

Reductions in lateral shoot growth and tillers has been shown to occur in various plants as a result of N, P, Cu, and Mn. Limitations in N is especially effective in reducing the number of tillers in grasses such as barley and wheat (Figure 3.29). Plants grown in 800 μM N had four times more tillers than plants grown in 50 μM N. It has also been demonstrated that in apple trees, the form of N applied can influence lateral shoot (spur) formation. As an example, N applied as NH_4 was more effective than N supplied as NO_3 or as a combination of NH_4 and NO_3. Low levels of Mn and Cu may either increase or decrease tiller formation. These effects may possibly be due to cultivar, species, or developmental differences. As an example of the last, increasing concentrations of Cu at the early booting stage in wheat resulted in fewer tillers but straw and grain yield increased. Such shifts in growth and developmental priorities probably reflect changes in sink strength, but may also suggest a regulatory function for Cu in controlling sink allocations, since plants grown under low Cu developed little grain yet maintained substantial tiller numbers and vegetative growth (Table 3.18). This response relates also to the effects of nutrient deficiencies on reproductive development, which will be addressed later.

Other nutrient deficiencies that have been observed to cause increased branching include Mn deficiency in lupins, Zn deficiency in citrus, and B deficiency in rape seed. Deficiencies in the latter two elements may relate to the reduction in IAA levels in plants as a consequence of the proposed requirement of Zn in IAA synthesis and the role of B deficiency in the production of IAA oxidase. Thus, reductions in IAA

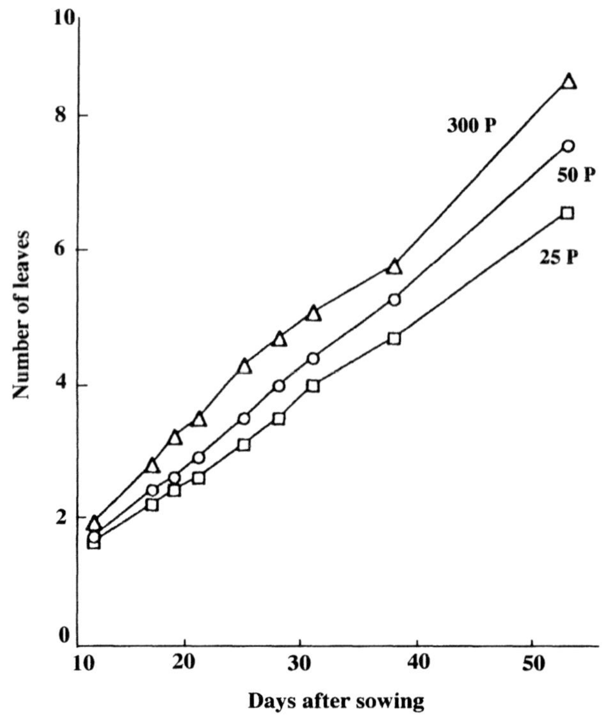

Figure 3.28. Effects of differing P levels on the rate of leaf emergence in barley. (Reprinted from Longnecker 1994, p. 151, by courtesy of Marcel Dekker, Inc.)

could release the correlative inhibition (apical dominance) of lateral growth as a result of deficiencies in these two elements.

D. Nutrient Deficiencies and Reproduction

Reproduction can be greatly affected by nutrient deficiencies. For example, low levels of N and N source can affect the number of flowers produced in apple trees. Nitrogen provided as NH_4 is more effective in flower induction than is NO_3 or combinations of NO_3 and NH_4. Low N and P levels can also promote early senescence of flowers and fruits in some legumes such as soybean, resulting in low seed yield and viability. As indicated earlier, low Cu results in reduced yields in wheat, while deficiencies in Mn, Zn, B, and Mo can result in lower yields in corn. Copper deficiency can also influence grain set in wheat (Table 3.19). Cross-pollination suggested that pollen is more sensitive to Cu deficiency than are female gametes (eggs), since pollen derived from Cu-deficient plants used to pollinate female reproductive parts resulted in lower grain yields than when the reverse cross was made. These effects are a result of several factors including reduced pollen production and viability, inhibition of anther formation, decreased pollen tube growth, self-incompatibility, preharvest germination of seed, and reduced germination and abnormal seedling development.

Much of the foregoing discussion related to the impact of mineral deficiency stress on growth and development has presented a view suggesting that mineral elements may have a direct effect on these processes in plants. However, the distinction between direct effects and indirect effects of mineral deficiency is not always clear, and the

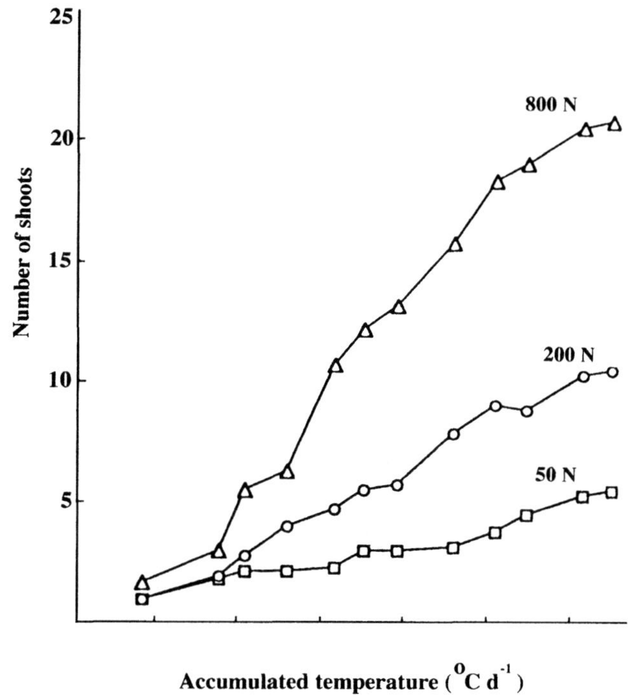

Figure 3.29. Effects of differing N levels on shoot development in wheat. (Reprinted from Longnecker 1994, p. 153, by courtesy of Marcel Dekker, Inc.)

resulting effects could be hormone mediated or could impact metabolic activity via vital pathways that affect growth and development. Whether deficiencies influence growth and development processes directly or indirectly may not be the most important question to ask. It is clear that mineral nutrients can influence some aspects of growth and development processes directly, but higher plant cells do not function in isolation, and thus are part of a continuum of physiological and metabolic activity that is constantly monitoring the external and internal environment of the plant, part of which includes the mineral nutrient status of the plant.

X. BENEFICIAL ELEMENTS AND ELEMENTS REQUIRED BY ONLY A FEW PLANTS

A. Functions and Types of Beneficial Elements

Essential elements are elements that are required for the completion of the normal life cycle of the plant, that cannot be completely replaced by another element, and that have a defined metabolic function as a component of an organic molecule or are involved as a cofactor to or are a required component in enzyme reactions. In contrast, *beneficial elements* may enhance growth or, in a limited sense, be required for only a few species of plants, but, in general, they are defined as elements that are not essential for plant growth.

Two elements that meet the criteria of beneficial elements are Na and Si. These elements are known to be essential for a limited number of plant species, however.

TABLE 13.18 Effect of Cu Levels on Tiller, Straw, and Grain Production in Wheat

	Copper Supply (mg pot^{-1})			
	0	0.1	0.4	2.0
Number of tillers	22	15	13	10
Straw yield (g)	7.7	9.0	10.3	10.9
Grain yield (g)	0.0	0.5	3.5	11.8

Source: From Nambiar 1976.

Sodium has been shown to be required for species of blue-green algae (bacteria) and also is required for several, but not all, C_4 plants. In the latter case, it is not clear why Na is required, but it may function at different physiological and metabolic levels depending on plant species. In some cases, Na may simply serve as a alternative osmotic ion to maintain turgor pressure within cells, particularly plants that are considered to be *halophytes* or *natrophiles*. In such cases, Na can replace K, which may be needed for more critical functions other than turgor maintenance. An example of a halophyte that requires Na is *Atriplex vesicaria*.

In the case of C_4 plants, Na has been demonstrated to stimulate the activity of PEPcase (phosphoenol pyruvate carboxylase), an enzyme essential for the initial fixation of CO_2 in the mesophyll cells of some C_4 plants. Another possible metabolic control point of Na in C_4 plants is the reduction of nitrate to form NH_3, which can influence the level of amino acids produced in the plant. This is particularly important in C_4 plants that utilize aspartate to shuttle carbon from the mesophyll to the bundle sheath cells, where it is decarboxylated to release CO_2 to be incorporated into the Calvin-Benson cycle. Also, in some C_4 species there may be specific Na requiring transport carrier proteins in the membranes that utilize Na as a co-transport ion in moving aspartate and malate between mesophyll and bundle sheath cells. Finally, there is some evidence that Na deficiency may cause damage to the light-harvesting complex in some C_4 species, thus limiting the amount of ATP being produced. The observed effects of Na deficiency on C_4 reactions in higher plants may possibly be associated with the Na requirements of blue-green algae, since some species have been demonstrated to exhibit C_4 photosynthetic reactions.

Silicon has been shown to be beneficial to growth for a number of species of plants, but has only been shown to be required for diatoms and some species of *Equisetum* (horsetails). The role of Si in plants may be similar to B in that it forms complexes with pectins and polyphenols and may function to strengthen cell walls and give rigidity to stem and other support structures in plants. This function of Si is demon-

TABLE 3.19 Effect of Cu Deficiency on Grain Set in Cross-Pollinated Wheat Plants

Cross			No. of	Total No.	
Female	×	Male	Grain Set	Florets	Heads
−Cu	×	−Cu	0	76	3
+Cu	×	−Cu	2	76	3
−Cu	×	+Cu	47	157	7
+Cu	×	+Cu	86	86	3

Source: Graham 1975.

strated in reduced lodging in grass crops and decreased shading as a result of more upright positioning of leaves in the canopy structures of plants. The accumulation of Si in the endodermal and other plant tissues may also act as a deterrent to invasion by pathogens and reduce the palatability of tissues to grazing animals. High Si levels have also been associated with reducing toxic levels of Fe and Mn in the tissues of bean and some grass species. In addition, deficiencies in Si have resulted in reduced incorporation of P into ADP, ATP, and phosphorylated sugars. Thus, Si interacts with other nutrients in beneficial and detrimental ways to affect plant growth.

Other elements that may have a beneficial function in plants include Co, Va, Se, and Al. Cobalt is required as a component of the vitamin B_{12} coenzyme required as a cofactor in N_2 fixation by *Rhizobium* and other nitrogen-fixing organisms. Thus, the beneficial effect for higher plants is derived indirectly through the symbiotic association of the microbial–higher plant interaction.

Vanadium has been shown to be required for the growth of two species of green algae, *Scenedesmus obliquus* and *Chlorella pyrenoidosa*. It is hypothesized that the requirement for Va may be related to chlorophyll synthesis, particularly, the synthesis of porphyrins. It is unusual that such a requirement has not been demonstrated for higher plants.

Selenium is accumulated to very high levels by some species of plants in the genus *Astragalus* and *Stanleya*. There is no evidence that there is a requirement for this element in these plants, but the beneficial effect may be related to their accumulation of selenocystine and selenomethionine, which are highly toxic to potential herbivores such as grazing mammals.

For the most part, Al is usually toxic to plants, particularly in acid soils. However, growth of some species of plants is enhanced by Al, including such plants as legumes, sugar beet, corn, *Eucalyptus,* and tea. Usually, enhanced growth is a consequence of low levels of Al. Tea and species of *Hydrangea* and *Pinus* have been shown to accumulate 16,000 ppm Al (leaves), 40,000 ppm (petals), and 2000 ppm (needles) Al in various plant tissues. Even so, Al does not appear to meet the requirements for essentiality in these plants. In some instances, Al may exhibit beneficial effects through preventing the uptake of toxic levels of such elements as Cu, P, and Zn.

XI. IMPROVING PLANT NUTRITIONAL EFFICIENCY

A. Genetic Potential for Increasing Nutrient Use Efficiency

Most agricultural plants have been bred for high yields and fast growth, and, as a result, many characteristics related to the ability of such plants to survive environmental stress, such as low soil nutrient levels, have been bred out of the genetic pool. It would appear that increasing concerns relative to the application of synthetic fertilizers for agricultural purposes, their potential for soil and groundwater pollution, and the need to bring marginal lands into agricultural production to feed the evergrowing population of people in the world should lead to an area of research that will ultimately identify the specific biochemical and physiological mechanisms involved in efficient nutrient utilization by plants. Nutrient requirements for plants are quite different, as exemplified by differences that occur in the concentration of nutrients in plants and the ability of plants to adapt to and grow in a variety of environments. This suggests, as do screening studies designed to identify differences in plant species, cultivars, and varieties exhibiting variability in nutrient use efficiency, that the ability to utilize nutrients efficiently is under genetic control. Thus, it is conceivable that plants may be bred or possibly developed through genetic engineering techniques to be more efficient in nutrient utilization under both high and low nutrient concentrations.

B. Morphology and Biochemistry of Nutrient-Efficient Plants

If one is to develop nutrient-efficient plants, then the ability to recognize morphological, physiological, and biochemical attributes that contribute to nutrient use efficiency is important. Much can be learned about nutrient use efficiency in plants by studying nutrient utilization of natural plant communities that typically have adapted to low nutrient soils. Plants that normally grow in such environments are characterized by slow growth, high root/shoot ratios, mycorrhizal associations with the roots, low turnover rates of leaves and roots, low root absorption capacity, and high luxury consumption of nutrients (i.e., storage of nutrients above the level needed for growth when nutrients sporadically become available). Agricultural plants bred for high yields require high fertilizer inputs and in general exhibit essentially the reverse characteristics. Table 3.20 outlines some physiological and morphological characteristics associated with genotypic adaptations of plants to nutrient deficiency stress. A better understanding of the molecular mechanisms associated with the acquisition of nutrients from the environment, the movement of nutrients into the root and ultimately into the xylem, the distribution and remobilization of nutrients between the root and shoot, and the way nutrients are utilized in metabolism and support of growth will enable scientists to breed or engineer new plant selections that can tolerate and produce effectively in low nutrient environments. Perhaps of particular importance will be research that focuses on control and regulation of specific ion and nutrient carriers that function in uptake, compartmentation and remobilization of ions within various plant tissues.

XII. SUMMARY

In this chapter we have attempted to explore nutrient requirements of plants from a more holistic view that ranges from how mineral elements move from the soil to the plant surface and into the apoplastic environment of the root, to the mechanisms of nutrient absorption into the symplast, to cellular compartmentation of ions and the importance of ionic homeostasis in the cell, to nutrient circulation within the whole-plant system, to the biochemical and physiological functions of the essential elements, and finally to how mineral nutrients can impact growth and developmental processes. Mineral nutrition in plants is very complex and quite an amazing biological accomplishment. How plants have adapted to the various nutrient niches that exist on the Earth and are able to survive is equally as impressive. Considering the diversity that exists in the plant kingdom, it is wondrous how and why plants selected only 17 or so of the over 100 known elements to sustain their growth and reproductive processes.

STUDY-REVIEW OUTLINE

Introduction

1. A good basis for understanding mineral nutrition in vascular plants must take into consideration plant-soil interactions, growth and the structure and function of roots, ion uptake

TABLE 3.20 Physiological and Morphological Characteristics Potentially Amenable to Genetic Manipulation to Alter Mineral Use Efficiency in Plants

A. Nutrient Acquisition from the Environment

1. Morphological factors
   a. Root/shoot ratio, particularly the increases that characteristically occur under deficiency stress
   b. Extent of the root system, both vertical and lateral
   c. Density of the root system within a unit of soil as related to:
      (1) Density of lateral roots
      (2) Frequency and lengths of root hairs
2. Root modification of the microenvironment (e.g., by proton efflux and organic exudate production)
3. Root ion exchange capacity
4. Root-microorganism interactions (e.g., with mycorrhizal fungi and N-fixing bacteria)
5. Biochemical ion uptake mechanisms

B. Nutrient Movement Across Root to Xylem

1. Transfer across endodermis
2. Release to the xylem
3. Control of rate nutrient uptake and distribution
   a. Types and numbers of membrane carriers
   b. Structure and composition of membrane lipids
   c. Number and control of symplastic connections (plasmodesmata)
   d. Sink types and phytohormone control
   e. Signaling mechanisms involved in balancing nutrient levels

C. Nutrient Distribution and Remobilization in Shoot

1. Retransport from older to younger leaves and from vegetable to reproductive parts
   a. Control mechanisms in timing and rates of senescence initiation
2. Chelates involved in xylem transport of trace element cations (e.g., Fe)
   a. Role of shoot transfer cells in removal and storage of ions
3. Variations in nutrient distribution among subcellular storage compounds and compartments (e.g., in vacuoles)

D. Utilization in Metabolism and Growth

1. Capacity for normal metabolism at reduced tissue concentration of a nutrient
2. Lower element concentration in supporting structures, particularly the stem
3. Element substitution (e.g., partial substitution of Na for K functions)

Source: Modified from Gerloff 1987.

and transport mechanisms, cellular and tissue distribution of elements, physiological and biochemical functions of elements, the role of mineral elements in plant developmental processes, and the reasons some plants are more efficient utilizers of nutrients than others.

Soil Characteristics and Mineral Stress

Soil Factors and Nutrient Acquisition

1. Ion exchange and water-holding capacity of soils along with soil mineral concentrations impact the transfer of nutrients from the soil to the roots of plants.
2. Texture and pH of soils will impact the availability and movement of mineral elements to root surfaces.

Mass Flow, Diffusion, and Nutrient Uptake in Roots

1. Several factors influence the movement of ions to root surfaces; these include mass flow, diffusion, and root interception.
2. Ions exhibit different rates of diffusion in soil solution depending on the *diffusion coefficient* of the element.
3. Mass flow and diffusion are thought to supply plants with the majority of mineral nutrients required by plants, while root interception only supplies about 1%.

Nutrient Absorption by Roots

Root Structure and Ion Uptake

1. The structure and tissue composition of roots may influence the ability of different plants to absorb nutrients from soils and successfully compete with other plants.
2. Generally, the tissues that ions must traverse in a root to reach the vascular tissue in the center of the root include (from outside to inside) the rhizodermis (epidermis), exodermis (hypodermis), cortex, endodermis, and pericycle.
3. The concepts of *apoplast* (nonliving free space of cells and tissues) and *symplast* (everything inside the plasmalemma of cells including plasmadesmatal connections) are important in understanding nutrient and water transport in plants and are the basis for explaining root pressure.
4. Roots of different plant species exhibit different *cation exchange capacities* (CEC).

Ion Transport Across Membranes

Transport Proteins

1. Ion transport across membranes involves both *active* and *passive* transport processes including the use of transport proteins such as antiports, symports, and uniports as well as gated protein carriers and ATP-driven proton pumps.

Cellular and Whole-Plant Nutrient Distribution

Plant Homeostasis and Ionic Balances

1. Plants exhibit a type of internal ionic homeostasis that is important in regulating ion balances and pH in various cellular compartments.
2. The vacuole and plasmalemma are important in regulating ion balances within the cell and also controlling growth and turgor.
3. Several elements may be involved with ionic balances in plants, but K plays a particularly important role in osmotic adjustment and maintaining proper charge balances in cells.
4. Organelles and tissues maintain and have different ion and pH requirements for proper functioning which is reflected in tissue, cellular, and organelle differences in ion composition.
5. Changes in ionic balances within the plant tissues could act as signaling and/or feedback mechanisms for controlling the uptake and redistribution of mineral elements in plants.

Ionic Compartmentation and pH Control

1. The tonoplast (vacuole) and plasmalemma help to maintain a constant pH (7.3–7.6) in the cytoplasm by pumping protons to the outside of the cell or into the tonoplast compartment.
2. Proton gradients generated in maintaining cytoplasmic homeostasis serves to provide the necessary gradient that drives various transport carriers.

Ionic Concentrations of Plant Cells and Tissues

1. Plant cell cytoplasm has a total inorganic concentration of 100–500 mOsm.
2. The vacuole contains a variety of organic and inorganic ions that are used in regulating osmotic balance in cells. These include Ca, Mg, NaCl, NO_3, malic acid, sugars, and amino acids.

3. Potassium is one of the most abundant and important cations in cells and is highly regulated in the cytoplasm.
4. Tissue and specialized cells also can be selective relative to accumulation of specific ions.

Ionic Signaling and Regulation of Ionic Balance

1. Signaling mechanisms that are concentration and ion specific are thought to control the balance of ions in the cytoplasm, vacuole, and the apoplast of the stele.

Long-Distance Transport of Mineral Elements

Xylem and Phloem Transport

1. The rate of movement and the composition of ions in the xylem and phloem differ and can be affected by differences in charge characteristics of ions and in the pH that occurs in the two tissues.
2. Transport in the xylem is under a negative tension as a result of leaf transpiration.
3. Movement in the phloem can be bidirectional, and movement between the xylem and phloem can occur readily, since these tissues are separated by only a few cell layers.
4. Transport in the phloem is under positive pressure generated according to the "pressure flow" hypothesis.
5. Some evidence suggests that some ions may move to sink areas in actively growing tissues in response to elevated levels of phytohormones.
6. Specialized transfer cells exist in shoots and remove ions from the xylem stream and may serve to store nutrients or remove toxic levels of nutrients before they reach the leaves.
7. Accumulation of nutrients in tissues can vary with transpiration rates. Fruits may exhibit nutrient deficiency symptoms as a result of low transpiration rates, resulting in reduced movement of ions to these tissues.
8. Ions are more concentrated in the phloem, and both organic and mineral nutrients are transported in the phloem, while mostly mineral elements are transported in the xylem.
9. Mobility of micronutrients is generally greater in xylem tissues compared with the phloem but may increase in the latter during reproduction.
10. Composition of the phloem varies with age, developmental stage, diurnal and seasonal cycles, soil nutrient availability, and plant tissue.
11. A type of circulatory system exists in the xylem and phloem with respect to the transport of K and malate, and K and NO_3.
12. The relative importance of xylem versus phloem in the transport of specific mineral elements depends on tissue type and mobility of the element in question. Nitrogen is transported in the xylem and phloem, but the amounts and forms transported depend in part on the N source and where in the plant the nitrogen is reduced.

Interactions of Mineral Elements

Antagonisms and Synergisms

1. Ion interactions in the soil depend on types of ions, concentrations, soil type, and pH.
2. Ions can interact by competing for exchange sites and transport carriers, and through the activation and deactivation of transport carriers, the substitution of ions in metabolic functions, the interaction of ions during transport processes, and the precipitation of ions as insoluble salts.
3. Examples of ions that can partially substitute for one another include Mn/Mg and K/Na.
4. Elements that are antagonistic include Ca/Na, Fe/Zn, and Cu/Fe.

Physiological and Biochemical Functions of Essential Elements

Essential and Required Levels of Elements

1. Seventeen elements are considered to be essential for normal plant growth and development: C, H, O, P, K, N, S, Ca, Fe, B, Mn, Cu, Zn, Mo, Cl, Mg, and Ni.

2. Mineral elements can be classified relative to general functions, such as structural, osmotic, ionic, transport, and as cofactor or enzyme constituents.

Structural Functions

1. Carbon, H, O, P, N, and S are macronutrients with important functions as structural components of numerous organic molecules, including sugars, proteins, amino acids, nucleic acids, and phospholipids.

Osmotic Functions

1. Osmotic relations are primarily controlled by K, but Cl, Na, NO_3, and Ca, as well as some organic molecules such as polyamines, are known to have some function in osmotic control under certain conditions.
2. Potassium is particularly critical in osmotic functions since it is involved in turgor control in such responses as stomatal opening and closing.
3. Plants deficient in K tend to produce polyamines that may substitute for K in regulating the osmotic concentration in cells.

Ionic Balances

1. Maintenance of ionic balances is important in establishing the electropotential gradients involved in nutrient and ion transport, regulating cellular compartmental pH, and controlling osmotic balances within the cell.
2. Ionic balances are regulated at the cell and at the whole-plant level.
3. Ions such as OH, H, K, Na, Cl, NO_3 H_2PO_4, SO_4, and some organic acids such as malic and aspartic acids are known to be involved with controlling osmotic balances in plants.

Transport

1. Boron forms complexes with hydroxyl radicals in compounds such as sugars, sugar alcohols, phosphate esters and *o*-diphenols. Such complexes may facilitate transport of such molecules across membranes and through the vascular system.
2. The exact function of B is unclear, but many physiological functions are affected, such as membrane permeability, phenol metabolism, lignification, respiration, indoleacetic acid synthesis, carbohydrate metabolism, and RNA metabolism.

Cofactors and Enzyme Constituents

1. Iron, Mn, Cl, Zn, Cu, Mo, and Ni are considered microelements and are either components of enzymes, required for enzyme activation, or components of cofactors needed for enzyme function.
2. Calcium is important in membrane stabilization and activates the enzyme calmodulin, which initiates other events leading to several developmental and physiological responses in plants such as gravitropism.

Functions of Nutrition in Plant Growth and Development

Mineral Nutrients and Cell Division and Expansion

1. Mineral nutrients can have considerable effects on the synthesis of phytohormones. Conversely, phytohormones such as indoleacetic acid and cytokinins have been demonstrated to regulate nutrient mobility in plants.
2. Mineral elements can affect growth directly by limiting functions such as photosynthesis or respiration, but also can affect processes indirectly by affecting hormone balances and the efficiency of metabolic processes.
3. Nitrogen is an example of an element that can influence growth processes such as leaf area, growth of tillers in grasses, lateral branching, and leaf number. In some cases, limitation of N can affect growth and development prior to affecting photosynthetic processes.

4. Phosphorus deficiency can result in reduced hydraulic conductivity. Ca deficiency can reduce both cell division and expansion, and cell expansive growth can be limited by deficiencies in K, Na, and Cl.

Nutrient Deficiencies and Root/Shoot Growth

1. Deficiencies in elements frequently result in increased root/shoot ratios.
2. Low N, P, and Cu have been observed to cause a reduction in leaf number in some plants.
3. Rates of development of leaves, tillers, and branches are also influenced by the availability of P.

Nutrient Deficiencies and Initiation of Lateral Shoots, Tillers, and Leaves

1. Deficiencies in N, P, Cu and Mn have been shown to reduce the numbers of lateral shoots and tillers in grasses.
2. The form of nitrogen (NO_3 or NH_4) applied to plants can affect the number of lateral branches formed. Ammonium nitrogen is more effective in apple trees.
3. Low Cu levels may control sink strength in grasses resulting in more vegetative growth than seed production.
4. Reports of B and Zn deficiencies causing increased branching may be a result of changes in apical dominance in plants since these elements are involved in indoleacetic acid (IAA) synthesis.

Nutrient Deficiencies and Reproduction

1. The form of nitrogen can affect flower formation in some plants, with NH_4 being more effective in apple trees.
2. Deficiencies in N and P can influence rates of flower senescence.
3. Low Cu levels can influence yields in wheat, possibly by affecting pollen production, viability, anther formation, reduced pollen tube growth, self-incompatibility, preharvest germination of seed and reduced germination, and/or abnormal seedling growth.

Beneficial Elements and Elements Required by Only a Few Plants

Functions and Types of Beneficial Elements

1. Beneficial elements are those that may enhance growth but are not required for completion of the life cycle of the plant.
2. Some elements may be required for the completion of the life cycle of a limited number of species and so are not generally viewed as being essential for most vascular plants.
3. Elements that have been included in the beneficial category include Si, Na, Co, Va, Se, and Al.
4. Some beneficial elements may function by preventing the uptake of otherwise toxic elements, facilitating photosynthetic processes, and helping in cell wall synthesis; they may also be required in symbiotic relationships, and could be deterrents to herbivores.

Improving Plant Nutritional Efficiency

Genetic Potential for Increasing Nutrient Use Efficiency

1. It is highly possible that plants can be bred or bioengineered for greater nutrient use efficiency, thus creating additional opportunities to bring marginal soils into productivity.
2. The use of marginal soils will undoubtedly result in a trade-off resulting in reduced productivity.

Morphology and Biochemistry of Nutrient-Efficient Plants

1. Much can be learned about breeding or bioengineering successful agricultural plants to be grown in nutrient-deficient areas by observing the morphological, biochemical, and physiological attributes of natural plant communities already adapted to these environments.

2. Plants growing in such environments are characterized by slow growth, high root/shoot ratios, mycorrhizal associations, low leaf turnover rates, low root absorption capacity, and high luxury consumption of nutrients.
3. Research that focuses on control and regulation of specific ion and nutrient carriers that function in uptake, compartmentation, and remobilizaiton of ions within and between tissues may yield the best results for improving nutrient-efficient plants.

SELF-STUDY QUESTIONS

1. Why is it important to study the mineral nutrition of plants?
2. Outline some of the soil characteristics that can influence the acquisition of mineral elements by plants.
3. Define mass flow, diffusion, and root interception, and indicate the relative importance of each to plant nutrition.
4. Mineral elements have different diffusion coefficients. What does this mean, and why is it important to plants?
5. Draw the cross-section of a root, and label all of the tissues that an ion would have to traverse to reach the xylem in the stele.
6. What is the Casparian strip, where is it located, and why is it important to a plant?
7. What is meant by the terms *symplast* and *apoplast,* and what is the significance of these areas to understanding nutrient movement in plants?
8. Describe the structure and function of plasmadesmata in plants?
9. What is root pressure, how does it form, and what role does it play in mineral nutrition in plants?
10. Roots exhibit cation exchange. Why is this important to the nutrition of a plant, and how do roots differ among plants with respect to CEC?
11. What are the different types of transport proteins that occur in plants?
12. What is the significance of ATP-driven proton pumps to nutrient transport across membranes?
13. What are gated carriers, and what function might they serve?
14. What is meant by the statement that plants have a type of "ionic homeostasis," and why is this important in plants?
15. What elements and structures or organelles are involved in maintaining ionic homeostasis at the cellular level?
16. Describe the distribution and concentrations of ions at the cellular level.
17. How might plants regulate ion balances throughout the whole plant?
18. How do ion concentrations differ among tissues?
19. What may regulate changes in the tissue concentrations of ions?
20. Indicate the types of substances that are transported in the xylem and phloem.
21. Describe the mechanisms by which substances move in the xylem and phloem.
22. Can nutrients be exchanged between the xylem and phloem? Why would this be important in a plant?
23. What are transfer cells, and why are they important in plant nutrition?
24. What factors may control the rate of movement of ions in the xylem and phloem?
25. How might phytohormones influence nutrient movement in plant tissues?
26. Describe a type of circulatory system that involves the transport of malate, K, and NO_3 in plants.
27. What factors govern the relative importance of the xylem and phloem in ion transport in plants?
28. What are the 17 essential elements needed for plant growth?
29. What factors determine whether an element is essential for plants or not?
30. What is the difference between a macronutrient and a micronutrient?
31. Name five general functions of essential elements in plants.
32. Give two specific examples of the physiological or biochemical role of each essential element in plants.

33. Give examples of how specific mineral nutrients may affect developmental processes in plants.
34. Discuss native plants relative to physiological, biochemical, and morphological adaptations that allow them to survive with limited nutrient resources. How might this information be used to improve nutrient use efficiency in crop plants?
35. Why is it important to understand mineral nutrition by using a holistic approach (soil and plant factors) rather than by focusing only on the functions of mineral elements in plants?

SUPPLEMENTARY READING

Basra, A. S. (ed.). 1994. Mechanisms of Plant Growth and Improved Productivity: Modern Approaches. Marcel Dekker, New York.

Chapin, F. S. III. 1980. The mineral nutrition of wild plants. Annual Reviews of Ecological Systems 11:233–260.

Chapin, F. S. III. 1991. Effects of multiple environmental stresses on nutrient availability and use. Pp. 67–88 in: H. A. Mooney, W. E. Winner, and E. J. Pell (eds.), Response of Plants to Multiple Stresses. Academic Press, New York.

Christiansen, M. N., and C. F. Lewis (eds.). 1982. Breeding Plants for Less Favorable Environments. John Wiley & Sons, New York.

Clark, R. B. 1983. Plant genotype differences in the uptake, translocation, accumulation and use of mineral elements required for plant growth. Plant and Soil 72:175–196.

Clarkson, D. T., and M. J. Hawkesford. 1993. Molecular biological approaches to plant nutrition. Pp. 23–33 in: N. J. Barrow (ed.), Plant Nutrition—from Genetic Engineering to Field Practice. Kluwer Academic Publishers, Boston, MA.

Epstein, E. 1994. The anomaly of silicon in plant biology. Proceedings of the National Academy of Sciences 91:11–17.

Fernandes, M. S., and R. O. P. Rossiello. 1995. Mineral nitrogen in plant physiology and plant nutrition. Critical Reviews in Plant Sciences 14:11–148.

Gabelman, W. H., and B. C. Loughman (eds.). 1987. Genetic Aspects of Plant Mineral Nutrition. Martinus Nijhoff Publishers, Boston, MA.

Nilsen, E. T., and D. M. Orcutt. 1996. The Physiology of Plants Under Stress: Abiotic Factors. John Wiley & Sons, New York.

Porter, J. R., and D. W. Lawlor (eds.). 1991. Plant Growth: Interactions with Nutrition and Environment. Cambridge University Press, New York.

Rendig, V. V., and H. M. Taylor. 1989. Principles of Soil-Plant Interrelationships. McGraw-Hill Publishing Company, New York.

Shkolnik, M. Y. 1984. Trace Elements in Plants (Developments in Crop Science, Vol. 6). Elsevier, New York.

Shuman, L. M. 1994. Mineral nutrition, Pp. 149–182 in: R. E. Wilkinson (ed.), Plant-Environment Interactions. Marcel Dekker, New York.

Welch, R. M. 1995. Micronutrient nutrition of plants. Critical Reviews in Plant Sciences 14: 49–82.

4 Mycorrhizae and Plant Stress Physiology

Outline

OBJECTIVES

1. Describe the diversity of types of mycorrhizal associations found in natural systems.
2. Differentiate these mycorrhizal associations based on structural characteristics and taxonomic relationships of the fungi.
3. Explain how mycorrhizal association affects plant productivity and allocation patterns.
4. Describe fungal requirements for reduced carbon.
5. Explicate the effects of mycorrhizae on plant photosynthesis.
6. Describe the mechanisms by which mycorrhizae can influence plant nutrient accumulation.
7. Explain the relationships between mycorrhizal associations and the nitrogen nutrition of plants.
8. Focus on how mycorrhizal association can influence phosphorus nutrition of plants.
9. Describe if and how mycorrhizae may influence the water relations of plants.
10. Explain the interrelationships between mycorrhizal fungi and plant growth regulators.
11. Describe how mycorrhizae could affect pathogen resistance and heavy-metal tolerance.
12. Describe the theory and evidence for the flow of carbon and nutrients from plant to plant through mycorrhizae.

I. INTRODUCTION

Mycorrhizae (Gr. *mykes* = mushroom + *rhiza* = root) are intimate, mutualistically beneficial symbioses between the absorptive organs of plants (usually the roots) and the hyphae of certain fungi. Usually both the plants and the fungi can benefit. However, the symbiosis can range from mutualism to commensalism or even become slightly parasitic, depending upon the species of plant and fungus involved, the age of the mycorrhiza, and the type of mycorrhiza. For example, orchid mycorrhizae are characteristically parasitic associations for a donor plant and occasionally for the host orchid. The mycorrhizal fungus connects the orchid to a donor plant that provides carbon to the orchid and the fungus. Classically defined, the functional symbiosis provides a suitable and sufficient carbohydrate source for the fungal symbiont. Benefits for the plant can be numerous and include improved nutrient and water uptake, additional carbon acquisition, increased sink strength for photosynthate translocation, increased production of phytohormones, improved resistance to pathogens, and heavy-metal tolerance. Mycorrhizae are critically important organs for resource uptake by most terrestrial plants. In the absence of an appropriate fungal symbiont, many terrestrial plants suffer from resource limitations and ultimately reduced growth, and fitness.

Mycorrhizal assemblages are some of the most important yet least understood components affecting ecosystem structure and functioning. As an integral part of ecosystem function, mycorrhizae are necessarily an important aspect of plant stress physiology. Plants that form symbioses with fungi include over 2500 different species, ranging from primitive seedless nonvascular plants to evolutionarily advanced members of the Anthophyta. Conservative estimates are that approximately 95% of all vascular plant species belong to families that are characteristically mycorrhizal (Trappe 1977). Only a few families of flowering plants characteristically lack mycorrhizae (e.g., Brassicaceae, Cyperaceae, and some members of Proteaceae). Furthermore, mycorrhizae also form between hyphal fungi and the underground structures of plants that do not form true roots such as the gametophytes of bryophytes and pteridophytes, as well as sporophytes of pteridophytes. Fungal symbionts are represented by members from three of the major divisions in the kingdom Fungi: Zygomycota, Ascomycota, and Basidiomycota. Mycorrhizae are found in a variety of ecosystems and plant communities including wetlands, deserts, deciduous forests, lowland tropical rain forests, high latitude as well as high-elevation plant communities, aquatic systems, prairies, and even canopy epiphytes (Figure 4.1). The only major terrestrial setting that is not characterized by mycorrhizal plants is the extreme Arctic. One predominant mycorrhizal type can often characterizes an ecosystem (e.g. ericoid mycorrhizae in heath vegetation, and vesicular-arbuscular mycorrhizae (VAM) in sand dune vegetation). Mycorrhizae are diverse, prolific, ubiquitous among ecosystems, and an integral part of plant physiology and ecology.

II. CLASSIFICATION OF MYCORRHIZAE

Root-fungi symbioses are spatially and temporally dynamic, resulting in remarkable variation in mycorrhizal function (mutualistic to parasitic) and structure (ecto- versus endo- forms). Mycorrhizae can be classified into roughly seven major groups based on key morphological features of the root-fungus symbiosis (Table 4.1). We will present the classification of all seven types of mycorrhizae, but generally we will focus on functional processes of ecto- and endomycorrhizae in relation to plant stress physiology.

A. Vesicular-Arbuscular Mycorrhizae

1. Diversity of Vesicular-Arbuscular Mycorrhizae. Approximately 70% of all plant families contain species that develop specialized endomycorrhizae called *vesicular-arbuscular mycorrhizae* (VAM) or just *arbuscular mycorrhizae* (AM) (Sieverding 1991) on their roots. In this chapter, "VAM" refers to the symbiosis between the plant root and the mycorrhizal endophyte, "VA fungus" refers to the fungal symbiont only, and "VAM plants" refers to plants that are capable of forming VAM on their root system. Vesicular-arbuscular mycorrhizae are spatially and temporally variable, often showing variation in form and function between different plant-fungus combinations. It is believed that VAM are ancient symbioses. The earliest plants that colonized land (e.g., Rhyniophyta) during the early Devonian had no true roots, but the aerial stems were colonized by hyphal fungi that formed arbuscules very similar to the modern VA fungi, or grass endophytes (Taylor et al. 1996). It is thought these endophytic associations could have been critical for successful establishment by early land plants because Devonian soils were probably poorly developed. At that time, the VAM-like association formed under high light–low nutrient availability. Currently, VAM are very common on plants colonizing new soils, such as early successional

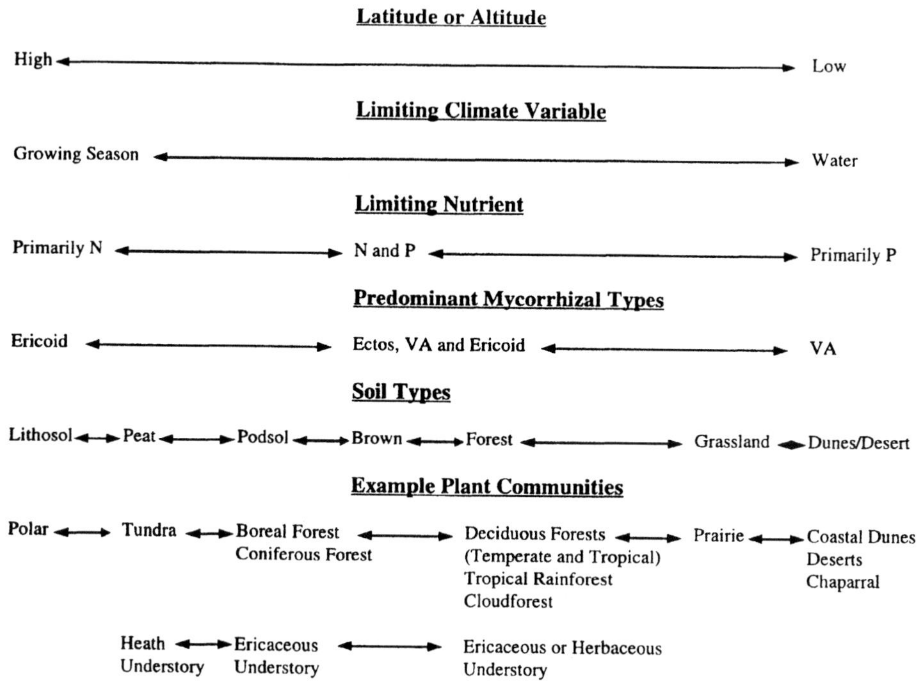

Figure 4.1. A simplified overview showing predominant mycorrhizal types along elevation and latitudinal gradients in conjunction with habitat types, limiting resources and common plant communities. N = nitrogen, P = phosphorus. (Redrawn with permission from Read 1984.)

habitats, newly formed volcanic soils, sand dunes, or even mine spoils. In fact, VAM are essential for dune ecosystems because early successional plants colonized by VA fungi subsequently help to agglutinate sand particles via external mycelia and thereby stabilize dune structure.

VA fungal species are extremely common soil fungi, classified in the phylum Zygomycota, order Glomales. Currently recognized are six widespread fungal genera that form VAM (*Glomus, Gigaspora, Entrophospora, Acaulospora, Scutellospora,* and *Sclerocystis*) (Schenck and Perez 1990), with fewer than 200 species involved. Glomalean fungi form mycorrhizae globally with many of the angiosperms, including most agronomically important plants, some gymnosperms, certain bryophytes, and pteridophytes, as well as a few algae. Spore dispersal occurs numerous ways including by wind, water, or animal vectors. Glomalean fungi, like most mycorrhizal fungi can colonize roots by germ tubes originating from spores or by hyphae growing from other propagules such as vesicles or colonized root fragments present in the soil. These modes are called *primary infections. Secondary infection* occurs when hyphae connected to active VAM invade the same root or even another root. The function of these associations can be highly variable in space and time. Consequently, plants that can harbor VA fungi range from being facultatively to obligatory mycorrhizal.

2. *Structure of VAM.* Unlike some mycorrhizal fungi, VA fungi cause little change to the external morphology of the root upon colonization (Figure 4.2). Consequently, only through appropriate staining and clearing procedures can the mycorrhizal structures be viewed with the aid of a light microscope. Following penetration of the root

TABLE 4.1 A Brief Summary of the Main Mycorrhizal Types and Potential Bidirectional Resource Exchange

Type	Host Diversity	Symbiont Range	Nutritional Interactions	
			Symbiont → Host	Host → Symbiont
Vesicular-arbuscular mycorrhizae	Many plant species, including representative bryophytes, pteridophytes, gymnosperms, and many angiosperms	Four genera of Zygomycota	Mineral nutrients	Carbohydrate
Ectomycorrhizae	Many trees and shrubs, especially in temperate regions	Many fungi, including species from at least 25 families of Basidiomycota, 7 families of Ascomycota, and 1 genus of Zygomycota	Mineral nutrients	Carbohydrate
Ericoid	Members of Ericales with fine hair roots, especially Ericoideae, Vaccinoideae, Rhododendroideae, Epacridaceae, and Empetraceae	Many isolates form steriles in culture: one has been identified as *Hymenoscyphus* (Ascomycota)	Nutrients, especially ammonia and organic nitrogen	Carbohydrate
Arbutoid	Members of Ericales with sturdier roots including *Arbutus, Arctostaphylos*, and Pyrolaceae		Probably mineral nutrients	Probably carbohydrates
Monotropoid	Achlorophyllous members of Ericales such as *Monotropa, Sarcodes,* and *Plerospora*	Those identified so far mostly Basidiomycota and also form ectomycorrhizae	Carbohydrates and mineral nutrients	None known
Orchid mycorrhizae	All members of Orchidaceae	Many isolates from sterile mycelia referable to "form" genus *Rhizoctonia* induced to form sexual stages, referable to about 8 genera of Basidiomycota including some pathogens	Juvenile achlorophyllous hosts: carbohydrates and mineral nutrients	None known
			Adult photosynthetic hosts: probably mineral nutrients	None known

Source: Adapted from Smith and Douglas 1987 with permission.

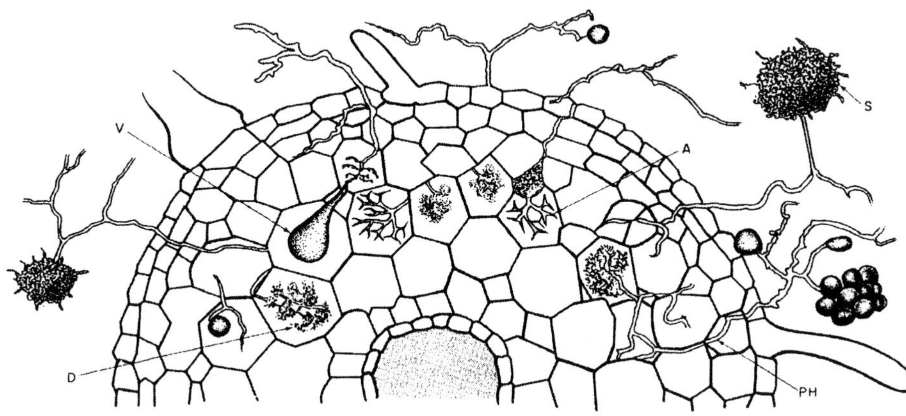

Figure 4.2. Cross section of root colonized by VAM. Note the penetrating hyphae (PH), dichotomously branched arbuscules (A), globular vesicles (V), digested haustoria, and external sporocarps (S) with spores. (From J. L. Harley 1965. Copyright © 1965 The Regents of the University of California.)

epidermis via a swollen hyphal segment called an appressoria, VA fungal hyphae can grow, both inter- and intracellularly, within the root cortex, where the vesicles and arbuscules, characteristic structures of this type of mycorrhiza, are formed. VA fungi are most active in newly formed, nonsuberized, secondary roots. Root endophytes do not colonize root meristems, nor is the vascular cylinder penetrated.

The overall morphology of many VAM roots can be similar in appearance, especially after staining and clearing the roots. As a result, identifying a VA fungal symbiont is challenging. Spore morphologies and the spore murograph (a characterization of the spore wall texture) are needed to correctly identify the fungal symbiont involved. Once again, this proves difficult, because the spores, when produced, are produced underground, and often VA fungi reproduce by a variety of asexually formed propagules (e.g., clamydospores or azygospores) that are not much help for identification purposes. As a result, wet sieving of the soil and decanting or other separation methods have to be employed to retrieve spores. Finally, complicating matters more, VA fungal spores do not germinate or grow in the absence of suitable host roots for reasons that will be discussed later. Consequently, VA fungi form no free-growing mycelia and cannot be cultured, so pot cultures are used to maintain viable fungal inocula.

Once VA fungal spores germinate, hyphae grow individually or as loose aggregates around the roots before penetrating. Following root colonization, the hyphae radiate into the soil forming the critical external mycelium or extraradical hyphae. The external mycelium extends like a fan into the soil, and is responsible for the uptake and transport of soil resources to the mycorrhizal interface as well as soil stabilization in some systems. External mycelia are delicate and easily severed making the study of intact VAM difficult. In the process of harvesting VAM plants, the external hyphal fan is often torn from the root.

Traditionally, a functional VAM was defined by the presence of arbuscules in the root. *Arbuscules* are dichomously branched, ephemeral structures, lasting only a few days before disintegrating and being digested by the colonized cells of the plant roots. Due to the ephemeral nature of arbuscules, a particular root sample may not show arbuscules even though a VAM association is present. The highly branched nature of the arbuscule increases the surface area of contact between the fungal hyphae and the cell plasma membrane. In addition, the plant plasma membrane in these sites is rich in H^+-ATPase complexes, indicting an active site for nutrient uptake. Arbuscules have been treated as the major sites of bidirectional resource transfer between the symbionts, and for this reason were thought to be critical to the functional association.

However, current research using *Arum*-type VAM suggests that arbuscules are not the exclusive site for resource exchange (Smith et al. 1994a, 1994b). Instead, the exchange interfaces for sugar and nutrients between the plant and the fungus are very likely spatially separate. Carbohydrate transport to the fungus may be associated more with nonarbuscular fungal hyphae than with arbuscules. Intracellular hyphae and tightly coiled hyphae such as *pelotons,* which are common to many endomycorrhizae, can function in much the same manner as arbuscules (i.e., a site of resource transfer). Like arbuscules, these structures increase the surface area of contact between the plant plasma membrane and fungal hyphae. Plus, the fungal plasma membranes of intracellular hyphae and pelotons are rich in H^+-ATPase complexes, which are energized and capable of generating the proton-motive force necessary to support co-transport mechanisms of sugars as well as other organic sources of carbon. The intracellular hyphae remain alive for the duration of an infection unit even as the number of arbuscules will decline with the age of the root, and the hyphae maintain contact with the external mycelium. Therefore, it is thought that intracellular hyphae and pelotons are most likely major sites of carbohydrate transfer to the fungus. These structures are also ephemeral, eventually resulting in breakdown followed by cellular digestion.

3. Modes of Plant Colonization by VA Fungi. Communication between partners in a symbiosis is critical for the initiation and later physical association of organisms involved. For VAM, it is now apparent that this communication system is controlled by both the fungal and plant symbionts involved. Careful observations of precolonization events indicate that communication is a two-way process, with each symbiont receiving and sending messages. Previously, it was thought that the plant symbiont entirely controlled precolonization, and formation of the active mycorrhiza was under strict plant control. It is now apparent that the fungal partner likely has the capacity to select or even reject a particular plant partner. Despite countless difficulties involved with studying communication between symbionts, it is clear that any confusion in communication will prevent the formation of the symbiosis. Because of the obligate nature of the symbiosis for VA fungi and the long evolutionary history of VAM, effective mechanisms of communication regulating all aspects of the symbiosis most probably exist and should be highly conserved.

VA hyphae, as with most biotrophic interactions, are chemotropically attracted to the plant root. It is becoming increasingly apparent that, although chance contact between symbionts cannot be ruled out as a plausible mechanism allowing for the formation of VAM, the functional symbiosis is under greater control than chance alone. Communication between symbionts begins with the accumulation of root exudates in the soil. It is not known whether location and growth responses are under the regulation of one or more exudate compounds from the plant. Roots produce a wide variety of organic chemicals that can and are released into the surrounding soil. These compounds fall into three categories based on their mobility in the soil: (1) diffusable–water-soluble, (2) diffusable-volatile, and (3) nondiffusable (e.g., mucilages and sloughed root cells). Some of these compounds may act as chemical messengers to initiate and coordinate the formation of VAM. However, research on the importance and function of these exudates is limited because of difficulty isolating, separating, and characterizing exudates (particularly diffusable-volatile compounds), as well as problems with introducing experimental artifacts different from natural conditions anytime these experiments are conducted in greenhouse settings or with sterile substrates. Yet evidence is accumulating that certain events can be mediated by these root compounds, including fungal spore germination, directional growth of germ tubes and hyphae, hyphal growth rate and branching, and positioning of fungal hyphae near entry sites on the root surface (Figure 4.3).

As hyphae approach the roots, exudates likely function as chemical messengers by directing them to accessible regions of the root. Hyphae select specific regions on the root surface where penetration hyphae can be formed. This response appears to be under the control of specific signal molecules that are attached to the host root cell walls as well as exudation of mobile root compounds. Once germ tubes or hyphae contact the root, surface recognition factors (including topography) likely assume the major regulatory function of the forming mycorrhiza. The exact mechanisms of recognition are still unknown and left mostly to speculation, but it is possible that extracellular polysaccharides located on fungal cell walls and on the plant root cell walls mediate the contact interactions between symbionts. Numerous plant and fungal genes are probably involved to insure appropriate compatibility at all stages of development. After fungal hyphae or germ tubes contact the roots and recognition occurs, growth rate and branching of the fungal symbiont might increase. In addition, fungi can stimulate young colonizable roots to increase branching and thereby approach spores and colonizing hyphae more readily. This stimulation of branching and attraction of

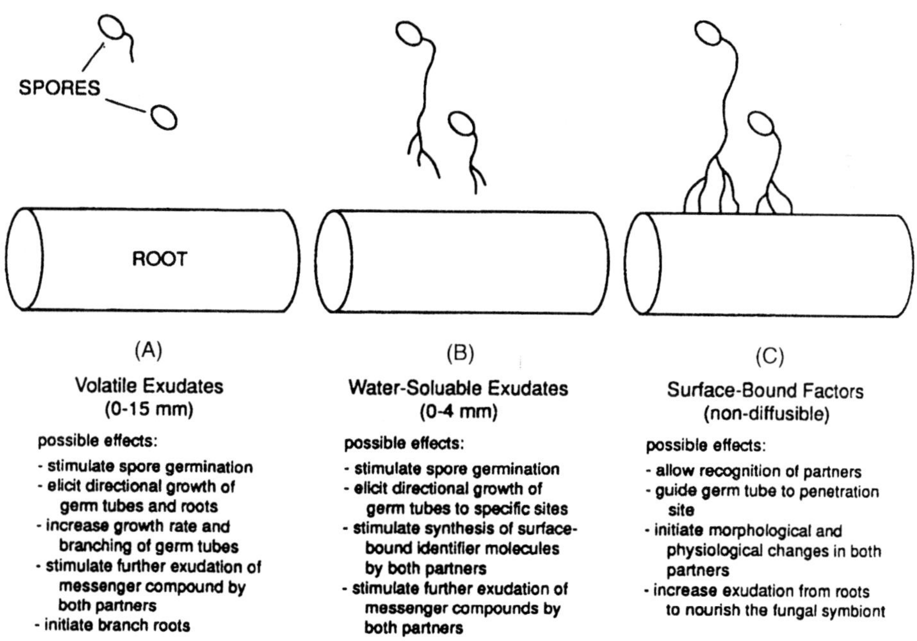

Figure 4.3. Possible communications between mycorrhizal fungi and plant roots prior to formation of mycorrhizae. Spores are shown at different distances from the root surface. (A) Possible effects of volatile compounds. (B) Effects of both water-soluble and volatile compounds. (C) Potential effects of water-soluble and volatile exudates plus surface-bound messengers. (From Koske and Gemma 1992, with permission.)

root tips to spores could be the result of increased phytohormone production that is either of fungal origin or plant origin but is induced by the fungus.

Penetration of the roots by the fungus occurs in areas near the root tip that are not suberized. Hyphae penetrate intercellularly within the epidermal layer, but intracellularly within the cortex. Hyphae penetrate root cortical cells by forming cellulolytic enzymes that dissolve a portion of the cell wall and subsequently induce an invagination of the cell plasma membrane. Production of such cellulolytic enzymes is thought to be under tight regulation by both the plant and fungus. For example, it is known that plant root cell walls are associated with proteins that inhibit the pectin-degrading enzymes of many plant pathogens. Thus, penetration of the root cell walls must involve upregulation and synthesis of wall-degrading enzymes and the down-regulation of plant proteins that might inhibit cell wall degradation. Once a cortical cell has been penetrated, VA fungi can produce the dichotomously branched haustoria like structures, arbuscules, hyphal coils, and, in some cases, terminal swellings of the hyphae or vesicles. Vesicles (formed between or within host cells) are thought to function as energy storage sites (e.g., lipid deposits), possibly for use when host metabolite transfer is low or impaired. It is not uncommon for one or more of these identifying structures to be missing from the functional mycorrhizae. For example, an association that lacks vesicles is often called arbuscular mycorrhizae.

Finally, fungal symbionts will often attempt to colonize nonmycorrhizal plant species. In these cases, the fungus may pass through the first few stages of mycorrhizal development (e.g., penetration of the epidermis and first few layers of cortical cells), before the plant realizes the fungus as an intruder. The plant then prevents further spread of the fungus, perhaps by increased suberization in the area of colonization or by mobilizing a set of chemical defenses, such as phenolics, that discourage further mycorrhizal development.

B. Ectomycorrhizae

1. Diversity of Ectomycorrhizae. Ectomycorrhizae (an association between a fungus (mycobioint) and plant (phycobioint) roots in which the mycobioint is primarily external to the root tissues) is estimated to occur in association with 10% of the world's flora, are not as widespread in nature as the endomycorrhizae and are restricted to seed plants (Marx 1991). It has been estimated that 3% of all seed plants form ectomycorrhizae (Meyer 1973). The exact number of ectomycorrhizal fungi is still unknown. In North America alone, there appear to be over 2000 species of ectomycorrhizal fungi (Alexopoulos et al. 1996); however, worldwide it is estimated there are more than 5000 species, often forming mycorrhizae with only one plant species. Most of these fungi are members of Basidiomycota. Genera of Agaricales that often form ectomycorrhizae include *Suillus, Boletus, Gomphidius, Chroogomphus, Russula, Lactarius, Amanita, Tricholoma, Heboloma, Cortinarius, Hygrophorus, Inocybe*, and *Laccaria*. Certain genera within Sclerodermatales, Hymenogastrales, and the Aphyllophorales also readily form ectomycorrhizae. Among Ascomycota, *Cenococcom geophyllum* and *Piloderma bicolor* are very common ectomycorrhizal symbionts, while species of Tuberales (the truffles) and Pezizales form mycorrhizae too. Unusually, within Zygomycota, a few members of Endogonales are also known to form ectomycorrhizae.

Any estimate of the number of ectomycorrhizal fungal species is approximate because of difficulties inherent in isolating, culturing, and therefore identifying many fungi as ectomycorrhizal. For example, members of Russulaceae, a large ectomycorrhizal family, are for the most part virtually impossible to culture by themselves implying an intricate nutritional relationship with the host plant. Therefore, it is extra difficult to verify that a particular species of *Russula* or *Lactarius* and its plant host form a viable symbiosis. The difficulty in culturing ectomycorrhizal fungi could be due to the complex nutritional requirements of ectomycorrhizae in the natural ecosystem; these include (1) a soluble carbohydrate source, (2) an exogenous source of thiamine, (3) other vitamins (required by some fungi), (4) growth factors from root exudates, and (5) other essential amino acids. Consequently, fungi currently considered saprotrophic might actually be ectomycorrhizal at some point during their life cycle.

Although the diversity of plants that form ectomycorrhizae is comparatively narrow (compared with plants that form endomycorrhizae), they are extremely important in many ecosystems because of the disproportionate amount of terrestrial habitat a relatively small number of plant species occupies and the economic importance of many plants that form ectomycorrhizae. For example, Pinaceae are exclusively ectomycorrhizal and dominate the northern boreal forests around the world. Likewise, Fagaceae, whose members comprise a large component of the deciduous temperate forest trees are also ectomycorrhizal (Table 4.2). Ectomycorrhizae occur on the roots of plants in most genera in Pinaceae, Betulaceae, Fagaceae, Dipterocarpaceae—as well as on members of Salicaceae, Myrtaceae, Juglandaceae, and Ericaceae all of which can be dominant trees in terrestrial ecosystems.

2. Structure of Ectomycorrhizae. Unlike endomycorrhizae, ectomycorrhizal fungi alter the external appearance of plant roots such that distinct ectomycorrhizal short roots can be visualized using a dissecting microscope (Figure 4.4). Ectomycorrhizae form hyphae that usually do not penetrate the cell walls of the root, except in older, senescent, cortical cells, where intracellular penetration can occur. Hyphae colonize actively growing secondary roots and usually grow in an intercellular fashion. Root tips are usually covered by a mantle of hyphae that vary from being a loose web to a parenchyma-like layer of cells enclosing the root. In fact, 20–40% of the colonized root weight can be composed of the fungal mantle and sheath. Construction of the mantle is well conserved and under fungal control. For additional information on mantle structure and morphology, it is suggested the reader reference an atlas on mycorrhizal varieties such as Agerer (1987).

An important hyphal network, the *Hartig net,* penetrates from the mantle into the first few cell layers of the root cortex. The structure and depth of penetration of the Hartig net differs between gymnosperms and angiosperms. The majority of angiosperms possess an epidermal Hartig net (i.e., penetration is confined to the epidermal layer). However, there are two variants. The *paraepidermal* type is characterized by partial encircling of the epidermal cells, while the *periepidermal* type is characterized by hyphae that completely encircle the entire cell. Upon careful examination of many ectomycorrhizal associations, it has been recognized that there is a continuum between the two extremes. The structure of the Hartig net is determined by the symbionts involved, the age of the mycorrhiza, and other edaphic factors. Ectomycorrhizal gymnosperms differ because they possess a cortical Hartig net that penetrates beyond the epidermis to enclose several layers of cortical cells.

Intercellular penetration by the Hartig net can only occur at a specific stage of epidermal cell development. Fungal enzymes are likely involved in the dissolution of the middle lamellae of epidermal cells in advance of the hyphal tips. The adjacent plant and fungal walls become indistinguishable and fused to form the interfacial

TABLE 4.2 Genera Reported to Contain at Least One Species on Which Ectomycorrhizae Have Been Described

Family or Subfamily	Genus[a]	Family or Subfamily	Genus[a]	Family or Subfamily	Genus[a]
Aceraceae	*Acer* (B)	Ericaceae (cont.)	*Rhododendron*	Pinaceae	*Keteleeria*
			Vaccinium	(cont.)	
					Larix
Betulaceae	*Alnus* (B)				*Picea* (B)
	Betula	Euphorbiaceae	*Poranthera*		*Pinus* (B)
	Carpinus (B)				*Pseudolarix*
	Corylus (B)	Fagaceae	*Castanea* (B)		*Pseudotsuga* (B)
	Ostrya (B)		*Castanopsis* (B)		*Tsuga* (B)
	Ostryopsis (B)		*Fagus* (B)		
			Lithocarpus (B)	Platanaceae	*Platanus* (B)
Caesalpinioideae	*Afzelia* (B)		*Nothofagus* (B)		
	Aldina		*Pasania* (B)		*Coccoloba*
	Anthonota		*Quercus* (B)		*Polygonum**
	Berlinia		*Trigonobalus* (B)		
	Brachystegia			Rhamnaceae	*Cryptandra*
	Eperua	Gnetaceae	*Gnetum*		*Pomaderris*
	Gilbertiodendron (B)				*Rhamnus* (B)
	Intsia (B)	Goodeniaceae	*Brunonia**		*Spyridium*
	Julbernardia (B)		*Goodenia** (B)		*Trymalium*
	Monopetalanthus				
	Paramacrolobium	Hammamelidaceae	*Parrotia*	Rosaceae	*Chaembatia*
	Swartzia				*Cirocarpus*
	Tetraberlinia	Juglandaceae	*Carya* (B)		*Crategus* (B)
			Engelhardtia		*Dryas* (B)
Caprifoliaceae	*Sambucus* (B)		*Juglans* (B)		*Malus* (B)
			Pterocarya		*Prunus* (B)
Casuarinaceae	*Casuarina* (B)				*Pyrus* (B)
	Allocasuarina (B)	Mimosoideae	*Acacia* (B)		*Rosa* (B)
					Sorbus (B)
Cistaceae	*Helianthemum* (B)	Myrtaceae	*Angophora* (B)		
	Cistus (B)		*Callistemon* (B)	Salicaceae	*Populus* (B)
			Campomanesia (B)		*Salix* (B)

Cupressaceae	*Cupressus* (B)		*Eucalyptus* (B)		
	Juniperus (B)		*Leptospermum* (B)	Sapotaceae	*Glycoxylon*
			Melaleuca (B)		
Cyperaceae	*Kobresia** (B)		*Tristania* (B)	Sterculiaceae	*Lasiopetalum* (B)
					Thomasia
Dipterocarpaceae	*Anisoptera* (B)	Nyctaginaceae	*Neea* (B)		
	Balanocarpus		*Torrubia* (B)	Stylidiaceae	*Stylidium* (B)
	Cotylelobium		*Pisonia* (B)		
	Dipterocarpus			Thymeliaceae	*Pimelia* (B)
	Dryobalanops	Papilionoideae	*Brachysema*		
	Hopea		*Chorizema*	Tiliaceae	*Tilia* (B)
	Monotes		*Daviesia*		
	Shorea (B)		*Dillwynia*	Uapacaceae	*Uapaca*
	Valica		*Eutaxia*		
			Gompholobium (B)	Ulmaceae	*Ulmus* (B)
			Hardenbergia (B)		
Elaeagnaceae	*Shepherdia*		*Jacksonia*		
			Kennedya		
Epacridaceae	*Astroloma*		*Mirabelia* (B)		
			Oxylobium (B)		
Ericaceae	*Arbutus*		*Platylobium*		
	Arctostaphylos		*Pultenaea*		
	Chimaphila		*Viminaria* (B)		
	Gaultheria				
	Kalmia	Pinaceae	*Abies*		
	Ledum		*Cathaya*		
	Leucothoë		*Cedrus* (B)		

Note: This list is not exhaustive, but illustrates the wide range of families and genera of angiosperms and gymnosperms in which ectomycorrhizae have been observed. A record of a mycorrhizal species within a genus does not mean that all species within that genus are ectomycorrhizal; nor does it imply that ectomycorrhizal colonization is consistently or normally present in any species within the genus.

[a]Those marked * are herbaceous plants, and those marked with a (B) may form both ecto- and VA mycorrhizae, with VAM the more common type of mycorrhiza observed.

Source: Modified from Harley and Smith 1983, with permission.

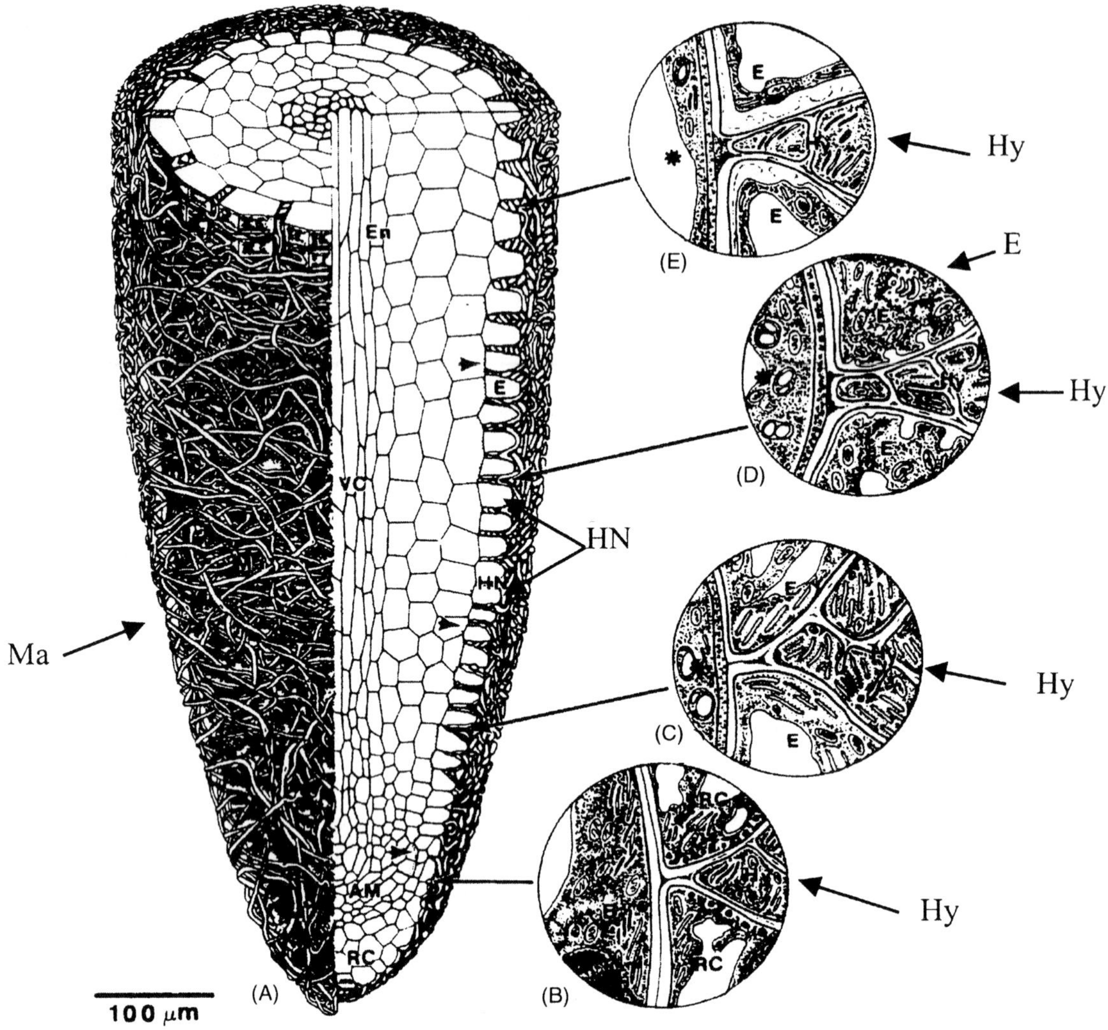

Figure 4.4. Schematic drawings of an ectomycorrhiza formed between *Alnus crispa* and *Alpova diplophloeus*. (A) Three-dimensional drawing of a mycorrhizal root, showing the mantle (Ma), paraepidermal Hartig net (HN) and epidermal cells (E), which become progressively more radially enlarged from the apex back. (B) Root cap region. Hyphae have begun to penetrate between the cells of the root cap (RC). The densely cytoplasmic epidermal cell (E) is also shown. (C) Hartig net region. Fungal hyphae (Hy) with rough endoplasmic reticulum and labyrinthine wall branching are penetrating between the epidermal cells (E). (D) Mature Hartig net region. Fungal hyphae (Hy) have reached as far as the modified wall of the hypodermis (*). Epidermal cells (E) show wall modifications including ingrowths. (E) Older Hartig net region. Fungal hyphae (Hy) show reduced numbers of cysternae, endoplasmic reticulum, and mitochondria. Epidermal cells (E) show modifications including deposition of wall material, vacuolation, and a decrease in the number of mitochondria. The hypodermal cell (*) is also more vacuolated. (From Massicotte et al. 1986 with permission.)

matrix. These are possible sites for resource exchange, and are usually found between zones on the root with less mycorrhizal development and senescent areas where hyphae actually penetrate the plant cell walls.

Evidence for cell wall degrading enzymes in ectomycorrhizae are still somewhat circumstantial. Cellulolytic enzymes have been found in *Pisolithus tinctorius* including three distinct isoforms of β-glucosidase, which indicates a tightly regulated cellulolytic capability in this fungus (Cao and Crawford 1993a, 1993b). Endoglucanases have also been found, possibly implicating their involvement in this process. In addition, some ectomycorrhizal fungi can be cultured and grown on pectin as a sole source of carbon indicating the presence of pectinase activity. In fact, wall-bound pectinase and endopolygalactonurase activity has been detected in many ectomycorrhizal fungi.

Hyphal strands, varying in structure among ectomycorrhizae, emanate from the mantle exploring the soil and subsequently conducting limiting soil resources (e.g., phosphorus and water) through the mycelial network to the root cortex. These hyphal strands are often connected to a larger mycelium present in the soil or even to fruiting bodies of the ectomycorrhizal fungi. It is also common in some fungal species for these hyphae to agglutinate together to form larger hollow tubes or rhizomorphs, which are also thought to function in long-distance resource transfer. There has been speculation that these strands may create connections between ectomycorrhizal plants, possibly allowing for reciprocal transfer of nutrients, water, and carbon. For a more detailed description of ectomycorrhizal root morphology, see Smith and Read (1997).

As mentioned, it is not uncommon for ectomycorrhizal fungi to assume a weakly pathogenic form and penetrate cortical cells in senescent portions of the plant root. However, should penetration of healthy root cells occur by the fungus, whether as part of the Hartig net or the sheath complex, then the association is usually defined as an *ectendomycorrhiza*. Some fungi are capable of forming both ectomycorrhizae and ectendomycorrhizae. For example, there are fungi that form ectomycorrhizae with members of Fagaceae and Pinaceae, and also form ectoendomycorrhizae with plants in the Ericaceae. These fungi form Hartig net mantles while colonizing the root epidermal cells intracellularly, and are called *arbutoid mycorrhizae*.

While three structural elements (Hartig net, sheath or mantle, and extraradical mycelium) routinely define the ectomycorrhiza, there is often a great deal of variation in their formation and morphology. Plant roots colonized by ectomycorrhizal fungi typically take on a unique if not distinct morphology (i.e., *morphotype*). Colonized root tips are shorter and cease to produce root hairs. Some mycorrhizal roots are bifurcated, and others have a bristle-brush appearance because stiff hyphae protrude from the mycorrhizal root tip. Such distinct morphotypes can be used to identify specific fungal symbionts in the absence of fruiting bodies (Agerer 1987), which is normally a difficult task. These techniques are improving field identification of ectomycorrhizal partners, and thus improving our understanding of the structural and functional relationships associated with symbioses between particular species of fungi and plant. The identification of fungal partners in symbioses is improving further with more modern techniques, including the use of molecular fingerprinting of the fungi associated with individual root tips (Gardes et al. 1991a, 1991b; Gardes and Bruns 1993, 1996).

3. Ectomycorrhizal Fungal Modes of Colonization. Like VA fungi, germination of ectomycorrhizal fungal spores and growth of mycelium in the viscinity of a suitable host is stimulated by exudates from the root, which contain a variety of sugars, amino acids, and vitamins. In fact, exudates from both mycorrhizal and nonmycorrhizal roots have been shown to induce ectomycorrhizal spore germination. Some of these exudates have homologous functions to flavonol compounds that also are responsible for inducing the formation of nitrogen-fixing associations between *Rhizobium* or *Agrobacterium* and plants. Root exudates also may include phytohormones such as cytokinins (Gogala 1991) and IAA (Gay and Debaud 1987), which subsequently lead to increased hyphal branching. Unidentified water-soluble volatile compounds are also thought to be important for colonization. Occasionally, root exudates promote the formation of enhanced microbial communities in the soil, which in turn stimulates the formation of ectomycorrhizae. For example, the rate and extent of ectomycorrhizae formation can be increased under certain conditions by the presence of fluorescent *Pseudomonas* (i.e., mycorrhizal helper bacteria) (Garbaye 1994). Consequently, we now know that trigger compounds can be produced by organisms other than the fungal symbiont and plant involved. Chemotrophic responses of fungi to root exudates enable fungal mycelia to orient their growth toward roots, to contact a root surface, and possibly to adhere to it. Upon contact with root hairs, ectomycorrhizal hyphae also can alter the orientation of their growth toward the surface of the root, eventually partially enveloping the root hairs. After contacting the root hair, the fungal hyphae change morphologically, showing increased branching and fusion of hyphal tips prior to complete mantle formation (Jacobs et al. 1989).

The exact mechanisms that allow the attachment of compatible hyphae to the root resulting in a successful symbiosis and rejection of incompatible hyphae remain unclear. However, several lines of evidence point to the involvement of phenolics, fibrillar proteins, and lectins. Changes in plant phenol production and localization in roots have been demonstrated and are likely involved with rejection of incompatible fungal hyphae. Increased deposition of phenol compounds at the hyphal point of contact with the root is likely the most important mechanism for rejecting incompatible strains of fungi. Ultrastructural studies have demonstrated the presence of fibrillar material, made up of glycoproteins, extending from the cell walls of the fungal mycelium to the plant root surface (Lei et al. 1990a). In fact, a layer of extracellular fibrillar polymers is present on the surfaces of the free-living mycelia of *Laccaria bicolor* and *Pisolithus tinctorius* even before there is contact with the plant root. These fibrillar polymers are possibly involved in recognition or attachment because they reorient toward the root of compatible hosts during mycorrhizal formation (Lei et al. 1990b).

Plant-specific binding sites for lectins, isolated from the cell walls of *Lactarius deterrimus,* have been found on the surfaces of *Picea* root cells. Binding of fungal lectins to plant-specific polysaccharides on the cell walls of the root is likely critical for initiating the symbiosis (Peterson and Farquar 1994). Subsequently, the fungal symbiont forms a loose aggregate, or weft, of hyphae over the root surface, which slowly thickens and forms a sheath. Some hyphae penetrate the root by mechanically penetrating between cells of the cortex. More extensive invasion of the root is usually inhibited, possibly by chemical defense mechanisms such as endogenously produced volatile compounds. Ectomycorrhizal roots stay in a juvenile condition, failing to suberize or form root hairs (Table 4.3).

Lectin receptor sites are most likely present on both the fungal and plant symbionts involved in the symbiosis, but it is thought that unreactive materials may mask these sites. Enzymes such as laminarase or protease are necessary to unmask these sites in the fungal cell wall. It is possible that these enzymes would unmask critical receptor sites on the surface of both partners allowing for symbiont recognition and subsequently mycorrhization (Lapeyrie and Mengden 1993). A similar mechanism has been

TABLE 4.3 Sequence of the Formation of *Eucalyptus* Ectomycorrhiza Obtained Using the Cellophane-over-Agar Technique

Time	Developmental Stage	Anatomical Features
0–12 h	Preinfection	Hyphal contact with root
12–24 h	Symbiotic initiation	Fungal attachment to root epidermis
24–48 h	Fungal colonization	Initial layers of mantle formed
		Hyphal penetration between epidermal cells
48–96 h	Symbiotic differentiation	Rapid buildup of mantle hyphae
		Hartig net proliferation
96 h–7 days	Symbiotic function	Mantle well developed and tightly appressed to epidermal cells
		End of Hartig net growth

found to occur during the formation of the *Rhizobium*-legume symbiosis. Following activation of nodulation genes by flavonols, proteins are released that unmask critical receptor sites on the legume root hairs allowing the functional symbiosis to form (Krishnan and Pueppke 1993).

Recent work has demonstrated that the establishment of the ectomycorrhizal symbiosis may be under the regulation of proteins produced by inducible genes from both partners (Hilbert et al. 1991). Hilbert and Martin (1988) established *Eucalyptus globulus–Pisolithus tinctorius* associations in paper sandwiches and used two-dimensional electrophoresis to compare the protein profiles of mycorrhizal and non-mycorrhizal plants over time. In established mycorrhizae, there was a marked downregulation of some polypeptides present only in the free-living partners. Of the 520 polypeptides screened, 10 were unique to the mycorrhizal condition. These symbiosis-related proteins (SR proteins) have been called *ectomycorrhizins.* Major changes in polypeptide profiles and initiation of mycorrhizal synthesis occur within hours of contact with the fungus. Hilbert et al. (1991) followed the incorporation of ^{35}S-methionine label into protein during mycorrhization, and found intense labeling of ectomycorrhizins. Seven ectomycorrhizins are upregulated during the earliest stages of synthesis, which is indicative of accelerated biosynthesis of SR proteins during mycorrhizae formation. At the same time, polypeptide cleansing, which is a downregulation of several plant- and fungal-specific polypeptides, simultaneously accompanies ectomycorrhizin synthesis. It is important to note that such changes were only evident when compatible races of *Pisolithus tinctorius* challenged compatible roots, indicating that the recognition event is specific.

Similar results have been found using other eucalypt mycorrhizae (Burgess et al. 1993, 1994, 1995). *Eucalyptus* seedlings were challenged with one of three strains of *P. tinctorius:* one aggressive, one moderate, and one that would not form compatible mycorrhizae. Only during colonization by the most aggressive isolate, which actually increased seedling growth, was there a dramatic upregulation of fungal protein biosynthesis within 4 days of contact. Again it was found that during mycorrhization a downregulation of plant polypeptides occurred and was accompanied by enhanced accumulation of certain fungal polypeptides as well as those recognized as SR proteins. The most apparent changes occurred in a group of fungal acidic polypeptides that increased in concentration as the symbiosis developed up to 4 days after initial contact. In particular, one protein with an apparent molecular mass of 32 kD was strongly upregulated and accumulated in the fungal cell walls (Figure 4.5). The upregulated protein is composed of isoforms of an acidic polypeptide (AP 32) that accumulates during the early stages of mycorrhization. The enhanced production and localization of a specific SR protein as an insoluble component of the cell wall indicates that it could be involved in recognition or adhesion processes.

There is increasing support for the hypothesis that SR proteins act as hydrophobins (Martin and Tagu 1995). *Hydrophobins* are cysteine-rich proteins that become polymerized into the cell walls of hyphae giving them a waterproof coat. SR proteins change from a soluble state in submerged conditions to an insoluble form in the

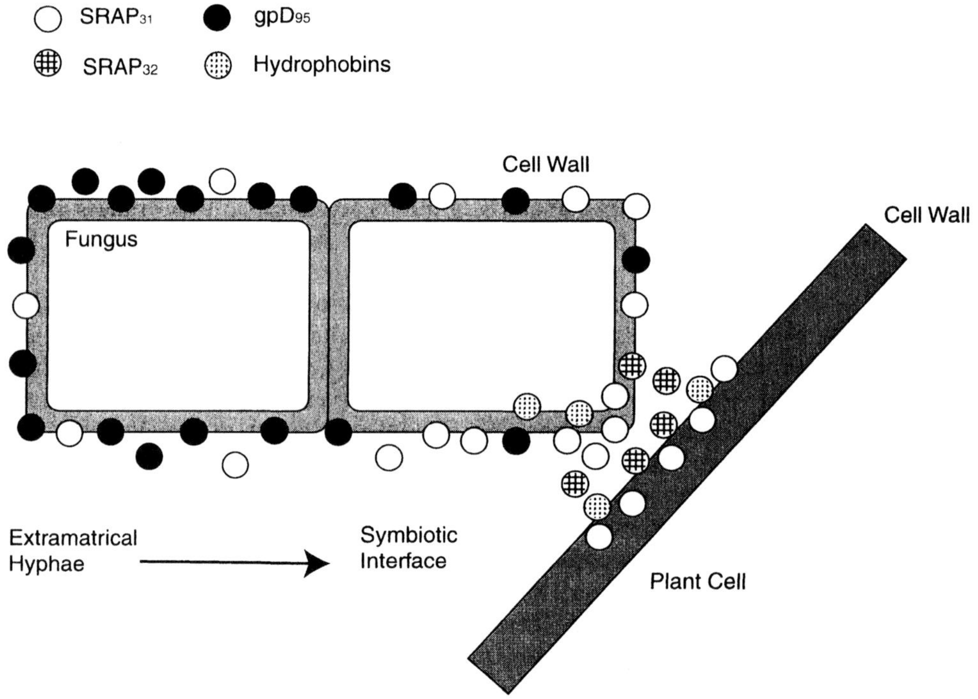

Figure 4.5. Hypothetical model summarizing current knowledge of the changes in protein composition of *Pisolithus tinctorius* cell walls. *Left:* In the free-living mycelium, the various isoforms of 31 and 32 kDa acidic polypeptides and the mannoprotein gp95 are located on the surface of the cell wall. An isoform of the alanine-rich 32 kDa acidic polypeptide is secreted into the extracellular medium. gp95 is a covalently linked polymer that undergoes extensive downregulation during ectomycorrhizae development; the disappearance of this protein and its associated glycans may be critical for early recognition (e.g., avoidance of plant defense reaction by suppression of surface glycans). *Right:* The mycelium interacting with *Eucalyptus* roots, synthesis of some isoforms of 32 kDa acidic polypeptides (i.e., $SRAP_{32}$) is upregulated; presumably they are components of the fibrillae bridging the symbionts. Mannoprotein is barely detectable in the interface. Hydrophobins are accumulated onto the hyphal surface and into the symbiotic interface. (Redrawn with permission from F. Martin and D. Tagu 1995. Ectomycorrhiza development: a molecular perspective. Pp. 29–58 in: A. Varma and B. Hock (eds.), Mycorrhiza: Structure, Function, Molecular Biology, and Biotechnology. Springer-Verlag, Berlin, Germany.)

mycorrhizal wall that is characteristic of a hydrophobin. This helps explain why the extracellular matrix of the mycorrhizal mantle has hydrophobic properties. Moreover, Rayner et al. (1994, 1995) have suggested that the deposition of hydrophobins in the hyphal walls behind the growing apices restricts nutrient absorption or loss to the apical region of the mycorrhizal root. Hydrophobins are likely important for hyphae-root recognition in the formation of the mycorrhizae.

Two popular theories have been developed and investigated concerning the physiological processes involved in the initiation and regulation of the ectomycorrhizal symbiosis. These theories have been researched extensively and critically reviewed. Both are flawed methodologically and conceptually. The first theory developed was the *carbohydrate theory* (Björkman 1949). The basic tenet of this theory is that mycorrhizae develop only under circumstances where the roots of the plant contain a surplus of soluble carbon. Björkman acknowledged that mineral deficiencies, especially N limitation, were an important prerequisite for mycorrhization, but his theory maintains that carbohydrate surplus is the main factor governing the formation of a functional symbiosis. Experiments using ergosterol extractions and regression analyses have concluded that variations in nitrogen concentration of the plant shoots alone could explain differences in the percentage of root colonization. Therefore, total N has an effect on mycorrhizal formation that is not necessarily mediated by surplus carbohydrate (Figure 4.6). Additional research has shown that carbon is important for mycorrhizae formation, but the requirement of a surplus supply of carbohydrate for synthesis may not be correct.

The second theory is the *auxin theory*, first proposed by Slankis (1973). According to the auxin theory, mycorrhizae formation is regulated through controlled production of auxins by the fungal symbiont. This theory is better supported now than in 1983, when Harley and Smith originally reviewed it. It has been demonstrated that some ectofungal symbionts form IAA in monoxenic culture. Yet the analyses of IAA concentrations in mycorrhizal roots have produced conflicting results. For example, a study by Mitchell et al. (1986) showed significantly more IAA in roots of *Pinus echinata* colonized by *Pisolithus tinctorius* than in uncolonized controls. Another study, by Wallander et al. (1994), found that *Pinus sylvestris* colonized by *Laccaria bicolor* experienced a 40% reduction in the IAA content of colonized roots relative to nonmycorrhizal roots.

Researchers have developed mutant strains of *Hebeloma cylindrosporum* that produce elevated levels of IAA from endogenously added tryptophane. Mutant strains were produced from wild-type monokaryons by a two-step procedure using fluoroindole resistance (FIR) as a selection factor. One mutant strain had an overproduction of IAA resulting from a single gene mutation (Durand et al. 1992). Additional research has shown that these mutants have an advantage in mycorrhizae formation, presumably because of higher levels of IAA. In fact, three to six times more *Pinus pinaster* roots are colonized by FIR mutants than by the wild type *Hebeloma* (Gay et al. 1994). Still, most of the evidence for the auxin theory is somewhat insufficient because the correlation between IAA production and subsequent mycorrhizal activity has not been well established.

Despite complications of the study systems, fungi that overproduce IAA develop thicker multiserrate Hartig nets and often form intracellular hyphae that are surrounded by the plasma membrane and tonoplast. These intracellular hyphae are atypical for ectomycorrhizae and indicate reduced or no pathogenicity. Interestingly, mutant plants forming these mycorrhizae exhibit no increase in growth upon mycorrhization. Are the mycorrhizae still considered functional? Of course, no increase in plant growth

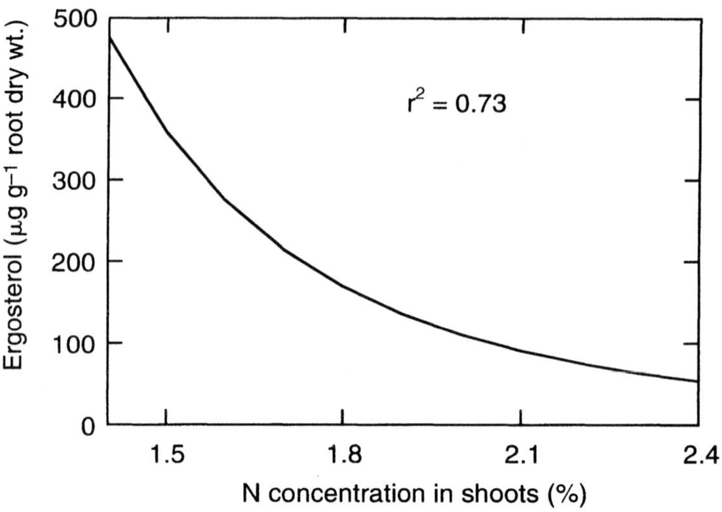

Nitrogen Treatment (mg N l^{-1})	Nitrogen (%) Needles before Inoculation	Root Sugar (%) (before inoculation)		Ergosterol ($\mu g\ g^{-1}$ of root) (after inoculation)
		Whole Roots	Lateral Roots	
0-10 N-b	1.5 (0.2) a	4.6 (0.1) a	3.6	214 (38) a
200 NH_4	2.7 (0.1) b	4.9 (0.3) a	3.2	112 (16) b

Figure 4.6. The general relationship between nitrogen content of the shoots in *Pinus sylvestris* and ergosterol concentration of the roots, which is an indicator of mycorrhizal colonization by *Laccaria bicolor*. Along with the table, these data demonstrate that nitrogen has an effect on mycorrhizae formation that is not mediated by carbohydrate content alone. Standard errors are in parentheses. Lowercase letters in the table denote significant differences. (From Salmanowicz and Nylund 1988 with permission.)

could be a result of increased sink demand and consequently greater fungal biomass being supported by the plant. We must keep in mind that these theories are a gross oversimplification of the process of ectomycorrhizae formation, but they serve as a foundation to which additional information can be added. The use of biochemically created mutants will certainly be important for future research on the mechanism of ectomycorrhizae synthesis.

C. Ectendomycorrhizae

As mentioned previously, these are mycorrhizae that exhibit characteristics of both ectomycorrhizae and endomycorrhizae. Macroscopically, ectendomycorrhizae are very similar to the ectomycorrhizae; however, upon closer inspection, they show increased levels of intracellular penetration beyond what normally occurs with ectomycorrhizae in cells that are senescing. Functionally, the symbiosis is essentially equivalent to that of ectomycorrhizae. Ectendomycorrhizae occur mainly on conifers (e.g., *Picea* sp.) and are formed by the ascomycete *Wilcoxina*. Two species (*W. mikolae* and *W. rhemii*) have distinctive habitat preferences (Egger et al. 1991). Originally, the fungi that form this association were all classified as the same unidentified strain of fungus called the E-strain (Mikola 1965). However, it was later discovered that E-strain fungi had strong ascomycete affinities when they formed operculate discomycetes, a characteristic of the Pezizaceae, in pot culture (Yang and Wilcox 1984).

D. Arbutoid and Monotropoid Mycorrhizae

Similar to ectendomycorrhizae upon initial inspection, mycorrhizae formed by Arbutoideae and members of Monotropaceae can and have been confused with ectomycorrhizae. Ultrastructural studies using electron microscopy have allowed for improved understanding and differentiation of the arbutoid and monotropoid types of mycorrhizae. Basidiomycete fungal symbionts and plants in Arbutoideae (e.g., *Arbutus* and *Arctostaphylos*) form arbutoid mycorrhizae. *Arbutoid mycorrhizae* are characterized by a thick mantle or sheath, intercellular development of mycelium within the root cortex to form a Hartig net, and intracellular penetration of some of the epidermal cells by fungal hyphae (Massicotte et al. 1993). The fungal symbionts involved can often form ectomycorrhizae with woody canopy plants or are closely associated with the ectomycorrhizae of canopy plants. Only upon close anatomical examination can these mycorrhizae be distinguished from similar-looking ectomycorrhizae.

Similar to arbutoid mycorrhizae, *monotropoid mycorrhizae* are formed by ectomycorrhizal fungal symbionts in association with achlorophyllous plants of the Monotropaceae (*epiparasites;* see Chapter 10). Phylogenetic studies have demonstrated that the closest relatives of the achlorophyllous monotropoid plants are likely the chlorophyllous members of the genus *Arctostaphylos* (i.e., which also form arbutoid mycorrhizae). Monotropoid mycorrhizae form a well-developed, multilayered mantle similar to the ectomycorrhizal sheath that is formed on many *Fagus* spp. However, the mycelium rarely penetrates the cortical layers of the root, and as a result, the Hartig net is usually located around the epidermis. From the Hartig net, individual hyphae penetrate the root epidermal cells in much the same manner as the arbutoid mycorrhizae. However, the invading hyphae are encapsulated by the epidermal cell walls forming haustoria-like structures called hyphal pegs. Hyphal pegs, for reasons not discussed here, are not true haustoria, but these specialized structures are the key defining character for monotropoid mycorrhizae in relation to the phylogenetically similar arbutoid and ericoid mycorrhizae.

Physiologically, monotropoid and arbutoid mycorrhizae differ from other endo- and ectomycorrhizae, because they are classified as *mycoheterotrophic*. These mycorrhizae often link the host plant symbiont to a second host plant, presumably using the second host as a carbon source and thereby assuming a somewhat parasitic role (acting as epiparasites). In addition, the seeds of *Monotropa* are similar to orchid seeds because they are very small and contain virtually no endosperm. The embryo is highly reduced, and ceases development shortly after germination unless the appropriate fungal symbiont is present. The embryo has a limited ability to absorb nutrients in the absence of the fungus. Furthermore, these mycorrhizae function well in acidic soils characteristically present around ericaceous plants (members of the Ericaceae) and likely play a large role in the liberation of nitrogen from these organic soils for subsequent mobilization, uptake, and transport to a host plant.

E. Ericoid Mycorrhizae

Several families of the Ericales (i.e., Ericaceae, Epacridaceae, and Empetraceae) are important components in Northern and Southern Hemisphere plant communities, often forming distinct heathland plant communities. Except for a small number of ericaceous genera that form arbutoid mycorrhizae (i.e., *Arbutus* and *Arctostaphylos*), all other genera in these families develop distinctive mycorrhizae associated with strains of the dark pigmented ascomycetes *Hymenoscyphous ericae* or *Oidiodendron* sp. The soils that are typically inhabited by ericoid plants are usually acidic and relatively depauperate of nutrients. Colonization of plant roots by the ericoid endophyte results in increased yield associated with increased uptake of P and N, and concentration in the plant tissues (Read and Stribley, 1975).

Ericoid fungi are challenging to study for several reasons. First, when grown on nutrient agar, these fungi produce slow-growing, sterile, dematiaceous hyphae. The obvious lack of reproductive structures originally made classifying the ericoid fungal symbiont a challenge. Second, there is increasing evidence that a large amount of genetic diversification exists between isolates of ericoid fungi. Isolates taken from epidermal cells on the same root can even be genetically different (Perotto et al. 1995). Hence, the exact number and classification of ericoid symbionts are still not clearly understood.

Ericoid roots are distinctive, with up to 80% of the root being colonized by the ericoid fungus. The diameter of the ericoid mycorrhizal root is usually less than 100 μm, causing them to often be refered to as "root hairs" (Smith and Read 1997). The ericoid root epidermal cells do not produce root hairs. Instead epidermal cells are colonized intracellularly by the ericoid fungal symbiont, which forms dark intracellular hyphal complexes. There can be one or more entry points per cell, and entry can be preceded by the formation of appressoria, but these structures are not always well developed. Epidermal layers in the ericoid root are short lived. This outer cell layer is routinely sloughed away, exposing cortical cells that are usually suberized and thickened, preventing further colonization of that root region. As a result, the formation and maintenance of ericoid mycorrhizae are both spatially and temporally variable.

The dark-pigmented ericoid hyphae form a loose hyphal weft over mature epidermal regions of the root. As the root elongates and new epidermal cells are produced, hyphae advance forward and colonize the newly formed cells. Many strains of *H. ericae* produce an extracellular fibrillar sheath that is rich in polysaccharides. The sheath is more strongly developed in infective strains when compared to effective strains of the fungus (Bonfante-Fasolo and Gianinazzi-Pearson 1982, Gianinazzi-Pearson and Bonfante-Fasolo 1986). The fibrillar sheath is thought to anchor the fungus to the plant, as part of the first step of mycorrhizae formation. See Smith and Read (1997) for more extensive information of ericoid morphology.

The mechanism of cell wall penetration by ericoid fungi is not well understood. There is evidence from work using periodic acid thiocarboxydrezide silver proteinate (PATAG) that the hyphae of the ericoid fungal symbiont are capable of producing hydrolytic enzymes that dissolve the epidermal cell wall allowing for cellular penetration (Bonfante-Fasolo and Gianinazzi-Pearson 1982, Duddridge and Read 1982). When grown in pure culture, *H. ericae* is capable off using plant cell wall–related mono-, di-, and polysaccharides including carboxymethylcellulose as a sole source of C (Pearson and Read 1975). The enzymes believed to be involved with the penetration of host cell walls (i.e., β-(1,3)- and β-(1,4)-glucanase) are the same enzymes that are used by plant pathogenic fungi to penetrate cell walls. Their production, along with a weakening of the cell wall prior to hyphal penetration, strongly suggests the involvement of hydrolase enzymes (Varma and Bonfante-Fasolo 1994). These same enzymes could be responsible for the breakdown of soil organic matter and allowing for the subsequent release of carbon or other nutrients contained within the substrate. This ability to degrade complex sugars in the soil is somewhat unique compared with other mycorrhizae and may provide ericoid plants benefits that ecto- or endomycorrhizal plants do not receive.

F. Orchid Mycorrhizae

It has been known for over 100 years that plants in the Orchidaceae are mycorrhizal. Species of the imperfect fungus *Rhizoctonia* are the main fungal symbionts thought to be responsible for the formation of orchid mycorrhizae. These fungal strains are readily isolated and can be grown in pure culture (Smith and Read 1997). Regardless of the fungal symbiont involved in the orchid mycorrhizae, all orchid mycorrhizal fungi are capable of obtaining an external carbon source for the plant and the fungus. Many of the orchid fungal symbionts are fast-growing saprophytes that can decompose complex polymeric sugars such as pectin, cellulose, starch, and occasionally lignin. Others, such as *Rhizoctonia solani* and *Armillaria mellea,* are better known as plant parasites.

Fungal colonization of the minute orchid seeds is a requirement for germination *in situ.* Subsequently, orchids form one or more specialized endomycorrhizae (i.e., orchid roots can host more than one fungal symbiont species simultaneously) that infect the protocorm at an early stage. If an appropriate fungus does not colonize the orchid protocorm (i.e., forming a functional mycorrhiza), then the protocorm ceases to grow. Intracellular penetration and development of highly coiled hyphal complexes called pelotons characterize these mycorrhizae (Figure 4.7). Fungal hyphae may penetrate cells separately from the soil or may also colonize cell to cell. As in other endomycorrhizae, the coiled hyphal pelotons are thought to be a site of resource exchange between the orchid and the fungal symbiont. Again, it is recommended the reader see Smith and Read (1997) for further information on orchid mycorrhizae morphology.

Some orchid seeds can successfully germinate asymbiotically. The undifferentiated embryos may absorb water, slightly swell, and possibly produce epidermal hairs, but the embryo does not develop unless it receives an exogenous supply of carbon, which

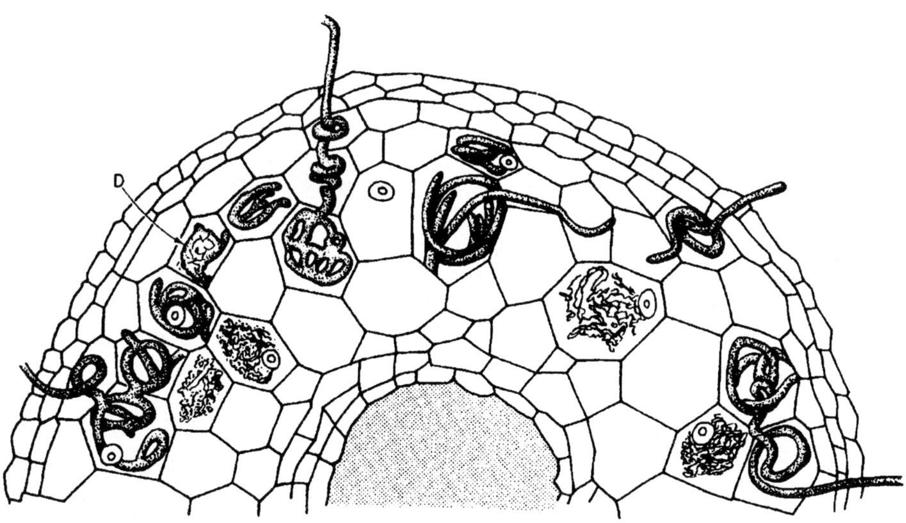

Figure 4.7. Cross section of an orchid mycorrhiza showing intracellular penetration, hyphal coils called pelotons and intracellular digestion of the hyphal coils by the plant host. (From Harley 1965. Copyright © 1965 The Regents of the University of California.)

in nature means being colonized by a compatible mycorrhizal fungus. Furthermore, in some species of orchid, an exogenous supply of vitamins and other growth factors may be required for continued growth following germination. Fungal colonization is required for glucogenesis stimulation, mobilization of starch reserves, and ongoing nutritional support until the plant becomes photosynthetic. In some cases, orchid species never form any photosynthetic tissue except that in the root endodermis (i.e., the ghost orchid).

Orchid mycorrhizae are unusual because, like monotropoid and some arbutoid mycorrhizae, they are usually epiparasitic or mycoheterotrophic on ectomycorrhizal plants. In parasitic fashion, the subterranean as well as achlorophyllous orchid species are obligately dependent upon their fungal symbionts to provide them with an adequate supply of carbon, presumably from a donor plant. Other nearby photosynthetic plants serve as carbon donors. Growing evidence demonstrates mycelial links between orchid mycorrhizae and ectomycorrhizal plants (Warcup 1985, 1991). Both the orchid and the fungus can subsequently use the carbon available from the photosynthetic plant. In fact, many orchid mycorrhizae are also formed by members of ectomycorrhizal fungi (e.g., *Thelephora* and *Tomentella*). Furthermore, all orchids including the chlorophyllous adult species pass through a post-germination stage, during which they are photosynthetically limited and dependent upon an exogenous supply of carbon to survive. Whether these are true mutualistic associations is debatable because there is little evidence showing any benefits derived by the fungus from the orchid in the association. There is no supporting evidence for the reverse flow of C from the orchid to the fungus in these relationships, even when the orchid matures and becomes photosynthetic. Any reverse transfer is considered negligible.

Orchid mycorrhizal associations are less stable than other mutualistic mycorrhizae (Figure 4.8). In many cases, the symbiosis borders on parasitism, and can even become highly pathogenic for the orchid. At any stage of the mycorrhizal development, the fungal symbiont can become pathogenic, stunting orchid growth or leading to orchid death. Conversely, for unknown reasons, the orchid might completely reject the formation of an intact mycorrhiza. Control over the nature of the symbiosis in orchid mycorrhizae is still not understood, but likely depends upon the orchid species involved as well as the fungal symbiont and on the prevailing edaphic conditions at the time of mycorrhization. Likewise, the association is not viable if the fungal symbiont is pathogenic initially or becomes pathogenic after colonization and mycorrhizal formation.

III. GROWTH AND CARBON ECONOMY OF MYCORRHIZAL PLANTS

A. Mycorrhizal Influences on Growth

It has been realized for some time that mycorrhizal plants usually exhibit increased growth relative to nonmycorrhizal plants of the same species. In fact, that is part of the definition of a functional mycorrhiza. It is largely for experimental convenience that vegetative growth or biomass accumulation has been used as a primary indicator of the benefits associated with mycorrhizae formation. However, changes in plant growth and yield in response to mycorrhizae formation might not be the best direct measures of fitness. For example, in long-lived ectomycorrhizal species, such as many woody tree species, biomass parameters can be difficult to determine.

Very few studies have provided equivocal evidence of improved growth from mycorrhizal formation in the form of nutrient response curves (Figure 4.9). Notice that

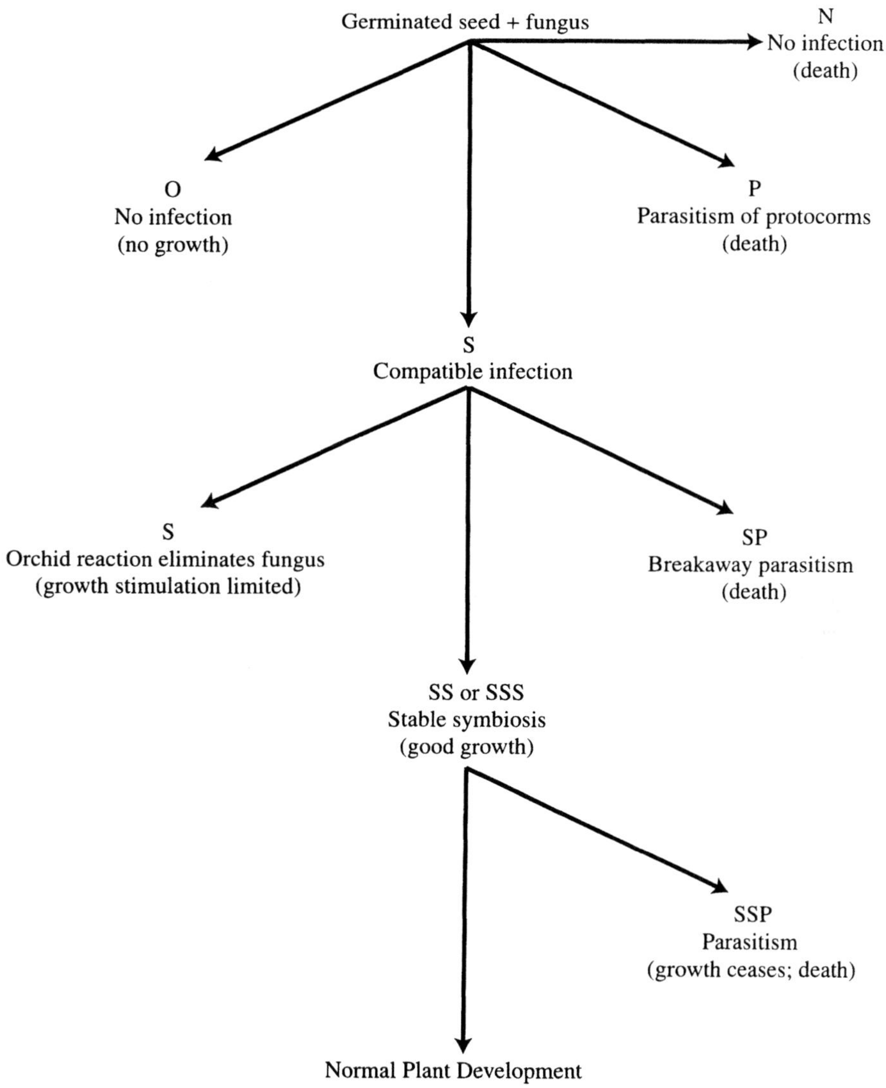

Figure 4.8. Diagram illustrating the possible outcomes of symbiotic development in orchids. O = noncolonized, no growth; S, SS, SSS = greater numbers of protocorm development; S* = initial colonization followed by hypersensitive reaction or complete fungal disintegration; P = fungus becoming an aggressive parasite, killing cells; N = no infection, protocorms dying for other reasons. The proportion of protocorms in each class depends upon the species of symbionts as well as prevailing environmental conditions. Little is known about what controls the balance of the symbiosis, but the mycorrhiza is clearly much less stable than other mycorrhizae. (Redrawn from Hadley 1970, with permission.)

the growth response of nonmycorrhizal plants is sigmoid in shape. There is an initial lag phase, due to a minimum threshold of P that must be overcome before growth will occur. Colonization by an appropriate mycorrhizal fungus eliminates this minimum threshold requirement and therefore allows plants to grow with greater success in nutrient-deficient soils. The removal of this threshold limitation is thought to be due to the external mycelium of the fungus being able to explore a soil volume that extends beyond the P depletion zones that occur within the local rhizosphere of the

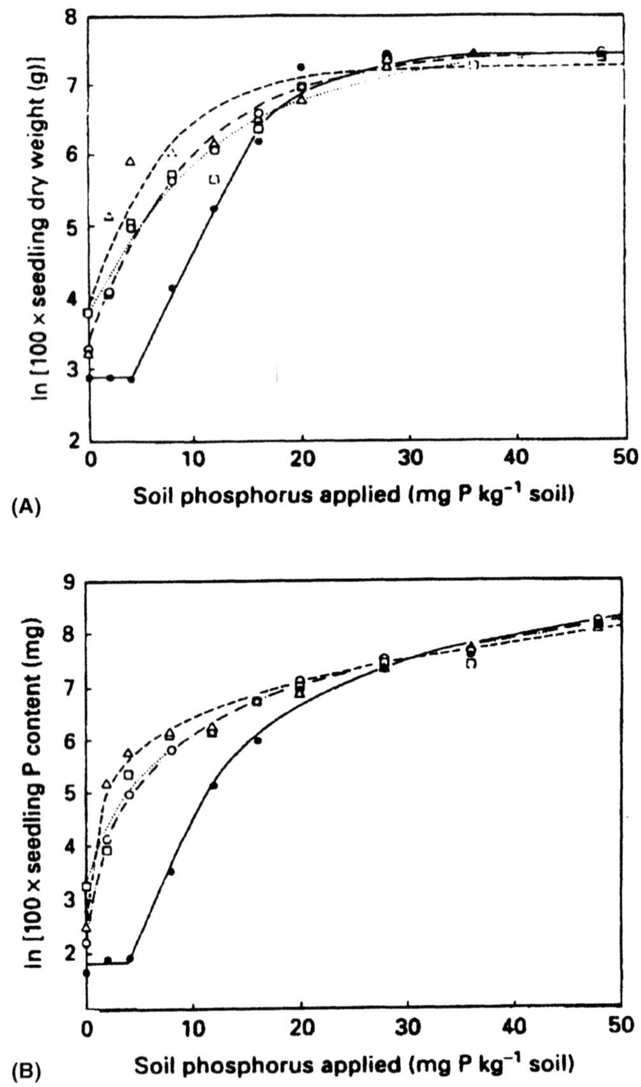

Figure 4.9. Growth and P uptake of *Eucalyptus diversicolor* in responses to P application and mycorrhizal colonization by two strains of *Descolea maculata* (□) A and (○) B or by (△) *Laccaria laccata* compared with (●) uninoculated controls. (A) Dry weight. (B) P content. Notice the sigmoid shape of the control response curves as a result of initial lag. (From Bougher et al. 1990, with permission.)

nonmycorrhizal plant. Fungal symbionts may also have a lower threshold concentration for absorption of P and might also have the ability to release P associated with complex substrates, which would also allow them to have greater uptake in these particular soils. Any of these mechanisms of improving P acquisition ultimately should increase plant growth. Now that we understand the importance of the external mycelium to the mycorrhiza, more interest is being devoted to the cost benefits of the mycorrhizal association, and there is less interest in the consequences to growth (discussed in a later section).

In mycorrhizal plants, growth continues until dry weight reaches a plateau, but unlike nonmycorrhizal plants, P content continues to increase linearly after the dry weight plateau has been reached. Consequently, mycorrhizae formation can change the growth response curves of plants that are growing in nutrient-poor conditions, and luxury consumption of P may occur. Although this pattern holds true most of the time, there is considerable variation induced by different fungal symbionts both between and within species. Also, it is important to realize that this example is illustrative of a response to P when other nutrients are supplied in adequate amounts. In natural systems, growth responses can be dependent upon other limiting nutrients, particularly N, and certainly other essential nutrients and resources such as water could be co-limiting the growth response to mycorrhization.

Root/shoot ratio has been used to determine the efficacy of mycorrhizal associations. If there is an increase in resource uptake from the mycorrhiza and a subsequent increase in carbohydrate production, then this should increase the root carbon sink and the root/shoot ratio. This parameter is difficult to determine and interpret in mycorrhizal plants because of the obvious intimate fungal association with the root system. How are the fungal and root components separated in order to obtain a true root/shoot ratio? Increases in the root/shoot ratio often can be attributed to increases in fungal biomass following colonization or to a change in the root system architecture (i.e., increased number of short roots). Alterations such as these cannot be disassociated from other mycorrhizal associated traits. For example, along with the increased root/shoot ratio, surface area and nutrient absorption dynamics are necessarily changing. Hence, it is difficult to determine whether the benefit to plants derives from the increased root/shoot ratio alone or whether increased root/shoot ratios are the result of other benefits from the association.

Although there is evidence of improved growth by mycorrhizal plants under nutrient-poor conditions, there has been little work to investigate the effects of mycorrhizae on plants under supraoptimal nutrient conditions. Under these conditions, it is thought that the benefits accrued by the plant from increased nutrient uptake through a mycorrhiza would be lost. Yet the cost to the plant in terms of carbon required by the fungus or respiration would still remain the same or would increase. Under such conditions, does the association become parasitic to the plant? In fact, many studies have demonstrated that there is no or very little growth reduction under adequate or supraoptimal nutrient conditions; however, under slightly lower than supraoptimal nutritional status of the substrate, such reductions in growth upon mycorrhization can be evident. Hence, the evidence for growth reductions are difficult to interpret.

Plants in moderately high nutrient regimes generally have lower root/shoot ratios relative to those under a low-nutrient regime. As a result, plants with a lower root/shoot ratio have more tissue to gain carbon and less tissue to consume C. Therefore, if equal quantities of root tissue were colonized in seedlings with different root/shoot ratios, those with a higher ratio would have proportionally higher mycorrhizal development per plant than the same species with a lower root/shoot ratio. If a mycorrhizal fungus constitutes 20% of the root system of a seedling with a 1:1 root/shoot ratio, it would make up 10% of the dry weight. In cases where the root/shoot ratio was 3:1, the fungus would only comprise 5% of the dry weight. The relative cost of supporting the fungus in a seedling with a root/shoot ratio of 1:1 would be double that of a seedling with a 3:1 root/shoot ratio, even though the fungal biomass was the same in both cases. There is additional evidence that growth reduction in mycorrhizal plants under higher-than-optimal nutrient conditions is further constrained by inhibition of the extraradical mycelium. Ultimately, the effects on root/shoot ratios may be related less to colonization that to overall plant size.

There have been studies that demonstrate decreased plant dry weight as a result of ectomycorrhizal colonization. These studies demonstrate that ectomycorrhizal development does not always result in increased plant growth in terms of size and weight, and there can be a decrease in plant dry weight shortly after colonization that is similar to the growth depression evident in many VAM associations. An initial lag phase can be prolonged, during which time growth is slower or equivalent to nonmycorrhizal control plants. Various plant-fungus combinations affect this relationship differentially (Table 4.4). Such decreases in plant growth could be expected in systems such as temperate deciduous forests where PAR, and not some other resource such as N or P, limits understory plant carbon gain. Under these conditions, the amount of photosynthate produced will be reduced because of reduced light availability, but the intensity of mycorrhizal colonization may not be affected.

B. Effect of Mycorrhizae on Plant Photosynthesis

Mycorrhizal symbioses can influence the net photosynthesis (P_n) of host plants. For example, ectomycorrhizal plants usually have higher P_n rates relative to nonmycorrhizal plants. Though this higher P_n may not always be evident in the long-term growth and carbon gain of the plant (Figure 4.10). It is important to remember that the carbon cost of maintaining the fungal symbiont is partly offset by increased photosynthesis by the host (Eissenstadt et al. 1993, Brown and Bethlenfalvay 1988, Fredeen and Terry 1988).

In mycorrhizal plants, the involvement of the fungal symbiont in nutrient acquisition may directly or indirectly increase the ability of the plants to fix CO_2. For instance, the increase in belowground resource acquisition by mycorrhizae is thought to be a major mechanism causing an increase in the net photosynthesis of the host plant. Increased photosynthesis in plants with VAM is likely mediated directly by increased P_i (inorganic phosphate) availability in the leaves, increased specific leaf area, increased specific leaf phosphorus, or increased leaf hydration (Fredeen and Terry 1988, Sivak and Walker 1986, Snellgrove et al. 1982).

However, Eissenstadt et al. (1993) showed that VAM citrus plants growing in low phosphorus conditions had higher photosynthetic rates than nonmycorrhizal plants

TABLE 4.4 Growth of *Quercus rubra* Seedlings Colonized by Different Strains of *Pisolithus tinctorius*

Fungal Strain	Height (cm)[a]	Fresh Weight (g)	Ectomycorrhizal Percent of Short Roots[b,c]
138	13.3	8.5a	72a
136	13.5	6.5c	15b
145	11.5	7.0b	1c
Control	12.5	6.7c	0

Note: Table illustrates that different combinations of plant and fungus are differently effective in growth. Variation could be due to differences in fungal species or (in this case) strains, or to differences in plant genotypes that react differently to fungal colonization.

Temperatures: day, 28°C; night, 22°C. Day length: 14.5 h. Growth period: 4 months.

[a] Height differences insignificant (p = 0.05).

[b] Within columns, common letters denote insignificance (p = 0.05).

[c] Estimated visually.

Source: From Smith and Read 1997, with permission; data from Marx 1979.

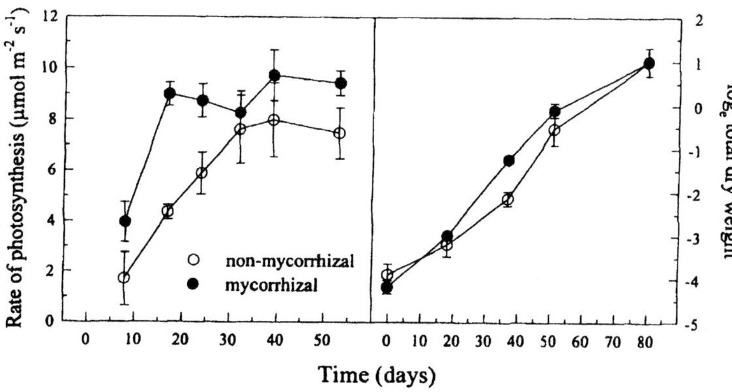

Figure 4.10. The steady-state photosynthetic rate of the youngest fullest-expanded leaf of mycorrhizal and nonmycorrhizal clover (*Trifolium repens*) (*left*), compared with rates of dry matter production (*right*). Significant increases in carbon assimilation of the mycorrhizal plants are not manifest in biomass accumulation relative to nonmycorrhizal plants over the course of the experiment. (Redrawn from Wright et al. 1997, with permission of Blackwell Science Ltd.)

amended with P fertilization, indicating an effect not mediated by P concentration in tissues. The most dramatic increase during this study occurred between weeks 7 and 8 of growth, when fungal colonization was most rapid and tissue phosphorus was the greatest. Relationships between mycorrhiza development, net photosynthesis, and foliar P indicated that with low and medium colonization, assimilation of mycorrhizal seedlings responded to increased foliar P concentration in the same way as uncolonized plants, suggesting a response to increasing P. However, at high levels of colonization, net rates of photosynthesis were significantly higher in mycorrhizal plants at the same foliar P concentration as nonmycorrhizal plants (Figure 4.11), thus indicating a possible nonnutritional indirect effect of mycorrhizae on P_n. The influence of increased sink strength on plant photosynthesis is one of the nonnutritional effects of mycorrhizae on plants.

It is generally thought that the rates of CO_2 assimilation by leaves are affected by increased demand for carbohydrate by mycorrhizae because the mycobiont is a strong sink for carbon (Sonnenwald et al. 1994). Several experiments have shown, at least for heavily colonized mycorrhizal plants, that carbon assimilation is significantly higher than noncolonized plants, and the observed effect is often not nutrient (i.e., phosphorus) mediated (Dosskey et al. 1990, Conjeaud et al. 1996). Eventually (at relatively high P content in the soil), both nonmycorrhizal and mycorrhizal plants have the same foliar concentrations of P but a different photosynthesis rate. Alternative explanations for the effect of mycorrhizae on photosynthesis have been proposed, including altered phytohormone production affecting photosynthesis, and the possibility that some nutrients other than P or N might be limiting photosynthesis in nonmycorrhizal plants. The data on this topic are conflicting and inconclusive. For example, occasionally plant growth does not increase, even though plant CO_2 assimilation and tissue nutrient concentrations do. In these cases, it is thought that increased photosynthesis in response to increased sink strength is not sufficient to account for the costs of mycorrhizae production and carbon allocation to external mycelium. In addition, different combinations of symbionts can have different effects on photosynthesis, such that some fungal symbionts stimulate increased photosynthesis while

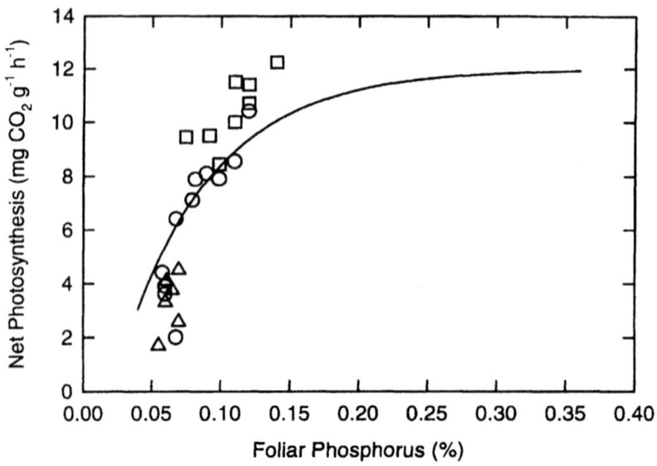

Figure 4.11. Net photosynthesis versus foliar P concentration (percent dry weight for ectomycorrhizal seedlings. The different symbols represent seedlings with different levels of glucosamine (i.e., different intensities of colonization), derived from the fungal symbiont associated with their roots: △ = low; ○ = medium; □ = high. The solid line shows the regression obtained from nonmycorrhizal seedlings. Notice the apparent nonnutrient-elicited effect on photosynthesis at the highest levels of colonization. (From Rousseau and Reid 1990 with permission.)

others do not. In general, mycorrhizal plants have higher photosynthesis than nonmycorrhizal plants, a difference most likely due to improved mineral nutrition, water uptake, and an increased sink demand by the mycorrhizal roots for carbohydrate.

C. Carbon Supplies for Ectomycorrhizal Fungi

Harley and Smith (1983) recognized that most ectomycorrhizal fungi have a limited capacity to degrade lignin and cellulose and subsequently use them as a carbon source for growth. Despite reports of slight lignase and cellulase activity in some ectomycorrhizal fungi, their capacity to degrade complex structural plant polymers is much lower than that of ericoid mycorrhizal fungi or wood decomposers that possess the laccase enzyme pathway (Cairney and Burke 1994). There is considerable inter- and intraspecies variability in fungal capacity to utilize oligo- and disaccharide sugars. Most fungi can only utilize simple monosaccharides, while some fungi can use pectins and others cannot. Potential sources of C vary among species of fungi, and in many cases we still do not know to what extent some mycorrhizal fungi can utilize sources of reduced carbon other than those provided by a host symbiont.

Certainly, enzymes allowing for the degradation of complex sugar polymers may be important for survival in the soil, but cellulose- and pectin-degrading substances are probably more important for penetration of root tissues. In ectomycorrhizae, a localized enzyme production allowing for degradation of polysaccharides could be important for softening the root epidermal and cortical cell walls during root invasion and Hartig net formation. A much higher enzyme activity would be required to release monosaccharides from complex polymers in the soil organic matter in sufficient amounts to affect fungal growth significantly. Therefore, the limited production of cellulose- and pectin-degrading enzymes by ectomycorrhizal fungi more likely serves the purpose of enabling penetration of the root tissues rather than of utilizing soil organic matter as a carbon source.

Do mycorrhizae utilize reduced carbon from plants? Melin and Nilsson (1957), provided the first proof that ectomycorrhizal fungi can derive carbon from plant roots. They fed $^{14}CO_2$ to the needles of axenically grown *Pinus sylvestris* plants in combination with the mycorrhizal fungi *Suillus variegatus* and *Rhizopogon roseolus*. Photosynthetic products were traced through the plants and subsequently found in the roots and the sheaths of the mycorrhizae. This study conclusively demonstrated the translocation of photosynthate from leaves to roots, but did not verify that the fungal symbiont was dependent upon carbohydrate from the plant or that root exudates were not the source for fungal carbon. Before we discuss the most accepted model for carbohydrate accumulation from plants to mycorrhizal fungi, some background information is necessary.

Mycorrhizal fungi do not have the capacity to accumulate sucrose and therefore must depend upon monosaccharide accumulation for their energy source. Monosaccharides, such as glucose and fructose, are readily absorbed from aerated solutions by excised mycorrhizal roots, with glucose being preferentially absorbed over fructose from the mixture. There is competitive inhibition between glucose and fructose uptake by fungi; in fact, glucose inhibits fructose uptake. Thus, it is probable that glucose and fructose have a common co-transporter located on the fungal plasma membrane. Furthermore, the rate of hexose absorption is temperature sensitive, oxygen dependent and is inhibited by the same metabolic inhibitors of the cytochrome oxidase pathway and of oxidative phosphorylation. Consequently, an active transport system requiring ATP is central to the accumulation mechanism. The glucose transport system is highly efficient and is capable of driving the extracellular glucose level close to zero. Once glucose and fructose are accumulated, mycorrhizae store those products as fungal-specific carbohydrates (Table 4.5). Monosaccharides such as glucose and fructose are common to both plants and fungi. Glycogen, mannitol, and trehalose are specific to fungi, whereas sucrose and starch are stored only in plants. Glucose is mainly converted to the fungal-specific sugar trehalose, while fructose is converted to mannitol. These fungal sugars are only very weakly reabsorbed by the plants (Table 4.6).

There are fungi (e.g., *Amanita muscaria*) that do not have the capacity to grow in a culture containing sucrose as the reduced carbon source. They are unable to hydrolyze sucrose to glucose and fructose because they lack the ability to excrete acid invertase (Schaeffer et al. 1995). As in many fungi, the protoplasts of *A. muscaria* can absorb both fructose and glucose, and glucose inhibits the uptake of fructose. Thus, this fungus has all the machinery needed to accumulate the products of sucrose hydrolysis, but does not have the enzyme to hydrolyze sucrose. The protoplasts of *A. muscaria* represent a common situation for many mycorrhizal fungi. The only way these fungi can utilize sucrose as an energy source is in symbioses when the disaccharide is hydrolyzed by an apoplastic or cell wall–bound invertase derived from the host plant.

Cell suspension culture has been used to determine that there are two forms of acid invertase produced by plants and moved into the apoplast. The two forms of invertase have different pH optima. There is an ionically bound form with a pH optimum between 3.5 and 4.5, and a tightly bound form that has a sharp pH optimum at 4.5 (Salzer and Hager 1993). Above a pH of 6, neither form has any significant activity. Both have a high K_m (K_m = Michaelis-Menten constant; the lower the value, the higher the enzyme efficiency) value for sucrose (16 and 8.6 m*M* for the ionically and tightly bound forms, respectively), and are competitively inhibited by fructose. These studies

TABLE 4.5 Carbohydrate Content of Excised Mycorrhizae and Changes after Storage for 20 h at 20°C in Water or in 0.5% (w/v) Solutions of Glucose, Trehalose, Sucrose, Fructose or Mannitol

	Initial	Changes after Storage In[a]					
	Sample[a]	Water	Mannitol	Fructose	Sucrose	Trehalose	Glucose
Total soluble	14.49	−5.36	−5.66	−1.53	−0.10	+2.02	+2.91
Total reducing							
Sugars	4.86	−1.02	−0.22	+0.34	−0.43	−0.69	+0.06
Sucrose	5.07	−2.34	−3.20	−0.99	−0.8	−0.06	+0.28
Trehalose	4.74	−2.00	−2.24	−0.88	+1.13	+2.7777	+2.57
Mannitol	Nil	Nil	+7.69	+6.36	+3.86	+2.27	+3.86
Insoluble (glucose units)	25.94	−2.84	+3.87	+6.83	+5.20	+8.10	+11.14
Total carbohydrate present	40.63	32.43	38.84	45.93	45.73	50.75	54.68

[a] Values as mg g^{-1} fresh weight (= 150 mg dry weight approximately).

Source: Data from Lewis and Harley 1965a. Table reproduced from Smith and Read 1997, with permission.

TABLE 4.6 Relative Rates of Carbohydrate Uptake by Mycorrhizal and Nonmycorrhizal Roots of *Fagus*

Sugar Supplied	Ratio of Uptake Rate, Mycorrhizal/Nonmycorrhizal
Glucose	3.1
Fructose	2.7
Sucrose	2.3
Trehalose	11.8, 10.8
Mannitol	19.6, 23.6

Source: Data from Lewis and Harley 1965b, 1965c; table from Smith and Read 1997, with permission.

suggest that carbohydrate transfer to the fungus of a mycorrhiza would be enhanced by acidification of the apoplast.

The current model for the movement of carbohydrate from plant to mycorrhizae is the *biological valve model* (Figure 4.12). This hypothesis maintains a net flux of carbon to the fungus from the plant and not in reverse. This model is dependent upon the rapid conversion of hexoses to fungal-specific storage compounds. It is generally accepted that a net flux of C to the fungus must be maintained. Trehalose, mannitol, and glycogen likely comprise a sink in the mycorrhiza, into which carbohydrates are accumulated in a form not readily available to the adjacent plant tissues even when present in the apoplast.

Sucrose produced by the plant is transported to roots, where it effluxes to the apoplast and is hydrolyzed by plant extracellular invertase into glucose and fructose. The apoplastic space must be acidified for optimum invertase activity. ATPase activity has been cytochemically demonstrated on both fungal and plant plasma membranes in the Hartig net region of *Pinus sylvestris–Laccaria laccata* mycorrhizae, and that activity has been shown to be associated with a H^+-ATPase transporter (Smith and Read 1997). In the acidified apoplast, glucose is absorbed by high-affinity fungal hexose transporters. Fructose, also released by invertase, exerts feedback inhibition on invertase activity, whereas glucose inhibits fructose uptake by the fungus. As glucose concentration in the apoplast declines, the fungus switches to absorbing fructose, which then releases the feedback inhibition on invertase, allowing renewed sucrose hydrolysis. Central to the biological valve hypothesis is the tenet that glucose and fructose do not diffuse back into the apoplast from the fungal mycelia because they are converted into fungal-specific carbohydrates.

There are some issues that the biological valve model does not take into account. First, it is still unlcear whether the plant may reabsorb sucrose and hexose sugars from the interfacial apoplast. Proton symports are the transporters for both fungi and plants; thus, there could be competition for the carbohydrate resources between the plant and the fungus. Net flux in the direction of the fungus, which we now know occurs, could only be maintained by mechanisms favoring net efflux of sucrose from the plant and net uptake as monosaccharides by the fungus. Consequently, monosaccharide uptake by the plant would likely not favor net flux to the fungal symbiont. Second, it has been suggested that hexose uptake by the fungus could be passive, with a concentration gradient being maintained by metabolic conversion within the fungus. Still, it seems most plausible that active proton co-transport would drive the inward

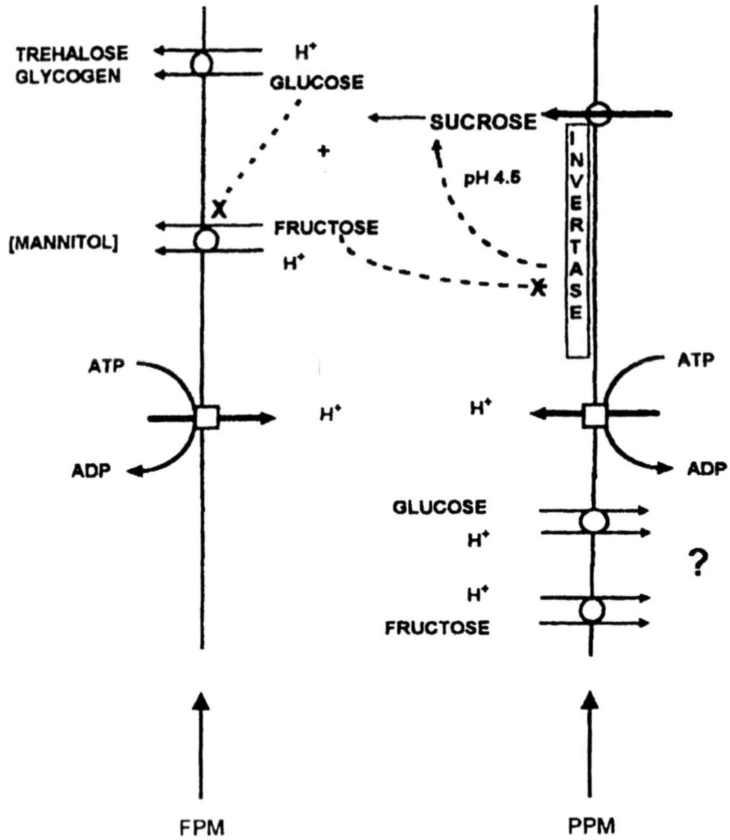

Figure 4.12. Diagrammatic representation of the plant-fungus interface in an ectomycorrhiza, showing the processes that might be important in sugar transfer from the plant to the fungus. Hydrolysis of sucrose effluxing from the plant occurs via the activity of a plant invertase, when the pH of the apoplast is at or below 4.5. Glucose is preferentially absorbed by the fungus and also inhibits (X) uptake of fructose by the fungus. Fructose might also inhibit invertase activity (X). Mechanisms for reducing or preventing reabsorption of sugars by the plant (?) are not understood. □ = H^+-ATPases; ○ = sugar transporter. (From Smith and Read 1997, with permission.)

flux of sugars, because low apoplastic pH, which is controlled by the membrane-bound H^+-ATPase complexes, is required to maintain invertase activity (Smith and Read 1997).

Contrary evidence to the biological valve model was found by Schaeffer et al. (1995) when using entire *Picea* mycorrhizae. These authors found a plant-derived acid invertase with a pH optimum around 4.0 and a K_m for sucrose of 5.7 mM. But, when comparing segments of roots that were mycorrhizal and nonmycorrhizal, it was found that the invertase activity was lower in mycorrhizal root segments. When the weight of the fungal tissue was taken into account, there was no effect of mycorrhizal colonization on invertase production by the plant or activity in the apoplast. In addition, 75% of the invertase activity was associated with the root cortical cells, which is tissue relatively inaccessible to the fungal symbionts for carbohydrate accumulation in ectomycorrhizal associations. Schaeffer et al. (1995) could find no evidence indicating that plant invertase activity increased with mycorrhization.

Of course, there are other plausible methods by which C might get from the plant to the fungus:

1. Cells sloughing off the root cap and epidermis, together with the mucigel surrounding the mycorrhizal root tip, might provide a colonizing fungus with a plant-derived carbon source.
2. As fungal hyphae penetrate the root to form the Hartig net, they might alter plant cell wall production at the *rosettes* (enzyme complexes that synthesize cellulose), and utilize plant cell wall precursors being delivered to the rosettes. For example, in areas around VAM haustoria, plant cells lose the capacity to organize a recognizable cell wall, but retain the capacity to secrete carbohydrates.
3. When hyphae penetrate senescent cortical cells, it is assumed that the fungus absorbs the original contents of the cells.

Although the mechanisms may not be clear yet, the establishment of the mycorrhiza does result in carbon flow from the plant to the fungus. However, it is also known that the movement of reduced carbon across the mycorrhizal interface is not exclusively toward the fungus. In fact, carbon compounds can return to the plant as the carbon skeletons of amino acids such as glutamine. The movement of glutamine originating in the fungal sheath to the plant has been demonstrated using ^{14}C labeling. It is unlikely that the amount of C transfer from the fungus to the plant by this route is significant to total plant carbon gain. However, it has been estimated that as much as 8% of the total weight of *Betula pendula* seedlings could be derived from their mycorrhizal fungus associate (Abuzinadah and Read 1989a,b).

Even though there is substantial evidence that C is translocated across the mycorrhizal interface, the amount of C transferred from the plant to the external mycelium is still mostly conjecture. It is only possible to make rough estimates for natural systems because of a series of complications in such analyses. For example, the extraradical mycelium provides connections between plants and between short roots on the same plant making determination of net flux from a particular plant to the mycorrhizal fungus difficult. Also, one plant can host a number of different fungal associations at the same time. There can also be flow of carbon from one plant to another via a mycorrhizal connection, so that the final destination is not the fungal mycelium. Sugars can be lost to the soil by root exudates. Finally, much of the fungal mycelium is lost with attempts to harvest host plants.

One way used to quantify the carbon cost of mycorrhization is to use a mass balance approach. The amount of carbon accumulated by photosynthesis above ground can be compared with that allocated to belowground organs for plants with and without mycorrhizal associates. Jones et al. (1991) studied C transfer in terms of P uptake per unit of C expended below ground using cuttings of *Salix viminalis* either colonized by the ectomycorrhizal *Thelephora terrestris* or not colonized. Efficiency was calculated by the formula from Koide and Elliot (1989):

$$\text{Efficiency} = \frac{\Delta P}{\Delta C_b} \tag{4.1}$$

where ΔP = amount of P taken up by the plant over a defined period; and ΔC_b = amount of C allocated below ground during the same interval (including C incorporated into root or fungal tissue, lost in respiration, and deposited in the soil). C_b is calculated for both mycorrhizal and nonmycorrhizal plants that were pulse labeled prior to each harvest for the intervals up to the first harvest (0–50 days) and from the first harvest to the final harvest (50–98) days. The amount of carbon allocated below ground was consistently greater for mycorrhizal plants relative to nonmycorrhizal plants (Figure 4.13). When the data were integrated, it was found that, over the interval leading to the first harvest, mycorrhizal plants allocated 2.5 times more C ($\Delta C_{b(P_n)}$) below ground than did those plants not colonized. On a weight basis ($\Delta C_{b(Wt)}$), carbon allocation was 1.75 times greater in mycorrhizal plants. Because colonized plants absorbed three times more P from the soil over the time period, they had a higher P use efficiency (Table 4.7). Over the second half of the experiment, the differences in P uptake between mycorrhizal and nonmycorrhizal plants was less; nonmycorrhizal plants were more efficient at acquiring P. Cost-benefit analysis, in terms of the amount of C required to produce and maintain nutrient absorbing structures, shows that the costs of maintaining the symbiosis are lower during the earlier stages when the benefits (i.e., greater P acquisition) are higher. Therefore, ectomycorrhizae are a significant drain on plant carbon gain.

The evidence for carbon flow to the fungal symbiont is both indirect and direct. Indirect evidence of carbon flow to the fungal symbiont includes increases in extraradical mycelium, increased fruiting body or spore production once mycorrhization occurs, correlation between increased irradiance and increased root colonization, as well as the magnitude of the mycorrhizal growth response (Harris and Paul 1987). Direct evidence of carbohydrate transfer from plant to fungus has been attained using highly controlled $^{14}CO_2$ labeling experiments and VAM symbioses. In these experiments, $^{14}CO_2$ is either fed to mycorrhizal and nonmycorrhizal plants separately with matched shoot P concentrations, or more often, commonly to one plant with a split root system (i.e., separate hyphal and root compartments) in pulse-chase-style experiments. These studies have demonstrated that the transfer of labeled photosynthate to the mycorrhizae is rapid, but it is difficult to quantify the amount of sugar translocated.

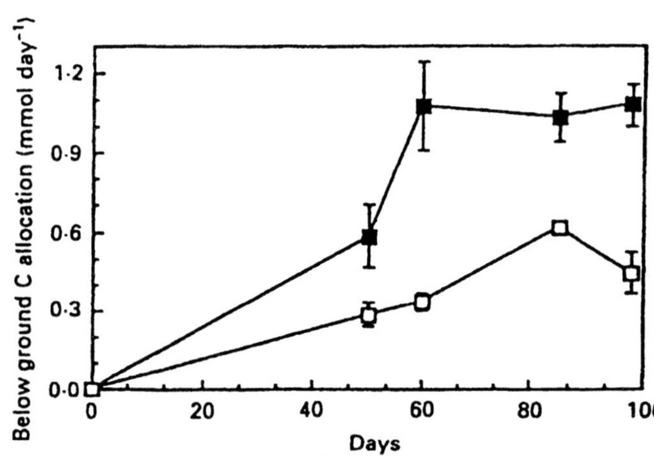

Figure 4.13. Belowground carbon allocation, $C_{b(Pn)}$ (mmol C d^{-1}) of mycorrhizal (■) and nonmycorrhizal (□) *Salix viminalis* during a 98-day growth period. Values are means of three replicates at 50, 60, and 85 days, and 4 (■) or 5 (□) at 98 days ± 1 standard error. Notice greater allocation below ground for mycorrhizal plants. (From Jones et al. 1991, with permission.)

TABLE 4.7 P Uptake, Translocation of C Below Ground, and P Acquisition Efficiency of Mycorrhizal *Salix viminalis* over a 98-Day Growth Period[a]

Mycorrhizal Treatment	Growth Period (days)	ΔP (μmol)[b]	$\Delta C_{b(P_n)}$ (mmol)[b,c]	$\Delta C_{b(Wt)}$ (mmol)[b,d]	$\Delta P/\Delta C_{b(P_n)}$ (μmol P mmol^{-1} C)[b,c]	$\Delta P/\Delta C_{b(Wt)}$ (μmol P mmol^{-1} C)[b,d]
Mycorrhizal	0–50	50.1	14.2	17	3.52	2.94
Nonmycorrhizal	0–50	17.1	5.7	9.7	2.97	1.75
Mycorrhizal	50–98	25.6	46.6	69.8	0.55	0.37
Nonmycorrhizal	50–98	17.9	22	21.2	0.82	0.84

[a] Data are based on 3–5 plants labeled with ^{14}C prior to each harvest.

[b] ΔP = phosphorus taken up during a given interval. ΔC_b = carbon translocated below ground during a given interval, including that deposited in root and fungal tissues, that respired by roots and soil, and that deposited in the soil.

[c] Values were calculated using photosynthesis measurements.

[d] Values were calculated using weight measurements.

Source: From Jones et al. 1991, with permission.

In a study using *Cucumis sativus* and *Glomus fasciculatum,* Jakobsen and Rosendahl (1990) found that as much as 20% (i.e., 7.4 mg $^{14}C\ d^{-1}$) of the total $^{14}CO_2$ fixed by the plant was transferred and used by the fungus. By feeding labeled $^{14}CO_2$ to mycorrhizal and nonmycorrhizal plants, and controlling for phosphorus nutrition effects, most data indicate that mycorrhizal roots receive 4–20% more of the plant photosynthate than do nonmycorrhizal roots. These data include mycorrhizae formed on a diversity of species including herbaceous and woody plants. In general, the most accepted estimate is that the mycorrhizal fungus uses up to 20% of total plant photosynthate (Smith and Read 1997).

It is also known through direct studies of C transfer that different VA fungal species acquired different amounts of plant carbohydrate. In one study using *Cucumis* sp. colonized by three different VA isolates, *Glomus caledonium, Glomus* spp., and *Scutellospora calospora,* plants respired 16.3, 17.3, and 26.2% of the labeled $^{14}CO_2$ respectively, compared with uncolonized control roots that only respired 9.6% of the labeled photosynthate. Uptake of ^{32}P was also monitored in the experiment. The data indicate that *S. calospora* was the least efficient fungal symbiont and *G. caledonium* the most efficient in terms of the amount of ^{14}C used by the hyphae per unit of ^{32}P transported (Pearson and Jakobsen 1993).

The significance of the carbon cost of mycorrhizae also depends on the nutrient resource availability in the soil. If phosphorus is readily available in the soil, then nonmycorrhizal roots are more efficient at nutrient absorption than are mycorrhizal roots, because of the carbon cost of maintaining the fungal symbiont. At high P concentrations, the cost of maintaining mycorrhizae would be greater than that of maintaining uncolonized roots (Eissenstadt et al. 1993). If phosphorus is limiting in the soil or temporally unavailable, then nonmycorrhizal roots cannot absorb nutrients at sufficient rates to maintain photosynthesis and growth. In order to increase P accumulation, plants must grow roots into new soil space. This is because P is relatively immobile in the soil solution and is not carried in the water that plants accumulate for transpiration. In these instances, the cost of producing roots can be greater than the cost of stimulating mycorrhizal growth. The carbon cost of producing a given length of hyphae (2–20 μm in diameter) can be two orders of magnitude less than that for producing the same length of root (200 μm in diameter). Under nutrient-limiting situations, plants might be able to "afford" hyphae, but not roots. Plants may not be able to develop roots fast enough (because of carbohydrate demands) in order to overcome rapidly forming depletion zones. In regions where phosphorus availability is limiting, plants may be dependent upon the fungal symbiont because mycorrhizal roots can obtain a resource that roots alone cannot.

IV. IMPROVED MINERAL NUTRITION

A. Uptake and Translocation of Nutrients

Before the uptake and translocation of nutrients is discussed, it is important to review soil factors that regulate the availability of macronutrients such as phosphorus and nitrogen, particularly in relation to plant uptake processes. Plants accumulate phosphorus as $H_2PO_4^-$ by an anion co-transport mechanism. A proton-motive force created by H^+-ATPase provides the energy and the co-transport ion. There is no other form of phosphorus accumulated by plants. The concentration of H_2PO4^- in the soil solution is primarily regulated by soil pH and the anion exchange capacity (CEC). Less than 1% of inorganic soil phosphorus will be in solution at any given pH. At relatively low pH (6 and below), phosphate ions are complexed with iron and aluminum to form insoluble iron and aluminum rock phosphates (e.g., varasite and strengite). At moderately high pH (7.5 and above) phosphate ions are complexed with calcium and magnesium to form insoluble precipitates (e.g., fluorapatite and bubierrite). Phosphorus in these rock salts is called *occluded phosphate* and is unavailable to plants. Thus, $H_2PO_4^-$ is soluble in the soil solution only between a pH of 6.5 and 7.5. High amounts of organic acids in the soil, such as oxalic, citric, or malic acid, can reduce the propensity of forming occluded phosphates at low pH. At neutral pH, phosphate ions are not necessarily abundant in the soil solution because of the anion exchange capacity of soils. Phosphate ions are strongly attached to the anion exchange sites of the soil (soil minerals). In fact, phosphate has the highest attraction affinity to these sites among common ions in the soil solution (Schlesinger 1991). The high affinity of phosphates to anion exchange sites makes phosphorus very immobile in the soil. Soils containing a high amount of minerals (lower quantities of organic matter) can have a significant amount of anion exchange capacity. Consequently, exposed anion exchange sites will remove phosphorus from the soil solution. Due to the constraints on phosphate availability in the soil solution, a large proportion of phosphorus is in organic forms (99% in arctic tundra, and 75% in temperate forest soils).

The nitrogen availability to plants depends upon the concentrations of both an anion (nitrate) and a cation (ammonium). Both of these ions have independent factors regulating their abundance in the soil solution. Nitrate is a very mobile ion in the soil with low affinity to the anion exchange sites. It is easily leached from the soil solution and is carried in transpiration water through the soil solution. Nitrate abundance is also regulated by environmental factors that affect the activity of microbes involved in *nitrification* (the conversion of ammonium ion to nitrate), and *denitrification* (the conversion of nitrate into dinitrogen gas). Ammonium ion is moderately attracted to cation exchange sites, so the soil can store some ammonium ion on these sites. Ammonium is less mobile than nitrate, and its concentration is dependent upon *ammonification* (the microbial release of ammonium from organic matter) and *ammonia volatilization* (the conversion of ammonium ion into volatile ammonia gas).

Plants have the ability to accumulate both nitrate and ammonium ion. There are both high-affinity antiports for accumulating ammonium ion as well as low-affinity ammonium channels in plant roots. Like phosphorus, rhizospheric depletion zones of ammonium ion develop around roots because ammonium ion is not very mobile. Processes regulating diffusion are more important for ammonium ion translocation to the root surface than nitrate. Nitrate is accumulated by a high-affinity, H^+-dependent, anion co-transport system. Depletion zones of nitrate ion do not develop around roots because nitrate is so mobile in soil solution. Nitrate carried in water that will be transpired by plants is a dominant mechanism for translocation of nitrate from the soil to the root surfaces. Some plants accumulate ammonium ion preferentially, while others accumulate nitrate preferentially. The flexibility in nitrogen accumulation and the greater mobility of these ions make them more available to plants than the ions of phosphorus. However, nitrogen requirements are 10 times those of phosphorus, which makes nitrogen an important limiting factor.

1. Nutrient Uptake by Mycorrhizal Plants. There are several ways that mycorrhizae enhance the uptake of nutrients to the plant (Allen and Allen 1990). First, mycorrhizal roots have a greater surface area, increasing the root absorption area. For example, a dense mantle that effectively increases the surface area, of the root usually covers ectomycorrhizal roots. Also, there is increased production of ectomycorrhizal short roots, which subsequently increases the overall surface area of the root. Likewise, VAM hyphae have a greater surface area/volume ratio than do bare roots. If the hyphae and extensions are considered, the absorptive surface can be increased up to 1800% (Okeefe and Sylvia 1991). Second, external mycelia of the fungus form an extensive network of anastomosing hyphae that explore a larger soil volume than do normal root hairs. Plus, fungi can increase the ability of root systems to access nutrients in narrow spaces (small soil pores), because hyphae are 1/10 the diameter of roots. This ability of hyphae to enter small soil pores can increase root absorption of nutrients 60-fold in some cases (Bolan 1991). Third, mycorrhizae provide faster movement of phosphorus to the roots by increasing the affinity of the root for phosphorus ions, thereby decreasing the threshold concentration needed for phosphorus to diffuse into the root cells (Bolan 1991). Thus, mycorrhizae essentially mobilize previously unavailable phosphorus. Fourth, many mycorrhizal fungi can chemically alter the rhizosphere. For example, VAM fungi solubilize soil phosphorus through the release of organic acids. This process enhances the absorption of poorly soluble rock phosphorus sources such as aluminum phosphates and iron phosphates (Gianinazzi-Pearson and Gianinazzi 1983, Bolan 1991, Alexander and Hardie 1991). Fifth, some research has suggested that fungal mycelia can connect adjacent mycotrophic plants, and nutrients (phosphorus) may be directly if not indirectly transferred between companion plants, based upon a phosphorus gradient established between the plants (Chiarello et al. 1982, Francis and Read 1984, Allen 1991). Sixth, there is some evidence that mycorrhizal roots live longer than nonmycorrhizal roots, which increases the total absorbing capacity of the root over its lifetime.

Are mycorrhizal associations critical to nutrient accumulation by plants that are normally mycorrhizal in nature? Although usually colonized in the field, many VAM plants grow well in the absence of appropriate fungal symbionts, as long as adequate mineral nutrition is provided. For many VA fungi, increased soil nutrient levels can actually depress mycorrhizal colonization. In contrast, fungal symbionts are obligately linked to the host plants. Propagules of VA fungi rarely develop mycelia in the absence of root colonization. As a result, development of mycorrhizal roots and subsequent effects on plant performance are usually only manifest in soils of low or imbalanced nutrient status (particularly if P is in short supply). Mycorrhization leads to increased growth of roots and shoots, often reduced root/shoot ratios, luxury consumption of phosphorus, increased N_2 accumulation and fixation leading to higher tissue N, and an in some cases increased root/shoot ratio. These associations can lead to increased flower production and seed yield thereby directly affecting the fitness of the plant. There is no question that plant fitness in low-nutrient sites is improved by the presence of mycorrhizal associates.

*2. **Translocation of Nutrients in Mycorrhizal Fungi.*** Rates of translocation of nutrients in the mycorrhizal fungus are expressed as mass transfer mol $m^{-2}\ s^{-1}$, and are highly variable depending upon the plant symbiont, fungal symbiont, season, temperature, experimental protocol, and sensitivity of equipment, among other factors. Furthermore, because intact mycorrhizal systems are difficult to manipulate, experimenters must often work with excised roots that create different conditions than the natural environments in which mycorrhizae are normally associated. Although it can be difficult to quantify, the rate of resource translocation is critical for determining the efficiency and importance of resource translocation to plant nutrition.

In order for translocation in hyphae to be effective, nutrients must be accumulated by hyphae distant from the root and transferred to the root without any leakage along the way, or at the root-hyphae interface. It is known that translocation of nutrients is faster within fungal hyphae connected to transpiring plants. This relationship is particularly true for ions that are relatively immobile in the soil solution, such as phosphorus. For ectomycorrhizae, the apoplasm of the fungal sheath is impermeable, and it serves as a barrier to leakage of water and nutrients from the fungus to the soil. As mentioned previously, during the development of the fungal sheath in ectomycorrhizae, a fungal hydrophobin gene is upregulated. The products of this gene are critical for binding together the hyphae of the sheath and for forming an impermeable barrier to resource leakage at the root interface (Martin and Tagu 1995). As a result, in order for nutrients to enter the root at the mycorrhizal root tip, they must pass through the fungal symplast. Hence, accumulated resources are first transported within the external mycelium and then effluxed into the interfacial apoplast in the Hartig net region. At the same time, resources passing from the plant to the fungus are forced to enter the fungal symplast and therefore will also not leak to the soil. The impermeable fungal layer surrounding the Hartig net is critical for the symbiosis because it allows for the maintenance of resource concentrations in the cortical apoplast and Hartig net area (Figure 4.14).

It is important to mention that although VAM do not form an impermeable sheath, most intracellular colonization occurs below a relatively impermeable layer of the root, the hypodermis. Also, the suberization of the hypodermis increases with age after mycorrhization, becoming increasingly impermeable and thereby insulating the cortex of the root from the soil. Entry or exit of resources must occur in very young sections of the root, through the symplast of passage cells that are nonsuberized, or via the hyphae of mycorrhizal fungi that themselves enter through nonsuberized passage cells. Furthermore, upon penetration of the host cell wall by a VA fungal hypha, a thick collar of plant cell wall material envelops the hypha initiating the haustoria. This collar gets thinner as the haustoria forms away from the cell wall periphery, and

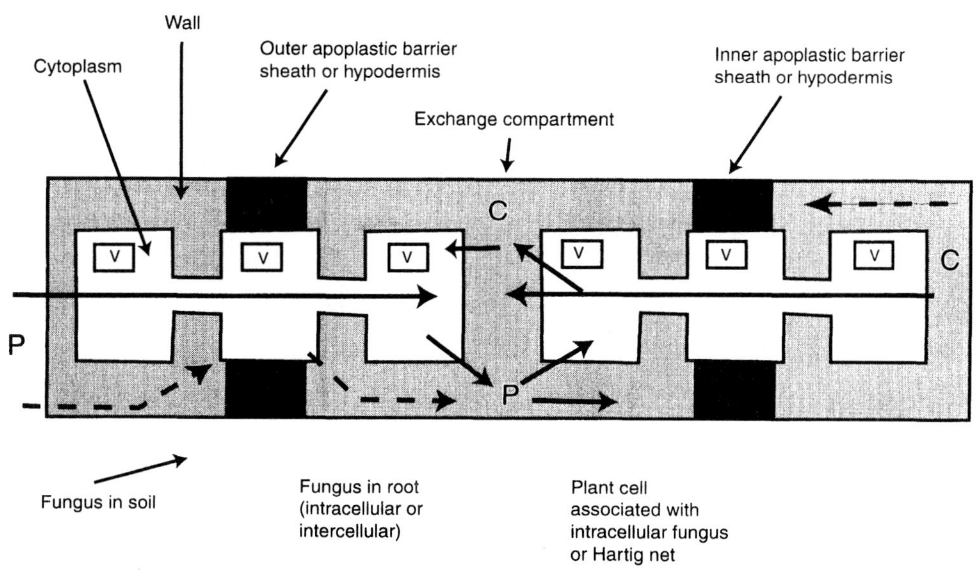

Figure 4.14. Diagrammatic representation of the proposed apoplastic exchange compartments in mycorrhizal roots. V = vacuolar compartments; white areas = cytoplasm (symplast); gray areas = apoplast; black areas = impermeable apoplastic barriers. Transport of P and C through the apoplast (*dashed lines and arrows*) through the symplast (*solid lines and arrows*). (Redrawn from Smith and Read 1997, with permission.)

is thin or absent over most of the intracellular fungus. Once again, this allows for tight regulation of the materials that enter the mycorrhizal interface zone.

Once the nutrients arrive at the fungus-root interface, translocation from fungus to plant must occur if the fungus is to alleviate nutrient limitations for the plant. In mycorrhizal symbioses, the interfaces generally have the same basic structure corresponding to their unique function (Figure 4.15). The fungal hyphae always colonize the root apoplast and remain outside the plant protoplast. The fungus can be located in the intercellular spaces of the root, closely associated with the walls of the epidermis and the cortical cells (Smith and Read 1997). In the mature symbiosis, the interface contains two membranes (one plant, one fungus), which are separated by the interfacial zone or apoplast. The interface regions can differ with regard to detail; in particular, wall and membrane components and structure can vary greatly.

The first step in the transfer of materials across the symbiotic interface must be the efflux of some resource from the donor organism into the interfacial apoplast. Mechanisms of nutrient efflux from the fungus are still unclear. It is likely that mechanisms similar to phloem loading and unloading in plants are responsible for the efflux of sucrose from the plant. See the companion *Plant Stress Physiology,* volume 1 (Nilsen and Orcutt 1996) for a better description of these mechanisms.

It is often assumed that bidirectional transfer of resources (nutrients and carbohydrates) between the plant and the fungus occurs across a common mycorrhizal interface at the same time, implicating that resource flow occurs across one specialized functional interface. However, in some mycorrhizal types (e.g., many of the VAM associations), there are several different interface structures that perform distinct trans-

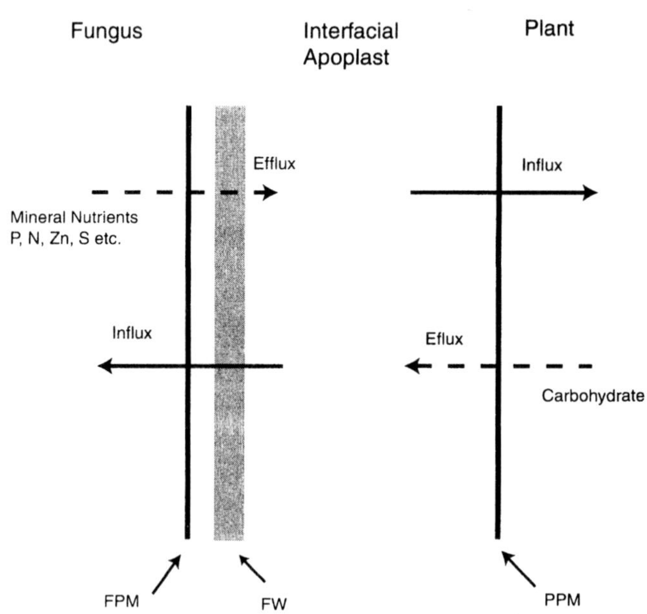

Figure 4.15. Simplified diagram of the symbiotic interface, indicating the key components including the interfacial apoplast bounded by the fungal plasma membrane (FPM) and plant plasma membrane (PPM). The arrows indicate the membrane transport processes involved in the transfer of carbohydrates and soil-derived nutrients. Dashed lines = efflux, solid lines = influx. The fungal wall (FW) is indicated by shading, but plant wall and wall-like components are not shown. (Redrawn from Smith and Read 1997, with permission.)

port functions (Table 4.8). In addition, the function and structure of the interface is spatially and temporally variable depending upon the type of mycorrhiza.

For VAM, there is mounting evidence that resource transfer mechanisms are spatially separated. Sugar is transferred from the plant to the fungus in the intercellular interface. Here, sucrose could efflux from the cortical cells of the root, be hydrolyzed to hexose, and be actively absorbed by the fungus. Fungal membranes in the intercellular spaces have a high activity of H^+-ATPase. At the arbuscular interface, plant cells have a high activity of H^+-ATPases on the plasma membrane, and efflux of minerals from the fungus would be followed by active uptake by the plant. The absence of H^+-ATPases on one membrane of each interface suggests that donor plants lack the ability to reabsorb resources at that site, hence polarizing transfer across the respective interfaces (Smith and Read 1997). Therefore, the position of the H^+-ATPases and whether they are on the plant or fungal membrane can help determine the function of a particular membrane interface (Figure 4.16).

B. Improved Nitrogen Nutrition

Low nitrogen constraints to plant growth are characteristic of many ecosystems in which ectomycorrhizal plants grow. For example, in temperate deciduous forests, nitrogen is likely the main mineral resource limiting the productivity of canopy trees. Therefore, assimilation, mobilization, and transport of nitrogen are critical for plant success, and ectomycorrhizae may enhance those processes. Recall that edaphic conditions such as pH, aeration, temperature, and moisture all can affect the processes of ammonification, nitrification, and denitrification leading to temporal and spatial variation in the amounts and proportions of NH_4^+, and NO_3^- within and among forest sites. Plus, the dynamics of ionic movement differ between these two nitrogen ions.

1. Inorganic Sources of N. The importance of mycorrhizae to NO_3^- absorption is not well described. This is because interest in this topic is lacking. The mobility of NO_3^- in the soil solution makes enhanced diffusion insignificant to accumulation. Also, nitrification is inhibited in acidic soils, which are most commonly inhabited by mycorrhizal species. Both of these factors reduce the importance of mycorrhizae for the uptake of nitrate. We will focus our discussion on the ways that mycorrhizae may enhance ammonium accumulation from the soil solution.

Ammonium is preferentially absorbed from solutions by mycorrhizae where both NH_4^+ and NO_3^- are available; in addition, NH_4^+ inhibits NO_3^- uptake and assimilation in the external hyphae. Ames et al. (1983), using mesh-compartmentalized VAM systems, found considerable $^{15}NH_4^+$ transfer from the fungus-only compartment to the mycorrhizal root compartment. The amount of transfer was correlated with percent colonization of the roots and the hyphal length density in the fungus-only compartment, as well as the number of hyphae crossing the mesh. The transfer rate of nitrogen was calculated to be 7.42×10^{-4} mol N m^{-2} s^{-1}, which is roughly the same flux rate that has been measured for phosphorus. Considering the high-nitrogen requirements of the plant, this rate is very low.

Studies of the importance of mycorrhizal fungi to ammonium assimilation, uptake, and metabolism have been carried out either axenically in culture or at the whole-root level (Figure 4.17). Research with axenic cultures using inorganic nitrogen has shown that most of the nitrogen from NH_4^+ enters plant metabolism of ectomycorrhizal fungi as either an amide group of glutamine formed by the glutamine synthetase (GS) pathway or as the amino group of glutamate through the glutamate dehydrogen-

TABLE 4.8 Types of Interface Present in Different Mycorrhizal Types, Illustrating Types of Intra- and Intercellular Interfaces, and the Presence of ATPases on the Interfacial Membranes

	Intracellular			Intercellular		
		ATPase[a]			ATPase[a]	
Mycorrhiza	Structure	Plant	Fungus	Structure	Plant	Fungus
VAM	Arbuscules	+	(+)	Hyphae	(+)	+
	Coils	ND	ND			
Orchid	Coils	+	+	None	NA	NA
Ericoid	Coils	ND	ND	None	NA	NA
Monotropoid	Pegs and membranous sac	ND	ND	Hartig net	ND	ND
Ecto-	None	NA	NA	Hartig net	+	+
Ectendo-	Hyphal complexes	ND	ND	Hartig net	ND	ND
Arbutoid	Hyphal complexes	ND	ND	Hartig net	ND	ND

[a]NA = not applicable, ND = no data, + = ATPase activity demonstrated cytochemically and always present; (+) = ATPase activity but not always present.

Source: Adapted from Smith and Read 1997, with permission.

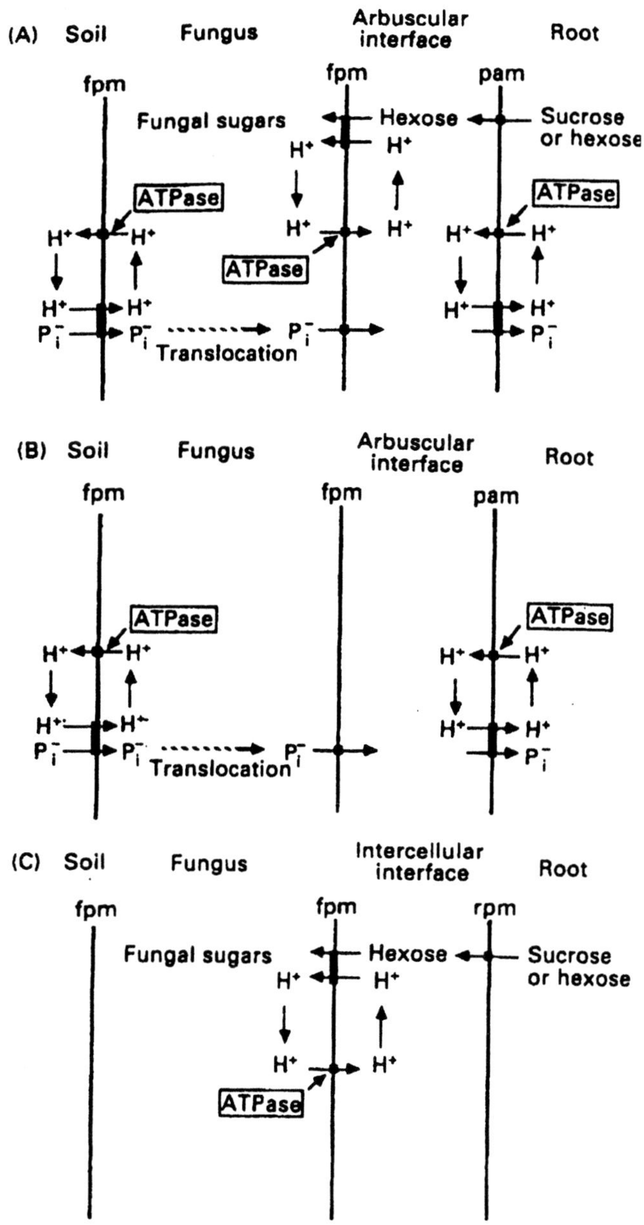

Figure 4.16. Possible spatial distribution of H^+-ATPases and the associated transport processes in VAM interfaces based on data for molybdate-sensitive ATPase activity. (A) Bidirectional transfer of sugar and PO_4^- across the same arbuscular interface. Both plant and fungal membranes are shown with H^+-ATPase activity. (B) Transport of PO_4^- from fungus to plant across an arbuscular interface. Absence of H^+-ATPase on the fungal plasma membrane could be associated with the passive loss from hyphae to the interfacial apoplast, and the presence of active uptake on the plant plasma membrane. (C) Transport of sugar from plant to fungus across the interface between cortical parenchyma cells and intercellular hyphae. Little or no H^+-ATPase activity on the plant plasma membranes could be associated with the passive loss of sugar to the interfacial (intercellular) apoplast and the presence of H^+-ATPase activity on the fungal membranes indicates active uptake into the hyphae. fpm = fungal plasma membrane; pam = periarbuscular membrane; rpm = plasma membrane of the root cortical cell. (From Gianinazzi-Pearson et al. 1991, with permission.)

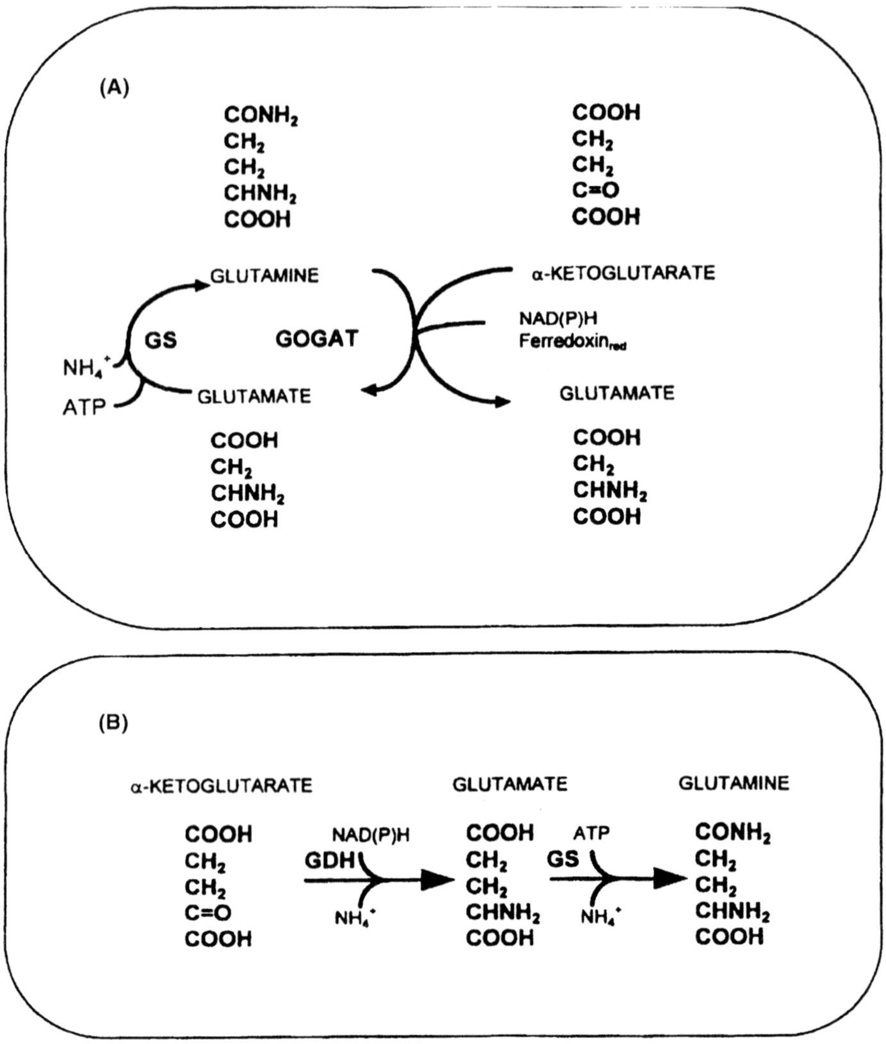

Figure 4.17. Pathways of NH_4^+ assimilation in plants and fungi. (A) The glutamine synthetase–glutamate synthase (GS-GOGAT) pathway. (Adapted from Martin and Botton 1993, with permission.) (B) The NADP–glutamate dehydrogenase–glutamine synthetase (GDH-GS) pathway. (From Smith and Read 1997, with permission.)

ase (GDH) pathway. These pathways, along with glutamine synthetase–glutamate synthase (GS-GOGAT), differ in relative importance to the symbiosis. In some mycorrhizal fungi, the GS pathway has been shown to be most important, while in other species, the GDH pathway seems to be used preferentially by the fungi involved. It now appears that the GS pathway is the pathway most frequently used by the greatest number of ectomycorrhizal fungi. However, glutamate, aspartate, and alanine accumulate in the mycelium of the ectomycorrhizal fungi when the GS pathway is inhibited by the presence of methionine sulphoximine, confirming that the GDH and GS-GOGAT mechanisms are also present in ectomycorrhizal fungi that use the GS pathway predominantly. It is most probable that nitrogen assimilation in mycorrhizal fungi involves a combination of all three mechanisms.

Studies that evaluate ^{15}N-labeled NH_4^+ accumulation by intact root systems also have demonstrated that glutamine is a major sink for absorbed N and that alanine, arginine, and aspartate-asparagine are also important (Botton and Chalot 1995). These data are consistent with experiments using ^{14}C and excised roots showing that glutamine is the major form of translocated N within the fungal mycelium and subsequently across the symbiotic interface to the plant (Smith and Read 1997).

The GS pathway for ammonium accumulation is most prevalent in fungi, but NH_4^+ assimilation in plants occurs primarily by the GS-GOGAT pathway. Because fungi and higher plant cells utilize different pathways of nitrogen assimilation, the question becomes, how is nitrogen assimilation controlled in the mycorrhizal condition, particularly within the mycorrhizal interface? It is believed that in many cases, fungal enzymes controlling nitrogen assimilation are downregulated in proximity to plant tissues, which has led to the *metabolic zonation model.* A good example of metabolic zonation occurs in *Picea* mycorrhizae that form with *Hebeloma.* It has been found that NADP-GDH activity is greatest in the extraradical mycelium and mantle but reduced in the Hartig net region (Dell et al. 1989). As a result, the quantity of GDH polypeptide decreases progressively from the peripheral cells of the mantle to the Hartig net, presumably as a result of the downregulation of fungal enzyme expression. Even more extreme in *Fagus* mycorrhizae, both the activity and the amount of GDH are strongly suppressed throughout the mantle regardless of the fungal symbiont involved (Chalot et al. 1990). There is expression of some enzyme activity as the mycelium moves away from the root. In other examples, enzyme activity is unaffected regardless of proximity to the plant. Finally, there is some evidence that the nitrogen source and concentration can affect nitrogen assimilation in *Laccaria bicolor* by regulating NADP-GDH biosynthesis and activity through altering GDH mRNA (Lorillou et al. 1996). Such research suggests that soil nitrogen availability could regulate nitrogen assimilation by fungi.

The pathways of NH_4^+ assimilation and the distribution of the enzymes involved have important implications for the mechanisms of transport of N across the symbiotic interface. Figure 4.18 summarizes the three basic models that have been suggested for the regulation of nitrogen flow in the fungus-root interface. Realize that the transfer of hexoses in these models is an oversimplification because transfer of organic C probably occurs as sucrose efflux from the plant (Smith and Read 1997).

2. Organic Sources of N. It is becoming more apparent that various ectomycorrhizal fungi have the capacity to utilize organic forms of nitrogen. In fact, many ectomycorrhizal fungi grow in culture better on organic forms of N than on inorganic forms. From studies using organic humus, it has been shown that ectomycorrhizal fungi have a limited capacity to acquire organic N, probably in the form of amino acids. This could be an important source of nitrogen for fungal symbionts because amino acids can comprise a large part of the soluble N pool in acidic soils, such as those occupied by ericaceous plants. Some amino acids that are predominant in the soil solution, such as glutamine and alanine, can be used by ectomycorrhizal fungi in pure culture and support relatively large yields compared with fungi just supplied with reduced carbon (sugars) and NH_4^+. Furthermore, alanine also serves as a major sink for organic C, and can be translocated by the fungus to the roots of plants resulting in translocation of nitrogen as well as the bidirectional flow of carbon. Therefore, it is likely that mycorrhizae can accumulate amino acids from acidic, organic soils and translocate them to the root-fungus interface.

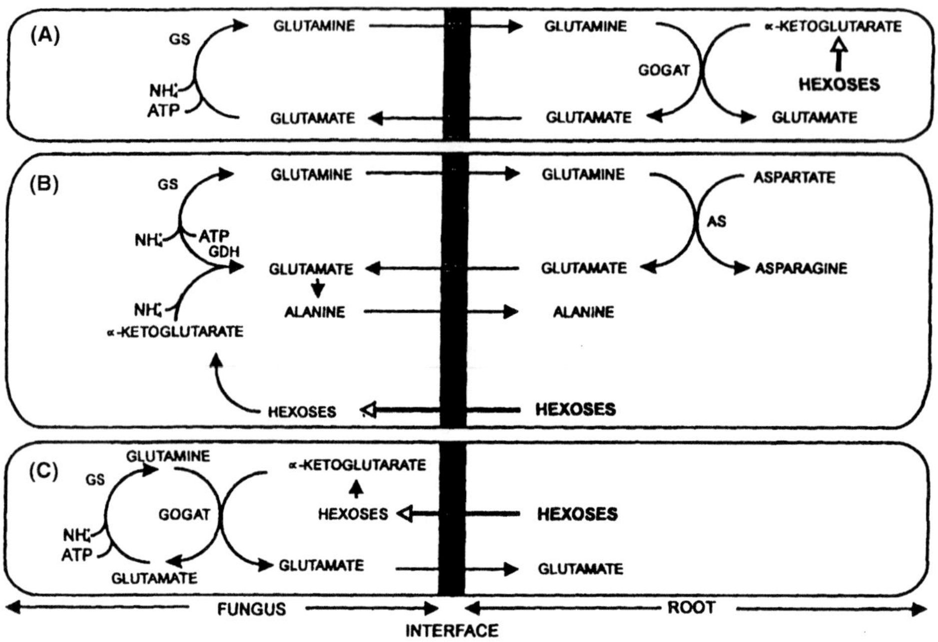

Figure 4.18. Three basic patterns of nitrogen assimilation recognized, and the possible location of enzymes of NH_4^+ metabolism in different types of mycorrhizae. (A) In *Fagus* mycorrhizae, GS is localized in the fungal sheath and GOGAT in the plant root. (B) In *Picea* mycorrhizae both GDH and GS occur in extraradical mycelium and sheath, respectively, with asparagine synthetase (AS) in the root. (C) A third pattern can be demonstrated using mycorrhizae formed by *Pisolithus tinctorius* in which GS and GOGAT operate in the fungus. (Modified from Martin and Botton 1993, with permission.)

Several lines of evidence suggest that mycorrhizae are important for amino acid assimilation by plants. It has been demonstrated that high concentrations of amino acids in the soil can be harmful to both the plant and the fungus. For instance, additions of aspartic acid can cause depressed growth in nonmycorrhizal plants. Yet accumulation of amino acids (alanine, arginine, and aspartic acid) has been demonstrated in the mycorrhizae and roots of *Pseudotsuga menziessii,* and the usual inhibitory effect of aspartic acid is diminished in mycorrhizal plants. When uncolonized by appropriate mycorrhizae, birch seedlings are incapable of using alanine as a sole nitrogen source and develop nitrogen deficiency. However, when colonized by ectomycorrhizal fungi (*Amanita muscaria, Hebeloma crustuliniforme,* or *Paxillus involutus*), the seedlings grew vigorously and their tissue nitrogen concentrations were higher and more representative of those seen in normal healthy plants (Abuzinadah and Read 1989b) (Figure 4.19).

The capacity to use organic forms of nitrogen varies greatly depending upon the type of mycorrhizae, the symbionts involved, and edaphic conditions prevalent. For VAM, there is no unequivocal evidence that extraradical hyphae are directly involved in the mineralization of organic forms of nitrogen. However, it is known that the ericoid mycorrhizal endophyte *Hymenoscyphus ericae* can degrade complex organic substrates and provide host plants with nitrogen that might otherwise be unavailable to them. Ericoid fungi produce an acid proteinase that appears to be structurally and functionally similar to other acid proteinases found in saprotrophic fungi (e.g., *Peni-*

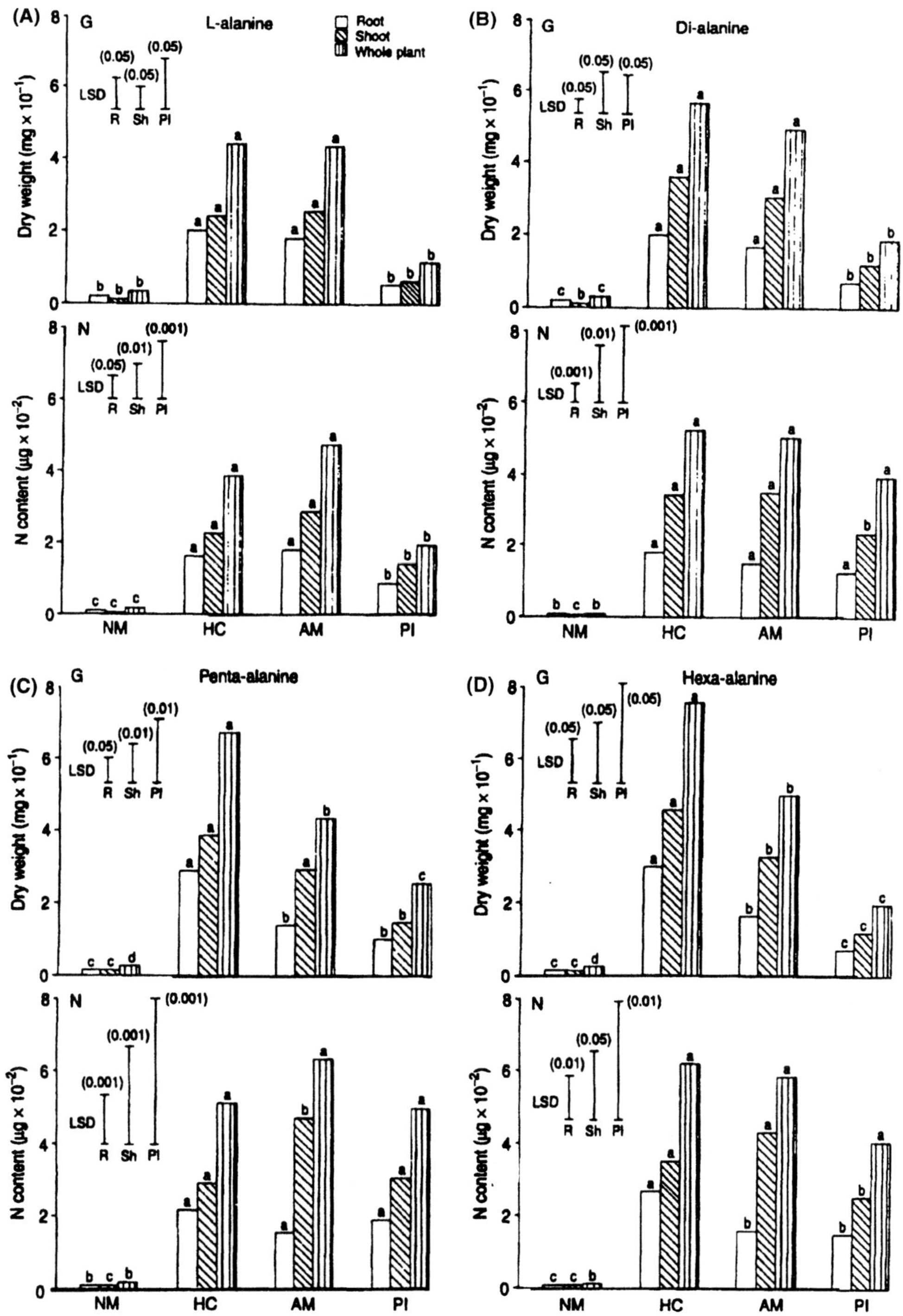

Figure 4.19. Growth (G) and nitrogen concentration (N) of roots; shoots and whole plants of *Betula pendula* colonized by *Hebeloma crustuliniforme* (HC), *Amanita muscaria* (AM), *Paxillus involutus* (PI), or noncolonized (NM) grown with different amino acids and peptides as sole N source for 75 days. Vertical bars represent least significant digit (LSD) for roots (R), shoots (Sh), and whole plants (Pl). (A) Alanine, (B) Di-alanine, (C) Penta-alanine, (D) Hexa-alanine. Notice that mycorrhizal plants have a greater capacity to utilize organic forms of N than non-colonized plants. Also, the capacity to use organic N differs depending upon the fungal symbiont used. (From Abuzinadah and Read 1989, with permission.)

cillium, Aspergillus) and in some plant pathogens such as *Fusarium* and *Alternaria*. Interestingly, these fungal proteinases are structurally and functionally similar to animal pepsin. It is thought that this has ecological importance because many ericoid plants inhabit soils in which a large amount of the N is accumulated in organic forms.

3. Relationship Between Dinitrogen Fixation and VAM. One of the ways that plants can improve their nitrogen accumulation rate is by performing nitrogen fixation through a symbiotic association with a prokaryote (i.e., *Rhizobia* or *Frankia*). Nitrogen fixation can be limited by carbohydrate flow from the plant and by nutrients such as phosphorus, iron, or molybdenum. Many nitrogen-fixing plants also host VAM associations, which has led to studies of the *dual mutualism* or *tripartite association*. Rhizobial plants growing in nutrient-deficient soils show increased rates of nitrogen fixation once they are colonized by appropriate VA fungi. In most cases, improved rates of N_2 fixation can be attributed to relief from phosphorus stress, and possibly from increased uptake of other essential nutrients.

Nitrogen fixation in field situations is often measured using the natural abundance of stable isotopes of nitrogen. This technique is dependent upon the fact that plant tissues containing nitrogen accumulated from the soil show a different $^{15}N/^{14}N$ signature than plant tissue containing a significant amount of nitrogen derived from the atmosphere through dinitrogen fixation. However, recent evidence suggests that the $^{15}N/^{14}N$ ratio in plant tissues may differ between mycorrhizal and nonmycorrhizal plants, due to different fractionation processes in plant and fungal nitrogen assimilation (Handley et al. 1993). Both absorption and assimilation of inorganic nitrogen by mycorrhizal fungi rather than by plant roots may contribute to this difference. This isotopic effect of mycorrhization lends support to the theory that mycorrhizae ultimately affect nitrogen assimilation processes in plants. In addition, differences in $^{15}N/^{14}N$ due to mycorrhization may confound the measurement of dinitrogen fixation by natural-abundance, stable-isotope techniques.

C. Improved Phosphorus Accumulation

1. Phosphorus in the Soil and Its Significance to Mycorrhizae. Inorganic phosphate ($H_2PO_4^-$, often abbreviated P_i) is absorbed by plants from the soil solution, where it is normally present at very low concentrations. Earlier in the chapter, we explained that phosphorus is a relatively immobile ion and that diffusion and translocation in water moving to roots are not important mechanisms for phosphorus accumulation. The availability of P in the soil solution can be enhanced by an abundance of chelating compounds in the soil solution, such as citrate, malate, and oxalate (Marschner 1995). Inorganic phosphorus also can be liberated from organic P-containing compounds (e.g., inositol phosphates, phospholipids, and nucleic acids) by enzymatic action (phosphatases). Organic P can be hydrolyzed by soil microorganisms, or it can be autolyzed by plant-exuded phosphatases. In many ecosystems, most of the phosphorus is contained in the organic matter. Plant roots are able to exude phosphatases that can liberate inorganic P from inositol phosphate (the largest fraction of organic P). This was elegantly shown by an experiment utilizing a double labeling of inositol phosphate with C^{14} and P^{32}. When plants of the arctic sedge *Eriophorum vaginatum* were exposed to double-labeled inositol phosphate, only the labeled P (not the labeled C) was found in the root tissues (Kroeler and Linkins 1991). The exudation of phosphatases can result in a decrease of organic P-containing molecules near the root surface and an increase in inorganic P (Tarafdar and Jungk 1987). There is some evidence of phosphatase activity on the root surface of VAM plants that could subsequently affect P hydrolysis, but this evidence is weak.

Soluble phosphate that enters the soil through mineralization or as fertilizer amendment results in localized short-term nutrient pulses. Most of the P_i in these nutrient pulses is rapidly lost from the soil solution because of fixation processes such as, adsorption of ions to charged particle surfaces, precipitation of mineral phosphates, and fixation by free living soil microorganisms. Inorganic phosphorus ions remaining in solution are supplied to the absorptive surfaces of roots by diffusion. Diffusion of P_i in the soil (10^{-8} to 10^{-11} cm^{-2} s^{-1}) is slower than that in an aqueous solution (10^{-5} cm^{-2} s^{-1}). Slow movement of P_i through the soil coupled with rapid absorption of P by plant roots results in the formation of depletion zones around the roots. Uptake of P_i by roots is primarily limited by the rate of $H_2PO_4^-$ diffusion across these depletion zones, instead of the rate of transport across cell membranes of the root or mycorrhiza.

2. Phosphate Metabolism in the Fungus. External hyphae of VA fungi must absorb P_i via active transport after the roots have been colonized. Phosphorus absorption has to be active because the electrochemical gradient favors efflux from the hyphae to the soil. Uptake occurs via a proton co-transport system, with the necessary proton-motive force generated by membrane-bound H^+-ATPase complexes. Fungal hyphae have a higher activity of H^+-ATPase complexes along the plasma membranes than do roots. As a result, P_i is accumulated to higher concentrations within VA hyphae than in plant roots (Lei et al. 1991). Most (80–90%) of the phosphorus accumulated by the fungal symbiont remains in the fungal hyphae. However, phosphorus is stored in the fungal hyphae in the form of polyphosphate granules, and can be mobilized and transferred to the plant symbiont when P is deficient.

A high-affinity phosphate transporter, GvPT, has been cloned from the cDNA of *Glomus versiforme*. The gene is expressed in the external hyphae of VAM roots. When the gene is used to transform yeast cells with a low-affinity for PO_4^- transport, wild-type properties are restored. It is interesting to note that germ tubes of *Gigaspora margarita* show no H^+-ATPase activity along the fungal membranes unless they are in the presence of root volatiles and CO_2, in which case they exhibit H^+-ATPase activity and an affinity for P. This helps provide reasoning for why VA fungi fail to grow in the absence of a suitable host, because hyphae alone lack the capacity to absorb phosphorus.

Rapid absorption of P_i by VA fungi is followed by the synthesis of inorganic polyphosphate in fungal vacuoles. Recent work using an ectomycorrhizal fungus *Pisolithus tinctorius* has shown that polyphosphate is present in fungal vacuoles as short-chain 15-mers, stabilized by K^+ ions (Ashford et al. 1994). Formation of polyphosphate prevents the accumulation of orthophosphate in fungal cells, when the external supply of P_i is plentiful and the uptake is rapid. Storage in polyphosphate form also reduces osmotic stress that accompanies high concentration of orthophosphate. Phosphorus is most likely transported as polyphosphate, but the actual transport of polyphosphate granules in the cytoplasm is unlikely.

Currently, there are two generally accepted mechanisms for the movement of polyphosphate and other solutes in mycorrhizal hyphae. The first mechanism, based on the net movement of vacuoles containing polyphosphate, requires the following:

1. Translocation occurs down a concentration gradient of phosphorus between a source and a sink. Cytoplasmic streaming in fungal hyphae, either unidirec-

tionally or bidirectionally, could lead to a net movement of vacuoles containing polyphosphate along this gradient.

2. A well-organized vacuole loading and unloading system must be present at the source and the sink. Ultimately, the rates of loading and unloading of phosphate into or out of vacuoles will determine the rate of translocation (Smith and Read 1997).

The second mechanism is contingent upon vacuolar movement along tubular cysternae in the fungal hyphae. These tubules connect among vacuolar clusters and potentially move material from cluster to cluster. Consequently, cytoplasmic streaming is not required for the movement of phosphate along long distances in the hyphae. Movement of resources in both directions, at a faster rate than diffusion alone, could occur by this mechanism because the tubules operate peristaltically.

Research on the dolipore septa of basidiomycetes (the septal pores between cells of basidiomycete hyphae) may help elaborate the actual mechanism of polyphosphate movement within mycorrhizal fungi. It is now known that the symplasm is continuous, not only through the septal pore, but also through small pores present in the parenthosomes (parenthetic-shaped, perforated membrane on each side of the dolipore septa, also called the septal pore cap). The pores in the parenthosomes are small enough to exclude organelles, but allow the passage of smaller materials such as organic and inorganic compounds. Furthermore, a series of tubular cysternae pass through both the septal pore and the parenthosomes and are connected continuously to the vacuoles. These tubules have been shown to transfer fluorochrome 6-carboxyfluorescein between adjacent cells at the tips of hyphae (Shepherd et al. 1993a, 1993b). The tubules appear to be similar to endosomes in cultured animal cells. They move in both directions across the septal pore independent of cytoplasmic streaming, and peristalsis within the tubules transfers materials between hyphal compartments (Shepherd, 1993a, 1993b). In addition, the direction and behavior of the tubules appears to be under the control of the pores in the parenthosomes. It is important to note that the same sort of tubular cysternae have also been found in zygomycetes and ascomycetes and always in conjunction with the hyphal vacuole system. They are likely the most important structure for the movement of substances within fungal hyphae

This system of pleiomorphic (flexible morphology), vacuolar tubules observed in many fungi contains polyphosphate and thus may be involved in the transport of P in hyphae. Peristaltic action within these tubules certainly can move the polyphosphate short distances. Movement of polyphosphate over distances measured in external hyphae of VAM, for example, is still an open question. In VAM, polyphosphate is stored in the external hyphae, transported to the internal root hyphae as polyphosphate, and broken down to P_i in the internal root hyphae. This is consistent with the observation that polyphosphate enzymes are absent in the external hyphae. However, the absence of polyphosphate enzymes in the external hyphae also may indicate that polyphosphate turnover and unloading in the tubules could be very low, which would argue against the mechanism for loading polyphosphate into tubules for transport in the hyphae.

One problem that plagues the aforementioned mechanisms is that both depend upon spatial localization of polyphosphate in impermeable fungal vacuoles. However, there is little evidence that spatial localization of polyphosphate occurs in vacuoles in a form that does not rapidly equilibrate with the fungal cytoplasm. At present, there is a small amount of evidence that spatial separation of polyphosphate metabolism occurs between the intraradical and extraradical mycelium in some VAM. Consequently, the vacuoles and attached tubules would have to be impermeable to allow long-distance translocation of polyphosphate.

3. General Effects of Mycorrhizae on P Accumulation. There are several reasons why the formation of extraradical hyphae can alleviate the limitation due to low mobility of P_i across depletion zones to roots (Figure 4.20). Hyphae can reach and grow into soil not colonized by roots and subsequently translocate P back to the roots. Therefore, mycorrhizal roots have a much greater area of contact with the soil than nonmycorrhizal roots. Translocation of nutrients back to roots through the hyphae is faster than the rate of P diffusion across soil depletion zones. Like roots, fungal hyphae also will create depletion zones that have to be compensated for through increased hyphal growth. The carbon cost of producing hyphae per unit absorbing area is less than that of roots. Thus, the production of new hyphae, or rapid hyphae population turnover, constitutes a smaller demand on carbohydrates from the plant in comparison with a similar growth and turnover of fine roots. Fungal hyphae have a significantly smaller diameter than fine roots. Consequently, hyphae can access P from soil spaces to which roots would not have access, increasing the volume of soil explored by the absorbing surfaces. Also, because of their size and spatial distribution relative to roots, hyphae are more competitive with soil microbes for recently mineralized or solubilized P. Hyphae have a higher affinity (lower K_m) for P than do roots. This leads to a more

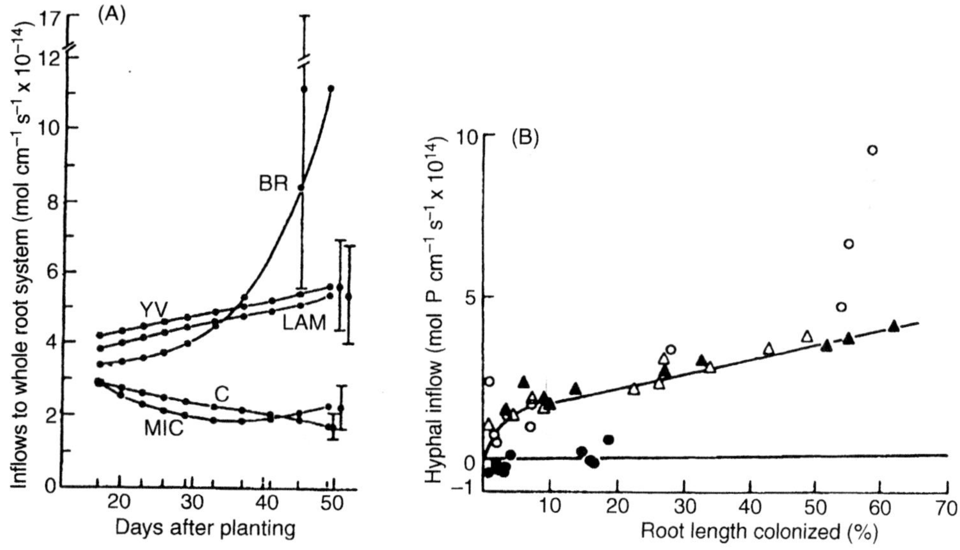

Figure 4.20. Effects of mycorrhizal colonization on flow of P into roots of *Allium cepa.* (A) Inflows of P into whole root systems computed for each harvest date. C = nonmycorrhizal control, LAM = *Glomus macrocarpus* var. *geosporus,* BR = *Scutellospora* (*Gigaspora*) *calospora,* MIC = *Glomus microcarpus,* YV = *Glomus mossea.* (B) Inflow attributable to the hyphae of ▲ = *Glomus mossea;* △ = *Glomus macrocarpus* var. *geosporus,* ○ *Scutellospora* (*Gigaspora*) *calospora,* ■ = *Glomus microcarpus.* Values calculated by subtracting inflow into non-colonized roots from inflow into mycorrhizal roots. Note the ineffective fungus *Glomus microcarpus,* which colonized roots very slowly and did not increase the P inflow over that of the controls. (From Sanders et al. 1977, with permission.)

effective uptake of P from low concentrations in the soil solution and probably lowers the threshold level at which P absorption normally stops. Finally, mycorrhizal roots can utilize sources of P in the soil (organic and rock phosphates) that would not be available to non-mycorrhizal roots.

Compartmental pots can be used to directly measure P uptake. Using these, researchers have conclusively demonstrated the importance of fungal hyphae for the uptake of P. Fine-mesh material generally separates roots in the root compartment and hyphal compartments containing the fungal inoculum. The hyphal distribution and contribution to uptake by plants or removal of nutrients from the soil can be studied in the absence of the roots themselves. Depletion of bicarbonate water extractable hyphal distribution, and uptake of ^{32}P from hyphal compartments have all been measured in different experiments and confirm the importance of external hyphae for P translocation; the mycorrhizal roots exhibit increased uptake of P relative to nonmycorrhizal roots, especially at greater distances from the roots. The evidence is overwhelming that mycorrhizae are ecologically important, particularly during later stages of plant development when depletion zones are near maximal and much of the P initially available to the plants has been absorbed. During these periods, hyphae are capable of reaching spatially isolated patches of P and translocating it to the plant.

4. Phosphorus Accumulation by Mycorrhizae. Much of the information concerning phosphorus accumulation by VAM and ectomycorrhizae is similar. More is known about VAM, so we will focus on VAM in this section. The importance of VAM hyphae for nutrient transfer has been directly demonstrated using compartmental chambers. ^{32}P, ^{35}S, and ^{45}Ca were injected at different distances from mycorrhizal onion plants (VAM), and the appearance of the tracer in the plant was followed. When hyphae between the root and the ^{32}P source were severed, P was not translocated from the isolated root chamber to the plant (Rhodes and Gerdemann 1975, 1978a, 1978b). Since the earliest experiments, continued research has shown that rates and fluxes of nutrients from the soil to the root interface are highly variable in space and are influenced by the particular fungal symbiont and plant symbiont. Some fungal symbionts transfer nutrients from greater distances and at a faster rate than others do, whereas some transfer nutrients faster at short distances. The variation in fungal symbiont effectiveness depends in part upon the length and arrangement of extraradical hyphae in the soil. Flux rates of nutrients to VAM plants have been measured in pots with different-volume compartmental chambers with soil and in split agar plates. These experiments show flux rates one to two times lower in split agar plates when compared to experiments using compartmental pots. This difference has been attributed to the lower transpiration rates of plants growing in agar relative to plants growing in soil such as those in compartmental pots. Also, plants growing in agar have much lower growth rates than plants grown in soil and therefore present a lower demand for P.

Mechanisms of translocation in VA fungi involve processes based on mass flow of protoplasm and cytoplasmic streaming. Several lines of evidence suggest that cytoplasmic streaming may be important for P_i translocation in VA fungi, For example, biirectional movement of nutrients occurs, nutrient translocation is temperature sensitive, and translocation can be inhibited by cytochalesin b (an inhibitor of cytoplasmic streaming). Most likely again, pleiomorphic, motile vacuolar tubules (common to the Eumycota, including the members of Glomales) that are continuous with cellular vacuoles are also important for translocation of nutrients within the fungal hyphal system.

The importance of VAM to the P nutrition of a particular plant depends on (1) the concentration of P_i in the soil, (2) the relative affinities of the root and fungal systems for P_i, (3) the characteristics of the root system extending into the soil, and (4) the P requirements of the host (Koide 1991). Among these four factors, mycorrhizal fungi may have the biggest impact on the second. Mosse et al. (1973) suggested that mycorrhizal colonization might lower the threshold concentration above which plants are capable of absorbing P, subsequently increasing the effectiveness of absorption. This concept is dependent upon the fungal symbiont having a greater affinity (i.e., lower K_m) for P_i than either the plant symbiont or the mycorrhizal association does. Tomato and cassava were used to test this hypothesis by other researchers. Mycorrhizal roots had a K_m of 1.6 mmol $H_2PO_4^-$, whereas nonmycorrhizal roots had a K_m of 4.0, even though the V_{max} of the uptake system across all ranges of nutrient concentration was the same for colonized and noncolonized plants. Hence, the fungal component of mycorrhizal roots had increased the affinity for P_i in comparison with that of uncolonized roots of the same species (Cress et al. 1979, Howeler et al. 1979). However, contrary evidence also has been observed. Thompson et al. (1990a) found that the kinetics of P_i accumulation of VAM germ tubes was similar to that of colonized roots and other fungi. As mentioned before, these contradictory results could be due to an inability of germ tubes to assimilate P in the absence of a suitable host symbiont. It remains questionable whether VAM hyphae have the capacity to lower the threshold concentration for P_i absorption by mycorrhizal roots.

*5. **Use of P Normally Unavailable to Plant Roots.*** It has been suggested that mycorrhizal roots might be able to utilize sources of P in the soil normally unavailable to plant roots. These sources include insoluble P_i such as rock phosphate (RP) as well as iron and aluminum phosphates and sources of organic phosphorus such as phytate (inositol phosphate). For example, the growth of ectomycorrhizal plants responded to RP or Ca_3PO_4, whereas these fertilizers had no effect on nonmycorrhizal plants (Wibawa et al. 1995). An increase in P_i availability in this case is most likely due to a higher release of chelating agents from mycorrhizal plants than from nonmycorrhizal plants. In contrast, there is no evidence that VA fungi produce chelating agents in mycorrhizal association. However, VAM are capable of reducing the pH of the hyphal compartment up to 1.0 pH unit concurrent with P depletion in a system where P_i was supplied as $Ca(H_2PO_4)_2$ and N as $(NH_4)_2SO_4$. The reduction in pH is most likely due to the extrusion of H^+ following uptake of ammonium by the fungal hyphae. In order for fungi to utilize organic P, they would need to exude phosphatases into the soil. Plant roots do exude phosphatases, but there is little evidence to support the ability of mycorrhizae to secrete phosphatases.

D. Uptake of Other Nutrients: Cu, Zn, and Others

There is plenty of evidence indicating that mycorrhizae are important for the uptake of nutrients other than P and N. For instance, mycorrhizae contributed up to 62% of copper accumulated by *Trifolia repens* regardless of the phosphorus uptake (Li et al. 1991). Increased Cu uptake by mycorrhizal plants has been confirmed for a number of plant-fungus combinations, but the transfer from fungus to plant is often very small, and the mechanisms of translocation have not been investigated. Uptake of other micronutrients by mycorrhizal hyphae is not well established. In fact, some nutrients such as Mn are often found to be lower in mycorrhizal plants than in noncolonized controls.

Another nutrient affected by mycorrhizae is zinc. Aside from phosphate, zinc may have the lowest mobility among required elements. Thus, we would predict that zinc accumulation is diffusion limited, and that mycorrhizae would have a large impact on zinc accumulation. It is difficult to demonstrate mycorrhizal effects on Zn because of the strong interaction with P accumulation. Interactions of P and deficiencies of trace elements are well known in several species of typically mycorrhizal plants (Wallace et al. 1978). For example, moderate P fertilization increases mycorrhizal colonization, which subsequently results in an increase in the uptake of P, and an increase in the uptake of other essential elements (e.g., Cu, Zn, and K). Conversely, heavy P amendments can have exactly the opposite effect, causing a decrease in mycorrhizal colonization, a decrease in P uptake, and therefore deficiencies of other essential nutrients (e.g., Cu, Zn, and K). The rate of Zn uptake is 2.1×10^{-8} mol m^{-2} s^{-1}, which is considerably lower than the rate of phosphorus uptake (Cooper and Tinker 1978). This low uptake rate is most likely appropriate for the low Zn requirement in plants. However, there is some evidence that mycorrhizae can improve Zn accumulation. Zinc deficiency symptoms in peach plants disappeared with an increase in mycorrhizal colonization (Gilmore 1971). In fact, large differences in Zn uptake could be attributed to the specific VA fungal species, distribution of mycorrhizae on roots, length of hyphae, and density of hyphae in the soil.

Potassium also is relatively immobile in the soil, and mycorrhizae may improve the uptake of K^+ by the plant. However, in many investigations, the concentration of K^+ has been found to be lower in shoots of mycorrhizal plants than in those of nonmycorrhizal plants when the effect of improved P nutrition is removed. Confounding effects of mycorrhizae on P accumulation make it difficult to determine whether the increased uptake in K^+ is a direct or an indirect effect of mycorrhizae. In addition, K^+ accumulation is strongly influenced by the form of nitrogen available to the plant. If ammonium is the primary source of N available, K^+ accumulation is enhanced relative to if nitrate is the predominant ion available for accumulation. Other soil cations (particularly Na) can also influence K^+ accumulation due to their effects on membrane permeability. (Some K^+ accumulation occurs by diffusion across the plasma membrane.) There is also some evidence that the synthesis and storage of polyphosphate granules in hyphae can affect the uptake of K^+ because it is thought that polyphosphate is stabilized by K^+. Therefore, the factors that regulate K^+ accumulation by both the fungal symbiont and the plant symbiont are complex, and it is still very difficult to determine whether mycorrhizae significantly improve the K nutrition of plants.

V. WATER RELATIONS OF MYCORRHIZAL PLANTS

A. Effects of Mycorrhizae on Transpiration

Numerous studies have demonstrated that transpiration rates of mycorrhizal plants are generally higher than those of nonmycorrhizal plants. For example, Huang et al. (1985) showed that pots with mycorrhizal plants had lower soil water potential than pots with nonmycorrhizal plants, thus indicating greater H_2O uptake by mycorrhizal plants. Mycorrhizae could increase whole plant transpiration by several general processes:

1. The resistance to water flow in root and shoots could be lowered (an increase in root hydraulic conductance or an increase in root surface).

2. An increase in transpiration could occur by an increase in stomatal conductance.
3. Water use could increase by an increase in plant leaf area.

Safir et al. (1971, 1972) were the first to demonstrate that mycorrhizal soybean plants had a lower root resistance to water transport than nonmycorrhizal plants, while shoot resistance was small and similar in mycorrhizal and nonmycorrhizal plants. In such studies, it is important to distinguish between direct effects of mycorrhizae on water relations, and indirect effects on water relations due to improved mineral nutrition. Safir et al. (1971, 1972) concluded that the differences they measured in root resistance were primarily due to improved nutrition because the effect could be eliminated with either fertilization or the addition of fungicides. In general, studies have shown that mycorrhizae have little effect on the hydraulic conductance of the root system (other than that due to increased P nutrition). However, it has been demonstrated that mycorrhizal hyphae can accumulate water and transfer it to the plant (Ruiz-Lozana and Azcon 1994). Thus, the increased surface area of mycorrhizal root systems (including the hyphae) could be an important mechanism by which mycorrhizal plants increase the amount of water removed from the soil. It is generally believed that the major effect of mycorrhizal colonization on plant-water relations is an increase in stomatal conductance, and a subsequent increase in transpiration (Figure 4.21).

Stomatal conductance could be influenced by hormonal changes in the plant elicited by mycorrhizal colonization. For example, a decrease in abscisic acid (ABA) or an increase in cytokinins transported from root to leaves would induce a higher stomatal conductance rate. For example, Sanchez-Diaz and Honrubia (1994) demonstrated that a lower level of ABA occurs in mycorrhizal plants than in nonmycorrhizal plants, which could subsequently impact stomatal conductance. Ebel et al. (1997) found that stomatal conductance in cowpeas was higher for mycorrhizal than for nonmycorrhizal plants, even when leaf P and tissue water potential were equalized. They postulated that lower ABA concentrations in the transpiration stream of colonized plants enabled them to maintain a high stomatal conductance. In the presence of mycorrhizae, the

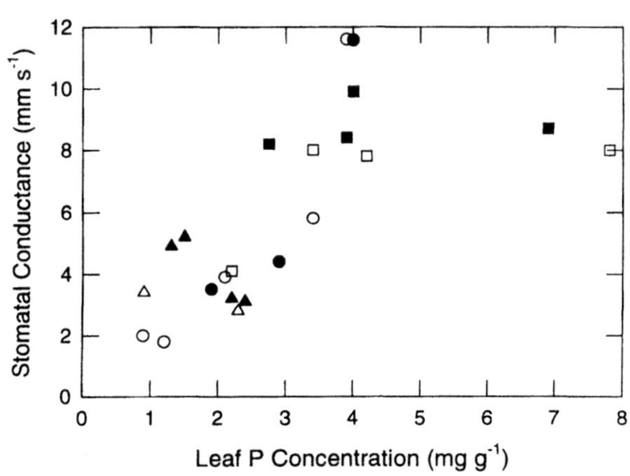

Figure 4.21. Positive relationship between stomatal conductance and leaf tissue P concentration. Symbols represent *Rosa* (▲,△), *Helianthus* (●,○), and *Trifolium* (■,□); solid symbols = mycorrhizal plants; open symbols = nonmycorrhizal plants. (Redrawn from data compiled by Fitter 1988, with permission.)

stomata remained open for a longer period of the day. Much of the data are still inconclusive, and additional work in this area needs to be conducted.

B. Effects of Mycorrhizae on Drought Tolerance

There are numerous studies indicating that mycorrhizae improve the drought tolerance of host plants. Mosse and Hayman (1971) demonstrated that onion with VAM did not wilt, whereas nonmycorrhizal onions did wilt when transplanted. VAM roots had a greater capacity to uptake water and enable the host plant to withstand water stress better under dry conditions. For example, VAM plants recovered from wilting better than non-VAM plants. Duddridge et al. (1980) found that specialized hyphae from 20–60 μm diameter formed within hollow fungal rhizomorphs when the transverse walls of the rhizomorph broke down. Water moved through the vessellike rhizomorphs into the roots and young leaves and was eventually lost to transpiration. Fungi that form rhizomorphs might be especially important in dry habitats.

Most VAM studies are controlled studies conducted in greenhouses. In greenhouse studies, VAM associations have been shown to increase the drought tolerance of cultivated crops such as wheat, soybean, red clover, onion, and pepper as well as many native plant species. Improved drought resistance could be a result of improved nutritional status, increased leaf elasticity (Augé 1987a), maintenance of stomatal conductance because of increased leaf turgor (Auge et al. 1987b), as well as improved water acquisition because of increased root length and rooting depth (Kothari et al. 1990). In contrast, there are reports that VAM fungi have no effect or even reduce the drought tolerance of certain host plants (Bildusas et al. 1986, Graham et al. 1987, Simpson and Daft 1990).

C. Salinity Tolerance and Osmotic Stress

Certain fungi can tolerate extreme osmotic stress (>−65 MPa) as well as halophytic plants. (See Chapter 5 for a discussion of halophytic plants.) Many plants growing in extremely arid environments with high soil-salinity support VAM associations. VAM fungi in coastal communities have been studied for their importance in dune formation and stabilization because most of the dune-forming grasses (e.g., *Ammophila breviligulata*) are VAM. Sand aggregation, which is required for dune formation and stabilization, is enhanced by VAM root systems. Young et al. (1994) found that the level of VAM formation on a barrier island shrub *Myrica cerifera,* contrary to coastal grasses, decreased along the island soil chronosequence. The decrease in colonization levels coincided with a marked increase in soil chloride levels, suggesting that soil salinity inhibited mycorrhizal colonization. It is possible that plants and fungi have coadapted to tolerate saline environments together. This is suggested by the fact that increasing soil salinity is associated with a change in the species composition of VA symbiont fungi (Stahl and Williams 1986). No generalities can be made about physiological effects of osmotic stress on VAM associations because very little work has focused on this topic to date.

VI. NONNUTRIENT ELICITED EFFECTS

A. Pathogen Resistance

VAM colonization reduces the stress that plants incur from pathogens (Davis and Menge 1980), especially root pathogens such as *Pythium, Fusarium, Phytophthora,* and *Verticillium*. It has been suggested that the benefits of improved pathogen resistance are the result of enhanced nutrition or increased production of phenolics and isoflavonoids. Much of the original work has been completed in horticultural settings where the use of moncultures increases the potential for epidemic attack by pathogens. Recent work has demonstrated a beneficial effect of VAM colonization on the annual grass *Vulpia ciliata* based upon enhanced resistance to pathogenic fungi (Newsham et al. 1994, 1995). Application of benomyl to field populations of the grass demonstrated that removing the fungal symbiont had no effect on P uptake or growth. However, a factorial study investigating the effects of different combinations of P and fungal symbiont showed the presence of *Glomus* protected the plants from the deleterious root pathogen *Fusarium,* enabling greater biomass production and ultimately greater fitness in the absence of a nutritional effect. For ectomycorrhizae it has been thought that the thick mantle provides a mechanical barrier to invasion by root pathogens. Recent work indicates that ectomycorrhizal fungi have antibiotic properties. Duchesne et al. (1988a, 1988b) observed that the presence of *Paxillus involutus* in the rhizosphere of *Pinus resinosa* significantly reduced the pathogenicity of *Fusarium oxysporum,* and seedling survival was increased. This was attributed to an antibiotic that was produced by *P. involutus* when in the presence of pine root exudates. Still very little is known about the cellular, molecular, or physiological basis of the mechanisms responsible for improved pathogen resistance following mycorrhizal colonization.

B. Heavy-Metal Tolerance

There are several VAM species that are adapted to soils with heavy metals. In fact, VAM are capable of providing host plants with metals at nutritionally required levels in both high and low P environments. Numerous metals can be toxic when present in high concentrations, such as Zn, Cu, Fe, and Co. VAM may provide plants with increased protection from heavy-metal toxicity. This could result from dilution of accumulated heavy metals due to the effect of increased P accumulation on increased growth. In addition, some heavy metals might be sequestered in the fungal hyphae and not transferred to the plant. That sequestering may occur by complexing with polyphosphate in fungal vacuoles, which would minimize the transfer of heavy-metal ions to the plant roots. Prolonged exposure to heavy metals could damage the fungal symbiont's ability to store the heavy metals and subsequently have a negative impact on the plant. In contrast, prolonged exposure to a high concentration of heavy metals could select for metal tolerant lines of fungal symbiont. Research has shown that prolonged exposure to Cd can lead to Cd-resistant strains of *Glomus* spp. (Weissenhom et al. 1994). Arsenate tolerance in mycorrhizal plants depends on modifications to the phosphate transport system, which also transports arsenate (Meharg, 1994). Again much of the research conducted has left unclear conclusions regarding the effects of mycorrhizae on heavy-metal tolerance.

VII. RECIPROCAL RESOURCE TRANSLOCATION

A. Carbon Transfer Through Mycelial Linkages

Different plant symbionts can be compatible with the same species of fungal symbiont, and consequently can be connected by a common mycelium. The movement of labeled ^{14}C fixed in photosynthesis can be traced as it moves from the root into the fungal sheath. In ectomycorrhizae, plant sugars are converted into nonplant polysaccharides (e.g., mannitol, trehalose, and glycogen). The fungal-specific sugars are not readily recoverable by the plant; however, they are mobile in the hyphal system. An important attribute of ectomycorrhizae is that the belowground mycelium can link several host trees together. It is thought that carbon as well as other nutrients may flow from one plant to another via the mycelial network. The magnitude and ecological significance of this translocation is poorly understood. A dominant tree may provide nutrients to other understory plants within the viscinity of the belowground network. In addition, a tree in a high light gap of a forest may provide carbon to trees located under a forest canopy. There is no doubt that shading of the receiver plants changes the magnitude of unidirectional carbon transfer indicating a source-sink mechanism of regulation (Figure 4.22). Newman (1988) outlines the existence of a pathway between plants through a mycelial network and the potential for carbon movement along the pathway, but this does not mean that carbohydrate follows this pathway under natural conditions, nor does it indicate ecological significance from this potential carbohydrate flow.

Although theoretically possible, it seems unlikely that plants share photosynthate from interplant connections formed by mycelial links to any great extent, except in the case of achlorophyllous plants. Instead, it is most likely that the mycorrhizal network is supported by larger, more photosynthetically active individuals, and other individuals unable to support a mycorrhiza alone utilize the network for improved mineral nutrition.

If the potential receiver plants are nonphotosynthetic, then carbohydrate flow from one plant to another is highly possible. Achlorophyllous members of Polygalaceae,

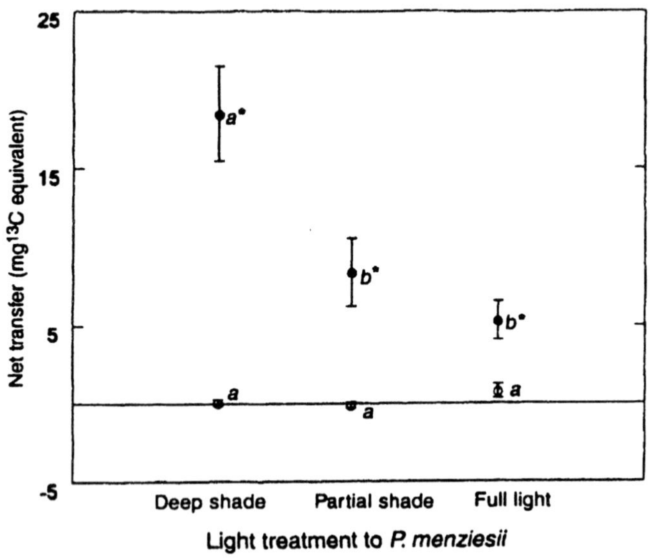

Figure 4.22. Mean ± (one standard error) net transfer of nitrogen from *Betula papyrifera* to *Pseudotsuga menziesii* in deep shade, partial shade, and full ambient sunlight in the first (two-year-old seedlings = open circles) and second (three-year-old seedlings = closed circles) experiment years. Within years, means denoted by the same letter (a or b) do not differ significantly ($p < 0.05$). Asterisks indicate treatments in which the net transfer was greater than zero ($p < 0.05$). (From Simard et al. 1997, with permission.)

Gentianaceae, and Burmannianceae are associated with glomalean fungi, which form mycorrhizae characterized by aseptate hyphae and extensive intracellular coils. It is believed that these plants (mycoheterotrophs) obtain their entire organic C via the fungal symbiont, which must come from a photosynthetic donor plant because glomalean fungi have little or no saprophytic capacity. The mycoheterotrophs are therefore epiparasites on a photosynthetic species (see Chapter 10).

B. Nutrient Transfer Across Mycelial Linkages

Also, it is possible that hyphal links among plants are potential pathways for the transfer of soil-derived nutrients between donor and receiver plants. Although this phenomenon of resource sharing is possible, controlled studies that demonstrate direct hyphal transfer of nutrients are infrequent.

Mycorrhizal hyphae within senescent roots likely redistribute and transfer P released from autolysis or the action of microorganisms to other plants in the mycelial network. Demonstrating that nutrient transfer occurs between living plants is more problematic. There are data indicating that transfers of ^{32}P and ^{15}N do occur, but net movement is lacking. Studies using isotopic tracers demonstrate that the magnitude of one-way transfer can be influenced by shading of the receiver plants, fertilization of donor plants with phosphorus, or use of nitrogen-fixing donor plants with nonnitrogen-fixing receiver plants. Most studies have concluded that changes in competitive balance between large nutrient-sufficient donor plants and small nutrient-deficient receiver plants is likely more important that the direct transfer of nutrients.

VIII. SUMMARY

Mycorrhizae are mutualistically beneficial symbioses between the absorptive organs of most terrestrial plants and basidiomycete, ascomycete, or zygomycete fungi. Mycorrhizal formation is under the control of both the plant and fungal symbionts and involves the upregulation of symbiosis-related proteins and the downregulation of genes for plant defenses. Mycorrhizae are highly variable, ranging in structure from ectomycorrhizae that cause considerable change to the external appearance of the root but do not normally penetrate root cortical cells, to endomycorrhizae that cause little change to the external appearance of the root but penetrate the root cortex inter- and intracellularly. Mycorrhizae also exhibit considerable variation in function, ranging from mutualism or commensalism to parasitism (see Chapter 1) in many of the monotropoid and orchid mycorrhizae.

Mycorrhizal plants benefit from improved acquisition of limiting soil resources, especially N, P, and H_2O as well as nonnutritional effects such as altered phytohormone production and increased sink demand by the roots. As a result, mycorrhizal plants are generally larger, have higher photosynthetic rates, improved drought, pathogen resistance, and heavy-metal resistance, and ultimately have greater fitness than nonmycorrhizal plants of the same species. In return for these benefits, the fungal symbiont receives an adequate source of reduced sugar. Mycorrhizae are ubiquitous and imperative for the establishment and survival of most terrestrial plants in almost all ecosystems.

STUDY-REVIEW OUTLINE

Introduction

1. *Mycorrhizae* are mutualistically beneficial symbioses between the absorptive organs of plants and the hyphae of appropriate fungi.
2. The functional symbiosis provides the plant with limiting edaphic resources such as N, P, and H_2O, while the fungal symbiont receives carbon.
3. Mycorrhizal assemblages are important to ecosystem functions such as nutrient cycling and carbon flow.
4. It is estimated that 95% of all terrestrial plant species belong to mycorrhizal families including bryophytes, pteridophytes, gymnosperms, and angiosperms.
5. Fungal symbionts are representative of three major fungal divisions (i.e., Basidiomycota, Ascomycota, and Zygomycota).

Classification of Mycorrhizae

1. Mycorrhizae exhibit tremendous variation in function (mutualism to parasitism) and structure (*ecto-* versus *endomycorrhizae*).

Vesicular-Arbuscular Mycorrhizae

1. Seventy percent of all plant families form endomycorrhizae called *vesicular-arbuscular mycorrhizae* (VAM).
2. VAM are probably ancient symbioses because fungi similar to modern VAM colonized the earliest land plants.
3. VA fungal species are extremely common soil fungi classified in the phylum Zygomycota, order Glomales.
4. There are six genera of VA fungi, with 200 species worldwide.
5. VAM cause little change to the external appearance of the root upon colonization, and fungal structures can only be seen following appropriate staining and clearing procedures.
6. VAM hyphae can grow both inter- and intracellularly within the root cortex but never penetrate the vascular cylinder.
7. VA fungi spores rarely germinate and do not grow in the absence of appropriate host plant roots, so VA fungi form no free-living mycelium and cannot be cultured.
8. After spore germination and while the hyphae penetrate the root, VAM fungal hyphae radiate into the soil forming a delicate external mycelial fan that is critical for resource uptake and translocation.
9. Arbuscules are dichotomously branched haustoria-like structures that penetrate intracellularly and are thought to be a site of resource transfer between the symbionts.
10. The plant plasma membrane around the arbuscule is rich in H^+-ATPase complexes indicating an active site for nutrient uptake by the plant.
11. Carbohydrate transport to the fungus is likely associated more with nonarbuscular intracellular hyphae and pelotons that, like arbuscules, increase the surface area of contact between the fungus and the plant plasma membrane.
12. The plasma membranes of these intracellular hyphae are rich in H^+-ATPase sites indicating an active site for resource acquisition by the fungus.
13. Intracellular hyphae and arbuscules are ephemeral structures, eventually breaking down followed by cellular digestion.
14. Communication between symbionts in a VAM association is critical for the formation of a functional symbiosis and is apparently controlled by both symbionts.
15. VA hyphae are chemotropically attracted to the plant roots, probably in response to root exudates.
16. Root exudates can induce VA fungi spore germination, affect directional growth of germ tubes and hyphae, impact growth rate and branching of VA hyphae, and attract fungal hyphae to acceptable sites on the root surface for colonization.

17. Hyphae select sites on the root where appressoria and penetration hyphae can be formed.
18. Penetration of roots by VA fungi occurs near the root tips, where there is little suberization.
19. Hyphae penetrate cortical cells intracellulary by producing cellulolytic enzymes and inducing invagination of the plasmalemma.
20. Cellular invasion is under tight regulation of both symbionts involving the upregulation of fungal specific genes and downregulation of plant specific genes.
21. Vesicles are terminal swellings of hyphae formed inter- or intracellularly by VAM and likely function as carbon storage sites for things such as lipids.
22. Because VAM are so variable and many structures are ephemeral, the association can be lacking any of the identifiable features upon examination.

Ectomycorrhizae

1. Ectomycorrhizae occur in association with 10% of the world's flora and are restricted to seed plants only.
2. It is estimated that there are currently 5000 species of fungi that form ectomycorrhizae worldwide, with 2000 species in North America alone.
3. Many ectomycorrhizal fungi are difficult or impossible to culture because complex nutritional requirements are probably affected by the host plant, so confirming a viable symbiosis is often impossible.
4. Ectomycorrhizal plant diversity is relatively small, but the plants that are ectomycorrhizal occupy a disproportionately large amount of the worlds terrestrial habitat when compared with other types of mycorrhizal plants (e.g., conifers in the northern boreal forests are largely ectomycorrhizal).
5. Ectomycorrhizae alter the external appearance of the root, forming unique root phenotypes called *morphotypes,* which can sometimes be used to identify the fungal symbiont involved.
6. Ectomycorrhizae do not form hyphae that penetrate root cells except in senescent regions of the root.
7. Hyphae colonize actively growing regions of secondary roots intercellularly.
8. The Hartig net is a hyphal network that emanates inward from the mantle, penetrating the epidermis and first few cell layers of the cortex.
9. The involving layer of the symbiosis is the mature interface between partners in which the adjacent plant and the fungal walls become indistinguishable and fused to form the interfacial matrix.
10. Cellulolytic enzymes are involved in intercellular penetration of the epidermis.
11. Hyphal strands also project from the mantle into the soil and subsequently are capable of transporting resources from the soil to the root cortex.
12. External hyphae are usually connected to a more expansive mycelium in the soil or even to fruiting bodies of the ectomycorrhizal fungus.
13. If hyphae penetrate root cortical cells that are not senescent, then the association is called an *ectendomycorrhizae.*
14. Germination of ectomycorrhizal spores is stimulated by root exudates that contain sugars, vitamins, amino acids, and phytohormones such as IAA and cytokinins.
15. Root exudates can also promote the formation of other soil microbial communities that release trigger compounds stimulating ectomycorrhizal formation.
16. Chemotrophic responses of fungal hyphae allow the mycelia to orient growth towards the roots, contact the root surface, and potentially adhere to it.
17. Mechanisms of attachment by hyphae include changes in phenolic deposition, fibrillar proteins, and lectins.
18. Phenolic deposition in the root is likely a mechanism for rejecting incompatible fungal hyphae.
19. Fibrillar polymers located on the surface of free-living mycelium are possibly involved in the recognition and attachment of hyphae to the root.
20. Initiating symbiosis, fungal lectins bind to plant-specific polysaccharides on the cell walls of the root cortex.

21. Unreactive materials mask lectin receptor sites that are present on both root cells and fungal hyphae, so these sites have to be unmasked by enzymes such as laminarase or protease.
22. *Ectomycorrhizins,* which are upregulated during mycorrhizae formation, are symbiosis-related proteins that are produced by inducible genes from both partners.
23. Some symbiosis-related proteins act as *hydrophobins,* which are cysteine-rich polypeptides that become polymerized into the cell walls of the hyphae giving them a waterproof coat.
24. The *carbohydrate theory of mycorrhizal formation* states that mycorrhizae develop under conditions where the root contains a surplus of soluble carbon.
25. According to the *auxin theory,* mycorrhizal development is regulated through the controlled production of auxins by the fungal symbiont.

Ectendomycorrhizae

1. Functionally and structurally, ectendomycorrhizae look similar to ectomycorrhizae, but upon closer inspection, they exhibit levels of intracellular penetration beyond what normally occurs for ectomycorrhizae.

Arbutoid and Monotropoid Mycorrhizae

1. Basidiomycete fungal symbionts and plants in Arbutoideae form specialized mycorrhizae called *arbutoid mycorrhizae,* which are characterized by a mantle, Hartig net formation, and penetration of epidermal cells by fungal hyphae.
2. Monotropoid mycorrhizae are formed by ectomycorrhizal fungal symbionts in association with achlorophyllous plants of the Monotropaceae.
3. Monotropoid mycorrhizae are characterized by a Hartig net that encapsulates epidermal cells, and haustoria-like structures called hyphal pegs within epidermal cells.
4. Monotropoid and arbutoid mycorrhizae are classified as *mycoheterotrophic.*

Ericoid Mycorrhizae

1. Ericoid mycorrhizae are mycorrhizae formed between ascomycete fungal symbionts and several plant families within Ericales.
2. Ericoid roots are distinctive, with up to 80% of the root being colonized by the dark pigmented ascomycete *Hymeoscyphus ericea* or *Oidiodendron* sp.
3. Ericoid roots do not produce root hairs, and the epidermal cells are colonized intracellularly by the ericoid symbiont that forms dark pigmented coiled hyphal complexes.
4. Ericoid fungal symbionts are capable of producing hydrolytic enzymes, similar to enzymes produced by pathogenic fungi, that dissolve the epidermal cell wall allowing for cellular penetration.
5. The same enzymes that allow ericoid fungi to dissolve plant cell walls during mycorrhization are also thought to be important for liberating carbon and nutrients from soil organic matter.

Orchid Mycorrhizae

1. *Orchid mycorrhizae* are mycoheterotrophic mycorrhizas formed between plants of the Orchidaceae and mainly species of the imperfect fungus *Rhizoctonia.*
2. Many of the orchid symbionts are fast-growing saprophytes that can utilize complex sugars, while others are better-known plant parasites.
3. Orchid seeds must be colonized by an appropriate symbiont for germination to occur *in situ.*
4. The orchid protocorm must host one or more specialized endomycorrhizae or the protocorm will cease to grow.
5. Intracellular penetration and development of highly coiled hyphal complexes called *pelotons,* which are thought to be a site of resource exchange, characterize these mycorrhizae.
6. Orchid mycorrhizae are atypical because they are frequently epiparasitic or mycoheterotrophic on ectomycorrhizal plants.

7. Orchid mycorrhizae are less stable than other mycorrhizae because the symbiosis can become parasitic or even pathogenic for the orchid.

Growth and Carbon Economy of Mycorrhizal Plants

Mycorrhizal Influences on Growth

1. Mycorrhizal plants usually exhibit increased growth relative to nonmycorrhizal plants.
2. Colonization by an appropriate mycorrhizal fungus eliminates the minimum threshold requirement of P that must be overcome before growth occurs.
3. P depletion zones occur in the local rhizosphere of nonmycorrhizal plants, but these depletion zones are not as problematic for mycorrhizal plants because the mycelium is able to explore a larger soil volume.
4. In mycorrhizal plants, growth increases linearly until a plateau is reached, but P content of the tissues continues to increase linearly.
5. Under natural conditions, plant growth responses can be dependent upon other limiting resources such as N or H_2O.
6. *Root/shoot ratio* has been used as an indicator of the efficacy of mycorrhizal associations, but the parameter is difficult to determine because of the intimate nature between the root and fungal symbiont.
7. Under slightly lower than supraoptimal nutrient conditions, mycorrhizae can cause a reduction in growth because the carbon cost of maintaining the mycorrhizae becomes high relative to the benefit to nutrient accumulation.
8. Plants under moderately high nutrient regimes generally have lower root/shoot ratios relative to those under low nutrient conditions.
9. There are documented cases of decreased rates of dry weight accumulation in ectomycorrhizal plants relative to plants that are not colonized.

Effect of Mycorrhizae on Plant Photosynthesis

1. Mycorrhizal symbiosis can affect the net photosynthesis of plants.
2. Ectomycorrhizal plants usually have a higher photosynthetic rate than nonmycorrhizal plants, but higher photosynthesis might not always be reflected in higher plant dry weight.
3. Improved mineral nutrition for mycorrhizal plants might directly affect photosynthesis.
4. There appears to be a nonnutritional effect of mycorrhizae on plant photosynthesis such that colonized plants have a higher photosynthetic rate than noncolonized plants under similar P concentrations.
5. CO_2 assimilation can be affected by the increased sink demand imposed by mycorrhizal roots.
6. Alternative explanations for increased photosynthesis in mycorrhizal plants have been proposed, including altered phytohormone production, as well as increased acquisition of some limiting resource other than N and P.

Carbon Supplies for Ectomycorrhizal Fungi

1. Most ectomycorrhizal fungi have a limited capacity to degrade lignin and cellulose, but this capacity is much lower than ericoid fungi or wood decomposers.
2. Cellulose- and pectin-degrading substances are probably important for the penetration of root tissue by the hyphae.
3. Studies using $^{14}CO_2$ have demonstrated that reduced plant sugars are transported from the plant leaves to the roots and fungal symbiont.
4. Mycorrhizae cannot accumulate sucrose but can accumulate certain monosaccharides.
5. Glucose and fructose are readily absorbed from aerated solutions by mycorrhizal roots, with glucose being preferentially absorbed over fructose.
6. Hexose absorption is temperature sensitive, O_2 dependent, and inhibited by the same metabolic inhibitors of the cytochrome oxidase pathway and oxidative phosphorylation.
7. ATP is required for the fungus to accumulate sugars.

8. Accumulated hexose sugars are stored as fungal-specific carbohydrates.
9. Some fungi do not have the enzymes to hydrolyze sucrose to monosaccharides and therefore are dependent upon a plant cell–derived invertase that can hydrolyze sucrose.
10. There are two forms of acid invertase produced by plants that have different pH optima, suggesting that carbohydrate transfer to the fungus requires acidification of the apoplast.
11. The *biological valve model* is the current theory explaining carbohydrate transfer to the fungal symbiont, and maintains a net flux of carbon to the fungus and not in reverse.
12. Sucrose produced by the plant is transferred to the roots, where it is effluxed into the interfacial apoplast where it is hydrolyzed by a plant extracellular invertase into glucose and fructose.
13. High-affinity fungal hexose transporters absorb glucose, while fructose exerts feedback inhibition on invertase activity.
14. Glucose inhibits fructose absorption, but as glucose concentration drops, fructose is absorbed, releasing the feedback inhibition on invertase activity.
15. It is unclear whether plants are capable of absorbing monosaccharides from the interfacial apoplast.
16. Hexose transport could be passive, with a concentration gradient being maintained by rapid conversion of sugars within the fungus.
17. There are other ways in which C might get from the plant to the fungus including the following: (1) Cells sloughing off the root cap and epidermis, together with the mucigel surrounding the mycorrhizal root tip, might provide a colonizing fungus with a plant derived carbon source. (2) As fungal hyphae penetrate the root to form the Hartig net, they might alter plant cell wall production at the rosettes and utilize plant cell wall precursors being delivered to the rosettes. For example, in areas around haustoria of VAM, plant cells lose the capacity to organize a recognizable cell wall but retain the capacity to secrete carbohydrates. (3) When hyphae penetrate senescent cortical cells, it is assumed that the fungus absorbs the original contents of the cells.
18. Carbon compounds can return to the plant as the carbon skeletons of amino acids such as glutamine.
19. The amount of carbon transferred from the plant to fungus is difficult to determine, but can be roughly quantified using the mass balance approach.
20. Most data indicate that mycorrhizal roots receive 4–20% more of the plant photosynthate than nonmycorrhizal roots.
21. Different species of fungal symbiont can acquire different amounts of the plant photosynthate.
22. The significance of carbon cost to the plant of maintaining a mycorrhiza will differ depending upon the nutrient status of the soil.
23. The carbon cost of producing mycelium can be two orders of magnitude less than producing the same amount of root tissue.

Improved Mineral Nutrition

Uptake and Translocation of Nutrients

1. Plants acquire P as $H_2PO_4^-$ by an anion co-transport mechanism.
2. Less than 1% of the inorganic soil P will be in solution at any given pH.
3. $H_2PO_4^-$ is only in soil solution between a pH of 6.5 and 7.5; otherwise, it is complexed with rock salts and is considered occluded.
4. Phosphorus is very immobile in the soil because of the anion exchange capacity of the soil particles.
5. Nitrogen availability depends upon the availability of ammonium cation and nitrate anion, which have independent factors affecting their abundance in soil solution.
6. Nitrate is a very mobile anion, while ammonium is less mobile in soil solution.
7. Plants have the ability to accumulate both nitrate and ammonium ions via different mechanisms in the roots.
8. Mycorrhizae enhance the uptake of nutrients to the plant by several mechanisms.

9. Mycorrhizal roots have a greater surface area increasing the root absorption area.
10. Fungal hyphae explore a larger soil volume than to roots alone.
11. Mycorrhizae increase the affinity of roots for P.
12. Mycorrhizae can chemically alter the rhizosphere.
13. Fungal mycelia can connect adjacent plants, allowing for the possibility of nutrient translocation from donor plants.
14. Mycorrhizal roots live longer than nonmycorrhizal roots.
15. Mycorrhizal development and beneficial effects on plant performance are greatest in soils with low or unbalanced nutrient status.
16. Translocation of most nutrients is faster within fungal hyphae connected to transpiring plants than in soil solution.
17. Ectomycorrhizal fungi form an impermeable fungal layer around the Hartig net that is critical for the symbiosis because it allows for materials in the apoplast to be actively regulated by the symbionts involved.
18. Although VAM do not form an impermeable layer around the root, most colonization occurs below the hypodermis of the root, which is impermeable.
19. In the mature symbiosis the interface consists of two membranes (one plant and one fungus) that are separated by an interfacial zone, or *apoplast*.
20. Mechanisms of sucrose efflux from the root are similar to those of phloem loading and unloading, while mechanisms of efflux from the fungus are still unclear.
21. Bidirectional transfer of resources does not occur across a common interface because the interface for nutrient acquisition by the plant is spatially separate from the interface for sugar acquisition by the fungal symbiont.

Improved Nitrogen Nutrition

1. Mycorrhizal roots preferentially absorb ammonium when ammonium and nitrate are both available, and ammonium inhibits nitrate uptake and assimilation in the external hyphae.
2. Research with axenic cultures using inorganic nitrogen has shown that most of the nitrogen from NH_4^+ enters plant metabolism of ectomycorrhizal fungi as either an amide group of glutamine formed by the glutamine synthetase (GS) pathway or as the amino group of glutamate through the glutamate dehydrogenase (GDH) pathway.
3. It is most probable that nitrogen assimilation in mycorrhizal fungi involves a combination of the GS, GDH, and GS-GOGAT (glutamate synthase) pathways, which results in considerable variation.
4. Ammonium assimilation in plants occurs primarily by the GS-GOGAT pathway.
5. Fungal enzymes for nitrogen assimilation are downregulated in the vicinity of the plant root, leading to the *metabolic zonation model*.
6. There are three basic models (summarized in Figure 4.18) describing ammonium assimilation and transport across the symbiotic interface.
7. Ectomycorrhizal fungi have the capacity to acquire organic nitrogen, probably in the form of amino acids.
8. Because alanine is a major source of C, translocation of it to the plant via the mycorrhiza can result in a bidirectional flow of carbon.
9. High concentrations of amino acids in the soil can be harmful to plants, but appropriate fungal symbionts can alleviate the harmful effect by accumulating amino acids in the external mycelium.
10. The ability to use organic N varies depending upon the type of mycorrhizae and the fungal symbionts involved.
11. Many nitrogen-fixing plants also host VAM that can increase the rates of N_2 fixation by alleviating P stress as well as by improving the uptake of other limiting nutrients.

Improved Phosphorus Nutrition

1. Chelating compounds released in the soil (e.g., citrate, malate, and oxalate) can enhance the availability of P in soil solution.

2. Inorganic P can be released from organic forms, the predominant form in most ecosystems, by the action of phosphatases.
3. Phosphorus absorption by fungal hyphae has to be active because the electrochemical gradient favors efflux to the soil.
4. Phosphate is stored in the fungal hyphae as polyphosphate granules and can be mobilized and transferred to the plant symbiont when P is deficient.
5. Rapid absorption of P_i is followed by the synthesis of inorganic polyphosphate in fungal vacuoles that is stabilized by K^+.
6. Storage as polyphospate prevents the accumulation of orthophosphate in fungal cells and reduces osmotic stress associated with orthophosphate accumulation.
7. P is transported as polyphosphate.
8. There are two accepted theories on the movement of polyphosphate and other nutrients within fungal hyphae.
9. The first mechanism is based upon the net movement of vacuoles containing polyphosphate, and requires that (1) translocation occurs down a concentration gradient of phosphorus between a source and a sink. Cytoplasmic streaming in fungal hyphae, either unidirectional or bidirectionally, could lead to a net movement of vacuoles containing polyphosphate along this gradient. (2) A well-organized vacuole loading and unloading system must be present at the source and the sink.
10. The second mechanism is contingent upon vacuolar movement along tubular cysternae in the fungal hyphae.
11. The tubules provide the means whereby resources could be transported long distances bidirectionally.
12. Compartmental chambers have been used to demonstrate the direction and rate of transport of nutrients in VAM hyphae.
13. It has been suggested that mycorrhizal roots can use sources of P in the soil normally unavailable to the plant root.
14. Mycorrhizae are important for the uptake of nutrients other than P and N, such as Cu, Zn, and others.

Water Relations of Mycorrhizal Plants

Effects of Mycorrhizae on Transpiration

1. Transpiration rates of mycorrhizal plants are generally higher than nonmycorrhizal plants.
2. Mycorrhizae could increase whole-plant transpiration by several general processes.
3. The resistance to water flow in roots and shoots could be lowered (an increase in root hydraulic conductance or an increase in root surface).
4. An increase in transpiration could occur by an increase in stomatal conductance.
5. Water use could increase by an increase in plant leaf area.
6. Stomatal conductance could be influenced by hormonal changes in the plant elicited by mycorrhizal colonization.

Effects of Mycorrhizae on Drought Tolerance

1. Mycorrhizae improve the drought tolerance of plants.
2. Improved drought resistance could be a result of improved nutritional status, increased leaf elasticity, maintenance of stomatal conductance because of increased leaf turgor, as well as improved water acquisition because of increased root length and rooting depth.

Salinity Tolerance and Osmotic Stress

1. Many plants growing in saline environments host VA fungi, but no generalities can be made as to the effect of VAM on plant physiology and osmotic stress.

Nonnutrient Elicited Effects

Pathogen Resistance

1. VAM colonization reduces the stress plants incur from pathogens, especially root pathogens.
2. Ectomycorrhizae provide a barrier to root pathogens by forming a thick mantle around the plant root; plus, many ectomycorrhizal fungi probably produce antibiotics.

Heavy-Metal Tolerance

1. There are plants with VAM that are adapted to soils with elevated concentrations of heavy metals.

Reciprocal Resource Translocation

Carbon Transfer Through Mycelial Linkages

1. Plant symbionts can be connected by a common mycelium from the same species of fungal symbiont.
2. It is thought that carbon and nutrients can flow from one plant to another across these mycelial linkages.

Nutrient Transfer Across Mycelial Linkages

1. Hyphal links among plants are potential pathways for the transfer of soil-derived nutrients between donor and receiver plants.

SELF-STUDY QUESTIONS

1. What are mycorrhizae?
2. Name seven different types of mycorrhizal associations.
3. What structures characterize a VAM association?
4. What is the function of a vesicle? What is the function of arbuscules? Are there any other sites for resource transfer in VAM associations?
5. What makes orchid mycorrhizae atypical when compared with ectomycorrhizae or VAM?
6. What mycorrhizal associations are considered to be ancient?
7. Can VAM fungi be cultured? Why or why not?
8. Can ectomycorrhizal fungi be cultured?
9. What sugars are fungal specific?
10. What are the primary benefits to the plant in hosting a mycorrhiza?
11. Describe the biological valve model for sugar transport from plant to fungal symbiont.
12. What is the basic tenet of the carbohydrate theory with regards to mycorrhizae formation?
13. According to the auxin theory, what regulates the formation of mycorrhizae?
14. How is phosphorus stored in fungal vacuoles?
15. What is used to stabilize the phosphorus storage product?
16. Give two mechanisms that can explain bidirectional phosphorus transport in fungal hyphae.
17. What are the flaws associated with the mechanisms of phosphorus transport?
18. In what form can carbon definitely return to the host plant?
19. What forms of organic nitrogen can mycorrhizae use? Are certain mycorrhizae better at using organic forms of nitrogen that others. If so, which mycorrhizae?
20. What monosaccharide is preferentially absorbed by mycorrhizae in aerated solutions?
21. What form of nitrogen is preferentially absorbed by mycorrhizae?
22. What is a phosphorus depletion zone?
23. Give six ways that mycorrhizae can improve the uptake capacity of plants in soils with limiting resources?
24. In what form is phosphorus absorbed by plants?

25. What are hydrophobins?
26. What are symbiosis-related proteins?
27. What are the functions of root exudates in relation to mycorrhizae formation?
28. What compounds can be found in root exudates that may affect mycorrhizae formation?
29. Describe three models of nitrogen assimilation and transport across the mycorrhizal interface.
30. What is the metabolic zonation model of nitrogen assimilation?
31. How might ectomycorrhizae improve the pathogen resistance of host plants?
32. Do mycorrhizae influence the net photosynthesis of host plants?
33. Are effects on photosynthesis always nutrient mediated?
34. Why are changes in photosynthesis not always manifest in the dry weight of the plant?
35. What is the effect of mycorrhizae on the water relations of host plants?
36. Do ectomycorrhizae penetrate cortical cells intracellularly?
37. What is the Hartig net, and in which mycorrhizal associations is it common?
38. Which is more soluble in soil solution—ammonium or nitrate?
39. Can you name an ecosystem or plant community characterized by a predominant mycorrhizal association?
40. Why are VAM important in sand dune environments?

SUPPLEMENTARY READING

Smith, S. E., and D. J. Read. 1997. Mycorrhizal Symbiosis. 2nd Ed. Academic Press, London, UK.

Varma, A., and B. Hock. 1999. Mycorrhiza: Structure, Function, Molecular Biology and Biotechnology. 2nd Ed. Springer-Verlag, Berlin.

Allen, M. F. 1992. Mycorrhizal Functioning: An Integrative Plant Fungal Process. Chapman & Hall, New York.

Alexopoulos, C. J., C. W. Mims, and M. Blackwell. 1996. Introductory Mycology, 4th Ed. John Wiley & Sons, New York.

5 Salinity Stress

Outline

- I. Introduction
  - A. Scientific Interest in Salinity Tolerance
  - B. Diversity of Salt-Tolerant Species
  - C. Salinity Effects on Plant Growth
  - D. Scientific Level of Understanding
- II. Saline Environments
  - A. Units of Salinity
  - B. Saline Environments
    - 1. Salt Marshes
    - 2. Coastal Bluffs
    - 3. Mangrove Swamps
    - 4. Desert Salt Flats and Alkali Sinks
    - 5. Agricultural Fields
- III. Detrimental Effects of Salinity
  - A. Direct Effects
    - 1. Enzyme Activity
    - 2. Nutrient Imbalance
    - 3. Membrane Dysfunction
    - 4. General Metabolic Processes
  - B. Indirect Effects
    - 1. Water Relations
    - 2. Collateral Nutrient Limitation
    - 3. Oxidative Stress
  - C. General Growth Inhibition Model—Biphasic
- IV. Mechanisms of Adaptation to Salinity
  - A. Salt Exclusion: Ion Homeostasis
  - B. Salt Reabsorption and Retranslocation
  - C. Salt Extrusion: Glands and Bladders
  - D. Salt Dilution: Succulence
  - E. Salt Compartmentation
  - F. Salt Tolerance
    - 1. Compatible Solute Accumulation
    - 2. Membrane Stability
    - 3. Growth Hormone Relationships with Salinity
  - G. CAM Photosynthesis Induction
- V. Salt Tolerance Genes and Genetic Engineering
  - A. Signal Perception and Transduction
  - B. Salinity-Inducible Genes and Proteins
    - 1. Ion Homeostasis
    - 2. Genes Involved in Solute Accumulation
    - 3. Genes Associated with Water Homeostasis: Aquaporins
    - 4. Dehydrins
    - 5. Osmotin
    - 6. Genes Associated with the Switch-to CAM Photosynthesis
  - C. Summary of Inducible Proteins and the Potential for Biotechnology

VI. Summary
Study-Review Outline
Self-Study Questions
Supplementary Reading

Objectives

1. Discuss the reasons why the study of plants that grow in saline habitats has important implications for plant physiology, ecology, and agriculture.
2. Cover the diversity of saline habitats as well as the diversity and classification of plants that grow in these habitats, in order to establish the context for plant response to salinity.
3. Define the terms *halophyte, glycophyte, obligate halophyte,* and *facultative halophyte.*
4. Describe the various processes by which physiological dysfunction occurs in plants when salts in the environment increase to a stressful level. Include direct and collateral affects.
5. Discuss interactions between salinity stress and other biotic stressors, particularly nitrogen limitation, elevated CO_2 levels, drought, and cold temperatures.
6. Explain the biphasic model for salinity-induced growth inhibition.
7. Expound upon the various mechanisms by which halophytes are adapted and can acclimate to high soil salinity.
8. Discuss the relationship between the salinity stress signal and the transduction of that information in relation to gene regulation.
9. Consider what is known about genes that are induced by salinity. Include the metabolic function of gene products and the regulation of these genes.
10. Review the potential for genetic engineering of salinity tolerance in crops. Include examples of specific genes, quantitative trait loci, and theoretical aims.

I. INTRODUCTION

A. Scientific Interest in Salinity Tolerance

Spread across the landscape, in relatively small patches, are regions with unusually high soil salt content (Figure 5.1). In these saline patches, the diversity of plants and productivity of communities are reduced compared with similar ecosystems that have lower concentrations of salt in the soil (Greenway and Munns 1980, Tanji 1990, Unger 1991). As a result of the reduced productivity in these sites and the salinization of agricultural soils, scientists have been studying the impacts of salinity on plants for many decades. Intensive research began in the 1950s, when the U.S. salinity lab showed that plants with low salt content in their tissues grew faster and that shoot salinity was controlled by roots. In fact, the number of citations on salinity tolerance in plant science journals is second only to that on photosynthesis (Garfield 1987). If the high-salinity patches are relatively infrequent in the landscape, why is there such a growing interest in the mechanisms by which plants can occupy those sites? The answer to this question lies in understanding the significance of three factors: (1) the ecology of these sites in relation to broader regional processes (Waisel 1972), (2) the physiology of halophytes in relation to comprehending the mechanisms of stress tolerance (Bohnert and Jensen 1996), and (3) the physiology of these species in relation to genetic engineering in agriculture (O'leary 1984).

Figure 5.1. The dark areas on this map represent areas of the world that are characterized by saline conditions. Included are salt marshes, saline deserts, and mangrove swamps. Other, smaller-scale salty areas (agricultural fields, salt bluffs, salt ponds, etc.) are not included. (Reproduced by permission from Chapman 1974.)

Several ecological attributes of saline ecosystems are important to regional processes. First, the productivity of salt marshes is among the highest for all ecosystems (Whittaker 1970, Turner 1976). Although this ecosystem type is a relatively small proportion of the world's terrestrial ecosystems, it may have a disproportional significance to regional productivity and carbon dioxide sequestering from the atmosphere. Second, many communities of high salinity are located at the interface between marine and terrestrial ecosystems (salt marsh, mangrove swamp). These coastal communities constitute a protective alcove for the young of many marine species (Hutchings and Saenger 1987). Third, due to their pivotal position at the interface between marine and terrestrial ecosystems (salt marshes, dunes, and mangrove swamps), they regulate the exchange of resources between the upland and marine ecosystems. That exchange of resources is critical to proper functioning of both ecosystems. Moreover, these coastal communities stabilize coastal habitats against the erosive forces of the sea.

Theoretical developments in ecological and physiological studies on stress tolerance have been well served by studies on halophytes. Salt marshes and mangrove swamps are characterized by sharp changes in species composition over short distances. This is an ideal situation for studying the factors that determine species distribution, community borders, regulation of community integrity, and other important ecological processes. On a physiological scale, adaptations to salt tolerance are similar to those for cold, oxidative, and water stress tolerance (Bohnert and Jensen 1996). Several proteins induced by salt stress are also induced by water stress and chilling, and the signal transduction pathways may be very similar for these three stressors (Wu et al. 1997). Therefore, studying the physiology of salt tolerance will shed light on a number of different aspects of plant stress physiology and may help to develop a unified model of stress tolerance in plants.

There is a rich complexity of environmental heterogeneity in saline environments. The ions involved in creating saline conditions can be diverse (sodium, calcium, potassium, sulfate, chloride, carbonate, etc.) and may contain toxic levels of boron or selenium. The combined effects of these specific ions can induce synergistic or antagonistic responses in plants. Along with the high ion concentration, plants may experience limitations in nutrient availability, high pH, low oxygen availability, mechanical damage, and flooding in saline ecosystems. In addition, there is wide spatial and temporal variation in salinity. Due to this complexity of environmental stresses in saline habitats, adaptations are also complex and diverse. The complexity of adaptation to saline environments and its interaction with other important abiotic stressors provides a worthy and timely research topic in plant stress physiology.

Through evolutionary time, many indigenous plants from a wide variety of families have adapted to these salt-rich regions. The same cannot be said for most agricultural species. However, the largest increase in acreage of saline habitat is occurring through salinization of agricultural fields (San Pietro 1982, Gorham 1992). The salt concentration increases in soils that experience long-term irrigation and poor drainage, particularly in arid regions. Agricultural management plans that involve repeated cycles of water use and lying fallow in the summer further increase soil salinity. Although these programs result in increased salinity and dramatically reduce crop productivity and yield (Greenway and Munns 1980, Ashraf 1994), they are necessary due to the marginal nature of the land in use. The problem of salinity in agriculture will only increase as populations increase and larger tracts of presently unused or marginally used lands are brought into service. The countries most affected by agronomic salinization are often those that have a critical need for agricultural products and a relatively small proportion of arable land. Sacher and Staples (1984) explain that the integrated biological and physical processes associated with saline land agriculture need to be understood in order to improve crop yield without amplifying the salinization problem. The development of new cultivars or the selection of new species with salt tolerance is important for solving this problem. Therefore, understanding the mechanisms by which native plants have adapted to saline environments, discovering the genes that determine those mechanisms, and transferring the genetic traits into agricultural crops are goals of many research labs around the world.

B. Diversity of Salt-Tolerant Species

Salt tolerance has evolved in many orders of vascular plants. In fact, Aronson (1990) lists more than 100 plant families that contain salt-tolerant species. Most of the families contain less that 10 salt-tolerant species, but Chenopodiaceae contains over 350 (Table 5.1). Based on the diversity of evolutionary lineages of salt-tolerant species, salt tolerance traits have most likely evolved many times over the history of vascular plants. It is thought that salinity tolerance probably evolved in several families of higher plants growing in estuaries and spread to inland areas (O'Leary and Glenn 1994). The result of that evolutionary process is that more than one-third of extant plant families have halophytic members, and about 2500 species of halophytes are found in the world's salt marshes and desert flats (Glenn 1997).

In general, the term *halophytes* refers to salt-tolerant plant species. Although there is no generally accepted definition of this term, we will use it to classify all those species that grow on saline soils and have some tolerance of elevated salinity. Ungar (1991) defined halophytes as those species that tolerate relatively high soil salinities and are capable of accumulating relatively high quantities of Na^+ and Cl^- in their tissues (Table 5.2). Glycophytes, on the other hand, are those species that show little

TABLE 5.1 Major Families Containing More Than 10 Salt-Tolerant Species

Family[a]	Example Genera	Number of Species[b]	% of Family[c]
Aizoaceae	*Mesembreanthemun*	52	2.2
Apiaceae	*Crithmum*	18	0.6
Arecaceae	*Cocos, Phoenix*	28	1.1
Asteraceae	*Aster, Parthenium*	67	0.3
Avicenniaceae	*Avicennia*	12	80.0
Boraginaceae	*Cortesia*	10	0.5
Brassicaceae	*Cakile, Lepidium*	30	0.9
Caesalpinaceae	*Caesalpinia, Intsia*	12	0.6
Caryophyllaceae	*Honkenya, Spergularia*	18	1.0
Chenopodiaceae	*Salicornia, Atriplex, Sueda*	368	28.3
Combretaceae	*Lumnitzera*	12	2.0
Convolvulaceae	*Cressa, Ipomoea*	15	0.9
Cymodoceaceae	*Cymbodocea, Halodule*	17	100.0
Cyperaceae	*Eleocharis, Scirpus*	49	1.2
Euphorbiaceae	*Euphorbia, Exoecaria*	22	0.4
Frankeniaceae	*Frankenia*	15	16.7
Hydrocharitaceae	*Halophila*	10	12.5
Juncaceae	*Juncus*	14	3.5
Malvaceae	*Pavonia, Thespesia*	19	1.9
Mimosaceae	*Acacia, Prosopsis*	20	0.7
Myoporaceae	*Eremophila, Myoporum*	12	13.3
Mytaceae	*Eucalyptus, Melaleuca*	22	0.7
Pandanaceae	*Pandanus*	12	42.9
Papiloneaceae	*Lotus, Dalbergia*	46	0.4
Plantaginaceae	*Plantago*	14	5.2
Plumbaginaceae	*Limonium, Aegialitis*	57	7.4
Poaceae	*Spartina, Leptochloa*	135	1.5
Polygonaceae	*Coccoloba, Polygonum*	11	1.4
Potamogetonaceae	*Cymadocea, Potamogeton*	10	10.0
Rhizophoraceae	*Bruguiera, Rhizophora*	26	21.7
Scrophulariaceae	*Bacopa, Gerardia*	22	0.7
Solanaceae	*Lycium*	24	1.2
Tamaricaceae	*Tamarix*	39	32.5
Zosteraceae	*Phyllospadix, Zostera*	18	100.0
Zygophyllaceae	*Nitraria, Zygophyllum*	27	11.3

[a]Not listed are 80 other families that contain less than 10 salt-tolerant species.
[b]Based on a very broad interpretation of salt tolerance.
[c]Halophytes, as a percentage of the family, are calculated based on Airy-Shaw 1973.
Source: According to Aronson 1989 and presented by Gorham 1992.

tolerance to elevated root-zone salinity, and do not accumulate high concentrations of ions in their growing tissues. The most tolerant halophytes such as *Salicornia europea* and *Sueda maritima* can tolerate salinities well above that of seawater (0.5 *M* NaCl), while glycophytes commonly are intolerant of salinities greater than 10% that of seawater. Obligate halophytes reach their maximum growth under moderate salinity (Figure 5.2) and have reduced growth in low salinity conditions (Ungar 1991). However, most halophytes reach their optimal growth in low salinity conditions, but can tolerate higher salinity than glycophytes can. There are also a number of facultative

TABLE 5.2 Best-Supported Definitions of Salt Tolerance Functional Types

Functional Type	Definition
Glycophyte	A plant whose growth and fitness decreases when exposed to any level of salinity greater than approximately 10 m*M*.
Intolerant glycophyte	Plants that cannot tolerate a soil salinity greater than approximately 10 m*M*.
Halophyte	Plants that tolerate higher salinity than glycophytes (up to approximately 50 m*M*) before a reduction in growth or fitness occurs.
Obligate halophyte	Plants that do not attain their highest fitness at low soil salinity (less than 10 m*M*). A moderately high soil salinity (50 mM) is required to attain maximum fitness.
Facultative halophyte	Plants that attain a halophytic nature only after they experience moderate salinity in the soils or as an aerosol.

halophytes such as the ice plant (*Mesembryanthemum crystallinum*) that gain the properties of halophytic metabolism when exposed to a pretreatment of moderate salinity.

A few agricultural crops have moderate salt tolerance (Table 5.3), and there can be a wide variation in salt tolerance among varieties or genetic lines of one crop species. For example, tomato cultivars have a diverse salt tolerance (Pasternak et al. 1979), and breeding programs have had some degree of success in promoting salt tolerance of tomato cultivars. Because of this and other successful breeding programs, there is extensive information on the salinity tolerance of crop varieties (Maas 1987). Even though there is a broad variation in salt tolerance among crop species and cultivars, that tolerance is very limited compared with many native species from saline habitats. For example, wild varieties of tomato possess many traits that confer a degree of salt tolerance and that are not found in saline-sensitive cultivated varieties (Sacher and Staples 1984).

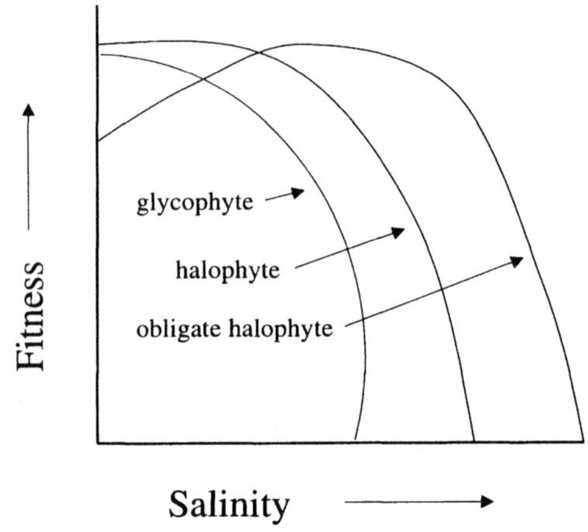

Figure 5.2. The relative fitness of three different functional types is plotted against salinity. Note that halophytes and glycophytes reach their highest fitness at the lowest salinity; however, obligate halophytes attain their highest fitness at moderate salinity.

TABLE 5.3 The Few Crops with a Relatively High Tolerance of Salinity

Common Name	Species	Salinity at 50% Reduction in Yield dS m^{-1}	Salinity at 50% Reduction in Yield mM	Tolerance Rating
		Crops		
Barley	*Hordeum vulgare*	18	200	Tolerant
Cotton	*Gossypium hirsutum*	17		Tolerant
Sugar beet	*Beta vulgaris*	15		Moderately tolerant
Sorghum	*Sorghum bicolor*	14		Moderately tolerant
Safflower	*Carthamus tinctorius*	14		Moderately tolerant
Wheat	*Triticum aestivum*	13	140	Moderately tolerant
Beet, red	*Beta vulgaris*	9.9		Moderately sensitive
Cowpea	*Vinga unguiculata*	8.9		Moderately sensitive
Alfalfa	*Medicago sativa*	8.9		Moderately sensitive
Tomato	*Lycopersicon lycopersicum*	7.6		Moderately sensitive
Cabbage	*Brassica oleraceae capitata*	7.0		Moderately sensitive
Corn	*Zea mays*	5.9	60	Moderately sensitive
Lettuce	*Lactuca sativa*	5.2		Moderately sensitive
Onion	*Allium cepa*	4.3		Sensitive
Rice	*Oryza sativa*	3.6	40	Sensitive
Bean	*Phaseolus vulgaris*	3.6	40	Sensitive
		Native Species		
Glasswort	*Salicornia biglovii*		>720	Tolerant
Salt bush	*Atriplex canascens*		700	Tolerant
Salt grass	*Distichlis palmeri*		600	Tolerant
Salt bush	*Atriplex canescens* sp. *canascens*		350	Tolerant
Beetle grass	*Diplanche fusca*		313	Tolerant

Source: Crops list is modified from Maas 1987; native plants list is from Glenn et al. 1997, with permission.

C. Salinity Effects on Plant Growth

As stated above, an increases in soil salinity commonly causes a reduction in growth and yield. The relationship between salinity and growth reduction is not uniform among ions; some are more toxic than others. It is also true that a combination of ions has a less detrimental effect on growth than does an equamolar concentration of one ionic species (Bernstein 1975). Originally, it was thought that reduced growth was due to direct effects of ions on photosynthesis that led to reduced growth. However, the current theory is that the specific influence of salinity on photosynthesis alone is not large enough to account for the magnitude of reduced growth. Munns (1993) explains that growth is reduced before photosynthesis decreases and growth decreases more than photosynthesis does. Allocation patterns and leaf turnover patterns in response to salinity are associated much more with reduced growth than with reduced photosynthesis.

One immediate response of plants to elevated salinity is a decrease in the rate of leaf expansion. Consequently, whole-plant leaf area is reduced. In fact, the decrease in growth during the initial buildup of salinity in soils can be attributed to a decrease in the leaf area ratio. Ball (1988) found that the diminution in growth of a mangrove species could be attributed entirely to a decrease in the leaf area ratio until very high salinities were reached. The common decrease in leaf expansion is not related to a loss in turgor pressure and is most likely a result of a change in hormonal signaling from roots to leaves. Even though salinity can induce a rapid reduction in root growth (Neumann 1995), shoot growth decreases proportionally more than root growth, causing an increase in the root/shoot ratio. A general pattern in the plant response to salinity is that the larger the change in the root/shoot ratio, the greater the effect of salinity on productivity (Cheeseman 1988).

D. Scientific Level of Understanding

The scientific understanding of salinity tolerance mechanisms in plants has matured extensively over the past 50 years. Preliminary studies on halophytes concerned the influence of salinity on plant productivity and plant distribution in the field. Following close on the heels of these ecological studies came a diversity of studies that investigated the morphological and basic physiological differences between halophytes and glycophytes. The differences that were found were used to propose mechanisms by which plants have adapted to the saline environment. It was not until agriculturists realized that salinization was a significant problem to agriculture that the real explosion in understanding the molecular basis of adaptation to salinity was initiated.

Today, we have a good idea of the diversity of physiological, morphological, and behavioral mechanisms used by plants to tolerate several different detrimental affects of salinity. Scientists are on the verge of understanding several mechanisms, at the gene regulation level, that determine salinity tolerance in halophytes. In particular, research with *Arabidopsis, Mesembryanthemum* (ice plant), and yeast have recently yielded significant breakthroughs. It is clear that there is no qualitative metabolic difference between halophytes and glycophytes (with the exception of salt glands) and that adaptations mainly concern quantitative differences in metabolic pathways related to ion homeostasis (Yeo 1998). Attempts at bioengineering salinity tolerance through the modification of one gene have had limited success, and the bioengineering of multiple genes is just beginning. In fact, there has been a shift in the way stress tolerance (including salinity) is studied. Molecular research on stress tolerance has found that only a few mechanisms at the biochemical level (often quantitative trait loci) may be associated with stress tolerance at the physiological level. Thus, salinity tolerance research is now focused on understanding the biochemical pathways that determine stress tolerance (of salinity, drought, low temperature) and understanding their molecular regulation.

The purpose of the rest of this chapter is to provide both the ecological context and physiological processes associated with the interaction between plants and salinity. The affects of salinity on plant metabolism and the mechanisms for tolerance and avoidance will be covered. The bioengineering of salinity tolerance in agricultural plants will be discussed in relation to its agricultural significance as well as its importance to defining the molecular regulation of salinity tolerance in plants.

II. SALINE ENVIRONMENTS

A. Units of Salinity

One of the most basic issues in the study of salinity is the choice of units used to define the magnitude of salinity. Salinity can be measured on the basis of the ionic concentration of a specific ion species (in seawater: Na^+ = 10.77 g l^{-1}, CL^-; = 19.37 g l^{-1}), or molecules (molality of Na^+ = 0.5 mol l^{-1} in seawater), or milliequivelents (in seawater: Na^+ = 480 mEq l^{-1}, Cl = 560 mEq l^{-1}). These kinds of units are useful when experimenting with a known salt in solution or when comparing the individual effects of various ions. However, in most natural situations, salinity is determined by a mixture of many different ions (seawater = 3.3–3.7% salt by weight), and mixtures containing different ions can have unlike effects on plant growth and metabolism. In order to determine the total salinity of a solution, a measurement of osmolality can be made; the unit osmoles (or milliosmoles, abbreviated mOsm) indicates the total amount of all osmotically active particles in solution (1.0 molal NaCl = 2000 mOsm NaCl kg^{-1} water; seawater = 1100 mOsm kg^{-1}).

Secondary measurements of salinity have been used. Electrical conductivity (E_c), measured in siemans per meter (1 dS m^{-1} = 1.0 mmho cm^{-1} = ($\approx$0.01 mol l^{-1} NaCl $\approx$ 0.06% NaCl), can be used because, as the amount of total salts in a solution increases, E_c increases in a consistent manner. Thus, regressions between E_c and salinity can be formulated (seawater = 46 dS m^{-1} at 20°C). This unit (E_c) is preferred by soil scientists, while molarity is the preferred unit of plant physiologists. Soil solution osmotic potential (in megapascals: seawater = −2.4 MPa) can also be measured to indicate the influence of all soluble ions on solution water potential. Freezing point depression has been used to indicate the osmolality of the solution (for seawater, −1.9°C). The greater the freezing point depression, the higher the salinity. In addition, the chlorinity of the water can be used to indicate salinity (salinity = 1.81 × chlorinity). Chlorinity is utilized in saline regions created from seawater in which the major anion is chlorine. Chlorinity may not indicate total salinity in regions with high concentration of sulfates. The diversity of units and methods used in studies of salinity makes it very difficult to compare studies from the literature. We recommend that all studies of salinity utilize osmolality or molality as the measure of salinity because these units reflect the total salt content in complex solutions, and this is most representative of the natural and agricultural environments as well as of tissue salinity.

Where do the soluble salts in the environment come from? The answer to this question includes both natural and anthropogenic factors. The primary source of salts in soils or waters is the weathering of geologic materials. Ions released from rocks and minerals will accumulate over time to form saline conditions. The ions flow in surface or underground water and eventually are concentrated by evaporation. Salinity of a location is also affected by plant transpiration, dilution from rainfall or snowfall, salt spray from the ocean, and seawater intrusion from tidal action or belowground basins. Anthropogenic causes of elevated salinity include irrigation with evaporation, soil and water amendments, animal manure, fertilizers, sewage sludge, and oil and gas field brines. Both natural and anthropogenic factors combine to produce a wide array of chemically and physically different saline environments.

B. Saline Environments

In a very general sense, regions that are characterized by soils with a high soluble salt content (halomorphic soils) can be divided into saline, saline-sodic, and sodic locations (Brady 1974). The majority of sites with high ion concentrations in the soil are of the saline category. They have high concentration of neutral soluble salts with an $E_c > 4$ mmohs cm^{-1}, and less than 15% of their cation exchange capacity is occupied by sodium ions. Soils that are saline-sodic have high concentrations of soluble salts ($E_c > 4$ mmohs cm^{-1}) and greater than 15% of exchange capacity occupied by sodium. However, both saline and saline-sodic soils have a pH below 8.5. An important difference between these two soil types is that when leaching occurs, the pH and the Na^+ concentration of saline-sodic soils will sharply increase while that of saline soils will remain fairly constant. Sodic soils do not contain a high concentration of neutral soluble salts but do contain a significant amount (enough to cause plant toxicity) of sodium. These sodic soils often have a high pH (as high as 10.0), largely due to the hydrolysis of sodium carbonate. The high pH and the excessive alkalinity resulting from the $NaCO_3$ cause the surface of sodic soils to be discolored. Consequently, such soils are frequently called "black alkali." In this chapter we will be focusing on saline soils and their influences on plant function.

1. ***Salt Marshes.*** There is a diversity of different types of saline environments, each of which has its own spatial and temporal patterns of salinity. The largest areas of natural saline habitats are salt marshes, which are located at the interface of ocean and land in the temperate zone. The range of salinity in salt marshes is between 0.1 and 1.2 *M*. (Chlorinity varies from 500 to 40,000 $\mu g/g^{-1}$ soil) depending on spatial and temporal factors (Figure 5.3). Site salinity is increased by tides, salt spray, and evaporation and decreased by precipitation. Salinity increases from shore toward the mean high tide level (high marsh) and begins to decrease thereafter, until it reaches its lowest level in upland sites that do not experience salt spray or tides. Following precipitation, this gradient is wiped out and salinity is relatively constant along the gradient from lowest tide level to mean high tide level. The interaction among the opposing forces of evaporation and precipitation results in a dynamic temporal variation in salinity (Figure 5.4). Salinity also varies spatially through the soil profile. Salts can be concentrated in surface soils by evaporation, but salts also increase in the root zone because salt retained around roots, due to ion exclusion by roots, can only leave by diffusion to upper soil layers (Passioura et al. 1992). Topographical variation creates added complexity to spatial variation in salinity. Water that collects in depressions can have high salinity because of evaporation, and rivulets of fresh water traverse the marsh creating zones of low salinity.

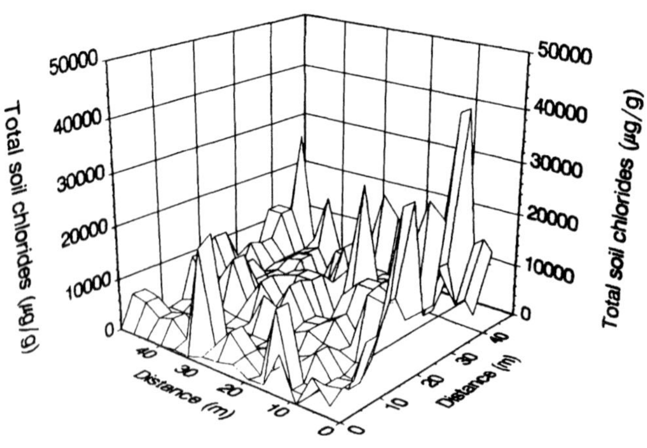

Figure 5.3. This figure shows the extent of spatial variability that can be found in saline environments such as a sandy shore environment. The high and low salinity locations in this area can change on a temporal cycle. (This figure is reproduced with permission from unpublished data collected by D. Young and his students at Virginia Commonwealth University.)

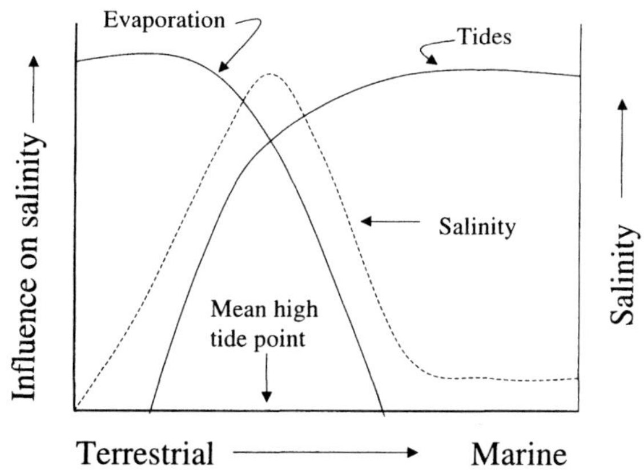

Figure 5.4. The influences of evaporation and tides on salinity are plotted against position along a gradient from terrestrial to marine (*solid lines*). The relative salinity that results from the interaction of these two forces has a peak near the mean high tide point (*dashed line*). The pattern of salinity represented in this figure becomes more evident as time since precipitation increases.

*2. **Coastal Bluffs.*** Coastal bluff communities are located on the crest of steep ocean boundaries where plants are not influenced by tides but are affected by salt spray. These regions are often dominated by cushion plants such as members of Aizoaceae. Salinity in coastal bluffs can become high during the summer because of constant salt spray off the ocean and extensive evaporation, particularly in Mediterranean regions that have limited summer precipitation. Salinity is located in the top few centimeters of soil and is easily washed out by precipitation. Probably the most stressful characteristic of the coastal bluff environment is the accumulation of salt from salt spray on leaves. This is easily observed as the phenomenon *salt pruning,* which describes the dwarfing of trees near coastal bluff communities. In some regions, the effects of salt spray on trees can be measured miles from the coast.

*3. **Mangrove Swamps.*** Mangrove swamps are located at the ocean-land interface in tropical regions, where tropical and subtropical forest is replaced by mangroves. Originally it was thought that the expansion of coastal land into the tropical ocean in some cases was due to growth of mangrove species. However, it is now thought that the mangroves are closely following the development of *siltation beds* (deposition of silt in the intertidal zone). There is often a clear zonation of mangrove species normal to the shore with prop-rooted species such as *Rhizophora* at the outermost edge of the swamp in the ocean, and other species with *pneumatophores* (root extensions into the air serving to absorb oxygen) further inland. Each vegetation band contains one to a few species of mangrove. In fact, the diversity of mangrove forests is particularly low in these tropical regions. A survey of the mangrove forests of the world identified only 114 species, which is less than that found in 1 ha of tropical forest (Tomlinson 1986). In stark contrast to other regions of the tropics, mangrove forests have very few palms, epiphytes, and vines and no conifers.

Similarly to salt marshes, tidal inundation is the primary environmental factor regulating vegetation zonation. The correlation between tidal frequency and duration with species zonation has been long established in mangrove swamps (MacNae 1968).

However, the correlations are site specific, which indicates that tidal inundation is the primary limiting factor, but other environmental factors are also important regulators of vegetation patterns. The three most important factors influencing vegetation in mangrove forests are nutrient availability and nutrient type, waterlogging, and salinity. Among these environmental factors, salinity is the most important (Ball 1996).

Salinity in mangrove swamps is primarily regulated by tidal inundation, precipitation, and evaporation. The seasonal and spatial variation in salinity can vary from freshwater to hypersaline conditions, and the salts in solution are dominated by NaCl. Most mangrove species require some salt for growth, with the optimal salinity varying between 25 and 250 m*M* Na^+ among species (Ball 1988); however, all mangroves exclude salts at the roots. Some species, such as *Sonneratia lanceolata,* grow optimally in freshwater conditions (Ball 1996), and some species, such as *Rhizophora mangle,* are unable to grow to maturity in fresh water (Werner and Stelzer 1990). In fact, it is very difficult to determine if mangrove plants are utilizing fresh water or salt water because of the complex spatial layers (*lenses*) of fresh and salt water below ground. Using an elegant isotope technique, Sternberg and Swart (1987) were able to show that mangroves in one swamp were using water with a wide variation in salinity. In fact, some plants that were bathed in seawater at the surface, were gaining most of their water from a freshwater lens below ground.

Tidal inundation transports nutrients into and out of the mangrove forest, and also influences the type of nutrients available. In frequently inundated sites, the sediments become reducing and phosphorus becomes more soluble, and therefore more available. Thus, in sites with a low frequency of flooding (aerobic-oxidative sediments), phosphorus availability limits growth (Boto and Wellington 1984). The ability of mangrove species to accumulate nitrogen is inhibited in flooded soils (Mendelssohn et al. 1981), which may explain why nitrogen fertilization increases the growth of mangroves in frequently flooded sites, even though ammonia is more abundant in these sites compared with infrequently flooded sites (Boto and Wellington 1984). Mangrove plants have several mechanisms to avoid oxygen deprivation in roots when flooded. The root system is close to the surface, *aerenchyma* (specialized tissues for gas transport) is present, many *lenticels* (fractures in the epidermis where gases can enter) occur on roots, and *pneumatophores* (protrusions from roots above ground that enhance gas transport) are often present. In fact, 70% of the root volume in *Avicennia marina* is occupied by aerenchyma (Curran 1985), and pneumatophores are present. Other structural adaptations to flooding in mangroves roots include root knees, prop roots, and root planks (McKee 1993).

4. Desert Salt Flats and Alkali Sinks. There are two main reasons why desert regions can become saline. In one case, salt flats can result from the evaporation of inland bodies of water. The Salton Sea of California is surrounded by a great salt flat that resulted from evaporating most of a large landlocked portion of the Pacific Ocean. The Great Salt Lake is a similar situation in Utah. In other cases, saline desert sites can result from evaporation of ground or surface water coming from surrounding mountains. Precipitation occurs at the higher elevations by orographic means, and the water travels down on the surface (flash floods) or under ground to the flat valleys, where the water will evaporate leaving the salts behind (Figure 5.5). Death Valley in California has saline soils because of this process. An extreme case is the Pampa de Tamarugal, located in the Chilean Andes, where a salt crust 1 m deep covers the soil. Sodic soils are also characteristic of desert and semiarid regions.

Spatial variation in salinity is low in desert salt flats because there is minimal topographic variation and no streams are present. During the rare precipitation events

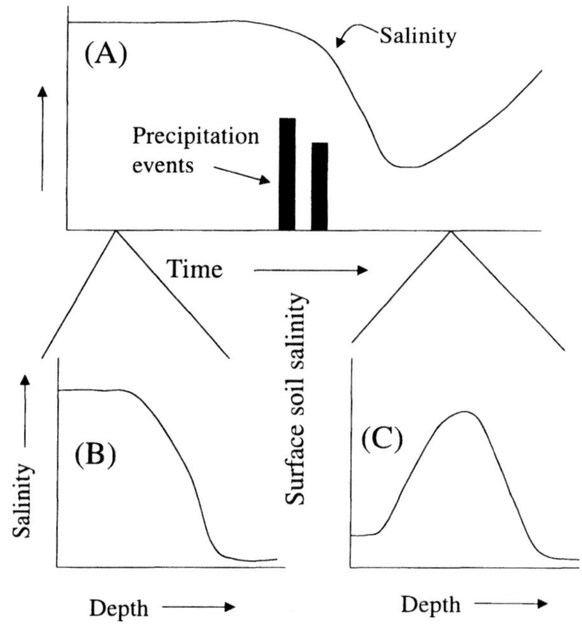

Figure 5.5. This figure represents the patterns of salinity found in desert salt pan environments. (A) Salinity is high in surface soils until a precipitation event occurs. Following precipitation, there is a short-term reduction in salinity at another point in time. (B) Before the precipitation event, salt is concentrated in surface soils as a consequence of evaporation. (C) Following precipitation, some salts can be washed down into the soil profile for a short time period. Evaporation will cause the salts to return to the surface over time.

the salts may be washed down into lower soil layers. However, over time evaporation of groundwater will wick the salts back to the surface again (Figure 5.5). Unlike salt marshes or coastal bluffs, there is no natural mechanism to remove the salts from these desert salt flats, and there is little opportunity for plants to avoid the places and times of high salinity. Therefore, salinity becomes most extreme in these desert sites compared with other saline habitats.

*5. **Agricultural Fields.*** Agricultural fields that receive irrigation can become salinized. In fact, historical records over the last 6000 years indicate that several societies based on irrigated agriculture have failed (Tanji 1990). Gelburd (1985) explains that irrigation in ancient Mesopotamia (now Iraq), from 4400 to 3700 years ago, led to excessive soil salinity. Crop records of the Sumerians indicate a switch from wheat (salt sensitive) to barley (salt tolerant), but barley yields also decreased with time. Armillas (1961) suggest that the Peruvian culture of the Viru valley, which peaked 1200 years ago, moved from the valley to the highlands 800 years ago partly because of the salinization of their agricultural fields. These are just a few of many cases in which salinization of irrigated agriculture has led to the relocation or demise of populations in the past.

The problem of salinization and its management in agriculture has not gone away in current-day agriculture. For example, salinization was revealed as a problem to agriculture in the San Juaquin valley of California in the mid-1890s. In 1905, tile drainage systems were first employed to reduce salinity in this valley. However, problems from subsurface drainage waters and selenium toxicity have kept this issue at the top of the list of agricultural problems in California. Contemporary problems of salinity also exit for agriculture elsewhere in the United States (e.g., the Colorado River basin and the northern Great Plains). In fact, increased salinity in irrigated soil is practically universal (Kovda 1976), and constitutes a major impediment for achieving crop yield necessary for the global population (Arbol et al. 1988).

As in other natural saline sites, salinity is located in the top few centimeters of agricultural fields as a result of evaporation. Farmers are unable to rid their fields of salinity by simply overwatering because the salts return to the surface again by evaporation. Agricultural practices, which include fertilization schedules, further enhance the buildup of salts. Special irrigation programs, involving leaching periods in combination with horizontal or vertical drainage, are required to limit salinization in agricultural fields.

The problem of salinization is most extreme in semiarid regions, where precipitation is infrequent, soils have some natural salinity, irrigation is required, and evaporation is relatively high. In the early 1970s between 70 and 80 countries and 35–50% of the world's population were in semiarid areas where salinization was a problem (Kovda et al. 1976). The proportion of the world's population in these countries is increasing, and most of the projected world population growth will occur in these countries. Although the problems of salinity are well known and the management of salinity is well established, salinity continues to be a widespread problem leading to decreased yields in irrigated agriculture, which counteracts the purpose of irrigation (Scheumann 1997). Moreover, because management of agricultural salinization involves periodic leaching of salts, this creates poor-quality drainage water (high salts, fertilizer, pesticides), which enters natural bodies of water or reservoirs of drinking water. Maintaining a balance between agricultural productivity in relation to salinization and protecting the quality of water resources for the population is a critical issue for our times.

III. DETRIMENTAL EFFECTS OF SALINITY

A. Direct Effects

Among all plant organs and developmental stages, salinity most directly affects seeds and shallow roots because salts are commonly concentrated in the upper layers of soils. In some cases, aboveground parts are directly affected by salt spray or by inadequate filtration of water moving from the soil into the xylem. Xylem water normally contains few ions (5–30 m*M*), even in halophytes that concentrate salts in leaves (Glenn et al. 1994). However, even though the xylem stream water has a low ionic concentration, the amount of water flow will make salt regulation in leaves a problem. In glycophytes, small increases in soil salinity (50 m*M*) can rapidly lead to damaging salt buildup in leaf tissues, reaching 500 m*M* in a few days. The direct effects of excessive ions on plants vary with the species of ion, such that sodium ions are more toxic than calcium, and sulfate ions are more toxic than chloride ions. The direct effects of ions can be observed at the level of enzyme activity, membrane function, and several important metabolic processes.

1. Enzyme Activity. As a general rule, enzymes are inhibited by salt *in vitro* equally in glycophytes and halophytes (Greenway and Osmond 1972, Greenway and Munns 1980). In fact, salinization of an enzyme extraction is an effective method of precipitating proteins. Only extremely halophilic bacteria have enzymes that are tolerant of high salinity. The proteins of the halophylic archaebacterium *Haloarcula marismortui* do not form functional higher-order structure until they are in the presence of 3 *M* NaCl (Cendrin et al. 1993). Although the enzymes of glycophytes and halophytes are equally sensitive to salinity, the effects of soil solution salts on plant metabolism vary among ionic species and between halophytes and glycophytes. In glycophytes, Na^+ salts are more toxic than K^+ salts, while in halophytes, toxicity follows the general pattern: $K_2SO_4 > KCL > Na_2SO_4 = NaCl$ (Egan and Ungar 1998). It is critical to separate the discussion of direct effects of salinity (intracellular) on enzymes from effects of external salinity on enzyme activity. The former is covered here, and the latter is covered in later sections of this chapter.

Enzymes associated with basic energetic pathways, such as glycolysis, in all plant species studied are sensitive to inhibition by salinity (Blum 1988). This inhibition of enzyme activity by salinity is not due to osmotic influences, because equamolar concentrations of compounds such as glycine-betaine or proline are much less toxic than NaCl (Cavalieri 1983). It has been hypothesized that *in vitro* inhibition of enzyme activity by salt most likely infers that elevated salt concentration (greater than 150 m*M*) *in vivo* would result in similar inhibition (Greenway and Munns 1980). This idea is supported by the NADH–nitrate reductase activity of rice, which decreases linearly as salinity increases (Figure 5.6) in a similar pattern both *in vivo* and *in vitro* (Richharia et al. 1997). The mechanism of salt inhibition of enzyme activity most likely involves an inhibition of active site function by noncompetitive inhibition.

2. Nutrient Imbalance. The interactions between other nutrients and salinity can be measured by looking for nutrient deficiency indices in plants growing in high salinity conditions. Many studies have shown that limitations in potassium, calcium, magnesium, and phosphorus can occur in glycophytes exposed to high salinity (Gorham

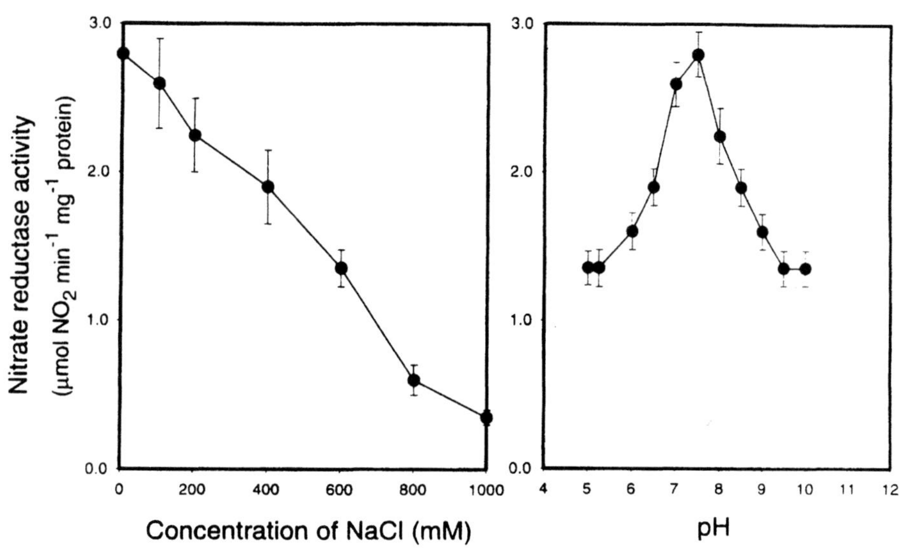

Figure 5.6. The activity of many enzymes is directly affected by salinity. In this case, the activity of nitrate reductase decreases linearly as salinity increases. This can be contrasted with the response of enzyme activity to other environmental factors such as pH. Redrawn with permission from Richharia et al. 1997.)

1992). Most of these deficiencies and toxicities are discussed below in the section on indirect salinity effects. Here we concentrate on direct effects of Na^+ and Cl^- on nutrient transport proteins in the root membrane.

Sodium is transported across membranes by a protein complex that moves Na^+ in one direction and H^+ in the other direction (Na^+/H^+ antiport). The energy source for this antiport system is a H^+ gradient established across the membrane by H^+-ATPase pumps. ATPase-dependent Na^+/H^+ antiport transporters are inhibited by NaCl. For example, in the salt-tolerant *Plantago* species, high concentrations of Ca^{2+} or Mg^{2+} stimulate H^+-ATPase activity, but H^+-ATPase is inhibited by an equamolar concentration of Na^+ (Kuiper 1984). However, long-term exposure to external NaCl can inhibit internal Na^+ stimulation of H^+-ATPase activity, which indicates that long-term exposure to salinity causes a conformational change in the H^+-ATPase complex, or the Na^+ transporter protein in the cell membrane. Because many root membrane transport proteins are dependent upon the electrochemical gradient established by H^+-ATPase, those transporters will be inhibited by the effect of NaCl on H^+-ATPase activity, leading to Na^+-dependent nutrient limitation.

3. Membrane Dysfunction. Some basic structural components of membranes are affected by salinity. Commonly associated with increased salinity is an increase in vesicularization of the plasma membrane (Flowers et al. 1972). There can also be a significant increase in endoplasmic reticulum. The increase is vesicles and endoplasmic reticulum may be a mechanism of compartmentalizing or exporting Na^+ ions.

Under saline conditions, plasma membrane leakage increases in glycophytes. In fact, there is a linear relationship between external salinity and membrane leakage rate (Figure 5.7). The mechanism for this increase in leakage is not due to osmotic factors because membrane leakage does not change when sorbitol concentration in the medium around the root is increased (Leopold and Willing 1984). It is clear that

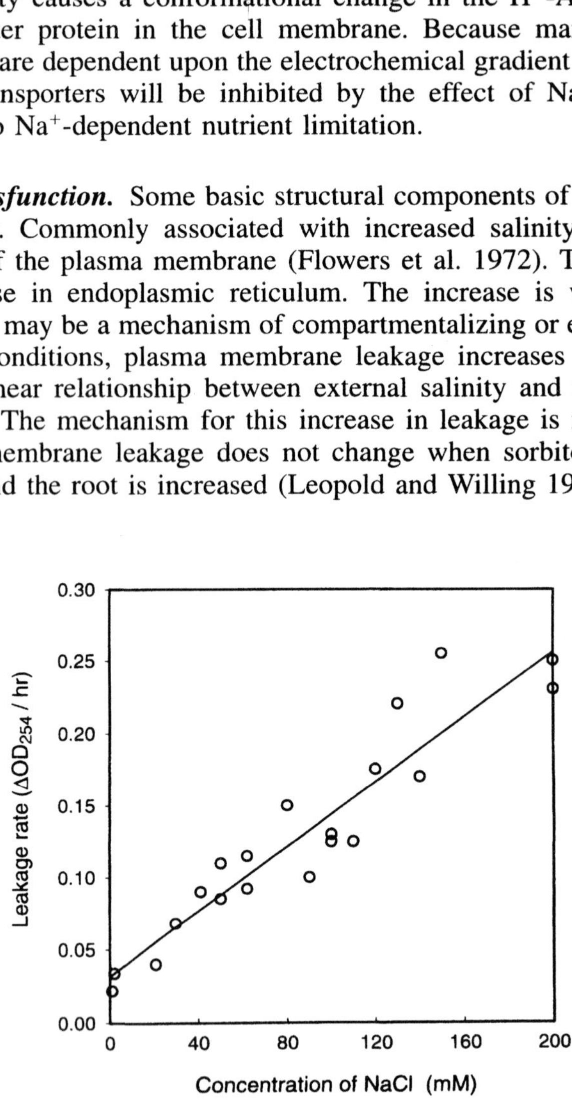

Figure 5.7. The leakiness of membranes (an indicator of their integrity and differential permeability) increases linearly with an increase in salinity. (Redrawn with permission from Leopold and Willing 1984. Copyright © 1984 by John Wiley & Sons. Reprinted by permission of John Wiley & Sons, Inc.

Na^+ replaces Ca^{2+} in membranes, and this ionic change disrupts the differential permeability of membranes (Renegal 1992).

Along with the changes in structure and permeability, membrane chemical composition is affected by salinity. Under hypersaline conditions, the total lipids of cells decrease (Kuiper 1984). In halophytes, there is often no change in the phospholipid/lipid ratio, while in some glycophytes, this ratio increases when exposed to salinity. The saturation level of phospholipids increases with increasing salinity in glycophytes, which reduces membrane fluidity and increases leakage. In addition, salt stress inhibits linoleic acid desaturase, which leads to a lower quantity of linoleic acid in chloroplast membranes and reduces chloroplast membrane function.

4. General Metabolic Processes. As described above, basic metabolic pathways such as respiration and photosynthesis are inhibited by salinity. Inhibition of respiration is primarily associated with the direct effects of salinity of enzyme function. However, elevated salt content in tissues directly influences photosynthetic enzymes and secondarily influences gas exchange and light reactions.

Originally, scientists assumed that salinity was inhibiting photosynthesis by directly inhibiting biochemical reactions (Long and Baker 1986). However, with the results of an increasing number of studies, it became clear that photosynthesis is inhibited by both stomatal and nonstomatal factors. Elevated salinity reduced photosynthesis in *Spartina* primarily by stomatal mechanisms (Longstreth and Strain 1977). In a study by Robinson et al. (1983), photosynthesis was inhibited 65% by salinity, and stomatal conductance was inhibited by a similar amount, while there was no change in chlorophyll concentrations. Zisca et al. (1990), showed that photosynthesis was inhibited by salinity through effects on mesophyll conductance only. These and other studies led many researchers to believe that salinity inhibits photosynthesis primarily by its influence on stomata.

However, many other studies found different results. Photosynthesis in *Atriplex patula* was inhibited by a decrease in both mesophyll and stomatal conductance (Longstreath and Nobel 1979), as was that of the gray mangrove (Ball and Farquhar 1984). Moreover, Pearcy and Ustin (1984) demonstrated that when exposed to high salinity, photosynthesis was primarily inhibited by nonstomatal factors in several salt marsh species. When all these experiments are considered together, the relative control of mesophyll and stomatal conductance by salinity varies among species without a general pattern among halophytes or glycophytes.

Other research has shown that light reactions may be the component of photosynthesis most sensitive to salinity. For example, photosystem II is quickly damaged in salt-sensitive species when exposed to 500 m*M* NaCl (Ball and Anderson 1986). This idea leads into a consideration of the interaction between light intensity and salt inhibition of photosynthesis. Salinity is more toxic to the photosynthesis of plants adapted to low light conditions than for plants adapted to high light conditions. High light–adapted plants have a relatively large capacity to generate ATP and reducing power through electron transport processes. Accordingly, a small inhibition of electron transport processes by salinity in high light–adapted plants may not be as important as the same inhibition in low light–adapted plants. Furthermore, C_4 plants are normally high light adapted, and they have a greater tolerance of salinity than C_3 plants (DeJong 1978). This may be another indication that a high ability to utilize adenolates by photosynthesis may protect the photochemical reactions from damage by the secondary effects of salinity. These relationships between adaptation to light environment and sensitivity to salinity damage suggest that the production of free radicals of oxygen may be an important mechanism of photosynthetic inhibition under hypersaline conditions (discussed later in the chapter).

B. Indirect Effects

*1. **Water Relations.*** The most observable indirect effect of salinity on plant performance is that of reduced soil water availability, because as salinity increases, soil water potential decreases. Over 100 years ago, Schimper (1898) realized that the osmotic problem of salinity is nearly equivalent to that of water stress. In fact, it has been difficult to separate physiological responses to low soil water potential from those in response to salinity. Only two types of experiments can be used to separate the individual effects of salinity and water stress on plant performance. One is to compare the results between equamolar concentrations of salts and organic osmotica (osmotically active substances). In this case, the water stress effect is the same, but the effect of salinity does not occur in the solution with organic osmotica. Another example is experiments that pressurize the soil solution containing high salinity in comparison to unpressurized, equimolar solutions. The pressurized solution has an osmotic effect but no water stress effect, while the unpressurized solution has both the salinity and water stress effects.

In agricultural systems, growth is typically suppressed when a threshold of salinity is reached. The threshold value varies with each specific crop, but yield reduction is often mostly associated with soil water potential, and specific ion effects are secondary (Maas 1990). The salt tolerance of many crop species can be defined by equation (5.1) (Maas and Hoffman 1977):

$$Y_r = 100 - b(E_C - a) \tag{5.1}$$

where Y_r = the yield percentage of the crop under saline conditions compared with nonsaline conditions, a = the threshold level of soil salinity at which yield begins to decrease, b = the percentage yield loss per increase of salinity above a, and E_c = the measure of salinity in units of dS m^{-1}.

Yield response to salinity also can be defined effectively by a linear function expressed in terms of soil osmotic potential alone:

$$Y_r = 100 - B(\mathrm{Op}_{fc} - A) \tag{5.2}$$

where Y_r = the yield percentage of the crop under saline conditions compared with nonsaline conditions, A = the salinity threshold expressed in bars, B = the slope of the relation between yield and soil osmotic potential (expressed in % change per bar), and Op_{fc} = the osmotic potential of the soil at field capacity.

The results of yield predicted with equation (5.2) are within 1% or 2% of those calculated with equation (5.1), which supports the concept that the effects of osmotic potential on yield dominates other aspects of the salinity response in crop species. The three most universal alterations in plant water relations that occur under hypersaline conditions are osmotic adjustment, turgor reduction, and decreased cell wall elasticity. Osmotic adjustment occurs both in the apoplast (salt accumulation) and the symplast (sugar, nonprotein amino acid, or quaternary amine accumulation). Osmotic adjustment in halophytes often occurs by the accumulation of inorganic ions. For example, *Salicornia europa* accumulates salts in the central vacuole to a concentration of 1000 m*M* (Gorham 1992). In contrast, glycophytes often accumulate organic osmotica. The significance of differential osmotic adjustment among species to salinity tolerance is covered in a later section of this chapter.

Stomatal closure is often a rapid response to the initiation of hypersalinity (DeJong 1978, Seeman and Sharkey 1985, Plaut et al. 1989). This rapid closure could be due to a reduction in tissue turgor pressure due to the low water potential of the medium, a specific effect of sodium on guard cell wall plasticity, or an increase in a root hormonal signal (most likely abscisic acid, or ABA). The evidence favors a specific ion effect or a hormonal signal rather than the turgor response. When epidermal strips of *Aster trifolium* are exposed to KCl solutions, stomata open, but when exposed to equimolar NaCl solutions, stomata close (Perera et al. 1994). Termaat et al. (1985) showed that there is no association between turgor pressure and leaf growth under hypersaline conditions.

For many years, it was thought that the reduction in turgor pressure for plants exposed to hypersaline conditions was the major cause of growth inhibition in addition to (and associated with) stomatal closure. Reduced turgor pressure and its influence on tissue growth is one of the most immediate responses to hypersalinity (Neumann 1997). However, reduced turgor is not a long-term regulator of growth in hypersaline conditions. Lloyd (1987) showed that turgor pressure in plants under saline conditions, when growth is inhibited, is equal to or greater than that of plants in nonsaline conditions with higher growth rates. In several studies that show reduced stomatal aperture and reduced growth in hypersaline conditions, turgor pressure was greater than in the nonsaline growth conditions (Kingsbury et al. 1984). When the low water potential of a saline soil is reversed by soil pressurization, stomatal conductance increases independently from leaf turgor potential (Gollan 1986). Therefore, hypersaline conditions may cause a short-term transient reduction in turgor pressure that may have an influence of stomatal aperture and plant growth, but in the long term the absolute amount of turgor is not the only or the most important determinant of plant growth or stomatal aperture.

Cell wall elasticity also is influenced by salinity. For example, cell wall extensibility rapidly decreases when corn seedlings are exposed to increased salinity (Chazen and Neumann 1994). It is well known that cell wall extensibility is closely related to growth rates (Cosgrove 1993); consequently, a part of salinity-induced growth depression is caused by a decrease in cell wall extensibility. It is thought that Na^+ replaces Ca^{2+} in the cell wall and that this reduces cell wall extensibility. This results in stiffer walls, slower growth, and a greater reduction in turgor pressure with any decrease in tissue water content.

2. Collateral Nutrient Limitation. As stated above, plants can experience limitations in other nutrients due to the effects of salinity, even in situations where the nutrients are not limiting in the soil solution. Trace elements are not particularly limiting in saline soils. Those most likely to be limiting are copper, iron, manganese, and zinc because these elements have low solubility in saline or alkaline soils (Tanji 1990). By contrast, boron, molybdenum and selenium have higher solubility in alkaline soils; therefore, plants may accumulate these elements to a higher level in saline soils. An increase in boron concentration in plant tissue could lead to phytotoxicity because plants have a limited tolerance to this trace element. This does not usually occur for Se^{2+} or Mo^{2+} because plants have a relatively high tolerance for these trace elements (compared with other heavy metals). However, the quality of the plant tissue for forage animals is reduced when Se^{2+} or Mo^{2+} are accumulated by plants in saline soils.

Macronutrients can become deficient in saline soils because of interference by sodium (Na^+) ions. In barley, the transport of Ca^{2+}, Mg^{2+}, K^+, and N from the root to shoot is reduced when plants are treated with NaCl (Termaat 1986). This inhibition of macronutrient accumulation is particularly true for potassium (K^+) ions. For example, some mutants of *Arabidopsis thaliana* are very sensitive to salt (20 times more sensitive to salinity than wild types) because salinity induces a reduction of K^+ accumulation by roots (Wu et al. 1996). In a saline environment where Na^+ predominates over K^+, the paramount nutritional requirement for plants is K^+ (Rains and Epstein 1967). Even when the transport mechanisms are highly selective for K^+ over Na^+ (high-affinity transporters), K^+ can become deficient in saline conditions (Ball et al. 1987). In fact, the ability to maintain metabolically significant concentrations of K^+ may be essential for salt tolerance in glycophytes (Zhu et al. 1998).

Calcium uptake dynamics also are influenced by salinity. As stated above, Na^+ can replace Ca^{2+} in cell membranes (Lynch et al. 1985). The first case showing a causative relationship between salinity and calcium deficiency was that of Elzam and Epstein (1969), in which a sensitive variety of wheat had reduced growth at 5m*M* NaCl, while a tolerant variety could continue growth until 100 m*M* was reached. The Ca^{2+} concentration in the roots of each species dropped precipitously at the same NaCl concentrations that reduced growth. When using isoosmotic solutions in which the Na^+/Ca^{2+} ratios were altered, Maas and Grieve (1987) showed that when the ratio dropped below 5.7:1, there was no inhibition of Ca^{2+} accumulation. Also, as the Na^+/Ca^{2+} ratio decreased (on a molar basis) from 34.6:1 to below 5:1, there was an increase in photosynthesis and water use efficiency (Plaut and Grieve 1988).

Perhaps the most interesting interaction between salinity and macronutrient uptake is that between salinity and N accumulation. Botella et al. (1997) reported that NaCl reduced the accumulation of nitrate and ammonium when provided together to the plant. In addition, when the sole source of nitrogen was nitrate, nitrogen accumulation was inhibited by salinity. In contrast, when ammonium was the sole source of nitrogen, there was no inhibition of nitrogen accumulation by NaCl. Ammonium ion protected nitrate accumulation transporters in plant roots against damage by salinity.

Some ions can serve to buffer the effect of salinity on the accumulation of other ions. For example, when excess Ca^{2+} or NH_4^+ is added to the growth medium containing high salinity, growth and nutrient accumulation can be stimulated (Glenn 1997) compared to the control. Ammonium ion is more effective at protecting ion discrimination by root membranes than is Ca^{2+}. Most soils already have a high enough level of calcium to be protective against some NaCl buildup, but this is not the case for NH_4^+. Therefore, the addition of NH_4^+ should increase growth and improve the accumulation of a broad spectrum of nutrients when added to soils high in NaCl sites, a concept that may be particularly applicable to agricultural fields.

3. Oxidative Stress. Several environmental characteristics common in saline habitats can lead to an increase in reactive oxygen species (ROS) and an inhibition of photosynthetic performance. Soil salinity rises to its highest value during warm and relatively dry periods. At these times, soil water potential is low, because of both the high salinity and the low water availability that causes a reduction in stomatal aperture. Saline habitats are also characterized by relatively high light intensity, which in combination with low stomatal aperture can result in the production of ROS in chloroplasts and mitochondria. This increase in ROS for plants experiencing high salt conditions is most likely ubiquitous among plant species (Smirnoff 1993). The importance of light intensity to the buildup of ROS during periods of high salinity is well demonstrated with mangrove species. Under low light conditions, the quantum yield of mangrove leaves is similar to that of nonmangrove species, and energy conversion (as indicated by fluorescence kinetics) is unaffected by high salinity. However, under high light and saline conditions, the quantum yield of photosynthesis decreases and fluorescence is quenched, indicating a reduction in photosystem II activity and potentially leading to the production of ROS (Björkman et al. 1988). Furthermore, stomatal conductance in mangrove species is relatively low, predisposing leaves to the potential production of ROS under high light conditions.

ROS accumulate in the chloroplast because light is stimulating electron flow, in the presence of a limited supply of NADP, to accept the electrons flowing through the photochemical reactions. Therefore, electrons are diverted into ROS through the Mahler reaction (Alscher et al. 1997). An initial oxyradical product, superoxide (O_2^-), cannot cross biological membranes, which restricts it to the site of production (chloroplast and mitochondria). ROS such as superoxide can severely disrupt enzyme activity and membranes associated with photosynthesis. Free-radical scavenging systems such as superoxide dismutase (SOD), and the peroxidase cycle, can be a critical component of salinity tolerance (Olmos et al. 1994, Olmos and Hellin 1996, Bohnert and Jensen 1996) due to their protection of chloroplast function during high light and high salinity.

C. General Growth Inhibition Model—Biphasic

A general model for the inhibition of growth by salinity was proposed by Munns and Termaat (1986). The first phase of growth reduction in this model is due to the osmotic effect of soil salinity (Figure 5.8). Early in phase 1 of growth reduction, turgor potential has an influence on growth, but the majority of phase 1 growth inhibition is maintained and regulated by hormonal signals from roots. Also, phase 1 of growth reduction depends on salt outside the plant rather than salt in tissues, and growth inhibition is not salt specific. This phase of growth inhibition corresponds to the equations developed between yield and salinity for crops shown earlier (equations 5.1 and 5.2).

Under extended periods of hypersalinity, salt begins to accumulate in older leaves and phase 2 salt injury becomes apparent (Figure 5.8). The rate of growth reduction in phase 2 depends on the rate of leaf turnover. If a few of these older leaves die and many new leaves are still produced, then the rate of phase 2 growth inhibition is minimized. However, if the number of older leaves affected approaches the rate of new leaf production, then there will be changes in the flow of assimilates or hormonal balance, leading to a more rapid phase 2 growth inhibition. The biphasic growth inhibition model also predicts that differences in growth response to salinity among varieties of a crop should only be evident during the second phase of inhibition, as was found for wheat, barley, and corn varieties (Munns et al. 1995, Fortmeier and Schubert 1995). However, in some published reports, variation in salinity response among genotypes of a species did not correspond to differential phase 2 effects (O'Reagan et al. 1993, Cramer et al. 1994a, Neumann 1997).

The two-phase model of salinity effects on growth depends on salt accumulation due to transpiration, and the diversion of accumulated salts to older leaves. This could occur by an active process whereby salts are pumped into older leaves, or it could simply be because older leaves have transpired water for a longer time than young leaves have. If salt arriving in the transpiration stream is sequestered in the apoplast,

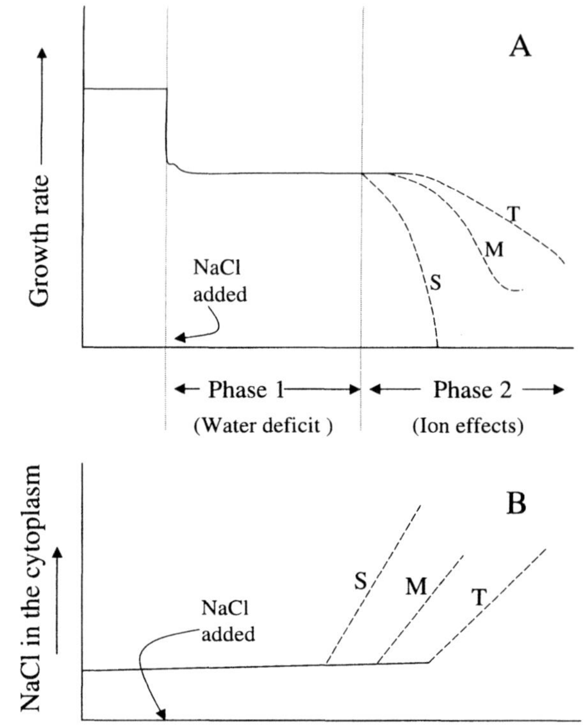

Figure 5.8. (A) The two phase model for salt inhibition of plant performance. T = a tolerant genotype, M = a moderately tolerant genotype, S = a sensitive genotype. The initial decrease in growth is due to a water stress effect, and the second decrease in growth is due to ion effects. (B) Cytoplasmic salt concentration increases sooner (early phase 2) in sensitive genotypes while tolerant genotypes can retain a lower cytoplasmic salinity for a longer period. (Panel A is redrawn from Munns 1993, with permission from Blackwell Scientific Ltd.)

it could be held in the cell wall or moved into the central vacuole. It would be much easier for cells to hold salts in the cell wall because there is no energy cost for transporting the salts across the plasma membrane and tonoplast. However, if salt accumulates in the cell wall of older leaves, the cell cytoplasm would eventually shrink and water stress could become severe. If the salt were concentrated in the vacuole, there would be a considerable energy requirement; eventually the vacuole would become saturated and would stop sequestering salt. Following vacuole saturation, cytoplasmic salinity would increase rapidly and soon become toxic for most enzymes. As we noted above, the rate of phase 2, growth inhibition depends on the ratio of older leaf death and young leaf production. The rate of older leaf death (and phase 2 growth inhibition) depends on the particular ions that constitute the high salinity, their location within tissues (cell wall, vacuole, cytoplasm), and the amount of salt that can be stored in various apoplastic compartments.

The biphasic model of growth inhibition by salinity describes glycophyte responses to salinity well. However, growth response to salinity in halophytes does not follow the same model. Halophytes are able to avoid the development of phase 2 growth inhibition by a number of physiological, morphological, anatomical, and behavioral mechanisms.

IV. MECHANISMS OF ADAPTATION TO SALINITY

A. Salt Exclusion: Ion Homeostasis

The most basic underlying attribute of salt tolerance is the maintenance of ion homeostasis in the cytoplasm. Glycophytes generally maintain ion homeostasis by salt exclusion, while halophytes accumulate salts as a mechanism of maintaining ion homeostasis (Gorham 1994). For example, in glycophytes such as sorghum (Weimberg et al. 1982), beans (Awada et al. 1995), wheat, barley (Gorham 1993), and corn (Alberico and Cramer 1993), salt tolerance is associated with Na^+ exclusion. However, insensitivity to salinity in glycophytes is not always associated with Na^+ exclusion. A salt-tolerant cultivar of maize transports Na^+ to leaves at twice the rate of tolerant cultivars (Cramer et al. 1994b). Many halophytes such as *Mesembryanthemum crystallinum* (Adams et al. 1998, Ungar 1991) do not exclude Na^+ but do maintain a homeostatic ion concentration in their tissues at relatively high external salinity. Salt tolerance, whether related to exclusion or not, is associated with the ability to maintain a homeostatic ion concentration in the cytoplasm.

The transport of ions in and out of the cytoplasm is one major factor that regulates ion homeostasis. Ion transport into the cytoplasm is determined by an electrochemical gradient across cell membranes (plasma membrane and tonoplast), carrier proteins, and ion ports. The cytoplasm is negatively charged compared with the apoplast, due to H^+-ATPase pumps in the plasma membrane and H^+-PPase pumps in the tonoplast that move H^+ ions into the apoplast (Figure 5.9). Sodium and calcium ions diffuse across the plasma membrane into the cell based on the electrochemical gradient, resulting in concentrations of about 10 and 0.2 m*M*, respectively, in the cytoplasm (Bush 1995). The high internal concentration of K^+ (100–200 m*M*) relative to outside

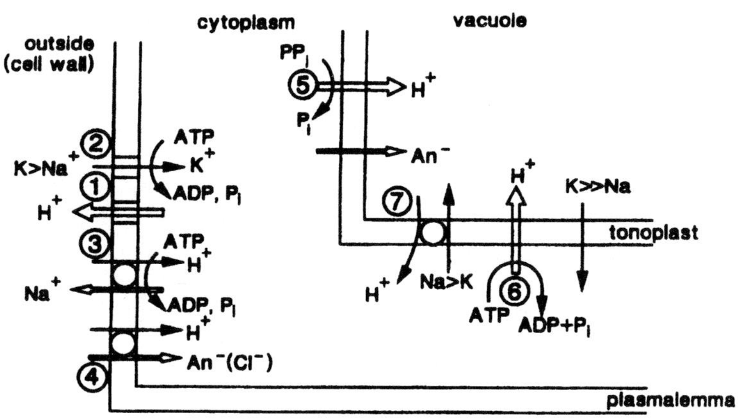

Ion transport mechanisms for Na^+ and K^+

1 = Proton pump	5 = H^+/PPase
2 = K^+ uniport	6 = H^+/ATPase
3 = H^+/Na^+antiport	7 = H^+/Na^+antiport
4 = Anion symport	

Figure 5.9. The major transport mechanisms associated with Na^+ balance among the external solution, cytoplasm, and vacuole are represented. (The upper portion of this figure is reproduced with permission from Glenn et al. 1997).

the cell is maintained by K^+ carriers and ports. When apoplastic Na^+ concentration is high, internal Na^+ homeostasis is regulated by H^+/Na^+ antiporters that couple the movement of Na^+ out of and H^+ into the cytoplasm (Dupont 1992). The antiporters work in conjunction with the H^+-ATPase pumps that are regulating the electrochemical balance of the cytoplasm. Therefore, under saline conditions, the level and activity of H^+/Na^+ antiporters and H^+-ATPase pumps should be critical to maintaining cytoplasmic ion homeostasis. There is significant variation in the response of H^+/Na^+ antiporters to salinity among species. Some salt-tolerant species such as barley increase H^+/Na^+ antiporter activity when in saline solution, while other species such as rice do not (Fukuda et al. 1998). This difference may be an important adaptation to salinity.

The discrimination between Na^+ and K^+ by cells may be directly related to salinity tolerance in glycophytes. Most species discriminate against Na^+ relative to K^+, and this discrimination results in a relatively low Na^+/K^+ ratio (0.1–0.5) in the cytoplasm. The genetic locus that determines Na^+/K^+ discrimination in wheat (*Triticum*) has been identified (Dvorak et al. 1992). When this locus is inserted into the genome of a salt-sensitive cultivar of *Triticum,* salt tolerance is enhanced (Dvorak et al. 1994). *Phragmites australis* is able to tolerate 180 m*M* salinity and maintain a low Na^+/K^+ ratio (0.13) in its tissues due to effective Na^+ exclusion (Glenn 1997, Matsushita and Matoh 1991, 1992). However, Na^+/K^+ discrimination is not required for salt tolerance in glycophytes, as evidenced by wild barley varieties that are salt tolerant and lack the Na^+/K^+ discrimination locus (Gorham 1993).

In contrast, halophytes radically alter the Na^+/K^+ ratio in tissues at high salinity (Glenn and O'Leary 1984). All halophytic monocots that survived at external salinity of 180 m*M* had a relatively high ratio of Na^+ to K^+. All 14 monocot species that survived full-strength seawater (540 m*M*) had very high Na^+/K^+ ratios except for *Phragmites,* which excludes Na^+. The Na^+/K^+ ratio in 20 halophytic species changed from less than 1 to between 2.5 and 5.8 when exposed to only 180 m*M* NaCL (Glenn and O'Leary 1984). In extreme halophytes, the Na^+/K^+ ratio increases and the water content decreases, while in salt-tolerant glycophytes, Na^+/K^+ increases and water content increases. This suggests that a certain amount of succulence is acquired in glycophytes for salt tolerance in order to maintain an appropriate Na^+/K^+ ratio, but such succulence is not required for extreme halophytes (see later section).

There is no question that ion homeostasis is required for salinity tolerance. It is also unequivocal that a constant increase in salt exclusion to maintain the same ion homeostasis (electrochemical gradient) as salinity increases is not energetically feasible (Schnapp et al. 1991). Therefore, salt tolerance depends upon being able to establish and maintain a new cell membrane electrical potential at a tolerable cytoplasmic ion homeostasis (Niu et al. 1995).

B. Salt Reabsorption and Retranslocation

If Na^+ and Cl^- ions are not strongly discriminated against at the root membrane, salinity of the xylem stream will increase under saline conditions. Some of the ions in the transpiration stream could be selectively accumulated by parenchyma cells in the xylem and retranslocated back to the roots via the phloem. Particularly in legumes, some parenchyma cells differentiate into transfer cells and reabsorb Na^+ from the transpiration stream (Jeschke and Wolf 1993). Perhaps the best case of reabsorption of Na^+ from the xylem stream is that of *Phragmites communis,* in which the phloem cells in the shoot base actively reabsorb Na^+ from the xylem stream (Matsushita and Matoh 1992).

Reabsorption of Na^+ and translocation back to the roots is found only in species with relatively low salinity tolerance. A comparison of three syndromes of retranslocation versus salinity tolerance is presented by Jeschke and Wolf (1993). *Lupinus albus* (salt sensitive) transports half of its total accumulated Na^+ to the leaves, of which 75% is retranslocated back to the roots, leaving 12% in the leaves. The castor bean (moderately salt tolerant) excludes a greater proportion of salt from the root, stores 80% of accumulated Na^+ in the root, retranslocates only 10% of that transported to the leaves back to the roots, and retaines only 2% of accumulated Na^+ in the leaves. In contrast, the salt-tolerant crop barley, has the highest total salt accumulation of the three species; 50% of the accumulated Na^+ is translocated to leaves, and only 10% of that transported to leaves is translocated back to roots, leaving 37% of accumulated Na^+ in the leaves. Results such as these suggest that halotolerant plants do not utilize efficient retranslocation of Na^+ back to the roots as a defense against Na^+ toxicity in leaves. In fact, the continued retranslocation of Na^+ back to the roots would not be effective in response to long-term salinity because Na^+ toxicity in roots would rapidly ensue.

C. Salt Extrusion: Glands and Bladders

Halophytes frequently have anatomical structures designed for extruding salt from tissues. These structures were originally classified as active hydathodes specialized for the secretion of ions and not directly connected to the vascular tissue (Hamberlandt 1914). These specialized excretion structures are particularly important since halophytes tend to accumulate rather than exclude salts. Three main classes of salt-extruding structures can be identified in plants. Salt glands and salt bladders (Figure 5.10) are found in at least 10 dicotyledonous plant families and 1 monocotyledonous family (Table 5.4). The grass family is particularly rich in species containing salt glands (33 genera), and many members of the Chenopodiaceae have salt bladders (200 species of *Atriplex* possess salt bladders). Salt glands and salt bladders both derive from epidermal tissue and have similar physiological function, but are anatomically distinct. The third type of salt gland is associated with insectivorous plants such as the venus flytrap, and serves both to regulate ion homeostasis and to excrete digestive enzymes.

Salt glands are embedded in the surface of leaves, are about the size of stomata, and can be abundant, reaching as many as 1000 cm^{-2} leaf surface (Hill and Hill 1976). Two types of salt glands can be identified. In many dicots such as *Tamarix, Limonium,* and *Frankenia,* they are comprised of 4–10 cells usually almost entirely covered in cuticle that functionally seals the glands from the rest of the plant except for small gaps on the leaf mesophyll side. These gap areas (transition gaps) of the secretory cells have a high density of plasmodesmata and are connected to the leaf mesophyll. This is the region of ion transfer into gland cells from the symplast of the mesophyll. The gland cells have relatively high concentrations of mitochondria and vacuoles, a characteristic trait of transfer cells (Thomson 1973). Salt and water are secreted out through small holes in the cuticle covering the gland.

In *Spartina* and some other monocots, there are only two cells per salt gland (basal and cap). Cell wall protrusions extend down into the basal cell from the cap cell. In association with these protrusions, the plasma membrane forms extensive involutions (called partition membranes) that occupy a large fraction of the cytoplasm of the basal cell. The internal sections of the partitions become distended (Levering and Thomson 1971) when placed in saline water (Levering and Thomson 1972). Moreover, the internal sections are tightly closed when the mesophyll is placed in distilled water.

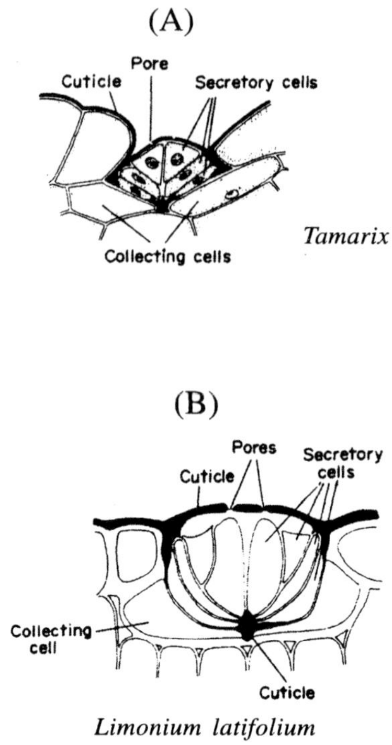

Figure 5.10. Salt glands are a mechanism of eliminating salt from tissues in some halophytes. (A) The salt gland found on the leaves of *Tamarix* trees. (B) The salt gland found on the leaves of *Limonium latifolium*. (Line diagrams are reprinted with permission from Fahn 1990.)

TABLE 5.4 Salt-Tolerant Plants Utilizing Salt Glands and Salt Bladders for Salt Balance

Family	Example Genera	Number of Species[a]
	Salt Glands	
Poaceae	*Cynadon*	25
Plumbaginaceae	*Limonium*	6
Tamaricaceae	*Reaumuria*	2
Avicenniaceae	*Avicennia*	1
Rhizophoraceae	*Ceriopps*	4
Primulaceae	*Glaux*	1
Frankeniaceae	*Frankenia*	2
Combretaceae	*Laguncularia*	1
Acanthaceae	*Acanthus*	2
Convolvulaceae	*Impomea*	2
Mysinaceae	*Aegiceras*	1
	Salt Bladders	
Chenopodiaceae	*Atriplex*	10
Aizoaceae	*Mesembreanthemum*	1

[a] Conservative estimates.

Source: From Gorham 1992.

These and other results indicate that the extracellular channels formed by the partition membranes are sites of ion and water accumulation. Such long protrusions in which ions accumulate suggest the *standing gradient model* for water and ion excretion proposed for animal secretory systems (Diamond and Tormey 1966). Ions are accumulated at the closed tips of the channel resulting in a standing gradient of ion and water diffusion from the tips to the open ends where secretion occurs.

The glands concentrate NaCl against a diffusion gradient resulting in up to 50-fold increases in ion concentration in relation to the mesophyll. The excretions can contain a number of different ions and organic solutes, but over 90% of the exudates are Na^+, K^+, Ca^{2+}, and Cl^-. When the mangrove *Aeluropus litoralis* was exposed to 300 m*M* NaCl for 3 days, 85% of the net Na^+ accumulated by the plant was excreted by salt glands (Pollak and Waisel 1970) onto the leaf surface. The exudates also have a high Na^+/K^+ ratio compared with the leaf cell sap.

Probably the best known ion pump in salt glands is the Cl^- pump in *Limonium vulgare*. The Cl^--ATPase transport pump is induced in a sigmoidal manner, with a lag of 1 hour, when the gland is exposed to saline conditions (Shachar-Hill and Hill 1970). Both transcription and translation are required to synthesize the Cl^- transporters during the lag phase.

In contrast to salt glands, salt bladders (sometimes called salt hairs) such as those of *Mesembryanthemum crystallinum* (Adams et al. 1998) are specialized trichomes (Fahn 1990). Salt hairs of the Chenopodiaceae (particularly *Atriplex*) have been extensively studied (Figure 5.11). These structures are comprised of two cells (stalk cell and bladder cell). The stalk cell has an ultrastructure that resembles the transfer cells of salt glands, while the bladder cell has an enormous central vacuole reaching up to 200 μm in diameter. In some cases, the bladder cells contain a high concentration of salts. For example, *Atriplex halimus* bladders contain 60× the salt content as the leaf (Mozafar and Goodin 1970). In other cases, the bladders serve only as a salt holding tank. For example, the salt concentration in salt bladders of *Mesembryanthemum crystallinum* is equal to that of the leaf (Adams et al. 1998). Eventually, the bladder cells die and rupture, depositing the salt on the leaf surface. In some cases, leaves can produce several populations of bladders in their lifetime (Hill and Hill 1976).

Certainly, salt bladders and glands are specialized for ion regulation in leaf tissues and are important mechanisms by which halophytes cope with elevated salinity. However, it is equally evident that similar salinity tolerance can occur in halophytes with or without salt glands. For example, one mangrove species, *Avicenna marina* has high salt tolerance and has salt glands, while another mangrove, *Rhizophora mangle*, has equal or greater salt tolerance than *Avicenna marina* but does not have salt glands or hairs.

D. Salt Dilution: Succulence

If Na^+ and Cl^- are not excluded, retranslocated, or extruded can they be diluted enough in leaf mesophyll to keep cytoplasmic salinity below a toxic level? Succulence, a high proportion of water to dry weight (or a high ratio of vacuole/cytoplasm), is thought to be an important aspect of halotolerant plants because of the number of halophytes that express some form of succulence (Gorham 1992). In some extreme halophytes (such as *Sueda maritima*) expansion and growth of succulent tissues requires relatively high soil salinity, and low salinity limits plant growth (Yeo and Flowers 1986). Cytoplasmic toxicity rapidly develops when the central vacuole becomes saturated with salt (Cheeseman 1988). Therefore, plants that have very large vacuoles

(A)

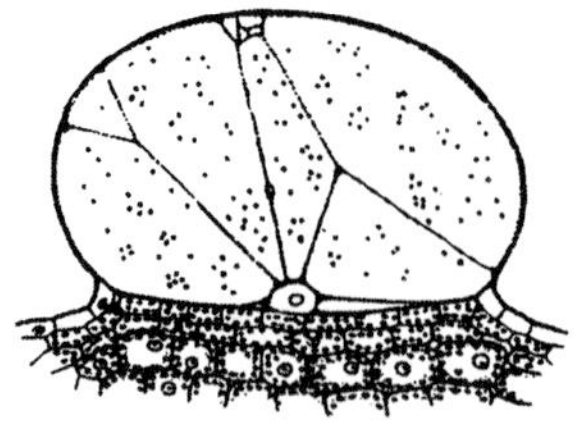

Mesembryanthemum crystallinum

(B)

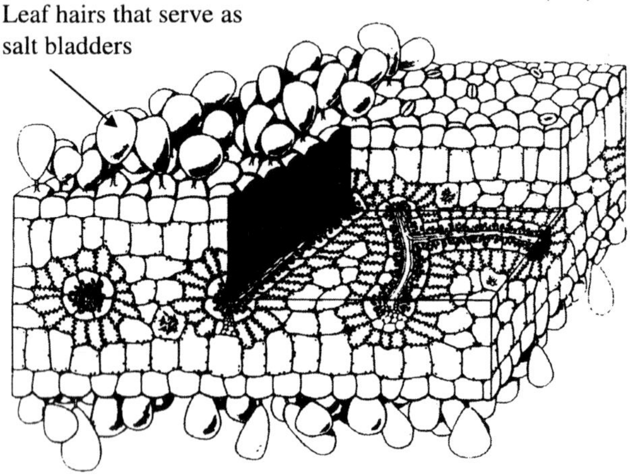

Atriplex spongiosa

Figure 5.11. (A) The salt bladder of *Mesembryanthemum crystallinum*. (Reprinted from Adams et al. 1998.) (B) This cross section of the *Atriplex spongiosa* leaf clearly shows the abundance and morphology of leaf hairs adapted to serve the purpose of salt bladders. (Reprinted with permission from Lüttge and Pitman 1976.)

(high succulence) can store more salt before the vacuole is saturated and toxicity ensues.

Only a few mangrove species have salt glands, 90% of external salt is excluded from entry into the transpiration stream, and transpiration decreases with an increase in soil salinity. However, intracellular salinity of mangroves can be highest in newly developing leaves. As leaves expand, the concentration of ions on a dry-weight basis increases (Popp et al. 1993), but that of a bulk leaf water is maintained relatively constant by an increase in leaf succulence (Popp et al. 1993). Therefore, increased succulence may be important for the regulation of ion concentration in cytoplasm during leaf expansion in mangroves and other halophytes (Adams et al. 1998). Of course, succulence is not utilized by all halophytes and is not normally induced in mature plants or tissues of glycophytes by increased salinity (Jeffries et al. 1979). Also, many other species in desert systems have succulence in stems, leaves, and roots, but are not in saline habitats (Nilsen et al. 1990). Consequently, succulence can be a contributing factor to bulk leaf water ionic balance during leaf development in some extreme halophytes, but succulence cannot be the only mechanism of salt tolerance.

E. Salt Compartmentation

Throughout this chapter, we have noted that many halophytes accumulate rather than totally exclude salts. In fact, halophytes utilize Na^+ as a mechanism of osmotic adjustment in order to maintain turgor at low soil water potential. An important characteristic of halophytes that do not have salt glands is the large quantity of Na^+ and Cl^- in leaf tissues and parenchyma cells. This suggests that many halophytes can tolerate a high concentration of salts in their cytoplasm, but this is not the case. For example, *Sueda maritima* maintains high tissue salinity, but enzymes and chloroplast function are inhibited by salt in a manner similar to that in glycophytes. Therefore, this species must be able to compartmentalize ions away from the cytoplasm (Hajibagheri et al. 1984). There are many other similar cases in which high bulk tissue salinity can be attained in halophytes but their cytoplasmic function is intolerant of the same salinity (Ungar 1991). Ions could be retained in the cell wall spaces by discrimination at the plasma membrane, accumulated in the vacuole after diffusion across the cytoplasm of parenchyma cells, or come to equilibrium in the apoplast and symplast. Traditionally, it has been thought that halophytes are able to compartmentalize Na^+ in the apoplast, particularly the central vacuole.

There is strong evidence that Na^+ and K^+ are accumulated in the central vacuole of cells (Pitman et al. 1981), and that the ratio of Na^+/H^+ to K^+/H^+ antiport systems is higher in halophytes (*Atriplex*) than in glycophytes such as cotton (Hassidim et al. 1990). *Salicornia europa* accumulates salts in the central vacuole of the mesophyll up to a concentration of 1000 m*M* (Gorham 1992). Furthermore, an increase in NaCl induces the transcription of H^+-ATPase genes in elongating roots and expanded leaves (Niu et al. 1993), but this induction is transient. The induction of transporters by NaCl and the relative abundance of Na^+/H^+ versus K^+/H^+ transporters in halophytes are strong indicators that ion compartmentation in the central vacuole is important for salinity tolerance.

However, there are several lines of evidence that limit the long-term significance of storing salt in the vacuole as a salt tolerance mechanism. First, the ability to concentrate salt in the central vacuole is limited to the point when the vacuolar salt concentration reaches saturation for the intrinsic membrane ion transporters. Following this point, all ions will be retained in the cytoplasm and the cytoplasmic ion concentration will increase very rapidly. Second, several studies show equimolar concentrations of Na^+ in cytoplasm and the central vacuole in halophytes growing in saline conditions. For example, Seeman and Critchley (1985) found that an intracellular concentration of 150 m*M* salt induced a 70% reduction in growth, but there was no difference between the salt concentration in the vacuole and that in the cytoplasm. Third, the vacuolar concentration of Na^+ and Cl^- only exceeded cytoplasmic concentration in maize root cells when external salinity increased to values that maize plants cannot tolerate (Hajibagheri et al. 1987). Therefore, strong compartmentation into the apoplast may be important to salt tolerance only when tissue salt levels increase above those tolerated by the cytoplasm, and such compartmentation has limited effectiveness for extended salinity conditions.

In most halophytes, salt increases in concentration more rapidly in older mature tissue than in developing tissue; *Mesembryanthemum crystallinum* is a notable exception (Adams et al. 1998). In addition, glycophytes, such as rice and many grasses, compartmentalize salts in older tissues (Cheeseman 1988, Gorham 1992). When these older tissues reach apoplastic saturation of ions, they will suffer cytoplasmic toxicity and die. Therefore, halophytes with a relatively fast growth rate will have a relatively low salinity in younger tissues and a relatively high salinity tolerance. This discussion points toward an interesting trade-off between growth rate and the amount of salt comparmentalized in the central vacuole. The larger number of ions (particularly Na^+ and Cl^-) that the central vacuole can hold (effective discrimination and volume), the greater the salt tolerance and the lower the growth rate necessary to tolerate extended salinity. However, if the central vacuole is able to develop a high ionic concentration, the cytoplasm must be able to balance the resulting low vacuolar water potential in the symplast, or the cytoplasm will suffer water limitation. It is this water relations concept that stimulated a large interest in studying the factors regulating cytoplasmically compatible solutes in halophytes and glycophytes.

F. Tissue Salt Tolerance

1. Compatible Solute Accumulation. Solutes that are compatible with the cytoplasm are accumulated by all groups of organisms. Plants accumulate these solutes in response to low water availability, low temperature, and salinity (Bohnert et al. 1995). Both glycophytes and halophytes have been found to accumulate a number of different solutes in the cytoplasm when they experience saline conditions. Accumulation of cytoplasmically compatible solutes cannot be the only method of osmotic adjustment for plants in long-term hypersaline conditions because much of the total plant photosynthate would be required for osmotic adjustment (Wyn-Jones et al. 1984). Therefore, both inorganic ions (Na^+, K^+, Ca^{2+}) and organic solutes are involved in osmotic adjustment of halophytes. Accumulation of organic solutes in hypersaline conditions occurs in response to the osmotic effects of the soil solution and is not a response to specific ions (Jeffries et al. 1979). Also, in any one species, several different organic solutes may accumulate in response to salinity. For example, reducing sugars, amino acids, proline, and quaternary ammonium compounds all accumulate when *Halimione portulacoides* is exposed to hypersaline conditions (Jeffries et al. 1979). In general, under saline conditions, low soil osmotic potential is associated with osmotic adjustment in tissues by an increase in organic solute concentration in glycophytes, and an increase in both inorganic ions and organic solutes in halophytes.

Proline. Proline (Figure 5.12) is accumulated by a number of different plants (e.g., *Arabidopsis,* alfalfa, bean, halophytes, tobacco, spinach, potato, soybean, barley, rice) in response to both water stress and salinity (McCue and Hanson 1990, Delauney and Verma 1993). Such accumulation is not unique to plants and has been observed in eubacteria, marine invertebrates, protozoa, and algae (Yoshiba et al. 1997). Therefore, this trait most likely appeared early in the evolution of plants (Bohnert et al. 1995).

Is the production of proline an adaptive response to osmotic stress, or is it a consequence of protein degradation during stress? In order to answer this question, it must be shown that increasing proline concentration by itself will increase osmotic protection. The amount of accumulation by proline and glycine-betaine is enough to account for all the osmotic adjustment of *Spartina alterniflora* during high salinity (Cavalieri 1983). A mutation in *E. coli* (proB74) results in the overproduction of proline and endows increased resistance to osmotic stress (Csonka et al. 1988). Also, the overproduction of proline in tobacco leads to an increase in tolerance of osmotic stress (Kishor et al. 1995). All of these studies suggest that the accumulation of proline in tissues confers some salinity tolerance.

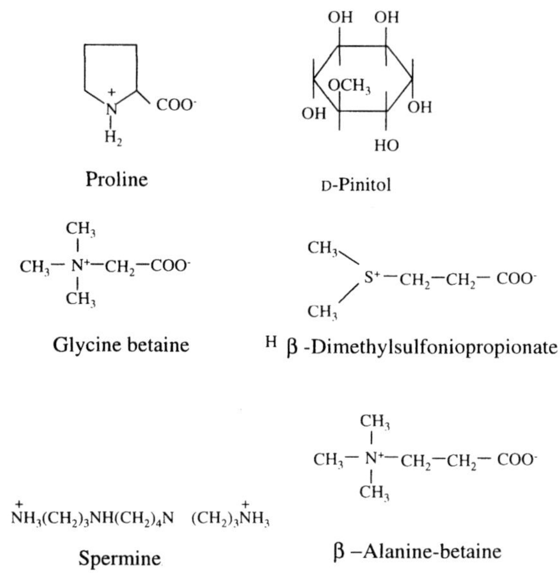

Figure 5.12. A selection of compounds commonly involved with osmotic adjustment in the cytoplasm of plants experiencing salinity or water stress.

In many cases, the concentration of proline alone cannot account for the required osmotic adjustment needed to maintain turgor under hypersaline conditions. Proline may have significance to metabolism other than just osmotic regulation. It has been suggested (Hare and Cress 1997) that a small increase in proline concentration can alter the redox balance of NADP/NADPH (increases at high proline) which may have a large effect (stimulation) on the oxidative pentose phosphate pathway. There is strong evidence to suggest that proline metabolism has a significant effect on cellular redox potential, which may be important for signaling processes associated with salinity tolerance as well as general stress tolerance traits (Hare and Cress 1997).

Polyols. Several polyols (Figure 5.12) have been found to accumulate in tissues experiencing either water or salinity stress. Such accumulation has been found in animals, plants, algae, yeast, and bacteria, making the accumulation of polyols as ubiquitous as that of proline. Straight-chain polyols such as mannitol and sorbitol, cyclic polyols such as myoinositol, and their methylated derivatives (e.g., pinitol) all have been found to increase in tissues when plants experience water or salinity stress. The polyols have two functions in osmotic stress tolerance that are difficult of separate (Bohnert et al. 1995). An increase in polyol concentration increases osmotic potential in the cytoplasm (osmoregulation) and allows Na^+ sequestration in the vacuole. In addition, polyols can act as an osmoprotectant of membranes and proteins, most likely by scavenging free radicals of oxygen (Sminoff and Cumbes 1989, Bohnert and Jensen 1996). Polyols, particularly nonreducing sugars, may also serve as a repository of carbon during periods of stress when photosynthate is not being rapidly used (Vernon et al. 1993).

The synthesis of polyols is closely related to the carbon reduction cycle of photosynthesis because most polyols are derived from glucose-6-P. For example, myoinositol is synthesized by a two-step process involving inositol-1-P synthase and inositol-1-P phosphatase. Similarly, mannitol and sorbitol are produced by a one-step process from fructose-6-P. Therefore, polyol biosynthesis can be intimately related to carbon reduction by photosynthesis and can easily be upregulated because few enzymes (outside of photosynthesis) are required for their synthesis.

It is important to recognize the significance of the inositol synthesis pathway. The synthesis of inositol is the first step in biochemical pathways leading to the synthesis of compounds required in the cell wall, mucilage, gums, and glycoproteins, many of which are important for plant stress responses (Bohnert et al. 1995). This intimate relationship between inositol and cell wall function places this compound in an important position relative to cell wall synthesis during salinity stress. Furthermore, an inhibition of cell wall synthesis induced by salinity stress may result in a buildup of inositol, and the inositol can act as a reservoir for cell wall synthesis when the salt stress ends.

Quaternary Amines. Quaternary amines (QAMs), in particular glycine-betaine (Figure 5.12), are accumulated during salinity stress in some slightly salt-tolerant glycophytes and halophytes. These compounds also serve multiple purposes related to both osmotic adjustment and osmoprotection. Glycine-betaine can protect enzymes from heat damage (Nash 1982), which indicates its capability to protect enzymes from stress damage (Gorham 1992). Increasing the concentration of glycine-betaine in cyanobacteria stabilizes ribulose-1,5-bisphosphate carboxylase/oxygenase to inactivation by NaCl (Nomura et al. 1998). In the Plumbaginaceae, various species accumulate glycine-betaine and other QAMs under salinity stress and the QAMs serve as an osmoprotectant (Hanson et al. 1994). Because glycine-betaine accumulates in many members of the Poaceae but does not accumulate in rice or tobacco, there has been a lot of effort to engineer the production of glycine-betaine in plants that do not normally produce it.

The synthesis of glycine-betaine is a simple pathway originating with choline and involving two main enzymes (choline monoexygenase and betaine aldehyde dehydrogenase). The simplicity of the biosynthesis pathway is reflected in straightforward genetics. In crosses between a glycine-betaine-deficient mutant and the wild type of sorghum, the ratio of genotypes in the F_2 population was 1:2:1, which is consistent with a single responsible gene (Grote et al. 1994). The fact that few enzymes involved (see later section) and the absence of glycine-betaine synthesis in some important crop species have led authors to speculate that genetic engineering of this trait may be a successful avenue for increasing salt stress tolerance in crops (McCue and Hanson 1992). However, others have noted that successful transformation and expression of glycine-betaine synthesis has not resulted in any benefit for plants in the field (Bohnert and Jensen 1996).

Other Methylated Organic Osmolytes. Several other groups of methylated compounds increase during osmotic stress, which also may have osmoprotective properties. For example polyamines such as spermine (Figure 5.12) and putracine, and tertiary sulfonium compounds such as dimethylsulfoniopropionate, can increase during osmotic stress (McCue and Hanson 1990, Bohnert and Jensen 1996). It is interesting to note that many of the compounds that increase during osmotic or water stress are methylated (e.g., methylated inositols, QAMs, tertiary sulfonium species, and polyamines). Central to the methyltransferases involved in the production of these compounds is the cosubstrate S-adenosylhomocysteine (SAM). During osmotic stress, the mRNA transcripts for SAM synthase increase in the tomato (Espartero et al. 1994), suggesting an importance of methylation for salinity stress tolerance. It is possible that the source of the methyl groups required for the SAM-mediated pathway is photorespiration (Bohnert and Jensen 1996). Photorespiration increases during salinity stress because stomata close and photosynthesis must rely more heavily on oxygen than on carbon dioxide. In photorespiration one of the resulting products, glyoxylate, is catabolized to CO_2 and methylene tetrahydrofolate which can be reduced to N^5-methyltetrahydrofolate (N^5MeTHF). Two enzymes, homocysteine S-methyltransferase and SAM synthase, are required to synthesize SAM from N^5MeTHF. Thus, the up-regulation of SAM synthase and increased photorespiration will both increase the rate of SAM-mediated methylation, which can result in the increase of several different osmoprotective compounds.

2. Membrane Stability. In order to maintain metabolic activity during high salinity, there must be mechanisms for ion regulation, osmoprotection of enzymes, and osmostability of membranes. Ion permeability of membranes in part depends on the amount and composition of sterols. Plants with relatively high salinity tolerance have higher quantities of sterols. The sterol/phospholipid ratio is high in halophytes (1.0–2.0), moderate in weakly tolerant glycophytes (about 0.5–1.0), and very low in intolerant glycophytes (0.2–0.5). The significantly higher sterol/phospholipid ratio in halophytes emphasizes the importance of sterols to membrane permeability and ion homeostasis in saline environments (Wu et al. 1998).

Various authors have suggested that salt tolerance is related to an increase in the ratio of cholesterol + camesterol to stigmesterol + sitosterol (also called the *more planar to less planar ratio*), which may relate to the ability to retain membrane fluidity at high levels of external NaCl. The larger the planar to less planar ratio of sterols, the more salt tolerant were the varieties of citrus root stock (Douglas and Walker 1984). Salinity induces an increase in more planar sterols in wheat (Mansour et al. 1994). In contrast, soybean plants exposed to 25 m*M* NaCl (Surjus and Durand 1996) have no changes in sterol composition of the plasma membrane fraction, and *Spartina pratens* plasma membranes increase in less planar sterols when they experience salinity (Wu et al. 1998). *Spartina pratens* has a high quantity of less planar sterol (sitosterol), which increase in abundance when plants are exposed to saline conditions. Less planar sterols, particularly sitosterol, are effective in reducing water permeability of membranes (Schuler et al. 1991); thus, the increase in less planar sterols in this halophyte may be important for retaining water in root cells when exposed to a high external osmotic concentration. Whatever the structural cause, there is direct evidence for reduced membrane fluidity when exposed to salinity (Wu et al. 1998), which may be an important factor for maintaining discrimination against Na^+.

Membrane permeability is also in part determined by glycolipid abundance. One function of glycolipids is the stabilization of the interaction between proteins and lipids (Williams 1988). This facilitates a more effective sealing of the protein in the membrane and reduces ion leakage across the membrane. In some species the concentration of glycolipids increases during salinity (Wu et al. 1998, Mansour et al. 1994), while in other species there is no change (Brown and Dupont 1989). There is not enough information on changes in glycolipid content to suggest a definite role of this molecule in salinity tolerance, but an increase in glycolipid content during salinity stress could help maintain ion and water homeostasis.

The phospholipid content of a membrane and its saturation state also influence the differential permeability and fluidity of membranes. Furthermore, *phase transitions* (changes in differential permeability and fluidity) are influenced by external salt concentration (Hauser 1991). In several glycophytes, there is either no change or a decrease in phospholipid content of the plasma membrane when treated with high external NaCl (Surjus and Durand 1996, Mansour et al. 1994). However, in halophytes there is either no change or an increase in phospholipid composition (Blits and Gallagher 1990, Wu et al. 1998). It is possible that halophytes are able to maintain and increase the phospholipid composition of the plasma membrane as a mechanism of salinity tolerance. More research is needed before this can be stated unequivocally.

Halophytes such as *Spartina pratens* have a relatively high proportion of short-chain fatty acids (C_{14}) compared with most glycophytes that have a preponderance of longer-chain fatty acids (C_{18}, C_{16}). A high preponderance of short-chain fatty acids and sterols results in a thinner membrane. Furthermore, short-chain fatty acids "fit" sterols better, allowing for tighter packing of sterols. Since sterols have the most significant influence on membrane rigidity among membrane components, it may be advantageous for halophytes to have a higher proportion of short-chain fatty acids. The saturation level of fatty acids in the plasma membrane increased for soybeans when exposed to 25 m*M* NaCl (Surjus and Durand 1996) and for *Spartina pratens* (Wu et al 1998). The 16:0 and 18:0 fatty acid ratios in the plasma membrane increased by 56% under salt treatment (Surjus and Durand 1996). The influence of fatty acid saturation increases membrane packing and decreases membrane permeability to water and ions. In addition, the degree of saturation and length of the acyl chain of fatty acids may be important membrane properties for the regulation of membrane-bound processes such as H^+-ATPase activity (Kasamo 1990).

The activity of H^+-ATPase in the plasma membrane may be one of the most important adaptations to salinity in both glycophytes and halophytes. The best chloride excluders among citrus varieties are those that have relatively high H^+-ATPase activity in the root plasma membrane (Douglas and Walker 1984). The activation energy of the H^+-ATPase activity of the plasma membrane is also inversely related to the ratio of phospholipids to sterols. *Salicornia europa* (an extreme halophyte without salt glands that does not accumulate organic osmolytes) grows optimally at 200 m*M* NaCl, and growth is much reduced at low external salinity. Both plasma membrane and tonoplast H^+-ATPase activity are increased dramatically in 200 m*M* NaCl compared with 50 m*M* NaCl (Ayala et al. 1996). Therefore, increases in H^+-ATPase activity in glycophytes is an effective way of excluding Cl^-, and in halophytes it is an effective way of increasing Na^+ accumulation. Higher Na^+/H^+ antiporter activity in the plasma membrane is also associated with salt tolerance (Glenn 1997) in halophytes. However, the upregulation of H^+-ATPase pumps or Na^+/H^+ antiporters is not exclusively required for salinity tolerance. Many factors regulate ion homeostasis and salt tolerance including membrane composition, Ca^{2+} and K^+ concentration, the composition and abundance of membrane surface proteins, root cell developmental stage, plant growth rate, and so on (Cheeseman 1988).

3. Growth Hormone Relationships with Salinity. In general, increasing salinity is associated with a decrease in auxin, cytokinin, and gibberellin levels in plant tissues, and an increase in ABA. Such changes in hormone levels are thought to be a primary process regulating the reduction in growth associated with salinity. In fact, the NaCl-induced reduction in growth can be partly counteracted by the application of gibberillic acid (Prakash and Prathapasenan 1990). Also, pretreatment of salt-sensitive seed with cytokinins or gibberellins will increase salt tolerance of wheat seedlings (Datta et al. 1998). Just before the induction of crassulacean acid metabolism (CAM) photosynthesis by NaCl treatment in *Mesembryanthemum crystallinum,* proline and phosphoenol pyruvate carboxylase (PEPcase) accumulate to high levels (Adams et al. 1998). The accumulation of these compounds (which are critical to salt tolerance and CAM photosynthesis) can also be induced by cytokinin treatment at low salinity

(Thomas et al. 1992), as can osmotin and pinitol accumulation (Thomas and Bohnert 1993). Cytokinin treatment could also improve the growth rate of *Sorghum* at high salinity (Amzallag et al. 1992). It was proposed that cytokinins improve the balance of nutrients in these plants causing an increase in growth at high salinity.

There is considerable evidence that ABA levels increase during salinity stress in both halophytes and glycophytes (Nilsen and Orcutt 1996), and that increase is positively correlated to the salinity of the growth medium. The activity of enzymes associated with pathways leading to osmolyte accumulation (such as BADH, an enzyme associated with glycine-betaine biosynthesis) are increased by salinity, and the increase correlates with a rise in ABA concentration. Also, exogenous application of ABA at low salinity treatments induces an increase in BADH activity (Ishitani et al. 1995). The reverse is not true; exogenous application of glycine-betaine influences stomatal conductance irrespective of ABA concentration (Makela et al. 1998). In response to increased salinity, ABA levels increase rapidly and then decrease after a few days to values similar to those before treatment. The short-term induction of ABA may constitute a signal for tissues to increase osmolality and induce the synthesis of ion transporters, but a long-term increase in ABA is not necessary to keep the new metabolism operating and may inhibit appropriate salinity response. In fact, salt-tolerant species such as *Brassica napus* have lower concentrations of ABA than salt-sensitive species such as *Brassica carinata* (He and Cramer 1996). The ABA signaling mechanism is most likely related to calcium biochemistry, and this concept will be covered in a later section. Although ABA has been implicated in the signal transduction pathway between salinity (also water stress) and gene transcription processes, not all salinity responses are induced by ABA. For example, ABA is not the singular factor inducing accumulation of proline and PEPcase in *Mesembryanthemum crystallinum* (Thomas et al. 1992).

Ethylene (C=C) also increases in the presence of salinity. Several studies with tomato have shown an increase in ethylene by leaves, particularly older leaves (Feng and Barker 1992). Ethylene is known to induce senescence, and its increase during saline conditions is likely a general stress response. Studies of citrus genotypes showed a biphasic response of ABA, and aminocyclopropane-1-carboxylic acid (ACC) which is the last substrate in ethylene biosynthesis. The initial transitory ACC increase was due to water stress effects, but the subsequent accumulation was due to ionic effects (Gomez-Cadenas et al. 1998). Severe leaf abscission (92%) was correlated with the increase in intercellular Cl^- and ACC accumulation. Therefore, specific ion effects of salinity may be the ultimate cause of ethylene increase and senescence induction. Inhibition of ethylene production under saline conditions by salt-tolerant plants could reduce the amount of tissue senescence and allow the accumulation of salt in the tissues before they die. Proline was able to overcome the effect of NaCl on ethylene concentration in *Allenrolfia occidentalis* (Crominski et al. 1988), suggesting that organic osmolytes can inhibit ethylene production during water stress and slow senescence in salt-tolerant plants.

G. CAM Photosynthesis Induction

A few extreme halophytes switch from the C_3 photosynthetic pathway to the CAM pathway when confronted with high salinity or water stress (Ting et al. 1996). The best-demonstrated example of this phenomenon is *Mesembryanthemum crystallinum,* first identified in the early 1970s (Winter and von Willert 1972). How is it that stomata that open during the day in low salinity are changed to opening at night in the same leaf surface at high salinity? The induction of CAM (and nocturnal stomatal opening) may be stimulated by the effect of water or salinity stress on stomatal conductance. It is generally held that an increase in salinity (or drought) causes a midday depression in photosynthesis. This depression can extend to include most of the day as the stress continues. This time period (of low water potential and low midday conductance) corresponds to a transient period of lower turgor potential. When this period extends for long enough, a genetic response is induced to initiate CAM (about 10 days under stress), stomata stay open only at night, and turgor returns to a value greater than the prestress level (Winter and Gadermann 1991). The induction of CAM in field situations has been elegantly demonstrated using carbon isotope and malate accumulation information (Bloom and Troughton 1979), indicating that CAM induction is an important adaptation to salinity and water stress in the field.

The induction to CAM is developmentally controlled and will not occur in juveniles (Cushman et al. 1990). The capacity to induce CAM coincides with the development of mature leaves and the ability to produce flowers and seed. In fact, seed production in *Mesembryanthemum crystallinum* depends on CAM (Winter and Ziegler 1992). CAM in these extreme halophytes will develop slowly upon maturity in unstressed situations; however, when plants are stressed by high salinity or drought, the rate of CAM induction increases (Winter and Gadermann 1991). The induction of CAM occurs after roots are influenced by salinity, and that induction is due to a stimulus conveyed to leaves from roots (Eastmond and Ross 1997). Therefore, CAM is a constitutive aspect of ontogenetic development in these halophytes, but high ambient salinity or drought stress accelerates the rate of CAM expression.

V. SALT TOLERANCE GENES AND GENETIC ENGINEERING

A. Signal Perception and Transduction

The overlap of physiological responses to salinity, drought, and chilling temperature cause difficulty in discovering and interpreting the signal transduction pathways for salt-induced responses. On one hand, researchers are looking for unifying aspects of signal transduction among these three stressors. On the other hand, researchers are trying to determine what unique signaling mechanisms are used to regulate metabolic responses to each specific stress. For example, water stress, salinity, and cold stress all stimulate ABA synthesis, and ABA causes an upregulation of many genes resulting in proteins that have significance to general aspects of all stress tolerances. The upregulation of proline synthesis enzymes will induce an accumulation of proline that can serve for osmotic adjustment (water stress), and osmoprotection (salinity), and that lowers the freezing point of cytoplasm (cold stress). However, the induction of one pathway leading to the accumulation of one solute only explains one mechanism of salinity tolerance, and cannot account for all aspects of salinity tolerance in any individual species. The integration of metabolism for salinity tolerance of any species must involve a regulatory signal cascade or transcriptional activation of many genes. In the case of salinity, a general stress response may occur rapidly after saline conditions are experienced (osmotic stress, phase I responses), while long-term mechanisms of salinity tolerance (ion homeostasis, phase II responses) may be the result of salinity-specific signal transduction pathways.

It has been well documented that the general stress-related signaling pathway, which includes salinity, is manifest through ABA. Many of the genes that are upregulated by high salinity are also upregulated by ABA (ABA-sensitive genes) and by water or cold stress. This generalized stress signal transduction pathway is associated with an upregulation of protein kinase genes such as PKABA1. In *Arabidopsis* a root-specific gene (ARSK1, *Arabidopsis* root-specific kinase) is highly expressed toward the meristem of roots, and its expression decreases upward along the root. The expression of this kinase is radically increased when roots are exposed to air or experience salt stress (Hwang and Goodman 1995). The ARSK1 protein product may be involved in signaling to shoots that desiccation (high salinity) is occurring, or it may be involved in root-specific responses to osmotic stress. The ABA signaling pathway (salt, cold, drought, and ABA-induced) is intimately related to a cytosolic Ca^{2+}-dependent salt stress signal cascade involving Ca^{2+}-dependent protein kinase (Sheen 1996). Two related Ca^{2+}-dependent protein kinases (CDPKs) induce a stress and an ABA-responsive HVA1 promoter that regulates a number of stress-induced genes. The link between ABA and the Ca^{2+}-dependent cascade is most likely cyclic ADP–ribose (Figure 5.13). The ABA signal is received by an ABA receptor protein in the plasma membrane, which when in contact with ABA stimulates the release of cyclic ADP–ribose that initiates the Ca^{2+}-dependent kinase cascade. The same signaling pathway is involved in the ABA-induced closure of stomata (Pennisi 1997). These protein kinase signal cascades are a major factor in the generalized stress response of plants (and animals) but are not specific to salinity (Sopory and Munshi 1998).

Salinity-specific signal transduction may be a response to long-term exposure (phase 2 of the biphasic growth model), and may relate to apoplastic ion concentration. Such signaling pathways are most likely related to regulating those proteins that de-

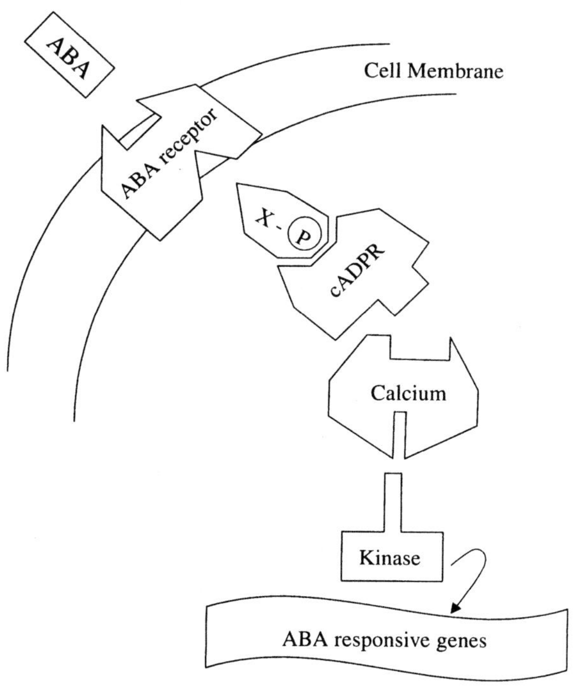

Figure 5.13. The signal transduction pathway for abscisic acid–induced genes, many of which are related to salinity tolerance. ABA = abscisic acid, X-P = xylose phosphate, cADPR = cyclic adenosine diphosphate ribose. (Patterned after a different diagram in Pennisi 1997.)

termine ion (and water) homeostasis. In yeast, a salinity-specific regulatory cascade depends on calcineurin (CaN), a protein phosphatase. CaN regulates the K^+/Na^+ differential selectivity to have high affinity for K^+ uptake and low affinity for Na^+ uptake. Furthermore, CaN in yeast induces the transcription of a P-type ATPase (located in the plasma membrane) of yeast that regulates Na^+ efflux. Therefore, the CaN-regulated responses determine ion homeostasis particularly related to the Na^+/K^+ ratio and Na^+ antiporters in yeast. There is biochemical and functional evidence that plants also have a CaN-mediated regulation system (Allan and Sanders 1995) that determines ion homeostasis of K^+ and Na^+. In addition, guard cell behavior is regulated by CaN by its influence on ion channels (Allan and Sanders 1995), and CaN has a critical importance for the salinity tolerance of roots (Pardo et al. 1998). These results suggest that CaN is an indispensable intermediate in one salt-specific signal transduction pathway in both root and shoot tissues.

B. Salinity-Inducible Genes and Proteins

A large number of genes are induced and accumulated in response to salinity treatment (osmotic or specific ion effects), many of which have no known function. Below we cover a selection of those salinity-inducible genes or proteins, and focus on those for which we have the most information on putative function.

1. Ion Homeostasis

H^+-ATPase Transporters. As reported earlier in this chapter, changes in membrane-bound (plasma membrane, organelle, and tonoplast) ion transport protein activity are an important part of salinity tolerance mechanisms in both glycophytes and halophytes. The H^+-ATPase complex that creates the membrane electrical potential and provides the energy base for Na^+/H^+ antiporters is the best known. Several different H^+-ATPase genes are known in plants (Niu et al. 1995). The expression of the various members of the ATPase gene family in plants is dependent upon tissue location and development (Harper et al. 1990), and it is possible that some of the gene family members are produced constitutively and others are induced by salinity (Niu et al. 1996). Salinity causes an increase in H^+-ATPase activity (and protein content) in the plasma membranes of the ice plant (Reuveni et al. 1990). In addition, the mRNA for the 70 kD subunit of the plasma membrane H^+-ATPase increases in intact tomato plants in response to salinity (Binzel 1995). H^+-ATPase genes are much more sensitive to induction by external salinity in halophytes than in glycophytes (Niu et al. 1993). During salinity shock, the mRNA for H^+-ATPase increased in *Atriplex,* then rapidly decreased, indicating that continued transcription is not required for the maintenance of ion homeostasis under prolonged salinity (Glenn 1997). It is not yet known whether H^+-ATPase activity is regulated by gene family transcription (Palmgren and Christensen 1994), enzymatic properties of the native protein (posttranscriptional regulation), or protein-membrane interaction in salt-tolerant plants experiencing salinity. However, it has been suggested that the response of the H^+-ATPase genes to salinity may be a good indication of salinity tolerance in plants (Perez-Prat et al. 1994), and may be a valuable trait for bioengineering salt stress tolerance in crops.

A recent study showed a strong relationship between H^+-ATPase activity in the tonoplast and salt tolerance in barley (Zhang et al. 1998). These results are consistent with the idea that salt tolerance in barley is partly dependent upon root cells accumulating Na^+ in the vacuole, thereby reducing leaf tissue Na^+ content. The activity of H^+-ATPase in barley may depend upon calcium because calcium may increase membrane differential permeability, or Ca^{2+} may induce protein modification of H^+-ATPase by a Ca^{2+}-dependent protein kinase (CDPK). For example, the C terminus of the plasma membrane H^+-ATPase from *Dunaliella acidophila* contains three phosphorylation sites (Sekler et al. 1994). Therefore, there is general agreement that increasing the activity of H^+-ATPase in membranes, by increasing the protein complement and by modulating protein activity, is a component of salinity tolerance in a wide group of species.

Na^+/H^+ Antiporters and Na^+/K^+ Selectivity. The greatest body of literature available on the regulation of cytosolic Na^+ has been done with yeast. Several different genes for Na^+ regulation have been identified. In the yeast *Saccharomyces cerevisiae,* the gene ENA1 (also called PMR2) codes for a putative Na^+ pump in the plasma membrane. The high salt-induced expression of the ENA1/PMR2 gene is regulated by the genes HaL1 and HaL3, and by calcineurin. If HaL1 is overexpressed in transgenic yeast, salt tolerance is improved by increasing intracellular K^+ and decreasing intracellular Na^+. The expression of HaL1 in transgenic melon causes increased tolerance of salinity in *in vivo* cultured shoots (Bordas et al. 1997). SOD2 codes for a Na^+/H^+ antiporter in *Schizosaccharomyces pombe,* and there is an analog in a salt-tolerant *Saccharomyces cerevisiae* called Z-SOD2. SOD2 is induced by salinity shock, but Z-SOD2 is constitutive and independent of salt shock. The predicted amino acid sequences of these two genes are not the same, the sequence for Z-SOD2 has an extra hydrophilic stretch in the C-terminal region (Watanabe et al. 1995). Further study is needed with plants (rather than yeast) to see whether genes for Na^+/H^+ antiporters are more effective and expressed at higher levels in halophytes than in glycophytes.

In addition to sodium exclusion accomplished by H^+-ATPase pumps and Na^+/H^+ antiporters, cells must accumulate K^+. The accumulation of potassium has a "low affinity" component that is determined by inward-rectifying potassium channels, which allow potassium to enter the cytoplasm along an electrochemical gradient (established by ATPase pumps) when K^+ is available in the soil (Smart et al. 1996). Increasing the numbers of K^+ channels in the plasma membrane would increase membrane permeability to K^+, but would not influence the cell's ability to concentrate K^+ relative to the external environment. Low-affinity K^+ transport is thought to be determined by regulating channel activity rather than by regulating the expression of channel protein genes. High-affinity K^+ transport is regulated by K^+/H^+ symports such as the symport coded by the HKT1 cDNA isolated from wheat roots. Under toxic concentrations of Na^+, K^+ transport by this protein is blocked and low-affinity K^+ uptake dominates K^+ transport (Gassmann et al. 1996). Although there is some leakage of Na^+ through voltage-regulated K^+ channels (low-affinity channels) and through the Na^+/H^+ symport at high external concentrations of Na^+, this leakage is not likely to account for differences in Na^+/K^+ discrimination among tissues, genotypes, or species (Tyerman et al. 1997).

2. Genes Involved in Solute Accumulation. Biosynthetic pathways for many organic solutes are upregulated by salinity; however, that of glycine-betaine, proline, and pinitol are best known. The pathway to glycine-betaine in plants is short. Choline monooxygenase (CMO) converts choline to betaine aldehyde, and betaine aldehyde dehydrogenase (BADH) converts betaine aldehyde to glycine-betaine. Among plants that contain the genes for BADH and CMO, BADH is more responsive to salinity than CMO. In sugar beets, BADH activity increases two- to fourfold upon salt treatment, and an eightfold increase in BADH mRNA occurs in barley when exposed to salt. The BADH transcript remains high until the external salinity is decreased (Ishitani et al. 1995). There are at least two copies of the BADH gene in the sugar beet haploid genome, and they do not have identical sequences. When BADH and CMO genes are inserted into the genome of species without these genes (such as tobacco), glycine-betaine synthesis increases and glycine-betaine is transported into the chloroplast, where it protects photosystem II.

The pathway for proline synthesis (Figure 5.14) is regulated by both proline synthesis and degradation enzymes. Synthesis of proline is thought to originate at glutamic acid or ornithine, but the glutamic acid pathway may be most abundant during osmotic stress or nitrogen-limited conditions (Delauney et al. 1993). Proline synthesis from glutamic acid is thought to involve two enzymes: pyrroline-5-carboxylase synthetase (P5CS), and Δ^1-pyrroline-5-carboxylase reductase (P5CR). Proline concentration is also regulated by a degradation pathway, which involves proline dehydrogenase (PDH) and pyrroline-5-carboxylase dehydrogenase (P5CDH). Following osmotic stress (drought), the mRNA population of P5CS increases and that of PDH decreases in *Arabidopsis* (Yoshiba et al. 1997), following which proline concentration increases. The reverse occurs during recovery (Peng et al. 1996). In contrast, the genes for P5CR and P5CS are not enhanced by dehydration, high salinity, or ABA concentration (Yoshiba et al. 1995, Peng et al. 1996). These results indicate that regulation of at least two genes associated with proline synthesis by external salinity and drought determine proline accumulation, and this accounts for a portion of osmotic tolerance in plants.

Polyols such as pinitol, sorbitol, and mannitol were identified earlier in this chapter as important organic osmolytes in halophytes such as *Mesembryanthemum crystallinum;* they also accumulate during saline conditions in other plants, bacteria, yeast, marine algae, and animals (Bohnert et al. 1995). The biosynthesis of these polyols originates with glucose-6-phosphate and is intimately related to photosynthesis and photosynthate use. When the plant is not growing, or is under stress such as salinity, the products of photosynthesis can be diverted into other carbon-containing molecules such as polyols. In several halophytes, pinitol concentration increases during salinity in accordance with mRNA populations of several enzymes involved in their synthesis. Pinitol is synthesized by a methylation pathway originating with myoinositol. One of the enzymes in this pathway, inositol o-methyltransferase (Imt1), is an excellent ex-

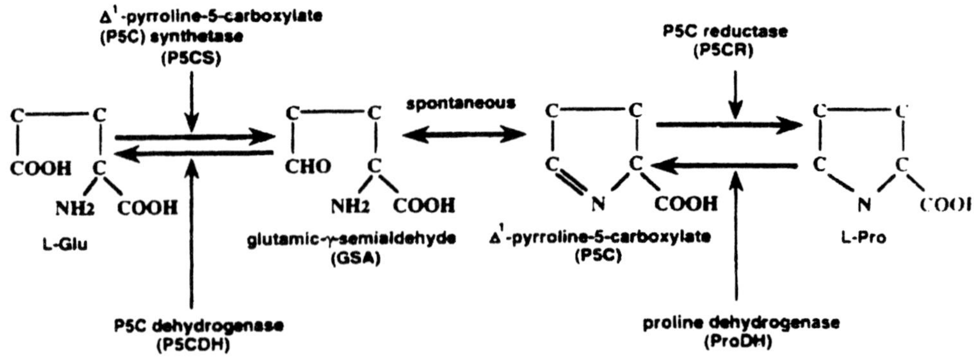

Figure 5.14. Synthesis and degredation pathways for proline (Redrawn and modified from Yoshiba et al. 1997, with permission.)

ample of a salinity-induced gene response. The transcription of the gene for Imt1 is induced only by salinity and low temperature (Vernon and Bohnert 1992). Following the salinity induction of Imt1, the flow of reduced carbon to pinitol increases, resulting in concentrations up to 700 m*M* in the cytosol and chloroplasts (Adams et al. 1998). Such increased activity of pinitol biosynthesis genes does not occur in glycophytes.

3. Genes Associated with Water Homeostasis: Aquaporins. *Aquaporins* are a group of membrane-bound proteins that regulate membrane water permeability in cells and may influence water movement among tissues and organs. First identified as a major intrinsic protein (MIP) in bovine eye lens (Gorin et al. 1984), these proteins have been extensively studied in animal systems because of their critical role in kidney function. Some aquaporins are highly conserved and are found in species ranging from bacteria, to plants, to mammals. In fact, aquaporins are abundant in plants, are associated with the regulation of water balance in plant cells, are located in the tonoplast and plasma membrane, and are found in a number of different isoforms (Chrispeels and Maurel 1994). These proteins are extremely abundant in plant membranes, for example the α-tonoplast intrinsic protein (α-TIP) accounts for 2% of total extractable protein in bean cotyledons. Also, several polyclonal antibodies raised against total purified plasma membranes were found to mostly react with aquaporins (Marty-Mazars et al. 1995).

The aquaporins are part of the major intrinsic protein (MIP) family, which can be divided into plasma membrane intrinsic protein (PIP), tonoplast intrinsic protein (TIP), and nodulin 26 from the peribacteriod membrane of legume root nodules. All cDNAs for aquaporins encode 25–30 kDa polypeptides that have six putative membrane-spanning segments with hydrophilic spans on each side of the membrane. Although aquaporins assemble into various organized structures *in vitro,* it is thought that each polypeptide monomer is able to form a functional water channel. Aquaporin water channels are blocked by mercury derivatives, possibly by the oxidation of cysteine residues near the channel pore, and this fact has been used to study their functional attributes. This assay of aquaporin activity is a minimal indication of the protein effectiveness because some aquaporins are insensitive to mercury. The most important test for identifying an aquaporin gene is injecting the mRNA from the gene into a *Xenopus* oocyte and measuring oocyte swelling. When an aquaporin mRNA is injected into an oocyte, it is rapidly translated, influences membrane permeability, increases water diffusing into the cell, and causes oocyte swelling.

The permeability of membranes to water depends on the number of aquaporins in the membrane and their effectiveness at transporting water (Maurel 1997). It is important to note that aquaporins selectively transport water and do not co-transport any other ions (Maurel et al. 1994). Therefore, these proteins are important for turgor maintenance and water distribution within and among cells, but not so important for ion homeostasis. In contrast, some other ion transport channels also transport water (Alexandre and Lassalles 1992). If aquaporins are important to water balance in cells and among tissues, do plants utilize this protein as a component of cell-based tolerance of salinity?

Before this question can be answered, it must be known how important aquaporins are to cell osmotic maintenance and water movement in plants. Water permeability of tonoplast vacuoles is blocked up to 80% in the presence of 1 m*M* mercuric chloride, but plasma membranes from the same cells have a low indication of aquaporin channel activity (Maurel 1997). In addition, the membranes of *Chara* and *Nitella* are highly permeable to water, and permeability is inhibited by 70% and 30%, respectively, when exposed to mercuric chloride. Therefore, the influence of aquaporins on membrane permeability varies among membranes and species. In general, aquaporins are predominantly located in the tonoplast, resulting in high tonoplast water permeability relative to the plasma membrane. In tobacco cell culture and onion epidermal cells, the tonoplast can be 100 times more permeable than the plasma membrane (Maurel et al. 1993). This high permeability of the tonoplast allows cells to use the vacuole (and ions in the vacuole) as a buffer against osmotic fluctuations in the cytoplasm.

Transport of water vertically in plants is obviously not associated directly with aquaporins because xylem elements have no membranes. However, radial movement of water through tissues from the xylem conduits occurs through both an apoplastic and a symplastic route. The proportions of water that travel through these differing paths vary with physiological state, but it is safe to say that a significant amount of radial water transport occurs through a symplastic route. It has not been possible to differentiate between water moving though plasma membrane channels and that moving through plasmodesmata. However, some lines of evidence suggest that aquaporins are involved in the radial movement of water. The endodermis functions to force water through a symplastic route. The expression of PIP1 in *Arabidopsis* is higher at the endodermis than elsewhere in the root tissue (Maurel 1997). The parenchyma cells surrounding the xylem in roots of maize have a high expression of the aquaporin ZmTIP1 (Barrieu et al. 1998). The high expression of aquaporins in xylem parenchyma cells would facilitate the control of water moving in and out of xylem elements and may be an important factor in xylem refilling following embolism. It is clear that aquaporins are important in maintaining osmotic balance in cells by their high expression in the tonoplast, and that aquaporins are important to symplastic radial movement of water to and from the xylem. Both of these processes are important in plant adaptation to the desiccation and ionic effects of saline conditions.

Although some part of water permeability may be determined by changes in membrane fluidity, the most powerful mechanism to mediate rapid and large changes in membrane permeability is changes in aquaporin activity. The activity of water channel proteins in membranes is regulated at the gene transcription, posttranslation, and cell localization levels. The rd28 gene in *Arabidopsis* codes for an aquaporin and is expressed during several forms of stress (Yamaguchi-Shinozaki et al. 1992). The MIP genes in the ice plant produce aquaporin-type proteins that serve as water channels, and the expression of these genes is associated with turgor recovery after salinity stress (Yamada et al. 1995). The halophytic sea aster increases the transcription of SaMIP A, B, and C when exposed to salinity (Uno et al. 1998). In addition, various isoforms and various genes are expressed in specific tissues at specific developmental times, which may be related to tissue-specific hydrological relationships and salinity tolerance. The function of TIP proteins can be regulated posttranscriptionally by phosphorylation. A calcium-dependent protein kinase can phospholylate α-TIP in bean seeds (Johnson and Chrispeels 1992). It has been proposed that phosphorylation of aquaporins can provide a mechanism to control *in situ* opening and closing of water channels.

Acclimation to saline conditions may involve the synthesis of aquaporins, their differential tissue location, and their phosphorylation. All of these possibilities have been found in both halophytes and glycophytes. The regulation of membrane permeability by these transmembrane proteins may be important for determining osmotic potential, increasing water accumulation, and directing water movement among tissues during salinity stress. However, our current knowledge of processes related to aquaporin activity among halophytes and glycophytes under salinity stress is too poor to make any general statements about their importance to salinity tolerance in plants.

4. Dehydrins. This class of proteins was named *dehydrin* because many of its members are induced by dehydration and high salinity (Close 1996). The first dehydrin identified was rab21 in rice (Mundy and Chua 1988), which is described below. The dehydrins are synonymous with the LEA (late embryogenesis activated) D11 gene family, which includes 18 distinct groups of proteins with diverse and sometimes unknown functions. All of the dehydrins have a consensus sequence called the K sequence, which may be repeated up to 11 times in one protein. A second sequence, the S segment, is found in many dehydrins and may be the binding site for nuclear locating polypeptides. Dehydrins function in a diverse way for improving desiccation and salt tolerance, including processes in the cytoplasm and the nucleus. This group of proteins has broad significance to osmotic stress response and a large fraction of metabolic functions, but this group is not related to salinity-specific metabolic responses.

Embryos, roots, and shoots of rice accumulate mRNA transcripts of a dehydrin gene identified as rab21 when exposed to 200 m*M* NaCl, desiccation, or ABA. The genomic clone of this gene encodes a glycine-rich polypeptide. A cDNA clone isolated from tomato (TAS14) encodes a single 16 kD glycine-rich protein with four structural domains similar to rab21 (Godoy et al. 1990), LEA D11, and some dehydrin genes. Consequently, rab21 is thought to be closely related to the LEA genes and dehydrins. Rab21, as with other dehydrins, is a desiccation-related gene product that is induced by a general stress signal transduction pathway that is not specific to salt. Thus, the induction of dehydrins is a general stress response rather than a salt-specific response.

5. Osmotin. Osmotin was one of the first proteins shown to increase in cells experiencing a gradual increase in salinity not in response to salt shock; hence the name. When plants are exposed to high salinity, this 25 kD protein increases dramatically (Ericson and Alfinito 1984), reaching 10% of total cellular protein. Two 26 kD polypeptides are formed in response to salinity, and only one of these accumulates in cells preadapted to saline conditions (LaRosa et al. 1989). The transcripts of the osmotin genes are constitutive, but they are at higher concentration in preadapted cells. ABA stimulates the ability to synthesize osmotins, but does not regulate the NaCl-induced accumulation of osmotin. Interpreting the significance of osmotin to salt tolerance became problematic when it was determined that there is a high degree of similarity between osmotin, thaumatin (a tobacco pathogen-related protein), and the maize trypsin/alpha amylase inhibitor. These other proteins are classified as defensive proteins against pathogen attack. It is now thought that osmotin protein is primarily related to antipathogen function rather than to salt tolerance (Liu et al. 1993).

6. Genes Associated with the Switch to CAM Photosynthesis. Genes associated with the CAM pathway are induced on a developmental schedule in the ice plant. Thus, following maturity, CAM genes are constitutively expressed at a low level. Upon increased salinity or drought stress, the transcript levels of these genes increase dramatically. Therefore, osmotic or drought stress accelerates the development of CAM in facultative CAM species. This post maturity acceleration is best demonstrated for the CAM-specific PEP-carboxylase gene, Ppc1, but it is also found for several other genes involved in the CAM pathway. For example, transcription of the gene encoding PEP carboxykinase, the enzyme that regulates the activity of PEP-carboxylase (Adams et al. 1998) is accelerated under saline conditions. The mechanism of acceleration of CAM gene transcription is not well understood, but does relate to a complex interaction between DNA sequences of promoters and transcription factors (Cushman and Bohnert 1992). There are several regulatory elements that govern transcription of Ppc1 during salt stress that are similar to other elements that respond to changes in plant growth regulators (Thomas et al. 1995). The acceleration of CAM gene transcription during osmotic stress is a response to the water deficit induced by high salinity, and is most likely related to a signal transduction pathway common to other growth regulator–induced responses.

C. Summary of Inducible Proteins and the Potential for Biotechnology

High salinity induces a diversity of different gene products. During phase 1 inhibition (osmotic stress), a generalized stress response signaling pathway (probably the ABA signaling pathway) induces the synthesis of enzymes and leads to the accumulation of solutes and the adjustment of water flow (aquaporins). However, during prolonged salinity stress (phase 2 inhibition), a salt-specific signaling pathway induces the synthesis of ion-homeostasis proteins (membrane transporters).

Undoubtedly, salinity is a problem for agricultural programs worldwide, and saline environments have shaped the metabolism of a number of species in unique ways. Furthermore, the wild progenitors of many crop species have higher salt tolerance than the domesticated crop. This suggests a potential to utilize evolutionary adaptation in other species to increase salinity tolerance in crops. The greatest potential for engineering salinity tolerance in crops is by either increasing the synthesis of compatible solutes during saline conditions, or increasing the amount or activity of Na^+ transporters (also H-ATPase) to enhance storage of Na^+ in the vacuole. In general, genetic engineering of crop salt tolerance can best be accomplished at the cell physiology level.

Upregulation and overexpression of glycine betaine have had little significance to the salt tolerance of plants in the field (Bohnert and Jensen 1996). Even though this pathway is short and easy to manipulate, the production of glycine-betaine will only be able to benefit plant performance as part of the intracellular compartmentation package induced by salinity stress (Yeo 1998).

VI. SUMMARY

Salinity is an important stress in the natural and agricultural environment. Ion concentration is increased by tides, salt spray, geologic salt, and evaporation. High salinity in any environment is extremely heterogeneous in space and time, and many plants use windows of low salinity to survive in such habitats. However, there is no question that plants have adapted to salinity at all levels of organization. Crops do not express many of the traits for salinity adaptation that are expressed by many plants of wide evolutionary origin in natural ecosystems. In fact, many progenitors of crop species have much greater salt tolerance than the domesticated crop. The historical and continuing salinization of agricultural fields, in combination with low salt tolerance of crops, makes the understanding of salinity tolerance by natural plants a critical issue to agriculture.

Halophytes have compensated for saline conditions at morphological and physiological levels. Specialized morphological features—salt glands, and salt bladders—are present in some halophytes and serve to extrude salt. However, most halophytes do not have salt glands and still accumulate salt in saline habitats. The ions accumulated by halophytes are concentrated in leaves because they follow the transpiration stream. Halophytes utilize these salts as an osmoticum to lower the osmotic potential of the central vacuole and maintain some turgor pressure. The osmotic potential of the vacuole is balanced by cytoplasmically compatible organic solutes that can also serve as osmoprotectants of proteins and some metabolic processes. The osmotic relationships of tissues and organs can be regulated by the activity of aquaporins in tonoplast and plasma membranes. When the vacuole becomes saturated with salts, the cytoplasm will rapidly increase in salt content and suffer damage. The likelihood of vacuole saturation can be postponed by having a very large vacuole/cytoplasm volume ratio (succulence) or by using salt glands or bladders. If none of those options is available, then the plant can grow relatively fast in order to replace the dying tissues with new growth.

By contrast, glycophytes (which constitute most crops) do not accumulate ions for osmotic adjustment, and their vacuoles rapidly become saturated. Most glycophytes must exclude ions at the root membrane or compartmentalize the salts in root systems to prevent damage. Increasing the activity of Na^+/H^+ antiporters and H^+-ATPase pumps in the root systems of glycophytes can increase their salinity tolerance. Some glycophytes can accumulate organic osmolytes in the cytoplasm to provide a partial tolerance of salinity. However, this trait alone cannot provide a high degree of salt tolerance.

Adaptation to salinity involves a complex interaction between genetic regulation of ion homeostasis, water transport, desiccation tolerance mechanisms, and ontogenetic and seasonal regulation of metabolic processes. A large amount of information has been attained using cell and subcellular systems. However, the overriding issue with salinity is the concentration of ions due to transpiration, which is determined on a whole-plant basis. Processes that are important for single cells or cell cultures in a liquid medium may not reflect those for whole plants. In order to apply the knowledge gained from research on single-gene upregulation in cell culture systems, this information must be viewed in the context of whole plants. Understanding the tissue and whole-plant signal transduction pathways that are common to desiccation tolerance and unique to salinity may provide the link between individual gene induction by salinity and whole-plant response to salinity.

STUDY-REVIEW OUTLINE

Introduction

Scientific Interest in Salinity Tolerance

1. Locations of high salinity on the landscape are patchy and relatively infrequent.
2. Communities in saline habitats have a relatively low productivity and low diversity.
3. The productivity of salt marshes is very high compared with other terrestrial ecosystems.
4. Coastal saline marshes and swamps regulate the exchange of nutrients between the terrestrial and marine environments.
5. Studying the physiological adaptations of plants to saline habitats has shed light on studies of gradient analysis in plants, as well as physiological tolerance of other environmental stresses.
6. Environmental characteristics such as pH, aeration, and nutrition of saline habitats are very heterogeneous.
7. The complexity of adaptive mechanisms found in plants adapted to saline habitats, and the importance of salinity tolerance to agriculture, have fueled the interest in this subject.

8. Salinization of agricultural fields is increasing and is a problem for agriculture because few crop species are adapted to saline environments.

Diversity of Salt-Tolerant Species

1. Salinity tolerance traits have evolved many times in a wide diversity of plant families (over 100).
2. Very few crop species are considered tolerant to saline conditions. Barley and cotton are some of the most tolerant crops.
3. Wild progenitors of many crop species have greater tolerance of saline conditions than do the domesticated crop.
4. Halophytes are those plants that can tolerate higher levels of salinity before production or fitness decreases than can glycophytes.
5. Obligate halophytes are those plants that require moderate levels of salinity before they attain their highest fitness.
6. Facultative halophytes are those species that have glycophyte traits. However, when they are exposed to moderate salinity, they gain halophytic traits.

Salinity Effects on Plant Growth

1. The relationship between ion concentration and growth reduction is not uniform among ions. Solutions containing a combination of different ions have a weaker inhibitory effect than equimolar solutions of one ion.
2. One of the first responses of plants to high salinity is a reduction in leaf growth rate.
3. In many cases, salinity reduces growth by decreasing the amount of photosynthetic leaf area.
4. Salinity causes a greater reduction in shoot growth than root growth.
5. The larger the change in the root/shoot ratio in plants exposed to saline conditions, the larger the effect of salinity on growth.

Scientific Level of Understanding

1. This field began with the study of productivity in saline habitats and has grown to understanding the molecular basis of salinity tolerance in plants.
2. Research with model plants (*Arabidopsis, Mesembryanthemum*) has caused an advancement in understanding ionic regulation and salt stress tolerance in plants.
3. The most interesting new research on salinity tolerance concerns the signal transduction pathways that regulate gene action in response to environmental salinity, and their interaction with other stressors.

Saline Environments

Units of Salinity

1. Primary measurements of salinity involve quantification of the concentration of sodium and or chlorine.
2. Molar concentration values are the most uniform and usable among experiments because the nature of the salts changes in different experiments and environments.
3. Secondary or indirect measurements of salinity include electrical conductivity, osmotic potential, freezing point depression, and chlorinity.
4. Soluble salts in the environment come from weathered rocks, salt spray, or the result of long-term evaporation. Anthropogenic sources of salinity are sewage sludge, soil and water amendments, animal manure, fertilizers, and oil and gas brines.

Saline Environments

1. Environmental characteristics vary greatly among different saline environments.
2. Saline soils are those in which less that 15% of their cation exchange sites are occupied by sodium, but they still have a high concentration of soluble salts.

3. Sodic soils are those that have a high concentration of neutral soluble salts, but do have a high amount of sodium and a pH greater than 10.

Salt Marshes

1. Salt marshes represent the largest area of natural saline habitat, and have salinity in the range from 0.1–1.2 *M*.
2. Evaporation, tides, and topography work together to regulate the salinity in salt marshes. This results in an overall gradient of salinity interrupted by many factors causing high spatial and temporal heterogeneity.

Coastal Bluffs

1. Coastal bluff communities receive most of their salinity from salt spray off the ocean.
2. Many species are shaped by the action of salt pruning, when winds deposit salt on their surfaces.

Mangrove Swamps

1. Mangrove swamps are the tropical equivalents of temperate salt marshes, but different factors regulate their salinity.
2. Species form a distinct zonation in mangrove swamps primarily due to their tolerance of salt, anaerobic soils, and nutrient availability.
3. Tidal inundation is the primary factor regulating salinity because evaporation is relatively low under the mangrove canopy.
4. Most mangroves require some salinity for optimal growth, with most requiring 25–50 m*M* NaCl in the growth medium.
5. In sites with infrequent tidal inundation, sediments become anaerobic and phosphorus becomes limiting.
6. Many mangrove species have structural adaptations, such as pneumatophores, in response to flooding stress.

Desert Salt Flats and Alkali Sinks

1. Desert salt flats can result from the evaporation of inland bodies of water, or from leaching of salts from mountainous regions around the salt flat.
2. Salt buildup in desert salt flats can be extensive, reaching a depth of more than 1 m of crystallized rock salt, such as in the Atacama desert.
3. Spatial variability of salinity in desert salt flats is less than in salt marshes or mangrove swamps.
4. Salts can be temporarily washed into greater soil depths by precipitation.

Agricultural Fields

1. Agricultural fields can become saline because of evaporation, irrigation, and fertilization.
2. In order to remove salts from agricultural fields, the field must be flushed, often with tiles to drain the salts from the field.
3. Salinization of agricultural fields has been a cause of the demise of cultures in historical times.
4. Problems of saline soils are frequent in developing countries, but are by no means isolated to these countries.
5. Salinization of agricultural fields is most problematic in semiarid regions. Such salinization has occurred in about 80 countries, including 35% of the world population.
6. Leaching agricultural fields to reduce salt causes a significant contamination of aquatic systems such as streams and estuaries.

Detrimental Effects of Salinity

Direct Effects

1. The greatest direct effect of salinity is on seeds and young plants, because of their connection with the surface soil environment.
2. Salts will accumulate in shoot tissues due to transpiration even if xylem water has low salinity. Therefore, all plants must have some salinity tolerance.

Enzyme Activity

1. Plant enzymes are generally intolerant of high salinity.
2. The enzymes from glycophytes and halophytes are equally inhibited by salinity.
3. Some bacteria, halophylic archaebacteria, have enzymes with high tolerance of salinity.
4. Even though all plants have sensitive enzymes, enzymatic reactions of whole plants are differentially influenced by different ions in the soil solution.
5. *In vivo* inhibition of enzyme activity in cytoplasm mimics *in vitro* inhibition of enzyme activity.
6. Nitrate reductase activity decreases linearly with an increase in salinity.

Nutrient Imbalance

1. Elevated salinity is associated with potassium deficiency in many plant species and crops.
2. Sodium is transported out of the cell across the plasma membrane by an antiport system, which also moves hydrogen ions into the cell. This transport is dependent upon a H^+-ATPase pump to create the high H^+ ion concentration outside the cell.
3. High concentrations of Na^+ outside of the root cells (saline soils) cause an inhibition of Na^+ antiport activity.
4. High Na^+ concentration in the soil also inhibits H^+-ATPase activity causing a change in the electrochemical gradient across the cell.
5. Since the activity of many ion transport systems for other ions, such as potassium, are dependent upon the electrochemical gradient across the plasma membrane, nutrient uptake is inhibited by a high external Na^+ concentration.

Membrane Dysfunction

1. High external Na^+ can be associated with an increase in internal membranes, particularly vesicular membranes. This may indicate increased compartmentation and export of cytoplasmic Na^+.
2. Plasma membrane leakage increases under high external salinity.
3. Na^+ substitution for Ca^{2+} in the plasma membrane may be the cause for increased leakage.
4. The saturation level of phospholipids in the cell membrane increases with an increase in salinity, which can decrease membrane fluidity.
5. Salt stress inhibits linoleic acid desaturase, which causes a reduction in linoleic acid in chloroplast membranes.

General Metabolic Processes

1. Both respiration and photosynthesis are inhibited by salinity at the level of enzyme activity.
2. Photosynthesis is further inhibited by influences of salinity on gas exchange properties and photochemical reactions.
3. Salinity influences photosynthesis by both stomatal and nonstomatal mechanisms.
4. In some plants (e.g., *Spartina,* stomatal conductance decreases upon high salinity treatment.
5. In some other plants (e.g., *Atriplex,* photosynthesis is inhibited by effects on both stomatal and mesophyll conductance.
6. In some plants adapted to low light conditions, high salinity directly affects photosystem II.

7. The production of free radicals under saline conditions, and their effects on photosynthesis, may be an important manner by which salinity reduces photosynthesis.

Indirect Effects

Water Relations

1. The most rapid and observable response of plants to salinity is wilting or other manifestations of water stress.
2. The high osmotic potential of saline soil solutions reduces water availability to plants.
3. In many crops, growth is suppressed when a threshold of osmotic concentration is reached.
4. Crop yield can be described by an equation that is based upon the soil osmotic potential alone.
5. The two most common responses of glycophytes to salinity are osmotic adjustment and stomatal closure.
6. Glycophytes accumulate cytoplasmically compatible osmotica, while halophytes often perform osmotic adjustment with inorganic ions.
7. When epidermal strips of some plants are exposed to KCl solutions, the stomata open. However, when they are exposed to equimolar concentrations of NaCl, the stomata close.
8. Although reduced turgor is an immediate response to increased salinity, this is not likely the cause of reduced growth and reduced stomatal conductance.
9. Cell wall extensibility is reduced in hypersaline conditions. This may be due to Na^+ replacing Ca^{2+} in the cell wall, which would reduce extensibility.

Collateral Nutrient Limitation

1. Only a few trace elements such as copper, iron, manganese, and zinc are likely to be limiting in saline or sodic soils. This is because of their decreased solubility in saline soil solution.
2. Trace elements such as boron, selenium, and molybdenum have a potential to become toxic in saline soils. Boron is highly likely to become toxic because plants have a low tolerance to boron.
3. Some macronutrients such as K, P, and N can become limiting in saline soils. This is particularly true of K because its accumulation by plants is severely affected by salinity.
4. NaCl inhibits nitrate accumulation, but not ammonium accumulation if nitrate is not present.
5. Calcium uptake is inhibited by salinity, and this is directly associated with reduced growth.
6. High concentrations of Ca^{2+} or ammonium in the soil can buffer the effects of salinity on nutrient relations in plants.

Oxidative Stress

1. Saline conditions, taken as a whole, can result in a buildup of *reactive oxygen species* (ROS) in plants.
2. An increase in ROS in plants experiencing salinity is most likely ubiquitous among plants in saline habitats.
3. The superoxide radical, produced under saline conditions, cannot leave the chloroplast or mitochondria.
4. Free-radical scavenging mechanisms, such as superoxide dismutase (SOD), and the peroxidase cycle, may be critical to survival in saline habitats.

General Growth Inhibition Model—Biphasic

1. Phase 1 of growth reduction is due to the osmotic effects of soil salinity on plant water relations.
2. Phase 2 of growth reduction is due to the effects of specific ions on cytoplasmic metabolism.
3. Halophytes and glycophytes respond similarly to phase 1 of growth reduction.
4. Halophytes are able to keep ion concentration lower in cytoplasm than glycophytes at higher soil salinity, which delays phase 2 of the growth inhibition model to later dates and higher salinities than glycophytes.

5. The turnover rate of old leaves may be an important mechanism for delaying the onset of phase 2 inhibition.
6. The biphasic model for growth inhibition fits well with patterns of glycophyte response to salinity. However, halophytes use many different mechanisms to slow the development of phase 2 inhibition.

Mechanisms of Adaptation to Salinity

Salt Exclusion: Ion Homeostasis

1. Salt tolerance in glycophytes is generally associated with salt exclusion, while salt tolerance in halophytes is generally associated with salt accumulation.
2. Ion homeostasis of the cytoplasm is dependent upon electrochemical gradients, carrier proteins, and ion pumps.
3. Hydrogen pumps regulate the electrochemical gradients across the tonoplast and plasma membrane.
4. Calcium and sodium ions diffuse across the membranes in response to the electrochemical gradients, causing internal concentrations of about 10 m*M*.
5. If external Na^+ concentrations are high and diffusion causes a high internal Na concentration, then Na^+/H^+ antiporters remove sodium from the cytoplasm.
6. The ability to increase the activity of H^+-ATPase pumps and Na^+/H^+ antiporters may be an important aspect of salinity tolerance.
7. The discrimination between K^+ and Na^+ accumulation by cells may be another important aspect of salinity tolerance in glycophytes.
8. A genetic locus for Na^+/K^+ discrimination in wheat was isolated and transferred into the genome of sensitive species. The result was increased salt tolerance.
9. Glycophytes normally maintain a constant or low Na^+/K^+ ratio when challenged with salinity. In contrast, halophytes may drastically increase their Na^+/K^+ ratio in saline conditions.
10. If glycophytes do increase Na^+/K^+, then succulence is required. However, succulence is not required for halophytes when they increase their Na^+/K^+ ratio during high salinity.
11. It is not energetically feasible to maintain the same electrochemical gradient when the cytoplasmic cation concentration increases; thus, a new stable electrochemical gradient is required for tolerating higher cytoplasmic salinity.
12. Exclusion of salt at the plasma membrane is important for ion homeostasis but cannot be the only mechanism for salt tolerance in halophytes.

Salt Reabsorption and Retranslocation

1. Salt exclusion at the root membrane will eventually cause increasing xylem salinity due to leakage past the root endodermis.
2. In some legumes, stem parenchyma cells differentiate into transfer cells and selectively remove salt from the xylem stream.
3. Phloem cells in the stem base of *Phragmites communis* reabsorb Na from the xylem stream.
4. Reabsorption and retranslocation of Na^+ is found only in species that have minimum salt tolerance.
5. In the most salt-tolerant species, most of the salt they are exposed to is accumulated and a large portion is transported to the shoots.
6. Continued retranslocation of Na^+ back to the roots is not a good strategy for tolerating salinity because it results in an excessive buildup of salt in roots, where the main salt problem already exists.

Salt Extrusion: Glands and Bladders

1. Some halophytes have specialized structures on their leaf and stem surfaces that serve to eliminate salts from their tissues: salt glands, salt bladders, and salt hairs.

2. Salt glands are found in at least 10 dicotyledonous families and one monocotyledonous family.
3. Both salt glands and bladders are derived from epidermal cells, but they differ in structure and function.
4. One type of salt gland, embedded in the epidermis, has 4–10 cells that are almost completely isolated from the mesophyll except for a few patches of "transition gaps" where salt transfer takes place.
5. The second type of salt gland, in *Spartina* and some other monocotyledons, has only two cells, and long protrusions of the cap cell wall extend down into the basal cell. The extracellular partitions formed by the convoluted cell wall are where ion and water exchange occur.
6. The glands serve to concentrate mostly Na^+, K^+, Ca^{2+} and Cl^- to values 50 times those in the cytoplasm.
7. The best-known salt pumping system in salt glands is that for Cl^- in *Limonium vulgare*.
8. Both transcription and translation are required to upregulate this Cl^--transporter protein following salt exposure.
9. Salt bladders are modified leaf hairs that can either concentrate salt at more than 60× the concentrations found in the leaf, or serve simply as holding tanks for salt at the same concentration as that of the leaf (an extended vacuole).
10. Eventually, bladder cells die and rupture depositing salt on the leaf surface. This is in contrast with salt glands, which continually excrete salt onto the leaf surface.
11. Similar salinity tolerance can occur in halophytes with or without salt glands. Thus, salt glands are not imperative for salt tolerance.

Salt Dilution: Succulence

1. Dilution of accumulated salts can occur by an increase in succulence.
2. Plants that have very large vacuoles can store more salt before cytoplasmic salinity increases.
3. In some halophytes, succulence is not manifest until soil salinity increases.
4. In many mangrove species, the bulk ion concentration of tissue is maintained at relatively constant levels due to changes in tissue succulence.
5. Succulence may be particularly important for the salt tolerance of developing and expanding tissues.
6. Succulence is not required for high salinity tolerance in many halophytes, and many nonhalophytes express succulence. Thus, succulence is not highly associated with salinity tolerance in plants.

Salt Compartmentation

1. Since halophytes generally accumulate salts rather than exclude them, and not many halophytes have succulence or salt glands, compartmentation of salts into nondetrimental areas may be very important for salinity tolerance.
2. It has generally been thought that halophytes are able to accumulate salts in their apoplastic space and vacuole, thereby keeping the symplast relatively free of salt accumulation.
3. There is very good evidence that Na^+ and K^+ are preferentially accumulated into the vacuole, and that this is one way of moderating salt accumulation into the cytoplasm.
4. When salinity reaches the saturation point for the Na^+/H^+ antiporters, salt compartmentation in the vacuole will stop, and salt will rapidly rise in the cytoplasm.
5. Several studies have shown that there is an equilibrium that develops for salt across the tonoplast membrane that cannot account for salinity tolerance.
6. In maize, vacuolar salt content increases above that in the cytoplasm only when intolerable salinity is reached.
7. In most halophytes, salinity increases faster in older tissues than in young tissues. (A notable exception is the ice plant.)

8. Salt increase in older tissues may simply be due to a longer duration of transpiration in these tissues, or it may be due to directed translocation.
9. A fast growth rate can be an effective salt tolerance mechanism because the bulk of the salt can be abscised in older tissues that are going senescent.
10. If the central vacuole is effective at accumulating salinity, there must be an adjustment of cytoplasmic osmotic potential to match that of the vacuole.

Salt Tolerance

Compatible Solute Accumulation

1. Compatible solutes are accumulated by all plants in response to a number of different environmental cues such as water stress, chilling temperature, and salinity.
2. Accumulation of these compounds cannot be the only method of osmotic adjustment because the cost in energy and carbon skeleton may be prohibitive.
3. Osmotic adjustment is accomplished by a combination of compatible solutes and inorganic ions in halophytes but only of compatible solutes in glycophytes.
4. Many compatible solutes can increase in concentration in the same plant when exposed to salinity.
5. The ionic composition of the soil environment does not influence the types or amount of compatible solutes accumulated. The response seems to be strictly associated with water potential effects of salinity.

Proline

1. Proline is accumulated in response to water stress and salinity by many plants, bacteria, protozoa, marine invertebrates, and algae.
2. Proline accumulation could be an adaptation to stress or a consequence of stress.
3. Increases in proline alone can confer greater water and salt stress tolerance in genetically engineered organisms.
4. Often, proline alone cannot account for the total osmotic adjustment that occurs in tissues in response to water and salt stress.
5. Another significance of proline accumulation can be a change in the redox balance of NADP/NADPH which may stimulate the pentose phosphate pathway and important signaling pathways.

Polyols

1. Polyol accumulation is equally ubiquitous as proline and occurs in animals, algae, yeast, bacteria, and plants when exposed to hypersaline conditions.
2. Polyols act in osmotic adjustment in several ways. They increase the osmotic concentration of the cytoplasm, they act as an osmoprotectant of membranes, they can scavenge free radicals of oxygen, and they can serve as a carbon storage pool.
3. Polyol synthesis can be easily upregulated in plants because the pathway from photosynthetically derived compounds to polyols such as mannitol and pinitol is short and not complex.
4. The synthesis pathway for inositol is central to the pathways leading to cell wall synthesis as well as those for glycoproteins and mucilage.

Quaternary Amines

1. Quaternary amines (QAMs) such as glycine-betaine are accumulated in response to water and salt stress in a few plant species.
2. QAMs are associated with osmotic adjustment in response to salt stress, but they do not account for all osmotic adjustment.
3. QAMs can protect enzymes from heat damage and high salinity.
4. Due to the multiple significance of QAMs to stress tolerance, the genes required for their accumulation have attracted a lot of interest.

5. Biosynthesis of glycine-betaine starts with choline and involves only two enzymes.
6. Although there are obvious reasons why transferring the ability to accumulate glycine-betaine into salt-sensitive species should increase salt tolerances, such experiments have not been particularly successful.

Other Methylated Organic Osmolytes

1. Polyamines such as putracine and spermine also increase during salt stress in some plant species.
2. Many of the compounds that increase during salt stress are methylated, which means that the activity of methyl transferases and their co-substrate SAM may be important to salt tolerance.
3. The mRNA message for SAM synthetase increases in some species such as tomato when exposed to salinity.
4. Photorespiration, glyoxylate catalysis to CO_2 and N^5MeTHF may be the sources of methyl groups for the methyl transferases involved in compatible solute accumulation during salinity.

Membrane Stability

1. The ability to maintain membrane differential permeability to ions is critical to salt stress tolerance.
2. Sterol abundance and composition affect membrane fluidity and differential permeability.
3. The ratio of sterols to phospholipids is high in saline-tolerant plants and low in salt-sensitive species.
4. The ratio of planar to nonplanar sterols may be important in salt stress tolerance. Generally, but not always, salt-tolerant species have a higher proportion of planar sterols.
5. Membrane fluidity generally decreases in response to salinity, which may be important for limiting Na^+ diffusion.
6. Glycolipids stabilize the seal between membrane and transport protein, which limits leakage around the protein. Therefore, higher glycolipid content may increase salinity tolerance. Evidence on this topic is not clear at this time.
7. In glycophytes, there is no change or a decrease in phospholipid content when exposed to salinity, while that for halophytes remains the same or increases.
8. It is possible that the ability to increase the phospholipid content of membranes is an important adaptation for salinity tolerance in halophytes.
9. If the proportion of short-chain fatty acids increases in a membrane, planar sterols can be accommodated better, and the membrane has the potential to become less permeable to ions.
10. An increase in fatty acid saturation and a decrease in fatty acid chain length can increase the activity of H^+-ATPase activity, which is critical to salt tolerance mechanisms.
11. Although many factors are associated with the ability to maintain ion homeostasis in halophytes and glycophytes, membrane attributes such as sterol content, fatty acid composition, fatty acid saturation, and phospholipid abundance may be intimately involved with salinity tolerance.

Growth Hormone Relationships with Salinity

1. When salinity increases in the environment, the amounts of auxin, cytokinins, and gibberellins in plant tissues decrease.
2. Pretreatment of seeds with gibberellins or cytokinins will increase salinity tolerance of seedlings.
3. Cytokinins can stimulate the accumulation of compatible solutes and can increase salt tolerance in plants. This effect may be due to the influence of cytokinins on ion homeostasis.
4. ABA increases in both halophyte and glycophyte tissues when exposed to salinity. This induces a number of genes, such as BADH, that are important for the accumulation of compatible solutes.

5. The increase in ABA under saline conditions is short lived, and salt-tolerant species generally have lower long-term concentrations of ABA.
6. Ethylene content and emission increases with an increase in salinity in most plants.
7. An ethylene increase in response to salinity is associated with an increase in ACC and an associated increase in the tissue abscission rate.
8. An inhibition of ethylene synthesis may allow higher concentrations of salt to occur in older tissues and result in higher salt tolerance for the plant.

CAM Photosynthesis Induction

1. A few extreme halophytes can switch from a C_3 metabolism to a CAM metabolism when they experience high salinity.
2. The induction of CAM is initiated by a deepening of the midday depression in stomatal conductance and an associated transient period of low turgor pressure.
3. CAM induction is also developmentally regulated so that juvenile plants are incapable of inducing CAM during salt stress.
4. CAM is a constitutive aspect of development in some halophytes, and stress hastens the normal induction of CAM in species such as *Mesembryanthemum crystallinum.*

Salt Tolerance Genes and Genetic Engineering

Signal Perception and Transduction

1. The water relations related responses to salinity are likely regulated by a general signaling pathway that is also associated with drought and cold stress responses. The ion homeostasis responses to salinity are likely to be regulated by a salinity-specific signaling pathway.
2. The general signal transduction pathway for water, salt, and cold stress is the ABA signaling pathway.
3. The ABA signaling pathway is directly related to cytosolic Ca^{2+}-dependent protein kinases.
4. The link (secondary messenger) between ABA and the calcium-dependent phosphorylation cascade is most likely cyclic ADP–ribose.
5. The salinity-dependent signaling pathways have not been elucidated in plants, but in yeast a salinity-specific signaling pathway involves calcineurin (CaN).
6. There is evidence that plants also have a CaN-regulated system of ion transporters.
7. CaN induces the P-type ATPase (located in the tonoplast) for increasing H^+ ion transport across the vacuole.
8. CaN maybe involved with the signaling pathway that regulates stomatal conductance in response to salinity and root-specific responses.

Salinity-Inducible Genes and Proteins

Ion Homeostasis

H^+-ATPase Transporters

1. The molecular genetics of H^+-ATPase are the best known among the ion transport complexes involved with salinity tolerance.
2. The nucleotide sequence of the ATPase gene family differs among species.
3. H^+-ATPase activity and protein content increase more rapidly in halophytes when they experience salinity than in glycophytes.
4. Salinity shock stimulates a short-term increase in H^+-ATPase mRNA, which indicates that continual transcription of H^+-ATPase genes is not necessary for long-term salinity tolerance.
5. Increased H^+-ATPase activity under saline conditions could be due to increased transcription, selective transcription of some gene family members, post-translational regulation of protein function, and membrane-protein interaction.
6. Increased activity of H^+-ATPase in the tonoplast of barley is directly related to root accumulation of Na^+ in the vacuole.

7. The activity of H^+-ATPase in membranes may be regulated by phosphorylation by CDPK at the C terminus of the protein.

Na$^+$/H$^+$-Antiporters and Na$^+$/K$^+$ Selectivity

1. The best information about the Na$^+$-antiporters and Na$^+$/K$^+$ selectivity is available for yeast.
2. The gene ENA1 (or PMR2) codes for a Na$^+$/H$^+$-antiporter that is localized in the plasma membrane.
3. The upregulation of ENA1 is determined by two genes named HAL1 and HAL3.
4. Two SOD genes in yeast code for Na$^+$/H$^+$ antiporters. One of the genes is only expressed in saline conditions (SOD2), while Z-SOD2 is a constitutive gene.
5. K^+ accumulation in cells has a low-affinity component and a high-affinity component.
6. The low-affinity component is due to inward-rectifying K^+ channel proteins, in which K^+ moves along the electrochemical gradient established by the H^+-ATPase.
7. Increasing the number of low-affinity K^+ channels increases K^+ transport but cannot increase cytoplasmic K^+ content above that of the soil solution.
8. High-affinity K^+ transport is accomplished by K$^+$/H$^+$ symports, which are coded by genes such as HKT1 isolated from wheat roots.
9. Under toxic levels of external Na$^+$, the activity of K$^+$/H$^+$ symports is inhibited and Na$^+$ leakage through the low-affinity K^+ channels increases, as does that along the K$^+$/H$^+$ symports.

Genes Involved in Solute Accumulation

1. Genes associated with glycine-betaine accumulation are sensitive to salinity.
2. Of the two genes involved with glycine-betaine biosynthesis, the BADH gene is more sensitive to salinity than the CMO gene.
3. Transcription of the BADH gene increases two- to eightfold in some plants when exposed to salinity, and the mRNA transcript for BADH remains high during the entire saline period.
4. Transferring the genes for BADH and CMO into tobacco causes the accumulation of glycine-betaine during salt stress and confers salinity tolerance.
5. During salinity stress, proline biosynthesis predominantly originates with glutamic acid.
6. Biosynthesis of proline through the glutamic acid pathway involves two enzymes (P5CS and P5CR).
7. The degradation of proline is regulated by two enzymes (PDH and P5CDH).
8. During osmotic stress, the mRNA of genes coding for P5CS and PDH increase while those for P5CR and P5CS are not enhanced. The reverse occurs during recovery from osmotic stress.
9. When plants are under salt stress, the products of photosynthesis (glucose-6-phosphate) are not being used by primary metabolism and can be diverted into polyol synthesis.
10. The mRNA transcripts of several genes that code for enzymes involved in pinitol synthesis increase during salinity stress in some halophytes.
11. In particular, Imt1 is upregulated during salt stress conditions in the ice plant.
12. The increase in Imt1 mRNA transcript results in a large accumulation of pinitol, which does not occur in glycophytes.

Genes Associated with Water Homeostasis: Aquaporins

1. *Aquaporins* are a group of membrane-bound proteins that affect water transport through membranes.
2. Some aquaporins are conserved from bacteria to plants, and others are kingdom specific. Aquaporins are abundant in plants, and they do affect the ease and direction of water transport in plant tissues.
3. Aquaporins are abundant in plant tissue, accounting for 2% of total protein in some cases.

4. Aquaporins are classified as major intrinsic proteins (MIPs) and can be subdivided into tonoplast and plasma membrane proteins.
5. MIPs are all 25–30 kD proteins that have six membrane spanning segments.
6. MIPs assemble into multiple polypeptide structures *in vitro,* but each ploypeptide is a water channel.
7. The best test for the aquaporin function of a particular protein is to inject the purified protein into a *Xenopus* oocyte and observe whether swelling occurs.
8. Aquaporins selectively transport water and not other ions. The activity of aquaporins is dependent upon their abundance and whether they are open or not.
9. The high abundance of aquaporins in the tonoplast makes them important for water and ionic balance between the cytoplasm and the vacuole.
10. Aquaporins are involved with radial movement of water because aquaporin expression is high in parenchyma cells that surround the vascular tissues.
11. The activity of aquaporin channels can be regulated by post-transriptional phosphorylation.
12. In beans, a CDPK phosphorylates α-TIP, which regulates water transport capability.
13. Phosphorylation of aquaporins while they are in the membrane may control water movement between cells.
14. Aquaporin activity and translation has been found in glycophytes and halophytes, but our knowledge of the differences in aquaporin activity between halophytes and glycophytes is not adequate to verify whether this constitutes a mechanism for adaptation to salt stress.

Dehydrins

1. The *dehydrins* are a class of proteins, (like the protein coded for by rab21), that are responsive to salt and drought stress.
2. The dehydrins are synonymous with the proteins coded by LEA D11 genes, which comprise 18 distinct families of proteins with many functions.
3. All the dehydrins have a consensus sequence named the K segment, which may be repeated up to 11 times in one polypeptide.
4. Dehydrins have broad significance for stress tolerance (salinity, water, cold) in the nucleus and the cytoplasm.
5. The abundance of the mRNA transcript from rab21 increases in rice plants under salt stress, drought stress, or ABA treatment.
6. Other dehydrins are upregulated in glycophytes and halophytes alike during all water stress–related times.

Osmotin

1. Osmotin, a 25 kD protein, was the first protein to be shown to be induced by salinity.
2. Upon salinity stress, osmotin can increase enough to account for 10% of the total protein in some plants.
3. There is a high similarity between the osmotin gene (which is constitutive) and genes related to pathogen attack such the one coding for as thaumatin in tobacco.
4. It is now thought that even though osmotin is upregulated by salinity it is most likely associated with pathogen defense rather that salinity tolerance.

Genes Associated with the Switch to CAM Photosynthesis

1. Genes associated with CAM photosynthesis, such as malate transporters and decarboxylases, are constitutively transcribed at a low level in *Mesembryanthemum crystallinum.*
2. The transcription of the CAM-specific PEP carboxylase gene is accelerated in mature plants when they are exposed to salinity.
3. The CAM-specific genes are all regulated by the same combination of elements.
4. CAM induction is a response of mature plants to water stress induced by saline conditions, and they all respond to the same growth-regulator-dependent signal transduction pathway.

Summary of Inducible Proteins and the Potential for Biotechnology

1. Salinity is a worldwide problem for agriculture due to the increase in salinized fields and the intolerance of most crops.
2. Many species in naturally saline habitats have evolved complex adaptations to salinity that include morphological, anatomical, behavioral, and physiological mechanisms.
3. The greatest potential for increasing salinity tolerance in crops is by increasing the responsiveness of solute accumulation mechanisms to saline conditions.
4. The two most important gene groups for salinity tolerance are those that regulate solute accumulation and Na^+, H^+, and K^+ homeostasis across the plasma membrane and the tonoplast.
5. Using osmotin or dehydrins in genetic engineering programs is not likely to improve salt tolerance traits in crops.
6. Much more has to be learned about aquaporins and their significance to salt tolerance in halophytes before a decision can be made about the potential of bioengineering these proteins for salt tolerance in crops.

SELF-STUDY QUESTIONS

1. Describe the reasons why the study of salinity has been a prominent aspect of the botanical sciences over the past 20 years.
2. What are the factors that determine the salinity of the soil solution in salt marsh and mangrove communities? How are they the same, and how do they differ?
3. Why would one expect that adaptations to salinity in plants occupying desert salt flats may be different than those on coastal salt bluffs?
4. Why is salinity a problem for agriculture? Under what conditions does agriculture have the greatest problem with salinity? What agricultural practices are best for controlling soil salinity?
5. Explain the differences between the terms *halophyte* and *obligate halophyte*.
6. What are three secondary methods of measuring salinity? Under what scenarios would it be appropriate to use these secondary measurements?
7. What is the difference between saline and sodic soils? Which is more abundant, and what geographical regions would be expected to have either of these soil types?
8. What are the three main environmental factors that cause stress for mangrove species in their natural habitat?
9. Are the enzymes of halophytic plants more tolerant of salt than those of glycophytes? How can you explain this in relationship to tissue salt concentrations in halophytes and in glycophytes?
10. What is the toxicity ranking for ion species in halophytes and in glycophytes?
11. What are the general reasons why high external sodium concentrations can causes cation deficiency?
12. How do the lipid properties (quantity and quality) change in glycophytes when exposed to high external salinity?
13. Glycophytes suffer reduced net photosynthesis per leaf area when exposed to high soil salinity. Can a mechanism for this inhibition be generalized among species? If so, what are the main causes of reduced photosynthesis?
14. Why is there an interaction between light intensity and the severity of salinity toxicity?
15. Salinity induces water stress–related reduction in yield. Why is this so, and what changes occur in glycophytes in response to this salinity-induced water effect?
16. Which ions can serve as a buffer against sodium inhibition of cation accumulation? Why is this so?
17. Why is it likely that reactive oxygen species will accumulate in plants when they experience salinity?

18. Describe the biphasic growth inhibition model. What are the critical aspects of this relationship for plant response to salinity?
19. How effective is salt exclusion as a salinity tolerance mechanism? What transport proteins regulate this process?
20. How do glycophytes discriminate against Na^+ versus K^+? How is this different from halophytes? What is the correlation between salinity tolerance in glycophytes and the ratio Na^+/K^+?
21. What is the relationship between salt sensitivity and the ability to reabsorb sodium or chlorine from the xylem stream before it travels to the leaves?
22. Differentiate between the structure and function of a salt gland and a salt bladder?
23. Why are halophytes with extreme salinity tolerance often succulent? Does succulence constitute a significant adaptation to salinity in glycophytes?
24. Why do the older tissues of halophytes and to some extent of glycophytes have a higher salt content that younger tissues?
25. What is the disadvantage of storing salt in the cell wall?
26. What are the limitations to long-term storage of salt in the central vacuole?
27. Why is it important for glycophytes to have a high relative growth rate when they are experiencing saline conditions?
28. What is the evidence that proline accumulation is an adaptive response to salinity rather than a consequence of stress?
29. Which solutes that accumulate during salinity are ubiquitous stress responses among all organisms, and which are specific to plants?
30. What basic metabolic process is closely tied to the synthesis of polyols? What two advantages can a high polyol concentration confer for plants under saline stress?
31. What is an example quaternary amine that accumulates during salinity stress? Why may this be an important molecule for increasing salt tolerance in crops?
32. What is the importance of the ratio of planar to nonplanar sterols in the plasma membrane to salinity tolerance?
33. What is the difference between the way that phospholipid abundance and composition respond to salinity in halophytes and glycophytes?
34. Why is the activity of H^+-ATPase so important to salinity tolerance?
35. What are the two most critical growth regulators for salinity tolerance? What is the significance of each?
36. What is the signal transduction pathway that determines phase I responses of plants to salinity?
37. What organism has provided the most information about a salinity-specific signal transduction pathway? What factor of ion homeostasis is regulated by this signaling pathway?
38. Describe the induction of H^+-ATPase transport activity in response to salinity, and relate this to the pattern of transcription activity.
39. What other transporter genes are upregulated during salinity, and what is the evidence for their significance to salinity tolerance in plants?
40. What enzymes are required for glycine-betaine synthesis, and how does the activity of transcription of these genes respond to salinity?
41. What biosynthesis pathways regulate the synthesis of proline? How do the genes regulating this pathway respond to salinity?
42. What are aquaporins? How could the upregulation of the genes associated with these proteins constitute an adaptation to salinity stress?
43. Differentiate between dehydrins and osmotin? Are these critical to salinity tolerance?
44. Which CAM-specific genes are responsive to salinity induction? Why is the induction of CAM an important mechanism of salt tolerance in some species?
45. Describe the general requirements for salinity tolerance found in halophytes. How does this differ from salinity tolerance in glycophytes? Do you think it is possible to transfer halophyte salinity tolerance traits to glycophyte crops in order to improve crop salt tolerance?

SUPPLEMENTARY READING

Arbol, I. P., J. S. P Yadav, F. I. Massoud. 1988. Salt-Affected Soils and Their Management (FAO Soils Bullitin, No. 39). FAO of the United Nations, Rome.

Basta, A. S. (ed.). 1994. Stress-Induced Gene Expression in Plants. Harwood Academic Publishers, Chur, Switzerland.

Cherry, J. H. (ed.). 1994. Biochemistry and cellular mechanisms of stress tolerance in plants. Springer-Verlag, Berlin.

Close, T. J., and E. A. Bray (eds.). 1993. Plant responses to cellular dehydration during environmental stress. 16th Annual Riverside Symposium in Plant Physiology. American Society of Plant Physiology, Rockville, MD.

Flowers, T. J. 1985. Physiology of halophytes. Plant and Soil 89:41–56.

Friedman, G. M., and W. E. Krunbein (eds.). 1985. Hypersaline Ecosystems. The Gavish Sabkha (Ecological Studies, No. 53). Springer-Verlag, Berlin.

Goodin, J. R., E. Epstein, C. M. McKell, J. W O'Leary (eds.). 1990. Saline Agriculture: Salt Tolerant Plants for Developing Countries (Panel on Saline Agriculture in Developing Countries). National Academy Press, Washington, DC.

Poljakoff-Mayber, A., and J. Gale. 1975. Plants in Saline Habitats. Springer-Verlag, Berlin.

Pomeroy, L. R., and R. G. Weigert (eds.). The Ecology of a Salt Marsh (Ecological Studies, No. 38). Springer-Verlag, New York.

Reimold, R. J., and W. H. Queen (eds.). 1974. Ecology of Halophytes. Academic Press, New York.

Staples, R. C., and G. H. Toenniessen. 1984. Salinity Tolerance in Plants. John Wiley & Sons, New York.

Tanji, K. K. (ed.). 1990. Agricultural Salinity Assesment and Management (ASCE Manuals and Reports of Engineering Practice, No. 71). American Society of Civil Engineers, New York.

Ungar, I. A. 1991. Ecophysiology of Vascular Halophytes, CRC Press, Boca Raton, FL.

Verma, D. P. S. (ed.). 1993. Control of Plant Gene Expression. CRC Press, Boca Raton, FL.

Wray, J. L. (ed.). 1992. Inducible Plant Proteins. Cambridge University Press, Cambridge, UK.

Yeo, A. R., and T. J. Flowers (ed.). 1994. Soil Mineral Stresses: Approaches to Crop Improvement. Springer-Verlag, Berlin.

PART II
Biotic Factors and Plant Stress Physiology

6 Influence of Plant Phytopathogens on Host Physiology

Outline

Objectives

1. Discuss the various types of plant pathogens and how they gain entry into plant cells and tissues.
2. Summarize what is known regarding the effects of pathogens on photosynthesis and respiration of host cells.
3. Describe how plants defend against pathogen attack with preformed and infection-induced chemicals.
4. Address how plant pathogens affect water relations, mineral nutrition, and translocation in host plants.
5. Establish how abiotic environmental factors influence disease development in host plants.

I. INTRODUCTION

This chapter addresses plant pathogens and their effects on the physiology of plants. Pathogens that influence plant growth and development include bacteria, fungi, viruses, viroids, fastidious bacteria—rickettsia-like organisms (RLOs)—mollicutes (mycoplasma-like [MLOs] organisms), nematodes, parasitic higher plants, and protozoa. The interactions of these organisms with their respective host plants are complex, (Sinclair and Backman, 1989) involving not only biotic interactions but also abiotic factors (Figure 6.1). When one considers the number of potential pathogens, varying degrees of pathogenicity, numerous differences in host susceptibility, and different ways in which host and pathogens respond to changing abiotic factors (light, temperature, water relations, soil conditions, pollutants, pH, and nutrition), the complexity of the interactions becomes enormous. Plant pathogens can attack virtually any part of the plant including leaves, stems, roots, flowers, fruits, and seeds. Age and relative vigor of the plant are also important for pathogen development, as are environmental conditions and various other factors associated with the ability of the pathogen to develop in the plant, such as stage of development, inoculum levels, dissemination factors, among others. In some cases, pathogens are tissue specific, invading only xylem or phloem or other localized tissues. In other instances, pathogens are very host specific, while some exhibit a broader host range.

Plant pests, including diseases, insects, and weeds, account for considerable economic loss annually. Approximate estimates of major crop losses on a global basis, due to these three groups of pests, are about 3319 million tons (1993 estimates) at a cost of about $500 billion. Additionally, it is estimated that $26 billion is invested in pesticides to control plant pests; otherwise, an additional $330 billion in losses would be incurred (Agrios, 1997).

II. PLANT INFECTION BY PHYTOPATHOGENS

A. Pathogen Dissemination

How do pathogens enter and come in contact with plants? Bacteria and fungi can be disseminated through the air by wind or by water splashing on the plant surfaces such as leaves or stems. Frequently, insects transfer bacteria and fungi on their external body parts. This is true for a number of animals that graze or feed on plants and move from one plant to the next. At other times, viruses, viroids, mollicutes, fastidious bacteria, and protozoa are spread by insect vectors that feed on plant parts or ingest plant juices. In some instances, the pathogens may live in the tissues or guts of the insect. Some viruses can also be disseminated by nematodes, mites, and fungal spores. Also, humans can transfer pathogens through pruning, grafting, and cultivation practices using contaminated tools and implements. Soilborne pathogens can come in contact with root surfaces as the roots grow through the soil. Motile forms of bacteria and fungi may also be attracted to roots by chemotaxis or possibly by electrotaxis. In either case, motile bacterial cells or fungal zoospores may be attracted to carbohydrates or amino acids exuded by roots or in the guttation fluids of leaves, which not only serve as attractants but can also be used as nutrients. As much as 25% of the carbon fixed in plants can be exuded into the rhizosphere of the soil. Thus, the rhizosphere is rich in microbial growth, consisting of both phytopathogens and free-living organisms.

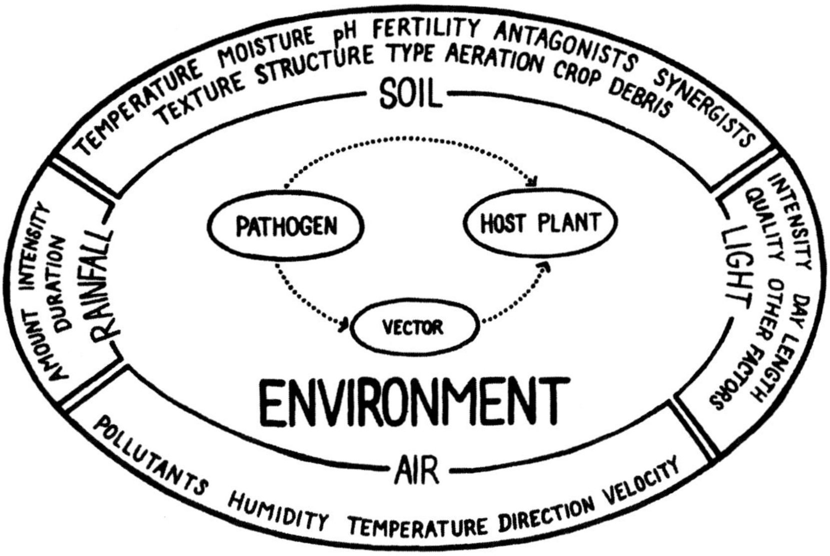

Figure 6.1. Abiotic environmental factors influencing host-pathogen relations in plants. (From Sinclair and Backman 1989, with permission.)

B. Pathogen Entry into Host Tissues

Once plant pathogens have made contact with a susceptible host, they employ various mechanisms to gain entry into host tissues. In the case of bacteria and some fungi, entry can be passive in that the organism gains entry via preexisting wounds (broken trichomes or root hairs, exudative glands, leaf traces, and exit wounds resulting from lateral root penetration) or through natural openings such as stomates, hydathodes, lenticels, and nectaries (Goodman et al., 1986). Figure 6.2 illustrates potential ports of entry represented by stomatal openings, lenticels, leaf traces, and hydathodes. Physiological conditions of the plant, such as water relations, may influence whether a specific infection port may be accessible by pathogens. Water status is known to affect stomatal opening, lenticel diaphragm rupture, leaf abscission, and guttation from hydathodes. In the case of lenticel diaphragm rupture and leaf abscission, the ability to repair damaged tissues quickly is important relative to whether infection occurs. Also, stomata represent important areas of pathogen entry, and may not necessarily have to be open for bacteria to penetrate them and establish infection sites in the stomatal cavities.

C. Growth Requirements of Phytopathogens in Host Tissues

In the case of bacteria, once entry occurs, environmental conditions, intercellularly, must be appropriate for reproduction of the bacterium. Sufficient moisture, nutrients, temperature, and possibly osmotic conditions are required to sustain growth. Once intercellular growth is established, the bacteria secrete a battery of pectolytic and cell wall–degrading enzymes that result in the destruction of cellular and tissue integrity. This, in turn, provides additional nutrients for growth and reproduction of the bacteria. Bacteria generally establish infection extracellularly but, once they invade the xylem, can be transmitted throughout the vascular system and proliferate readily in other plant tissues. Generally, bacteria do not reproduce well in the phloem, probably as a result of high osmotic potentials in the sieve cells.

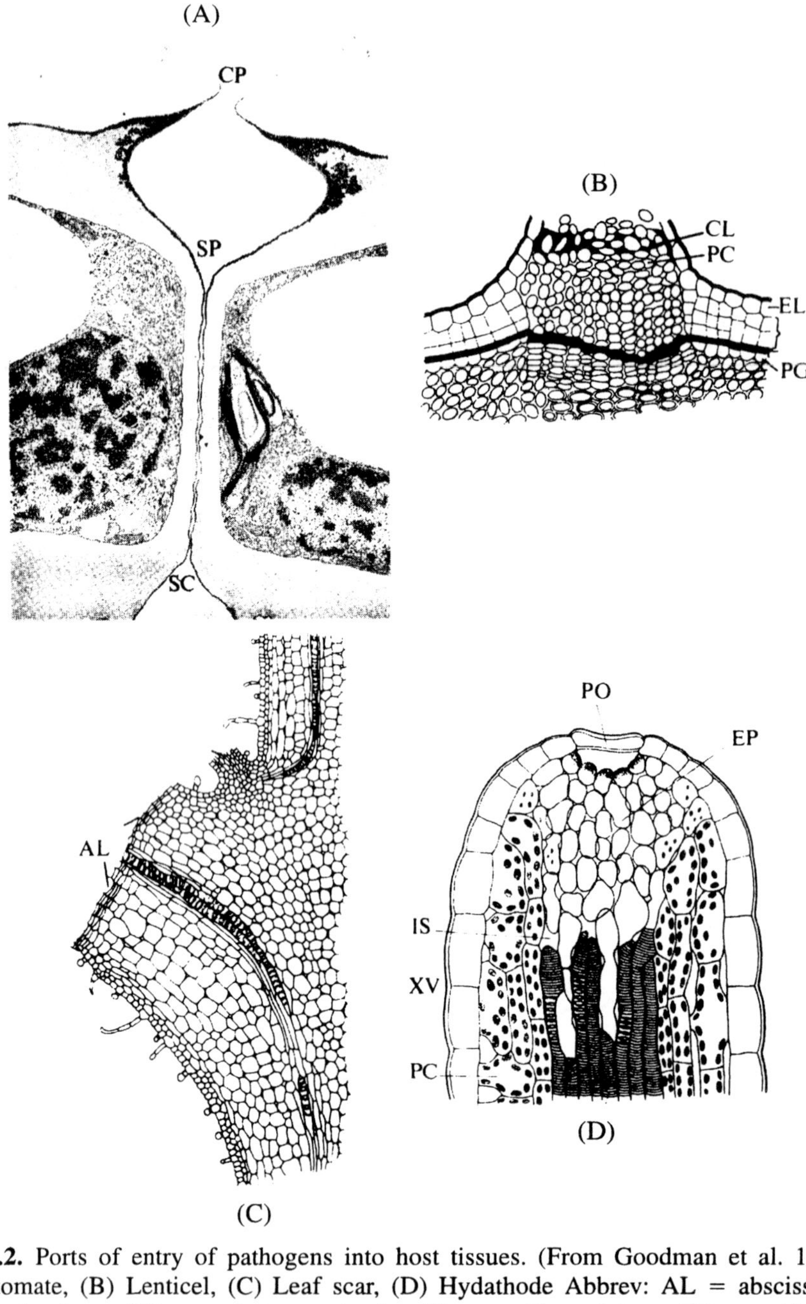

Figure 6.2. Ports of entry of pathogens into host tissues. (From Goodman et al. 1986.) (A) Closed stomate, (B) Lenticel, (C) Leaf scar, (D) Hydathode Abbrev: AL = abscission layer CL = closing layer CP = cuticular projection EL = epidermal layer EP = epithem IS = intercellular space PC = parenchyma cells PG = phellogen PO = pore SC = substomatal chamber SP = stomatal pore XU = xylem vessel

Fungal infection of the host can occur actively at the surface of healthy tissues, passively through natural openings as described before, or as a result of mechanical injury. Generally, the sequences involved with active fungal infection include spore germination, germ tube growth, appressorial formation, and, in some instances, haustorium formation. In some fungi, penetration of compatible host tissues starts with the germination of spores followed by germ tube growth across the surface of a tissue and, in the case of leaves, the formation of an appressorium over a stomate. The growth of the germ tube may be thigmotropic and/or chemically directed toward stomates. Hyphal growth can ensue down into the stomatal cavity with the formation of haustorial connections with specific host cells (Spencer-Phillips and Gray 1981). The haustorium apparently does not invade the host's cytoplasm, but an extrahaustorial membrane (of host origin) is formed that keeps the cytoplasm of the host separate from that of the haustorium. *Neck bands* form a seal between the membrane of the haustorium and the host membrane (extrahaustorial membrane), creating a matrix within a closed compartment that then allows exchange of nutrients from the host cell to the haustorium. It is presumed that the fungal parasite must control influx of nutrients from the host, which in essence creates a type of metabolic sink for the flow of nutrients (Figure 6.3). It appears that the extrahaustorial membrane is either free of ATPase or has limited activity compared to noninvaginated regions of the host cell and also the fungal haustorium. This suggests that the host cell may act as a type of transfer cell in which carbohydrates are loaded actively from the apoplast into the host cell and then transported by ATPase-mediated transport systems in the fungal pathogen. It is assumed that hyphal growth of infected tissues can also facilitate uptake of nutrients from surrounding tissues.

Some fungi may penetrate host tissues through mechanical force by the formation of *pegs* that grow from the base of the appressorium. It is thought that forces generated by the peg, as it grows, result in penetration of the host cell. In some cases, penetration may be facilitated by the production of hydrolytic enzymes that degrade various cellular components (cutinases, pectinases, cellulases) by the pathogen. Hyphae of some fungal species may also penetrate host cells directly by the production of cuticle- and

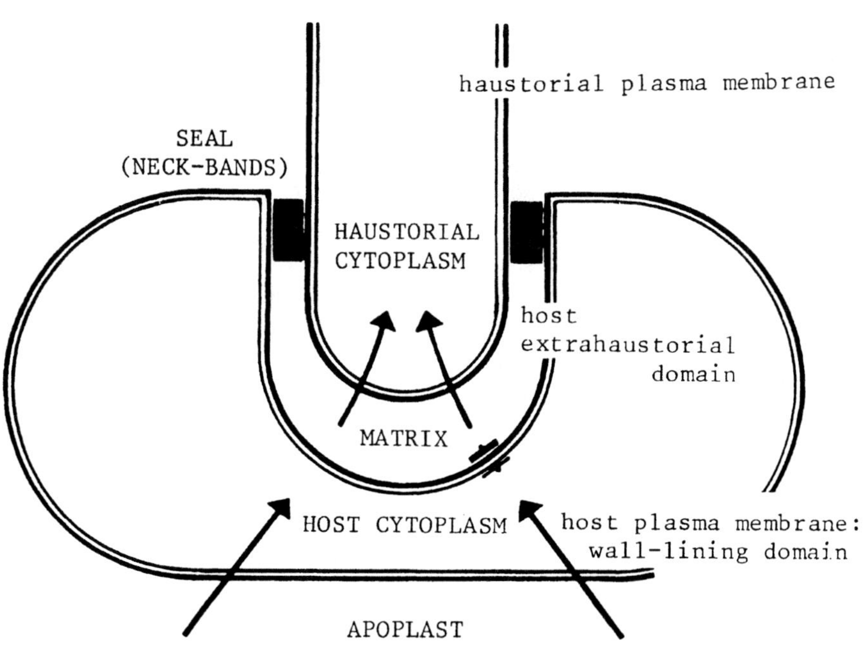

Figure 6.3. Haustorial connection between fungus and host cell. (From Spencer-Phillips and Gay 1981.)

cell wall–degrading enzymes. Other fungi require open wounds to be able to infect plant tissues. Some are quite pathogenic, while others are considered to be weak pathogens.

Viruses, fastidious bacteria, and mollicutes are considered to be *obligate parasites* since they cannot reproduce or grow in the absence of a host. These organisms infect plants via insect vectors and/or in some cases through mechanical wounding of plant tissues. All are considered to be *systemic* in that all enter the plant at a specific point and then invade susceptible tissues intracellularly. Viral insect vectors include aphids, leafhoppers, and whiteflies, while leafhoppers are the primary vectors for mollicutes, fastidious bacteria, and protozoa. Viruses can move systemically throughout the plant intracellularly via plasmodesmata and/or via the xylem and phloem. Some viruses occupy predominant the phloem, while others reproduce in the phloem as well as other tissues. Mollicutes are found exclusively in the phloem and have no cell wall, while fastidious bacteria have cell walls and are divided into two groups, one associated with xylem and the other with phloem.

Nematodes living in the soil come in contact with actively growing roots or are also attracted to root surfaces by exudates and elevated CO_2 levels generated by the respiration of the roots. Nematodes either can feed externally on roots (*ectoparasitic*) by the insertion of their stylets into root tissues or can penetrate plant tissues (*endoparasitic*) and move from cell to cell feeding on cellular contents.

Parasitic angiosperms are propagated by seeds that can be disseminated by wind, birds, and other animals. Those that parasitize crop plants develop considerable seed banks in the soil. Once a seed encounters a suitable host, chemical signals, in some cases emanating from the host root, trigger germination of the parasite, and once the germinating radical makes contact with the host, haustorial connections develop with the host, which ultimately leads to transfer of nutrients to the parasite. Vascular plant parasites will be addressed in Chapter 10.

III. HOST-PARASITE INTERACTIONS: PHYSIOLOGICAL EFFECTS

A. Response of Host Photosynthesis to Phytopathogens

1. Indirect Effects on Photosynthesis. One of the most sensitive physiological parameters of plants, with respect to biotic and abiotic stress, is photosynthesis. Attack by pathogens can have effects on photosynthesis that vary with pathogen, host, environmental conditions, and severity of the disease. Evidence that photosynthesis is affected by pathogens can be seen in infected plants through the development of chlorotic or necrotic tissues, loss of leaves, reductions in chlorophyll and numbers of chloroplasts, and stomatal closure. Not much is known about the effects of plant pathogens on the biochemistry of photosynthesis. Depending on pathogen and growth conditions, phytopathogens can increase, decrease, or have limited effects on photosynthesis.

When pathogens colonize plant tissues, in some cases the organism itself can affect photosynthesis indirectly by blocking out sunlight needed for photosynthesis. This is true of powdery mildews and sooty molds that colonize the surfaces and intercellular spaces of leaf tissues. In addition, the color of fungal pigments may affect leaf temperatures, which in turn can influence rates of enzyme activity associated with photosynthetic activity of leaves. Light-colored fungal hyphae may reflect light and cool leaf surfaces, while dark-colored sooty mold fungi may absorb light energy causing the tissues to increase in temperature. Surface-colonizing organisms can also influence photosynthesis by affecting gas exchange in leaves. Fungal hyphae growing into stomatal cavities can injure cells or block openings, thus preventing the exchange of CO_2, O_2, and water vapor. Other indirect means by which pathogens can affect photosynthesis include altering leaf shape, size, arrangement, and position. Disease can cause rolling of leaves, reduce leaf size, cause epinasty, result in abscission, and promote witches'-brooms. In the last case, leaves are reduced in size, irregularly arranged, and frequently chlorotic.

Necrotrophic pathogens are more detrimental to host plants with respect to photosynthesis than are biotrophic forms. At least, this is true relative to the rate at which photosynthesis might be affected. *Necotrophic* organisms generally result in necrotic tissues developing rapidly, thus affecting the ability of those tissues to photosynthesize. *Biotrophic* pathogens depend more closely on their host for continued supply of nutrients for their existence; thus, they usually affect photosynthesis at slower rates. Some of the more direct effects on photosynthesis involve chlorophyll, which, depending on the pathogen as well as other factors, can result in the degradation of chlorophyll, changes in the ratio of chlorophylls a/b, and inhibition of synthesis. Chloroplast integrity is also compromised by pathogen infection resulting in abnormal rounding of chloroplasts, disintegration of lamellar structure, and unusually high levels of starch accumulation. These functions may be impaired as a result of toxin formation by fungal and bacterial infection or by rendering the host cell incapable of repair or synthesis of proteins required for the formation of new chloroplasts as might occur with viral infection. As a result of toxin production, permeability changes can occur that could result in the decompartmentalization of function within the chloroplast, which subsequently leads to disfunction.

2. Phytopathogen Effects on Photosynthetic Reactions and Carbon Allocation. Photosynthetic activity can also be affected by influencing the Hill reaction, photophosphorylation reactions, and the carbon reduction cycle. Depending on disease severity, these processes can either increase or decrease. Usually, in early stages of disease development, stimulation may occur, which may be a response by the host to compensate for localized reductions in photosynthesis and/or a requirement for increased metabolic activity relative to repair of injured tissues. Later stages of disease result in reduced levels of activity that can cause up to 30% reduction in noncyclic photophosphorylation in the host plant (Magyarosy et al. 1976). Table 6.1 illustrates how powdery mildew affects noncyclic photophosphorylation in sugar beet chloroplasts. Effects on cyclic photophosphorylation appear to be less pronounced. Several enzymes involved in carbon reduction have also been shown to be affected by various pathogens. In the case of a beet virus, ribulose bisphosphate carboxylase/oxygenase

TABLE 6.1 Impact of Powdery Mildew Infection on Noncyclic Photophosphorylation by Isolated Sugar Beet Chloroplasts

Chloroplast Source	ATP Formed (μmol mg^{-1} Chl h^{-1})	$NAD(P)H_2$ Formed (μmol mg^{-1} Chl h^{-1})	Ratio $P/2e^-$
Healthy leaves	109	121	0.90
Infected leaves	80	64	1.25

Source: From Magyarosy et al. 1976. © ASPP. Reprinted with permission.

activity was reduced while viruses infecting cabbage and cowpea caused increases in phosphoenol pyruvate carboxylase (PEPcase) but a decrease in tobacco. Viruses as well as other pathogens can also influence how photoassimilates are partitioned among important organic compounds used in metabolism. It appears that several pathogens cause considerable shifts toward the formation of organic acids and amino acids and away from monophosphates, sugars, and glucose (Bedbrook and Matthews 1973). Table 6.2 shows the effects of turnip yellow mosaic virus–infected Chinese cabbage on the distribution of ^{14}C-labeled CO_2 among organic compounds in diseased versus healthy plants. Note that phosphoglyceric acid, monophosphates, diphosphates, sucrose, and glucose all are reduced in virus-infected cabbage. On the other hand, phosphoenolpyruvic acid, malate, citrate, aspartate and alanine increase. The amino acid alanine does not change. These changes occur in early stages of virus replication and levels return to the levels of healthy plants once replication rates are reduced. The increase in amino acid synthesis facilitates the replication of viral proteins.

B. Response of Host Respiration to Phytopathogens

1. Types of Respiration in Healthy Plants. Respiration is rather complex in higher plants and involves a number of processes. The functions of some of these processes remain somewhat speculative. Adding to the complexity of respiration in plants is that these processes occur in different parts of the plant cell and may function at different times during the life cycle of the plant and can be influenced by external abiotic factors. The most commonly described respiratory process in plants is *dark respiration*

TABLE 6.2 Allocation of ^{14}C Derived from $^{14}CO_2$ to Sugars, Sugar Phosphates, Amino Acids, and Organic Acids in TYMV-Infected and Healthy Chinese Cabbage Leaves after 5 Minutes of Photosynthesis

	Amount of ^{14}C Incorporated (cpm[a])	
Organic Constituent	Healthy	Diseased
Phosphoglyceric acid	420	340
Sugar phosphates:		
Monophosphates:	1422	1240
Diphosphates	192	80
Sugars:		
Sucrose	874	110
Glucose	152	136
Fructose	62	140
Phosphoenolpyruvic acid	5	40
Organic acids:		
Malate	10	100
Citrate	82	422
Amino acids:		
Aspartate	10	66
Alanine	192	528
Glutamine	56	54
Total cpm	3487	3322

[a] Counts per minute.

Source: From Bedbrook and Matthews 1973, with permission.

and involves both glycolysis and Krebs cycle reactions. The enzymes that catalyze these reactions are localized in the cytoplasm and mitochondrion, respectively. Glycolysis occurs under both anaerobic and aerobic conditions and terminates with the production of pyruvate, which is then transported to the mitochondrion where it is further oxidized to CO_2 and water. In the process, oxidative phosphorylation of ADP to form ATP occurs. Krebs cycle reactions will occur only under aerobic conditions. If molecular oxygen is not available, pyruvate produced in glycolysis is converted to ethanol and lactic acid. If anaerobic respiration occurs, considerably less energy is produced in the form of ATP, and growth can be substantially affected. Another interesting component of oxidative phosphorylation, associated with the Krebs cycle, is what is referred to as *alternative respiration* or *cyanide resistant respiration.* In this process, protons are shunted out of the normal electron transport pathway between coenzyme Q and cytochrome b and accepted by molecular oxygen to form water. This process also occurs in the mitochondrion and is coupled with little to no production of ATP, so it is very energy inefficient. It is thought that CN-resistant respiration occurs normally in plants for energy dissipation when excess nutrients are being produced.

Another respiratory process, which is unique to plants is photorespiration. This process occurs in C_3 plants and is thought to be a wasteful process in that when it occurs, oxygen is fixed rather than CO_2, and previously fixed carbon is lost as CO_2 in a series of reactions involving the chloroplast, mitochondrion, and peroxisomes. Consequently, the potential energy production of the plant is reduced considerably. Some hypothesize that under conditions of drought and high sunlight, photorespiration may act to dissipate high levels of O_2 being produced in the light reactions. The production of high levels of O_2 in such situations can lead to another type of "respiration" in the chloroplast in which electrons from NAD(P)H are passed to oxygen via ferredoxin to produce toxic H_2O_2. Thus, photorespiration may act to reduce the production of harmful oxidative substances such as H_2O_2, which could oxidize membrane lipids and affect enzyme function. However, during the process, some modicum of energy production is maintained until the stress condition is relieved.

Another respiratory pathway involved with the utilization of glucose is the *pentose phosphate shunt.* The enzymes involved in this process occur in the cytoplasm, and the pathway basically functions to produce NAD(P)H, which is used in support of the production of important molecules such as nucleic acids, hormones, cell wall materials, among others needed during early plant development and required during repair of injured tissues.

There are other oxidase enzymes (phenol oxidases, ascorbic acid oxidase, glycolic acid oxidase, peroxidase, and catalase) that at one time were thought to act as terminal oxidases, and thus could be considered yet another form of respiration in plants. However, the exact role of such oxidases is uncertain in plants but may be related to the oxidation of NAD(P)H such as occurs in the pentose phosphate shunt. Such oxidases appear to be organized into short nonphosphorylating electron transport chains located in the cytoplasm.

Thus, when one considers the interactions of plant pathogens with host plant respiration, the interpretation of the effects on respiration become difficult and must be done with the understanding of the complexity involved.

2. Respiration in Diseased Plants. Generally, diseased plants exhibit increased respiration. The only exception may be with viral infections in which host respiration increases only slightly and only after systemic infection has become established. The mechanism(s) of how respiration increase occurs and the extent to which it can be attributed to the host or the parasite are difficult to ascertain. Increases in respiration may result from reaction to host tissue injury, decompartmentalization due to membrane degradation, elevated levels of respiratory enzymes, increased availability of enzyme substrates, increased oxidase activity, uncoupling of mitochondrial electron transport from ATP production, and increased demand for ATP and NAD(P)H. Depending on susceptibility of the host, the type of pathogen, and the stage of disease development, it is possible that all or combinations of several of the above responses to infection could occur. The patterns of respiratory change can differ depending on whether the plant is resistant or susceptible to a pathogen (Smedegaard-Petersen and Støllen 1981). For example, resistant and susceptible barley cultivars to infection by *Erysiphe graminis* f. sp. *hordei* (virulent and avirulent races) show differences in respiratory response (Figure 6.4). In the case of the resistant cultivar, respiration increased rapidly the first day after inoculation and then started to decline to near normal levels after the second day. In the susceptible reaction, respiration did not increase until the third day after inoculation, reaching a peak after six days. The resistant response may have been a defense response in which the plant engaged early defensive metabolism mechanisms to ward off infection by the pathogen. Increased respiration reflected the need for additional energy to initiate such defensive mechanisms. In the case of the susceptible reaction, early defensive strategies were not employed resulting in a later rise in respiration that was most likely a result of invasion and destruction of the host tissue. This most likely involves respiration processes that are less efficient in terms of host metabolism.

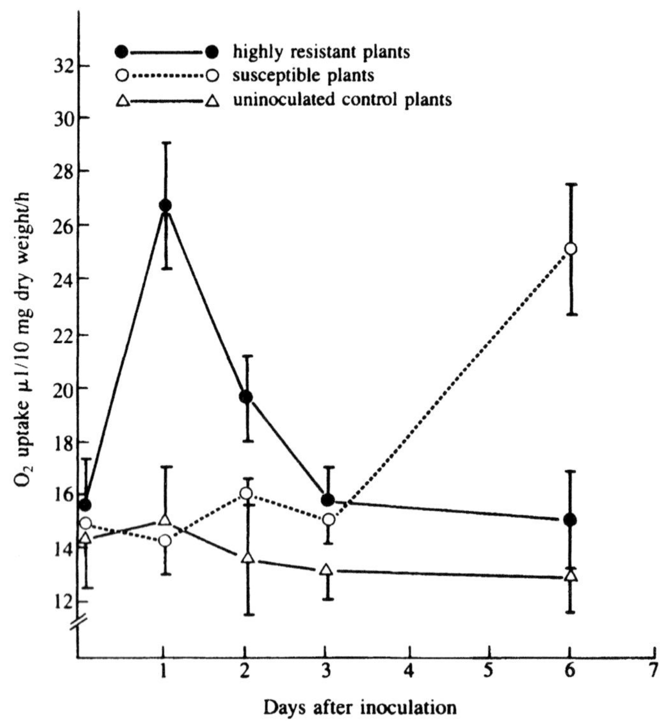

Figure 6.4. Influence of *Erysiphe graminis* f. sp. *hordeii* on respiration in susceptible and resistance barley cultivars. (From Smedegaard-Petersen and Støllen 1981, with permission.)

It has been proposed that almost all aspects of host respiration can be affected by pathogens. In some instances, interference with the Pasteur effect has been implicated, in that anaerobic fermentation processes are initiated that ultimately produce ethanol and less energy for other metabolic processes. Other evidence points to the uncoupling of mitochondrial electron transport from ATP production in the Krebs cycle, although no specific uncouplers have been detected from pathogens. Also, the pentose phosphate pathway (PPP) is activated in infected host plants leading to the production of NADPH and intermediates necessary for the repair of damaged tissues, and also to the production of phenolics thought to be used in the synthesis of chemicals associated with defense mechanisms. However, initiation of the PPP is thought to be a nonspecific stress response in plants and, if it represents the primary respiratory process, results in considerable reduction in the energy requirements of the plant. Although this pathway can be initiated by all pathogens, it appears that, in the case of viral infection, it occurs predominantly in only those that result in the formation of local lesions. Cyanide-resistant respiration appears also to be employed by diseased plants. Again, this alternate mode of respiration is inefficient and probably nonspecific. However, some evidence suggests that it is required for the production of defense compounds such as phytoalexins. Usually, oxidases come into play as electron acceptors in the final stages of infection when cellular degradation and tissue decomposition are occurring. They may also be operative in hypersensitive reactions in which rapid necrosis isolates the pathogen from surrounding healthy tissues. Oxidases may also function in the oxidation of NAD(P)H derived from the PPP. Finally, photorespiration may also be affected by rust fungi, in that the pathogen inhibits glycolic acid oxidase, an enzyme required in the oxidation of glycolic acid, which is a major substrate in the initial reactions of photorespiration. How this may impact the carbon and energy balance in the host plant is uncertain, but it could certainly have some implications for increased carbon fixation efficiency and/or potential for increased oxidative reactions in diseased cells.

C. Plant Defense Against Pathogen Attack

Plants have both preformed and infection-induced factors that help to defend against pathogen infestation. *Preformed factors* have been divided into morphological and preexisting chemical defenses. Evidence for morphological factors in plant resistance is weak but could include such factors as cuticle thickness, chemical composition, and surface wettability. Also, stomatal function, frequency, size, and morphology could impact the ability of pathogens to gain entry into healthy tissues. Cuticle and epidermal hairs may also contain preexisting chemicals that may inhibit pathogen growth and entry into tissues. Also, the nature of cell walls in epidermal cells may be extremely thick or have a composition that is resistant to the penetration of pathogens. The arrangement of cells surrounding vascular tissue—in particular, the presence of sclerenchyma cells that add strength to vascular bundles—also resist invasion of the vascular system by pathogens. Such factors most likely delay pathogen entry rather than provide a level of resistance to the plant.

1. Preformed Chemical Defenses. Preformed chemicals in plants probably do provide some degree of resistance to pathogen infection. In some cases, these chemicals are exuded from plant tissues as exudates that can influence the germination of fungal spores or the growth of bacteria or fungal hyphae. Root exudates specifically support the growth of organisms that could produce antagonistic substances that prevent phytopathogens from invading root tissues. In other cases, chemicals are not exuded but accumulate in high concentrations within plant tissues. Chemicals such as phenolics, tannins, short-chain fatty acids, glycosylated steroids, unsaturated lactones, sulfur-containing compounds (leek and mustard oils), and terpenoids are relatively simple molecules known to inhibit bacteria and fungi. Many of these types of molecules are capable of inhibiting hydrolytic enzymes produced by pathogens and thus of preventing or isolating their growth in the host. Other larger preexisting defensive chemicals produced in plants include proteins that can act as proteinases or cell wall–degrading enzymes such as glucanases or chitinases that can attack and break down pathogen cell walls. Although many of these occur naturally as preexisting chemicals in plants, they may increase considerably if the plant is infected with a pathogen. In may cases, pathogens are specific for a particular host or range of hosts. The ability of the pathogen to recognize a susceptible host is most likely chemically based and depends on whether a plant has specific recognition factors that allow the pathogen to attach and produce the necessary penetration and absorption structures and enzymes required to establish infection. Also, the pathogen may produce toxins that, depending on the susceptibility of the host, may or may not be toxic, and in turn the host may not produce necessary substances required for host survival.

2. Infection-Induced Physical Factors. Several physical factors can be induced at several different levels of organization in plant tissues. Individual cellular responses may suffice to inhibit or destroy invading pathogens or host tissues may respond by forming of thicker, more resistant cell wall material (lignification), which resists enzymatic degradation by the pathogen. Thickened areas called *papillae* are sometimes formed by the host in response to fungal infection beneath the *appressorium,* which is thought to prevent the penetration of the fungal infection peg. Sometimes, fungal hyphae that have penetrated the host cells are surrounded by a cellulosic sheath that becomes infused with phenolics that restrict further invasion of host tissues. Cork layers and abscission zones can also form to wall off diseased tissues or completely remove the diseased part. Also, tyloses formed from outgrowths of parenchyma cells adjacent to xylem vessel elements grow through xylem cell wall pits resulting in blockage of the vessels and restricting the distribution of the pathogen in the vascular tissue. Various gums and resins are produced by plants in response to pathogens, forming impenetrable barriers around the infection site and thus sealing off the organism to nutrients and restricting growth and distribution of fungal spores and propagules.

3. Infection-Induced Chemical Factors. A special type of resistance called the *hypersensitive response* occurs in some plants. It can be induced by obligate and non-obligate fungi, bacteria, viruses, mollicutes, and nematodes. This type of response occurs rapidly in the plant and effectively isolates the pathogen within localized regions of the tissue, resulting in necrotic areas that in some cases fall out of tissues, such as in leaves, giving them the "shot hole" symptomology associated with some diseases. The response is the culmination of a series of biochemical events that include rapid oxidative processes leading to membrane disruption and leakage as a result of the production of oxygen free radicals; the strengthening of cell walls by cross-linking with phenolics; the formation of antimicrobial quinones through the oxidation of phenolics; the production of toxic phytoalexins and cell wall–degrading enzymes such as chitinases, which degrade fungal cell walls; and the formation of signaling molecules such as jasmonic acid, which is formed from the hydroperoxidation of unsaturated fatty acids and is involved in the production of numerous proteins involved in the hypersensitive reaction. The whole process is initiated by pathogen-produced elicitors that trigger a resistance gene in the host that initiates the events that results in the hypersensitive reaction. This interaction between host and pathogen is considered to be an incompatible reaction, in that it is a very rapid response that results in minimal infection and damage to the host. Other chemical responses to pathogen attack include the formation of *pathogenesis-related proteins* (PRs), which include a diversity of proteins that are toxic to invading pathogens and include such proteins such as chitinases, lysozymes, osmotin-like proteins, β-1,3-glucanases, proteinase inhibitors, proteinases, chitosanases, peroxidases as well as a number of other proteins. The PRs occurring in the inter- and intracellular spaces are quite soluble, highly reactive, and induced by signaling compounds such as salicylic acid, jasmonic acid, ethylene, xylanase, and polypeptides.

Phytoalexins are pathogen-induced chemicals typical of incompatible biotrophic reactions in plants and are also produced during the hypersensitive reaction. Most act as toxins to fungi, but some are effective against bacteria and nematodes. Chemically they are a diverse group including terpenoids and isoflavonoids. About 60 different chemicals have been identified to have phytoalexin-like properties. These have been isolated from about 100 species of plants representing 21 families (Friend 1981). Figure 6.5 illustrates several chemical structures of terpenoid and isoflavonoid-type phytoalexins found in plants. Isoflavonoid phytoalexins are typically found in legumes and usually, but not always, induced by fungi. Figure 6.6 illustrates an incompatible reaction between a fungus and a resistant host and shows how the defensive mechanism of the host is triggered by the fungus. Initially, glucomannan elicitors are produced as a result of degradation of the fungal cell wall by glucanases produced by the host plant. The elicitors then bind with membrane receptors in the host membrane cells triggering second messengers that activate the production of new mRNA in the nucleus that then is involved with the synthesis of phytoalexins that are toxic to the fungus (Callow 1983).

Other chemicals induced by pathogens include simple phenolic compounds such as chlorgenic acid, caffeic acid, and ferulic acid, all of which may be inhibitory to pathogens. Also, pathogens may produce glycosidases that can hydrolyze nontoxic phenolic glycosides to the toxic-free phenolic forms. In addition, polyphenol oxidases and peroxidases are activated by pathogens resulting in the oxidation of phenolics to form quinones which are quite effective inhibitors of phytopathogens. Finally, plants may be able to detoxify toxins produced by pathogens either by metabolizing the toxin to a nontoxic chemical or by producing a nontoxic bound conjugate. Host plants respond to pathogen infection using a diversity of options including the formation of toxic peroxidation products, cell wall–hydrolyzing enzymes, toxic phytoalexins, and the formation of mechanical barriers to infection.

D. Water Relations, Translocation, and Nutrition in Diseased Plants

All of the areas under the above heading are related. They all involve movement of water and nutrients from cell to cell and tissue to tissue. The basic differences reflect the distances involved and how transport is achieved. Long-distance transport involves movement in the xylem and/or phloem and involves principles of passive diffusion, facilitated diffusion, and active transport. In the case of xylem, transport is driven by evaporative transpiration, while in the phloem transport is facilitated by bulkflow and active transport. Short-distance transport involves transport between cells via the plas-

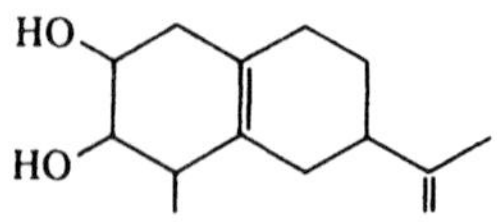

Rishitin (potato)

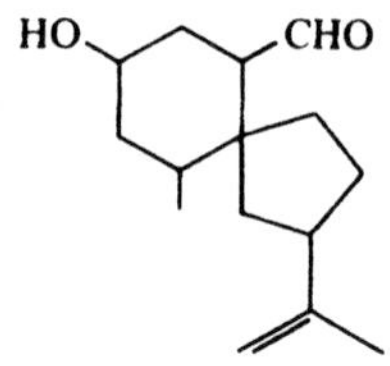

Lubimin (potato)

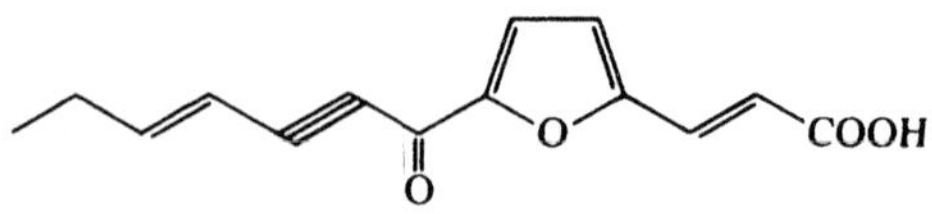

Wyerone Acid (broad bean)

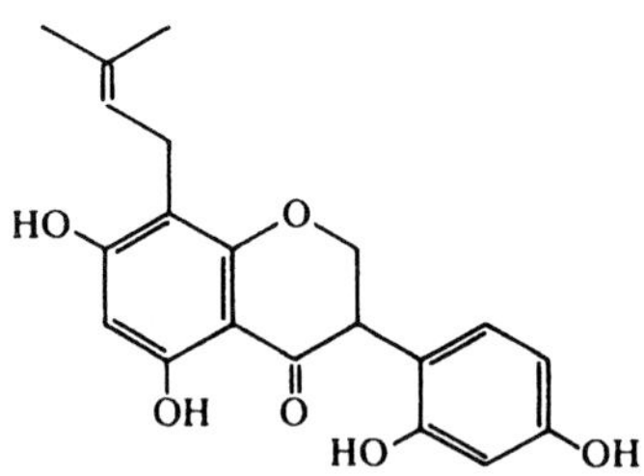

Kievitone (french bean)

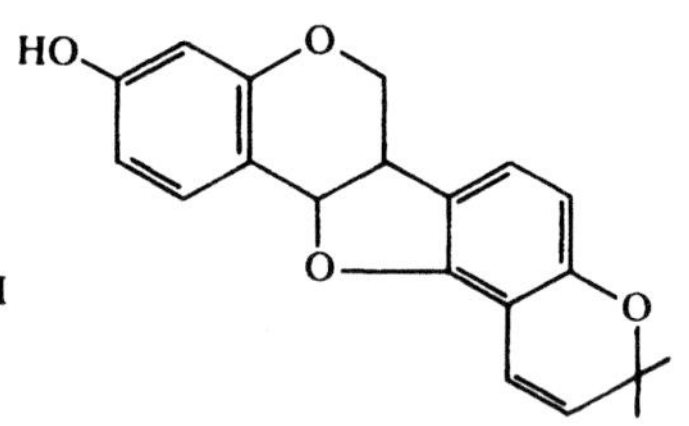

Phaseollin (french bean)

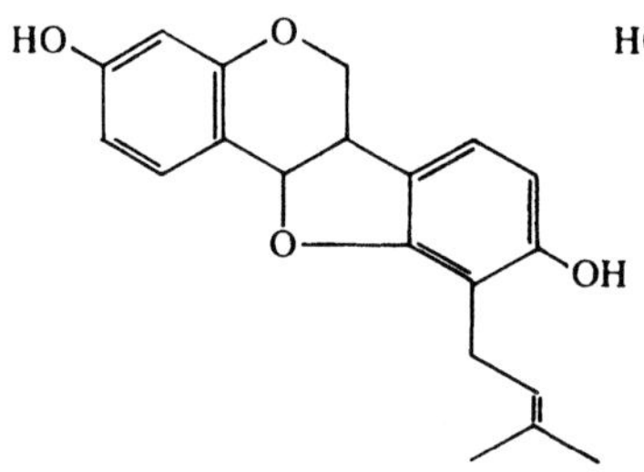

Phaseollidin (french bean)

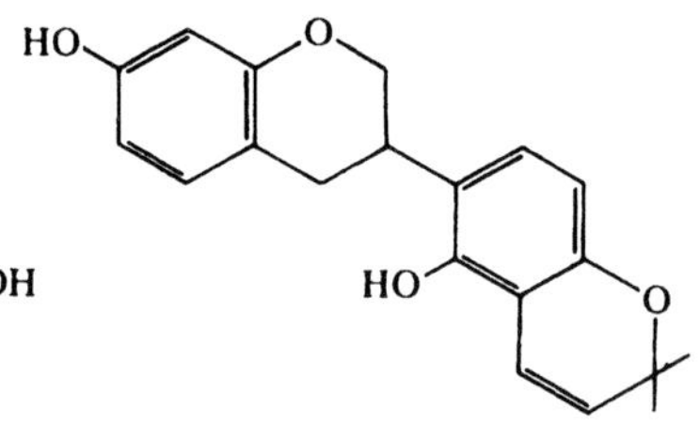

Phaseollinisoflavan (french bean)

Figure 6.5. Examples of terpenoid and isoflavonoid-type phytoalexins found in plants. (From Friend 1981. Copyright John Wiley & Sons Limited. Reproduced with permission.)

modesmata and pit connections or through intercellular spaces (apoplast). The transport system in vascular plants is continuous from the roots, through the shoots and into the leaves. Various resistances are encountered to the movement of water and nutrients through plant tissues. Such resistances are imposed via cell walls, membranes, sizes of conducting cells, cuticle on tissue surfaces, and the number, location, morphology, and degree of opening of stomates. Thus, anything that changes the resistance to the flow and transfer of water and nutrients through vascular tissues and

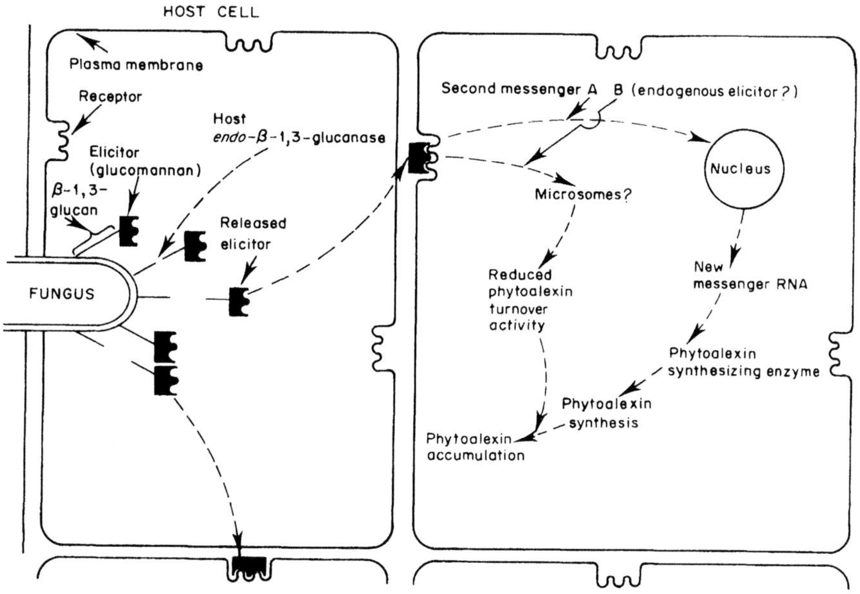

Figure 6.6. Plant defensive responses in an incompatible reaction induced by a fungus. (From Callow 1983.)

from cell to cell can have important ramifications relative to the distribution of water and dissolved nutrients throughout the plant.

1. Nutrient and Water Flow in Diseased Plants. Once pathogens invade plant tissues, resistances to the flow of water and nutrients can be altered in various ways. The way in which pathogens alter water and nutrient movement in plants depends, in part, on what tissues the pathogen invades. In the case of leaf pathogens such as rusts and powdery mildews, water relations and nutrient transport can be affected by alterations in stomatal function, degradation of leaf cuticle, formation of fungal mats on leaf surfaces, and invasion of stomatal cavities and mesophyll regions of the leaf. In addition, such pathogens can alter the leaf surface temperature depending on the color of the fungus (powdery mildews vs. sooty molds), thus affecting rates of transpiration and metabolism in the leaf. Soilborne pathogens that cause root rots may attack root tissues, affecting the ability of the root to take up water and nutrients from the soil. Some viral infections have been observed to cause root hair necrosis, which significantly alters the efficiency of nutrient and water absorption. Other pathogens invade vascular tissues and, by various mechanisms, decrease nutrient and water transport efficiency. Pathogens can occlude or increase resistance to water and nutrient flow in the xylem by various mechanisms. Such mechanisms include pathogen production of high-molecular-weight polysaccharides and glycopeptides, formation of vascular gels by pathogen-produced cell wall–degrading enzymes such as polygalacturonase, pathogen degradation of vascular elements, vascular occlusion by the pathogen itself, and pathogen-induced embolisms in vascular conducting elements. The host may also respond to infections by producing *tyloses* in the vascular tissue, which may restrict flow but may also function to isolate the pathogen and prevent its spread. Fungal spores, bacterial cells, and viruses are readily transported in the transpiration stream of the xylem.

The ultimate result of increased resistance to water and nutrient flow is to cut off vital water and nutrient supplies required for the growth and maintenance of plant tissues. The result is reduced water potentials in plants causing wilting of plant tissues and reductions in essential metabolic processes. This results from reduced uptake of water and nutrients by the roots, increased transpiration in the leaves as a result of stomates not functioning properly, and increased loss of water from tissue surfaces as a result of cuticular wax degradation and/or increased necrotic areas. Figure 6.7 illustrates how transpiration rates change as rust infection progresses in bean leaves. Up to six days after inoculation, infected plant transpiration in the light is lower than the controls, but after day 6, transpiration increases. In the dark, transpiration is the same in both infected and control plants up to about day 6 but dramatically increased in the dark in the infected plants, suggesting stomatal control in the dark has been compromised (Duniway and Durbin 1971).

In many cases, disease symptomology resembles nutrient deficiency symptoms, which frequently is the reason for the development of discoloration and chlorotic symptomology in diseased plants. These problems result, in part, from the inability of plants to reallocate nutrients from older leaves to newly developing regions of the plant as well as from the reallocation of nutrients to the site of infection. In the latter case, biotrophic pathogens create a type of metabolic sink in the region of infection resulting in the flow of nutrients and essential elements to the pathogen. It appears that cytokinin-like substances may be produced either by the pathogen or induced in the host by the pathogen that facilitates the accumulation of nutrients and ions near the infection site.

2. Nutrition and Plant Disease. Several inorganic nutrients have been associated with changes in disease development in plants. It is hard to predict how the nutritional status of a plant can influence disease establishment in a host since both nutrient-sufficient and -deficient plants can be susceptible to pathogens, all be it that different pathogens are involved in causing disease in nutrient-stressed plants compared with nonstressed plants. Nitrogen is one of the most severely limiting nutrient elements in

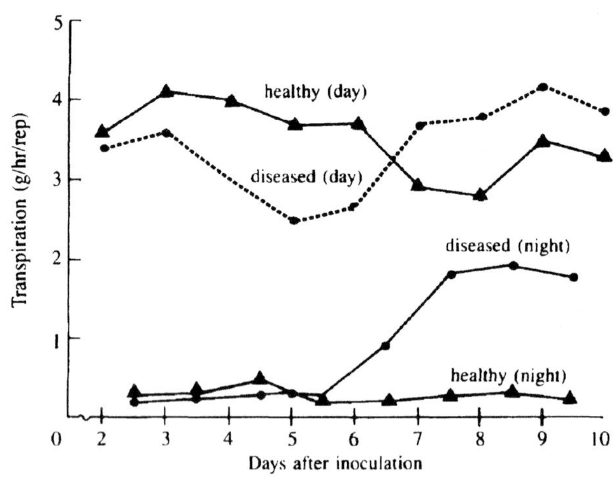

Figure 6.7. Influence of rust infection on the transpiration in bean leaves during the day and at night as a function of post-inoculation time. (From Duniway and Durbin 1971, with permission.)

the environment. Depending on the particular plant and specific pathogen, plants high in nitrogen produce healthy succulent tissues that may be susceptible to a particular group of pathogens, while in the case of N-deficient slow-growing plants, other pathogens may be involved. The form of nitrogen applied to plants (i.e., nitrate or ammonium) seems to play a part in what pathogens infect plants. This probably relates to changes in soil pH in the root rhizosphere associated with application of these fertilizers and the ability of the pathogens to grow and colonize plant tissues.

Calcium is one of the most studied essential elements relative to its effect on disease development in plants. In many instances, treating plants with increased levels of Ca has reduced disease severity or even prevented disease development. Resistance or reduced incidence to various vascular wilts, root rots, and stem pathogens has been reported as a consequence of treating plants with Ca. Several mechanisms may be involved with increased resistance, including chemical changes in pectin, making it more resistant to degradation by pathogen-induced enzymes. Also, Ca functions to maintain the integrity and permeability of cellular membranes and thus could prevent or reduce leakage induced by tissue invading organisms. Calcium may also interfere with the uptake of ions such as Fe and Zn, which appear to be required in the sporulation, growth, and virulence of some vascular wilt pathogens. Calcium, however, does not always increase disease resistance. Some reports suggest it may increase disease severity caused by some viruses and interfere with toxic phenolic synthesis in barley plants causing increased susceptibility to fungal pathogens. Elevated Ca levels have also been observed to cause increased disease incidence in some plants infected with viruses. Calcium may also interfere with the uptake from the soil of toxic ions that may actually be toxic to bacteria and fungi.

Other macronutrients reported to increase resistance to disease include P and K. Potassium has been effective in reducing "take-all disease" in barley and potato scab. However, P increases some viral diseases, which perhaps reflects the different strategies and growth requirements of viruses relative to invading and colonizing host tissues compared with other pathogens. Phosphorus is important in maintaining a proper balance of nutrients within the plant and also increases the maturity rate of the plant. In and of itself, this may be important in changing the window of opportunity for a pathogen to infect a potential host.

Potassium is very important in osmotic adjustment in plants and thus may have a critical role relative to the interaction of a developing pathogen and a host at the site of infection. For nutrients to flow from the host to the pathogen, the pathogen has to maintain a much more negative osmotic potential than the host. High osmotic pressures in host tissues tend to make them more resistant to infection by pathogens than those at low osmotic pressures. Thus, host K levels may be important in regulating nutrient flow between the host and the pathogen, which may be important in controlling infection and the spread of the pathogen. Also, K is important in stomatal function, and since some pathogens enter plant tissues via the stomatal pore and can alter stomatal function, K levels in the plant may affect this important avenue of entry. Potassium can increase freezing tolerance in plants, which could indirectly affect ports of entry for pathogens as a result of frost cracks. Also, K delays the maturation processes in plants, which could influence when and what pathogens may opportunistically invade a host.

The availability of Fe, an essential micronutrient in plants, also appears to influence disease development in some plants. Studies using fungal conidia from different species cultured with or without Fe have shown that those grown in Fe-deficient environments are more aggressive pathogens than those grown in the presence of Fe. Also, the presence of Fe induce conidia germination and appressoria formation, while deficiencies reduce this response. More extensive lesion development is frequently observed in plants inoculated with conidia that were cultured on Fe-deficient media. It has been proposed that Fe associated with conidia may have some role in the initiation of the resistance response in noncompatible reactions between host and pathogen. It has further been proposed that Fe is required to initiate toxin (phytoalexin) production in defense mechanisms associated with the host plant.

The essential micronutrients Cu and Zn can increase resistance in plants to pathogens. Copper appears to stimulate the production of peroxides, phenolics, and quinones that are all highly toxic to bacteria, while Zn has been shown to reduce the incidence of root disease in rice and wilt disease in cotton when applied as a root dip prior to transplant or as a soil amendment, respectively. Zinc has also been reported to be essential in some fungi for growth, sporulation, and virulence as well as the formation of pathogen-produced toxins. In some instances, the use of heavy metals may be directly toxic to the pathogen by interfering with enzyme function. Some Cu and Zn formulations are effective fungicides for the control of fungal diseases.

Silicon, although not considered to be an essential element for most plants, could have an essential role in some grasses relative to support tissues. It is essential for cell wall production in single-celled diatoms. Silicon has been shown to increase disease resistance to rice blast and rice brown spot in field-grown rice and also to root rots and mildews associated with some greenhouse crops such as cucumber. In some instances, plants respond to infection by translocating Si to the site of infection and perhaps forming organosilicon constituents in cell walls that are highly resistant to attack by pathogen-hydrolyzing enzymes.

Nutrition is very important to plants relative to plant health and vigor. Chapter 3 dealt with the physiological implications of elemental nutrition in plants. Complex interactions between host and pathogens disturb normal nutrient sinks in plants, the mechanisms of which are not evident at this time. However, it would appear that much could be learned regarding the resistance and establishment of disease in plants if more were known about the nutritional kinetics of host-pathogen interactions.

IV. ABIOTIC ENVIRONMENTAL FACTORS AFFECTING DISEASE DEVELOPMENT IN PLANTS

When considering host-parasite relations, one must consider three factors: a susceptible host, a pathogen that can infect the host, and the way the environment affects the host and the pathogen. Several abiotic factors can influence pathogen growth and disease development in plants. Such factors as temperature, moisture, wind, soil pH, light, soil nutrients, air pollutants, and pesticide usage can all influence pathogen and disease development. Temperature and moisture are two of the most important factors, however. Disease problems are prevalent in locations or times of the year when temperatures are warm and humidity is high. Frequently, pathogens will infect hosts when they are growing at less than optimum temperatures assuming those temperatures are adequate for growth of the pathogen. A single pathogen may even have different disease development optima (temperature) on different hosts. Figure 6.8 shows how the fungal pathogen *Gibberella zeae* affects seedling mortality in corn and wheat at differing soil temperatures. At a soil temperature of 8°C, disease development is greater in corn, while at 25°C, seedling mortality is higher in wheat. Interestingly,

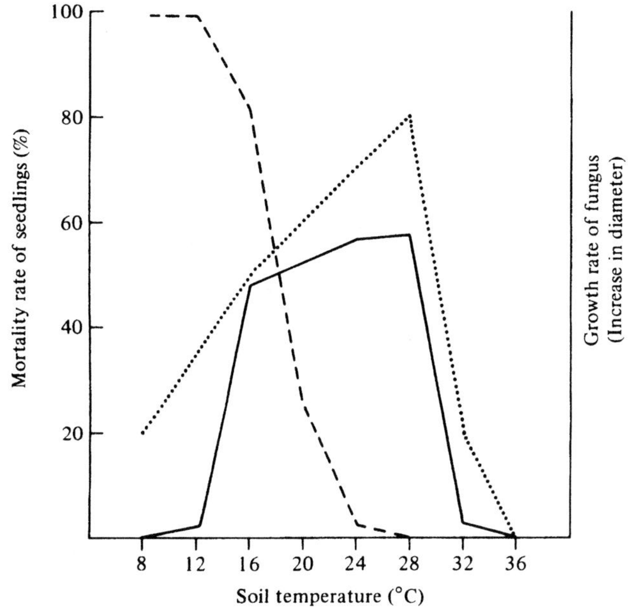

Figure 6.8. Influence of soil temperature on growth of *Gibberella zeae* (*dotted line*) and subsequent influence on seedling mortality of corn (*dashed line*) and wheat (*solid line*). (From Manners 1993, as developed from data by Dickson 1923.)

both hosts are susceptible at both temperatures when plants are subjected to drought stress (Manners 1993 and Dickson 1923).

Moisture is required for sporulation and germination of fungal spores. Also, water is required for the movement of motile zoospores and bacterial cells. Thus, high humidity resulting from rain, overhead irrigation, dew, fog, mist, or guttation generally will facilitate pathogen dissemination and growth. Some fungi, such as the powdery mildews, prefer dryer conditions for spore germination, and thus are more prevalent in dryer regions or during dry seasons of the year.

Light can also affect pathogen growth and possibly disease development in the host. In some fungi, sporulation and spore germination have a light requirement. Whether light directly or indirectly affects disease development in the host by affecting metabolic processes is debatable. In the case of viral infection, plants maintained in shade or low light environments appear to be more susceptible to viral infection and reproduction. Ultraviolet radiation may also have an impact on spore and cell survival, particularly over long distances. This may help explain why rust fungi appear not to traverse the Indian and Atlantic Oceans via wind currents but can traverse short distances across the North Sea and English Channel. Estimates based on laboratory studies suggest that fungal spores traveling 3000 m on a sunny June day in Britain will be rendered nonviable in seven days.

Wind can impact liberation and dispersal of spores and bacterial cells, as well as create infection ports in plants by causing branches and stems to rub against one another causing abrasions. Movements of air masses at upper altitudes represent important mechanisms for the spread of diseases. The ability of meteorologists to track such air masses has important implications relative to the study of the epidemiology of plant disease.

Soil pH can affect soil pathogen growth by affecting the availability of nutrients. Extremes in pH may either limit or result in toxic levels of nutrients in soils. Also, pH can affect the growth of other microorganisms that may be antagonistic to the growth of pathogens or be in competition for nutrients. In addition, pH can impact the host, directly affecting its overall vigor and ability to extract nutrients from the soil.

The effects of air pollution on host-pathogen interactions are very complex, in that pollutants can influence the pathogen directly, indirectly influence a potential pathogen by altering interactions with other microbes as might occur in a soil environment, or indirectly affect the pathogen by altering the host physiology. Examples of pollutants that could result in such interactions include CO_2, SO_2, O_3, NO_x, fluorides, dust, smoke, acid rain, trace metals, as well as others. Reports are mixed as to how pollutants affect pathogens, and range from having no effect, to increasing or decreasing the growth of the pathogens. It is difficult to separate the direct effects of pollutants from indirect effects that may support growth as a function of altering the host metabolism. An example of the latter can be seen with CO_2 and how it influences the germination, infection, and growth of powdery mildew (*Erysiphe graminis*) in barley (Hibberd et al. 1996). In this example, elevated CO_2 concentrations (700 ppm vs. 350 ppm) resulted in no difference in conidia germination, reduced colony formation in barley leaves, and increased growth of established colonies. The inability to establish new colonies was attributed to increased photosynthesis under elevated CO_2 levels and the ability of the plant to establish more efficiently papillae and silicon around the sites of appressorial penetration, thus preventing penetration of the organism into host tissues. Figure 6.9 illustrates the proposed mechanism for how the host and pathogen interact in elevated CO_2 environments. Interpretations as to how a pollutant impacts a pathogen must be evaluated on the basis of several aspects of the life cycle of the parasite as well as the host in order to accurately evaluate the impact of the pollutant.

Increasing atmospheric CO_2 levels may also influence soilborne pathogens indirectly by increasing photosynthesis in plants and subsequent exudation by the roots into the soil. This could have either beneficial or detrimental effects depending on the microbial composition of the soil as well as other environmental factors, in that pathogens may be increased or reduced as a consequence of competition for nutrients and/or possible antagonisms or synergisms between microbes.

It is likely that air pollutants most directly impact pathogens that infect leaves or aboveground organs, although heavy-metal pollution and acid rain could impact soilborne pathogens. Airborne particulates such as fluorides or heavy metals, which are frequently deposited directly on leaf surfaces, probably have little effect on pathogens unless the particulates are of high enough concentration, as would be experienced near a point source. In regions where gaseous air pollutants are high, several reports of both reduced and increased incidence of disease have been reported for several host-pathogen interactions. A higher incidence of disease most likely reflects the predisposition of the host as a consequence of the pollutant resulting in reduced vigor.

Herbicides and insecticides may increase or decrease disease development either directly by affecting the pathogen or indirectly by influencing the metabolism of the host plant. Such chemicals could also alter the soil microclimate or cause elimination or selection of alternate hosts for pathogens. More on this subject is covered in Chapter 11.

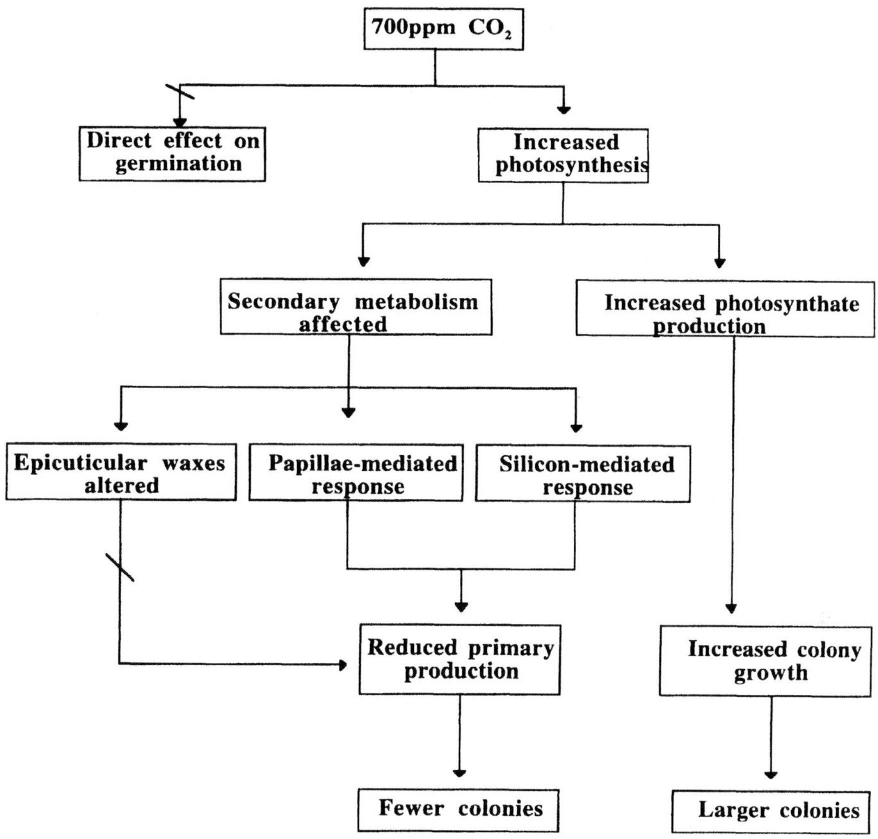

Figure 6.9. Proposed model for the interaction of barley with *Erysiphe graminis* at elevated CO_2 levels. Oblique slashes indicate the process is inhibited. (From Hubberd et al. 1996, with permission.)

V. SUMMARY

Considering the variety of plant pathogens and the range of plant hosts, it is not surprising that both hosts and pathogens have developed numerous offensive and defensive strategies to deal with the complex interactions that can occur. Morphological, biochemical, physiological, and environmental factors that result from, or influence, the interaction of the host and the parasite must be considered. Most studies regarding plant-pathogen interactions have dealt with the genetics of host-parasite interactions and mechanisms of disease resistance. Surprisingly little has been done to examine the effects of disease on the physiology of the host per se. It would seem that much could be learned from more extensive studies on the interactions of host and pathogen with respect to plant water relations, nutrition, carbon allocation, and general metabolic functions. Equally of concern is that little information is available concerning the disease and pathogens of native plants and plant communities. Physiological interactions are practically nonexistent. Both of these observations probably reflect economic and production considerations relative to the need for increased yields in important agricultural crops.

In view of the increasing levels of air pollutants, the potential for global warming, the predicted changes in rainfall patterns associated with global warming, and the increased levels of heavy metals and pesticides in the environment, it would seem important to have a better understanding of how these environmental changes will impact the physiology of plant-parasite interactions in both "natural" and agricultural ecosystems.

STUDY-REVIEW OUTLINE

Introduction

1. Plant pathogens include bacteria, viruses, viroids, fastidious bacteria, mollicutes, nematodes, protozoa, and parasitic higher plants.
2. Numerous factors affect the ability of a pathogen to infect a plant tissue, including plant age, tissue type, environmental factors, pathogen type, inoculum levels, and methods of pathogen dissemination.
3. Diseases, insects, and weeds cost growers and consumers billions of dollars per year in lost production.

Plant Infection by Phytopathogens

Pathogen Dissemination

1. Pathogens can be disseminated by grazing animals, insects, humans, wind, rain, growth of plant roots through the soil, and motile spores and cells.

Pathogen Entry into Host Tissues

1. Pathogens can gain entry into plants via stomates, wounds, lenticels, hydathodes, broken trichomes and root hairs, leaf traces, and lateral root outgrowths.

Growth Requirements of Plant Pathogens in Host Tissues

1. Generally, bacteria must establish infection sites intercellularly where suitable moisture, nutrients, temperature, and osmotic relations will more readily support growth.
2. Fungi can establish themselves on the surface of healthy tissues, enter passively through natural openings or wounded tissues, enter actively as a result of the production of cuticle and cell wall–degrading enzymes, or gain entry through the penetration of tissues by structures called pegs that develop from the base of the appressorium.
3. Some fungi form haustorial connections with host cells through which they then extract nutrients from the host by establishing a strong metabolic sink.
4. Viruses, fastidious bacteria, and mollicutes are all obligate, systemic parasites and enter plants via insect vectors or through wounds.
5. Nematodes are either endo- or ectoparasitic, actively feeding on plant roots by utilizing specialized feeding structures called stylets.
6. Parasitic angiosperms are disseminated by insects and birds via seeds that, in some cases, germinate in response to host signals and invade host tissues by establishing haustorial connections.

Host-Parasite Interactions: Physiological Effects

Response of Host Photosynthesis to Phytopathogens

Indirect Effects on Photosynthesis

1. Depending on host and pathogen, photosynthesis can increase, decrease, or not be affected.
2. Pathogens can indirectly affect photosynthesis by blocking light, altering leaf surface temperature, induce abscission, affect stomatal gas exchange, and alter leaf size, shape, arrangement, and position.

Pathogen Effects on Photosynthetic Reactions and Carbon Allocation

1. Pathogens can affect photosynthesis directly by altering chloroplast integrity, reducing chlorophyll concentration, and inhibiting the Hill reaction, carbon reduction, and noncyclic photophosphorylation.
2. Carbon allocation can be influenced by many pathogens and can cause shifts in carbohydrate, organic acid, and amino acid composition.

Response of Host Respiration to Phytopathogens

Types of Respiration in Healthy Plants

1. Healthy plants respire by several methods including dark respiration, photorespiration, CN-resistant respiration, the pentose phosphate shunt, and possibly the utilization of oxidative enzymes.

Respiration in Diseased Plants

1. Generally, respiration increases in diseased plants with the possible exception of viral diseases.
2. The mechanisms of how respiration is increased most likely varies with host and pathogen.
3. Proposed mechanisms of increased respiration in diseased plants include membrane degradation and decompartmentalization, increased levels of respiratory enzymes, increased availability of enzyme substrates, increased oxidase activity, uncoupling of mitochondrial electron transport from ATP production, and increased demand for ATP and NAD(P)H.
4. Diseased plants frequently exhibit reductions in dark respiration and increases in the pentose phosphate shunt, CN-resistant respiration, and oxidase enzyme activity.

Plant Defense Against Pathogen Attack

Preformed Chemical Defenses

1. Chemicals such as phenolics, tannins, short-chain fatty acids, glycosylated steroids, unsaturated lactones, sulfur-containing compounds, and terpenoids are known to be effective inhibitors of fungal and bacterial growth.

Infection-Induced Physical Factors

1. Some physical factors that can influence pathogen establishment in plant tissues include the development of lignified cell walls; the formation of papillae that resist the penetration of fungal pegs; the formation of cellulose sheaths impregnated with toxic phenolics that surround fungal hyphae; the formation of tyloses in xylem elements, which restricts the movement of pathogens; and the production of gums and resins that seal off the pathogen from nutrients needed for growth.

Infection-Induced Chemical Factors

1. Resistant species of plants may respond to pathogens by engaging the *hypersensitive reaction,* which is a series of rapid chemical responses induced in the plant by the pathogen that effectively isolates the pathogen from healthy tissues and produces substances that will inhibit or kill the invading organism.
2. Phytoalexins are a diverse group of chemicals consisting of terpenoids, flavonoids, and isoflavonoids that can be induced in the host by the pathogen and can inhibit the growth or kill the pathogen.
3. Other chemical responses of plants to pathogen attack include the formation of pathogenesis- related proteins (i.e., chitinases, lysozymes, osmotin-like proteins, glucanases, proteinase inhibitors, proteinases, chitosanases, and peroxidases).
4. Both host and pathogens can detoxify various chemicals by directly metabolizing such molecules or by forming nontoxic conjugates.

Water Relations, Translocation, and Nutrition in Diseased Plants

Nutrient and Water Flow in Diseased Plants

1. Leaf pathogens can alter water relations in plants by interfering with stomatal function, altering leaf cuticle and wax structure and composition, and forming fungal mats on leaf surfaces that affect leaf temperature and gas exchange.
2. Soilborne pathogens will interfere with nutrient and water uptake by disrupting membrane function and the capacity of roots to absorb water.
3. Other pathogens invade vascular tissues resulting in the physical occlusion of the transport cells, the production of polysaccharides or glycopeptides that block vessel elements, the formation of degradative enzymes that decompose the vascular tissues, and the formation of tyloses that help to restrict the movement of the pathogen but also restrict the movement of water and nutrients.

Nutrition and Plant Disease

1. Several elements that are essential for plant growth and development have been shown to increase resistance to plant disease, and include: Ca, K, P, Fe, Cu, Zn, and Si.
2. Mechanisms for increased resistance are for the most part uncertain, but could include factors related to the involvement of specific elements with membrane integrity (Ca), osmotic relations (K), nutrient balance and rate of maturation (P), phytoalexin synthesis (Fe), other toxin production (Zn and Cu), and development of enzyme-resistant host cell walls (Si).

Abiotic Environmental Factors Affecting Disease Development in Plants

1. Plant disease is most frequently associated with moist warm climates, which are conducive to the spore germination and growth of pathogens.
2. Little is known concerning the effects of light on disease development in plants, except that light can influence spore germination and survival and that plants kept under low light conditions are more susceptible to some viral diseases.
3. Wind can affect the liberation and dispersal of fungal spores and bacterial cells.
4. Changes in soil pH can impact the availability of nutrients (toxic and deficient levels) for the growth of soilborne pathogens as well as affect competition with other nonpathogenic microbes.
5. Air pollutants result in direct and indirect effects on both the host and the pathogen, causing difficulty in determining how air pollutants affect host-parasite relations.
6. The effects of pollutants such as CO_2, SO_2, O_3, NO_x, fluorides, dust, smoke, acid rain, and trace metals on host-pathogen interactions range from having no effect to increasing or decreasing disease development and severity.
7. Increasing atmospheric CO_2 levels could influence disease development through the production of increased sugars and other nutrients, which could impact the growth of leaf and soilborne pathogens.
8. Pesticide application can influence disease development, either by directly affecting the pathogen or by altering the metabolism of the host to increase or decrease disease development.

SELF-STUDY QUESTIONS

1. Name the various groups of plant pathogens that affect plant growth and development.
2. Discuss interactive factors that can influence disease development in plants.
3. What three groups of plant pests contribute most significantly to global reductions in crop yields?
4. Describe various means and mechanisms by which plant pests can be disseminated.
5. Discuss various ways in which pathogens can enter plant tissues.

6. Compare bacteria, fungi, and viruses relative to how they gain entry into plant tissues and their respective requirements for growth.
7. Discuss fungi relative to haustorium development and function.
8. Discuss pathogens relative to the types of tissues and cells they may colonize.
9. Indicate how pathogens can influence directly and indirectly photosynthetic responses in host plants.
10. Describe the various types of respiration that occurs in health plants.
11. Indicate how plant pathogens can influence the types of respiration described in Question 10.
12. What types of defense mechanisms are employed by host plants to defend against phytopathogens?
13. What are some preformed and pathogen-induced chemicals found in host plants that are used to defend against pathogen invasion?
14. Discuss various ways that water and nutrient flow can be altered by phytopathogens.
15. Name some plant nutrients (ionic elements) that can influence disease development and indicate the mechanism(s) by which they may affect resistance or susceptibility to pathogens.
16. Outline the various abiotic environmental factors that can influence disease development in plants, and indicate how each may influence the formation of disease.
17. Indicate how increases in atmospheric CO_2 levels could ultimately affect host-parasite interactions.

SUPPLEMENTARY READING

Agrios, G. N. 1997. Plant Pathology, Academic Press, New York.

Ayres, P. G. 1991. Growth responses induced by pathogens and other stresses. Pp. 227–246 in: H. A. Mooney, W. E. Winner, and E. J. Pell (eds.), Response of Plants to Multiple Stresses. Academic Press, New York.

Fowden, L. T. Mansfield, and J. Stoddart (eds.). 1993. Plant Adaptation to Environmental Stress. Chapman & Hall, New York.

Fritig, B., and M. Legrand (eds.). 1993. Mechanisms of Plant Defense Responses, Kluwer Academic Publishers, Boston, MA.

Heagle, A. S. 1982. Interactions between air pollutants and parasitic plant diseases. Pp. 333–348 in: M. H. Unsworth and D. P. Ormrod (eds.), Effects of Gaseous Air Pollution in Agriculture and Horticulture. Butterworth Scientific, Boston, MA.

Kohmoto, K., U. S. Singh, and R. P. Singh (eds.). 1995. Pathogenesis and Host Specificity in Plant Diseases: Histopathological, Biochemical, Genetic and Molecular Basis. Pergamon Press, New York.

Mills, D., H. Kunoh, N. T. Keen, and S. Mayana (eds.). 1996. Molecular Aspects of Pathogenicity and Resistance: Requirement for Signal Transduction. APS Press, St. Paul, MN.

Misaghi, I. J. 1982. Physiology and Biochemistry of Plant-Pathogen Interactions. Plenum Press, New York.

Patil, S. S., S. Ouchi, D. Mills, and C. Vance (eds.). 1991. Molecular Strategies of Pathogens and Host Plants. Springer-Verlag, New York.

Smith, W. H. 1990. Air Pollution and Forests: Interaction Between Air Contaminants and Forest Ecosystems, Springer-Verlag. New York.

Vidhyasekaran, P. 1988. Physiology of Disease Resistance in Plants, Vol. I. CRC Press, Boca Raton, FL.

Vidhyasekaran, P. 1988. Physiology of Disease Resistance in Plants, Vol. II. CRC Press, Boca Raton, FL.

Walters, D. R., J. D. Scholes, R. J. Bryson, N. D. Paul, and N. McRoberts (eds.). 1995. Physiological Responses of Plants to Pathogens. Association of Applied Biologists, Wellesbourne, Warwick UK.

7 Herbivory and Plant Stress

Outline

I. Introduction
  A. Abundance and Diversity of Herbivores
  B. Potential Influences of Herbivores on Plants
  C. Influence of Herbivory at an Ecological Level
  D. Evolutionary Relationships
  E. Applied Significances of Plant-Herbivore Interactions
II. Mechanisms of Adaptation Against Herbivory
  A. Nonphysiological and Indirect Defenses
    1. Morphological and Behavioral Defenses
    2. Compensatory Growth
    3. Indirect Plant Defenses
  B. Direct Physiological/Chemical Defenses
    1. Tannins and Lignins
    2. Phenolics and Resin
    3. Cyanogenic Glycosides
    4. Nonprotein Amino Acids
    5. Alkaloids
    6. Hormones and Antihormones
    7. Proteinase Inhibitors
  C. Physiological Basis for the Induction of Defensive Chemicals
III. Costs of Defense for the Plant
  A. Description of Costs
  B. Resource Allocation
  C. Autotoxicity and Biochemical Disruption
IV. Importance of Quantitative Variation
  A. Heterogeneity of Host Tissue Quality
  B. Effects on Insect Populations
V. Environmental Stress and Herbivory
  A. Insect Preference for Specific Host Plants
  B. Host Suitability and Insect Performance
  C. Herbivore Influence on Subsequent Abiotic Stress Tolerance
VI. Summary
  Study-Review Outline
  Self-Study Questions
  Supplementary Reading

Objectives

1. Describe the general diversity of herbivores as a ubiquitous stress on plants.
2. Explain in a broad sense the various impacts that herbivores can have on plants at the individual, population, and community levels.

3. Explain some of the morphological, anatomical, behavioral, and physiological defenses plants utilize to cope with herbivory stress.
4. Focus on physiological defenses by describing the diversity of defensive compounds, their compartmentation, and their efficacy.
5. Explicate the importance of quantitative variation to the plant-herbivore interaction, including the induction of defensive mechanisms.
6. Discuss resource costs for the defense of tissues against herbivores. In particular, consider the cost-benefit relationships between defense and growth rate.
7. Describe how environmental stresses influence herbivory. Include the influence of stress on plant susceptibility, attractiveness, and defense.
8. Consider the influence that herbivory has on plant response to subsequent environmental stress.

I. INTRODUCTION

The fact that plants are exposed to intense herbivory at all times during their life cycle is dogma in plant sciences. The intensity of the herbivory and the diversity of herbivores may change among species, seasons, and locations, but herbivory per se is a constant pressure on plant performance in the short term, and on population fitness and plant evolution in the long term. Due to this continuous selective pressure on plant evolution over the history of the plant kingdom, it is not surprising that plants have evolved a complex array of defensive mechanisms, and that plant structure, behavior, and physiology have been partly sculpted by the coevolution between herbivores and plants. The purpose of this chapter is to examine some of the interactions between herbivores and plants, and consider in particular the influence of insect herbivory on the physiological performances of plants and subsequent physiological responses to environmental stress.

The interaction between plants and herbivores has been churning for millions of years. However, a general theory about plant-herbivore interaction extends back only about 35 years, when research on this topic exploded in part due to the work of Ehrlich and Raven (1964) on butterflies and their use of specific host plant species. Since then, the literature on plant-herbivore interactions has increased in a logarithmic manner, resulting in a cornucopia of examples and theoretical concepts. This is not surprising given the inherent potential for scientific inquiry about plant-animal interaction. The short life span of insect herbivores provides a model system for the study of coevolution. Successful crop production is dependent upon managing the relationship between crops and phytophagous insects. Furthermore, there is an intricate web of interactions between environment, plant secondary chemistry, and insect population regulation that is tailor made for imaginative ecological and physiological research. In fact, this field of research has been referred to as a scientist's "candy shop." Compilation of the resulting diversity of examples and concepts into one umbrella theory has been difficult.

Recently, a list of important "messages" about plant–insect herbivore interaction has been compiled (Stamp, 1996). These messages relate to the interaction between plant and insect herbivore at an ecological scale (Table 7.1). A similar set of messages

TABLE 7.1 A Short Version of the 12 Messages of Plant–Insect Herbivore Interactions at an Ecological Scale

The Plant's ability to defend itself by physiological means:
1. Plants have a dynamic array of chemical defenses.
2. Chemical defenses can have a metabolic cost.

The herbivore's ability to exploit a plant:
3. Herbivores are capable of developing immunity to phytotoxins.
4. The physiological development of herbivores can be highly attuned to the seasonal and diurnal changes in defensive compounds.
5. Plant diversity plays a large role in the development of specific detoxification mechanisms developed by insects.

Lateral effects: herbivore vs. herbivore, plant vs. plant:
6. Competition among herbivores depends on type of feeding and life history traits.
7. Indirect (resource-mediated and enemies-mediated) interactions are common between herbivores.
8. Herbivory lessens a plant's ability to compete and, consequently, alters distribution and abundance of plants.

Bottom-up effects of plants on the herbivore's enemies:
9. Plant traits can affect the performance of enemies of herbivores.
10. Parasitoids are often more specialized than predators.

Top-down effects of the herbivore's enemies on plants:
11. Enemies can limit herbivores and improve plant productivity.

Community structure:
12. Much biodiversity is due to plant-herbivore interactions.

Source: Revised from Stamp 1996, with permission.

can be developed for this interaction at a physiological scale (Table 7.2). This chapter will include all the processes listed in Table 7.2.

A. Abundance and Diversity of Herbivores

Herbivores are virtually as diverse as the complexity represented in all animal phyla. This diversity exists because *herbivory,* which is broadly defined as the ingestion of live plant material, is a major link between producers and the rest of the energy flow pathway in most ecosystems. We will be classifying herbivores in this chapter on the basis of their feeding mechanisms. The majority of these mechanisms are covered in Table 7.3.

Each of the feeding mechanisms may be utilized by a wide diversity of different herbivores. For instance, chewing on plant parts is characteristic of many mammals, birds, nematodes, and insects. Some of the organisms that harvest seeds as an energy source are ants, birds, reptiles, and mammals. Insects have developed the greatest diversity of feeding mechanisms such that at least one group of insects can be assigned to each of the feeding mechanisms listed in Table 7.3. It is not possible to cover all types of herbivores in this chapter; therefore, we will focus on insect herbivores because they encompass the entire range of feeding mechanisms.

Herbivores can be further classified on the basis of the diversity of plants they utilize. Specialist herbivores (those that feed on one or a few plant species) may often be closely tied to the life cycle patterns of their host plant species. In some cases, the survivorship and/or fitness (reproductive success) of the host may also depend on the presence of the herbivore. These tight insect-plant relationships that assist the survival of both parties are classified as *mutualistic* and have been refined by coevolutionary

TABLE 7.2 Some Messages of Plant–Insect Herbivore Interactions at a Physiological Scale

The plant's ability to defend itself by physiological means:
1. Plants have a wide array of chemical defenses.
2. Chemical defenses have a metabolic cost.
3. Some physiological defenses are constitutive, while others are inducible.

The herbivore's ability to induce physiological responses in plants:
4. Herbivores have stimulated the development of chemical defenses in plants on an evolutionary scale.
5. Chemical defenses can be induced following herbivore attack.
6. Quantitative and qualitative variation in herbivore defensive chemicals is an important aspect of plant-insect interaction

Collateral effects: plant vs. plant, plant vs. pathogen:
7. Some chemical defenses against herbivores may also be involved in allelopathic effects on other plants.
8. When attacked by a pathogen, a degree of tolerance can be conferred to the plant herbivore defense chemicals.

Bottom-up effects of plants on herbivore's enemies:
9. Specific herbivore defense chemicals produced by plants can be consumed and stored by a herbivore to confer protection against its predators.

Top-down effects of the herbivore's enemies on plants:
10. Herbivore population regulation by predators enhances the quantitative variation of physiological defenses in plants.

processes over millennia. In contrast, generalist herbivores utilize a diverse array of plant species and have not formed those strong mutualistic relationships with any particular host plant species. Generalist herbivores are able to develop life history patterns that are relatively independent of the phenological patterns of any one host species. These herbivores are able to switch to equal or lower-quality alternative host species when the most favored species is unavailable. For example, the highest-quality host species for gypsy moth larvae is white oak. During a gypsy moth outbreak, the larval population may completely defoliate white oaks, after which the larvae will switch to using a diversity of other tree species. Considering the diversity of herbivorous insects, it may be too simplistic to divide them into generalists or specialists. In fact, this dichotomy among herbivores is useful only for theoretical considerations. Actually, there is a continuum of herbivorous insects, from obligate specialist, through facultative specialist, to extreme generalist.

Herbivores are found in every ecosystem, but their diversity varies among ecosystems. Herbivorous insects are most diverse in tropical systems because of the diversity of plant species as well as the vertical and spatial complexity of tropical vegetation. Herbivore diversity decreases with increasing latitude and increasing elevation in accordance with the changes in plant species diversity. However, most ecosystems, no matter how limited the plant diversity, will support more herbivores than plant species. For example, in a desert ecosystem that includes only one tree and a few shrub species (mesquite woodland), all classes of herbivores in Table 7.4 exist (Cates and Rhodes 1977). Also, there is a large diversity of herbivorous nematode species at a 5 m depth in desert soil, where there are roots of only one phreatophytic tree species (Freckman and Virginia 1989).

All plants experience an array of different herbivores during their lifetime. Embryos are susceptible to herbivory by seed-boring insects such as bruchid beetles. In the seed stage, plants are exposed to the herbivory of chewing or gathering herbivores.

TABLE 7.3 Various Feeding Guilds of Phytophagous Herbivores and Example Insect Groups That Utilize Those Feeding Mechanisms

Feeding Mechanism	Insect Example (common name)	Insect Order[a]	Insect Family[a]
Cambium bark or sapwood borers:			
Main stem	Bark beetles	Coleoptera	Scolytidae
Roots	Long-horned beetles		Cerambycidae
Sap feeders: leaves, twigs, branches	Aphids	Homoptera	Aphidae
	Leafhoppers		Cicadellidae
Ingesting leaves (folivores):			
Any age leaf	Common sawfly larvae	Hymenoptera	Tenthredinidae
Mid-season leaves	Tiger moth larvae	Lepidoptera	Arctiidae
Early season leaves	Inchworms, loopers		Geometridae
Feeding on young roots	Nematodes	Nematoda	
	White grubs (Scarab beetle larvae)	Coleoptera	Scarabaeidae
Root sap feeders	Cicada nymphs	Homoptera	Cicadidae
Consuming pith and wood	Stem-boring larvae (stem sawflies)	Hymenoptera	Cephidae
	Long-horned beetle larvae	Coleoptera	Cerambycidae
Mining leaf parenchyma	Leaf-mining beetles	Coleoptera	Chrysomelidae
	Leaf-miner flies	Diptera	Agromyzidae
Consuming seed in fruit	Seed beetles	Coleoptera	Bruchidae
Harvesting seed after dehiscence	Harvester ants	Hymenoptera	Formicidae
Inducing cancerous growth on stems	Pine and spruce aphids (adelgids)	Homoptera	Adelgidae
	Gall midges	Diptera	Cecidomyiidae
Inducing cancerous growth on leaves	Gall wasps	Hymenoptera	Cynipidae
Consuming nectar and pollen	Various bees	Hymenoptera	Apidae, Halictidae

[a]Listed in order of most to least likely to damage plant fitness; severity of herbivory is inversely related to a plant's ability to compensate for the damage.

TABLE 7.4 Diversity of Herbivores That Utilize Mesquite (*Prosopis glandulosa*) Simultaneously in the Desert Southwest of the United States

Feeding Class	Example Herbivores
Cambium or sapwood borers	Stem-mining beetles
Sap feeders: leaves, twigs, branches	Aphids
Leaf chewing	Lepidopteran larvae, cicadas
Root feeding	Many nematode species
Root sap feeding	Some nematode species
Stem-boring larvae	Long-horned beetles
Mining leaf parenchyma	Leaf-mining beetles
Seed chewing	Bruchid beetles
Consuming fruit	Coyote, jackrabbit
Consuming detritus	Kangaroo rat, wood rat
Consuming nectar and pollen	Various bees

Source: Compiled from information in Simpson 1977.

Once the seed germinates, herbivory pressure from small rodents, ants, and snails can be extreme. Once the plant has grown to maturity, a number of different herbivores may attack the plant simultaneously. On one tree, nematodes feed on roots, bark beetles feed in the phloem, aphids suck phloem sap, miners consume leaf parenchyma cells, lepidopteran larvae chew on leaves and flowers, and so on. One individual tree can harbor hundreds of different species of herbivores at any given time. Fumigation of individual tropical trees often records hundreds of phytophagous insect species, some of which have never been previously identified. Therefore, there can be no question that herbivore populations can constitute a significant stress to plant performance in all ecosystems.

B. Potential Influences of Herbivores on Plants

Within the context of population or community ecology, herbivory of some individuals may stimulate the growth or fitness of others, leading to overall net effects that are complex. However, most scientists would agree that herbivory generally has a negative influence on individual plant growth and fitness. In some cases, plants have the capacity to compensate for herbivory to such an extent (i.e., to overcompensate) that their fitness actually increases in the presence of herbivores (McNaughton 1986). A few studies have clearly demonstrated the mechanism behind a positive effect of herbivory on plant fitness (Crawley 1987, Dyer et al. 1995). Those mechanisms include stimulation of growth by chemical factors in saliva, and stimulation of growth by increased nutrient availability in the soil derived from insect frass. When all studies are viewed together, there is a continuum from strongly negative to moderately positive influences of herbivores on plant growth and fitness (Whitham et al. 1991).

One of the easiest ways to determine the influence of herbivores on plant performance is to fumigate plants with an insecticide (or otherwise remove herbivores with exclosures) and compare the growth and reproduction (i.e., the fitness) to controls that experience natural herbivory. These studies have been done on many species in different environments. In grassland systems, removal of large ungulates (wildebeasts) caused increased growth in savanna species (McNaughton 1985a). In chaparral systems, removal of seed herbivores resulted in increased seed recruitment and survival of *Haplopappus squarrosus* (Lauda 1982). When blue butterfly larvae were removed from lupines in Colorado, flower production increased by 85% with a corresponding increase in seed set and plant fitness (Breedlove and Ehrlich 1972). In agroecosystems, fumigation with insecticides increases plant growth, fruit production, and the quality of yield (Heinrich 1988).

The intensity of the negative impact that herbivory has on plants depends mostly upon six factors:

1. *Life Stage of the Plant.* A few bites out of the leaves of a mature tree has little effect on tree fitness, but the same number of bites out of a young seedling or a seed has dramatic effects on subsequent plant fitness.
2. *Type of Tissue Consumed.* Consumption of inactive tissue (senescent leaves, cork) has less of an effect on plant performance than does consumption of active tissue (mature leaves, cambium, apical meristems). Seed-consuming herbivores have a stronger effect on reproductive performance of a species than bark-mining herbivores.
3. *Timing of the Herbivore Attack.* Severe middle- to late-season defoliation has a stronger effect on canopy regrowth than does early-season defoliation (Gregory and Wargo 1986).
4. *Genome of the Individual Plant.* A mean tissue loss of 5–15% by herbivory is common in natural populations. However, within the population, particular genotypes suffer far greater percentage losses of tissues than other genotypes. Therefore, the greater the variation of genetic resistance to herbivores, the greater the impact of herbivory on plant evolution (Fox and Morrow 1986).
5. *Environmental Conditions.* When a plant experiences environmental stress, it may be more susceptible to damage by herbivores. For example, pines are much more likely to die by wooly adelgid infestation if they first experience water stress (Coulson 1979).
6. *Species-Specific Responses.* In some species, small amounts of specific tissue consumption can have dramatic effects on growth (Dirzo 1984), seed production (Archer and Tieszen 1980), and mortality (Clark and Clark 1985). In other species, large amounts of tissue can be consumed with little effect on growth (Nilsen et al. 1987).

These six factors clearly demonstrate that a simple measure of average mass consumed by herbivores will not be very useful as a description of the influence herbivores have on plant fitness.

In general, herbivores consume some of a plant's net production, and this is one of the factors that causes the increment in plant mass to be lower than that predicted from measurements of net photosynthesis. However, one of the responses plants exhibit to herbivory is *compensatory growth,* defined as an increase in plant growth rate following herbivory when compared with plants that did not experience herbivory. Compensatory growth allows plants to regain their competitive advantage following herbivory. Often the quality of tissues produced by compensatory growth differs from that produced before herbivory. Regrowth tends to have higher nutrition and greater defense against herbivores than the original tissue. Enhanced defense in compensatory growth can occur at a structural level (increased spines or trichomes) or a chemical level (increased concentration of toxic secondary substances).

Plants utilize substances called *allelochemicals* (chemical defense compounds) to defend themselves against herbivores (Whittaker 1970). There is a large diversity of chemical defenses that can be induced at different time scales by folivory (Baldwin, 1994). *Performed induced responses* occur immediately upon damage, such as HCN release, which occurs upon tissue crushing (previously separated cyanogenic glycosides and hydrolyzing enzymes are mixed). *Rapidly induced responses* occur within hours or days in the damaged tissues or throughout the plant. The induction of nicotine, proteinase inhibitors, or lechtin accumulation are rapidly-induced responses. *Delayed induced responses* are those that occur in regrowth tissue or in the next season's foliage. The delayed induced responses have long-term effects on herbivore populations (Haukioja and Neuvonen 1987).

Herbivory also can influence the reproduction of plants in a positive or negative manner. Many herbivores induce pollination or relieve endogenous seed inhibition mechanisms. The fruit of figs is pollinated by phytophagous wasps, and a large number of seeds require passage through the gut of birds before successful germination can occur. Most tropical plant species have fleshy fruit, the seeds of which require a fruit-consuming herbivore for dispersal (Terborgh 1986). However, it is clear that herbivores can also have a negative effect on plant reproduction by consuming or otherwise destroying the embryo in seeds. Insects feeding on flowers and fruits significantly reduce the recruitment of new plants in chaparral and desert ecosystems (Lauda 1982). The larvae of bruchid beetles bore into fruit pods and consume seeds inside the pod, reducing the production of viable seed. Flea beetles can severely reduce the aboveground crop of seeds in legumes. In fact, flea beetles are used as a biocontrol method for reducing the population of legume weeds in natural systems. Floral damage by herbivores can also reduce reproductive fitness of plants by reducing their attractiveness to pollinators. For example, pollen beetles, *Meligethes rufimanus,* damage flowers, which in turn causes a reduced visitation by pollinators, thus compounding the negative influence of the beetle on plant fitness (Krupnick et al. 1999). However, the indirect effects of floral herbivory on pollinator visitation were found to be limited to male reproductive effects, while direct influences of floral feeding by gamete destruction was the main mechanism of fitness reduction in *Isomeris arborea* (Krupnick and Weis 1999). Foliage-feeding herbivores also can reduce reproductive output. Pollen production (number of grains, and size of grains) of *Raphanus raphanistrum* (annual wild radish) was inhibited by leaf herbivory (Lehtilä and Strauss 1999). Leaf damage by herbivores also reduced the attractiveness of plants to pollinators in the same research system (Lehtilä and Strauss 1997).

Herbivores also may influence the sex ratio of plant populations and the reproductive maturity of plants. One of the effects of moth larvae (*Dioryctria albovitella*) feeding on pinyon pine (*Pinus edulis*) is to alter tree architecture from an upright form to a more prostrate form. In pinyon pine, female cones are born on apically dominant branches. Therefore, the number of female cones decreases as a consequence of herbivory (Whitham and Mopper 1985). Attack on the arroyo willow (*Salix lasiolepis*) by a sawfly (*Euura lasiolepis*) changes the willow tree architecture, which indefinitely maintains the trees in a juvenile condition and significantly reduces willow reproduction (Craig et al. 1986).

Once seeds have been successfully produced, herbivores can have a pronounced effect on seed and seedling survival. Mature seeds of some plants, such as species in Pinaceae and Asteraceae, can be consumed before being released into the environment. After seeds fall, they are gathered by seed predators such as ants, beetles, birds, and mice. Some of those seeds are cached for future consumption. Most of the seeds in the caches will be consumed by the herbivore or contaminated by fungi. However, in the case of blue jays, many of the cached acorns are not recovered, and the seeds germinate successfully. Also, ants often collect whole seed, bring it to the top of the nest, cut off the food body and leave the seed outside. The result is a concentric ring of seedlings around the nest the following year. Seeds that are resistant to the rigors of herbivore digestive systems are deposited in the herbivore's scat. In the Sonoran desert, coyotes consume most of the mesquite fruit and deposit a scat that is mostly mesquite seed. Young seedlings are consumed by snails, mice, and a multitude of insect foliovores. Many experiments that exclude herbivores from a natural ecosystem have shown that herbivory can have a substantial negative effect on seedling survival.

Herbivores can influence plant susceptibility to pathogens. Tissue damage due to the action of herbivores provides avenues for infection by fungi, bacteria, and viruses. In fact, insects are normally contaminated by a diversity of microorganisms that subsequently develop in the herbivore-affected host plant. Insects often serve as vectors for transporting fungal pathogens to plant hosts (Klepzig et al. 1991). In many cases, there is no advantage to the insect; however, in some cases, the fugal infection may aid the insect's ability to utilize the host plant. For example, termites increase wood digestibility by introducing fungal pathogens (Cowie and Wood 1989). Elm bark beetle adults feed on healthy elm trees and infect the elm trees with a pathogenic fungus (*Ophiostoma ulmi*). The breeding of these beetles occurs only in dead elm trees, and the fungus is likely to kill the host trees. Therefore, the elm bark beetles are increasing their breeding sites by infecting live trees with a lethal pathogen while feeding in the bark. Viruses and bacteria are also transported between plants, in or on insects. The overall result is that herbivory increases the incidence and virulence of pathogens in plants.

Herbivory also influences a plant's capability to compete. In any situation where a herbivore population restricts the foraging ability of plants for resources, the competitive ability of those plants decreases. Nematodes that reduce root biomass cause a decrease in plant nutrient acquisition, reduced aboveground growth, and decreased fitness. Leaf-chewing insects reduce plant leaf area, decrease photosynthesis, decrease growth potential, and decrease the plant's ability to compete for available light.

Moreover, herbivores have the capacity to influence the manner by which plants respond to environmental stress. In some cases, herbivores induce a change in plant morphology that preadapts the plants to certain stresses. For example, when herbivores reduce leaf area, this reduces a plant's demand for water, and the plant will subsequently be adapted to tolerate less soil water availability. In other cases, herbivory induces plants to become more susceptible to environmental stress (particularly biological stress).

The overall influence of herbivores on plants encompasses all temporal and spatial scales. Herbivores impact plant performance throughout the plant's life cycle at every possible location on the plant. Herbivores also elicit physiological, morphological, and behavioral responses from plants. Because of the powerful linkages between phytophagous insect and plant population fitness, ecological ramifications abound.

C. Influence of Herbivory at an Ecological Level

Although the bulk of this chapter concerns the interactions between the physiology of individual plants and their herbivores, it is prudent to establish an ecological and evolutionary context. *Ecological interactions* can be broadly defined as those occurring among species and between species and the environment. Therefore, it is indisputable that herbivory has important ecological implications. Many studies have focused on the influence that herbivory in general has on system productivity. In fact, the terminology used to describe productivity includes a herbivory component (i.e., *net primary production* = net increase in standing biomass + loss through mortality + loss through herbivory). In addition, any ecosystem model that concerns energy transfer must contain a herbivory function. The efficiency of energy transfer into herbivores is one of the primary regulators of ecosystem complexity. A good example of the extent to which a herbivore can influence ecosystem processes is that of wooly adelgids in eastern hemlock forest (Jenkins and Aber 1999). The structure (diversity and density) of the forest is dramatically altered, as is the function (productivity and nutrient cycling) by an infestation by one insect species.

The nature of community species composition (diversity) may be strongly influenced by herbivory. In many cases, the dominance of one species is dependent upon herbivore pressure on subdominant species. Removal of herbivores by exclosure often results in a change in species dominance, particularly in grasslands. The speed and direction of succession also can be severely influenced by herbivory. Following herbivore exclosure, many systems rapidly develop a new successional trajectory ending up with a different composition of species than in unexclosed sites. When herbivores are excluded from a system long enough, a new steady-state community structure will result (Bach 1994).

In some communities, herbivory may increase species diversity. Herbivores can reduce the competitive advantage of a potentially dominant species such that other species can acquire some of the available resources. The result is increased diversity. In fact, there may be a "play-counter play" interaction between plant diversity and insect diversity. Herbivores can stimulate diversity, which in turn can stimulate herbivore diversity. Processes such as plant defenses, species extinction, climate change, predation, disease, and competition among herbivores check this spiral of increasing diversity.

Implicit in the previous discussion on biodiversity is the assumption that herbivores can influence the competitive interactions between species. The outcome of competition between two plant species may be regulated by a third party, such as a specialist herbivore. A theoretical example: plant species A is a strong competitor and severely reduces the survival and fecundity of plant species B. In fact, species B is excluded when it is grown with species A in laboratory conditions. However, in the field, both species coexist as codominants. A specialist herbivore that preferentially attacks species A may be the reason why competitive exclusion of species B does not occur in natural communities.

The potential impact of herbivores on competitive ability has been applied in programs designed to control plant pests. Invasive exotic species (plants, animals, fungi, microbes) often cause significant alterations in community structure, productivity, and ecosystem function. In fact, invasive species now constitute a major component of global change research (Vitousek et al. 1997, Mack 1996). The transport and establishment of new plant species into disturbed lands can often result in a massive, unregulated population explosion of the exotic to the detriment of native species. Such occurrences are abundant in recent history, including annual grasses and brooms in California (*Cytisus scoparius* and *Genista monspessulanus*) and *Casurina, Maleleuca,* and *Schinus* in Florida. In all these cases, the exotic species has developed monocultures that exclude native species and change the nature of ecosystem processes. The most often cited reason for the unregulated population growth of the invasive species is the absence of population regulation by phytophagous insects or pathogens. Subsequent regulation of these invasive plant populations is difficult, but the most effective mechanisms may be the introduction of a specialist herbivore or pathogen. Potential biocontrol insects must be carefully screened in order to assure that they are highly specific for the invasive plant. Usually, the biocontrol insect is selected from specialist herbivores in the native range of the invasive species. At times, biocontrol of invasive species works extremely well (e.g., the regulation of invasive cacti in Australia), and at other times the biocontrol insect becomes a pest itself, such as the seven-spot lady beetle (*Coccinella seplempunctata*) introduced into the United States.

Population models that describe the interaction between grazers and producers have been developed by modifying Lotka-Volterra equations. In general, these models predict "top-down" control of plant productivity by herbivores unless carnivores relieve the herbivore pressure (Rosensweig 1973). However, a number of different formulations of these models have been studied. A good example is a study of the interaction between herbivory and primary productivity in a salt marsh (van de Koppel et al. 1996). Results from this study indicate that at high plant productivity, herbivore efficiency is decreased, and at very low plant productivity, there is not enough resource for extensive herbivory. This suggests that a stable state of high herbivore fitness would be attained at moderate plant productivity, and a stable state of low herbivore fitness would occur at high plant productivity, irrespective of the presence of predators. Also, it implicates a potential influence of nutrient availability on plant herbivore interaction. At high nutrient levels, plant productivity could increase to a point where a positive feedback between the increased standing crop and reduced herbivory becomes established, relieving the vegetation from control by herbivores.

One unique aspect of herbivory, in comparison with the interaction between pathogens and plants, is the interplay between plant, herbivore, and parasitoid. This *tritropic interaction* includes control of the herbivore population by both the plant and the parasitoids (both top-down and bottom-up interactions). This leads to the concept of direct and indirect defenses of plants against herbivores. Most of the chemical defensive mechanisms we will discuss are directed against the herbivore population. In contrast, plant defense against herbivory can be attained by stimulating the ability of parasitoids, predators, or pathogens to attack insect herbivores (Baldwin and Preston 1999). These indirect plant defenses against herbivores are as important as direct chemical defenses and may be much less costly to the plant. The effectiveness of direct and indirect defense mechanisms depends on the ecological context of the plant; thus, the relationship between the nature of plant defensive mechanisms and plant fitness is intimately tied to the ecological setting of the plant. The study of plant defensive mechanisms at a physiological level must be integrated with ecological theory.

D. Evolutionary Relationships

Generally, in a steady-state ecosystem, trees that die will be replaced with trees of other genotypes each 100 or more years. Contrast, the tree's life cycle with the fact that each phytophagous insect is replaced at least once or twice a year by many more than one individual. Furthermore, factors that regulate the rates of evolution often concern population size, reproductive rate, and population size stability. In general, plant populations are relatively small and relatively stable, particularly in the case of arborescent trees. In contrast, phytophagous populations can become extremely large very rapidly, and can have dramatic temporal and spatial variation in population size. Natural selection theory predicts that over a short time frame (in relation to the plant's life cycle), phytophagous insects will maximize their potential use of the host tree. If this is true, how do trees maintain their dominance of ecosystems? This important evolutionary paradox has been evaluated in many essays. Later in this chapter we will consider the importance of defensive heterogeneity as a mechanism to dampen phytophagous insect evolutionary rate.

Coevolution, in the classic sense, often pertains to the interaction between herbivore and plant. Results of this coevolution can be easily observed in nature. For example, some species of *Acacia* produce food bodies near their spines that attract ant species. The ants feed on the food bodies and also clear out competing seedlings from the base of the tree, thereby limiting the abundance of potential competitors with the *Acacia* tree (Cronin 1998). In other cases, the coevolution may not be so obvious. As a speculative exercise, imagine that a plant species develops an unusual defensive compound. One herbivore species subsequently develops a new and unique mechanism for detoxifying the compound, but by doing so becomes restricted to feeding on that specific host species. The result is a specialist herbivore, and mutualistic relationships are likely to occur.

Another important evolutionary concept for plant-herbivore interactions is the cost-benefit relationships of plant defensive mechanisms. There are many built-in costs to plant defense, such as the resources required to synthesize and maintain the defensive mechanism. Allocation of resources to defense may have a benefit to the plant if herbivores are present, but it will also result in a cost by reducing the resources that can go to other metabolic processes such as growth. Other costs involved with herbivory include self-toxicity, catabolic imbalance, catabolic repression, and ecological trade-off. Optimality theory defines resources used for defense as necessarily diverted from other fitness-enhancing attributes of the plant. In contrast, models concerning coevolution theory assume that plant defenses evolve in the context of coevolving herbivores, not of cost-benefit relationships to the plant alone. The cost-benefit relationships of herbivory will be expanded in a later section of this chapter.

E. Applied Significance of Plant-Herbivore Interactions

In an applied sense, herbivory is best known in agriculture. Every agricultural field around the world is exposed to herbivores that are potentially damaging. The use of single varieties of a crop in a field makes the potential for herbivore damage much greater than in a natural community. The herbivores severely decrease yield per acre and also spoil the attractiveness of the product to the customer. Therefore, it is critical to control these potentially explosive insect herbivore populations (*pests*) in agricultural fields. Agrobusiness has evolved, in part, due to the need for procedures that control insect pests. A cornucopia of chemicals has been discovered to control these insect herbivores (*pesticides*), and an enormous worldwide industry is the result.

Plants can tolerate a number of chemical applications without decreasing yield, particularly if the chemical is targeted at insect physiology rather than general physiological processes. However, there is a limit to the ability of crops to tolerate pesticides. Selective breeding programs have tried to identify varieties that have greater tolerance of pesticides so that greater application intensities could be used. In addition, genetic engineering has been used to increase crop tolerance of pesticides. But, there are some detrimental aspects of increasing the use of pesticides in agriculture.

Following the development of the Ames test and regulations for testing chemicals, it became clear that some pesticides can be toxic when consumed by people and therefore may pose a health risk. The greater the application of these pesticides, the greater the risk that the produce will be contaminated upon harvest. Also, the continual application of a pesticide on an insect population is likely to stimulate the evolution of resistance in the insect. Therefore, the greater the application frequency and intensity, the more likely that resistant populations of insects will occur, rendering the pesticide ineffectual in the future.

Due to the above problems with pesticide use, other agricultural technologies have been developed. One called *integrated pest management* utilizes pesticides and other mechanisms to regulate herbivore populations. The main purpose of this technology is to minimize pesticide use by utilizing other aspects of the crop biology and physiology to regulate herbivore damage. A second technology is the *engineering of natural defense mechanisms* in crops. Many crops, through artificial selection, have lost their defensive mechanisms. The genetic structure of the crops can be adjusted by molecular techniques to increase these natural defenses. However, such upregulation of natural defenses may come at a cost to production, as resources are diverted from processes that regulate yield to those that regulate defense. Alternatively, a defensive mechanism utilized by other organisms could be added into the genome of the crop to increase defense without disrupting pathways that determine yield. For example, the BT genes from the bacterium *Bacillus thuringiensis,* which produces proteins inhibiting insect digestion, have been incorporated into the genome of crops as a mechanism of increasing herbivore resistance. Technologies to inhibit herbivore damage to crops will have a long-term association with agriculture into the future.

The substances produced by plants in response to herbivores also have a large applied significance. Different cultures around the world have made extensive use of the defensive chemicals in plants. Seeds of spice plants such as pepper, dill, cumin, and coriander gain their flavors from compounds that serve to inhibit granivore digestion of the seed embryo. Leafy spices such as mint, oregano, rosemary, and sage derive their flavor and volatile fragrances from putative herbivory defensive compounds. Caffeine in tea and coffee, and nicotine in tobacco are plant secondary compounds that serve in defense against herbivores. Other plant products used in tanning, dying, and perfumes are also putative defensive compounds. Extractions of resins from the creosote bush (*Larrea tridentata*) were used to preserve food until 1965. Moreover, many pharmaceuticals are derived from or produced by plants for defense against herbivores.

II. MECHANISMS OF ADAPTATION AGAINST HERBIVORY

A. Nonphysiological and Indirect Defenses

1. Morphological and Behavioral Defenses. There are a relatively limited number of morphological, behavioral, or anatomical adaptations that inhibit the effectiveness of herbivores. The classic morphological adaptations are spines that serve the purpose of inhibiting relatively large herbivores (Cooper and Owen-Smith 1986). Spines are diverse in shape and size and are found on a diversity of taxa. Some of the largest linear spines (over 20 cm long) are found on *Prosopis ruscifolia.* The ecosystems with the largest diversity of spiny plant species are water-limited systems such as deserts, dry woodlands, and savannas. In semiarid tropical thorn woodlands, a majority of the species are armed with spines. However, these large spines have little effect on phytophagous insects.

Trichomes of all sorts have a greater impact on insect herbivores than do spines (Levin 1973). Stinging hairs or glands can influence both large mammalian herbivores and insects. For example, the stinging hairs of nettles (*Urtica dioca*) have antiherbivore properties. Other tissue surface structures, particularly stelate trichomes, can serve to inhibit insect herbivory. However, one must keep in mind that leaf surface trichomes (when dead) can also increase leaf reflectivity and reduce leaf temperature in high light environments (Ehleringer and Bjorkman 1978). Thus, the effect of trichomes on herbivory may not be the only selective factor determining the evolution of this plant trait.

Several behavioral characteristics of leaves may reflect antiherbivore adaptation. For example, the thigmotropic movement of leaves on the sensitive plant (*Mimosa* sp.) may serve to reduce the *apparency* (ability to be located by herbivore senses against the background of all other species) of leaves and expose spines to herbivores. Nyctinastic movements of leaves in legumes, where leaflets fold into one another at night, may also serve to reduce the *apparency* of leaf tissue to nocturnal herbivores. If rapid phenological development can be considered a behavior, then it may likely reflect an antiherbivory mechanism (Nilsen et al. 1987, Hunter 1992). Leaf tissues and meristems are most susceptible to herbivory during young stages of phenological development. If tissues mature rapidly, then they are not exposed to the possibility of herbivory for too long. This may be the reason why new branches and leaves of *Prosopis glandulosa* (a desert phreatophyte) develop extremely rapidly (Nilsen et al. 1987). Stochastic development of new tissues among individuals of a population may also be a mechanism for inhibiting the magnitude of herbivore damage by limiting the available resources for herbivores.

2. Compensatory Growth. Earlier in the chapter, the concept of compensatory growth was considered. This reaction to herbivores concerns both morphology and chemical traits of new tissues produced after a bout of herbivory. Compensatory growth has the greatest effect in grasses and young tree seedlings or saplings. In grasses, removal of a set of swards usually results in a new, more vigorous set of shoots. Anyone who maintains a mown lawn is aware of this phenomenon. The new tissue has an enhanced physiological capability. For example, in a study conducted by Richards and Caldwell (1985) defoliation in grass caused a significant increase in photosynthesis of the new tissues. In natural systems dominated by perennial grasses, compensatory growth is an adaptation against herbivory by large ungulates (McNaughton 1985b). Productivity in grasses increases because of grasshopper herbivory (Dyer et al. 1993, Dyer et al. 1995) or large ungulate herbivory (McNaughton 1979, Knapp et al 1999). A growth-stimulating substance in or induced by the saliva of herbivores may be one mechanism for increased compensatory growth in these systems (McNaughton 1985b, Detling and Dyer 1981, Dyer et al. 1995), however the action of large ungulates alter many ecosystem processes (Knapp et al. 1999).

Compensatory growth in tree seedlings may not always be beneficial to the success of the species. Oak seedlings in the California oak woodland are subject to intense herbivory by deer. The seedlings develop into small shrubs instead of large trees because the deer are pruning all the new meristems resulting in a bushy compensatory growth form. Only when the deer are excluded by fences or the deer population crashes, can the oaks break out of the shrub growth form and develop into trees.

In contrast to the case for oaks, compensatory growth is a successful adaptation for some other tree species. For example, *Prosopis glandulosa* (mesquite) is a desert phreatophyte that is successful growing and producing seed as a shrub or a tree (Sharifi et al. 1987). When a young sapling is clipped off by rabbits or other mammalian herbivores, it resprouts with greater vigor and more frequent larger spines than the original sapling. The increased growth rate and defense soon allows the sapling to develop into a successful large reproductive shrub at a faster rate than would have occurred without the herbivore damage.

3. Indirect Plant Defenses. Recall that indirect defenses are those that involve a third trophic level such as predators, parasites, and pathogens of the herbivore. Upon attack of plant material by a herbivore, the nature of the plant and its environment could change to increase the likelihood that a herbivore could be attacked by a predator, pathogen, or parasitoid. The best example of this is the herbivore induction of *volicitin,* which causes the release of volatile compounds that attract parasitoids (Alborn et al. 1997). Plants such as tobacco, cotton, and maize each produce distinct blends of volatile compounds in response to damage by different herbivores. Parasitic wasps are capable of differentiating among the volatile blends to locate their specific arthropod host (De Moraes et al. 1998). The key factor for eliciting plant volatile compounds that attract parasitoids and pathogens is *N*-(17-hydroxylinolenoyl)-L-glutamine (volicitin). This compound, which causes the release of plant volatile substances, was first isolated from beet army worm caterpillars (Alborn et al. 1997). However, it is clear that the fatty acid component of this molecule is plant derived, and the 17-hydroxylation reaction and the conjugation with glutamine are carried out by the caterpillar (Pare et al. 1998). Therefore, the caterpillar is consuming linolenic acid, converting it into volicitin. The volicitin induces the release of plant volatile compounds that attract caterpillar parasitoids.

Although these studies show the complex interaction between plant, herbivore, and parasitoid, they do not show that plants with greater infestation of herbivores actually do serve to increase the number of parasitoids or predators on herbivore-infested plants in the field. Although done in a greenhouse rather than in the field, a recent study addressed this limitation with spider mites and their parasitoids (Janssen 1999). The parasitoids were released in the center of a hexagon of plants, some of which had an infestation of two-spotted spider mites. Within 7 hr, more that half of the released parasitoids were recaptured on the plants, and about 80% of those were located on plants that had spider mites. Therefore, the blind parasitoids are attracted to plants that are infested with spider mites on the basis of herbivore-induced volatile substances.

In general, the tritrophic interaction between plant, herbivore, and parasitoid is complex and tightly interwoven suggesting strong coevolution. Chemical plant based cues are the main way that parasitoids can locate the widely dispersed and often cryptic hosts (De Moraes and Mescher 1999). The specific complex blend of chemicals released by plants during herbivore attack are important both for the parasitoids to sort among the diverse chemical characteristics of the environment and the parasitoids' ability to choose among different host insects.

B. Direct Physiological and Chemical Defenses

Chemical defenses of plant tissues are the most studied and perhaps the most important defenses among the mechanisms used by plants to limit the destructive power of herbivore populations. Originally called allelochemicals (Ehrlich and Raven 1964), this group of compounds is diverse in structure and efficacy. The term *allelochemical,* was first coined by Molish (1937) in reference to substances released by plants to inhibit neighboring plants. The concept of allelopathy is treated in Chapter 8, where some, but not all, of the same compounds considered to be natural pesticides in plants are discussed. We will restrict use of the term *allelochemicals* to the discussion of allelopathy. Most chemicals of plants associated with herbivore defense are grouped within the secondary chemistry of plants. Secondary chemicals were originally clearly defined by Sachs (1873) as by-products of metabolism that are no longer useful for the formation of new cells (Hartmann 1999). The function of these secondary chemicals, in relation to insect herbivores, was established by entomologists (Fraenkel 1959, Ehrlich and Raven 1964). Once the significance of these secondary chemicals was introduced, a rapidly evolving field of chemical ecology developed.

A large diversity of compounds has been implicated in defense against herbivores (Table 7.5). Some of these substances are part of the primary metabolism, a few are the result of specific pathways that have evolved exclusively for herbivore defense, but most are classified as secondary metabolites (potentially waste products of metabolism). Because there is a diversity of pathways leading to the production of compounds with antiherbivore action, we will use the term *defensive chemicals* to refer to those substances. There is no reason to conclude that these substances cannot have dual roles in natural systems; however, the evolutionary pressure of herbivory is more likely than allelopathy to have induced the diversity of defensive chemicals found in the plant kingdom (Whittaker 1975).

Putative defensive chemicals are synthesized by three major biochemical pathways (Figure 7.1). The *shikimic acid pathway* (named after an intermediate compound) begins from simple carbohydrates, associated with glycolysis or the pentose phosphate pathway, and proceeds to the aromatic acids. This pathway is found in plants, fungi, and bacteria, but it is absent from animals, which precludes animals from forming three aromatic amino acids (phenylalanine, tyrosine, and tryptophan). Many herbivore defensive compounds are derived from these three amino acids. Moreover, the biosynthesis of herbivore defensive compounds often can be linked to acetyl coenzyme A (acetyl CoA). This molecule is the main substrate for the *acetate-malonate pathway* leading to the synthesis of fatty acids and some aromatic defensive compounds (Figure 7.1). In addition, acetyl CoA is the substrate for the *acetate-mevalonate pathway* leading to the synthesis of steroid defensive compounds. Therefore, the synthesis of defensive compounds is complex and is most related to biosynthetic pathways associated with the synthesis of aromatic compounds and acetyl CoA.

Some general aspects of defensive chemicals, related to herbivory, should be considered before a discussion of particular substances can be untertaken. Defensive compounds can be generally divided into two classes:

1. *Broad-spectrum compounds,* which have a general inhibitory impact on herbivore physiology and are common to a relatively wide diversity of plant taxa. Examples are tannins, lignins, phenolics, terpenes, cyanogenic glycosides, and proteinase inhibitors.
2. *Narrow-spectrum compounds,* which are restricted to a small group of plant taxa and may negatively affect a small group of herbivores by a specific metabolic process. Examples of narrow-spectrum compounds are hormone facsimiles, alkaloids, and nonprotein amino acids.

Broad-spectrum defensive compounds are often constitutive and at relatively low concentration throughout the tissue, while specialist substances are often induced by herbivore attack and come to high concentrations in localized regions.

Variation in defensive chemical concentration is the rule in plants systems rather than the exception. Defensive chemicals vary in concentration among populations as influenced by environmental conditions, among genotypes within one population,

TABLE 7.5 Some of the Most Important Classes of Compounds Associated with Plant Defense Against Herbivory

Class of Defensive Compound	Some Example Compounds	Distribution in Plants
Alkanes, aldehydes, ketones	Plant surface waxes	Found in all plants
Lignins and tannins (phenolic polymers)	Lignin and tannic acid	Integral cell wall constituents
Iridoids	Irodial	Essential oils in 57 dicot families
Phytojuvenile hormones	Sesamin, juvenile hormone II	From seeds and other tissues in a diversity of plants
Antijuvenile hormones	Preconene 1 and 2	From one species in Asteraceae
Sesquiterpene lactones	Alantolactone, cumambranolide	Located in glandular hairs of species in Asteraceae
Saponins	Diosgenin, β-amyrin	Widespread in angiosperms
Phytoecdysones (steroid antihormone)	Ponasterone, β-ecdysone	Found in more that 100 plant families
Phenolics	Cinnamic acid, chlorogenic acid	University in leaves and often in other tissues
Coumarins	Furanocoumarines, umbelliferone	In roots and seed coats of plants in more than 70 families
Flavonoids	Anthocyanins, flavones	Universal in angiosperms, gymnosperms, and ferns
Quinones	Benzoquinone, antharquinone	Found in vascular plant families, not in ferns and mosses
Alkaloids	Reticuline, vinblastine, morphine, papaverine	Widely distributed in angiosperms' roots, leaves, and fruits
Nonprotein amino acids	Canavanine, L-hypoglycine	Widespread in plants but abundant in legume seeds
Cyanogneic glucosides	Dhurrin, amygdalin	In 2500 plants in over 140 families
Glucosinolates	Isothiocyanate, sinigrin	In brassicaceae and similar families
Lectins	Sugar-binding proteins	Universally in plants, up to 11% of total protein

Source: Adapted from table 3.1 in Panda and Khush 1995.

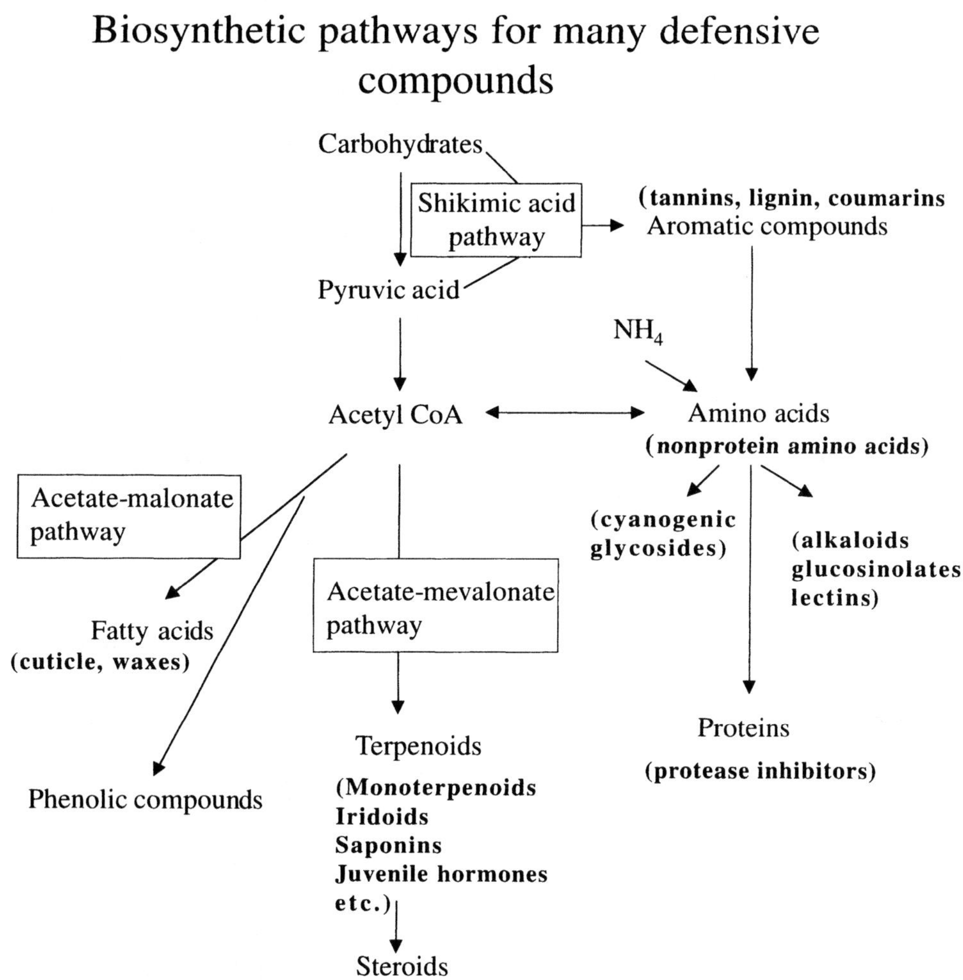

Figure 7.1. The three main biosynthetic pathways involved in the production of defensive compounds (identified in boxes). Notice that acetyl CoA is central to the synthesis of many of the compounds used in defense against herbivores. (Modified from Panda and Khush 1995, with permission.)

among branches on one individual, among times on one branch, and among times and locations on one leaf. Understanding this variation as it relates to defense against herbivores has been a central theme in research on plant chemical defense mechanisms, and is one of the most crucial factors defining plant-herbivore interactions.

A complete coverage of all the molecules associated with defense against herbivores is a large task for a book, let alone a chapter. Therefore, in the next section we discuss the most salient aspects of a selection of substances that represent the major types of compounds influencing the interaction between plant and herbivore. Substances such as lectins, ketones, terpenes, and several others that have antiherbivore action are covered in the supplementary readings or in Chapter 8. We discuss each substance by considering its structure, diversity in plants, tissue and subcellular localization, and effects on herbivores.

1. Tannins and Lignins. Both tannins and lignins are complex molecules based upon phenol monomers. These substances are ubiquitous in the plant kingdom but are not found in animals. Tannins obtained their name because these are the phenol-rich plant extracts that are used to convert animal hides into leather. This terminology was established before the chemical nature of these compounds was known, which often leads to confusion about the difference between tannins and lignins. Both are polymers of phenol-containing subunits, but the monomers are underivatized in tannins (Figure 7.2). Tannin molecular weights range from a minimum of 500 to values exceeding 20,000, and tannins are water soluble (except in the case of very large molecular weight tannins). Tannins fall into two basic categories: condensed and hydrolyzable.

The *condensed tannins,* the most widely distributed type among plant groups, are oligomers of flavonoid monomers held together by carbon-carbon bonds (Figure 7.2A). In natural systems, these compounds cannot be hydrolyzed, but the flavonoid units can be liberated when dissolved in a strongly acidic solution containing trace metals and heated. When hydrolyzed in this manner, the degradation products include anthocyanidins and a number of uncharacterized products. A single plant tissue may contain many different oligomers of anthocyanidins (Kolodziej 1990). The relative proportions of different cyanidins (e.g., procyanidin/prodelphinadin) in the tannin complement of tissues varies among species (Ayers et al. 1997).

In contrast, the *hydrolyzable tannins* have monomer units tied together by carboxylic ester linkages (Figure 7.2B,C). These compounds can be hydrolyzed in either a basic or an acidic environment or by esterase enzymes. The basic monomer is a polyol, commonly D-glucose with hydroxyl groups esterified by gallic acid (called gallotannins) or hexahydroxyphenic acid (called ellagitannins). The most common gallotannins are heavily esterified, having six to nine gallolyl groups per molecule (Figure 7.2).

Tannin concentration varies with plant developmental stage, environmental condition, and genotype. In addition, the amount of tannins in plants varies widely among both wild and domesticated species. In general, the lower the amount of tannins the more palatable the tissue. Thus, breeding programs have lowered tannin concentrations in crops. Herbivory pressure has induced the evolution of higher tannin concentrations in wild plants (Skogsmyr and Fagerstrom 1992).

Tannins have been traditionally thought to affect metabolism by binding to proteins thereby deactivating them or causing their precipitation. The types of proteins that bind well to tannins are relatively large, open, and flexible, with high concentrations of proline. Glycoproteins can either improve or hinder tannin binding. Tannin-protein bonds are based on hydrogen bonding and hydrophobic interactions. Thus, this binding is not permanent and can be reversed by the addition of another effective tannin-binding compound such as polyvinylpyrrolidone (PVP). This is the reason why most enzyme extraction protocols utilize PVP in the isolation medium to protect the target enzyme from inhibition by tannins. Tannin-protein complexes may cause protein precipitation in some cases, while in other cases the tannin-protein complex is soluble.

Tannins also bind to some other cellular components, which may suggest a possible *in vivo* activity for tannins. Tannins bind to several metals such as iron, thereby sequestering iron in the vacuole. Therefore, tannins may serve as a type of heavy-metal tolerance mechanism. Tannins also scavenge free radicals of oxygen which may serve as a method to minimize oxidative damage during high light intensity and low intercellular carbon dioxide concentration. Tannins also bind to membranes, and the tannin-phospholipid complex may serve to moderate membrane morphology and permeability (Johnson et al. 1986.). In fact, it is this general stickiness that makes these compounds effective antiherbivory agents (Zucker 1983).

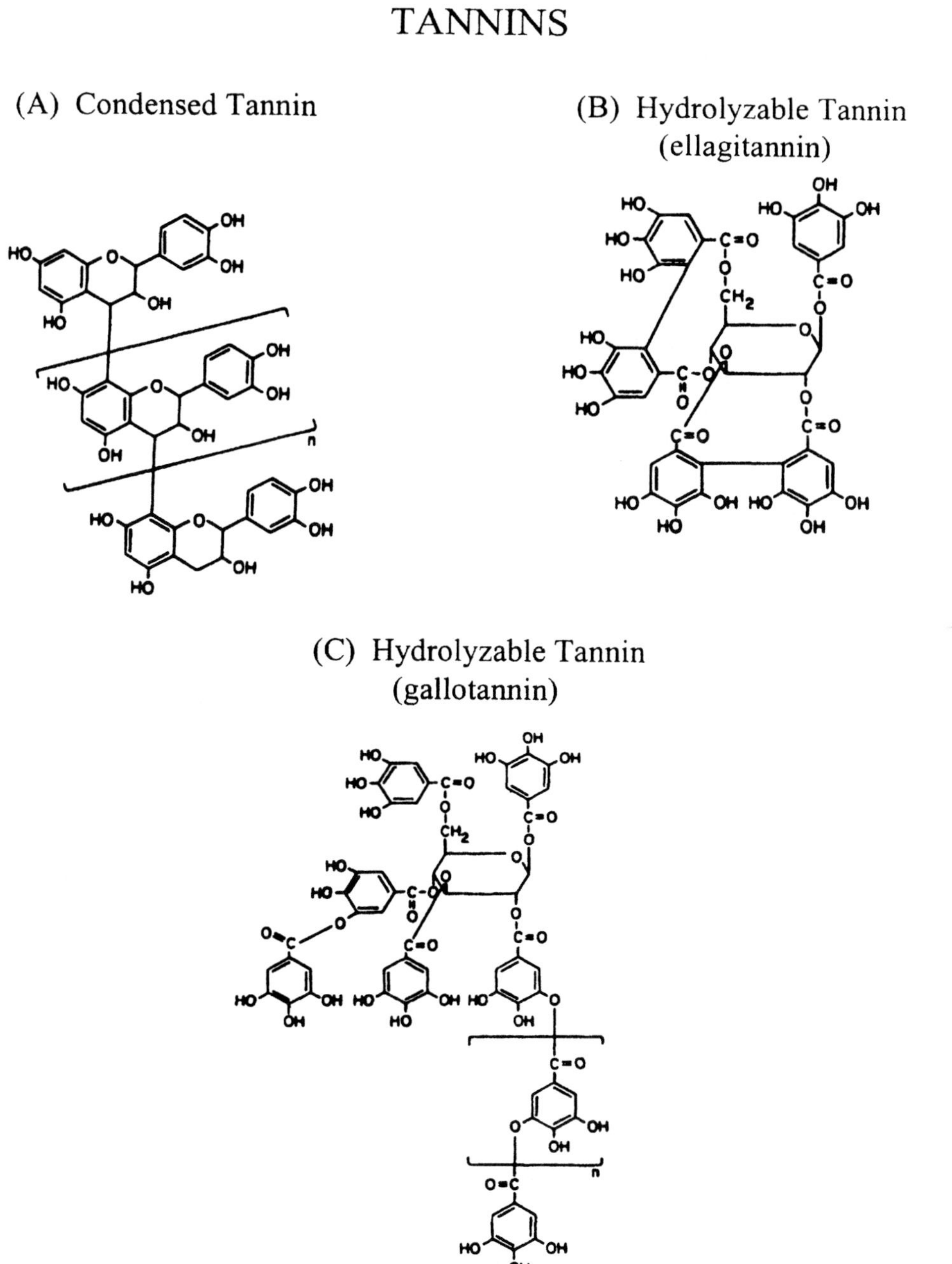

Figure 7.2. The two main classes of tannins, represented by three molecular diagrams. (A) Condensed tannins. (B, C) Hydrolyzable tannins from two groups. (From Hagerman and Butler 1991, with permission.).

The efficacy of tannins on herbivore fitness was originally thought to relate to the protein binding characteristics of this compound (Feeney 1970, Harborne 1988), which causes a reduction in protein assimilation by herbivores (Robbins et al. 1987). Other studies suggest that the direct toxicity of tannins to digestive cells is the most important antiherbivory process (Voltura and Wunder 1994). Some research on the tannin contents of oak leaves has raised significant questions about tannin toxicity. The concentration of condensed tannins increases dramatically during oak leaf senescence. In contrast, the protein complexing characteristics of oak leaves remain fairly constant throughout leaf development (Haslam 1988). These and other studies of wild plants with high tannin contents have led some authors to speculate that tannins do not serve an antiherbivory function, but rather are simply waste products of metabolism (Martin et al. 1987, Haslam 1988, Smith et al. 1992). In fact, tannins may simply make the plant tissue less palatable by increasing its toughness.

There are several reasons why it is hard to believe that tannins have no significance to plants except as a dumping ground of waste materials. A considerable amount of energy in the form of ATP and carbon resources is required to synthesize these compounds and maintain their metabolic pathway. Furthermore, tannins are known to be toxic. For example, hamsters given a diet of 4% tannins die within 3 days (Mehansho et al. 1987). One possible mechanism for the mortality of these hamsters relates to the astringency of tannins in the oral cavity which causes coagulation of mucoproteins. There is no doubt that animals tend to reject astringent food (Mole and Waterman 1987); however, there is no general agreement about what physiological dysfunction the animals are avoiding by not eating the tannin-rich food. When animals are fed only tannin-rich diets, they generally consume less than they would if they received tannin-free diets (Mole and Waterman 1987), and this may result in reduced growth and slow development (Clausen et al. 1990). Reduced consumption cannot be the only explanation because it cannot account for the entire reduction in growth of herbivores offered only tannin-rich feed.

A recent theory about the mechanism by which tannins inhibit the growth of insects relates to post-digestive use of amino acids rather than the inhibition of digestion enzymes, or precipitation of proteins. The effectiveness of this mechanism is dependent upon absorption of a tannin-associated component in the gut (Butler 1989). It was originally thought that tannins were too large to be taken up in the digestive system and were very resistant to degradation. However, studies have shown that in some cases (sheep), a large proportion of tannins in diets can be digested and absorbed, while in other cases (deer), no tannins are degraded or absorbed (Robbins et al. 1991). Whether tannins inhibit digestion efficiency at low protein content or inhibit post-absorptive amino acid use, herbivores must develop some mechanisms of resistance if they are to utilize the wild plant species that often contain relatively high tannin content (Appel and Schultz 1992).

The effects of tannins on growth efficiency are variable among different organisms and developmental stages. Carnivores are more susceptible than herbivores (Salunkhe et al. 1990), and ruminants are more susceptible than nonruminants (Barry and Manley 1986). The effect of tannins on the growth efficiency of herbivores also varies with developmental age, because juveniles are more susceptible than adults (Joslyn and Glick 1969). In addition, the efficacy of tannins in reducing growth efficiency may relate to the amount of protein in the feed. Tannins are most effective at reducing growth and development in birds when dietary protein is low (Garwood and Rogler 1987). The chemical composition of the tannins in plants and the particular herbivore species both influence the level of toxicity (Ayers et al. 1997).

Herbivores can utilize both behavioral and physiological mechanisms to avoid tannin toxicity. Behavioral mechanisms include avoiding high tannin-containing plants, pretreatment of food with an alkali, or concurrent consumption of tannin-binding compounds such as milk. Herbivores utilize a diversity of physiological mechanisms to detoxify tannins. Some examples include: (1) preventing protein precipitation in the gut with surfactants (Tugwell and Branch 1992); (2) increasing the pH of the gut to prevent interaction of tannins and proteins (Felton and Duffey 1991); (3) releasing enzymes, such as polyphenol oxidase (PPO), that degrade tannins in the gut (Nichols-Orians 1991); (4) increasing the gut redox potential (Appel and Martin 1990, Appel 1993); (5) inducing proline-rich tannin-binding proteins in saliva (Hagerman and Robbins 1993); and (6) reducing the resting metabolic rate to minimize intake of tannins.

Although there are many similarities in the chemical structure of tannins and lignins, lignins are an integral component of secondary tissue in plants and cannot be considered a secondary metabolite. Lignins are synthesized by one branch on the phenylpropanoid pathway (Hahlbrook and Scheel 1989). Two of the enzymes responsible for the synthesis of lignins are cinnamyl alcohol dehydrogenase (CAD) and peroxidase (POD). CAD is a lignin-specific enzyme that catalyzes the production of cinnamyl alcohol subunits that eventually form the complex three-dimensional lignin structure (Walter et al. 1988). POD functions in a number of different pathways that include polymerization of cinnamyl alcohol subunits into the lignin backbone and producing hydrogen peroxide or other phenoxy radicals that may interfere with herbivore digestion (Appel 1993). The end result is a complex molecule formed of highly methylated phenolics (Figure 7.3). There is no single structure that can be pointed to as representing all lignins. The phenolic compounds are linked in a random manner, unlike the directed polymerization of nucleic acids or proteins. These complex molecules are resistant to enzymatic degradation and are primarily located in the secondary walls of vascular tissue. In gymnosperms, lignins are composed primarily of conifryl alcohol subunits, while in angiosperms, cinnamyl and sinapyl alcohol-derived subunits are also present. In grasses, lignin is unique from other herbaceous or woody species because it can be solubilized by sodium hydroxide. This may be because the linkages which bind lignin to the cell wall in grasses differ from those in other species.

Lignin also has been shown to be toxic to herbivores (Appel 1993). All plant tissues contain some lignin because of its importance in secondary cell wall structure. The toughness of plant tissue is directly related to lignin concentration. Therefore, all herbivores will be exposed to lignin, and the tougher the tissue, the greater the lignin content. Tissue toughness is one factor that regulates the amount of herbivore damage a species may experience. Tissue with a high lignin concentration (and therefore tough) is less palatable than tissue with low lignin content. For example, the brown-midrib mutants of corn have been shown to be more digestible and more sensitive to pathogen infection than wild-type corn because they have lower amounts of, and chemically altered, lignin (Oba and Allen 1999). Furthermore, browning of the pericarp of maize is associated with increased lignin and increased insect resistance (Dowd 1994). The specific inhibition of lignin biosynthesis in wheat strongly reduces wheat resistance to rust (Moerschbacher et al. 1990). In addition, several other studies have shown that the digestibility of feed is directly related to lignin content (Chesson 1981, Foley and Hume 1987). In fact, mechanical damage due to wind induces the lignin branch of the phenylpropanoid pathway in leaves and this is associated with increase in resistance to insect damage (Cipollini 1997).

POD and polyphenol oxidase (PPO) are enzymes involved in the polymerization of phenolics leading to lignin synthesis in the cell wall. They are also involved in the

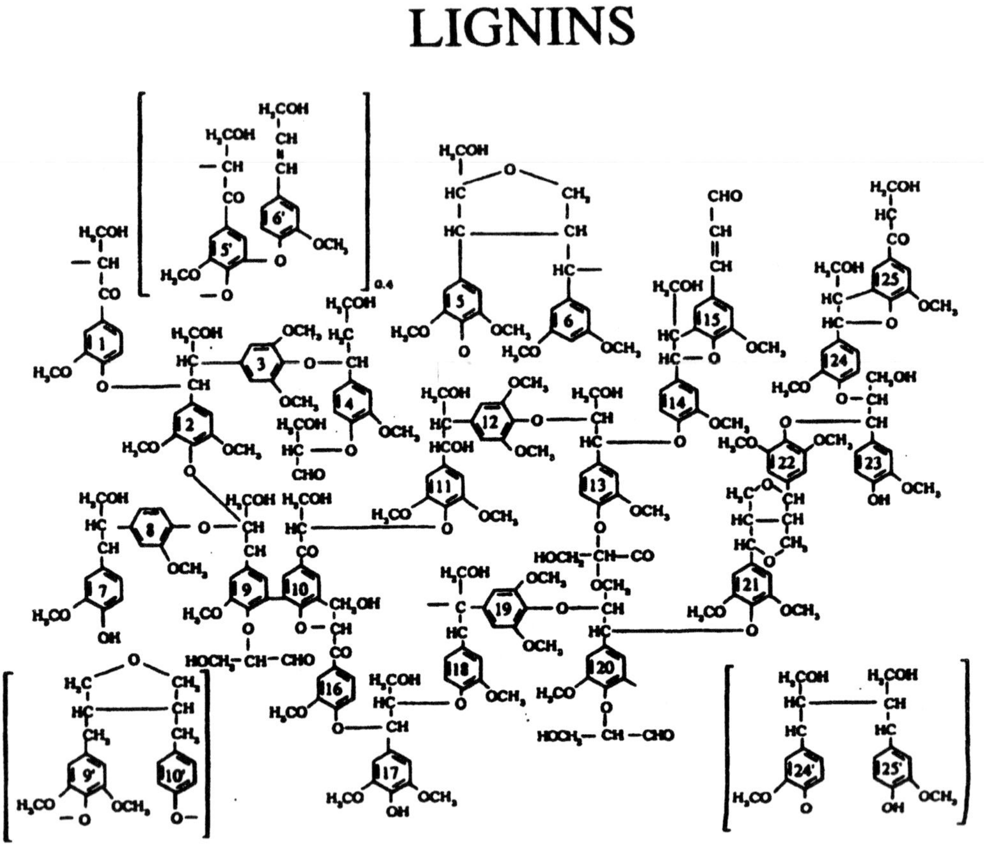

Figure 7.3. Lignins are complex structures consisting of cinnamyl alcohol subunits. This representation is based on the lignin in beech cell walls. (From Hagerman and Butler 1991, with permission.).

generation of reactive oxygen species, quinones, and hydroperoxides (Felton et al. 1989). Therefore, increasing the activity of POD and PPO for increased lignin synthesis also results in products that are known to inhibit insect performance (Appel 1993). A strong relationship has been found between increased activity of POD and PPO in plant tissues and decreased growth rates of herbivores feeding on those tissues (Cipollini and Redman1999). Age-specific induction of these two enzymes and the resulting effects on insect growth rates suggest that PPO is most associated with induced defenses in young (4-week-old) seedlings and POD is most important in older (8-week-old) seedlings (Cipollini and Redman 1999). Therefore, increasing the synthesis of lignin has negative effects on insect fitness because PPO and POD are important enzymes in lignin synthesis pathway and they produce toxic by-products. Increased foliage toughness, and increased toxic by-products of lignin synthesis (POD and PPO activity) are the most likely mechanisms for the lignin-associated protection of plant tissues from pests (Coley 1983, Appel 1993).

2. Phenolics and Resins. Phenolic compounds are aromatic in nature and are synthesized by the shikimic acid pathway. Each compound in this group contains one or more phenolic hydroxyl groups. One group of phenolics, the phenylpropanoids, are an abundant homogeneous group of substances found in a wide diversity of plant species. The phenylpropaniods are derivatives of cinnamic acid, which is produced by the action of phenylalanine ammonia lyase (PAL). This simple reaction scheme can produce many phenylpropanoids that may have antiherbivore action:

Phenylalanine → cinnamic acid → ρ-coumaric acid →

caffeic acid → ferulic acid → 5-hydroxyferulic acid → sinapic acid

Phenolic acids have both allelochemical and antiherbivore action. Chlorogenic acid is toxic to corn earworm, and is concentrated in nightshade fruits as well as the trichomes of tomato. Although phenolic acids can have some antiherbivore efficacy, these compounds are most often studied in relation to allelopathy. See Chapter 8 for a more extensive discussion of phenolics in plants.

Resins are found in leaves and stems in some species, particularly desert taxa and conifers. These resins are aromatic and are composed of a mixture of different phenolic compounds. The best-known plant resins in relation to herbivore defense are those of pines (*Pinus* spp.) and desert creosote bush (*Larrea tridentata*). Resins in pines are located in resin ducts in wood and leaves, so that mechanical damage of plant tissues by a herbivore causes the oozing of resins onto the plant surface, which inhibits feeding of many insects. Of course, there are insects, such as bark beetles, that are not inhibited by the resins and are actually attracted by the aromatic volatile substances emanating from pine resins. In contrast, the resins of creosote bush are exuded out on the surface of leaves and stems at all times. The relationship between creosote bush resins and specialized herbivores is particularly well known.

The resins on the surface of creosote bush are composed of saponins, phenolics, terpenoids, and wax ester (Rhodes 1977). Over 80% of the resin is composed of phenolic aglycones, and the major component of these is nordihydroguiairetic acid (NDGA). NDGA has significant fungicidal, bactericidal, and antiherbivore properties (Gonzales-Coloma et al. 1988). The concentration of these resins on the leaf surface decreases with developmental age. On average, the resins account for over 26% of juvenile leaf dry weight but only 10.5% of mature leaf dry weight. The amount of NDGA in the resins is regulated both by genetic and abiotic factors. There is a large amount of variation in NDGA content among creosote shrubs in the Sonoran desert, but high-producing shrubs are consistently high producing from year to year. Environmental factors such as latitude or atmospheric chemistry (ozone) can affect the concentration of NDGA on a seasonal or regional basis.

Most herbivores avoid eating creosote bush tissues. If nonspecialized herbivores are fed creosote bush tissues, they have severely reduced growth rate and slow developmental progression. The toxicity of the tissues is related to the resins because toxicity decreases with a decrease in resin content. The direct linkage between NDGA and toxicity is complicated because of the multiple other phenolic compounds in the resin (Rundel et al. 1994). However, there are some herbivore species that are able to feed on creosote bush, and some of these are specialist feeders utilizing only creosote bush. One species of specialist grasshopper (*Ligurotettix coquilletti*), which feeds only on creosote bush, chooses only creosote tissues that are relatively low in NDGA concentration. The growth rate of this grasshopper decreases linearly with increasing NDGA in its feed. However, another species of specialist grasshopper (*Bootettix argentatus*) does not avoid tissues with up to 20% NDGA, and there is no impact of

NDGA on insect growth rate (Chapman et al. 1988). The mechanism by which some specialist herbivores are not affected by NDGA is not known.

3. Cyanogenic Glycosides. Cyanogenic glycosides are soluble polar molecules that are accumulated in the central vacuole of plant cells. The general structure of these compounds (Figure 7.4) can be characterized as a glycoside of various hydroxynitriles (also called cyanohydrins). In the presence of β-glycosidase (which cleaves off the sugars) and hydroxynitrile lyase (which decomposes the cyanohydrins), cyanogenic glycosides release cyanide (HCN) which can be highly toxic to herbivores. This group of compounds may also include nitrile glycosides, nitro acids, nitro alcohols, and cyanolipids (Figure 7.4), because these compounds are derived from similar pathways and also release HCN upon enzymatic degradation. These nonglycosidic, cyanogenic compounds are restricted to a smaller diversity of plants than that of the cyanogenic glycosides (e.g., cyanolipids are restricted to the Sapindaceae and the Hippocastanaceae).

Cyanogenic glycosides are considered to be part of plant secondary chemistry, and their biosynthesis is linked to the production of glucosinolates, or other organic, nitro compounds in plants such as amino acids. There are about 60 known cyanogenic glycosides that are derived from the L-amino acids phenylalanine, tyrosine, leucine, isoleucine, and valine. A few others are derived from nicotinic acid and the nonprotein amino acid (2-cyclopentenyl)glycine. During the synthesis of cyanogenic glycosides from amino acids, the amino acid backbone loses its carboxyl carbon and the alpha carbon (which holds the amino group) is oxidized to the nitrile level. The beta carbon on the original amino acid backbone is oxidized to form a hydroxyl, and this is where the sugar of the glycoside is attached (Figure 7.5).

These substances have been found in more than 2500 species in over 500 genera and 150 families including all major vascular plant orders as well as some nonvascular plants families. Cyanogenesis is particularly common in members of Asteracea, Euphorbiaceae, Fabaceae, Papaveraceae, Graminaceae, and Proteaceae. In most cases, only one cyanogenic glycoside occurs in an individual species, but in some cases mixtures of two or more cyanogenic compounds can be found in a specific tissue of one species. For example, complex mixtures of di-, tri-, and tetraglycosides have been found in members of Asteraceae (Schwind et al. 1990). Also, mono- and diglycerides co-occur in the same tissue in flax (*Linum usitatissimum*) (Smith et al. 1980) and rubber (*Hevea brasiliensis*) (Lieberei et al. 1985).

The most-studied and best-known cyanogenic glycosides are amygdalin derived from L-phenylalanine, and linamarin derived from valine (Figure 7.4). Amygdalin is found in the seeds and seed coats of members of Roseaceae such as peaches, apples, cherries, and apricots. Linamarin is probably the most widespread cyanogenic glycoside, occurring in more plant species that any other cyanogenic compound.

Cyanogenic glycosides are nontoxic by themselves to plants or their herbivores; however, free HCN is extremely toxic to both plant and herbivore. Very small (micromolar) amounts of HCN may be found constitutively in plant cells and may serve to regulate nitrate reductase because HCN can reversibly inhibit nitrate reductase. However, plant tissues cannot tolerate higher quantities of HCN. The cytochrome *N*, a protein complex of oxidative electron transport, is inhibited by HCN. Plants do have an alternate electron transport pathway that is resistant to HCN, but his pathway alone cannot provide the energy resources for all of cell metabolism for any length of time. Therefore, if cyanogenic glycosides are to be used by plants as an effective defense against herbivores, plants must have a low constitutive HCN concentration and a

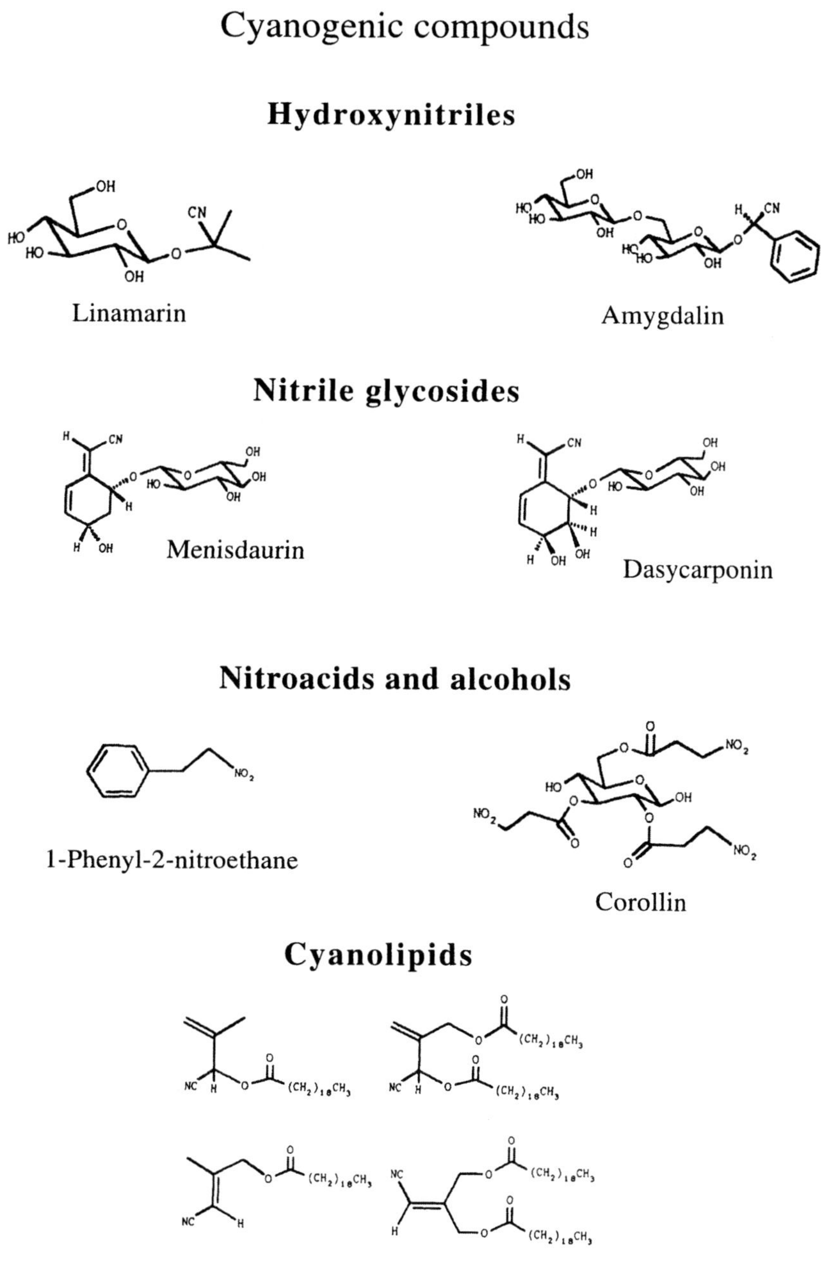

Figure 7.4. Four different classes of molecules with cyanogenic properties. (Molecular diagrams reprinted with permission from Seigler 1991.)

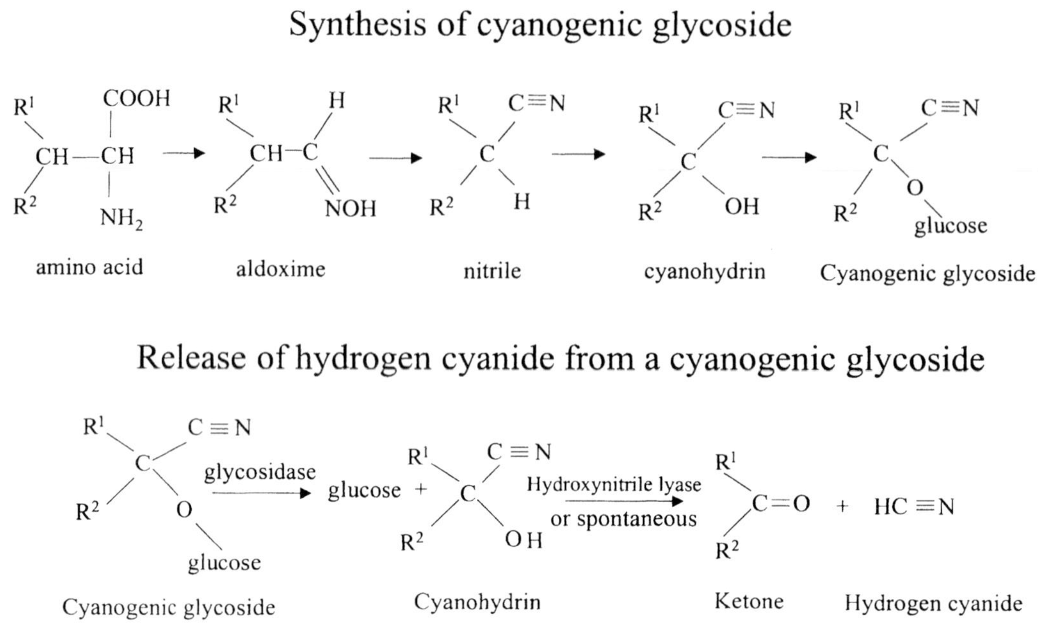

Figure 7.5. (*Upper panel*) Main method of cyanogenic glycoside synthesis from an amino acid precursor. (*Lower panel*) Two-step process required for the release of hydrogen cyanide from a cyanogenic glycoside. (Redrawn from Panda and Khush 1995, with permission).

mechanism for inducing the release of HCN from cyanogenic glycosides upon herbivore attack.

Recall that the enzymes β-galactosidase and hydroxynitrile lyase are required to release HCN from cyanogenic glycosides. These enzymes must be released upon herbivore damage of tissue in order for HCN to be generated from cyanogenic glycosides (Figure 7.5). In some cases, the hydrolytic enzymes are located in one cell type and the cyanogenic glycosides are in another. For example, in sorghum, dhurrin is predominantly located in epidermal cells, while the hydrolytic enzymes occur in mesophyll cells. Upon herbivory, the tissue is crushed and the contents of the two cell types merge, releasing HCN. In other plants such as lima bean or clover leaves and cassava roots, the cyanogenic glycosides are located in the vacuole while the hydrolytic enzymes are located in the apoplast (intercellular spaces).

The effectiveness of HCN as a poison in humans and domestic livestock is indisputable, but there is some controversy about the efficacy of HCN as a defense against herbivores in natural systems. Part of this doubt comes from the observation that many herbivores can detoxify small amounts of HCN. Moreover, some phytophagous insects have evolved mechanisms to detoxify large quantities of cyanide. Larvae of the southern army worm grow better when there is KCN in their diet than when the diet is free of cyanide (Brattsen et al. 1983). African army worms can consume grass containing enough cyanide to kill cattle without any detrimental effect on army worm growth (Georgiadis and McNaughton 1988). In fact, for nonadapted insects, the lethal dose of cyanide from cyanogenic glycosides is at least 10–20 times the amount found in most consumed plant tissues. Few studies have provided statistically sound data indicating that HCN release from cyanogenic glycosides during herbivory is an effective mechanism of defense against insect herbivores (Hruska 1988). There is also evidence that other compounds released during the hydrolysis of cyanogenic glycosides such as aldehydes and ketones may be the responsible herbivore repellent rather than cyanide itself (Spencer 1988).

4. Nonprotein Amino Acids A unique characteristic of plants is their ability to produce an extremely wide variety of amino acids that are not incorporated into a protein structure. In general, an amino acid has the formula R-CH(NH_2)COOH and about 20 different molecules are utilized in proteins. In plants, another 600 or more nonprotein amino acids have been identified. Most are free compounds (unbound to proteins, membranes, or carbohydrates); however, some are attached to a carbohydrate and a few occur as 4-glutamyl-linked peptides. Most nonprotein amino acids are aromatic or heterocyclic. In such cases, the R group is cyclic and may contain noncarbon elements such as O, S, and N. Aromatic nonprotein amino acids are often hydroxylated, and many have a phenyl group associated with alanine or glycine (Figure 7.6). For a complete discussion of nonprotein amino acid structure and biochemistry, see Rosenthal (1982).

Many of the nonprotein amino acids are nontoxic and can be accumulated in response to stress. For example, glycine-betaine and proline are accumulated by some plants in response to water, or salt stress. Other nonprotein amino acids are accumulated by plants during chilling stress. These cytoplasmic compatible substances can alter osmotic potential, conferring drought tolerance or decreasing tissue freezing point. In addition, some nonprotein amino acids are toxic to herbivores. Several ver-

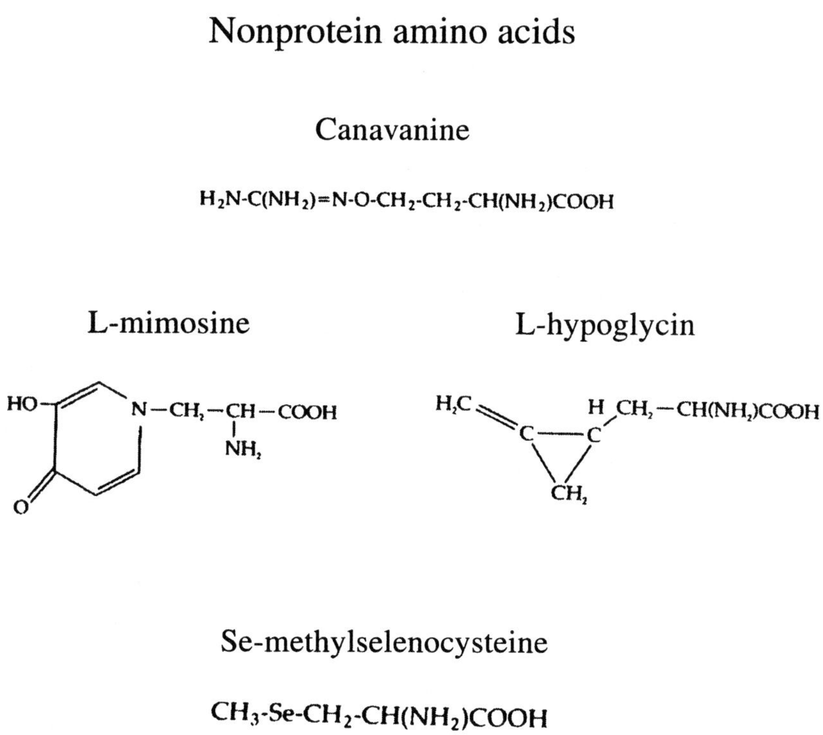

Figure 7.6. Several cyclic and noncyclic nonprotein amino acids involved in plant defense against herbivores. (Redrawn with permission from Rosenthal 1991.)

tebrate diseases can be directly tied to consuming plants that contain a particular nonprotein amino acid. For example, neurolathyrism is a disorder of the nervous system that results from eating *Lathyrus* species that contain toxic nonprotein amino acids such as methylselenocysteine. Canavanine in plant tissues can severely disrupt growth and development of herbivores. Consumption of other nonprotein amino acids can induce violent vomiting (hypoglycine), hair loss (mimosine), or other symptoms in herbivores.

The effect of canavanine on potential herbivores is arguably the best documented case of a toxic nonprotein amino acid affecting phytophagous insects. Canavanine is found in one subfamily (Lotoideae) of the Fabaceae. However, it is found in 1500 species of this group. In general, canavanine has powerful antimetabolic effects on a broad array of organisms; insects, bacteria, viruses, and fungal pathogens are deleteriously influenced by canavanine. Therefore, the effect of canavanine on metabolism must occur at a basic level that is uniform across biological kingdoms. The linkage between nucleic acids and proteins is the most basic and ubiquitous component of metabolism, and a likely site of action for this toxin.

Canavanine is an analog of arginine, which is an important protein amino acid. In fact, arginyl-tRNA synthetase, which normally links arginine to its appropriate tRNA, can not distinguish between arginine and canavanine. Therefore, this amino acyl-tRNA synthetase catalyzes the activation and attachment of canavanine to the tRNA that normally carries arginine. Therefore, canavanine will be incorporated into proteins instead of arginine. Canavanine is much less basic that arginine. Consequently, polypeptide chains that contain canavanine in the place of arginine will not have appropriate basic domains that are required for proper protein folding. Thus, the consumption of tissue rich in canavanine could disrupt normal structure and function of proteins in herbivores.

Canavanine-sensitive insects accumulate large amounts of this amino acid in their proteins when placed on a canavanine-rich diet. The substitution error frequency (SEF) represents the frequency at which canavanine replaces arginine in proteins. In canavanine-resistant insects such as the bruchid beetle *Caryedes brasiliensis,* the SEF ratio in total proteins is 1 in 365 (Rosenthal et al. 1987). In contrast, the tobacco budworm (*Heliotus virescens*), a canavanine-sensitive species, has an SEF value of 1 in 3.3 when fed on the same canavanine-laden diet as the bruchid beetle. Therefore, canavanine-sensitive species incorporate the toxin in most of their proteins when fed a canavanine-laced diet, while canavanine-resistant species are able to discriminate between canavanine and arginine when synthesizing proteins.

Nonprotein amino acids that contain a cyclic group attached to the alpha carbon of the amino acid backbone can be particularly toxic. For example, L-hypoglycine has a three-carbon ring and is concentrated in the seeds of various members of Sapindaceae (Figure 7.6). This nonprotein amino acid reaches 0.008% of the dry weight in arils (seed coats) of the ackee tree (*Blighia sapida*) and is particularly toxic to mammals (Hassall and Reyle 1955).

Ripe ackee seeds are consumed by people living in Jamaica. When the seeds are green, the levels of L-hypoglycine are 15 times those of the ripe seeds. Therefore, as long as ripe seeds are eaten, there is little harm. When food supplies become low, starving people consume the green seed. After consuming a large quantity of this green seed, humans experience vomiting and convulsions, followed by coma and death. These symptoms are due to a severe case of hypoglycemia. Blood sugar can decrease from 100 mg ml^{-1} to 3 mg ml^{-1} shortly after consuming the green seed. The hypoglycemic state results from an inhibition of fatty acid oxidation, which puts a large demand on glucose. At the cellular level, L-hypoglycine induces mitochondrial swelling, inner membrane clumping, and a severe decrease in phosphorylation (McKerns et al. 1960).

As stated above, members of the genus *Astragalus* also contain toxic nonprotein amino acids. These compounds differ from other nonprotein amino acids in plants because they contain selenium instead of sulfur. In most cases, the selenium is stored in conjunction with cysteine or methionine. Approximately 20 of the 500 species of *Astragalus* accumulate selenium amino acids in high enough doses to be toxic. Selenium accumulators produce Se-methylselenocysteine (Figure 7.6) and selenocystathionine and concentrate them up to 15,000 ppm (Virupaksha and Shrift 1965). The selenium-containing nonprotein amino acids are close enough analogs to cysteine and methionine that they are incorporated into proteins of the herbivore. The selenium amino acid will not be able to form disulfide bridges in proteins that are usually associated with cysteine. This will severely alter the shape of proteins and induce dysfunction (Shrift 1967).

Selenium toxicity is often seen in cattle or other animals that feed on the range in the American Southwest. Symptoms include neural disorders and hoof or leg dysfunction. Two conditions—alkaline disease and blind staggers—are attributed to selenium toxicity in range animals.

In general, nonprotein amino acids are toxic to many herbivores when they resemble the structure (but not the function) of an amino acid that is normally an important part of protein structure. The toxic nonprotein amino acid is not discriminated against by the appropriate aminoacyl-tRNA synthetase; thus, the toxin becomes activated for participation in translation. Why is it that plants accumulating these toxic nonprotein amino acids do not incorporate them into their own proteins? In fact, toxic nonprotein amino acids can be 50–200 times the concentration of the amino acids they resemble and not be incorporated into protein by plants. Aminoacyl-tRNA synthetase molecules in plants that concentrate nonprotein amino acids are capable of discriminating against the accumulated nonprotein amino acid in their tissues (Fowden et al. 1979). A few herbivores can develop resistance in much the same way that plants avoid toxicity. For example, *Dioclea megacarpa* (a tropical legume) concentrates canavanine to 8% of seed dry weight and is remarkably free of seed herbivores. However, there is one bruchid beetle that can consume the seeds and completely avoid toxicity (Bleiler et al. 1988).

5. Alkaloids. Alkaloids are the most diverse and abundant of the defense-related secondary compounds in plants. At least 10,000 different alkaloids have been described (Southon and Buckingham 1989), and 37 volumes on the chemistry and pharmacology of alkaloids were written between 1950 and 1990. In general, *alkaloids* are defined as cyclic compounds that contain nitrogen in a negative oxidation state and are produced by a limited number of organisms. It is this last characteristic of alkaloids that makes them specific rather than broad-spectrum substances. Even though the term *alkaloid* covers a vast array of different compounds, each compound is produced by a small number of plant species and influences a relatively small number of herbivores. Most alkaloids have been described from plant systems (20% of plants studied accumulate alkaloids); however, these compounds also are known to be produced by amphibians (Daly and Spande 1986), marine organisms (Christopherson 1985), and insects where alkaloids can serve as defensive agents or pheromones (Numata and Ibuka 1987).

Alkaloids can be divided into five broad classes based on chemical structure (Figure 7.7):

1. *Benzylisoquinoline alkaloids* (BAs) are derived from the union of two tyrosine molecules leading to (S)-norcoclaurine, which can be modified into a diversity of compounds. There are approximately 2500 benzylisoquinoline alkaloids including morphine, papaverine, and berberine.

Alkaloids

Benzylisoquinoline alkaloids

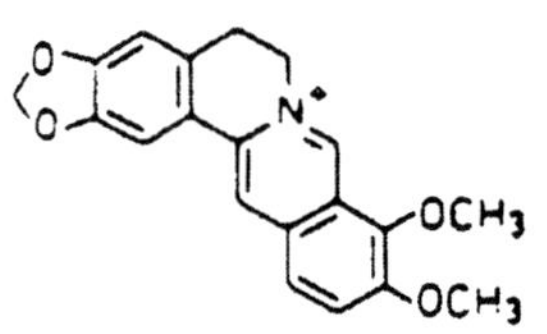

(Berberine)

Monoterpenoidindole alkaloids

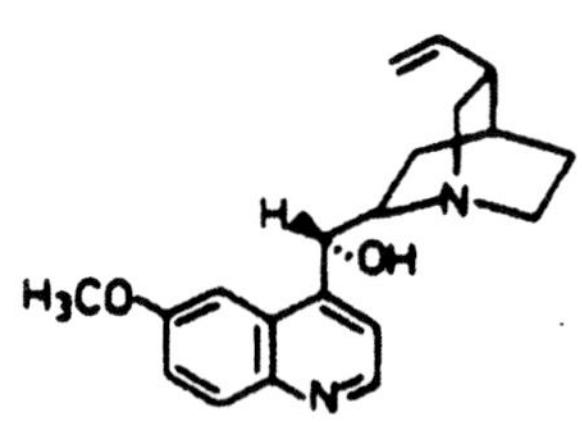

(Quinine)

Pyrrolizidine alkaloids

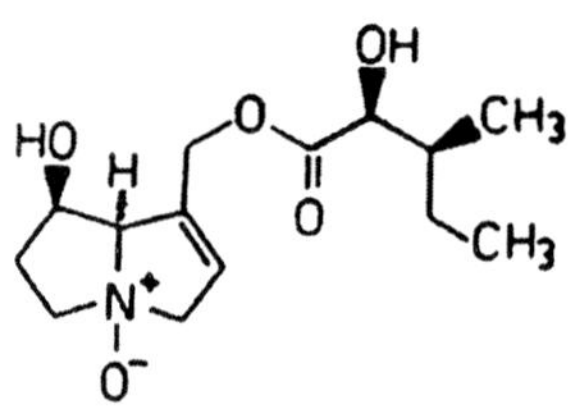

(Creatonotine *N*-oxide)

Quinolizidine alkaloids

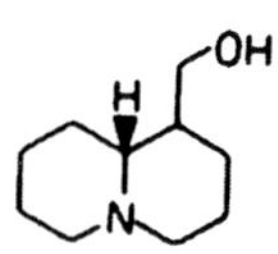

(Lupinine)

Polyhydroxindolizine alkaloids

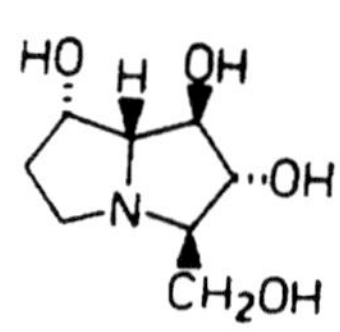

(Australine)

Figure 7.7. The five major classes of alkaloids, represented here by one compound each. (Molecular diagrams redrawn with permission from Hartmann 1991.)

2. *Monoterpenoid indole alkaloids* (MAs) are predominantly produced by tropical plant families and are usually derived from (S)-strictosidine. There are about 1700 of these compounds, of which several are important pharmaceuticals. For example, vinblastine and vincristine have been used as medicines for the treatment of leukemia and Hodgkin's disease. Other important alkaloids in this class are quinine (used to treat malaria) and strychnine (a potent poison for vertebrates).
3. *Pyrrolizidine alkaloids* (PAs) are based on necine, which is derived from a pair of putracine molecules. These alkaloids are most abundant in Asteraceae and Boraginaceae, and about 250 different molecules have been identified.
4. *Quinolizidine alkaloids* (QAs) are derived from lysine and are restricted to a few tribes of the legume subfamily Papilionoieae. About 170 of these compounds in four to six structural groups have been identified.
5. *Polyhydroxypyrrolizidine alkaloids* are polyhydrated compounds that mimic sugar molecules. These compounds (such as australine in Figure 7.7) interfere with sugar metabolism and the normal action of certain enzymes such as galactosidases. Most of these compounds are found in five plant families: Moraceae, Fabaceae, Polygonaceae, Aspidaceae, and Euphorbiaceae.

Alkaloids are generally considered to be a constitutive part of plant defense. This means that they are present at all times in some or all plant tissues, including the phloem, and their concentrations may vary among tissues. Only a few alkaloids can be considered to be induced by herbivore attack or wounding. For example, quinolizidine alkaloids (QAs) in the leaves of *Lupinus* species, indole alkaloids in *Catharanthus roseus,* and dianthalexin in *Dianthus* species are all induced upon tissue wounding.

The synthesis of alkaloids is commonly restricted to one organ system (leaf, stem, root, fruit, or seed), and they are differentially transported to other organs. In *Lupinus* species, QA synthesis is restricted to shoots, as evidenced by the localization of enzymes required to synthesize QAs in green stems and leaves (Wink 1987). QA synthesis is most likely restricted to green stems and leaves because the chloroplast is directly involved in QA synthesis. However, QAs can be found in all tissues of *Lupinus* species with the highest concentrations in fruits and seeds. Transport from the site of synthesis to nongreen tissues occurs in the phloem, as demonstrated by analyzing phloem fluid (Wink and Witte 1984). No QAs can be found in xylem sap. Pyrrolizidine alkaloids (PAs) in *Senecio vulgaris* are synthesized almost exclusively in roots, and these alkaloids are transported in the phloem to all *S. vulgaris* tissues particularly the inflorescence. The synthesis of PAs is highly controlled and tightly linked to root activity (Hartmann 1999). Hemiparasites, connected to the phloem of the host root system of PA-accumulating plants, accumulate PAs in their tissues (Stermitz and Harris 1987). In contrast, hemiparasites attached to *Lupinus arboreum* may be able to avoid accumulation of QA alkaloids (Marvier 1998b).

Most alkaloids are synthesized in vesicles derived from the endoplasmic reticulum. These vesicles eventually fuse with the tonoplast, depositing the final alkaloid form into the vacuole. The tonoplast membrane also has specific transport mechanisms for alkaloids in the cytoplasm. For example, reticuline and scoulerine are accumulated by an ATP-dependent transport system in *Fumeria capreolata*-isolated vacuoles (Deus-Neumann and Zenk 1986). Vacuoles that were isolated from plants that do not accumulate these alkaloids were not able to transport the alkaloids into the vacuole. In poppies (*Papaver bracteatum*), alkaloids are stored in vacuoles contained in the latex of laticifers (specialized structures associated with the transport of latex in the vascular tissues).

The mode of action of alkaloids is as diverse as their taxonomic complexity. PAs are nontoxic by themselves, although they may cause food to be rejected by herbivores on the basis of taste. However, when ingested, the PAs are transformed, by cytochrome P-450 monooxygenases of the liver, into highly reactive compounds that are carcinogenic and mutagenic (Mattocks 1986). Since PAs are nontoxic in themselves, many insects feed on PA-accumulating plants (such as aphids on PA-accumulating plants) in order to gain protection for themselves against predators. QAs are well known to cause a type of intoxication in mammalian herbivores. When sheep or cattle consume large amounts of Lupine containing Qas, they become uncoordinated, may suffer convulsions, experience hepatic damage, and death may follow. There are actually a large number of biological effects of QAs (Wink 1988), such as (1) inhibition of the reproduction of plant viruses and bacteria, (2) inhibition of seed germination by competing species, (3) multiple toxic effects on insects, particularly leaf minors, aphids, and beetles. In a general sense, most alkaloids are bitter and may deter feeding because of this characteristic. In fact, the chemicals used by the food industry as the standards of bitterness are alkaloids.

Are alkaloids defensive compounds in natural situations? In most cases, there is little evidence in support of the defensive qualities of alkaloids against herbivores in natural systems except for the fact that they show pharmaceutical activity. However, in the case of QAs and PAs, several lines of evidence point to their efficacy as antiherbivore substances. First, these compounds have been found to deter the feeding of herbivores at concentrations found in natural plant tissues. Second, these compounds tend to concentrate in locations where herbivory would have the greatest effect on plant fitness (seeds, inflorescence, and shoot tips). Third, their synthesis and translocation are specific processes restricted to alkaloid-producing plants. Fourth, insects that utilize alkaloid-accumulating plants have evolved specific mechanisms to cope with the detrimental effects of the specific alkaloid.

6. Hormones and Antihormones. The defensive compounds that have been covered so far in this chapter can be autotoxic to plants, which makes them difficult for plants to store at high concentrations. It is easy to argue that if plants were able to store a toxin that was specific to insects, this would be particularly effective because of its benign interaction with plant cytoplasm. However, insects can rapidly develop counterdefensive mechanisms to protect themselves from an insect-specific toxin. If, however, the stored compound is a natural hormone of insects, then it would be much more difficult for the insect to develop a counterdefense. One of the most important hormones in insect systems is *juvenile hormone* (JH), which, along with the ecdysones, regulates growth, reproduction, metamorphosis, and behavior in insects (Figure 7.8). Plant production of these hormones could strongly impact insect populations, and it would be difficult for insect populations to acquire resistance to a phytochemical that is also important to their own endocrine system. It is not surprising to discover that plants have substances with JH activity called *phytojuvenoids* (Bowers 1976) and steroids that are analogous to ecdysones called *ecdysteroids*.

There are several types of JH, among which the most common in insects is JH III (Figure 7.8), or *trans, trans*-10,11-epoxymethylfaresenate. The first phytochemical discovered to have JH activity was farnesol and its aldehyde extracted from yeast

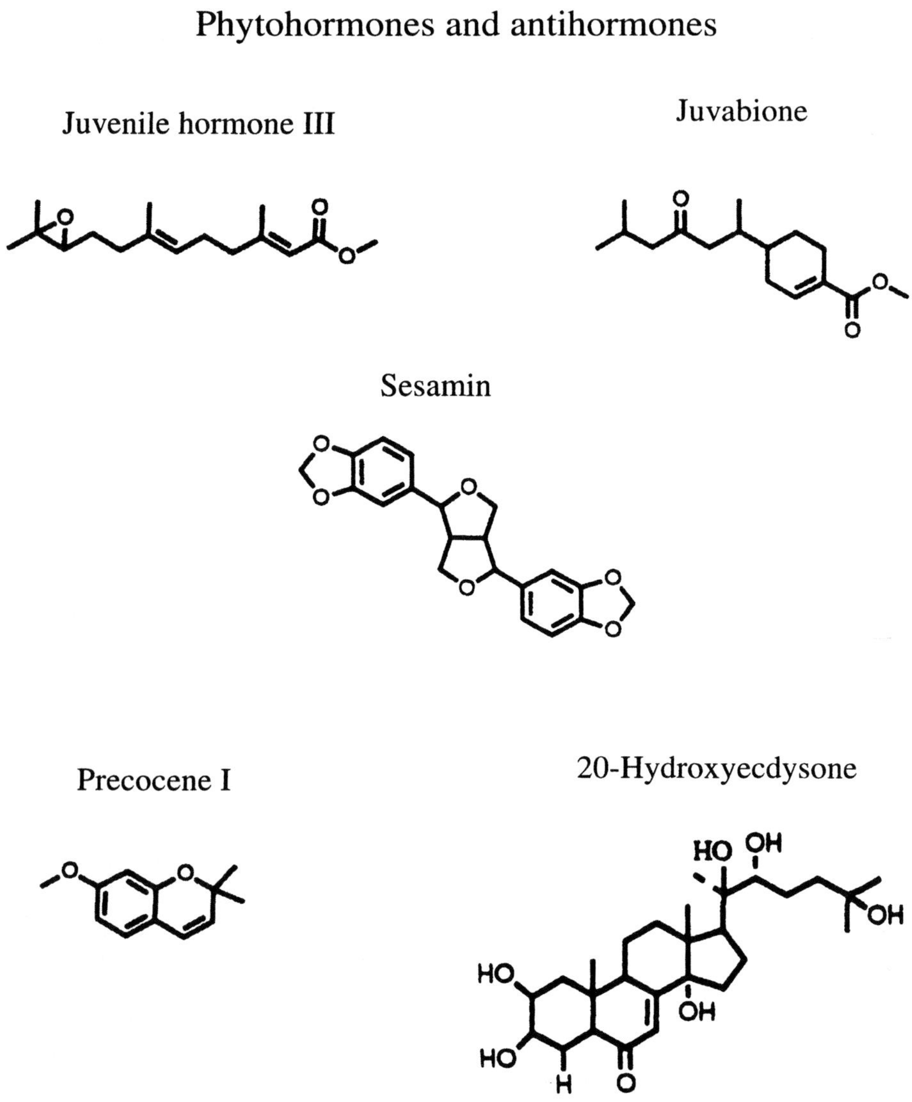

Figure 7.8. Several compounds produced by plants that are insect hormones (juvenile hormone III), or are mimics of insect hormones (sesamin). Precocene I is a compound with antihormone activity. (Molecular diagrams redrawn with permission from Bowers 1991.)

(Schmialek 1961). Later, it was noted that the European linden bug (*Pyrrhocoris apterus*) regularly molted into the nymphal-adultoid stage prematurely when rearing chambers were lined with paper towels (Sláma and Williams 1965). The "paper factor" was determined to be a compound with JH activity in the paper towel derived from balsam fir (*Abies balsamea*). The compound isolated was called *juvabione* (Figure 7.8). This compound had intense JH activity with a few insect species (cotton stainers and linden bugs) and reduced activity with a number of other species. Therefore, phytochemicals with JH activity can be selective against specific insects. One plant, *Cyperus irida,* has been found to contain JH III in the exact structural configuration as found in natural insects (Toong et al. 1988). Other compounds that are not analogous in structure to JH III can also have JH activity. For example, the aromatic compounds sesamin and sesamolin (Figure 7.8), extracted from sesame seeds, have JH activity.

Phytojuvenoids can affect the development and reproduction of phytophagous insects. For example, grasshoppers fed on *Cyperus irida* molted into intermediate forms expressing immature characteristics (Toong et al. 1988), and females fed the *Cyperus* diet produced few and infertile eggs. Elevated levels of phytochemicals, with JH activity, retard insect development and interfere with the reproductive activity of adults. However, damage to plants is often the result of juveniles not adults. Elongating the juvenile stages of insect life cycles may not be effective for reducing herbivory in the short term, but may have an effect on herbivore population size in the long term.

If plants were to produce compounds that interfered with the synthesis, secretion, transport, or action at a receptor cite of JH, this would cause premature development and shorten the feeding time of juveniles as well as reduce insect reproduction. In fact, some plants do produce compounds that interfere with particular insect hormones. Antihormones are substances produced by plants that interfere with the proper function of insect endocrine systems. Two compounds isolated from a member of Asteraceae (*Ageratum houstonianum*) have anti-JH activity. Precocene I (Figure 7.8) and precocene II were so named because they induced precocious metamorphosis in many insect systems. These compounds act by inhibiting neural secretions and may act to inhibit the enzymes responsible for the biosynthesis of JH. Many studies indicate that the precocenes act as specific cytotoxins destroying the insect gland (*Corpus allatum*) that is responsible for JH synthesis.

Another important hormone system of insects is the ecdysones. Approximately 30 different ecdysones have been isolated from insects, but the most common one is 20-hydroxyecdysone. Ecdysones are responsible for inducing molting and pupation (Williams 1956). When studying *Podocarpus nakaii* leaf extracts, Nakanishi et al. (1966) discovered a steroid constituent that was almost structurally identical to 20-hydroxyecdysone (Figure 7.8). Several members of Chenopodeaceae have substantial concentrations of ecdysteroids in seeds and newly developing shoots (Dinan 1995). In fact, 70 ecdysteroids in 100 plant families have been isolated (Alder and Grebenok 1995) and can be found at concentrations near 5% of dry weight, particularly in the bark of gymnosperms (Russell and Fenmore 1970). Although many insects can be affected by these endogenous plant ecdysteroids, some species are remarkably resistant to their hormonal affects (Blackford et al. 1996).

7. ***Proteinase Inhibitors.*** Proteinase inhibitors are low-molecular-weight proteins that bind to proteinases and deactivate them. The first to be isolated was the soybean trypsin inhibitor (Kunitz 1945). Since the first proteinase inhibitor was found, more than 20 different proteinase inhibitors have been isolated from plant tissues, and they are most frequent in members of Solanaceae and Fabaceae. The concentration of these inhibitors is highest in reproductive tissues (particularly seeds). In some cases, they can account for up to 10% of storage protein in seeds (Laskowski and Kato 1980). The main function of these inhibitors may be as antibiotics against bacterial and fungal pathogens (see Chapter 7). However, these compounds are thought to serve as antiherbivore agents because they (1) reduce the growth rate of aboveground phytophagus insects (Augustyniak et al. 1997, Boulter and Latoszek-Green 1997, Broadway 1996),

(2) reduce the growth rate and fitness of nematodes (Urwin et al.1998), (3) are induced by herbivore attack (Broadway et al. 1986), and (4) are more effective against animal and fungal proteinases than against plant proteinases (Richardson 1977). In addition, plants transformed with genes that code for a proteinase inhibitor have gained additional resistance to herbivores (Johnson et al. 1989, Urwin et al. 1998, Jouanin et al. 1998) compared with control plants.

The genes that produce these proteins have been designated as proteinase inhibitor (PIN) genes and are induced by wounding. Some are induced only in the local area of wounding, while other PIN genes are induced sytemically throughout the plant upon wounding (Pearce et al. 1991). Jasmonic acid (JA) activates transcription of PIN genes (Farmer et al. 1992, Titarenko et al. 1997). Other factors that may be involved in PIN gene expression are a small protein called systemin, the hormones ABA (stimulation) and auxin (inhibition), and electrical signals (Herde et al. 1995, Stankovic and Davies 1998). One signal that has been shown to travel from the wound site to other locals is systemin (Figure 7.9), which travels through similar channels as sucrose does when it is transported (Narváez-Vásquez et al. 1995). The signal transduction pathway mediated by systemin is discussed later in the chapter.

Many proteinase inhibitors are grouped as *serine inhibitors* because they all have the same mode of action. Proteinases of herbivores and pathogens recognize these inhibitors as substrate based on the phosphate residue in the reactive site of the inhibitor. The inhibitor then binds with high affinity to the proteinase active site, excluding water and forming a stable complex that inactivates the proteinase. One family of serine inhibitors, called Kunitz type, is comprised of 20–22 kD proteins found mostly in the seeds of plants in Fabaceae, although serine inhibitors are also found in potato and wheat. These proteins inhibit trypsin, chymotrypsin, and cathepsin. An-

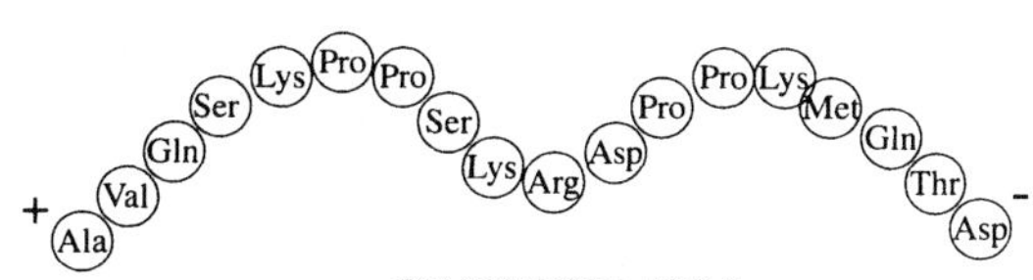

SYSTEMIN

Some of the proteins induced by Systemin

Defensive Proteins	Signal transduction pathway
Inhibitor I	Systemin
Inhibitor II	Calmodulin
Cystatin	Lipoxygenase
Aspartic protease inhibitor	Allene Oxide Synthase
Polyphenol Oxidase	

Proteinases	Others
Leucine Aminopeptidase	Threonine Deaminase
Carboxypeptidase	Acetyl CoA Binding Protein
Aspartic Proteinase	Nucleotide Diphosphate Kinase
Ubiquitin-like Protein	

Figure 7.9. (*Upper panel*) The 18 amino acid sequence for systemin. (From Bergey et al. 1996.) (*Lower panel*) Some of the proteins upregulated by the systemin transduction pathway. (Adapted with permission from Annual Review of Cell and Developmental Biology, Volume 14, © 1998, by Annual Reviews http://www.AnnualReviews.org.).

other family of serine inhibitors, the Bowman-Birk inhibitors, are smaller (8 kD) than the Kunitz type, and are found mostly in legumes. These inhibitors have two active sites, one for trypsin and a second for chymotrypsin (Wu and Whittaker 1991). Common in Solanaceae, proteinase inhibitors I and II are abundant (Pearce et al. 1988) and are another family of serine inhibitors. Inhibitor I can be found as a tetramer in potato tubers with a monomer weight of 8 kD. Inhibitor II is similar to those in the Bowman-Birk class because the latter have two active sites for trypsin and chymotrypsin. The squash serine inhibitor family consists of small (3.2kd) molecules in Cucurbitaceae. This group of inhibitors is very active against trypsin. The barley family of serine inhibitors are found in a number of grasses including corn and are very effective against trypsin. Several members of this family of inhibitors are effective against other enzymes such as α-amylase.

A second class of inhibitors is the *cysteine proteinase inhibitors* (Barrett 1987). These are low-molecular-weight (5.6 kD) proteins called *cystatins* that are active against endoproteinases in the papain class. Cystatins have been isolated from pineapple stem (seven different molecules), potato tuber (an 85 kD protein with eight active domains), rice (oryzacystatin I and II), *Wysteria* sp. (several proteinase inhibitors), and avocado (Dopico et al. 1993).

Aspartic proteinase inhibitors are Kunitz-type inhibitors that have been isolated from potato tubers. They inhibit trypsin and aspartyl proteinase cathepsin D (Mares et al. 1989). Aspartic proteinase inhibitors can also affect metalloexopeptidases. For example, carboxypeptidase A and B are inhibited by an aspartic peptidase inhibitor that accumulates in potato tubers during development (Hass and Ryan 1981).

C. Physiological Basis for the Induction of Defensive Chemicals

Several of the compounds discussed above increase in concentration following herbivore damage (*inducible defense*) including some alkaloids, proteinase inhibitors, tannins, and so on. There are several possible mechanisms for the induction of defensive chemicals following herbivory:

1. Increased synthesis (transcription and or translation)
2. Lower degradation rates of the defensive compound
3. Increased transport of the compound from the site of synthesis to the target tissues
4. Release of bound defensive compounds such as from membranes or vacuoles
5. Increased accumulation from the transporting tissues

Number 5 is the best explanation for alkaloid accumulation in wild tobacco leaves following herbivory (Baldwin 1989), and number 1 is the best explanation for proteinase inhibitor induction in tomato (Ryan 1990).

It is clear that the induction of some of these compounds occurs at the genetic level. For this reason, inducible defenses have been of considerable interest to biochemists because of the challenges to understanding their complex signaling systems, and because of their potential in bioengineering crops for herbivore tolerance. Early in the 1970s, it was found that transcription of PIN genes is upregulated after herbivore damage of tomato leaves (Ryan 1974), and the activity of enzymes associated with nicotine synthesis in the roots increases when tobacco plants are damaged (Mizusaki et al. 1973). Since that time, there has been extensive interest in the signal transduction pathways leading from herbivore (and pathogen damage) to defensive gene induction.

There are several different signal induction pathways that regulate defensive compound accumulation; however, the systemin pathway is best known. Most of the research on this topic has focused on tomato plants. After the initial discovery that there must be a systemic signal that induces defense-related genes throughout tomato plants following herbivore damage, the first signaling compound isolated was oligogalactouronides (Bishop et al. 1981). These molecules are derived from the plant cell wall when herbivores or pathogens lyse plant cells. It was soon found that oligogalactouronides have a low activity and are not transported far from the point of atttack, so they are thought to influence localized defense. In a similar manner, chitin and chitosan oligomers (products of fungal cell walls) were also found to be active inducers, and also are restricted to localized effects. It was not until the early 1990s that a molecule induced upon herbivore damage and transported throughout the plant was discovered (Pearce et al. 1991). This oligonucleotide, systemin, was almost 1 million times more effective at inducing defensive genes in tomato leaves (Pearce et al. 1993) than oligogalactouronides or chitin oligimers.

The systemin-induced wound response was first associated with the increase in proteinase inhibitors, but it has been found to influence a number of other protein systems as well (Bergey et al. 1996). The genes producing these protein products can be classified into four categories (Figure 7.9): (1) defensive genes producing proteinase inhibitors and phenol oxidase, directed toward reducing protein digestion efficiency in insects; (2) genes that produce signal transduction pathway proteins; (3) genes that produce proteinases that are active in plants; (4) genes that produce enzymes with unknown roles in plant defense.

Systemin is a low-molecular weight signaling protein with 18 amino acids (Figure 7.9). This molecule is active at the femtomole per plant level, which ranks it among the most powerful gene-activating proteins known (Ryan and Pearce 1998). The polypeptide is most likely a Z-like β-sheet, with no tertiary structure, which is similar to the structure of many DNA-binding proteins. The polypeptide is highly charged, and its sequence includes a palindrome with a proline-proline component (*xx*Q*x*-BPP*x*BB*x*PPB*x*Q*xx*) where B is either lysine or arginine (P = proline, Q = glutamine). The activity of the protein is determined by proline 13 (substitution resulted in only 2% activity) and threonine 17 (substitution resulted in an inactive protein). In fact, deletions starting at the N terminus progressively reduces activity, but the four–amino acid sequence (Met15–Gln16–Thr17–Asp18) retains some activity (3×10^5 less activity than the whole protein). Removing only the C-terminus amino acid completely inactivates the protein. Thus, residues near the C terminus are necessary for activity (possibly a phosphorylation of threonine 17), and the N terminus may be the location that interacts with a receptor (Ryan and Pearce 1998).

Systemin is processed from a larger protein (prosystemin; 200 amino acids) at the C terminus region. The prosystemin gene has 11 exons containing 5 repeated regions that arose by several gene duplication and elongation events (McGurl and Ryan 1992). Systemin is located at residue 179–196, and is not part of the repeated regions. Since there is no sequence suggesting a transit peptide or membrane-spanning domain in prosystemin, this protein is most likely synthesized in the cytoplasm. The prosystemin gene has been found in several members of Solanaceae such as nightshade, peppers, and potatoes. Although there are differences among the sequences of the various prosystemins, there are no differences in the systemin part of the genes. It is interesting that prosystemin from tomato does not induce defensive genes in tobacco, and vice versa. Therefore, the processing machinery required to release systemin is species specific. Systemin has not been found in families of plants outside Solanaceae, but similar patterns of wound-induced increases in proteinase inhibitors have been found in other families of plants (Saarikoski et al. 1996). The prosystemin gene has been directly related to wound response, because transgenic plants with an antisense gene for prosystemin have no wound response mechanism and are highly susceptible to herbivore damage.

Prosystemin expression following wounding is localized in the vascular bundles close to the phloem of petioles, leaf main veins, and stems. Therefore, upon tissue wounding by a herbivore, prosystemin is synthesized in cells where systemin would be released in the vicinity of the phloem, facilitating transport to other parts of the plant. When radioactive systemin is placed on fresh wounds, systemin was distributed throughout the plant in 30 minutes (Narváez-Vásquez et al. 1995). Within the first 15 minutes after application, the systemin was located primarily in the xylem and phloem tissues of leaf veins. The application of *p*-chloromercuribenzene (an inhibitor of apoplastic phloem loading) to a wound, just before applying radiolabeled systemin, completely blocked the transport of systemin out of the wound region. These results suggest that systemin moves from the wound site to the vascular traces through the apoplast, then into the phloem, where it is transported to target cells (sinks for sucrose).

The systemin signaling pathway is mediated through lipid-derived intermediates. At the site of action, systemin binds to a receptor that activates an intracellular cascade resulting in the release, from the plasma membrane, of linolenic acid which is subsequently converted into phytodienoic acid (PDA), and eventually jasmonic acid (JA) (Figure 7.10). Both PDA and JA are powerful activators of defensive genes in plants. It is interesting to note that the pathway from linolenic acid (LA) to JA (the octadecanoid pathway) is also the cascade that induces defensive genes in response to carbohydrates generated from fungal and pathogen attacks. In fact, the octadecanoid pathway is a general signaling pathway for defense, environmental stress, and cell development (Weiler 1997). The mechanism by which systemin induces the intracellular cascade of LA to JA is not known, nor has the receptor protein been identified, but the earliest phases of the mechanism may concern ion transport (Ryan and Pearce 1998).

Information derived from the signal transduction pathway of systemin can elucidate some general attribultes of the pathogen or herbivore attack response in plants. There is a localized response and a systemic response. The localized signals are products of cell wall interactions that induce the same pathway as the systemic signal, as well as an oxidative burst. The systemic signal induces a generalized stress response pathway that is mediated through PDA and JA. The plant defense system resembles that of the animal inflammatory response, except that the plant system is less complex (as far as we now know) and may not include wound healing signal pathways (Bergey et al. 1996).

III. COSTS OF DEFENSE FOR THE PLANT

A. Description of Costs

Early in the development of theory about herbivore defense, it was assumed that plant defensive mechanisms were costly to the fitness of plants. The rationale behind this theory is that many plants attain resistance traits only after induction following damage. If resistance traits are assumed to benefit plant fitness, why is resistance not

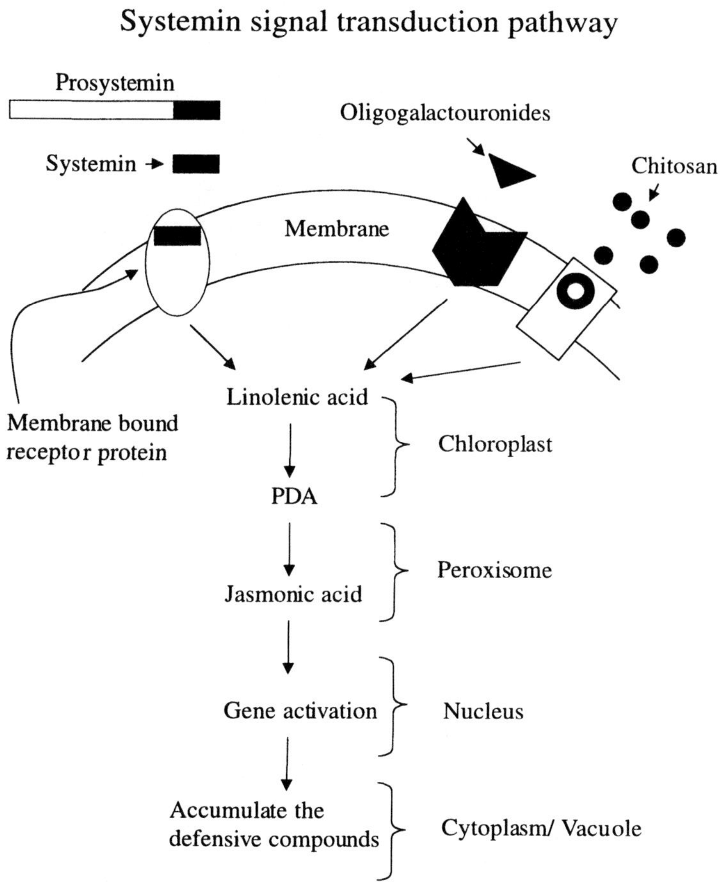

Figure 7.10. Signal transduction pathway involving jasmonic acid. This pathway is induced by attack of herbivores (systemin signal), and pathogens (oligogalactouronides and chitosan signals) alike. PDA = phytodienoic acid. (Modified from Bergey et al. 1996.)

always constitutive? Hence, the optimal defense theory; resistance has costs for plants, so resistance traits are expressed only when needed, and, in the absence of damage, allocation to growth and reproduction can be maximized (Feeney 1976, Clark and Harvell 1992). Also, optimal defense theory predicts that any defensive strategy of plants (chemical and/or morphological) diverts resources away from other fitness traits (photosynthesis, growth, reproduction). This concept is based predominately on examples of chemical defenses and includes a number of different biochemical and physiological costs associated with the allocation to defense.

The assumption that defense against herbivory has fitness costs seems logical, but it has been difficult to verify. Defensive compounds are often produced in response to stimuli other than herbivory, such as drought (Karban 1991), ultraviolet radiation (Lee and Lowry 1980), and nutrient deficiency (Lieberman and Lieberman 1984). In fact, it has only rarely been shown that the induction of a defensive compound by a herbivore results in an increase in plant fitness in the presence of a herbivore (Agrawal 1998). Thus, in many cases, the production of putative defensive compounds may be a general response to stress rather than a specific herbivory defensive mechanism. In addition, several putative defensive compounds are storage compounds. Given this dilemma, it is prudent to examine the methods by which the cost of defense is measured.

A physiological assay for the cost of defense is an analysis of the energy involved in the pathway used to synthesize the defensive compound. This method involves calculating the energy requirement for synthesizing a given quantity of the defensive compound by determining the biochemical structure of the defensive compound and knowing the energy required to form the bonds in the molecule (McDermitt and Loomis 1981). Add to this value the cost of any aspect of the synthetic pathway that is not a constitutive part of the metabolism and the cost of maintaining the concentration of the compound. The result is an estimate of the number of moles of ATP, CO_2, or glucose required to synthesize a given amount of the defensive compound. This can then be converted into an estimate of the amount of photosynthate that is required to synthesize the defensive substance. Not all defensive compounds have the same cost coefficient. For example, alkaloids have twice the construction cost per gram as tannins or lignins. In addition, the maintenance costs of alkaloids and terpenes are high because the longevity of the molecules is relatively short (Mooney and Gulmon 1982, Coley et al. 1985). The relative cost of synthesizing defensive substances may explain why alkaloids are normally low in concentration except when induced by herbivore damage while tannins are relatively high in concentration without the induction by herbivory.

Pathway estimates for the cost of defense normally indicate high costs for plants based on the need for energy and carbon resources. However, it is important to realize that glucose, carbon dioxide, and ATP are not the only important currencies for the estimate of cost (Chapin 1989). Similar pathway analysis techniques can be used to determine the cost in terms of nitrogen or sulfur. And these currencies may be more significant to plant fitness in regions where nutrient availability is low.

Does pathway analysis give an accurate estimate of the cost of defense? Recent studies indicate that determining cost in allocation units by pathway analysis cannot be used to deduce the total cost of defense (Simms and Rauscher 1987, Simms 1992). The technique can only determine the cost of known defensive compounds and cannot account for all defensive compounds since some are unknown. Moreover, the compound may have purposes other than defense against herbivores. If so, the calculation of cost should be reduced by advantages gained through those alternative uses. In addition, other costs to defense such as catabolic disruption, are not accounted for by the pathway calculation. The following section gives a more critical coverage of the defense cost theory for plants, which focuses on resource allocation and biochemical disruption.

B. Resource Allocation

Resources for the synthesis of defensive compounds include nitrogen, carbon, ATP, and sulfur. Further costs for nonconstitutive defensive compounds are attributable to the resources required for synthesizing all the other enzymes needed for the nonconstitutive metabolic pathway. As stated above, optimal defense theory is based on the assumption that allocation of these resources to defense compound synthesis causes a reduction in fitness. This could only be a valid assumption if resources are limiting for fitness traits (Gould 1988, Bryant et al. 1988,Zangerl and Bazzaz 1992).

The two resources that are most often considered as being limiting to defense and fitness are nitrogen and carbon. Plants grown in low nutrient environments have long-lived leaves with relatively high levels of constitutive carbon-based defense compounds (Bryant et al. 1983) such as tannins. The intrinsic slow growth rates of plants from low nutrient environments may reflect the elevated allocation of resources to defensive chemicals compared with plants growing in high nutrient sites. In fact, most plants with slow growth rates have high allocation to defense (Coley et al. 1985), and plants with high allocation to defense have slow growth rates (Coley 1986). Also, in the absence of herbivores, agricultural varieties that are resistant to herbivory have slower growth rates than varieties that are sensitive to herbivory (Pimentel 1976). All these observations support the concept of a trade-off between defense and growth. Yet these are correlative results that need to be treated with caution. For example, defensive compounds are also thought to be storage moieties for nitrogen and carbon, and slow-growing plants often have a greater proportion of resources in storage pools. As discussed above, verifying a cause-and-effect relationship between allocation to defense and decreased fitness has been difficult. One of the best tests of this trade-off is to study the effect of inducing defense on growth or fitness in the absence of herbivores (Coleman 1991, Simms 1992). If the induction of a defensive pathway causes a reduction in fitness, this is good support for a resource cost of the induced compound. Such relationships have been shown in only a few cases (Brown 1988, Baldwin et al. 1990).

One of the difficulties in proving the cause-and-effect relationship between defense and fitness is that the induction of defensive chemicals is regulated not only by herbivore damage (Figure 7.10); constraints on the capacity to induce defensive chemicals are due to genotype (a result of evolutionary processes), plant phenological stage, plant age, abiotic environment, and biotic influences. Plant physiological functions must operate within these constraints to acquire, allocate, and partition resources among organs and physiological pathways that determine growth, reproduction, maintenance, defense, senescence, and so forth. The result of that allocation process regulates the quality of plant tissues for herbivores (physical, chemical, and nutritional) and the fitness of the plant (growth, reproduction). Tissue quality regulates both the preference and performance of herbivores. Herbivore population size (and subsequent intensity of plant damage) are determined by tissue quality, environmental factors that influence herbivore fitness, and biotic factors (predators, parasites, pathogens). Therefore, the trade-off between defensive chemical induction and plant fitness is logical but complicated by a large number of interactive processes. Verifying this trade-off requires a thorough consideration of these interactive processes when designing experiments.

Another difficulty with determining the relative costs of defense is that the relationship between nutrient limitation and allocation to defense within a species is inverse to that between species. Previously, we stated that the allocation to defense among species increases at low resource availability. However, allocation to defense among individuals within a species decreases at low resource availability. For example, in plants that utilize a carbon-based defensive compound, allocation to defense is low in shaded sites (carbon limited) and high in sunny sites (excess carbon). In plants that utilize a nitrogen-based defensive compounds, allocation to defense also decreases under nitrogen limitation. The answer to the apparent discrepancy between intra- and interspecies variation in defensive allocation resides in the difference between time scales. Evolutionary adaptation selects for plants with increased tissue longevity and increased defense at low resource availabilities. However, during stressful periods of low resource availability, plants allocate resources to critical fitness traits such as reproduction, root growth (low levels of nutrients or water), shoot elongation, and leaf area (low light availability) rather than to defensive compounds.

C. Autotoxicity and Biochemical Disruption

Increased concentrations of defensive chemicals in plant tissues can disrupt metabolism not just in the herbivore but also in the plant, resulting in autotoxicity. For this reason, compounds such as cyanogenic glycosides are compartmented away from the mainstream metabolism. The compartmentation system itself must therefore be considered a cost of this defensive posture. In addition, if the synthesis of defensive compounds causes an imbalance or disruption of metabolism, this consequence must also be considered a cost to defense. If resistance traits arise by mutation, a biochemical pathway that normally produces an important enzyme may have been affected by the mutation, and a fitness reduction could occur. In situations where a number of copies of a gene are present in a genome, a mutation in one of the copies may result in the production of defensive compounds, while the other copies continue to support the metabolic pathway. However, decreasing the functional copy number supporting a metabolic process may also reduce the capacity of that metabolic pathway. Therefore, mutations that result in the synthesis of herbivory defensive compounds could result in catabolic disruption or catabolic imbalance (Berenbaum and Zangerel 1988).

Likewise, the production of chemicals against herbivores can allow disease agents to take hold in plants. For example, when *Hevea brasiliensis* increases hydrogen cyanide concentrations in tissues, this inhibits its ability to induce the defensive pathway against fungal pathogens (Lieberei et al. 1989). Although it is not clear how to quantify this cost, changes in susceptibility to disease in plants that express defensive compounds results in decreased fitness and should be considered in the cost calculations.

In summary, the calculation of the costs of defending tissues by chemical means is a difficult and complex issue. The theory that inducible defenses have evolved because of the high cost of producing the defensive chemicals for plant growth and fitness seemed logical at first. However, the interpretation of what exactly constitutes cost and what are alternative benefits of these compounds has made the quantification of the cost of defense arduous. The difficulties encountered in the interpretation of costs have led authors to consider a second possible selective factor for the evolution of inducible defense. The alternative theory is based on the heterogeneity of the defensive posture that results from inducible defenses among individuals in a population.

IV. IMPORTANCE OF QUANTITATIVE VARIATION

A. Heterogeneity of Host Tissue Quality

There are a number of reasons why the quality of host tissue for insect consumption is not homogeneous in the plant community. Some of the character traits that can vary among host species are defensive chemical quantity and quality, nutritional quality, toughness, mineral content (such as silicate or calcium carbonate), and the presence of sticky fluids that gum up insect mouth parts (Pennings and Paul 1992). In fact, variation in tissue nutrition, digestibility, and defense among host species can be a primary factor stimulating host specialization among insects. Host specialization can become complex enough to allow herbivore species composition to be used as a diagnostic trait for establishing plant taxonomic relationships (Abrahamson and Melika 1998). Moreover, tissue quality varies among individuals (genotypes) within a species. Within a population of *Trifolium,* there is significant difference in cyanogenic glycoside content among genotypes. Also, there can be significant variation in tissue quality among organs of an individual. For example, reproductive structures have the highest nutritional quality among plant organs. There is also temporal variation in tissue quality within a specific organ. For instance, early in the development of leaves, nutrient content per leaf area is high, and defensive mechanisms (particularly physical defenses) frequently are poorly developed. Later in leaf development (after maturation), leaf nutrient content decreases, digestibility decreases, toughness increases, and chemical defenses increase.

A large degree of variation in tissue quality and defensive posture is expressed in tree canopies. Tissue quality for herbivores varies among branches on a given individual at a given time. Upper canopy leaves have higher nitrogen content than lower canopy leaves (Field 1983). Specific leaf weight is lower and trichome development is less within a canopy compared with the outer, sun-exposed branches (Hardin 1979). Furthermore, there is a strong influence of phenological pattern on the heterogeneity of tissue quality in canopies. Short-flush species (those that produce new growth during a short period of time) have dynamic changes in tissue quality over time due to the even aged leaf population. In contrast, species with indeterminate growth have a greater degree of spatial heterogeneity of tissue quality due to the presence of different leaf age cohorts on the tree at the same time. In addition to these factors causing a constitutive variation in defense, induced defenses are a source of variability. If herbivores have recently attacked one particular branch of a tree, defensive compounds may increase throughout the tree, but compensatory growth that is highly defended will occur on the affected branch more dramatically than on other branches. The result is a further increase in spatial variation of tissue quality among branches in the canopy. Thus, spatial and temporal heterogeneity in tissue quality (nutrition and defense) are the rule rather than the exception.

The variation in defensive posture in a population and its impact on insect performance may be the evolutionary stimulus for inducible defenses (Karban et al. 1997). This is an alternative to the optimal defense theory described earlier (i.e., that inducible defenses have evolved in response to the cost of defense). Plants that vary in resistant phenotype as a result of inducible defenses have a selective advantage over plants with uniform defenses, because herbivores will have reduced performance when faced with variability and will avoid variable host species (Zangerl and Rutledge 1996). This concept is based on the fact that as defenses increase, the benefit of consuming the plant tissues to insect populations decreases. If one assumes that the relation between the benefit to the herbivore and the benefit to plant defense is concave, then a variable level of defense will always result in a lower benefit to the herbivore than will a mean level of defense produced by the plant constitutively (Karban et al. 1997). A model that represents this relationship can be plotted based on means and variance in toxin concentration. If the frequency of distribution of toxin concentration in tissues or plants is distributed normally with a mean defense level t and a variance σ^2, expected insect benefit (where benefit = fitness) curves for different levels of defense can be calculated (Figure 7.11). Increasing the mean levels of toxin leads to decreased fitness of the herbivore. However, very high mean toxin concen-

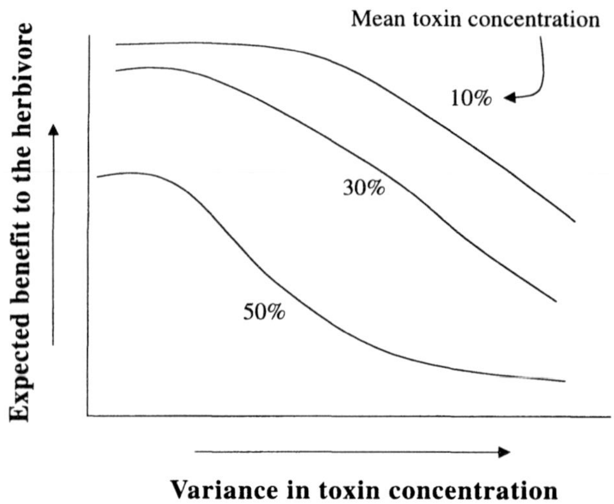

Figure 7.11. The expected benefit of consuming plant tissues for insects plotted against the variance in concentration of plant tissue defensive compounds (toxins). Notice that, as the mean toxin concentration increases, the benefit decreases. Also, as the variance in toxin concentration increases the benefit to herbivores decreases. (Modified from Karban et al. 1997, with permission).

trations can be a strong selective pressure for resistance in the herbivore population. Over the long term, producing very high levels of a toxin will not be a successful strategy for keeping herbivore fitness relatively low. Moreover, the fitness (benefit) of herbivores decreases as the variability (indicated by variance) in toxin concentration increases. Thus, plants with a moderate average toxin concentration and a high level of tissue toxin variation will provide the least benefit for the herbivores. Based on this analysis, selective pressure on plants by herbivores should favor an inducible defense that enhances variation in toxin levels of plant tissues and reduces the fitness of the herbivores. In this case, the selective pressure for inducible defenses is not related to the cost-benefit relationships between plant fitness and herbivore fitness as predicted by *optimal defense theory.* The relative significance of the optimal defense theory or the variability theory for the evolution of inducible defenses is still controversial.

B. Effects on Insect Populations

This variation in tissue quality of the host can result in both positive and negative effects on insect populations. In a positive sense, the insects can select high-quality tissues to maximize their growth rate. In addition, based on various sensing tools, insects can select tissues that are less defended. Consequently, insect populations can continue to survive on plants that have strong insect defenses without going through natural selection for more costly antidefensive mechanisms. Since many of the antidefensive mechanisms result in the reduced growth rate of the herbivore and its population, being able to feed on undefended tissues increases the intrinsic growth rate and fecundity of individuals (compared with feeding indiscriminately), and this is reflected in population growth. The mechanisms by which insects forage for tissue quality will be discussed in the next section.

Of course, there are important negative aspects to insects foraging for highest-quality tissues. If high-quality tissues are restricted in space or time, then predators and parasites can develop a search image, on the basis of the plant tissue, to locate potential insect prey. Moreover, plant tissues damaged by herbivores can release volatile compounds that attract parasites (discussed earlier). This fact suggests an important trade-off for insect populations in natural systems: locating host tissue of the highest quality versus avoiding predators and parasites. Another problem with insect selection of high-quality tissues is that there are a number of traits that define high quality. Tissues that are high in N, high in water, have high digestibility, and lack defensive structures or chemicals are optimal for insect population performance. Unfortunately (for the insects), these traits do not always covary in a positive manner. For example, the leaves with the highest nitrogen in a canopy are those with the lowest water content and the greatest amount of trichomes. Thus, high-quality traits often have a negative covariance, making the selection of high-quality tissue by herbivores difficult. The presence of tissues with low defense on plants weakens the selection for optimal antidefensive tactics. Since plants maintain a proportion of their tissues without defensive chemicals, a portion of the herbivore population will not experience the defensive posture of the plant. These individuals (feeding on the low-defended tissues) will have a higher fitness than those feeding on defended tissue and will contribute proportionally more to the next generation. As a result, natural selection for genetic antidefensive mechanisms in herbivores is diluted by the high fitness of herbivores feeding on undefended tissues.

V. ENVIRONMENTAL STRESS AND HERBIVORY

Environmental stresses can have a complex influence on herbivore populations. Included are positive, negative, and neutral influences on herbivore growth, development, fecundity, and population size. Since we are taking a phytocentric view in this chapter, direct influences of climatic stress on herbivores will not be considered; therefore, the host plant will be considered the interface between environmental stresses and the herbivore population (Figure 7.12). The common denominator for most of the concepts about environmental stress and phytophagous insect populations is the quality of the plant tissue. We have already discussed the fact that plant tissue quality (nitrogen, water, secondary chemicals, toughness, etc.) changes in response to environmental conditions. The logical next question is, how do these changes affect insect populations? Before discussing the answer, it is important to realize that most of our knowledge about the interactions among environmental stress, plants, and herbivores is derived from agroecosystems (Dale 1988). Consequently, the understanding of fertilizer effects on individual host-insect interactions is better developed than that of long-term evolutionary effects among a diversity of host-stress-insect interactions.

A. Insect Preference for Specific Host Plants

In order to broach this subject, preference and performance must be distinguished (Denno and McClure 1983, Thompson 1988). *Preference* refers to the manner by which insects select host individuals, which is often concerned with oviposition choice. *Performance* is defined as the growth, development, and fecundity of insects after they are located on a particular host. There can be direct relationships between

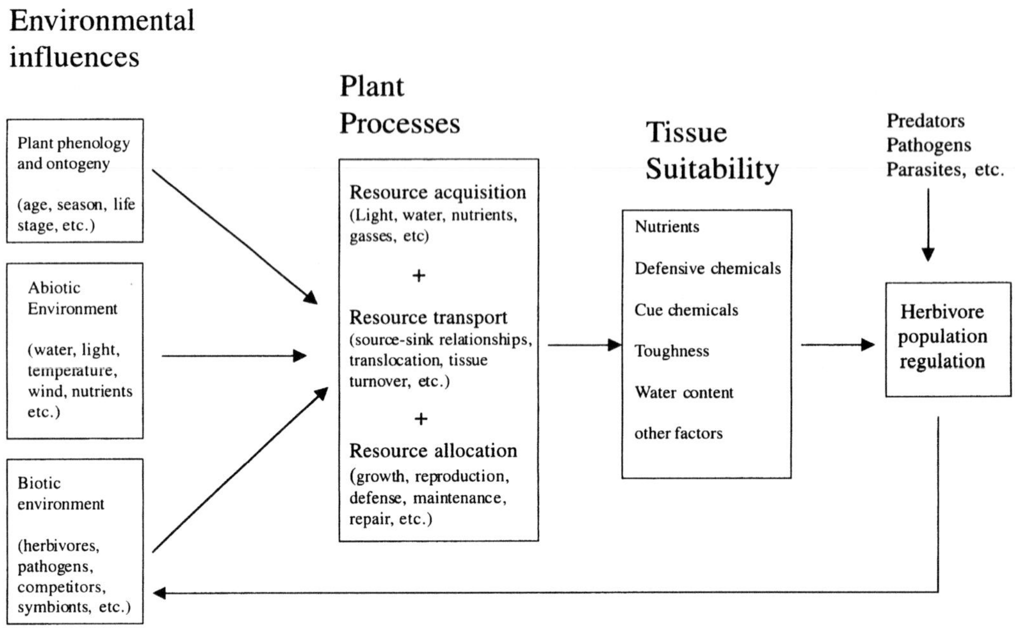

Figure 7.12. Flow diagram representing the interaction between environmental influences on plants and ways the resultant plant processes influence tissue suitability for insect herbivores. Herbivore populations are also regulated by external factors such as predators and parasites. (Modified from Coleman and Jones 1991. Copyright © 1991 Reprinted by permission of John Wiley & Sons, Inc.)

preference and performance (Valladares and Lawton 1991, Nylin et al. 1996), but this relationship is complex and, as yet, not well founded on experimental evidence.

There are many cases in which resource limitation in the environment can influence the manner by which phytophagous insects select hosts (Figure 7.12). Some herbivorous insects are able to sense the quality of tissues before they attack. Tissue nitrogen concentration, the forms of available nitrogen in plant tissues, tissue water content, and tissue toughness are some of the factors involved in insect host selection. There are also a complex group of volatile cues and surface chemicals that can be important in host selection.

Nutrient deficiencies in plant tissues can influence preference by phytophagous insects. Nitrogen seems to be the main factor regulating these choices. For example, in some situations, nutrient limitation in potassium, phosphorus, or sulfur increases insect selection of tissues (Benepal and Hall 1967), because these limitations result in an increase in free amino acids in the plant tissue. Many studies indicate that insect populations select tissues with relatively high nitrogen content in comparison with other neighboring species or tissues (Dale 1988). In fact, some insect species switch host plants when tissue nitrogen concentrations change (Prestidge 1982, Hunter et al. 1992). There are situations in which low nutrition protects host plants from being chosen by herbivores. A classic case is the efforts to regulate alligator weed in the southeastern United States. The biocontrol agent (a flea beetle) clearly selected individual host plants based on nutritional level, so that plants in low nutrient sites were protected from damage (Maddox and Rhyne 1975). Similarly, the biocontrol of prickly pear cactus in Australia could not have been complete without fertilizing some locations where prickly pear cactus had relatively low tissue nutrient content (Dodd 1940).

Tissue water content (relative water content) can also influence insect preference for particular host plants. For example, host selection for oviposition in *Pieri rapae* L. was dependent on tissue water content (Wolfson 1980). However, several authors have measured negative relationships between phytophagous insect abundance and leaf water content (Chiang and Norris 1983, Lewis 1984). In these cases, it is likely that the increase in free amino acids, characteristic of water-stressed tissues, may stimulate selection by phytophagous insects.

The diversity of insect species on a host plant can usually be predicted from the host species' geographical range, habitat range, tissue nutrition, and growth form or architecture (Strong et al. 1984). However, some insect species do not follow these traditional criteria and select specific host species based on chemical characteristics (Erlich and Raven 1964). For example, the black swallowtail butterfly feeds almost exclusively on species in the Apiaceae (Blau 1981). As originally demonstrated by Dethier (1941), these and other insects can select specific species on a behavioral basis (Jermy 1976, 1984; Bernays and Chapman 1994), which is usually associated with oviposition preference (Renwick and Chew 1994). Volatile and visual cues lead to the selection of an appropriate plant for oviposition by some butterflies (Bauer et al. 1993).

Subsequent to volatile cues, compounds on the surfaces of tissues may also be important for preference. Some female butterflies drum on the surface of the leaf with their foretarsi (front legs). Chemoreceptors on the surface of the foretarsi identify chemical cues from the plant which the females use to decide whether to deposit eggs (Feeney et al. 1983). Two compounds, *trans*-chlorogenic acid and luteolin 7-O-(6′-O-malonyl)-β-D-glucoside, have been identified as potential cues (Feeney et al. 1988), and chemoreceptor sites have been located of the foretarsi (Roessingh et al. 1991). These two compounds are not likely to be the only chemical cues. Instead, a complex chemical message is likely with a complex neurological response (Van Loon 1996). Identification of specific hosts is important for insects such as the swallowtail butterfly because larvae of these butterflies concentrate the secondary compounds from the host plant in their tissues as a defense against predators.

In general, environmental stress can either increase or decrease the concentrations of secondary compounds that regulate oviposition. During nitrogen limitation, secondary compounds decrease for plants adapted to high nutrient availability. Such decreases may reduce the ability of female insects to locate hosts. In legumes, nitrogen limitation increases the concentration of alkaloids that may make it easier for insects to locate hosts. Water stress can increase the intensity of volatile emissions from plants (such as ethylene), making them more likely to be located by insects. An excellent example of this is the ability of bark beetles to preferentially locate stressed trees (Coulson 1979).

B. Host Suitability and Insect Performance

A large impact of environmental stress on herbivory is its effect on the suitability of the host as a feed source for insects. Recall that our definition of insect performance is based on host tissue qualities such as nutrient content, toughness, protective spines, and the concentration of secondary chemicals. All of the factors that regulate tissue quality can be influenced by environmental stress (Hunter et al. 1992). Therefore, environmental stress, through its influence on plant tissue quality, can regulate insect growth, development, fecundity, and population size.

Many studies have shown an interaction between larval growth rate and tissue nutrient quality (Figure 7.12). Jansson and Smilowitz (1985) found that Colorado potato beetles were larger on plants that received fertilizer than those on unfertilized plants. The larger size was ascribed to both host tissue quality and other factors such as emigration. Increasing atmospheric carbon dioxide can result in increased insect performance (Lindroth et al. 1995). This response is not uniform among species or among soil nutrient levels (Kinney et al. 1997). Insect growth efficiency is also related to tissue nitrogen content (Slansky and Feeny 1977, Landsberg and Ohmart 1989). Water stress can induce an increase in free nitrogen compounds, particularly proline and glycine-betaine. These could lead to an increase in insect growth rate, although this is not well established. Therefore, the influence of plant water stress on insect performance can be both positive and negative depending on the specific species involved (Wearing 1972, Gershenzon 1984, Preszler and Price 1988). Temperature primarily regulates the growth rates of insects directly rather than through changes in host plant tissue quality. Separating the influence of temperature on plant and on insect physiology and its effect on insect performance is very difficult (Benedict and Hatfield 1988).

Any environmental stress that influences secondary chemistry can influence insect performance by inhibiting larval growth or development, or by preventing the development or survival of eggs or larvae (Rosenthal and Berenbaum 1992). Water stress can alter red oak defensive chemistry (low water potential causes an increase in tannin concentration) and thereby influences insect performance (Stowe et al. 1994). Also, recall that one of the problems with defining the role of secondary chemicals in defense is the fact that the chemicals are also induced by several different environmental factors.

C. Herbivore Influence on Subsequent Abiotic Stress Tolerance

There is abundant evidence that herbivores can have important effects on plant physiology and ecology. However, the evidence for prolonged effects of herbivores on stress resistance is minimal. Some relationships are clear. For example, the removal of leaf area by herbivores reduces water use by plants and may actually increase the plants tolerance of limited soil water availability. In opposition, if leaves are chewed but not completely consumed, the increase in exposed surfaces can act as a desiccating agent. Root-feeding nematodes can induce water stress (and nutrient stress) because they reduce the abundance of root hairs. Therefore, herbivory can relieve or accentuate water stress.

Herbivores that consume seeds may decrease the population size of a plant species over the long term and thereby conserve resources. This would reduce the probability that resources would become limiting for the host plant population (reduced intraspecific competition).

One critical factor regulating the effect of herbivory on secondary environmental stress is the timing of the damage in relation to plant development. In cotton, attack by tobacco budworm during the plant's establishment phase or the fruit maturation stage will have no effect on yield during future drought stress. However, attack during the fruit-filling stage of cotton will reduce yield by 20% during subsequent drought (Reynolds et al. 1982).

Collateral effects of insect damage can modify the plant's response to environmental stress. Herbivory can increase phenylpropanoid synthesis which leads to increased lignification (Chiang et al. 1986). Increased tissue lignification can reduce plant susceptibility to air pollutants. Insect damage to leaves can lead to increased carbohydrate or to nonprotein amino acid levels. Two-spotted mite damage to peppermint leaves causes an increase in leaf soluble-carbohydrate levels, which results in an increase in short-term drought tolerance (DeAngelis et al. 1983). However, a reverse effect is also possible. Feeding by green bugs on drought-tolerant wheat results in significant membrane dysfunction and a decrease in wheat's tolerance of low soil moisture (Dorschner et al. 1986).

In general, herbivory damage can both increase and decrease plant tolerance of any future environmental stress. This is the case for nutrient limitation, water stress, and air pollution. The type of interaction that occurs between herbivory and subsequent stress tolerance must be determined on a species-by-species basis.

VI. SUMMARY

The physiological and ecological interactions between plants and animals are vast and diverse. The complex processes provide a wealth of research topics and have profound significance to agricultural and natural ecosystems. It is no surprise that the written information on this topic is voluminous. However, even though many high-quality scientists have been working on this topic for many years, a generalized theory has been difficult to support and controversial. It is the diversity of insects, plants, and their interactions that is stifling the development of integrative models.

It is clear that plants have evolved defensive postures that can successfully reduce herbivory. Those traits are both structural and metabolic, and they are regulated by a complex set of genes. Plants that take or induce a defensive posture usually grow less than those which do not when free from herbivory. This evidence suggests that there are costs (plant growth, and fitness) to defense. It has been difficult to perform controlled experiments about the cost of defense, and few studies have conclusively quantified the cost of defense or its relationship to plant fitness under herbivore pressure. But, in a theoretical sense, allocation to defense should rob other growth functions of ATP energy, potentially disrupt metabolism, and cause autotoxicity. Minimizing the costs of defense in the absence of herbivory may be one of the selective forces for the evolution of inducible defenses.

Defenses against herbivory can be direct or indirect. Direct defenses are targeted at the herbivore population. Some of these direct defenses are structural (spines, trichomes), while others are chemical (tannins, alkaloids, etc.). The chemical defenses of plants are quite extensive and include at least eight classes of compounds. Some classes are constitutively produced, while others are induced, by herbivore damage. Constitutive defenses are often carbon based and broad-spectrum, while inducible defenses are often nitrogen based and narrow-spectrum. Indirect defenses involve the ecological setting of plants. These defenses concern enhancement of predator and parasitoid search image for the herbivore. Induced volatile compounds to be released from plants can attract specific parasitoids, which in turn reduces the growth and fecundity of herbivores.

Variation in host plant tissue quality can be very influential on the interactions between insects and plants. Host plant tissue variation induced by seasonal and spatial resource availability, developmental stages, phenology, genetic factors, environmental stress, and previous herbivore damage all can influence insect preference and performance. This variability in tissue quality also reduces the rate of insect adaptation and helps stabilize the interaction. In fact, variation in defensive status among individuals in a population may be another important selective force for the evolution of inducible defenses.

The physiological interactions between insect and plant do not happen in a sterile environment. Bacterial infection of plant tissues can be detrimental to insect physiology, insects are a common vector for plant fungal pathogens, host tissue variability provides search images for insect predators and parasites, and herbivore damage influences secondary plant tolerance of environmental stress. Thus, there a number of outside forces that can influence the physiological interaction between herbivore and plant. No plant-animal interaction has evolved without the influence of abiotic and biotic environmental factors. Studying the tritrophic or multisystem complexity in communities is critical to a thorough understanding of the physiological reactions between plants and herbivores.

The study of plant-herbivore interactions is complex, and research on this topic has led to significant insights into both insect population biology and the regulation of plant physiology. Bioengineering of these interactions is a fertile research topic because of its importance to agriculture and because studies of plant-animal interaction may lead to increased understanding of signal transduction mechanisms that regulate plant physiological processes.

STUDY-REVIEW OUTLINE

Introduction

1. Plants are exposed to a multitude of herbivores at all stages of their life history and on all organs.
2. Over the long term, herbivory has been a significant factor in the evolution of plants.
3. The origin of modern plant-herbivore interaction theory can be traced back to a publication by Ehrlich and Raven (1964), about butterfly evolution.

Abundance and Diversity of Herbivores

1. Herbivores are as diverse as the diversity of the animal phyla because ingestion of plant material (the definition of herbivory) is the foundation of energy support for animals.
2. Herbivores can be classified into guilds based on their feeding mechanisms.
3. At least one group of insects can be assigned to every guild of feeding mechanism.
4. Herbivores can also be classified on the basis of whether they utilize one species of plant exclusively as a food source (*specialists*) or a wide diversity of plants (*generalists*).
5. Specialist herbivores develop life history patterns that are highly coordinated with the physiological activity of the host plant.
6. Herbivores as a whole form a continuum from *obligate specialist,* through *facultative specialist,* to *extreme generalist.*
7. Herbivores are found in every ecosystem of the world, but the diversity of herbivores and the intensity of herbivory vary among ecosystems.
8. Herbivore diversity is greatest in the tropics and decreases generally with latitude and elevation.
9. During their lifetime, all plants will experience a diversity of herbivores in different feeding guilds.

Potential Influences of Herbivores on Plants

1. In many cases, the net effect of herbivory is a reduction in productivity.
2. When herbivores are excluded from plant populations, productivity and fecundity often increase.
3. However, in some cases herbivory can cause an increase in plant productivity.

4. Six factors regulate the intensity of herbivory on plant species:
   a. Life stage of the plant
   b. Type of tissue consumed by the herbivore
   c. Timing of the herbivore attack
   d. Genetics of the individual plant
   e. Environmental conditions
   f. Particular plant species
5. *Compensatory growth,* an increase in plant growth following herbivory, is a mechanism by which plants can regain their competitive advantage following herbivory.
6. Plants produce chemicals that can help defend against herbivores.
   a. Performed induced responses occur immediately upon damage.
   b. Rapidly induced responses occur within hours or days of the damage.
   c. Delayed induced responses are those that occur during regrowth following herbivory.
7. Herbivores can enhance seed production or reduce yield.
8. Herbivores feeding on seeds and flowers can reduce reproductive output of plants directly by reducing seed yield and pollination.
9. Herbivores feeding on leaves can reduce seed yield by reducing the leaf area, which produces the photosynthate required for floral and seed production.
10. Herbivores can influence the sex ratio of a population of plants as well as the architecture of the plants.
11. Seed-feeding herbivores can determine the frequency and location of seed germination.
12. A plant's susceptibility to pathogens can be influenced by herbivory.
13. Herbivore damage can provide avenues of pathogen infection, and many pathogens are carried on the bodies of herbivores.
14. Herbivores can influence a plant's ability to acquire resources and, as a result, the plant's competitive ability.
15. Herbivores can also influence a plant's response to environmental stress.

Influence of Herbivory at an Ecological Level

1. Ecosystem productivity is partly determined by herbivory.
2. Systems ecologists include herbivory as a transfer function in models of energy and nutrient flow.
3. In some communities, herbivory can increase species diversity.
4. Herbivores can reduce the competitive advantage of a potentially dominant species.
5. Naturalized species can become invasive species if there are no herbivores to control the reproductive capability of the introduced species.
6. The outcome of competition among species can be affected by the presence of a herbivore.
7. Models of herbivory predict top-down regulation of plant productivity by herbivores unless a predator relieves the herbivore pressure.
8. In other models, plant growth rate influences the way herbivores regulate community productivity. At low growth rates, herbivores do not have enough food to support a large enough population to influence plant productivity. At high plant growth rates, the efficiency of herbivory is decreased, and herbivores have a reduced effect on plant productivity.
9. The *tritrophic interaction* between plant, herbivore, and predator is central to the evolution of the interaction between plants and herbivores.
10. Plant defenses against herbivores can be direct against the herbivore, or indirect through stimulating the effectiveness of herbivore predators.
11. The relative effectiveness of direct and indirect defenses is dependent upon the ecological context of the particular plant population.

Evolutionary Relationships

1. The generation time for plants (trees in particular) is much greater that that of herbivores (insects in particular). Also, insect population sizes are much larger than that of plant populations.

2. Therefore (following number 1), the evolutionary rate of herbivores should be faster than that of plants, and herbivores should quickly overcome plant defensive mechanisms.
3. The evolutionary rate of herbivores in reference to antidefense mechanisms is dampened by variation in defense and inducible defenses in plants.
4. Coevolution between plant and herbivore can result in specialist herbivores and mutualistic relationships between herbivore and plant.
5. There is a trade-off between the benefit a defensive mechanism affords a plant and the cost (in resources) that the mechanism requires.
6. *Optimality theory* is based on the assumption that resources invested in defensive strategy are diverted away from processes that determine growth rate.
7. *Coevolution theory* assumes that plant defenses evolve in the context of coevolving herbivores and not in the context of cost-benefit trade-off relationships.

Applied Significances of Plant-Herbivore Interactions

1. Efforts to increase yield in agricultural situations have involved the eradication of herbivores. Herbivores are removed from agricultural crops because they decrease yield and damage the quality of the yield for the consumer.
2. An enormous global business has developed around the production of chemicals that are effective at reducing herbivore populations in agricultural fields.
3. Much genetic engineering is occurring in agriculture to enhance plant defenses against herbivores. Examples are engineering defenses against pesticides so that higher doses of pesticide can be applied, enhancing defensive mechanisms in the variety, and adding new defensive mechanisms into the crop varieties.
4. Naturally produced *secondary chemicals,* which may be associated with herbivore defense, are the raw materials for many pharmaceuticals.
5. Herbivore inducible pathways regulate the synthesis of important secondary chemicals such as caffeine and nicotine. By modifying the product of these inducible pathways, plants can become *bioreactors* (biological producers of specific molecules).
6. The regulation of exotic plant invasions can be accomplished by utilizing specialist herbivores.

Mechanisms of Adaptation Against Herbivory

Nonphysiological and Indirect Defenses

Morphological and Behavioral Defenses

1. There are only a limited number of morphological, anatomical, and behavioral defenses against herbivores.
2. Spines on plant stems serve to inhibit some large herbivores, but they will not inhibit insect herbivory, and many herbivores are capable of feeding on spiny plants.
3. *Trichomes* (leaf and stem hairs) have a greater impact on insect herbivores than do spines.
4. Trichomes can serve as a barrier to feeding by some insects, but the significance of trichomes is not limited to defense against herbivores: as a reflecting coating of the leaf surface, trichomes serve as a means for keepng the leaf cool.
5. Thigmotropic movements of leaves, for example on *Mimosa,* may serve to decrease the apparency of leaves and expose the defense mechanism (spines).
6. Rapid development of shoots and leaves may be a behavioral mechanism to limit herbivory of juvenile tissues.

Compensatory Growth

1. In grasses, removal of leaf tissue by herbivores causes an increase in growth of the remaining tissues.

2. Productivity in natural grasslands increases because of large ungulate herbivory or grasshopper herbivory.
3. One of the mechanisms by which herbivores can influence the growth rate of plants is through the release of growth-stimulating substances in the saliva.
4. Large ungulates in grassland systems alter many ecosystem processes that indirectly could influence the subsequent growth rate of plants.
5. The nature of the regrowing tissues and organs after herbivory may include greater investment in defensive structures.

Indirect Plant Defenses

1. Indirect plant defenses involve the stimulation of predation on herbivores by parasitoids.
2. Plants produce a distinct blend of volatile compounds after damage by herbivores that are used by parasitoids to locate herbivores.
3. Volicitin, a compound first isolated from army worms, causes plant tissue to release the volatile attractants of parasitoids.
4. The insects consume plant fatty acids that are acted on by insect enzymes to produce volicitin. Volicitin in turn initiates the release of volatile compounds that attract insect parasitoids.
5. Parasitoids of spotted spider mites are also able to locate the spider mite population by the volatile emissions from plants infected by spider mites.
6. Volatile attractants produced by plants are the main way for parasitoids to locate the widely dispersed insect population.

Direct Physiological and Chemical Defenses

Secondary Metabolism

1. *Secondary chemicals* are "by-products" of metabolism, which are no longer useful for the formation of new cells (original definition by Sachs).
2. The functions of secondary chemicals in herbivory defense were established by entomologists.
3. A wide diversity of compounds are involved in defense against herbivores. Some are part of the primary metabolism, most are secondary metabolites, while others are present only after herbivore damage.
4. The evolutionary pressures of herbivory are more likely than allelopathy to have been the cause of the diverse nature of secondary chemistry.
5. Three major pathways are utilized to synthesize the secondary chemicals referred to as defense chemicals:
   a. The *shikimic acid pathway* leading to a number of different defensive compounds based upon phenylalanine, tyrosine, and tryptophan
   b. The *acetate-malonate pathway* leading to fatty acids and some aromatic defense compounds
   c. The *acetate-mevalonate pathway* leading to steroid defensive compounds
6. Some defensive chemicals can be called *broad-spectrum substances* because they have a general inhibitory influence on many different herbivores (e.g.; tannins, lignins, proteinase inhibitors).
7. Other defensive chemicals are *narrow spectrum* because they are restricted to a small group of plant families and may affect only a narrower range of herbivore species (e.g., alkaloids, nonprotein amino acids).
8. Broad-spectrum defense substances are often constitutive, while narrow-spectrum defense compounds are often induced by herbivore attack.
9. Variation in the concentration of defensive substances in time and space is the normal situation in plants. Understanding the mechanisms that regulate that variation leads to an

understanding of the defensive mechanism and possibly a window into the evolution of these substances.

Tannins

1. Tannins and lignins are polymers of phenol monomer units; their molecular weight ranges from 500 to 20,000, and they are composed of underivatized monomers.
2. Tannins can be classified as either *condensed tannins* or *hydrolyzable tannins.*
3. Condensed tannins are polymers of flavonoid monomers and cannot be hydrolyzed in nature.
4. When the condensed tannins are hydrolyzed, the monomers produced are anthocyanidins. The proportion of different cyanidins within condensed tannins varies among species.
5. In hydrolyzable tannins, the monomers are held together by carboxylic ester linkages and can be broken down by natural means.
6. The monomers are either a polyol (gallotannins) or a hexahydroxyphenic acid (ellagitannins).
7. The amount of tannins in tissues varies with age, environment, season, and species. In general, the lower the tannin content, the more palatable the tissue.
8. Domestication of crops is associated with lower tannin concentrations, and herbivory in wild plants has stimulated the development of higher tannin concentrations.
9. Traditionally, tannins were thought to act by binding to proteins and inactivating them, and possibly by causing their precipitation.
10. Tannins bind well to heavy metals; therefore, tannins may serve a metal-sequestering function in plant tissues.
11. Tannins also bind effectively to free radicals and membranes.
12. The efficacy of tannins as toxins to herbivores was originally thought to relate to the binding of proteins, making them unavailable to the insect digestion.
13. Recent evidence suggests that tannins may be toxic to the digestive cells of the epidermis of the insect gut.
14. In other studies, there does not seem to be direct relationship between tannins and poor insect performance.
15. There are several reasons why tannins must be associated with a defense mechanism: (1) tannins are expensive (resources) to synthesize; (2) they have been shown directly to be toxic to mammals; (3) animals reject food with high tannin content; and (4) herbivores eating only tannin-rich food show slow growth rates compared with a control group.
16. The chemical composition of tannins in the feed has a different effect on the growth of different insects.
17. Herbivores can use behavioral mechanisms to avoid tannin toxicity: (1) they can avoid tannin-rich individuals, (2) they can pretreat their food with alkali, and (3) they can consume their food with a tannin chelator such as milk.
18. Several physiological defenses of insects limit damage to herbivores by tannins:
    a. Preventing protein precipitation with surfactants
    b. Increasing the pH of the gut
    c. Releasing polyphenol oxidase to decompose tannins
    d. Increasing the gut redox potential
    e. Inducing of proline-rich tannin-binding proteins in the saliva
    f. Reducing the resting metabolic rate to minimize the intake of tannins

Lignins

1. Lignins are very large complex molecules formed from cinnamyl alcohol subunits.
2. Two enzymes are involved with lignin synthesis: CAD catalyzes the formation of the cinnamyl alcohol subunits, and POD (a peroxidase) polymerizes the lignin and releases hydrogen peroxide.
3. Lignins are resistant to decomposition and are found in secondary walls of plants.

4. Lignins inhibit herbivory because of their effect on plant toughness.
5. The higher the concentration of lignin, the tougher the tissue and the less likely herbivory will be successful.
6. The higher the concentration of lignin, the lower the tissue palatability.
7. Mechanical damage of plants by wind can induce increased lignin for support and is associated with reduced herbivory.
8. Lignin also may serve as a protection against pathogen attack.
9. Polyphenol oxidase and peroxidase are required for lignin synthesis, and they may also produce by-products that are toxic to herbivores.
10. Lignin increases plant defense against herbivores, both by increasing tissue toughness and by increasing toxic compounds (by-products of lignin synthesis).

Phenolics

1. Phenolic compounds have one or two phenolic hydroxyls and are synthesized by the shikimic acid pathway.
2. Phenylpropanoids are the most abundant of the phenolic substances in plants.
3. Phenolics have both allelochemical and antiherbivory action.
4. Resins are found in conifers and many desert plants.
5. Resins are composed of a number of different aromatic compounds including phenolics.
6. Resins have antiherbivory activity, but they can also serve as attractants for some insects.
7. NDGA (nordihydroguiairetic acid) is a phenolic aglycone that is abundant in the resins of the creosote bush.
8. NDGA induces reduced growth rate and retarded developmental in many insects.
9. Some insects are able to consume creosote bush tissues. Some of these specialist insects avoid creosote tissues with high NDGA content.
10. One specialist grasshopper can consume creosote bush tissues with up to 20% NDGA without reduced growth rate. The mechanism of this ability to avoid toxicity is unknown.

Cyanogenic Glycosides

1. Cyanogenic glycosides are glycosides of various hydroxynitriles that accumulate in the central vacuole of some species.
2. In the presence of β-glycosidase and hydroxynitrile lyase, these compounds release cyanide into solution.
3. Other cyanide-releasing compounds such as nitro acids exist, but they are usually restricted to a small number of species.
4. The biosynthesis of cyanogenic glycosides is linked to the synthesis of nitrogen-containing amino acids.
5. These compounds have been found in 2500 species in 150 different families, including all vascular plant orders.
6. Usually, each species has only one cyanogenic glycoside compound, but occasionally one species will have two or three.
7. The cyanogenic glycosides known best are amygdalin and linamarin.
8. Linamarin is the most widespread cyanogenic glycoside compound among higher plants.
9. Cyanide (HCN) is very toxic to plant and animal cells, but it can be found constitutively at very low levels in plant tissues.
10. HCN inhibits cytochrome a activity, which severely reduces aerobic respiration. HCN also reversibly inhibits nitrate reductase activity.
11. Upon herbivore activity, HCN is released rapidly by the mixing of enzymes with the cyanogenic glycosides.
12. In some plants, the cyanogenic glycoside and hydrolytic enzymes are located in one cell type, while the glycoside is in another cell type.
13. In some plants, the cyanogenic glycoside is located in the vacuole, and the hydrolytic enzymes are located in the apoplast.

14. The efficacy of HCN as a poison to mammals is indisputable, but there is some question about its efficacy as a defense against insect herbivores.
15. Some insects, African army worms, have evolved the ability to detoxify large amounts of HCN.
16. Other compounds released during the hydrolysis of cyanogenic glycosides may be more toxic than the HCN itself.

Nonprotein Amino Acids

1. In the plant kingdom, more than 600 different amino acids are produced that are not incorporated into proteins.
2. Most of these nonprotein amino acids are unconjugated, but a few are attached to carbohydrates or occur as 4-glutamyl-linked polypeptides.
3. Many nonprotein amino acids are aromatic or heterocyclic where the R group is cyclic.
4. Many nonprotein amino acids are nontoxic and cytoplasmically compatible. They accumulate in response to stress such as drought and chilling (e.g., glycine-betaine and proline).
5. Some nonprotein amino acids are toxic and can be linked to specific disease symptoms.
6. The effect of canavanine on insects is the best documented case of toxicity against insect herbivores.
7. Canavanine is produced by one subfamily of the Fabaceae, but it has broad toxicity to organisms from viruses to mammals.
8. Canavanine is an analog of arginine, and can replace arginine in proteins.
9. In most organisms, arginyl-tRNA synthetase cannot distinguish between arginine and canavanine.
10. Canavanine is much less basic than arginine, so proteins formed from canavanine instead of arginine have a dysfunctional structure.
11. The *substitution error frequency* (SEF) represents the frequency of arginine substitution by canavanine. Canavanine-sensitive insects have a high SEF when fed canavanine-laced diets (1:3), while canavanine-resistant insects have a much lower ratio (1:365).
12. Nonprotein amino acids with a cyclic side chain such as L-hypoglycine can be particularly toxic.
13. Ripe seeds of the ackee tree are nontoxic, but juvenile seeds have a high concentration of L-hypoglycine and are very toxic.
14. Consumption of L-hypoglycine induces a severe case of hypoglycemia (low blood sugar).
15. Nonprotein amino acids containing selenium are concentrated by *Astragalus* species. Twenty species of *Astragalus* are known to accumulate a toxic level of selenium nonprotein amino acids.
16. The selenium nonprotein amino acids are analogs of cysteine, and are incorporated into the herbivore protein. The selenium nonprotein amino acid cannot form the disulfide bridges in proteins that normally form through cysteine. Thus, proteins can lose their shape and become dysfunctional.

Alkaloids

1. Alkaloids are an extremely diverse group of cyclic compounds containing nitrogen in a negative oxidation state.
2. Most individual alkaloids are produced by a relatively small number of organisms and inhibit a small number of herbivores.
3. Twenty percent of plants accumulate alkaloids, which are also found in amphibians, marine organisms, and insects.
4. Alkaloids are an extremely diverse group of substances, but they can be broadly classified into five large groups: benzylisoquinoline alkaloids (BAs), monoterpenoid indole alkaloids (MAs), pyrrolizidine alkaloids (PAs), quinolizidine alkaloids (QAs); and polyhyroxypyrrolizidine alkaloids.

5. Alkaloids are mostly constitutive defense substances; however, some are inducible.
6. Alkaloids are usually synthesized in one organ (leaves, roots) and are translocated in the xylem or phloem to the rest of the plant.
7. QAs of lupines are synthesized in the leaves, in association with chloroplasts, and transported in the phloem to the rest of the plant.
8. Alkaloids are synthesized in vesicles and transported to the plasma membrane, where they are released into the apoplast by endocytosis.
9. Alkaloids are also transported into the central vacuole by alkaloid-specific transport proteins, resulting in a buildup of alkaloids in the central vacuole of some species such as poppies.
10. PAs are nontoxic in themselves; however, in the liver of mammals, they are converted by P-450 monooxygnase into highly reactive mutagens.
11. Since PAs are nontoxic to insects, many insects feed on plants that accumulate PAs in order to gain some protection from mammalian predators.
12. QAs can have a wide diversity of effects, including disorientation and convulsions in mammals.
13. Most alkaloids are bitter and may deter feeding based on their bitterness.
14. QAs and PAs are the classes of alkaloids that have the best experimental evidence for antiherbivory action.

Hormones and Antihormones

1. If plants are able to store a toxin that is specific for insects but not plants (i.e., not autotoxic), the plants can accumulate this toxin to higher concentrations.
2. If plants are able to synthesize and concentrate an insect hormone, there is little chance for the insect to develop resistance to this defense.
3. Juvenile hormone (JH) and some of its analogous forms are accumulated by some plant species.
4. *Phytojuvenoids* are plant substances that have juvenile hormone activity in insect systems.
5. Some plants also accumulate steroids that have ecdysone activity.
6. Juvabione is a substance that occurs in the wood of balsam fir and has juvenile hormone activity.
7. *Cyperis irida* produces JH III in the exact configuration as that in insect systems. Insects that consume this tissue remain in the juvenile state for a longer period of time than insects that consume other leaves.
8. Female grasshoppers fed *Cyperis irida* leaves produced few and infertile eggs.
9. Most of the damage done to plants is done when insects are in a juvenile stage. Thus, slowing down the developmental stages of insects is not a very good defense against short-term herbivory, but is a good defense against long-term herbivory.
10. Some plants produce compounds that interfere with the action of hormones in insects.
11. These antihormones (precocene) cause premature development or metamorphosis, which reduces survival and reproduction.
12. *Ecdysones* are responsible for initiating molting and pupation in insects.
13. *Ecdysteroids* have been found in many members of the Chenopodiaceae. In fact, 70 different ecdysteroids have been isolated in 100 different plant families.
14. Some insects are affected by the ecdysteroids, but most have remarkable resistance to the plant ecdysteroids.

Proteinase Inhibitors

1. Proteinase inhibitors are low-molecular-weight proteins produced by plants that bind to animal proteinases and deactivate them.
2. About 20 different proteinase inhibitors have been found, and they are most abundant in the Fabaceae and the Solanaceae.

3. Proteinase inhibitors reach their highest concentration in reproductive tissues such as seeds.
4. Although the main function of these compounds may be for defense against fungal and bacterial infections, they also may have other defensive qualities.
5. Plants transformed with genes for proteinase inhibitors have gained additional herbivore resistance.
6. The genes for proteinase inhibitors (*PIN genes*) can be induced upon wounding by herbivores.
7. The signal transduction pathway for PIN genes operates through jasmonic acid.
8. *Serine proteinase inhibitors* bind to the proteinase active site with high affinity.
9. Particular proteinase inhibitors work most effectively on specific proteinase enzymes. For example, the Kunitz type of serine inhibitor most effectively inhibits chymotrypsin and trypsin.
10. A second class of inhibitors is *cysteine proteinase inhibitors,* which are effective against endoproteinases in the papain class.

Physiological Basis for the Induction of Defensive Substances

Generalities on Induction

1. Many narrow-spectrum defensive substances are not constitutive, and are induced by herbivore attack.
2. There are several possible mechanisms for the induction of defensive substance concentration following herbivory damage:
   a. Increased synthesis involving increased transcription and translation of proteins required for the synthesis of the substance
   b. Decreased destruction rate of the defensive compound
   c. Increased transport of the substance from the site of synthesis to the target site
   d. Release of bound (inactive) defensive compound
   e. Increased activity of the accumulating mechanism from the transport tissue.
3. Some of the defensive compounds are induced at the genetic level (mechanism a above).
4. The transcription of PIN genes is upregulated after herbivore damage in tomato; genes coding for enzymes needed for nicotine synthesis are upregulated following herbivore attack.
5. Upon herbivore attack of one leaf, the synthesis of defensive compounds increases throughout the plant due to a hormonal signal.
6. The first defense signaling pathway discovered was that for fungal infection and involved oligogalactouronides, which are products of the fungal cell wall.
7. Oligogalactouronides are not very mobile, so they are thought to induce localized defense against infection.
8. A small polypeptide, systemin, has been isolated and has very high induction capacity for defensive compounds.
9. Systemin induces the upregulation of a number of genes in basically four classes:
   a. Genes producing proteinase inhibitors that are active in animal digestion
   b. Genes that produce signaling pathway proteins
   c. Genes that produce proteinases that are active in plants
   d. Genes that produce proteins with unknown roles
10. Systemin is an 18–amino acid–long polypeptide with the highest known gene activation capability.
11. The protein has a shape similar to other DNA-binding proteins, and its activity is dependent upon proline 13 and threonine 17.
12. Systemin is processed from a larger protein prosystemin. Systemin is located at residue 179–196 of prosystemin.
13. Prosystemin genes have been located in a number of different Solanaceae species, but the prosystemin of one species cannot be processed into systemin by another species.

14. The prosystemin gene is important for wound response because transgenic plants with an antisense gene for prosystemin are very susceptible to wounding and herbivory and have no biochemical response to wounding.
15. Systemin is synthesized in the cells near the vascular traces, and systemin is rapidly placed into the phloem for transport. Within 30 minutes after a wound, the systemin has been transported to other parts of the plant.
16. Systemin signaling operates through lipid-derived intermediates: linolenic acid (LA) and jasmonic acid (JA).
17. Systemin induces the octadecanoid pathway (involving JA and LA), which is associated with herbivore defense, pathogen infection, and cell development.
18. The plant wound response is similar to the inflammatory response in animals, in which there is a localized response and a systemic response.

Costs of Defense for the Plant

Description of Costs

1. The induction of plant defensive compounds following herbivore damage suggests that there are costs to producing these compounds in high enough concentration to be effective against herbivores.
2. *Optimal defense theory* is based on the concept that any resources allocated to producing defense related substances will divert those resources from growth and reproduction.
3. There are several difficulties in proving the optimal defense theory:
   a. Defense compounds are induced by other stresses besides herbivory.
   b. Only rarely has it been shown that the upregulation of the defensive compound results in an increase in plant fitness.
   c. Several putative defense compounds such as alkaloids also are storage compounds for nitrogen or carbohydrates.
4. One method for determining the cost of defensive substance production is an analysis of the energy consumed during the production of the substance.
5. The synthesis cost estimate includes all enzymes associated with the biosynthesis of the substance that are not otherwise associated with primary metabolism as well as the cost of the substance itself.
6. The cost of synthesizing defense substances is usually reported in moles of ATP, $CO_{2,}$ or glucose required.
7. The cost of synthesizing a mole of each specific defensive compound is different. For instance, alkaloids have a much higher cost coefficient than tannins.
8. The difference in biosynthesis cost among defensive compounds may explain why constitutive defense substances have a low cost coefficient.
9. Similar pathway analysis can be used to estimate other cost currencies (N or P investment).

Resource Allocation

1. Resources required for the synthesis of defensive compounds are N, C, S, and P.
2. Optimal defense theory is based on the concept that any resources allocated to producing defense-related substances will divert those resources from growth and reproduction. This is only valid if the resources are limiting for fitness.
3. Plants growing in low nutrient sites have relatively long-lived leaves with large amounts of carbon-based defense compounds.
4. The slow growth rate of plants in low nutrient environments may reflect the cost of synthesizing the defense substances.
5. There is a broadly based correlation between a high allocation to defense and a slow growth rate.
6. It is difficult to verify a cause-and-effect relationship between the allocation to defense and the growth rate.

7. If induction of a defensive compound causes a reduction in fitness in the absence of herbivores, this indicates a fitness cost for synthesizing the compound. Such examples are rare.
8. One complication to proving a cause-and-effect relationship between the synthesis of defensive compounds and fitness is that many other aspects of the environment influence resource allocation patterns, induction of defensive compounds, and growth rate.
9. Furthermore (in relation to number 8), tissue quality (allocation of resources) determines herbivore feeding. Thus, low allocation of resources to a tissue may inhibit herbivore damage and complicate the herbivore-allocation correlation.
10. The relationship between the relative cost of defense and the allocation to defense within one species is opposite to that between species.
    a. Allocation to defense among species decreases for those species in low resource environments.
    b. Allocation to defense among individuals in one species increases with the addition of nutrients.
11. Evolutionary processes select for a functional type in low nutrient environments that has relatively long-lived leaves with relatively high defense.
12. On a shorter time scale, plants under stress allocate resources to critical fitness traits such as growth and reproduction, and under excess resource availability more resources are allocated to defensive compounds.

Autotoxicity and Biochemical Disruption

1. High levels of defensive compounds can disrupt plant metabolism as well as herbivore metabolism, resulting in autotoxicity.
2. Therefore (in relation to number 1), an additional cost to defense is the compartmentation system that keeps the defense substance away from plant metabolic processes.
3. The evolution of the pathway leading to the defense compound could disrupt normal metabolic processes.
4. If a gene mutation results in the synthesis of the defense compound, then the metabolic pathway originally regulated by that gene has been altered, which may result in a loss of fitness.
5. If one of the copies of a gene in a genome is altered to synthesize the new defense substance, then the functional copy number of the original gene has been reduced, which may place a constraint on a metabolic pathway.
6. The induction of defensive compounds can interfere with other nonherbivore defensive pathways resulting in increased susceptibility to pathogens. It is difficult to calculate this cost of defense against herbivores.

Importance of Quantitative Variation

Heterogeneity of Host Tissue Quality

1. The quality of host tissue for insect consumption is not homogeneous among individuals in a population of plants.
2. Tissue quality (toughness, nutrition, defensive compounds, etc.) varies among genotypes within a species.
3. Tissue quality differs among organs on one individual, and among developmental stages of each organ.
4. As leaves mature, tissue quality for herbivores decreases based upon nutrition, toughness, water content, and structural defenses.
5. Tissue quality in a canopy of an individual has a large amount of heterogeneity due to location and developmental stage of the leaves.
6. Short-flush species have more homogeneous tissue quality on a spatial scale, but more rapid variation on a temporal scale than species with indeterminate growth.
7. The induction of defense is another cause of the heterogeneity of defensive compounds in plants. Leaves that are attacked will express localized defenses and send a systemic message

to other organs, inducing widespread defense. The result is a mosaic of defensive conditions among leaves on plants that are experiencing herbivory.

8. The heterogeneity of defensive posture in a population can be an evolutionary stimulus for inducible defenses.
9. Herbivores will avoid species that have greater variability in defensive posture, because the herbivore will have reduced fitness on variable plants compared with homogeneous plants.
10. Increasing the mean toxin level leads to a decrease in herbivore fitness, but also selects for herbivore resistance.
11. As the variance in toxin concentration increases, herbivore fitness decreases.

Effects on Insect Populations

1. Heterogeneity in host defense can lead to positive and negative effects on insect populations.
2. Insects are able to select tissues with relatively low defense, and have relatively high fitness by consuming these tissues.
3. Insects that select low-defended, high-quality tissues can have relatively high fitness because there is no selection for resistance traits, which usually reduces herbivore fitness.
4. Herbivores feeding on tissue with low defense in small locations make the herbivores localized and more susceptible to predators.
5. An important trade-off for herbivores is: Locating the highest-quality tissues versus avoiding predators and parasitoids.
6. Tissue quality traits do not often correlate with each other. Selection for high N content does not always correlate with selection for high water content.
7. Variation in defense also weakens selection for resistance traits in herbivores.

Environmental Stress and Herbivory

1. In a phytocentric view, the host plant is an interface between the environment and the herbivore population. Keep in mind that environmental effects also regulate herbivore populations to some degree.
2. As earlier stated, environmental effects alter host tissue quality, which will influence herbivore populations.
3. Most of the information about the environmental effects on insect populations is derived from agricultural systems. Therefore, long-term relationships in natural systems are not well represented in this database.

Insect Preference for Specific Host Plants

1. Insect preference refers to the manner by which insects select host species.
2. Insect performance relates to the growth, development, and fecundity of insects on a particular host species.
3. Environmental resource limitations can change insect preference.
4. Host preference can be a result of tissue quality (nitrogen, water) or particular volatile or tactile substances.
5. Limitations in potassium, phosphorus, or sulfur in plant tissues can increase insect preference for the tissues because these limitations result in reduced protein synthesis and an increase in free amino acids.
6. Herbivores prefer species or individuals with high nitrogen concentrations, and shun plants with lower nitrogen.
7. The effectiveness of biocontrol insects is enhanced by fertilizing sites that are occupied by invasive plants.
8. Herbivores may also prefer tissues with relatively high water content. However, water stress induces an increase in free amino acids which are favored by herbivores.
9. Some insects select species based on chemical cues rather than tissue quality cues. This is most often true of specialist herbivores that feed on one or a relatively few host plants.
10. Some butterflies utilize volatile and visual cues to find their preferred species.

11. Some butterfly species have chemical sensors on their front legs that identify chemicals found only on their preferred host. The result is that they lay eggs only on leaves that possess these chemicals.
12. It is often a defensive compound or compounds that are detected in this host selection process.
13. Environmental stress can cause either a reduction or increase in the chemical cues in plant tissue for insect oviposition.
14. During nitrogen limitation, the chemical cue concentration can decrease, making it hard for the herbivores to locate their preferred host.
15. Water stress can increase the release of volatile substances that can make it easier for insect herbivores to find preferred hosts.

Host Suitability and Insect Performance

1. Environmental stress, through its influence on tissue quality, can regulate herbivore growth, development, and fecundity.
2. Fertilization of a plant population can result in higher tissue nutrition and a greater growth rate and fecundity of herbivores.
3. Other stresses such as water stress and temperature stress can either increase or decrease insect growth rate depending on the host species.
4. Any environmental factor than changes the concentration of defensive compounds in plant tissues will influence herbivores negatively.

Herbivore Influence on Subsequent Abiotic Stress Tolerance

1. The evidence for prolonged effects of herbivory on plant stress resistance is not well established.
2. The removal of leaf area by herbivores causes reduced water loss, increased water conservation, and a reduced chance of experiencing water stress.
3. Partial removal of leaf tissues can induce greater water stress because of the increased evaporation surfaces.
4. The timing of herbivory can influence the effect of environmental stress on plant fitness.
5. Herbivory can lead to increased lignification. Higher lignification can reduce plant susceptibility to air pollution stress.
6. Herbivory can cause an increase in free amino acids or carbohydrates, which can result in increased drought tolerance.
7. The interactions between environmental stress and herbivory must be understood on an individual species basis because herbivory can increase or decrease tolerance or resistance to any stress.

STUDY-REVIEW QUESTIONS

1. Why is the diversity of herbivores so high?
2. What are the "important messages" about plant-herbivore interaction on a physiological basis?
3. Describe several feeding mechanisms utilized by herbivores. Identify several classes of arthropods that utilize these feeding mechanisms.
4. What is the difference in population stability for generalist and specialist herbivores? Why is this the case?
5. What is the relationship between the diversity of herbivores on an individual plant during the various phases of the plant's life cycle?
6. Does herbivory have a positive or negative influence on plant growth?
7. What are the six factors that determine the magnitude of herbivory effects on plants?

8. What is a definition of *allelochemicals?* What do you think is the best terminology for the substances that are associated with defense against herbivores in light of other plant-plant or plant-animal interactions?
9. What are the three basic time scales that are associated with the induction of secondary chemicals in response to herbivory?
10. In what ways can herbivores influence the reproductive fitness of plants? Why is this an important concept to long term interactions between herbivores and plants?
11. What are the potential interactions between herbivores, pathogens, and plants?
12. In what ways can herbivores influence community structure?
13. Can herbivores have an influence on ecosystem function? If so, what are the main ecosystem processes that herbivores influence?
14. Why would one predict that insect herbivores have a faster rate of evolution than plants? How does this relate to the defense mechanisms of plants?
15. What are some examples of coevolution between insect herbivores and plants?
16. Does the interaction between herbivores and plants have any applied significance? If so, what are some examples of the ways this interaction has been used by human cultures?
17. What are two morphological mechanisms that plants can use to reduce the possibility of herbivory? How effective are they?
18. What is compensatory growth, and why is it significant to plant-herbivore interactions?
19. What are the three metabolic pathways that lead to the synthesis of important substances in herbivore defense?
20. What are some indirect defense mechanisms in plants? Describe the tritrophic interaction between tobacco, herbivores, and parasitoids.
21. Describe the patterns of defensive chemical concentrations among plants and among tissues over time?
22. Differentiate between tannins and lignins.
23. What are the two classes of tannins, and what is their mode of action in defense against herbivores?
24. Differentiate between broad-spectrum and narrow-spectrum herbivore defense compounds. Give some examples of each.
25. Some scientists think that tannins have no significance to herbivore defense. What is the evidence against this idea?
26. What mechanisms do insect herbivores use to avoid the toxic nature of tannins?
27. Peroxidase and polyphenol oxidase are involved in lignin synthesis. What is the relationship between these enzymes, lignin concentration, and defense against herbivores?
28. How can tissues with a high lignin concentration inhibit herbivory?
29. What are cyanogenic compounds (four classes), and how are they synthesized?
30. Why do cyanogenic glycosides need a special compartmentation scheme in plant tissues, and how do these compounds release cyanide?
31. Describe the resins of the creosote bush (*Larrea tridentata*), and explain how these serve as a herbivore defense mechanism.
32. What are nonprotein amino acids, and how are they toxic to insects?
33. How can insects become immune to the effects of nonprotein amino acids?
34. What type of nonprotein amino acids does *Astragalus* have, and which herbivores do these compounds primarily affect? What are the symptoms of *Astragalus* poisoning?
35. Describe the special case of canavanine as a defense against herbivores? Are their any resistant herbivores? If so, why are they resistant?
36. What are the four major classes of alkaloids, and what is the basic unit of each?
37. Where are alkaloids often synthesized? Are alkaloids constitutive or induced? How are alkaloids transported in plants?
38. What is the mode of action of alkaloids on insects?
39. What is the evidence that QA and PA alkaloids may be defensive substances? If they are not defensive, what is the significance of alkaloids to plants?
40. Which insect hormones are produced by plants?

41. Why is the production of insects hormones by plants an effective defense strategy against herbivores?
42. What are proteinase inhibitors, and what is the evidence that they act as a defensive substance?
43. Are proteinase inhibitors inducible? If so, what is the signaling mechanism for their induction?
44. Describe the Kunitz-type proteinase inhibitor.
45. What are five ways by which herbivore defense compounds can be induced to higher concentration?
46. Diagram the systemin signal transduction pathway. How is this pathway different for herbivore attack and pathogen attack?
47. What types of genes are induced by the systemin signaling pathway?
48. What are three of the main costs of defense against herbivores for plants?
49. Differentiate between the cost-benefit model and the variability model for the evolutionary significance of inducible defenses.
50. What is a pathway analysis for the cost of defense? Why is this technique an underestimate of costs?
51. What is an important trade-off for insects foraging for the highest-quality plant tissues?
52. What is the common denominator in all interactions between environmental stress, plants, and herbivores? Why is this so?
53. What factors determine insect preference for plant tissues? What factors determine insect performance on plant tissues?
54. How can environmental stresses influence subsequent tolerance of herbivores? How can herbivory influence subsequent abiotic stress tolerance?

SUPPLEMENTARY READINGS

Bernays, E. A. 1992. Insect-Plant Interactions. CRC Press, Boca Raton, FL.

Bernays, E. A., and R. F. Chapman. 1994. Host-Plant Selection by Phytophagous Insects. Chapman & Hall, New York.

Denno, R. F., and M. McLure (eds.). 1983. Variable Plants and Herbivores in Natural and Managed Systems. Academic Press, San Diego, CA.

Fritz, R. S., and E. L. Simms. 1992. Plant Resistance to Herbivores and Pathogens: Ecology, Evolution, and Genetics. University of Chicago Press, Chicago.

Gregory, R. A., and P. M. Wargo. 1986. Timing of defoliation and its effect on bud development, starch reserves, and sap sugar concentration of sugar maple. Canadian Journal of Forest Research 16:10–17.

Heinrichs, E. A. (ed.). 1989. Plant Stress-Insect Interactions. John Wiley & Sons, New York.

Karban, R., and I. T. Baldwin. 1997. Induced Responses to Herbivory. University of Chicago Press, Chicago.

Palo, T., and C. T. Robbins. 1991. Plant Defenses Against Mammalian Herbivory. CRC Press, Boca Raton, FL.

Price, P. N. 1984. Insect Ecology, 2nd Ed. John Wiley & Sons, New York.

Price, P. W., T. M. Lewinshon, G. H. Fernandes, and W. W. Benson. 1991. Plant-Animal Interactions: Evolutionary Ecology in Tropical and Temperate Regions. John Wiley & Sons, New York.

Raupp, M. J., and D. W. Tallamy. 1990. Phytochemical Induction by Herbivores. John Wiley & Sons, New York.

Rosenthal, G. A., and M. Berenbaum (eds.). 1991. The Chemical Participants. Herbivores: Their Interaction with Secondary Plant Metabolites, 2nd Ed., Vol. I. Academic Press, San Diego.

8 Allelochemistry as a Plant Stress

Outline

Objectives

1. Define *allelopathy* as a form of plant-plant interaction via chemical means.
2. Explain the ecological and physiological significances of allelopathy.
3. Assess the debate on the existence of allelopathy and its role in the natural environment.
4. Describe the difficulties in experimentation with allelopathy.
5. Provide an overview of biosynthesis and structure of the most common secondary metabolites associated with allelopathy.
6. Discuss the biochemical and physiological bases of allelopathic interference including modes of release and modes of action.
7. Explain the mechanisms by which plants can detoxify allelochemicals.
8. Expound on the various effects of environmental stresses on allelopathy.
9. Describe the applied significance of allelopathy to agriculture and forestry.
10. Consider an appropriate experimental design for allelopathy research, particularly related to separating effects of allelopathy and competition.
11. Evaluate the current state and future direction of allelopathy research

I. INTRODUCTION

A. Presence of Allelopathy

Plant fitness depends on acquiring sufficient resources for growth and reproduction. When the availability of resources is limiting or finite, it is advantageous for a plant to exclude its competitors (Harper 1977). Akin to territorial protection in animals, plants possess chemical means to interfere with intruding neighboring plants. The interference between plants through the release of chemicals is known as *allelopathy.* Molisch (1938), who first coined the term *allelopathy,* defined it more broadly as interactions between plants, and between plants and soil microorganisms, that result in either a stimulatory or an inhibitory response. While Rice (1974) initially considered allelopathy to be only the inhibitory effect of one plant on another, he later (1984) concurred with Molisch's broader definition. Soil microorganisms, although still central to allelopathy, have in recent assessments assumed the role of an intermediary by modifying the chemical environment between a source and a receptor plant.

Chemical interference between plants is achieved through the release of secondary metabolites. The term *allelochemicals,* which refers to these secondary compounds, was also adopted by scientists studying plant-insect interactions. Many of the same chemicals are involved in plant-plant and plant-insect interactions. Therefore, it was natural to use the same term to define the chemicals associated with both interactions. Secondary chemicals that have antiherbivory efficacy were discusses in Chapter 7. In this chapter, we will consider the allelopathic efficacy of secondary substances, and we will focus on the phenolic acids and terpenoids.

Allelopathy could be an unintended consequence of discharging secondary compounds as waste products (Muller 1966). Most of these substances released from plants occur as volatiles in arid regions or as aqueous leachates in humid regions (Del Moral and Cates 1971). The amount of secondary metabolites (allelochemicals) reaching a receptor plant is greatly modified by above- and belowground environmental factors such as precipitation and soil microbes. For example, decomposition in soil can rapidly degrade allelochemicals in the field. Given the great diversity of these substances and their temporal and spatial variations in the environment, there is no strong selective pressure for the evolution of defense by a particular target plant to a specific allelochemical. Detoxification of allelochemicals occurs by general metabolic pathways that also serve to detoxify other chemicals such as herbicides. Moreover, as originally defined, any given secondary compound with allelochemical activity can cause both stimulatory and inhibitory effects. This pattern is common and is concentration dependent: stimulation at low concentrations and suppression at higher concentrations (Figure 8.1).

Chemical interference is also observed in aquatic environments. The benefits of preempting local resources by chemical means in aquatic systems are less easily attained than they are on land (Lewis 1986). Nevertheless, many of the processes linking toxins from source to target plants, mediated by the growth environment, are the same as those in terrestrial environments and present the same challenges in demonstrating such links (Inderjit and Dakshini 1994a). Chemical interference has been reported in freshwater macrophytes and phytoplanktons (Gopal and Goel 1993) and in marine systems such as coral reefs (Lovett 1989). In this chapter, however, we focus our discussions only on terrestrial ecosystems.

Earlier experiments aimed at demonstrating allelopathy had been suggestive but never conclusive (Harper 1977). To address deficiencies in those experiments, recent work has become a painstaking process excluding all potential sources of interference except the phytochemical one in question. This type of scientific rigor is rarely required even for the study of competition often cited as a confounding factor in the study of allelopathy. Though difficult to demonstrate unequivocally, compelling evidence suggests the participation of allelopathy in plant-plant interactions and in ecosystem processes. However, because allelopathy may function largely indirectly through an intermediate agent (e.g., soil microbes) or by altering the available resources in the environment (Wardle et al. 1998b), there is a controversy over its

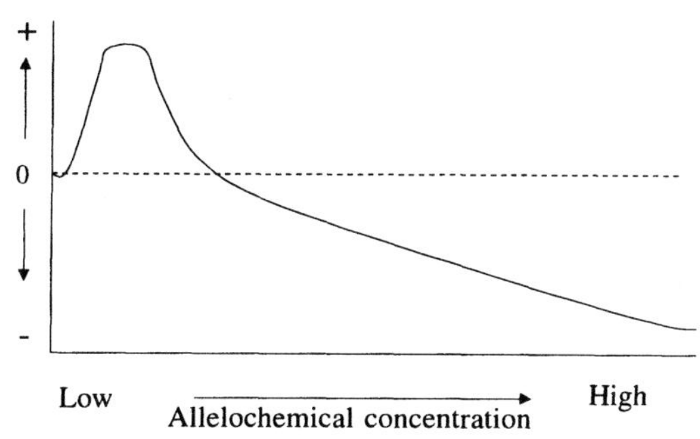

Figure 8.1. Biological response to the presence of allelochemical (measured as percent control). Notice that the response is stimulatory at low concentrations and inhibitory at increasing concentrations. (Redrawn from An et al. 1993.)

significance in field situations. In the following section, we examine the basis for either affirming or dismissing the role of allelopathy in plant-plant interactions.

1. The Case in Support of Allelopathy. A compelling case for exclusion of plants by allelochemicals exists when a bare zone occurs around a plant and no other apparent cause for its occurrence is evident. Examples of these bare zones are found in the Florida scrub (Fischer et al. 1994), in California soft chaparral (Muller 1966) or in a temperate forest (Lodhi 1976). The bare zones in California chaparral (e.g., McPherson and Muller 1969) may be particularly critical for "repelling" potential competitors for resources. The advantage of repelling other plants of arid lands such as Mediterranean regions is an increase in light availability, which is critical because plants of this habitat are intolerant of shade. The comprehensive study of bare zones in the chaparral by McPherson and Muller (1969) established an ecological imperative for broad validations of allelopathy in the natural and agricultural environment for the next three decades.

One of the best known examples of a plant producing allelochemicals is black walnut, which is referenced and discussed in almost all reviews and chapters on this subject. In fact, the phytotoxic potential of black walnut has been recognized for more than 1500 years because it was noted by ancient Romans (Pliny, cited in Willis 1985). The toxic compound that black walnut releases is *juglone* (a quinone), which is capable of inhibiting physiological processes and growth of test species (José and Gillespie 1998b). Juglone can disrupt mitochondrial activitiy at less than 10 μM, and photosynthesis of target plants declines linearly as juglone concentrations increase from 0 to 5 μM. The presence of a walnut tree has long been known to inhibit crop production (Massey 1925). Juglone is not the only substance that can create zones of inhibition below the source tree. Substances such as phenolics, glycosides, terpenoids, and resins also create zones of suppression beneath other species such as *Eucalyptus baxteri*. The routes of chemical interference resulting in bare zones (or zones of inhibition) vary, and include direct foliar contact, volatile transmission, root uptake, or indirect inhibition of mycorrhizal formation of target plants (Del Moral et al. 1978).

While unequivocal evidence of allelopathy in natural and agricultural systems remains elusive, the presence and participation of secondary chemicals in plant-plant interactions is indisputable. The key question is, to what extent is the inhibitory response observed in target plants attributable directly to secondary chemicals from source plants? Many of the same chemicals also defend a plant against herbivory (see Chapter 7), where the direct physiological cause and effect of defensive compounds can be clearly demonstrated. Allelochemicals in contrast, must traverse a circuitous path involving complex intervening processes before reaching the target plant (Figure 8.2). The challenge is to verify allelopathy by tracing the chemical link from the affected target plants back to the source plants.

2. The Case Against Allelopathy. The difficulty for scientists studying allelopathic plant-plant interactions is to experimentally demonstrate the action of specific compounds from source to target plant (Seigler 1996). This issue raises continued doubts about the direct contribution of allelochemicals in plant-plant interactions. In fact, most researchers agree that it is nearly impossible to isolate an inhibitory response observed in the field solely to a single or a small group of allelochemicals (Einhellig 1987). Views critical of the methodology of allelopathy research and evidence supporting its existence abound, beginning with Harper (1977) and gaining more support recently (Wardle et al. 1998a, Weidenhamer 1996, Foy 1999).

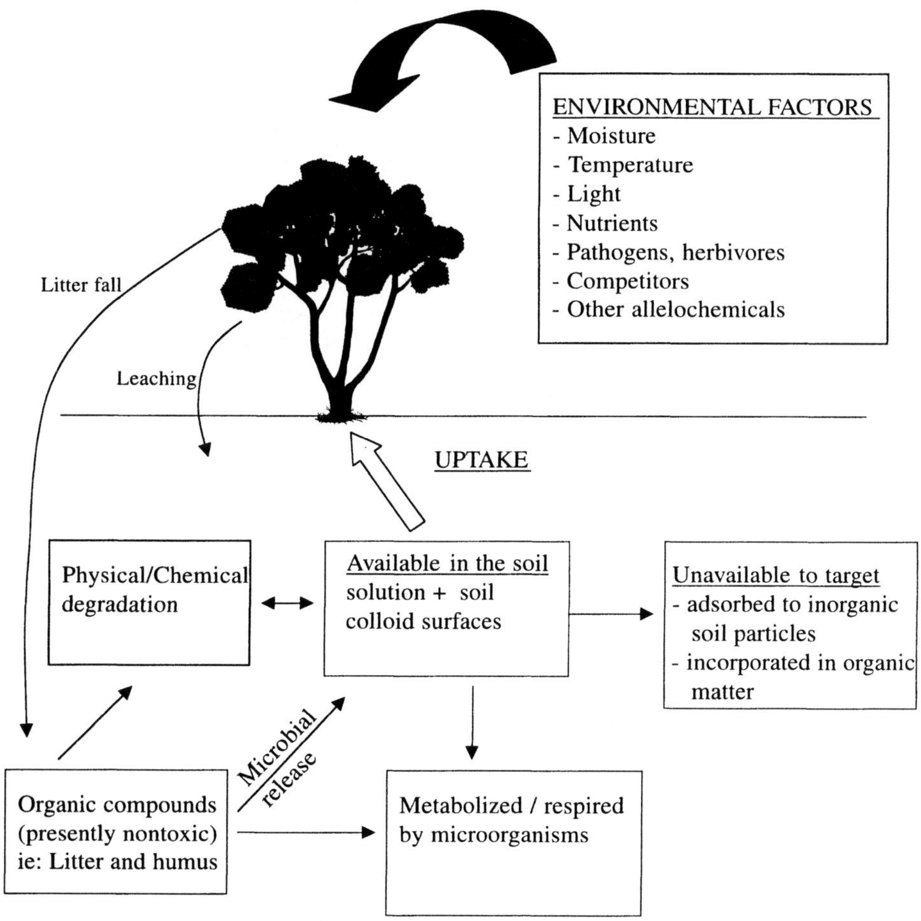

Figure 8.2. Allelochemical pathways in the soil environment. Notice the central significance of microbes to allelopathic processes. (Redrawn from Weidenhamer 1996, with permission.)

In order to address the criticism of methodology, experimenters have attempted to systematically isolate the source of allelopathic suppression by eliminating all other possible factors. An example of this type of study is that done with the inhibition of herbs beneath *Adenostoma fasciculatum* (McPherson and Muller 1969). The origin of the toxicity associated with *A. fasciculatum* was found to be in the soil beneath the shrub (i.e., a product of soil activity) and not from the plant canopy. Although the substrate for microbial activity may have been the litter of *A. fasciculatum*, it is the *de novo* synthesis of microbial toxins that caused the apparent allelopathy. Therefore, the zone of suppression is the result of the particular *A. fasciculatum*–soil microbe association (Kaminsky 1981) rather than a specific allelochemical released by *A. fasciculatum*. Many of the difficulties in demonstrating the direct effect of allelopathy are given by Stowe (1979): (1) realistic concentration used for bioassay, (2) the effect of soil physicochemical properties and microbial activities, (3) equivocal findings from field experiments of allelopathy, (4) autotoxicity, and (5) the lack of correspondence between plant community structure and a presumed allelopathy. Often, the expected zone of suppression around plant species that show strong allelopathy in bioassays is absent or ephemeral.

Clearly, the isolation of a toxic compound from plant tissue or the demonstration of its inhibitory effect in bioassay does not imply allelopathy in the field (Friedman and Waller 1985, Heisey 1990). For example, even though juglone shows a clear toxic effect in bioassay, those concentrations used in bioassay have not been well established in the field (Hejl et al. 1993). Among the several possible reasons why bioassay and field results do not match, one is the extraction method. The extraction of allelochemicals using chloroform is not realistic because it can remove compounds from plant tissue not normally leached by rain; moreover, it can extract from the soil and microbial fraction those compounds not available for uptake by target plants (Schmidt 1990). Therefore, such extraction methods do not produce realistic types and concentration of phytotoxins present in the soil solution (Kaminsky 1981). An alternative to chloroform extraction may be the use of chelators such as EDTA (ethylenediaminetetraacetic acid) (Kaminsky and Muller 1977).

Even though allelopathy requires the presence of chemical agents released by one plant into the environment (Rice 1974), the identity of the active compound may not be readily apparent, and in some cases, it may not even be phytotoxic. For example, organic carbon such as carbohydrates, not considered allelochemicals, indirectly can suppresses the growth of neighboring plants by elevating microbial activity that depletes available soil nutrients (Michelsen et al. 1995, Schmidt et al. 1997a, 1997b). Nilsson (1994) also found nutrient immobilization following the addition of leaf extract of arctic woody species like *Empetrum hermaphroditum*. Wardle and Nilsson (1997) later presented evidence that showed allelopathy at work in the suppression of graminoids not caused by the nutrient immobilization of microbes. This apparent contradiction may be explained by temporal variations of soil toxicity. In the early spring, during the growth initiation stage of target plants, *E. hermaphroditum* was found to be most toxic (in undecomposed litter). Later in the season, toxins are neutralized by soil microbes (Zackrisson and Nilsson 1992). The effect of substances released by a source plant on neighboring plants is clearly influenced by temporal and spatial characteristic of the environment and the plants themselves. Moreover, any source plant may influence neighboring target plants by a number of different allelopathic mechanisms. Very different conclusions can arise if environmental and phenological variations are not properly considered.

Even the activity of the well-known allelochemical juglone has come under scrutiny. Despite its suppressive effects in countless bioassay experiments, there is no irrefutable evidence that juglone persists at toxic levels in the soil. This can be attributed to the presence of native bacteria found in soils under walnut trees that can utilize juglone as their only source of carbon (Schmidt 1988). The breakdown of juglone by soil microbes can be so rapid that a 10-min incubation can reduce the original concentration to 1% (De Scisciolo et al. 1990). Diffusion of juglone in the soil is also restricted such that its concentration is reduced by 80% at 4.25 m from a source black walnut (José and Gillespie 1998a). Coupled to the short residence and poor diffusion of juglone in the soil, understory vegetation also fails to show consistent patterns that support a strong allelopathic influence (De Scisciolo et al. 1990). This study concludes that juglone is a constituent of the suppressive regime surrounding a walnut tree, and the extent of its role in that suppression remains to be determined.

It is widely acknowledged that allelochemicals must operate within a complex environmental system that often obscures their activity. Moreover, recent studies have recognized that inhibitory interactions between plants often involve a group of allelochemicals interfering with several physiological processes (Einhellig 1996). It will require a rigorous and critical experimental approach to convincingly assign allelopathy to a plant-plant suppressive interaction in the field. More frequently, studies refuting earlier claims of allelopathy are reported in the literature. For instance, Kolb (1988) reexamined a report of allelopathy by Kentucky bluegrass (*Poa pratensis*) on northern red oak (*Quercus rubra*), but failed to substantiate the earlier conclusion and attributed the suppression to competition.

3. The Special Case of Autotoxicity. If allelochemicals confer an advantage to the source plant by excluding neighboring plants of another species, then why do conspecific individuals often show similar vulnerability to the same substances? Such *autotoxicity* is quite prevalent when this effect is specifically examined (Table 8.1). But in most cases, toxicity on the source plant itself is not considered and not assessed. Rice, sugarcane, asparagus, and *Digitaria* are all found to be autotoxic. The effect is more acute under waterlogged situations, where toxins such as phenolics and aliphatic hydrocarbons cannot be degraded.

Stowe (1979) suggests that autotoxicity may be a cost of defense associated with the multiple activities of secondary compounds that evolved as defense against herbivory and pathogens. Friedman and Waller (1985) concur because caffeine autotoxicity is an unintended consequence of the protection against predation and competition conferred by this allelochemical. In this study, the benefits of protection by caffeine, particularly during coffee plant reproduction, outweighed the harm caused by autotoxicity. Among plants in an old-field community, autotoxicity is as strong as *allotoxicity* (toxicity against other species). In a different interpretation, Seigler (1996) suggests that autotoxicity is a form of density regulation within a population causing a reduction in the spread of pathogens and a promotion of outcrossing. However, autotoxicity as a means of self-thinning has not been experimentally substantiated. Autotoxicity of ragweed appears to lead to its demise (death of the whole population) rather than a portion of the population (thinning) in an early successional environment. Jackson and Willemsen (1976) speculated that this may have been an unfortunate consequence of a seed dormancy mechanism. The allelochemical from the aboveground tissues enforces a prolonged dormancy for the seed of the source plant in preparation for the next disturbance event. An excessive buildup of the allelochemical kills the parent plants as well as enforcing the seed dormancy.

As with allotoxicity, even if autotoxicity can be shown in the laboratory, it may not necessarily occur in the field. By having deep roots, an established plant can avoid the contaminated shallow soil layer caused by its own litter. Also, the timing of seed germination may be regulated to avoid seasonal pulses of allelopathy by emerging when sufficient degradation of its own toxins has taken place (Zackrisson and Nilsson 1992). This, however, still leaves young seedlings vulnerable to autotoxicity (e.g., eastern hemlock; Ward and McCormick 1982). When the exocarp is removed from seeds of *Schinus* species (pepper trees), dormancy is relieved, much as when the seeds are soaked (Nilsen and Muller 1980a). This is due to an allelopathic substance in the exocarp of the seed. The toxins leach out of the pepper tree seeds into the soil below the source tree canopy creating a bare zone devoid of pepper tree seedlings. However, the source trees can grow vigorously because they have deep roots. Only when the pepper tree seeds travel away from the host plant will the toxins be leached from the seed and germination occur. Moreover, the efficacy of the toxins of pepper trees varies among species, and this factor may be important to their ability to become naturalized (and invasive) in different environments (Nilsen and Muller 1980b). Without detailed

TABLE 8.1 Examples of Plant Species Showing Autotoxicity from Allelochemicals They Release into the Environment

Species	Source of Toxin	Toxin	Concentration	Toxic Effect	Reference
Tomato	Shoots	—	—	—	Qasem and Hill 1989
Coffee	—	Caffeine	—	Seedling growth	Friedman and Waller 1983
Picea abies	Litter and humus	—	—	Root extension	Gallet 1994
Engelmann spruce	—	Stilbene	—	—	Taylor and Shaw 1983
Alfalfa	Shoots	Coumarin *trans*-Cinnamic acid	60 μg mL^{-1}	—	Hegde and Miller 1992
Arctostaphylos uva-ursi		—	—	—	Hobbs 1984
Artemesia princeps var. *orientalis*	Leaf material	—	—		Yun and Kil 1992
Osmunda cinnamonea	Litter	—	—	Germination and gametophyte growth	Munther and Fairbrothers 1980
guayule (*Parthenium argentatum*)	*trans*-Cinnamic acid	—	6.7 × 10^{-3} m*M*	—	Bonner and Galston 1944
Thymus vulgaris	Essential oils and phenolics	—	—	Germination	Tarayre et al. 1995
Salsola kali root	—	—	—		Schmidt and Reeves 1989
Ailanthus altissima	—	Ailanthone	0.7 ml L^{-1}	Radicle extension	Heisey 1996
Sorghum	—	Sorgoleone	<10 u*M*	Photosynthesis (intact leaf)	Heisey 1996
			0.2 u*M*	Photosynthesis (chloroplast isolates)	Einhellig et al. 1993
Ragweed	Leaf material	Phenolic acids	1 m*M*	Tomato leaf [N] and [P]	Einhellig et al. 1993
		p-Coumaric and ferulic acids	0.1 mM	Reduced P	Mersie and Singh 1988
		p-Hydroxyacetophenone	—	—	Mersie and Singh 1988
		Caffeic acid	50 μ*M*	Spruce growth	Gallet 1994
		Flavonoids	10–80 μ*M*	Electron transport	Moreland and Novitzky 1987
		Phenolics	1 m*M*	Spruce seedlings	Zhu and Mallik 1994
		Quinoline alkaloid	0.3 m*M*	—	Aerts et al. 1991

information on the prevalence of autotoxicity among species and its physiological effects, it may be best to assume that phytotoxins affect conspecifics in the same manner and undergo the same mediating processes as they do against other species.

*4. **Allelopathy Versus Competition.*** Allelopathy is a form of interference among plants that also includes competition (Muller 1969; see Chapters 9 and 10 for further discussion of plant-plant interactions). Competition occurs when plants share a finite amount of resources; therefore, by definition, it can never result in a stimulatory response. But when suppression of one species by another is observed in the field, it is exceedingly difficult to ascertain whether the interaction is the result of allelopathy or competition (Inderjit and Del Moral 1997, Weston 1996). To further complicate the matter, Einhellig (1995) has suggested that some allelochemicals prevent nutrient uptake of target plants, in effect functioning in the same realm as resource competition. This phenomenon may, in fact, be common if we consider the frequency of suppression of radicle extension by phytotoxins in bioassay. By inhibiting the normal development of roots, and by reducing mycorrhizal infection in some cases, a decrease in nutrient uptake of the affected plant will likely follow. Also, changes in soil resource availability can alter the toxicity of source plants. For instance, some plants respond to phosphorus limitation by producing higher amounts of phytotoxins (Koeppe et al. 1976), which can be interpreted as a form of induced allelopathy by nutrient limitation.

Direct evidence for plant competition has been as rare as that for allelopathy, and few studies of competition consider allelopathy as a probable factor (Muller 1969, Fuerst and Putnam 1983). Recent studies (e.g., Seigler 1996) have made an attempt to factor out the role of competition when inferring chemical interference. In natural plant communities, both allelopathy and competition act in a density-dependent manner, but they are driven by different mechanisms (chemical interference and resource limitation respectively). Separation of these two processes should be possible based on their mode of action. Conceptually, if competition for nutrients is the agent of suppression, then the amendment of fertilizer should alleviate the observed outcome. However, if allelopathy is the mechanism for inhibition, fertilization should have no effect. In nature, however, separating the two processes of suppression by alleviating resource limitation may not be simple due to the mediating microbial flora. Nutrient amendment may stimulate the activity of microorganisms that hasten the degradation of allelochemicals. The subsequent amelioration of inhibition is consistent with competition, leading to a misinterpretation of the interaction. Also, as stated above, nutrient limitation can increase source plant release of allelochemicals that causes inhibition. Relieving the nutrient limitation would reduce allelochemical production and increase target plant growth, producing a result that resembles competition. Furthermore, the sensitivity of target plants to allelochemicals may vary with the degree of resource competition. Even if a direct allelopathic relationship can be demonstrated, the intensity of allelopathy will vary among populations experiencing different levels of resource availability. In fact, there is no reason to assume that a negative interaction between plant species must result from either competition or allelopathy—both processes could affect the interaction. Techniques developed to partition the respective influences of competition and allelopathy on growth are available and will be discussed later in the chapter.

B. Ecological Significance of Allelopathy

*1. **Community Structure and Succession.*** In natural communities, co-occurring plants are by definition tolerant of the allelochemicals that others release. Are allelochemicals only significant to suppression of alien species that have not coevolved a defense (detoxification mechanism) or tolerance against allelochemicals in the environment? This appears to be true in situations where weeds and crop species of divergent origins mingle in an agricultural setting. Yet Wardle et al. (1998a) found evidence of strong allelopathic interactions in a natural ecosystem with a small number of co-occurring species. This is because the chemical environment is dominated by a single species (with the participation of soil microbes), and suppressed co-occurring species show no signs of tolerance. Only after catastrophic disturbances, such as periodic fire, can plants reinvade zones of suppression. The episodic reversion of chemical dominance by disturbance events may be the driving force that sustains subordinate species without evolving a tolerance to allelopathy. In a species-rich community, one is less likely to find strong allelopathic interactions because the number of co-occurring species acts to disperse source plants and limit their chemical influence on the environment (Wardle et al. 1998a). Consequently, allelopathic interactions should be most common in regions with low species diversity. Despite these generalizations, there are examples of succession in diverse communities that are mediated, at least partly, by allelopathy.

In one such case, phytochemical interactions were originally thought to have played a role in the replacement of an early successional species, *Salsola kali* (Russian thistle), in disturbed sites by grasses such as *Agropyron smithii* Rydb. The decline of *S. kali* could have been due to several factors: autotoxicity from its own litter, the enhancement of mycorrhizal formation of *A. smithii,* and competition. Schmidt and Reeves (1989) found no autotoxicity originating from *S. kali* litter. Moreover, root leachate of *S. kali* was autotoxic but did not affect mycorrhizal infection of *A. smithii,* and *A. smithii* was found to have no influence on the mycorrhization of *S. kali* or itself (Schmidt and Reeves 1989). Hence, in this case, succession was the result largely of nonchemical processes.

Fire has been identified as a key disturbance factor maintaining the community structure of the evergreen Florida scrub and the California chaparral. When adjacent to a shrub community in Florida, a sandhill community can resist the invasion of the shrub species because it experiences much less frequent fires. Fischer et al. (1994) have implicated allelopathy as the process that prevents the sandhill community from becoming dominated by shrubs, a vegetation type that is fuel-rich and fire prone. The allelochemicals identified are a group of monoterpenes and sesquiterpenes soluble in water that show strong inhibition of native grass germination. In many cases, fire can remove the source plant and its litter from the environment, allowing a window of opportunity for sensitive target plants to germinate and establish. Fire, however, also acts to eliminate competition (Fisher 1987) and to release a pulse of mineral nutrients; both are known to contribute to a flush of shrub regeneration. The challenge of distinguishing among these chemical changes following fire and their significance to plant-plant interaction remains.

There is considerable controversy over the significance of allelopathy to dominance, diversity, and succession in communities. McPherson and Muller (1969) advanced the idea that allelopathy has a role to play in the "dominance and diversity" of plant communities. They considered that an examination of community dynamics would not be complete without due consideration of the chemical interference between species sharing a site through time. Yet Keever (1950) and Muller and Muller (1956) found no correlation between bioassay results and successional changes in the old-field environment. This led Stowe (1979) to conclude that "any species can be shown to have allelopathic properties in bioassays, and that bioassays may, for many communities, have no ecological meanings." Seigler (1996) echoed the sentiment in a recent review in which he found no conclusive evidence for allelopathy mediating successional changes. But clearly, chemical interaction does occur in every successional stage and in every plant community. The key issues are the intensity of the chemical interaction (allelopathy), the source of the toxins, and the impact of those substances on community structure and succession processes.

Wardle et al. (1998b) suggest that allelopathy is more compatible as a factor regulating ecosystem-level processes than as one acting at the level of populations or between individual plants. They reason that the amount of secondary metabolites in plant litter contributes to litter quality and microbial activities. Allelochemicals determine important aspects of plant litter composition; thus, they can alter ecosystem processes such as nutrient cycling and have a significant impact on community structure. Perhaps, future research should be directed toward studying the contribution by emergent properties of allelopathy, at the scale of single plants and populations, to the regulation of ecosystem processes.

2. Evolution of Allelopathy. When considering the origin of secondary compounds in plants, it is necessary to explore beyond the original assumption that they are waste materials to be disposed of in the most efficient manner. Several groups of plant secondary metabolites serve important functions such as providing flower color, defending against herbivores, attracting pollinators and parasitoids (Thaler 1999), and allelopathy. It is not known whether allelopathy arose as an associated property of the diversity of herbivore defensive chemicals that plants produce or as an independent evolving entity. However, evidence does suggest a dual function (herbivory and allelopathy) of several secondary compounds. For example, nonprotein amino acids that are toxic to seed predators can also inhibit competing plants. In many cases, the production of allelochemicals or their toxicity increases with an increasingly stressful biotic environment, suggesting that these secondary metabolites have a broad-spectrum protective function. Gliessman (1976) believes that allelopathy and antiherbivory defense not only evolved to benefit the source plant by protecting its tissue and resources, but also to redirect herbivores or competitors toward other plants. More recent reports indicate that allelochemicals also play a role as chemical signals in the coevolution between plants, animals, and pathogens. For example, salicylic acid (a known phytotoxin) was recently found to function, in the form of methyl salicylate, as a chemical signal. Tissue infected by tobacco mosaic virus releases methyl salicylate (a volatile substance) that induces healthy tissue of neighboring plants to upregulate antipathogenic protein synthesis (Shulaev et al. 1997, Gaffney 1993). Methyl jasmonate in Solanaceae and Fabaceae functions in a similar fashion to induce proteinase inhibitors even from one species to another (Farmer and Ryan 1990; See Chapter 7). Clearly, the allelopathic activity of any particular secondary metabolite constitutes but one of its myriad functions.

If phytotoxins confer an advantage to the source plant in general, then incidents of a stimulatory or a beneficial effect of allelopathy on the target plant would be a secondary consequence. Phytotoxins at low concentration often stimulate seed germination and growth of target plants. Parasitic angiosperms are another example of allelochemicals conferring a stimulatory effect on other species. Just as some herbivores ingest plant toxins for their own defense, the hemiparasitic plant *Pedicularis* sp. can acquire a host plant's alkaloids without apparent ill effect (Schneider and Stermitz 1990) and deploy the alkaloids for the parasite's own protection (see Chapter 10).

Moreover, the allelochemical sorgoleone stimulates the germination of parasitic plant seeds. Very little published information is available to assess the extent of the benefits that an allelopathic environment has for target plants.

Jones and Firn (1991) observed that the high diversity–low activity nature of plant secondary compounds is favored in the evolution of plant protection. By retaining a large variety of inactive compounds synthesized via different pathways, the chances of a novel compound arising with defensive or allelopathic activity improve. By keeping within the metabolic cost limits maintaining a high diversity of secondary chemicals, a plant is better endowed with the potential to defend or repel pathogens, herbivores, and competitors.

II. CLASSES OF ALLELOCHEMICALS

A. Synthesis and Biochemical Pathways

Plant allelochemicals, a term coined by Whittaker and Feeney (1971), are secondary metabolites that "do not function directly in the primary biochemical activities that support the growth, development, and reproduction of the organism" (Conn 1981). But, as mentioned earlier, many of these compounds have an important role in primary plant functions. A large number of secondary metabolites are involved in antiherbivory (see Chapter 7) and other physiological processes such as maintaining seed dormancy (Aerts et al. 1991a). Secondary metabolites are synthesized by the further modification of primary metabolic products such as sugars, fats, proteins, amino acids, purines, and pyrimidines, along several major biosynthetic pathways illustrated in Figure 7.1. The most abundant and most studied allelochemicals are synthesized from the shikimic acid and mevalonic acid pathways (Figure 8.3). For example, the biosynthesis of phenolics is derived from the shikimic acid pathway. Flavonoids contain structural groups originating from both the shikimic (phenylalanine) and acetate pathways. Other biosynthetic pathways are also associated with allelochemicals. Steroids and terpenoids originate from acteyl CoA and mevalonic acid (Figure 8.3). All of the known alkaloids including cyanoglycosides are derived from amino acids. Nonprotein amino acid synthesis depends on aminoacyl-tRNA synthetase that determines whether toxic or nontoxic species of amino acids are selected for protein synthesis (Friedman and Waller 1985).

Because of their direct biochemical link to several key plant products, secondary compounds are diverse and numerous. They are likely to occur in all species and in most plant parts (e.g., root, stem, leaves, flowers, and seeds). Classes of secondary compounds, their occurrence, and their effects are summarized in Table 8.2 and Figure 8.4. Even though closely related species share a chemical signature of secondary metabolites making chemotaxonomy possible, two congeners can differ significantly in the diversity of secondary compounds with allelopathic activity they produce. Seigler (1981) presents an in-depth discussion of the biosynthetic pathways and phylogenetic variations of secondary compounds in plants. In general, allelochemicals can be grouped into those that are nitrogen based (alkaloids, nonprotein amino acids, cyanogenic glycosides), those that are carbon based and have an aromatic hydrocarbon ring (phenolics), and those that are carbon based without the phenolic structure such as terpenoids (Figure 8.4). Most secondary compounds implicated in allelopathic interactions are either terpenoids (mono- or sesquiterpenes) or phenolics (phenols, phenolic acids, cinnamic acids, hydroxyquinones).

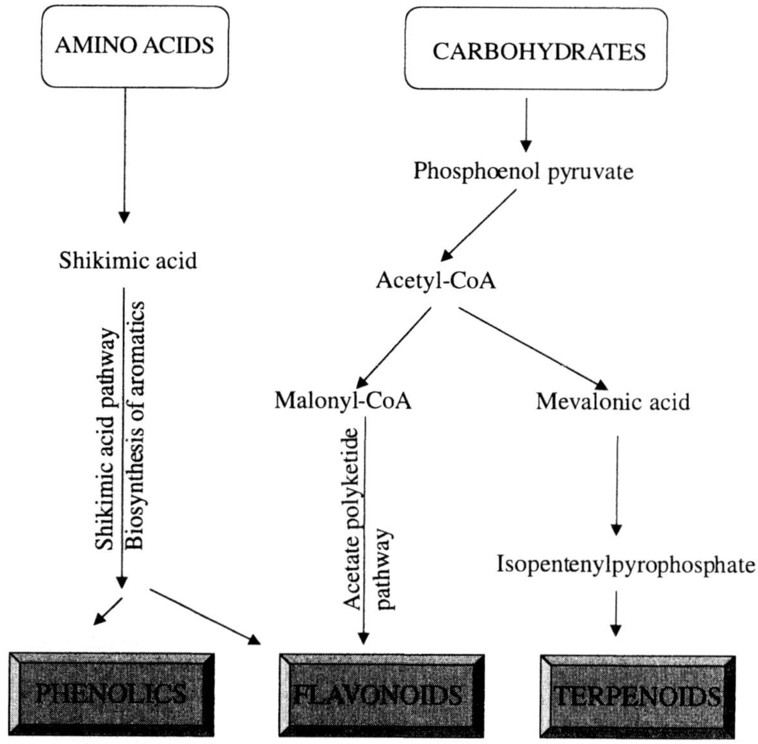

Figure 8.3. General scheme of the biochemical pathways involved in the synthesis of the main classes of secondary substances with allelopathic activities. (Redrawn and adapted from Larcher 1995, with permission.)

B. Nitrogen-Based Compounds

Alkaloids are a very large group of compounds, only rare members of which have been studied in relation to allelopathy; these include caffeine, nicotine, racinine, cocaine, codeine, and strychnine. There is also a group of aromatic nitrogen-based compounds, of which aristolochic acids are the principal representatives, that may have some allelopathic activity. Nitrogen-based secondary chemicals have mostly been implicated in antiherbivory mechanisms (see Chapter 7). In fact, it is surprising (because of their diversity) that alkaloids have only rarely been implicated in allelopathy (Harborne 1988, Aerts et al. 1991).

Nonprotein amino acids (of which there are about 600) are toxic to other organisms as analogues to amino acids, replacing them in protein synthesis and thereby disrupting protein function (Figure 8.5). These compounds are not autotoxic because the plant's own tRNA synthetase is capable of discriminating against the analogue (Friedman and Waller 1985). These compounds are commonly found in legume seeds and seedlings. Azetidine-2-carboxylic acid, an analogue of proline, is found in lily-of-the-valley (*Convallaria majalis*), in rhizomes of Solomon's seal (*Polygonatum multiflorum*), seeds of legume (*Bussea*), and in low concentrations in other plants (Conn 1981). Nonprotein amino acids, like alkaloids, are mainly known for their antiherbivory functions (see Chapter 7). One allelopathic example, however, is found in seed exudates of *Glycine wightii* containing canavanine (a nonprotein amino acid and an analogue of arginine) which inhibited the germination and growth of lettuce (Wilson and Bell 1978).

TABLE 8.2 Classes of Secondary Compounds with Allelopathic Potential

Group	Chemical Structure[a]	Biosynthesis Pathway	Occurrence[b]
		Nitrogen Based:	
Nonprotein amino acids	1	Citric acid, shikimic cycle	Common in Fabales, Cucurbitaceae
Alkaloids	2	Amino acids, acetate	Widespread in angiosperms; in root, leaf, and fruit
Anthraquinones	3	Polyketide, shikimic acid and mevalonic acid	Rhamnaceae, Lythraceae, Iridaceae
Cyanogenic glycosides	4	Amino acids and carbohydrates or fatty acids	Not common; Myrtaceae, Fabaceae; in fruit and leaf
Glucosinolates	5	Amino acids and carbohydrate	Brassicaceae, not common
		Phenolics:	
Simple phenols	6	Shikimic acid, mevalonic acid, acetyl-malonyl CoA	Universal; in leaf and cortex
Cinnamic acids	7	Shikimic acid	Universal
Flavonoids	8	Aromatic amino acids, acetyl-malonyl CoA	Universal
Tannins	9	Shikimic acid and acetate polyketide	Common
Coumarins	10	Shikimic acid	Hippocastaneaceae, Fabaceae, Asclepiadaceae
		Terpenoids:	
Monoterpenes	11	Mevalonic acid	Common, Myrtaceae, Juglandaceae; as essential oils and resins
Sesquiterpenes	12	Mevalonic acid	Asteraceae, Myrtaceae, Apiaceae
Diterpenoids	13	Mevalonic acid	Solanaceae, Asteraceae; as resins and latex
		Others	
Alkanes, alkenes, aldehydes, alcohols, ketones	14	Acetyl-malonyl CoA	Graminae, Amaranthaceae, Fabaceae; as volatile; in all plant parts

[a] Molecular structures in Figure 8.4 show the common root structure without R groups.
[b] Only representative plant groups containing the phytochemicals are given.

Source: Data compiled from many sources.

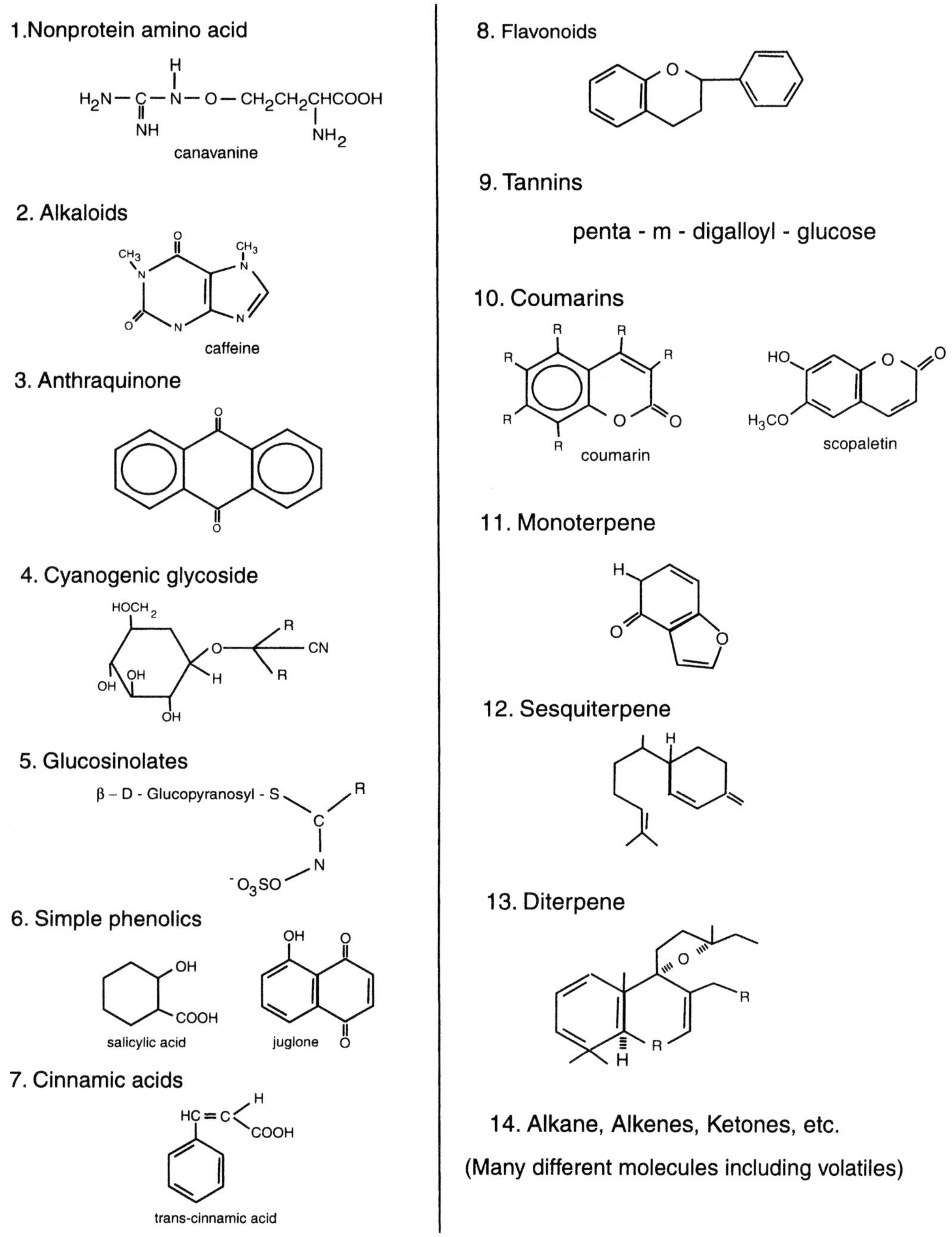

Figure 8.4. Representative chemical diagrams for the classes of allelochemicals listed in Table 8.2. R refers to a variable side chain.

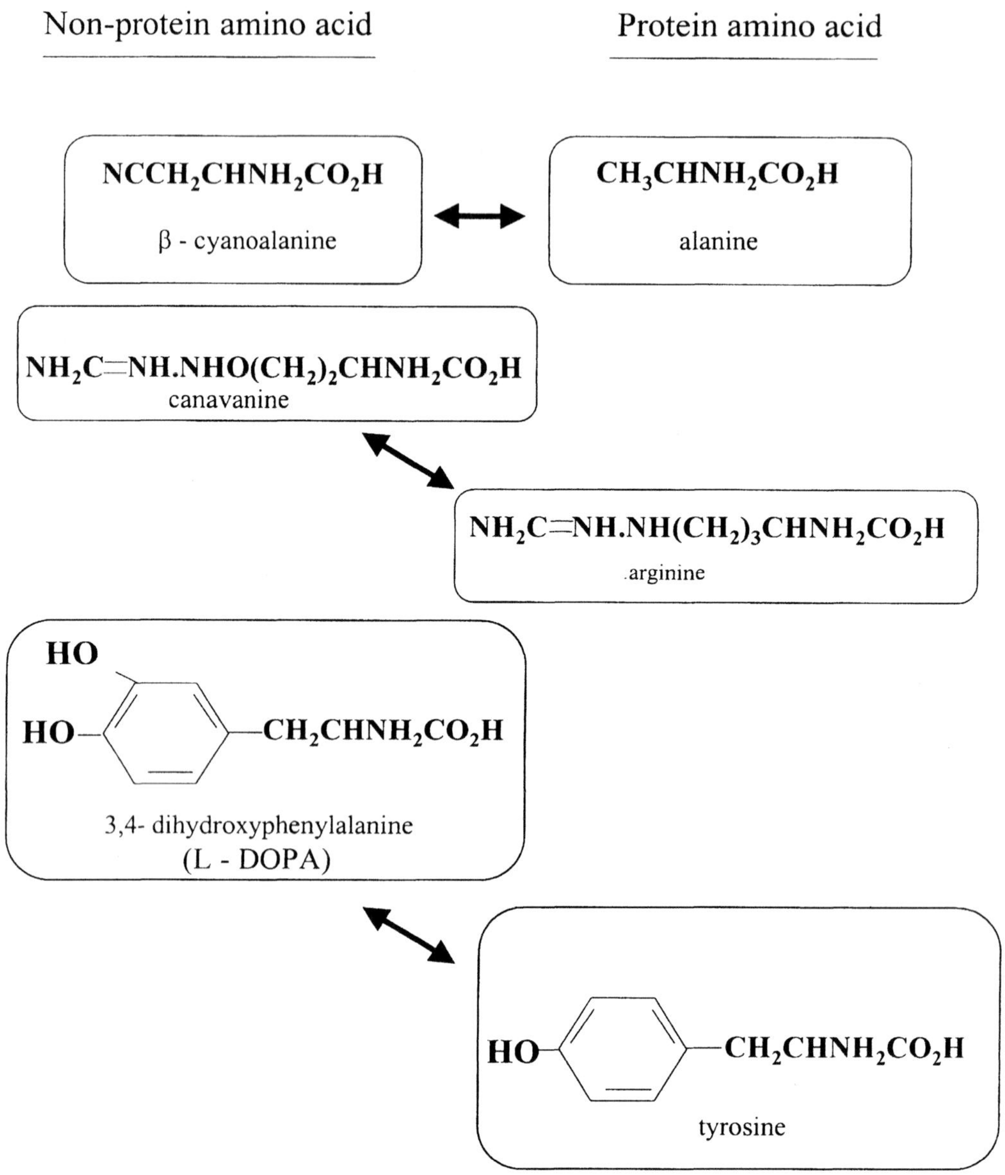

Figure 8.5. Toxic nonprotein amino acids and the amino acids they replace in target plants. (Redrawn and adapted from Harborne 1988, with permission.)

Hydrogen cyanide is a potent germination and growth inhibitor. It is liberated enzymatically from cyanogenic glucosides (see Figure 7.5). Glucosinolates (which contain sulfur) are closely related to cyanogenic glucosides (Figure 8.4) but the two types of compounds have never been found in the same species (Conn 1981). Cyanogenic glucosides are found in more than 2000 plant species, while glucosinolates are less widespread, commonly found in Brassicaceae (e.g., mustard oil). Glucosinolates are hydrolyzed to yield isothiocyanate or organic cyanide. Cyanogenic glycosides and glucosinolates have strong antiherbivory activities (see Chapter 7), but some also have been reported to inhibit other plants. Since HCN is usually released by reaction with β-glucosidase when tissues containing cyanogenic glycosides are crushed (Hegarty et al. 1991), allelopathic activity of these compounds is restricted to decomposed leaves. Black mustard releases allyl isothiocyanate that is adsorbed in the soil and could have contributed to the exclusion of plant competitors (Bell and Muller 1973). Dhurrin, a cyanogenic glycoside released from sorghum litter, has also shown allelopathic activities (Weston 1996). Cyano-group (C≡N) molecules, such as β-aminopropionitrile produced by sweetpea (*L. odoratus*), are very toxic to closely related species, but thay have no autotoxicity (Hegarty et al. 1991).

C. Carbon-Based Compounds

1. Phenolics. Phenolics are very abundant in plants, and their biosynthesis is well known for both seed plants and ferns (Munther and Fairbrothers 1980). Phenolics constitute the bulk of the structural matrix in plants (in cell walls, the cuticle, and lignin) and most of coloration of flowers (primarily flavonoids). Coinciding with their wide distribution in plants, phenolics also have many functions including: (1) defending against herbivores, (2) defending against microbe attack, (3) signaling molecules in plant interaction with pathogens, (4) influencing the development of mutualistic microbes (dinitrogen fixation), (5) influencing plant-parasitic plant interactions, and (6) regulating allelopathic processes. The most important group of phenolics to allelopathy—phenolic acids (Figure 8.4)—are relatively simple compounds, are easily identified, and are abundant in the soil solution. Usually, the term *phenolic acid* refers to simple molecules that are derived from benzoic acid or cinnamic acid. Benzoic acid and its derivatives have been found in all angiosperms currently tested (Harborne 1988). Cinnamic acid and its derivatives are thought to be universal in higher plants. Although these compounds are common to all plants, some species synthesize high concentrations of particular phenolic acids. In addition, decaying plant material is a rich source of phenolic acids because lignin (about 30% of the leaf mass) decomposes into phenolic acids and other aromatic substances. In fact, phenolic acids have been found in soils from almost all ecosystems (Dalton 1999).

Phenolics such as *p*-coumaric acid and quinones are produced by the well-known allelopathic chaparral shrubs (e.g., McPherson and Muller 1969, Friedman and Waller 1985). Phenolics are more soluble in water than are terpenoids, which allows them to occur at higher concentrations in leachates than terpenoids. Common, simple, aromatic acids such as *p*-hydroxybenzoic, cinnamic, *p*-coumaric, ferulic, caffeic, protocatechuic acid (3,4-dihydroxybenzoic acid), chlorogenic acid, and gallic acid are found in plant material and decomposing litter. Common phenolics found in the soil include 3-hydroxyhydrocinnamic, benzoic, phenylacetic and hydrocinnamic acids. The phenolic acids present in soil, originating from root exudates or other plant sources, can be trapped by using a XAD-4 resin column (Figure 8.6), developed by Tang and Young (1982). Chlorogenic acid is the main component of allelochemicals found in *Chenopodium album* shoot tissue (Mallik et al. 1994). Salicylic acid (hydroxybenzoic acid) is a water-soluble phenolic toxin occurring in small quantities in many plants (Figure 8.4). It is found in leachates of cherrybark oak (*Quercus falcata* var. *pagodaefolia* Ell.) and was shown to reduce radicle extension of radish (DeBell 1971).

Another group of phenolics, the coumarins (Figure 8.4), are found in epicuticular waxes on plant surfaces as well as on roots and seeds. These substances can come to relatively high concentration without autotoxicity because they are located on the outsides of plant cells. The first such compounds studied were furanocoumarins of *Daucus carota* L. (wild carrot) in relation to herbivory. These compounds may also

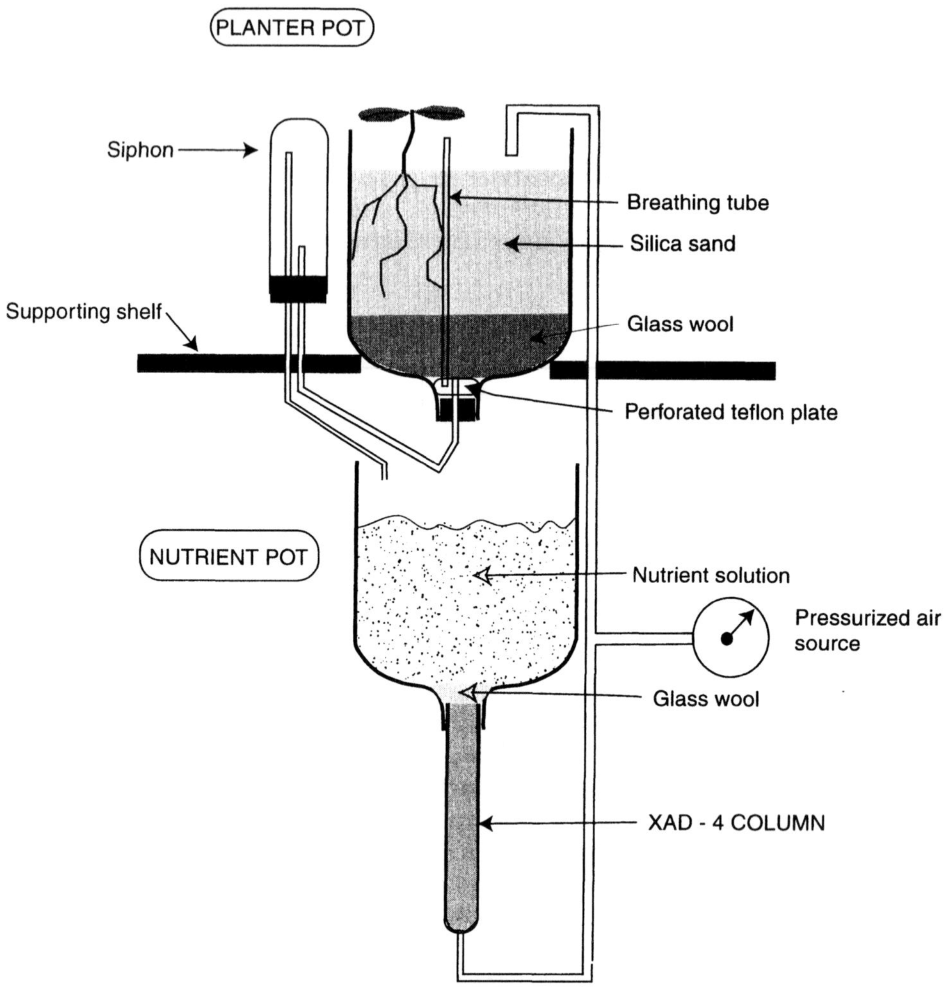

Figure 8.6. Experimental setup showing the collection of plant and root exudates through the base of container (XAD-4 resin column). (Redrawn from Tang et al. 1989.)

serve to protect leaf tissues from ultraviolet (UV) radiation because they absorb UV light and fluoresce the energy back as longer wavelengths (i.e., they autofluoresce). However, antimicrobial function may be the main role of coumarins in plants because they have antifungal, antibacterial, and antiviral properties. Furanocoumarins also act as potent inhibitors of seed germination. These compounds have been found as crystals on the embryo of *Psoralea* seeds (Zobel and Brown 1991). At 100 ppm, these compounds inhibit mitosis, but at 10 ppm, mitochondrial activity is stimulated. Thus, at high concentration, seed germination is inhibited, but when spring rains leach away the coumarins, they switch over to stimulating germination. Another coumarin, scopoletin and its glucoside scopolin, are synthesized from hydroxylated cinnamic acids (Fritig et al. 1970). The aqueous leachates of white oak leaves or litter contain scopolin (Lodhi 1978).

Juglone (5-hydroxy-1,4-naphthoquinone), a quinone found in black walnut leaves, is a water-soluble phenolic that can be washed into the soil from the canopy by rain or leached from roots (Figure 8.4). Juglone becomes toxic after being hydrolyzed in the soil. Stilbene, a phytoalexin, acts as an antiherbivory and antifungal defense chemical of Engelmann spruce (*Picea engelmannii* Parry), but it can also suppress the germination of conifer seed and seedling growth (Taylor and Shaw 1983). Stilbene in the leaves of *Empetrum hermaphroditum* shows inhibitory activities against tree seedling germination (Zackrisson and Nilsson 1992).

Other phenolics are flavonoids (Figure 8.4). These substances function in a variety of processes (including determining flower color and pathogen defense). Approximately 4000 structural types of flavonoid are known (Harborne 1988). The biosynthesis of plant flavonoids has been extensively studied and reviewed (Stafford 1990). These substances are synthesized from phenylalanine via the general phenylpropanoid pathway (Figure 8.7). This is the same biosynthetic pathway that produces coumarins, stilbene, phenolic acids, lignin, and others. Some flavonoids are constitutive; however, many are induced. Induction of flavonoid synthesis occurs by environmental stress and mechanical damage. Flavonoid synthesis is definitely induced by UV radiation, and the induction is linearly related to the intensity of UV-B radiation such that flavonoid concentration can be used as an indicator of UV-B radiation intensity (Beggs and Wellmann 1994). The flavonoids are synthesized in the first few millimeters of the upper leaf surface (Cen and Bornman 1993) and are capable of screening out 90% of the incident UV radiation before it reaches the mesophyll cells (Day et al. 1992). There are large variations in the quantity and quality of flavonoids among species, individuals, organs, and seasons (Chaves and Escudero 1999). Water stress is another factor that usually promotes the synthesis of flavonoids in many tissues of some plants (Chaves and Escudero 1999). Flavonoids also are produced in response to microbial infection (Jordan et al. 1994, Klepzig et al. 1995). They can be leached out of leaf surfaces by rain because they are close to the top surface and are induced in broken regions of leaves (pathogen or herbivore damage sites).

Flavonoids have wide antipathogenic and antiherbivore action, such that they have often been considered for possible biopesticide production (Dakora 1995). Other studies have examined the allelopathic activity of flavonoids (Baruah et al. 1994, Inderjit and Dakshini 1992) . Flavonoids can be leached from leaves, leaf litter, or roots into the soil solution. In most studies examined so far, flavonoids in the soil solution inhibit seed germination and radical elongation (Berhow and Vaughn 1999). The flavonoids also show antagonistic properties with plant hormones (IAA and gibberellins) and can exhibit cytokinin-like activities. The metabolic action of flavonoids on seed germination may be due to inhibition of the energy transfer system. Flavones have been shown to interfere with ATP formation by mitochondria (Stenlid 1976).

Phenolics vary in their toxicity among the various groups and within any particular group. Phenolic acids, coumarins, and aglycones have the greatest allelopathic activity. Within one species, bioassay results found a decreasing level of toxicity in the following order of acids: *o*-hydroxyphenylacetic, *p*-hydroxybezoic, vanillic, *p*-coumaric, gentisic, syringic, ferulic, and *m*-coumaric (Zhu and Mallik 1994). Some phenolics are also autotoxic. *Para*-hydroxyacetophenone (spruce-specific metabolite) and caffeic acid reduced spruce growth at 50 μM (Gallet 1994). The order of autotoxicity among phenolics in alfalfa is as follows: coumarin > trans-cinnamic acid = hydrocinnamic acid = o-coumaric acid > *p*-coumaric acid > 1,8-dihydroxyflavone = 3-hydroxyflavone (Hegde and Miller 1992, Zhu and Mallik 1994).

In summary, phenolics are a broad variety of compounds with a wide array of metabolic activities. Among all the phenolics, phenolic acids are most important to allelopathy. Phenolic compounds are water soluble and leach from leaves, stems, and

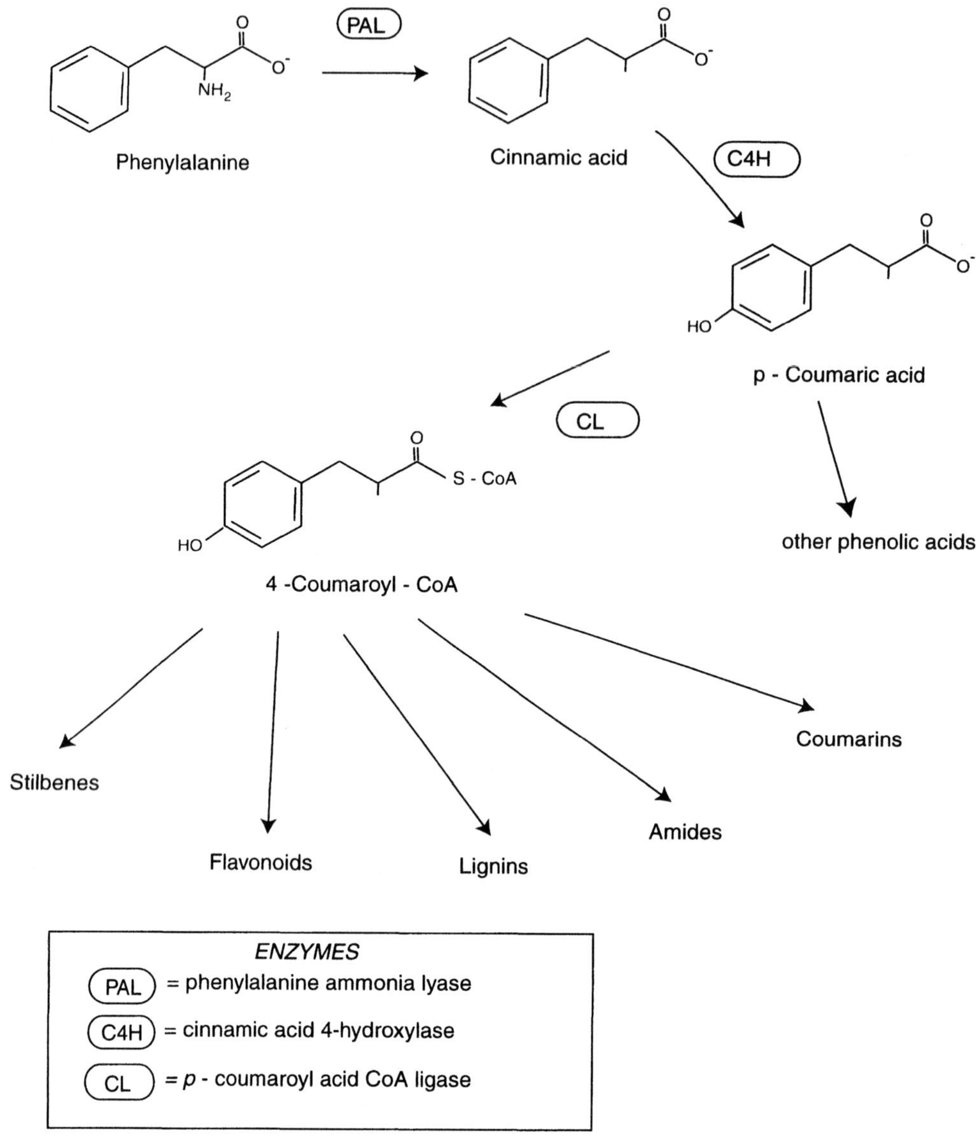

Figure 8.7. Beginning of the phenylpropanoid pathway leading to the synthesis of many allelochemicals is represented. (Redrawn and adapted from Berhow and Vaughn 1999.)

roots into the soil solution. In addition, decomposition of litter releases simple phenolic acids from lignin into the soil solution. The resultant soil solution is a complex mix of phenolic compounds, some of which may have phytotoxic activity. Soil microbes further modify phenolic acids, which can increase toxicity or degrade toxic compounds. The intermediate processing of phenolics by microbes is critical to allelopathy and makes it difficult to trace the phytotoxin to the source plant.

*2. **Terpenoids.*** Terpenoids are important allelochemicals because they have activity at relatively low concentrations (1–3 μM) and they diffuse a long distance because they are transmitted through the atmosphere. These compounds are less ubiquitous compared with the phenolics; however, they have been implicated in many cases of allelopathy. The basic structure of terpenes can be divided into monoterpenoids (10 carbons), sesquiterpenoids (15 carbons), diterpenoids (20 carbons), triterpenoids (30 carbons) and tetraterpenoids (40 carbons). Monoterpenes are the most common, and are relatively nonpolar. Some have low water solubility ($<$10 ppm, linolyl acetate), but others with ketone and aldehyde groups are highly soluble ($>$6000 ppm, verbenone; Fischer et al. 1994, Weidenhamer et al 1993). Monoterpenes such as α- or β-pinene, camphene, cineole, diol, and piquerol are essential oils often secreted by glandular hairs (Kelsey et al 1984). Camphor and cineole are volatile monoterpenes that are absorbed and concentrated by dry soils or seed coats, and can inhibit target seed germination or plant growth. Sesquiterpenes include over 2000 known types in the form of sesquiterpene lactones (ringed hydrocarbons). As with the sesquiterpenoids, the majority of diterpenoid compounds are unsaturated lactones (including nagilactones, inumakilactone, podolactone). Diterpenoid alkaloids can inhibit the biosynthesis or transport of gibberellin in cucumber and lettuce (Worsham 1989). The diversity of terpenoid compounds can be high in some species. In fact, grapes emit about 45 different volatile terpenes. Terpenes are commonly emitted from species inhabiting semiarid climates such as Mediterranean regions, where many species of *Salvia* and *Eucalyptus* release monoterpenes into the atmosphere.

The biosynthesis of terpenoids is based on isoprene (2-methylbutadiene). In fact, the terms *terpenoid* and *isoprenoid* are interchangeable because *isoprenoid* refers to the 5 C isoprene unit from which all terpenoids are derived. The classification of terpenoids (see above) is based on the number of isoprenoid units. The isoprene units are derived ultimately from the condensation of three acetyl CoA units into mevalonate, a 6-carbon unit (Panda and Khush 1995). The mevalonate is pyrophosphorylated and then undergoes decarboxylation to yield isopentanyl pyrophosphate (Figure 8.8).

Saponins are substances based upon a steroid or triterpene skeleton and substituted with carboxyl, carbonyl, and hydroxyl functional groups. The compounds can also be additionally substituted with sugar chains that include glucose, galactose, arabinose, and others (Oleszek et al. 1999). Saponins are a diverse group of compounds found in many organs, particularly the seeds of Fabaceae, Sapindaceae, Aceraceae, and Hippocastaneaceae (Seigler 1981). The diversity of structural types may elicit variable responses in plants, animals, and microbes. The mechanisms of action of these compounds are not as well studied as those of the phenolic acids or monoterpenes, but they have been shown to have herbicidal potential (Miller 1996).

Among all secondary chemicals, phenolics and terpenoids predominate in allelochemical studies. The path for release and uptake of terpenoids involves the complex participation of plants and soil microbes described by Langenheim (1994) and shown in Figure 8.2. These interactions stress the indirect role terpenoids play in allelopathy by modifying the soil microbial activity and nutritional status. Lipid-soluble monoterpenoids can also be concentrated on the epicuticular waxes of seeds and can inhibit seed germination. The mode of action of terpenoids is covered in a later section of this chapter.

III. ACTIVITY OF ALLELOCHEMICALS

A. Location of Allelochemicals

Since a great number of plant secondary products possess an allelopathic potential, and since they are derived from processes occurring in all living plant cells, it is likely that most plant tissues contain an array of such compounds. However, specific species

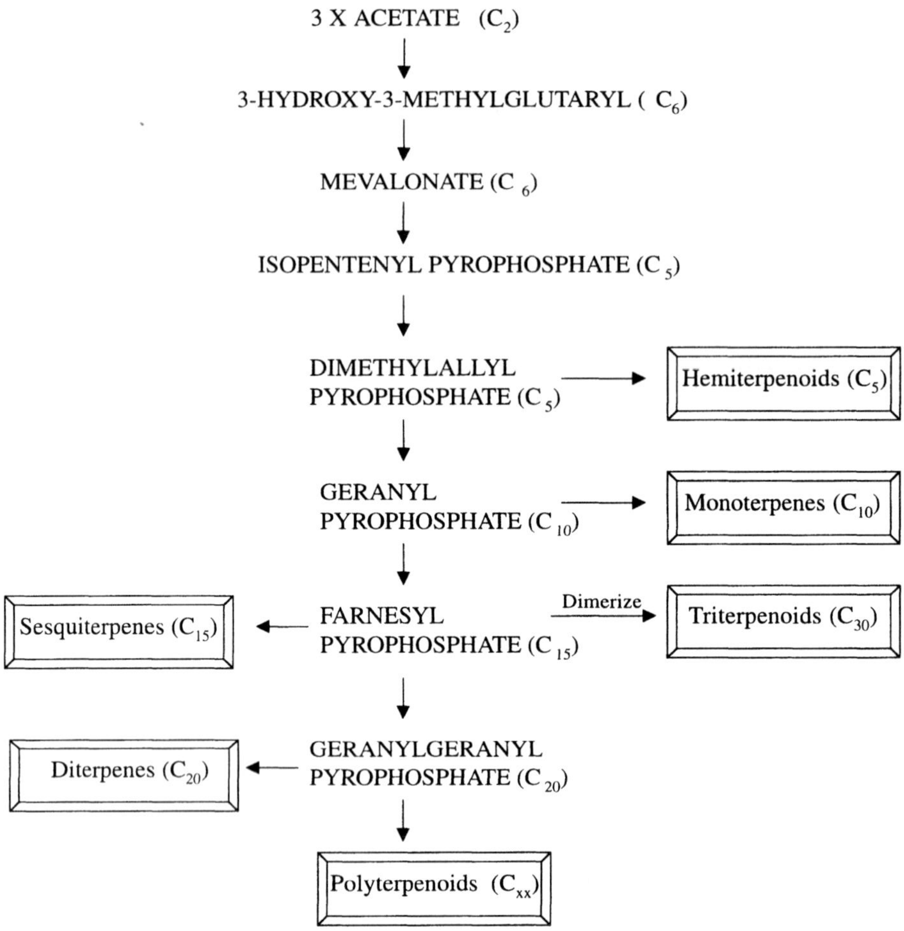

Figure 8.8. Biosynthetic pathway for the major classes of terpenoids. (Adapted from Panda and Khush 1995, with permission.)

or plant taxa are capable of producing phytotoxins at unusually high quantities or efficacy. The location for storage of allelochemicals within the plant may be determined by lipophilicity (an affinity for lipids) of the compound. Lipophilic compounds, such as many terpenoids, are found in secretory structures (e.g., glandular hairs and resin ducts) or in cuticular wax. Hydrophilic secondary compounds tend to be found in cell walls and vacuoles (Gershenzon 1984). Variation in toxicity among plant tissues may be determined by both the rate of synthesis and the storage capacity of secondary metabolites. The activity of alfalfa phytotoxins (autotoxicity) decreased in the order leaf, seed, whole plant, soil, root, flower, and stem. Some plant tissues tend to be more phytotoxic in the reproductive phase than in the vegetative stage (Chung and Miller 1995b), suggesting a phenological influence on toxin production or storage. Given the increased carbon and nutrient demand by developing fruit, a reduction in "waste" metabolite synthesis should be expected. The production of *Ailanthus* phytotoxins was also greater in young plants compared with mature plants (Lawrence et al., 1991).

1. Seed and Fruit. Allelochemicals such as HCN produced by germinating seeds of apple are believed to protect it from pathogenic attack; other allelochemicals may also have a role in maintaining seed dormancy. Away from the embryo, the outer dead skin of the fruit of *Ammi majus* (Bishop's weed, Umbelliferae), the husks of rice, the calyx of persimmon, and the seeds of several Cucurbitaceae species all contain phytotoxic compounds ranging from diterpenes to nonprotein amino acids (Friedman and Waller 1985).

2. Leaves and Litter. Plants in the Labiatae, Compositae, Lauraceae, and Malvaceae families commonly possess unicellular glands on leaves containing essential oils and other phytotoxins. Velvetleaf (*Abutilon theophrasti,* Malvaceae) exudes phytotoxins from glandular trichomes on the leaves. These are either volatilized or leached with precipitation (Sterling et al. 1987). Even though fresh leaves contain high concentrations of essential oils, leachate from intact leaves may not produce an inhibitory effect. Rather, phytotoxins are released from disrupting the leaf tissue during litter decomposition by microbial activity. Similarly, green foliage of *Rhus copallina* in the forest-prairie ecotone of North America is not toxic, but senescing leaves are (Petranka and McPherson 1979). It was believed that tannins released into the cell sap as cell degradation occurred were oxidized enzymatically to toxic gallic and tannic acids (Petranka and McPherson 1979). In many instances, litter contains more phytotoxin than fresh tissue (Del Moral and Cates 1971), suggesting that decomposition has initiated the conversion of some inactive secondary compounds to toxic ones. Similarly, litter was found to be more toxic than fresh leaves in the pine species of the Florida scrub community (Richardson and Williamson 1988). However, this trend is not universal, as Jobidon (1986) detected a more inhibitory effect from fresh leaf leachates of *Abies balsamea* and *Pinus resinosa* than from their litter. Extracts of the green phyllodes of *Acacia cyclops* are also far more toxic than litter. In fact, the inhibitory effect of living phyllodes was found to be reversed, to stimulation by dead phyllodes, suggesting a detoxification process in the litter environment (Rutherford and Powrie 1993). Alternatively, this change from inhibition to stimulation could simply be due to a reduction of toxin concentration from high to low (Figure 8.1).

3. Stem and Bark. Bark and litter of *Eucalyptus* species are inhibitory, but stemflow, especially from species with fibrous bark, is most inhibitory of seedling growth in *Acacia, Eucalyptus,* and *Lolium* (May and Ash 1990). Toxins released by stemflow are enhanced by those from leaf and bark litter, which also are accumulated in high concentrations at and around the trunk (May and Ash 1990). Even though the lateral spread of allelochemicals from stemflow may be restricted to the immediate region around the trunk, these compounds can sometimes reach considerable depths. For example, it was found that tannic acid leached into stemflow from *Rhus copallina* reached 70 cm below the surface (Petranka and McPherson 1979).

B. Mode of Release

The release of allelochemicals may occur as a continuous flux (e.g., as volatile substances from leaves or root exudates) or as pulses or flushes (e.g., leaching by rains or snowmelt). Seigler (1996) believes the latter may be as common as the former. We will examine the major means by which allelochemicals are delivered from a source plant to a target plant.

1. Soil Solution. Other than the special cases of pollen allelopathy, most chemical interference between plants takes place in the soil via an aqueous solution. Different soil textures and soil chemical composition can alter phytotoxcity. Leachates from *Pluchea lanceolata* (Asteraceae) were more toxic when applied to a sandy loam soil than to silty loam or to sand. Therefore, using soils in bioassays that are different from those occurring *in situ* could lead to spurious inferences about allelopathy (Inderjit and Dakshini 1994b). Phenolics could be removed from solution and rendered less suppressive through complexing with soil colloids. Also, phenolics in the soil can reduce available dissolved nutrients, thereby indirectly inhibit neighboring plants by limiting resources (Rice 1984, Inderjit 1998). Therefore, the activity of allelochemicals must be considered in context with the soil structure and chemistry. Spatial variabiliy in soil properties are the rule rather than the exception. Thus, even if allelochemicals are produced uniformly in a site, their activity will be spatially variable based on heterogeneous soil properties. Soil properties can change dramatically from the base of a source tree to the surrounding environment (De Scisciolo et al. 1990), causing spatial variation in the activity of stemflow allelochemicals. The fate of allelochemicals modified by soil properties and their activity on target plants cannot be easily predicted.

The modes of allelochemical release can vary from locality to locality. An interesting example of the locality-dependent means of allelochemical release is given by Gliessman (1976). He determined that bracken fern [*Pteridium aquilinum* (L.) Kuhn] maintains its dominance by allelopathy. In the Pacific Northwest of the United States, phytotoxins were found in dead fronds (less so in roots and rhizomes) of bracken fern, and their toxicity was most intense at the start of the rainy season (Gliessman and Muller 1978). But toxins of the same fern growing in the tropics (Costa Rica) are mainly leached from green fronds. It has been argued that in regions of high decomposition, and low litter accumulation, the effect of allelochemicals would be maximized by releasing them from green tissue. In contrast, in regions with slow decomposition and a semiarid climate, allelochemicals released from dead litter would give maximal selective advantage because, upon the first rain (when competing species are germinating), the highest dose of allelochemicals would occur.

2. Atmospheric Moisture. *Throughfall* (precipitation penetrating through a canopy) produces leachates containing a wide diversity of compounds. The amount of allelochemical leached through precipitation depends on the solubility of the compound. However, the phytotoxicity of a compound is not correlated with its solubility in an aqueous solution like rainfall. Among the monoterpenes, solubility in rainfall varies by three orders of magnitude, but their suppressive activity does not vary by the same amount. Most toxins, in fact, exhibit inhibitory activities at concentrations much lower than their saturation limit, which makes rapid solubility unimportant for toxicity. The release of some hydrophobic phytotoxins (e.g., monoterpenes) from leaves in the canopy is facilitated by the presence of detergent compounds, such as ursolic acid, found in the cuticular wax of leaves (Fischer et al. 1988). Ursolic acid acts by forming a shell around the water-insoluble substance, thereby providing solubility. Without ursolic acid, the concentration of hydrophobic substances in rainfall and their infiltration into target plant tissue would be much reduced.

The rate of phytotoxin production can be quite rapid in regions and time periods of frequent rains. For example, toxic secretions removed from the glandular trichomes of *Abutilon theophrasti* (velvetleaf) leaves were replenished in 2 days (Sterling et al. 1987). In addition, infrequent precipitation in arid regions can cause flushes of microbial activity that release toxins, an occurrence that simulates phytotoxic effects. In such situations, the only contribution by a source plant may be the carbon substrate that is required to sustain microbial metabolism (Kamisky 1981). In flooded situations, oxidation and decomposition of some phytotoxins is inhibited. Therefore, in sites or during periods of excessive moisture (flooding), saturated anoxic soils can accumulate phytotoxins to high, suppressive concentrations (Zackrisson and Nilsson 1992, Fisher 1987).

3. Physical Disruption. Disintegration of cells may be necessary to release certain allelochemicals. A well-known example is the release of hydrogen cyanide from cyanogenic glycosides. Cyanogenic glycosides are separated spatially from the enzymes necessary to hydrolyze them and release cyanide (see Chapter 7). Aspects of this one case have significance for many other allelochemicals. By disrupting the plant tissue, sequestered enzymes are liberated that hydrolyze nontoxic glycosides and produce toxic aglycones (Friedman and Waller 1985). Macerated tissue can be much more toxic than entire tissue. For example, fragments of leaves from ericaceous species (e.g., *Arctostaphylos uva-ursi*) and *Eucalyptus* species are more inhibitory than intact leaves. In most situations, phytotoxins are freed from within the cell wall when plant tissue is ruptured by saprophytic microbes in the soil. Since leaching of living tissue by precipitation and the extraction of decomposing litter often yield different suites of phytotoxins with different inhibitory effects, testing allelopathy from macerated green tissue is no longer considered a valid methodology. Extracts made from macerated tissue can also have sufficiently high osmotic potentials that alone can cause suppression in bioassay (Bell 1974). Bell reported that solutions that have greater than 100 mOsm (milliosmoles) inhibited radicle extension, but more than 200 mOsm was required to affect germination. Recent studies using bioassay have taken the precaution of eliminating high osmotic potential in extracts as a possible confounding factor. An overestimation of allelopathy resulted when bioassay tests evaluated against a distilled water control were compared with tests using a control adjusted to the same osmotic potential as plant extracts using mannitol (Wardle et al. 1992).

4. Oxidation. Nontoxic secondary metabolites can be converted into phytotoxic compounds by oxidation. However, oxidation can also degrade and inactivate toxic allelochemicals. Juglone released by walnut trees is in the nontoxic glycoside form (4-β-glycoside of the corresponding hydroquinone), but it becomes the toxic form (quinone juglone) when oxdized. Leaf leachate of the woody goldenrod (*Chrysoma pauciflosculosa*) in Florida contains a mild allelochemical: 17-hydroxygrindelic acid (HGA, a diterpene alcohol) active at 60–360 μM against native grasses (Fischer et al. 1994). Oxidative activation under sunlight of HGA, however, produces the more potent 17-oxogrindelic acid (OGA), active at 40–150 μM. Further oxidation will degrade OGA to a diterpene diacid that has no allelochemical activities at the same concentrations.

5. Volatilization. Dry soils absorb volatile chemicals (e.g., terpenes, terpenoids) more effectively than do wet soils. These chemicals are released in the dry season in semiarid zones and can be adsorbed on lipophilic dust particles that settle and accumulate in the rhizosphere (Muller and Del Moral 1966). Examples of plants releasing volatile products are the chaparral species of *Salvia leucophylla, S. apiana,* and *Artemisia californica* (Muller and Del Moral 1966) and *Eucalyptus* (volatile monoterpenes, (Del Moral et al. 1978, Fischer 1986). Because of the prevalence of wind and high soil temperature, condensation and accumulation of volatile allelochemicals are less likely to occur in the Florida scrub than leaching by dew (Fischer et al. 1994). Volatile compounds (such as monoterpenes) of a lipophilic nature can be dispersed in aqueous solution with the aid of surfactants, such as ursolic acid, found in large amounts in the leaf surface of Florida scrub species. However, it is not known whether the micelles that the surfactant form around the compound will eventually release the toxic substance. Fischer et al. (1994) speculated that leaf waxes and resins can help retain and accumulate volatile allelochemicals in the soil to inhibitory levels. The toxicity of volatile hydrocarbons appears to be related to their chemical structure (i.e., chain length and straight versus branched molecules. Although no consistent patterns emerged, some straight-chain hydrocarbons are less inhibitory than branched variants (Bradow 1991).

*6. **Root Exudation.*** The collection and isolation of root exudates require a special trapping system (Figure 8.6). Using such a system, Tang and Young (1982) demonstrated the presence of cinnamic acid produced by the intact root of bogalta limpo grass, which inhibits the radicle elongation of radish. Large quantities of sorgoleone (1% of seedling dry weight) were produced in the root exudate of sorghum seedlings (Nimbal et al. 1996). *Ailanthus altissima* (tree-of-heaven) is an early successional, shade-intolerant species that often forms a pure stand that persists by inhibiting neighboring species (Lawrence et al. 1991). It is believed that such exclusion can be attributable to root toxins if a continuous release occurs. *A. altissima* root bark contains a strong phytotoxin identified as ailanthone, a water-soluble quassinoid compound (Heisey 1996). However, this phytotoxin can degrade in a petri dish in less than 3 days and in soil within 4 weeks (Heisey 1990). In the field, ailanthone applied at 4 kg ha^{-1} lost its toxicity in less than 5 days (Heisey 1996). Both the observation of some tolerance to phytotoxins in target plants (Lawrence et al. 1991) and the lack of evidence for the persistent release of ailanthone in the field (Heisey 1996) obscure the importance of allelopathy by *Ailanthus* on community structure.

*7. **Decomposition.*** Allelochemicals produced by plants generally undergo chemical changes upon reaching the soil microbial environment. In the decomposition of bilberry (*Vaccinium myrtillus*) leaves, one phytotoxin, caffeic acid, is converted to its degradation products of *p*-coumaric and protocatechuic acids (Gallet 1994). Microbial decomposition plays a key role in the transformation of plant metabolites that can result in an increased or a decreased toxicity in the rhizosphere.

After incubating the shoot leachate of *Chenopodium album* for 5 days, the inhibitory effect was reduced by 40%, and only 20% of the original activity remained after 30 days of incubation (Mallik et al. 1994). Rapid degradation of juglone and ailanthone have also been reported (Schmidt 1990, Heisey 1990). Although some inhibitory compounds in the leaf and bark litter of *Eucalyptus* species degrade quickly, others remained even after 5 months in the dry climate (May and Ash 1990). Variable toxicity from the degradation of leaves of *Eucalyptus* and other species is attributed to three factors: species, duration of incubation, and soil type. Some of the interaction terms (species × soil) are also significant. Soils with high humus or clay content can complex with phytotoxins, removing them from solution but also preventing them from breakdown and leaching (Chou 1987). Blum (1998), however, found that the sorption by soil does not prevent microbial breakdown of phenolics (i.e., no accumulation of phenolics to toxic levels in clayey soil). Clearly, key edaphic factors and their interactions with microbial populations need to be considered in assessing the factors that determine the presence and stability of allelochemicals (Souto et al. 1994).

In the Florida scrub, litter and leachates of fresh leaves of the Florida rosemary (*Ceratiola ericoides*) contain the inert compound ceratiolin, which, when exposed to sunlight, is transformed to hydrocinnamic acid, which has an inhibitory effect on native grasses and pines (Tanrisever et al. 1987, Williamson et al. 1992). Hydrocinnamic acid is further degraded to acetophenone in the soil solution by microbial activities. Acetophenone is also chemically active against the same native grasses (Fischer et al. 1994). This demonstrates the complexity of chemical interactions mediated by both abiotic and biotic factors.

In the bioassay of allelochemicals, the length of incubation of leaf washes can have a significant effect on the microbial population size, and the amount of decomposition products in the final washing. Lovett and Jackson (1980) reported that a 12-hour incubation period was sufficient to produce an inhibitory effect by washings of *Camelina sativa* (L.) Crantz (false flax, Brassicaceae) leaf in bioassay. The 12-hour period was associated with an exponentially increasing bacterial population that caused the phytotoxicity. One way to experimentally create the interaction between plant and microbes in the rhizosphere is through the "continuous root exudate trapping" (CRET) system (Tang et al. 1989; Figure 8.6). This study, performed with or without the presence of soil microbes, did result in either a suppressive or a stimulatory effect, respectively, of allelochemicals on shoot growth of lettuce (Inderjit et al. 1997). In some situations, microbial breakdown of nontoxic nitrogenous plant compounds can produce strong nonallelochemical inhibitors (e.g., ammonia) of germination and growth (Kaminsky 1981).

C. Active Concentrations

In general, the response of plants to allelochemicals exhibits a concave pattern (Figure 8.1) in which it is stimulatory at low concentrations, but becomes suppressive at higher concentrations (Rice 1984). For example, phenolic acids applied at low doses can have both positive and negative effects on target plants, but are always inhibitory at high doses. Different concentrations of weed extract can have differing effects on the nodulation of soybean (acting on the N-fixing bacteria) in the same pattern as that found for the phenolic acids (Mallik and Tesfai 1988). It appears that the original definition of allelopathy encompassing both stimulation and suppression of one plant over another by chemical means is an inherent part of phytochemical-induced plant-plant interaction. Therefore, allelochemical concentration is critical because this is the factor that determines whether the compound will be stimulatory or inhibitory.

Because plant secondary chemicals tend to degrade rapidly in the soil (Heisey and Delwiche 1983), it is difficult to ascertain the concentration of putative allelochemicals present at the interface between soil and plants (Newman 1983). Often, we do not know whether the dose required for an allelopathic effect observed in laboratory experiments actually occurs under field conditions. For example, quinoline alkaloid is strongly inhibitory at 0.3 m*M* in bioassay, but the measured concentration in soil is only 0.02 m*M*, well below the active threshold (Aerts et al. 1991a). In the rhizosphere, for allelopathy to exert a continuous effect on neighboring plants, the source plant must be able to maintain a steady supply of phytotoxins to replenish declining levels (due to allelochemical degradation). Beyond knowing the soil concentration of phytotoxins, it is another challenge to determine their concentration at the physiologically active sites. The concentration at the physiological site is determined by substance concentration in the environment, substance diffusion, and plant accumulation of the substance by uptake or adsorption.

For most active compounds, predictions of stimulatory or inhibitory effect based on field measures of concentration is not possible, perhaps because of the range of doses applied in laboratory bioassay—10–1000 ppm—is much wider than concentrations in the field. Moreover, there is a wide variation in the threshold of inhibition of different allelochemicals and different species. For example, lactones at 250–300 ppm inhibit the germination of lettuce. Cineole, an essential oil and the main ingredient of bay leaf, has inhibitory activity at 5 m*M* (Lorber and Muller 1980). The inhibitory effect of long-chain alcohols, aldehydes, and ketones increased with increasing lipophilicity, perhaps due to their greater solubility through the cell membrane. At 0.1 m*M*, camptothecin (an alkaloid) can inhibit tobacco growth but will not affect beans and sorghum. The threshold concentration of allelochemicals can vary with environmental factors such as temperature. At 29°C, ferulic acid inhibits grain sorghum at 0.4 m*M*, but the same effect at 37°C required only 0.2 m*M* (Einhellig 1989). Environmental influences on the action of secondary compounds is discussed further in Section V. Therefore, there is no general concentration for the threshold of activity that can be applied to a wide range of allelochemicals.

Significantly, the response curve of allelochemicals (Figure 8.1) shows two points where no plant response is observed: at zero concentration and at the crossover point between being stimulatory and suppressive. Testing for activity at a number of concentrations is critical in order to understand allelopathic processes in the field. Clearly, without conducting a test at a series of concentrations, it would not be possible to distinguish between a compound having no bioactivity and one tested at the crossover concentration.

D. Presence of Multiple Allelochemicals

1. Bioassay and Germination Tests. Much of the research on phytotoxicity has relied on bioassays of common test species such as lettuce and cress because of their rapid rates of development. Bioassays in allelopathy research usually involve germination tests for standard species on an inert medium containing the putative allelochemical. Commonly, many concentrations of the putative allelochemical are tested against a control treatment. The rate of seed germination and the rate of radical elongation are used as indices of toxicity.

The use of a standard "test" species has some distinct advantages. The test species germinate to 100% in control solutions. Germination and radicle growth is rapid, which minimizes the chance for contamination by pathogens during the bioassay. The "test" species do not have any allelochemicals in their seed to interfere with the allelochemical being tested. There is no pretreatment required; thus, bioassays can be completed in a few days rather than the months required for many native species. Gopal and Goel (1993) also have used *Lemna,* a floating aquatic macrophyte, as a sensitive bioassay for low concentrations of allelochemicals in the growth medium. However, there are some disadvantages to using "test" species in bioassays of allelopathy. These quick-germinating and -growing test species may react to the putative allelochemicals quite differently than will the native target species. Also, the outcome of the bioassay may be misleading when a single standard set of environmental conditions is used for several test species, each of which requires different optimal germination and growth conditions (Leather and Einhellig 1988).

Techniques used to calculate germination indices on test species bioassays also vary in their reliability and sensitivity. Chiapusio et al. (1997) compared different methods of describing the germination process, and found the one known as "speed of germination" (S) was most sensitive to both total germination rate and the delay in germination caused by toxicity. This technique computes S as

$$S = (N_1) + \frac{N_2 - N_1}{2} + \frac{(N_3 - N_2)}{3} + \cdots \frac{(N_n - N_{n-1})}{n}$$

where N_1, N_2, N_3, . . . , $N_n - 1$, N_n represent the proportion of germinated seeds obtained the first, second, third, . . . , $n - 1$, and n days or hours (Einhellig et al. 1982, Chiapusio et al. 1997).

2. Toxicity Threshold. Determining the concentration threshold of allelochemicals in bioassay is only the first step. It is equally important to relate the concentration of these compounds to their total flux rate in the natural environment such as that in the rhizosphere. The one-time measure of concentration in the field (as reported by earlier workers) does not account for the rate of release of allelochemicals and the rate of microbial decomposition and modification of those chemicals. Ion-exchange resins placed in soil for a period of days could be a useful tool in assessing the flux of allelochemicals available to root uptake by target plants. Other materials that adsorb phytotoxins, such as polyurethane foam plugs and methyl silicon film, can be used and analyzed in a gas chromatograph after desorption (Skogley et al 1990).

Recent findings have indicated that the concentration of individual allelochemicals extracted from nature is often below the threshold required to elicit an inhibitory response in laboratory bioassays (Einhellig 1987). These studies show that target plants are suppressed by a combination of allelochemicals, but the concentration of any individual compound is well below (10- to 100-fold) the threshold of suppression (Einhellig 1987). Although many studies lack the rigor to adequately account for the presence and concentration of all phytotoxins, there is sufficient evidence to support the additive or synergistic effect of multiple allelochemicals at low concentrations (Mersie and Singh 1988). Moreover, whether a mixture of phytotoxins (e.g., phenolic acids) has an additive, synergistic, or antagonistic effect may depend on the concentration and composition of the constituent chemicals (Blum et al. 1985). Therefore, inhibition observed in the field may be due to the combined effect of several allelochemicals (see review in Einhellig 1989), potentially from several different source species. In any given soil environment where allelopathy occurs, we are likely to find a complex suite of chemical compounds of both plant and microbial origins. While several studies have shown the additive or synergistic effect of multiple compounds (Brown 1967, Einhellig 1987, Einhellig 1996), the allelopathic properties of complex mixtures are still poorly understood (Einhellig 1987).

E. Modes of Action

Even though secondary compounds with allelochemical potential are pervasive in plants, the expression of allelopathy requires both an appropriate abiotic environment and the presence of susceptible target plants. The activity of most allelochemicals appears to be broad spectrum against other plants, herbivores, and pathogens. But "clear insight into the precise physiological perturbations caused by allelochemicals has not been obtained" (Einhellig 1986). The site of allelopathic activity on the target plant is also quite variable, ranging from inhibiting mitosis in pollen development of target species (Kanchan and Jayachandra 1980, Murphy and Aarssen 1995a, 1995b) to disrupting cellular functions such as photosynthesis and respiration. Phenolic acids, coumarins, and flavonoids, for example, can affect mitochondria and chloroplast electron transport activities (Moreland and Novitzky 1987). Many allelochemicals behave like coumarins in that their inhibitory effects may be due entirely to nonspecific influences on metabolism. Coumarins affect the action of many enzymes through a reaction with the sulfydryl group of enzymes as well as interacting with plant hormones such as indoleacetic acid (IAA), cytokinins, and gibberellins. These reactions can inhibit organogenesis and callus growth, and exert a general effect on growth regulation such as inhibiting root growth (Brown 1981). Coumarins interfere with cell mitosis and elongation, water relations, and photosynthesis, possibly due to the disruption of chloroplast inner membrane (Einhellig 1996).

Quinones released by root exudates generally influence the energy relations of cells. Sorgoleone (a quinone found in sorghum root exudates) inhibits seedling growth of weeds and prevents photosynthetic oxygen uptake (Einhellig et al. 1993) by inhibiting electron transport in photosystem II (PSII) more strongly than does 3-(3,4-dichlorophenyl)-1,1-dimethylurea (DCMU) (Gonzalez et al. 1997). Sorgoleone and juglone interfere with ATP production at PSII, blocking electron flow between cytochrome b and c_1 complex (Einhellig et al 1993, Einhellig1995). Juglone acts directly on the chloroplast by reducing photosynthesis, but does not inhibit the chloroplast electron transport chain. Juglone also affects mitochondria metabolism by significantly altering normal oxygen uptake (Hejl et al. 1993).

Phenolics show a range of activities such as suppressing mitochondria metabolism, stimulating hypocotyl elongation and rooting, and affecting flowering. The mode of action is attributed to an interaction with IAA oxidase, but the effect could both inhibit growth (monophenolics stimulate IAA decarboxylation) and stimulate growth (*o*-dihydroxy phenolics inhibiting IAA oxidation). Salicylic acid can alter membrane function and mineral absorption, and affect plant-water relations (Einhellig 1996).

Terpenes also have a diversity of toxicities. Sesquiterpenes have antigibberellin and antiauxin activities. The inhibitory effect appears to be related to the polar nature of sesquiterpenes, becoming less inhibitory with less polar species (Fischer 1986). Sesquiterpene lactones are found in Asteraceae, Umbelliferae (stong inhibitors of germination) and Mangnoliaceae (Harborne 1988). Unsaturated lactones inhibit root growth and block mitosis. Lactones inhibit extension growth by disrupting auxin receptor sites (Fischer 1986). Cineole interferes with oxidative phosphorylation in mitochondria (Lorber and Muller 1980). Colchicine acts by disrupting microtubule formation, which halts mitosis and may induce nondisjuction of homologous chromosomes in meiosis. The diterpenoid-alkaloid delcosine inhibits tissue growth and, because of its similarity to gibberellins, may act as a competitive inhibitor of gibberellin action. Benzoxazolinone and related compounds inhibit auxin-induced growth and auxin receptor binding.

Bioassay studies have shown cases of germination being more sensitive to allelochemicals than radicle extension is (in native grasses of Florida sand pine scrub community; Richardson and Williamson 1988) and root extension being more sensitive to phenolics than germination (Hegde and Miller 1992). The physiological process of germination and radicle extension is sufficiently different to expect differences in their sensitivity to phytotoxins (Stowe 1979). If radicle elongation or root growth is affected by allelochemicals, then it is reasonable that nutrient uptake would be hindered as the root volume declines and root tissue deteriorates. Evidence consistent with this assumption was found where, responding to phenolics extracted from ragweed (*Par-*

thenium hysterophorus L.), tomato plants had lower shoot growth than was associated with reduced leaf N and P (Mersie and Singh 1988). However, there is no consistent pattern indicating which plant process—germination or root elongation—is more sensitive to a particular phytotoxin. It may depend largely on the concentration of the substance reaching the target tissue, the particular receptor species, and the growth environment of the plant.

One form of allelopathy not occurring in the rhizosphere is pollen allelopathy. In *Parthenium hysterophorus* L. (a tropical weed), pollen allelopathy may reduce fruit set of neighboring plants, thereby reducing competition from their progeny (Kanchan and Jayachandra 1980). A convincing study of interference by allelopathic pollen was presented by Murphy and Aarssen (1995b), where the reduction in seed set of the competitor was demonstrated in the lab and validated in the field. Pollen clusters on leaves can also reduce chlorophyll contents (Kanchan and Jayachandra 1980); however, this form of suppression may be a fortuitous one. Even when the pollen of *Hieracium* species (Asteraceae) inhibits the germination of pollen from noncompatible species, the direct impact of such suppression in the field was found to be inconsequential because pollinators did not cross-visit flowers of different species (Murphy and Aarssen 1995a).

In summary, the modes of action for allelochemicals are diverse, but there are some general patterns. Many allelochemicals interfere with metabolic processes in energy-regulating organelles (chloroplast and mitochondria) or with cell division processes (microtubule organization). Allelochemicals are most effective against seeds and newly growing radicles. In some target species, seeds are more sensitive (germination), and in other species, root growth is more sensitive. Allelochemicals also can influence growth regulator function.

IV. DETOXIFICATION OF ALLELOCHEMICALS

In order for plants to tolerate allelochemicals in their environment, they must have mechanisms of avoiding the toxicity, tolerating the toxins, or neutralizing the toxins. Avoiding toxicity could be accomplished by a phenological behavior in situations where allelochemicals are temporally distributed. In the case of bracken fern, seeds of other species that germinate late in the rainy season will not suffer the full dose of allelochemicals as will those seeds that germinate after the first rain. In the California soft chaparral, terpenes released by *Salvia* species are most abundant during the dry season, producing a strong bare zone during dry years. Species that germinate after several rain events (a prolonged period of moisture) will not be likely to suffer inhibition by terpenes released by *Salvia* and will be able to colonize the bare zone. Species that germinate exclusively in postfire environments will also avoid allelopathy, particularly in Mediterranean shrublands. Germination requirements for many ephemeral postfire species and woody dominant species are based on unique properties of the postfire environment (Keeley 1987). For instance, many species require heating, a flush of nutrients, contact with charcoal, or nitrous oxide to germinate. Germinating in the postfire environment compared with the mature community is advantageous because of the higher nutrient levels, and the reduction of suppressive chemical concentration in soils (Kaminsky 1981). Avoidance of soil phytotoxins can also be accomplished by deep rooting, because most phytochemical toxins are found in the upper soil layers. In fact, as stated earlier, this is one mechanism by which source plants can avoid autotoxicity.

There is a diversity of tolerances to allelochemicals among species. Differences in growth stimulation or inhibition have been found for many species in response to a wide variety of allelochemicals including phenolic acids, amino acids, benzoquinone, anthoquinone, and others (Schulz and Friebe 1999). For example, L-tryptophan released by oat (*Avena sativa*) roots inhibits root growth of other species. The LD_{50} (the concentration of L-tryptophan that causes 50% loss of root growth) for cockscomb (*Amaranthus caudatua*) is 0.14 m*M*, while the LD_{50} for wheat (*Triticum aestivum*) is 1.7 m*M* (Kato-Nogochi et al. 1994). Moreover, the effect of DIBOA [2,4-dihydroxy-(2H)-1,4-benzoxazolinone, which is released by the roots of a number of species], on seedling growth varies greatly among species (Figure 8.9). In fact, dicots tend to be more sensitive to DIBOA than monocots (Schulz et al. 1994). Therefore, species vary in their allelopathic tolerance in general and to specific compounds.

Tolerance of allelopathic compounds could be due to a number of different processes:

1. The ability to exclude toxins at the root or leaf surface
2. The ability to compartmentalize toxins in the apoplast
3. The ability to localize allelochemicals in glandular trichomes for exudation
4. The ability to withstand allelochemicals in the cytoplasm (differential sensitivity at the site of action)
5. The ability to detoxify the allelopathic compound

The capacity to exclude toxins at the plant surface is primarily due to morphological characteristics of the organ surface. For instance, the permeability of phenolic sub-

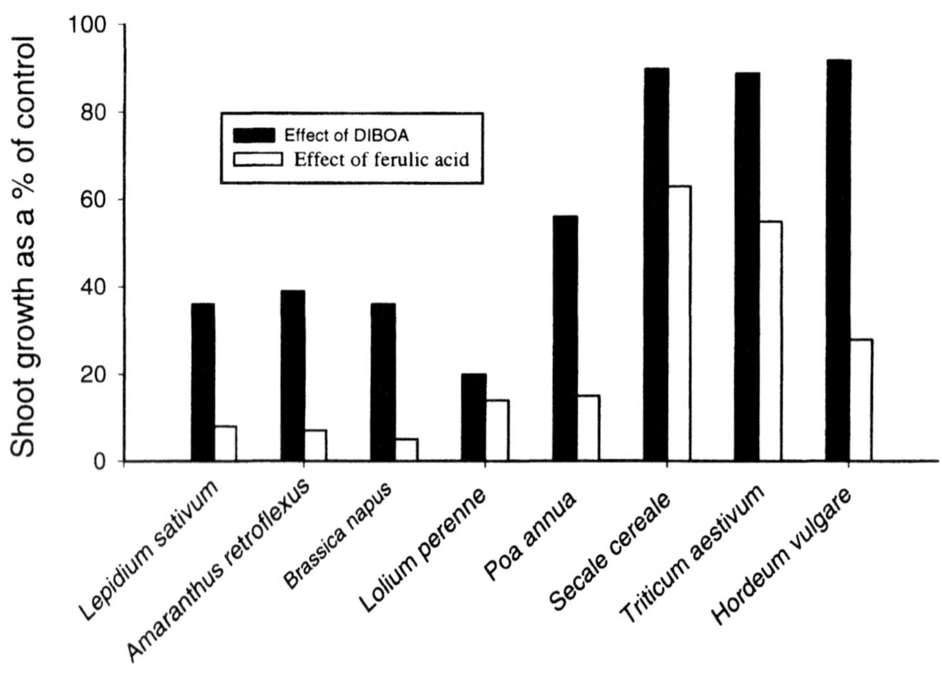

Figure 8.9. Effects of DIBOA [2,4-dihydroxy-(2H)-1,4-benzoxazolinone] and ferulic acid on the shoot growth of a diversity of species. (Redrawn from Schulz and Friebe 1999.)

stances into leaves varies by two orders of magnitude depending on the lipid composition of the cuticle (Schafer and Schönherr 1985). Plants that have glandular hairs (and exude toxins from those hairs) would be able to recycle toxins accumulated by roots back into the glandular hairs.

Some tolerance of allelochemicals can be attributed to the ability to deposit allelochemicals in nonmetabolic compartments. The central vacuole may be the most important depository of allelochemicals, particularly after conjugation. Vacuoles normally store compounds such as anthocyanins and other secondary chemicals. Esculin and scopalin are accumulated by wheat vacuoles (Werner and Maitile 1985). The aglycones are glycosylated before transport into the vacuole by an ATP-requiring proton-dependent antiport (Martinoia 1992). Such storage in the vacuole has limitations similar to those of storing salt in the vacuole (see Chapter 5). Once the transport mechanism is saturated, the concentration of allelochemical in the cytoplasm increases rapidly, causing severe metabolic disruption. Other allelochemicals, particularly phenolic acids, can be stored in compounds located in the cell wall. For example, ferulic acid (a compound that has allelopathic potential) is a precurser for lignin synthesis. Can plants that accumulate external ferulic acid incoporate it into cell wall lignin? This question was answered by studying cucumber (*Cucumis* sp.), which is known to continuously absorb ferulic acid from the growth medium (Shann and Blum 1987). When radiolabeled ferulic acid was applied to cucumber plants, lignin synthesis increased and the radiolabel was located in new lignin synthesis. In general, some plants can accumulate extrinsic allelochemicals and incorporate them into biosynthesis pathways for secondary chemicals. Some of these pathways lead to the production of polymers that are exuded into the apoplast, while others are stored in the central vacuole. Plants that grow rapidly or are naturally phytotoxic themselves may have the highest capacity for this type of defense against allelochemicals in the environment.

The excretion of allelochemicals that have been absorbed from the environment is most likely to occur in plants that already excrete allelochemicals. For example, *Urtica* species (nettles) that exude toxins from glandular hairs are likely to be able to exclude toxins accumulated from extrinsic sources. Hydroxyquinone absorbed by wheat and bean is glycosylated to form arbutin and is exuded into the growth media by roots (Winter 1961). This and a few other studies indicate that plants can exude allelochemicals (usually after modification) into the growth medium. However, little is known about this concept.

Detoxification of organic compounds accumulated by plants has been extensively studied in relation to herbicides and pesticides. Herbicide resistance is well known, and the mechanism for that resistance has been exhaustively studied. The fact that some species and varieties can detoxify herbicides indicates that plants have the capacity to detoxify absorbed extrinsic allelochemicals. Some of the concepts learned in studies of herbicide detoxification can be applied to understanding defense mechanisms against allelochemicals for plants in natural ecosystems.

A general concept for plant defense mechanisms against extrinsic organic toxins has been proposed by Cole (1994). Three phases are suggested. Phase I is a metabolic attack on the toxin. This usually involves increasing the polarity of the molecule by hydroxylation and dealkylation. These reactions often are catalyzed by cytochrome P450–dependent oxidases, or peroxygenases. Phase II of the detoxification pathway usually involves conjugation with sugars, amino acids, or malonic acid. Such conjugation increases the water solubility of the compound. Phenolic acids that are absorbed by plants become conjugated to monosaccharides, disaccharides, amines, amino acids, and peptides (Balke et al. 1987, Russness and Still 1977). The third phase of detoxification is the translocation of the compound to a storage location such as the vacuole or cell wall, or excretion onto the plant surface.

Although detoxification of herbicides is well studied, there has been little research on detoxifying allelochemicals. The lack of studies on this topic is primarily due to the general theory that allelochemicals are so diverse and variable in the environment that there is little selective pressure for the evolution of specific detoxification mechanisms. Clearly, that assumption is unfounded based on the evidence for differential sensitivity to herbicides. Some of the mechanisms associated with detoxifiaction of allelochemicals are constitutive metabolic processes, but other mechanisms can be induced or are specific to certain species. Therefore, detoxification of certain allelochemicals can be species dependent. Detoxification mechanisms for approximately 12 different allelochemicals have been studied (Table 8.3). The results indicate that there are differences among species in the way that certain allelochemicals are detoxified, and different species have different capacities to detoxify allelochemicals.

V. ENVIRONMENTAL STRESSES AND ALLELOPATHY

As the chapters in this volume illustrate, plants experience a multitude of stresses from both the biotic and abiotic environment. For convenience, we define *stress* broadly as a condition that prevents the plant from attaining its full growth and reproductive potential. At any given time in its life, a plant is likely to be limited in its growth (e.g., due to N deficiency) or reproduction (e.g., seed predation) by one or more factors. In many cases, stress factors can exert an additive or a synergistic effect on plants. Considering allelopathy, the susceptibility of a target plant can increase in the presence of other stresses such as drought and nutrient limitation. A clear example of multiple stress interaction is the combination of allelopathy and competition. As the foregoing discussion has emphasized, demonstrating allelopathy requires the removal of competition effects. In the field, it is rare to find situations where competition and allelopathy are not intimately connected. In essence, the two are inseparable since the degree of chemical interference between neighboring plants impinges directly on the intensity of competition.

In this section, we will consider the effect of abiotic and biotic stress factors on source and target plants. It has been reported that the synthesis of secondary metabolites changes in response to drought, shade, and nutrient deficiency. If abiotic or biotic factors reduce primary processes such as photosynthesis, it is reasonable to expect a reduction in secondary metabolites because their synthesis requires energy and precursors derived from primary processes. However, it is also possible that the synthesis of specific secondary compounds is triggered by stress events such as pathogen attack or herbivory (Chapter 10). Therefore, the rate of allelochemical synthesis will likely depend on both the nature and severity of the stress.

A. Abiotic Stresses

Because of drought, nutrient limitation, and grazing pressure, allelopathy in a desert ecosystem is likely to be more intense than in humid regions (Friedman 1987). There is also likely to be higher concentrations of phytotoxins in this environment, where volatile toxins can accumulate in the soil over long periods of time aided by low microbial decomposition and reduced leaching (Friedman 1987). Drought stress increases chlorogenic acid, phenolics, monoterpenes, and hydroxamic acids in some

TABLE 8.3 Representative Examples of Detoxification Mechanisms Used by Higher Plants Against Allelochemicals in Their Environment

Allelochemical	Allelochemical Class	Mechanism	Product
BOA[a]	Hydroxamic acid	Hydroxylation Glucosylation	BOA-6-*O*-glucoside
BOA[a]	Hydroxamic acid	Conversion into *O*-amino phenol, and oxidative dimerization	2-amino-3H-phenoxacin-3-one 2-acetylamino-3H-phenoxacin-3-one
Brassinolide	Steroid	Glucosylation	23-β-glucoside
Castasterone	Steroid	Brasinolide pathway	Brassinolide
Daphnetin, esculetin	Alkaloids	Glucosylation	Glucosides
Ferulic acid	Phenolic acid	Polymerization	Lignin
Hydroquinone	Quinone	Glucosylation	Arbutin
p-Hydroxybenzoic acid	Phenolic acid	Glucosylation Esterification	*p*-*O*-β-Glucose-hydroxybenzoic acid *p*-Hydroxybenzoic acid ester
Resorcinol		Glucosylation	Glucosides
Salicylic acid	Phenolic acid	*O*-glucosylation	Salicylic acid-*O*-β-*D*-glucoside
Salicylic acid	Phenolic acid	Hydroxylation	Gentisic acid-*O*-β-D-glucoside
Tryptophan	Amino acid	Conversion into IAA Conjugation with amino acids	IAA-aspartate, IAA-glycine

[a]BOA = benzoxazolinone.

Source: Adapted from Schulz and Friebe 1999.

plants (Einhellig 1996) but decrease phenolics in other species. For example, the toxicity of leaf exudates of velvetleaf was found to increase in plants under high temperature and water stresses. The increased toxicity however, was not the result of a higher volume of toxins released (no change), but rather to an increase in the concentration of toxins under drought stress conditions (Sterling et al. 1987). Moreover, different lines (varieties) of one species can have different responses to drought-induced accentuation of toxicity. Among four half-sib white ash families, higher growth rate under drought stress was found to correspond to their tolerance of fescue leachates. It is not known how prevalent this correspondence is among species in general (Rink and Sambeek 1987).

Temperature and light intensity in the field can alter the effect of phytotoxins from weeds on crop production (Bhowmik and Doll 1983). In some instances, high PAR (photosynthetically active radiation) and UV light can induce higher levels of phenolics in weeds (Einhellig 1996). But other studies found that phytotoxins such as mimosine in the leaf leachate of *Leucaena leucocephala* are degraded by sunlight (Chaturvedi and Jha 1992). Thus, the general stimulation of phenolics in plants by UV light may be negated because UV light also degrades these compounds when they are released from plants.

The inhibitory effect of phenolics released by sunflower on *Amaranthus retroflexis* increases with greater nutrient limitation (Hall et al. 1983, Koeppe et al. 1976). Similar responses in grasses under low nutrient levels have been reported by Chou and Lee (1988). By alleviating nutrient stress, target plants can become more tolerant of phytotoxins, but there are exceptions to this generality (Einhellig 1989, 1996). Nutrient amendment also indirectly affects neighboring plants by increasing microbial metabolism, which can break down and detoxify plant metabolites (Hall et al. 1983). However, decomposition can also lead to a more rapid release of toxins from litter (Chou 1987). If a plant is already being suppressed by allelopathy, it will not respond to fertilizer application. Moreover, interference with nutrient uptake by allelochemicals such as phenolic acids can occur when the electrical potential of root cell membrane is disrupted (Glass and Dunlop 1974, Balke 1985). When source plants experience abiotic stress such as deficiencies in N, K, Mg, and S, they produce more allelochemicals (e.g., scopoletin and chlorogenic acids) than those grown in high nutrient soils (Einhellig 1996). In combination with greater sensitivity in target plants, allelopathy is accentuated in nutrient-limited environments. Weidenhamer (1996) has also reported bioassay results showing an increase in allelochemical inhibition under low concentrations of soil N and P. The biochemical basis for higher phytotoxicity to target plants by allelochemicals in nutrient limited plants is unknown.

Under high soil moisture and warm temperatures, when plants show optimum growth, they are less susceptible to phytochemicals (Einhellig 1989, Weston 1996). This could be partly due to higher microbial breakdown of active phytotoxins in the soil under the same conditions. Seed germination of understory herbs in the *Adenostoma fasciculatum*–dominated community increased under higher soil temperature. It was found that this was due to indirect effect of higher rates of microbial degradation of allelochemicals rather than a direct link between germination and high temperature (McPherson and Muller 1969). But high temperature and moisture also favor greater decomposition of plant residues, potentially releasing a greater amount of derived phytotoxins (Einhellig 1989).

B. Biotic Stresses

The process of allelochemical transformation in the soil involving the participation of both the biotic and the abiotic environment is central in allelopathic interactions.

Current research attempts to better assess the indirect effect of a source plant on recipient plants through its action on nutrient dynamics via mycorrhizal formation, nodule-forming bacteria, soil decomposition flora, and diseases via pathogenic fungi and microorganisms. As an example, the combination of phytotoxins released by microbial decomposition of litter and pathogenic fungi can suppress Chinese fir (*Cunninghamia lanceolata*) seedling growth (Zhang 1993). Therefore, over time the Chinese fir changes the nature of the soil so that it becomes autotoxic. Hence, *C. lanceolata* does not grow well in repeated plantings. Extracts from such soils reveal the presence of phenolics that inhibit the growth of nonpathogenic fungi while promoting the growth of pathogenic fungi. The effect of this dichotomy is an increased likelihood of disease and a reduction in the rates of microbial detoxification of allelochemicals and soil mineralization. All of this contributes to the inhibition of growth when trees a planted a second time in the same area.

If the presence of allelochemicals can increase nutrient limitation experienced by target plants, then the added suppression of the toxins on mycorrhizal formation can only exacerbate the level of stress. (For a description of mycorrhizal function, see Chapter 4.) Phenolics (benzoic and catechuic acids) extracted from green foliage of aspen (*Populus tremuloides* L.) inhibit the growth of *Boletus* sp. (mycorrhizal fungi) and *Marasmius* sp. (decomposer fungi). But *Boletus edulis* and *B. elegans* show stimulatory effects on plant growth at low concentrations of phenolics (Olsen et al. 1971). Foliar leachates of raspberry (*Rubus idaeus* L.) at 2.5% concentration inhibit the growth of some black spruce ectomycorrhizal fungi (in decreasing severity: *Paxillus involutus, Laccaria proxima, L. bicolor, Thelephora terrestris,* and *Cortinarious pseudonapus*) but stimulated others (*Hebeloma cylindrosporum* and *Cenococcum geophilum*). Therefore, when raspberry is present, tree seedlings form mycorrhizae mainly with *C. geophilum* (Cote and Thibault 1988). It is possible that *C. geophilum* has some tolerance of allelochemicals because it is also particularly resistant to antibiotics (Goldner and Medve 1986). Clearly, similar to plant species, fungal species show vastly different sensitivities to plant secondary metabolites. In addition, the response of ectomycorrhizal fungi to phytotoxins exhibits a dose response pattern (i.e., stimulatory at low concentrations and inhibitory at high concentrations) (Rose et al. 1983) like that for plant responses to allelochemicals.

Soil and litter under *Kalmia angustifolia* inhibits seedling growth (but not germination) of black spruce (*Picea mariana*) through abnormal root tip development. This may be partly due to the observed suppression of mycorrhizal formation on spruce roots when grown in soil from under *Kalmia* plants, but the direct effect of lowered nutrient level in these soils and an increased susceptibility to pathogens are also implicated (Yamasaki et al. 1998). Black spruce seedling growth is also affected by altered soil chemistry associated with *Rhododendron (Ledum) groenlandicum* leaves and litter (Inderjit and Mallik 1997). Gallet (1994) has also suggested that autotoxicity in *Picea abies* could increase when mycorrhizal formation is inhibited. Since the suppression of mycorrhizae by allelochemicals can reduce plant growth indirectly, it may be possible to confer greater resistance to allelochemicals by inoculating vulnerable plants with mycorrhizal fungi. Indirect evidence shows that *Suillus* (an ectomycorrhizal fungus) could moderate the effect of allelochemicals from fern on red oak (Hanson and Dixon 1987).

The effect of allelochemicals on community structure through its interference with mycorrhizal formation has also been implicated. In the *Calluna* heathland, root washings and humus extract of *Calluna* inhibited mycorrhizal fungi (e.g., *Boletus elegans* and *Amanita muscaria*) and pathogenic fungal growth (*Phytophthora cinnamomi*) (Robinson 1972). The maintenance of the *Calluna* population, Robinson contends,

TABLE 8.4 Stress-Induced Changes in Allelochemical Concentration and Effect on Target Plants

Stress	Source Plant	Target Plant	Allelochemical[a]	Change in Concentration	Change in Response	References
			Herbicide			
2,4-D	Sunflower	—	Coumarins	+	—	See Einhellig 1989
Torson	Tobacco	—	Scopoletin, scopolin	+	—	Wendler 1970
			Minerals			
Low N/P/K	Sunflower	—	Scopoletin, chlorogenic acids	+	—	Various
Excessive (?) N	Digitalis	—	Phenolics	+	—	Chou and Lee 1988
N deficient	Various	—	Cyanogenic glycosides	—	—	Gershenzon 1984
N deficient	Various	—	Glucosinolates	+	—	Gershenzon 1984
N deficient	Various	—	Alkaloids	—	—	Gershenzon 1984
N deficient	Various	—	Phenolics	+	—	Gershenzon 1984
N deficient	Trees	—	Terpenoids	—	—	Gershenzon 1984
P deficient	Various	—	Phenolics	+	—	Gershenzon 1984
P deficient	Herbs, shrubs	—	Terpenoids	+	—	Gershenzon 1984
K deficient	Various	—	Alkaloids	+	—	Gershenzon 1984
K deficient	Various	—	Phenolics	+	—	Gershenzon 1984
S deficient	Various	—	Glucosinolates	—	—	Gershenzon 1984
S deficient	Various	—	Phenolics	+	—	Gershenzon 1984
P, K, Mg uptake	—	Grain sorghum	Ferulic acid	—	—	Kobza and Einhellig 1987
Mineral uptake	—	Several	Benzoic and cinnamic acids	—	—	Alsaadawi et al 1986

			Temperature			
Chilling	Tobacco	—	Scopoletin, chlorogenic acids	+	—	Koeppe et al. 1970
Low and high	—	Barley	Phenolic acids	—	+	Glass 1976
High	Barley	—	Alkaloid (gramine)	+	—	Hanson et al. 1983
High	Velvetleaf	—	ND	+	—	Sterling et al. 1987
High	Wheat	Weeds	ND	—	+	Steinsiek et al. 1982
High	*Amaranthus*	Corn	Plant residues	—	+	Bhowmik and Doll 1983
High+low precipitation	*Ailanthus altissima*	—	—	—	+	Lawrence et al. 1991
			Light			
High	Grain sorghum	—	Phenolics	+	—	Woodhead 1981
High UV	Tobacco	—	Scopoletin, scopolin, chlorogenic acids	+	—	Koeppe et al. 1970
			Moisture			
High Osmoticum	Sunflowers	—	Chlorogenic acids	+	—	Del Moral 1972
		Sorghum	Ferulic acid	—	+	Einhellig 1987
Low	*Pinus taeda*	—	Monoterpenes	+/−	—	Gilmore 1977
Low	Various	—	Cyanogenic glycosides	+	—	Gershenzon 1984
Low	Various	—	Glucosinolates	+	—	Gershenzon 1984
Low	Various	—	Alkaloids	+	—	Gershenzon 1984
Low	Trees	—	Terpenoids	—	—	Gershenzon 1984
Low	Herbs, shrubs	—	Terpenoids	+	—	Gershenzon 1984
Low leaf Ψ	Velvetleaf	—	Not determined	+	—	Sterling et al. 1987
			Disease			
Fusarium	—	Asparagus	ND	—	+	Hartung & Stephens 1983
			Mycorrhizae			
Ectomycorrhizae	—	Oak	ND	—	+	Hanson and Dixon 1987

[a]ND = not determined

may be partly due to its ability to prevent invasion of species such as birch by inhibiting their ectomycorrhizal formation. Ericaceous species, such as *Kalmia* and *Calluna,* create an acidic soil environment (pH 2–4) that alone is sufficient to suppress co-occurring plants but is further exacerbated by the release of phenolic acids from decomposing litter (Zhu and Mallik 1994). Low pH alone can also affect the local occurrence of mycorrhizal fungi. Some workers have suggested that the regulation of mycorrhizal infection by allelochemicals is dependent on competition. Koide and Li (1991) observed a depletion of soil P by *Setaria* through root competition with *Abutilon*. And in response, *Abutilon* suppressed its own mycorrhizal infection through autotoxicity. The authors reasoned that this occurred because the benefit of mycorrhizal association would be minimal in such a low resource situation.

For those plants with symbiotic nitrogen-fixing bacteria, the presence of phytotoxins can inhibit the formation of nodules. Growth of nodule-forming bacteria (*Bradyrhizobium japonicum*) was inhibited by extracts of foxtail, lamb's-quarters, and Johnsongrass at 1–10 mg ml^{-1} concentrations (Mallik and Tesfai 1987). Nodulation of several legumes was reduced by as much as 20-fold when grown in quackgrass (*Agropyron repens* L. Beauv.) soil (Weston and Putnam 1985). Toxic root exudate was implicated in these studies. Nodulation of soybean occurs in shallow roots, while that of navy beans occurs on deeper roots. Quackgrass roots are most abundant in deeper soil layers, and navy bean suffers greater suppression of nodulation by quackgrass than does soybean (Weston and Putnam 1985). Phenolics released from *Populus* species inhibited nodulation in *Alnus incana* (Younger and Kapustka 1983). However, there was a concomitant increase in nitrate levels with phenolics, making it difficult to resolve which of these two processes was responsible for the suppression of nodulation. Variable responses to allelochemicals are also seen in nodule-forming microorganisms. The switch-over point between stimulation and inhibition of soybean nodulation by *Bradyrhizobium japonicum* was 1% when shoot extracts of nutsedge and sunflower were used, but 0.5% when that of lamb's-quarters was used (Mallik and Tesfai 1988).

While it is expected that additional stresses such as nutrient limitation may exacerbate the effect of allelopathy (Table 8.4), it is also likely that because of reduced resource demand in stressed plants (e.g., slower growth rate requires fewer nutrients), the uptake of allelochemicals will also be reduced. When viewed in light of the synergism between competition and allelopathy, several important issues arise. One, a stressful environment may support lower densities of plants, which could result in higher phytotoxin availability to each plant (i.e., no dilution effect). Two, a rhizosphere of low nutrients, moisture, or temperature suppresses the allelochemical detoxification activity of soil microbes. In opposition, if phytotoxins were derived from microbial processes, then their concentrations would be lower in a suboptimum rhizosphere. Future research should focus on elucidating (1) the effects of allelochemicals on plant performance when other conditions for growth are limiting and (2) how allelopathy might alter the ability of a target plant to cope with other environmental stresses. Although previous studies do demonstrate that, when stressed, a source plant generally increases its phytotoxin production, and target plants generally show an increased sensitivity to allelochemicals (Table 8.4), it is important to note that the lack of such responses is usually not reported. Without a comprehensive study of the effects of environmental stresses on allelopathic interactions, we must view the cited examples as specific cases that may not represent a general rule.

VI. APPLICATIONS OF ALLELOPATHY IN AGRICULTURE AND FORESTRY

In a simplified community, typical of agriculture and agroforestry, using source plants as natural herbicides against target plants (weeds) has merited serious consideration (Worsham 1989). A natural herbicide, such as caffeine and hydroquinone, could act selectively against weed growth while not affect deep-rooted crop or tree growth. Although the potential for natural herbicidal application abounds, its efficacy depends on the right combination of crop, weed, and soil conditions (Shettel and Balke 1983).

Phytotoxins are produced by several common crop species such as sunflowers, rye, oats, wheat, and sorghum. These species can be grown as cover or mulch crop to suppress weed growth. Some crop species capable of inhibiting weed growth through their own secondary compounds are beans, corn, cassava (*Manihot esculenta* Crantz), sorghum, barley, oats, wheat, rye, sunflowers, sweet potato (*Ipomoea batatas*), soybean, squash (*Cucurbita pepo*), legumes (*Stizolobium deeringianum, Canavalia ensiformis*), and sudangrass (*Sorghum sudanense* L.).

Alfalfa germination and growth are also suppressed by extracts of shoot and root of a number of forage grasses, including tall fescue (*Festuca arundinacea* Schreb.) and grain sorghum [*Sorghum bicolor* (L.) Moench] (Chung and Miller 1995a). In pasture, foliage tissue of tall fescue, Italian ryegrass, and little barley can suppress alfalfa and Italian ryegrass seed germination and seedling growth (Smith and Martin 1994). To avoid the likelihood of one crop species interfering with the subsequent crop of the same species or another crop species, careful agronomic management of the rotation and cultivation is required.

One such agronomic practice is *intercropping*. Intercropping was found to be effective when combining allelochemical-producing squash with resistant crop species such as corn. The toxins from squash-inhibited weeds, and this released the corn from inhibition by weeds (Weston 1996). Wheat, oats, and other crops (Weston 1996) can also be used to provide allelochemicals that act as herbicides through rotational cropping, living mulching, and intercropping. Naturally produced allelochemicals can increase their efficacy when combined with compounds derived from microbial breakdown of plant tissue. While intercropping has been extensively practiced in regions such as South America and Africa, where the use of costly herbicides is rare, it has not always proven effective against weeds. Environmental conditions and poor crop-weed combinations are cited as possible reasons for the lack of weed suppression by intercropping.

In agriculture, demonstrating a direct link between specific allelochemicals and weed control is secondary to its effectiveness in the field. It is likely that microbial mediation of allelochemical release is as important in the agricultural setting as it is in natural communities. Effective weed control using natural herbicides may be maximized when combined with crop plants that can preempt resources available to the weed through competition. In many cases, it is that slight advantage in early season growth from allelopathy that allows the crop species to overtop the weed species by competition (i.e., it becomes a smother crop). Such serial "one-two punch" benefits conferred by allelopathy and competition should also be occurring in natural communities, but there is no information available.

One of the most promising allelochemicals that has been considered as a natural herbicide is ailanthone, extracted from *Ailanthus altissima*. It meets all of the requirements of an effective herbicide: broad-spectrum activity, an active concentration rivaling those herbicides currently in commercial use (glyphosate and Parquat), and rapid degradation in soil (Heisey 1996) that minimizes the hazards of residue to other organisms.

Many of the major invasive weeds of the world may owe their success to having an effective suite of allelochemicals (Ismail and Mah 1993). The proliferation of weeds could be enhanced by a lack of adaptation of native plants (and microbes) to their new phytotoxins. Willis (1985) thought that allelopathy from weeds would be most severe in an agricultural setting where coevolution has not taken place between target plants and weeds. But most plants retain a broad spectrum of secondary compounds with allelopathic potential (Jones and Firn 1991), and evolution for defensive mechanisms against that broad spectrum does not occur. Thus, it would be equally unlikely for crop species to develop specific tolerances to allelochemicals produced by introduced weeds as it would be to those allelochemicals produced by their native neighbors. There is a need to examine the degree and rate of coevolution in chemically mediated interactions between plants in native and agricultural settings.

VIII. CURRENT STATUS AND FUTURE DIRECTIONS

A. Can Allelopathy Be Separated from Competition?

By applying the *allelochemical dilution hypothesis* (Fuerst and Putnam 1983, Weidenhamer et al. 1989, Thijs et al. 1994), it is possible to experimentally distinguish allelopathy from competition (Choesin and Boerner 1991, Gentle and Duggin 1997). When allelopathy is responsible for suppression, an increase in target plant density results in improved growth through dilution of the toxins among a larger number of plants (Figure 8.10). When density-dependent competition is the dominant factor, individual plants performed worse with an increase in target plant density. Thijs et al. (1994) showed an increase in biomass of a sensitive target plant (soybean) to atrazine (an herbicide) when grown with an increasing number of nonsensitive neighboring plants (corn). This suggested that allelopathy was involved because the toxin concentration that the soybeans experienced was lower when more corn plants were present. If the interaction had been competitive, a reduction in soybean biomass with more corn neighbors would have been expected.

The dilution hypothesis assumes that plants passively take up phytotoxins present in the root zone. A variation of this hypothesis is the *replacement series design experiments* (Dekker et al. 1983) where the "relative" yield totals (of two mixed species) of less than 1 indicates allelopathy. Others have criticized this method for not considering density-dependent competitive outcome (reduction) in yield. Despite the methodology available, Inderjit and Del Moral (1997) still do not think it is possible to separate allelopathy from resource competition. In a community of mixed species of various patch sizes, both competition and allelopathy may be occurring, perhaps separated by time and space, producing a mosaic of interactions that cannot be easily sorted apart.

B. Separating Microbial-Origin from Plant-Origin Phytotoxins

Kaminsky (1981) raised an important question about what constitutes allelopathy. In the chaparral environment of California, the allelopathic activity of the shrub *Adenostoma fasciculatum* was attributed to toxins produced not by the associated soil microbes but by the plant itself. However, the close association of the microbial flora

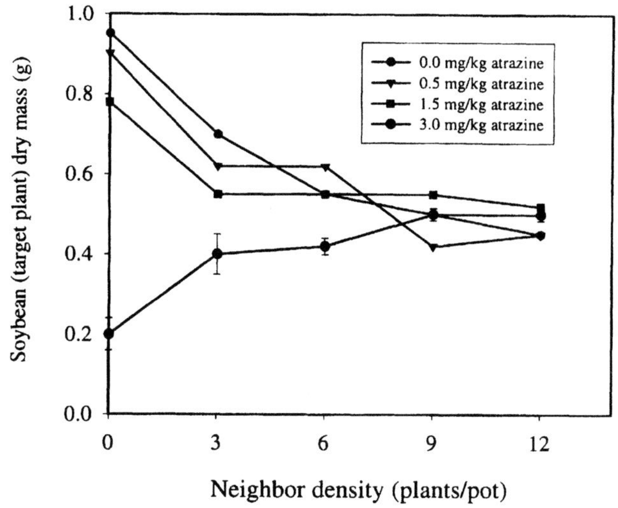

Figure 8.10. The effect of a toxin (atrazine) on the performance of target plants (soybean). At low atrazine concentrations, plant performance decreases as population density increases, but at high toxin concentrations, plant performance increases with increasing plant density. The negative slope of the nonallelopathic situation indicates competitive effects (intraspecific competition), and the positive slope at high toxin concentrations indicates allelopathy. At lower plant densities, individual plants take up more chemicals (greater inhibition), and at higher densities, the chemical effect is diluted and competitive inhibition dominates. This figure supports the dilution hypothesis and may be one way to separate the effects of allelopathy and competition. (Redrawn from Thijs et al. 1994.)

with the shrub suggests at least a plant-mediated process (e.g., due to the chemical composition of the litter), which could still meet the definition of allelopathy. But whether the active compound must be directly of plant origin, or can be produced by microbes in the presence of nonallelopathic plant substrate, remains unresolved. It is clear, however, that the action of most allelochemicals is probably indirect (Willis 1985). Without the intervention of soil, allelopathy would impact the natural community in a completely different manner.

C. Ecologically Relevant Methodology

It is now widely acknowledged that before assigning allelopathy to an observed effect, there is a need to partition out the possible effect not only of competition but also of herbivory, disease (pathogenicity), and nonresource-related abiotic stresses such as osmotic potential, aerobic condition, moisture, and temperature. Moreover, observed suppression in the field may shift from one factor (water stress) to another (allelopathy) at different times of the growing season (McPherson and Thompson 1972), raising the further need to isolate allelopathy within temporal and spatial dimensions. Weidenhamer (1996) points to a need for more research on distinguishing allelopathy from resource competition and other interference mechanisms, and on developing new analytical techniques to measure the flux rates of allelochemicals in soil. The following are some key steps required to demonstrate allelopathy:

1. Identify the allelochemicals involved in a system.
2. Determine the concentration flux of allelochemicals instead of using one-time measurements (e.g., using resin bags).

3. Extract suspected phytotoxins using an aqueous and polar solvent.
4. When making leachates, use plant tissue appropriate to its growth cycle in amounts and for a duration that mimics precipitation or other leaching events in nature (May and Ash 1990).
5. Determine the origin of chemical suppression, i.e., a chemical produced by the plant, a decomposition product of soil organisms, or a toxin that indirectly interferes with microbial activities such as mycorrhizal formation.
6. Demonstrate inhibitory effects on native target plants under conditions that preserve as much of the natural biotic and abiotic elements as possible. This calls for particular caution in making inferences from bioassay.
7. Take caution when measuring the treatment response proportional to a control. Using the ratio of treatment over control as an index of activity can artificially increase the effect of stimulation. The distribution of such ratio may not conform to statistical assumptions of normality and homoscedasticity (Williamson and Richardson 1988).
8. Consider plant density and its possible dilution effect on the uptake of allelochemicals; and consider an interaction between toxicity and density-dependent resource competition (Figure 8.11).

VIII. SUMMARY

Unlike defensive compounds against herbivory, the route of a phytotoxic compound from a source plant to a target plant involves many intervening processes. The likelihood of degradation of allelochemicals by microbes or abiotic (UV light, oxidation) factors is high, but low concentrations do not necessarily mean a stimulatory response or an absence of response. The additive effect of multiple phytotoxins from possibly multiple sources on target plants is difficult to assess in the field but should not be

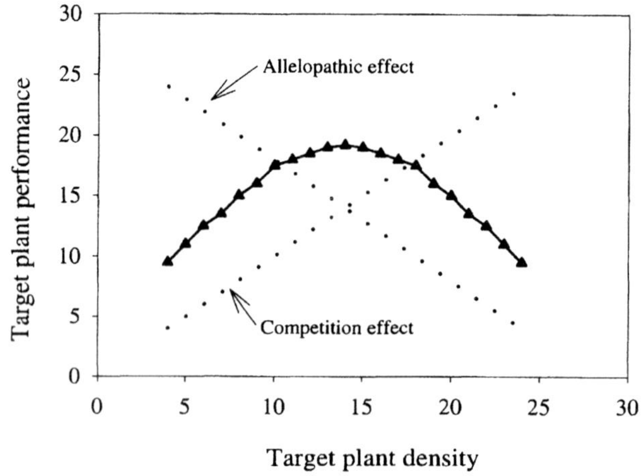

Figure 8.11. Change in the influences of allelopathy and competition in plant assemblages along a gradient of increasing target plant density (allelotoxin concentration is constant in the environment). Target plant performance is a result of the interactive effects of competition and allelopathy.

discounted. In assessing the benefit of allelochemicals to a source plant, we should not be limited to only the negative impact on the neighboring target plants. In a resource-rich environment, allelochemicals could confer greater fitness to a source plant by preventing pathogenic fungal infection (i.e., the target is fungi) than by suppressing neighboring plants (i.e., the target is competitors) (Langenheim 1994). The challenge in allelopathy research may well be in defining the participatory role of allelochemicals in interactions between neighbors, within a community, and within an ecosystem.

STUDY-REVIEW OUTLINE

Presence of Allelopathy

1. When resources are finite, plants that exclude their neighbors have a greater chance of survival.
2. Plants possess chemical means to interfere with the success of potential neighbors.
3. A term first coined by Molisch, *allelopathy* is a chemical interaction between plants that results in either an inhibitory or stimulatory response.
4. Soil microorganisms are critical to allelopathy, but they serve as intermediaries in the chemical interaction between plant species.
5. *Chemical interference* is due to one plant releasing *secondary chemicals* into the environment that subsequently inhibit seed germination or growth of seedlings.
6. Allelochemicals are discharged as volatile substances (in arid and semiarid environments) or liquid leachates (in humid environments).
7. The quantity of toxic substances reaching target organisms is dependent upon soil chemical processes, precipitation, and wind.
8. The variation in toxic substance quality and quantity does not result in strong selection for resistance in target species.
9. At low concentrations, allelopathic substances can stimulate germination and growth of target species, but at high concentrations, the allelopathic substances are inhibitory.
10. Chemical interference also is evident in aquatic systems, where similar constraints and processes in allelopathy are involved.
11. Experiments aimed at verifying allelopathy are usually suggestive but not definitive.
12. All other possible mechanisms of interference must be ruled out before allelopathy can be viewed as a significant mechanism regulating the interaction between species.
13. Allelopathy can influence ecosystem processes, although this has not been demonstrated unequivocally.

The Case in Support of Allelopathy

1. The presence of a bare zone or a zone of species repression around a source plant indicates chemical interference if no other apparent cause can be identified.
2. The comprehensive study of shrub inhibition of grasses in the California chaparral established the ecological imperative to validate allelopathy in agricultural and natural systems for over 30 years.
3. Zones of repression are also induced by black walnut (*Juglans* sp.), and *Eucalyptus* sp.
4. The pathway for chemical interference can be direct, or secondary through mycorrhizae or bacteria.
5. The presence of secondary chemicals in atmosphere and soil is indisputable, and chemicals are involved in plant-plant interactions.

6. A key question is whether a substance released by a source species is the factor inhibiting the target species.

The Case Against Allelopathy

1. Although many chemicals involved in defense against herbivores are the same as allelochemicals, the allelochemicals must traverse an environmental path fraught with complicated transformations before reaching the target organisms.
2. It is virtually impossible to attribute any plant-plant interference to only one cause.
3. Multiple criticisms have been levied at the experimental designs used in studies of allelopathy.
4. The method of choice for developing a case of allelopathy has become to systematically eliminate all other possible mechanisms of interference except allelopathy.
5. Several of the difficulties with verifying allelopathy are:
   a. A realistic concentration of toxin must be used.
   b. Soil physicochemical properties (including microbes) must be evaluated.
   c. Field experiments must be performed.
   d. Autotoxicity to the source plant must be determined.
   e. Plant community structure must be studied.
6. Carbohydrates (a microbial energy source) applied to soils induce the growth of microbes and sequestering of resources that reduces plant growth. The resultant plant response resembles allelopathy.
7. Chemicals released by a source plant also could inhibit neighbors by stimulating microbial populations.
8. Even the best-made cases for allelopathy (as with juglone or terpenes) are questioned due to complications of chemical transfer in the environment.
9. Isolation of compounds from ground tissues cannot represent toxicity in the field; therefore, such experiments have no validity.
10. Bioassays of compound toxicity can only suggest that a compound could be toxic; they do not indicate that toxicity occurs in the field.
11. It is highly possible that chemicals can have activity in a group since soil chemistry is complicated; it is unlikely that one substance can account for toxicity by itself in a natural system.

The Special Case of Autotoxicity

1. It is possible that a toxic substance released by a source species into the environment will be toxic for the seedlings of the source species (*autotoxicity*).
2. In most experiments, testing for toxicity to the source plant is not considered.
3. Autotoxicity is very prevalent in studies that do examine the possibility.
4. Autotoxicity is more evident in waterlogged situations where the toxic substance cannot be degraded by aerobic microorganisms in the soil solution.
5. Autotoxicity of caffeine to coffee plants occurs, but this cost is outweighed by the benefit to the reproduction of coffee plants.
6. Autotoxicity may be a cost of producing secondary chemicals for defense against herbivores.
7. In old fields (agricultural fields that have been abandoned), autotoxicity is stronger than *allotoxicity* (toxicity to other species of plants).
8. Autotoxicity also may serve to stabilize the dormancy of seeds from the source plant in the seed bank.
9. Autotoxicity demonstrated in the lab may not occur in the field. For example, plants with deep root systems will not experience autotoxicity because toxins have a low mobility in soil.

Allelopathy Versus Competition

1. When suppression of one species by another is observed in the field, it is difficult to separate competitive interactions from allelopathic interactions.

2. A further complication is the fact that allelochemicals can reduce the nutrient accumulation ability of target plants. The result of this interaction resembles competition.
3. Under nutrient limitations (conditions when competition may be important), phytotoxin production can increase, which enhances the possibility of allelopathy.
4. Many studies of allelopathy go to great lengths to eliminate the possibility of competition, but few studies on competition consider allelopathy.
5. The sensitivity of plants to allelochemicals may vary based on competition.
6. Allelochemicals may influence microbial populations that regulate resource availability, and resource availability may regulate competition.

Ecological Significance of Allelopathy

Community Structure and Succession

1. In communities dominated by one or a few species, allelochemicals may be a factor regulating that dominance.
2. Catastrophic events such as fire can return the ecosystem (described in number 1) to a more diverse state.
3. In a diverse community, it is not likely to find strong allelopathic effects because the source plants are at low density.
4. In low diversity communities, allelopathy is easier to identify because one or a few species dominate the soil chemistry.
5. Phytochemical interactions may be involved in the replacement of early successional species in some ecosystems.
6. Several studies have found no relationship between toxicity bioassays and successional change.
7. Wardle argues that allelopathic interactions may have more significance at the ecosystem scale than at the population or individual scale.
8. Phytotoxins in the litter may regulate litter decomposition processes and thereby influence nutrient cycling in ecosystems. Therefore, the phytotoxins released from one source species may have a proportionally larger effect on nutrient cycling than will litter from other species.
9. The sandhills of Florida may be prevented from being fire-prone by the allelopathic inhibition of fire-prone shrubs.

Evolution of Allelopathy

1. Phytotoxins that induce allelopathic interactions may have evolved as antiherbivore agents, and secondarily cause allelopathy.
2. Phytotoxins may have evolved as a mechanism of enhancing resource acquisition. The antiherbivory properties redirect herbivores to other plants, and the allelopathic properties limit the number of neighbors that might preempt the resources.
3. Allelopathic stimulation of target plant germination and growth when phytotoxins are at low concentration may be a cost to source plants. Stimulation by phytotoxins has rarely been shown in field studies.
4. A high diversity of potentially toxic secondary chemicals is a more successful adaptation against herbivores, pathogens, and neighboring plants than a high concentration of one phytotoxin.
5. The diversity of secondary compounds provides a greater chance that one of the substances will have allelopathic or antiherbivory action.

Classes of Allelochemicals

Synthesis and Biochemical Pathways

1. Many of the secondary chemicals called *allelochemicals* (not participating in growth metabolism) do influence some important traits such as flower color (flavonols) and pollinator attractants (terpenes).

2. Several secondary metabolites are involved in stress responses, particularly water and salinity stress. Moreover, some of the secondary chemicals serve in signal transduction pathways.
3. Many secondary compounds are synthesized by small modifications of molecules that are directly involved in important metabolic pathways such as amino acid synthesis.
4. Secondary compounds are diverse in all species, and they are produced by all organs of the plant.

Nitrogen-Based Compounds

1. Alkaloids and nonprotein amino acids are nitrogen-based compounds that can have allelopathic activity.
2. Alkaloids are an extremely large group of compounds, including a few that have allelopathic potential. In contrast, many have no allelopathic capability because they are rapidly degraded in the soil.
3. Nonprotein amino acids are commonly found in legume seeds.
4. Leachates of legume seeds that contains canavanine inhibit the seed germination of lettuce.
5. Other nonprotein amino acids exuded by roots of lilies can inhibit seed germination of test species.
6. Some cyanogenic glycosides and glucosinolates have been reported to inhibit other plant species.
7. Dhurrin, from sorghum, has been shown to be allelopathic.

Carbon-Based Compounds

Phenolics

1. Phenolics are abundant in plants and can be released in large quantities by rainfall.
2. Several of the most well established cases of allelopathy involve phenolic toxins.
3. Phenolics are very water soluble, and they range in complexity from simple compounds such as cinnamic acid to complex polymeric tannins.
4. Many phenolics from roots and leaves are found in soils, and they can inhibit seed germination and seedling growth.
5. Juglone is a quinone that has limited mobility in the soil solution. Juglone is exuded from roots and leached from leaves into the soil solution
6. Stilbene, commonly thought of as a phytoalexin, can be allelopathic. For example, stilbene from the leaves of *Empetrum* can inhibit tree seedling germination.
7. Some phenolics are much more toxic than others. Bioassay results indicate the following order of toxicity (highest to lowest) for these acids: *o*-hydroxyphenylacetic, *p*-hydroxybenzoic, vanillic, *p*-coumaric, gentisic, syringic, ferulic, and *m*-coumaric.
8. Phenolic acids are toxic to the seeds and seedlings of the plants that produce them (autotoxicity). The strength of autotoxicity varies among phenolics in the same manner as does their toxicity to target species.

Terpenoids

1. *Monoterpenes* are aromatic 10-carbon compounds. Some are very soluble in water, while others are hydrophobic.
2. Some monoterpenes are a major component of essential oils secreted by glandular hairs.
3. *Sesquiterpenes* are 15-carbon compounds that, when in ring form, are called sequiterpene lactones.
4. Phenolics and terpenoids predominate in allelopathy studies.
5. Some terpenoids are allelopathic because of their action on microorganisms and the subsequent action on plant function.

Activity of Allelochemicals

Location of Allelochemicals in Plants

1. Most plants have some secondary chemicals with potential allelopathic activity in their tissues.
2. Specific species are able to produce particularly toxic substances at particularly high concentration.
3. Lipophilic compounds such as terpenes are located in secretory structures such as glandular trichomes.
4. Some plant tissues are more phytotoxic in the reproductive phase rather than the vegetative phase.
5. Young plants may produce proportionally more toxin than mature plants.
6. The outer layers of seed often contain phytotoxins that may serve as a mechanism of maintaining dormancy.
7. Leaves of many plant families (particularly Lauraceae, Asteraceae, and Malvaceae) have glandular hairs on their leaves that contain high concentrations of essential oils.
8. Many of these leaves may not release these phytotoxins into water passing over the leaves. Rather, the toxins are released when fungi and insects decompose the fallen leaf litter.
9. Tannins released during leaf decomposition can be oxidized releasing toxic gallic acid.
10. In many cases, litter contains greater toxicity than fresh material, suggesting that decomposition converts relatively benign compounds into toxic substances.
11. In some cases, such as *Acacia,* leachates of green leaves are far more toxic than leachates of recently fallen leaves.
12. Bark can also release phytotoxins into stemflow down to the base of trees.
13. Toxins in stemflow are restricted to the region around the trunk of trees; however, the toxins can penetrate relatively deeply into the soil.

Mode of Release

1. The release of allelochemicals can occur continuously or in short flushes or pulses.
2. Most allelopathic interactions take place in the soil via an aqueous solution.
3. The nature of the soil can influence the efficacy of the allelochemical against a target species. This is a critical issue to consider when experimenting with allelopathy.
4. The toxicity of bracken fern (*Pteridium aquilinum*) is predominantly related to decomposing fronds, and toxicity is focused in a spring pulse at the start of the rainy season in Mediterranean climates.
5. Rainfall passing through the canopy (throughfall) leaches allelochemicals off of or out of leaves.
6. The amount of secondary substance that is found in throughfall depends upon the substance's solubility in water, but the toxicity of a compound is not related to its solubility in water.
7. Physical disruption of tissues may be necessary to release some allelochemicals.
8. Allelochemicals are freed from the cell wall when saprophytic microbes rupture the cells.
9. Allelochemicals released during decomposition of leaf tissue are different from those released by extracts of macerated tissue.
10. A complication in bioassay studies is the osmotic potential of the extract, particularly when macerated tissues are used.
11. Oxidation reactions can convert an inert substance into a toxic substance.
12. Juglone released by walnut trees is nontoxic until it is oxidized in the soil.
13. Extensive oxidation can render a toxic substance benign.
14. Some allelochemicals are released as volatile substances. These volatile substances are more effective in dry soil regions such as the Mediterranean and semidesert.
15. The volatile substances can be concentrated in lipophilic areas such as seed coats.

16. Lipophylic compounds can also be dispersed in water if they are associated with a surfactant such as ursolic acid.
17. Some roots exude allelochemicals such as cinnamic acid.
18. Sorgoleone is released in high quantities from sorghum roots.
19. *Ailanthus* root bark contains a strong phytotoxin identified as ailanthone, a water-soluble toxin.
20. Microbial decomposition of allelochemicals can cause heightened toxicity or neutralization.
21. Most often, microbial populations decrease the toxicity of allelochemicals fairly rapidly (in a few days or a week).
22. Soil particles and soil organic matter can complex with allelochemicals, taking them out of solution.

Active Concentrations

1. At low concentration, allelochemicals can stimulate growth and germination.
2. At intermediate concentrations, there may be no effect of the allelochemicals on target plant growth.
3. At high concentrations, allelochemicals are always inhibitory.
4. Phytotoxins degrade rapidly in the soil, so it is hard to determine the precise concentration of allelochemicals at plant surfaces.
5. The source must be continually supplying the phytotoxin to the soil to keep the concentration of allelochemical within the inhibitory range.
6. Physiological concentration of the phytotoxin at the active site depends upon (1) the rate of supply from the source, (2) the rate of microbial degradation, (3) the rate of diffusion, (4) the rate of accumulation by the plant, (5) the rate of translocation to the site of action in the plant, and (6) the rate of chemical modification in the plant tissues.
7. There is no hard-and-fast rule about what concentration of a phytotoxin will be stimulatory or inhibitory.
8. The concentration-dependent response of target plants to allelochemicals varies among different compounds and among species.
9. Threshold concentrations for inhibition can vary greatly based on environmental conditions.
10. Ferulic acid inhibits sorghum growth at 0.4 m*M* at 29°C, but at 37°C, the same inhibition requires a concentration of only 0.2 m*M*.
11. There are two points of neutral effect of chemicals on plants in a concentration response curve: extremely low concentration, and the switch-over point between stimulation and inhibition.

Bioassay and Germination Tests

1. Bioassay has been the mainstay of research on allelopathy.
2. Most bioassay experiments are based on germination and root elongation of test species.
3. In some studies, a floating aquatic angiosperm, *Lemna minor,* is used as a bioassay organism.
4. There are severe problems inherent in using test species in bioassay:
   a. The test species may not behave in a similar way to native species of the region.
   b. If more than one species is used in the test, standard testing procedures may cause different results for the species that have nothing to do with the allelochemical but are related to the basic phenology of the species.
   c. Techniques used to calculate results from a bioassay can produce different results with the same data.
5. The are also advantages of using test species in bioassay:
   a. The test species has a very dependable germination and growth. Any small deviation from the norm can be measured.
   b. Germination and survival of test species is 100% in control solutions.
   c. There is much less variation among individual responses within a test species seed lot.

d. Test species do not require special treatment to break dormancy.
6. Test species are useful for determining whether there is a potential for allelopathic influences in field situations.
7. The best index for germination bioassay tests is the "speed of germination" index.
8. Any bioassay must be carefully coordinated with the concentration expected in the field, but bioassays cannot address the rate of allelochemical delivery.
9. Field delivery rate (influenced by source plant, microbes, and diffusion) could be assessed using ion-exchange resins similar to those used in mineral nutrition studies.
10. Most measurements of allelochemicals in the field fall below the inhibition threshold (as determined in bioassay), even though target plants are inhibited in the field.
11. Inhibition of target plants in the field may be the result of synergistic effects among allelochemicals; thus, a measurement of one compound does not indicate allelopathy should be occurring.
12. Inhibition in the field may be due to a number of different allelochemicals possibly from several different source plants.
13. While some studies have shown additive effects of allelochemicals, the knowledge base on this topic is very small.

Modes of Action

1. The activity of most allelochemicals is broad spectrum, which means they are generally inhibitory to many other target species.
2. Allelochemicals also have activity against herbivores and pathogens.
3. The site of action of allelochemicals is quite variable.
4. Processes inhibited by allelochemicals include the following:
   a. Mitosis (microtubule organization)
   b. Photosynthesis (particularly, photosystem II)
   c. Respiration (particularly oxidative phosphorylation)
   d. Electron-transport chains
   e. General enzymatic reactions (coumarins)
   f. The action of plant hormones (particularly IAA)
5. Quinones such as juglone and sorgoleone inhibit energy-producing and -using systems.
6. Phenolic acids have a wide variety of actions, including inhibiting mitochondria, stimulating hypocotyl elongation, and flowering.
7. Phenolics also are thought to act through inhibiting or stimulating IAA oxidase.
8. Terpenoids have a wide diversity of actions, including inhibiting gibberellin activity, mitosis activity, and the proper function of auxin receptors.
9. The physiological activity of germination and radicle elongation is different enough to expect different responses of these processes to allelopathic substances. This is the case.
10. When root growth is inhibited, one would expect a decrease in seedling nutrient and water accumulation.
11. There is no consistent pattern as to which character—germination or radicle elongation—is more sensitive to allelopathy.
12. Allelopathy also occurs between pollen grains on a stigma.
13. Pollen allelopathy can determine seed set, and therefore can have significant impact on natural selection in plants.
14. The significance of pollen inhibition to species fitness has yet to be tested in the field.

Detoxification of Allelochemicals

1. In order for plants to have uninhibited fitness in an environment with a relatively large allelochemical complement in the soil, plants must either avoid the chemical, tolerate the inhibition of the chemical, or neutralize the toxicity.
2. Avoidance could be accomplished by a dormancy mechanism that regulates germination to occur at a time when allelochemicals are at their lowest.

3. Mechanisms of avoiding allelopathy upon germination are to germinate only after large disturbances (fires) or after extensive leaching (several consecutive rainfalls).
4. Avoidance by young seedlings could be attained by having deep rooting (a prominent taproot.
5. Tolerance of toxins varies greatly among species.
6. Tolerance of allelochemicals could be a result of the following:
   a. Excluding toxins at the root surface
   b. Compartmentalizing toxins in the apoplast
   c. Localizing toxins in glandular trichomes (excreting toxins)
   d. Nonsensitive cytoplasm
   e. The ability to detoxify the compound
7. Some toxins are translocated (at times by specific transport proteins) into the vacuole.
8. Storage in the vacuole is only a short-term tolerance mechanism.
9. Some toxins can be incorporated into cell wall constituents such as lignin.
10. Plants that grow rapidly or are naturally phytotoxic themselves have the greatest ability to compartmentalize and or excrete toxins.
11. Herbicide resistance is well established to occur in some plants; thus, plants have the ability to detoxify naturally occurring phytochemicals.
12. Three phases of plant defense against allelochemicals have been proposed:
    a. Metabolic attack
    b. Conjugation
    c. Translocation, storage, and excretion
13. Metabolic attack usually involves making the phytochemical more polar by hydroxylation or dealkylation.
14. Altered phytochemicals are usually conjugated to sugars, amino acids, or malonic acid.
15. Cytochrome P-450 oxidases are important in the metabolic attack of allelochemicals.
16. There have been very few studies of allelochemical detoxification because allelochemicals are thought to be so diverse and diffuse that there is little selection for defensive mechanisms.
17. The science of herbicides and plant detoxification of the herbicides may be very applicable to that of plant defense against allelochemicals.

Environmental Stress and Allelopathy

1. The susceptibility of plants to allelochemicals can increase under stress conditions.
2. Competition and allelopathy occur simultaneously in the field. In fact, allelopathy and competition are inextricably connected.
3. If environmental stress reduces carbon gain of a source plant, it is logical that the production of secondary chemicals would be reduced.
4. In contrast, some stress situations (herbivory) induce an increase in the synthesis of allelochemicals by source plants.

Abiotic Stresses

1. Deserts, because of their low water, low nutrients, and high heat, are likely to have more allelopathy than in humid regions.
2. Toxins in the soil may be higher in semiarid than in humid regions because of the volatile release of terpenes from plants and their absorption in soils.
3. Low microbial activity and infrequent leaching could cause relatively high concentrations of allelochemicals in semiarid regions (particularly Mediterranean regions where shrub biomass can be high.
4. Drought stress can increase the concentration of allelochemicals in volatile emissions from plants.
5. High light (particularly UV) can induce higher concentrations of toxins in source plants; on the other hand, high UV light increases the rate of degradation of toxins after they are leached or volatized from the plant tissues.

6. In general, nutrient limitation increases the release of allelochemicals from source plants.
7. Increased decomposition (microbial activity) is a two-edged sword for allelopathy. On one hand, allelochemicals will be degraded faster, but on the other hand, increased decomposition of litter can increase the delivery rate of toxins.
8. The combination of greater allelochemical production by source plants and greater sensitivity by target plants under nutrient limitations creates a much greater potential for allelopathy than under high nutrient conditions.
9. Under high soil moisture and relatively warm temperature, target plants have a greater tolerance of allelochemicals.

Biotic Stresses

1. Biotic activity in the soil (microbes, mycorrhizae, nodules, parasitic plants, etc.) is critical and central to allelopathy.
2. After Chinese fir has grown to maturity in a site, the resultant soil chemistry inhibits the growth of nonpathogenic fungi and promotes the growth of pathogenic fungi.
3. Allelochemical inhibition of microbe activity will decrease nutrient availability and cause the inhibition of plant growth.
4. Phenolic acids can both inhibit the activity of mycorrhizal fungi at high concentration and stimulate the fungi at low concentration.
5. There is wide diversity in the tolerance of allelochemicals among mycorrhizal species, and it is highly possible that allelochemicals can regulate the colonization and diversity of mycorrhizal fungi on target plants.
6. Allelopathic compounds can have multiple effects. For example, leachates from *Kalmia angustifolia* inhibit the mycorrhizal synthesis, nutrient uptake, and pathogen susceptibility of black spruce seedlings.
7. Source species in Ericaceae may enhance the activity of their leached allelochemicals by acidifying the soil.
8. Allelochemicals can inhibit the proper development of nitrogen-fixing nodules on legumes and other dinitrogen-fixing plants.
9. The nodulation of several legumes is inhibited by quackgrass, and a root exudate from quackgrass is suggested to be the cause.
10. Phenolic acids leached from the aboveground parts of a source plant can inhibit nodulation in *Alnus*.
11. Several important interactions between competition and allelopathy should be considered when studying plant-plant interactions.

Applications of Allelopathy in Agriculture and Forestry

1. The possibility that natural chemicals produced by plants could act as herbicides for shallow-rooted weeds has not been lost on the agricultural community.
2. Compounds such as caffeine or hydroquinone could inhibit surface-rooted seedlings and weeds, while the deep-rooted source plant could avoid autotoxicity.
3. A number of different crop species are capable of inhibiting weed growth.
4. Agricultural systems utilize rotation and cultivation methods to minimize the negative interference between successive crops. For example, a number of forage grasses inhibit the performance of alfalfa.
5. Intercropping is an agricultural technique where more than one crop species is grown in the same field, usually in alternate rows.
6. Allelochemicals produced by one crop can inhibit weed competition for the other crop in an agricultural field.
7. Intercropping with squash (which produces allelochemicals) and corn (which is resistant to those allelochemicals) was found to be an effective way of reducing weeds.
8. The effectiveness of weed suppression by intercropping is dependent upon the climatic conditions and the particular species used.

9. The combination of allelopathy (early in the growing season) and competition (later in the growing season) is the most effective natural method of weed prevention known.
10. It is very likely that microbial modification of secondary chemicals from the source plants is important in the success of intercropping as a weed inhibition technique.
11. One of the most promising natural herbicides is ailanthone from *Ailanthus altissima*. The efficacy of this compound rivals that of commercial herbicides such as Paraquat.
12. Many of the most invasive weeds of the world may owe their success in part to their production of allelochemicals.
13. It is possible that in agricultural systems, allelopathy will be magnified compared with natural systems because (1) the diversity is very low, (2) the species combinations have not evolved together. Both of these minimize the potential for evolution of defensive traits.
14. Understanding the microbial component of a potentially allelopathic system is an important key for the ability to manipulate the efficacy of the natural allelochemical inhibition of competing weeds.

Current Status and Future Directions

Can Allelopathy Be Separated from Competition?

1. Competition may be inextricably linked with allelopathy in natural systems, which makes it very difficult to measure the effects of allelopathy.
2. The *dilution protocol* is one possible mechanism for separating the effects of allelopathy and competition.
3. When allelopathy is the main cause of an interaction, an increase in target plant density results in improved growth of target plants.
4. When competition is the main cause of inhibition, an increase in target plant density causes a decrease in performance of target plants.
5. A replacement series design may also give some indication of the relative effects of allelopathy and competition.
6. In natural systems, it is impossible to separate allelopathy and competition because they co-occur in a mosaic of time and space.

Separating Microbial-Origin from Plant-Origin Phytotoxins?

1. There is a close association between the materials produced by plants and the actions of microbes.
2. *Adenostoma fasciculatum* (chamise) produces litter that is not allelopathic. However, the products of microbial decomposition of these materials are allelopathic.
3. The vast majority of allelopathic effects are indirect (mediated through soil microbes).
4. The source plant may simply provide the organic material with the appropriate chemistry for microbial conversion into allelochemicals.
5. Without an intervention by soil microflora, allelopathy would be an entirely different process.

Ecologically Relevant Methodology

1. Natural ecosystems are extremely complex systems, within which a wide array of chemicals are produced and transformed.
2. The ability to isolate any of these individual processes is a difficult task, and allelopathy may be the most difficult to study independently of other ecosystem processes.
3. Allelopathy must be isolated from competition and herbivory.
4. Allelopathy needs to be studied in a spatially and temporally explicit manner.
5. Dose and flux characteristics of allelochemicals are critical.
6. Allelopathy must be studied in a multidisciplinary manner in order to include the substance chemistry, soil physics, soil microbiology, plant physiology, and ecology.
7. Steps required to demonstrate allelopathy:

a. Extract substances with an aqueous solution.
b. Identify the allelochemicals.
c. Demonstrate the field concentration and flux dynamics of the substances.
d. Make leachates in a manner that mimics how they are made in the natural system.
e. Determine the origin of the substance (plant, litter, microorganism).
f. Demonstrate the inhibitory effects on the native target species individually.
g. Take caution with the statistics used in bioassay, particularly in relation to the control treatment.
h. Consider the density of target plants in an effort to evaluate the dilution effect.

SELF-STUDY QUESTIONS

1. What is the critical difference between the meaning of competition and the meaning of allelopathy?
2. Why is the action of secondary chemicals in herbivory different from the action of the same substances in allelopathy?
3. What are the main reasons why allelopathy is a controversial topic in ecology and physiology?
4. Is allelopathy restricted to terrestrial environments? If not, how is the action of allelopathy in aquatic environments different than in terrestrial habitats?
5. What are the two main types of secondary chemicals that are involved with most cases of allelopathy?
6. Explain why there is a difference in the types of secondary chemicals associated with allelopathy in semiarid and humid environments?
7. The intensity of allelopathy increases in habitats with low species diversity. Why is this so?
8. What is a "bare zone," and why is it important to the history and conceptual development of the theory of allelopathy?
9. What is the evidence in favor of allelopathy as an important plant-plant interaction in nature?
10. Why is allelopathy more likely to be an important process in agricultural systems than in natural systems?
11. What is juglone, and how does this compound cause allelopathic responses in target plants?
12. If a chemical substance is isolated from plants and inhibits seed germination and radicle elongation in bioassay, why cannot we say that this substance has allelopathic activity?
13. In what three ways can microbes be involved with the process of allelopathy?
14. What factors regulate the dose of allelochemicals to a target plant (in a soil environment)?
15. How can plants protect themselves from autotoxicity by the allelochemicals they produce?
16. Is autotoxicity rare or frequent? What is the evidence for this?
17. Can allelopathy be stimulatory, and if so, what factors regulate the difference between a stimulation and an inhibition?
18. Is allelopathy important for succession? If so, in what ecosystems is it important?
19. What is often the relationship between fire and allelopathy in the California chaparral?
20. What is the controversy about the interaction between allelopathy and dominance/diversity relationships?
21. Why is allelopathy thought of as a fortuitous benefit to the defense against herbivores?
22. What is the rationale for suggesting that allelopathy evolved in concert with competition?
23. What are two reasons why plants have a diversity of secondary chemicals with potential allelopathic activity?
24. Describe what a phenolic acid is. What is the main biosynthetic pathway for phenolic acids?
25. What are two nitrogen-based substances (secondary chemicals) that have some allelopathic potential?
26. Why are there only a few cases of nitrogen-based chemicals identified as putative allelochemicals?

27. What are some of the functions of phenolic acids in plant metabolism?
28. What are coumarins, where are they found in plants, and do they have allelopathic activity?
29. What are flavonoids, what is their main biosynthesis pathway, and are they involved in allelopathic processes?
30. How do phenolics—such as phenolic acids, coumarins, and flavonoids—enter into the soil solution?
31. What are terpenoids, and how are they synthesized in the plant?
32. How do sesquiterpenes leave plants and enter into allelopathic processes?
33. What is the relationship between saponins and terpenes? Do saponins have any allelopathic activity?
34. In what tissues are allelochemicals located?
35. Which allelochemicals are most likely to be found in fruits?
36. Why is it not advisable to use macerated tissue when extracting leaves for a bioassay?
37. What happens to the allelopathic activity of hydroxygrindilic acid when it is oxidized?
38. What is the significance of ursolic acid to the release of lipophilic terpenes?
39. Why is it that scientists cannot clearly identify the concentration at which allelochemicals are toxic in the field?
40. What is the most common type of bioassay used in allelopathy studies? What factors must one be aware of when doing bioassays in allelopathy studies?
41. What are the primary modes of action for allelochemicals? In particular, what is the mode of action for terpenes? Quinones? Phenolic acids?
42. Which is more sensitive to allelopathic inhibition, seed germination or radicle elongation?
43. What is the general three-step pathway used by plants to detoxify allelochemicals?
44. Are all species equally tolerant of particular allelochemicals?
45. What are the possible mechanisms for tolerance to allelochemicals in plants?
46. In what way has the study of herbicide resistance influenced our understanding of defense against allelochemicals?
47. How does water stress influence the quantity of allelochemicals released by plants? How does water stress affect the sensitivity of target plants?
48. What is the effect of nutrient limitation on source plant allelochemical release? How does nutrient limitation affect target plant sensitivity to allelochemicals?
49. How do phenolic acids in the soil influence mycorrhizal associations in target plants?
50. What are your thoughts about the significance of allelopathy to the ecology and physiology of plants in natural or agricultural systems?

SUPPLEMENTARY READINGS

Chou, C. H., and G. R. Waller (eds.). 1989. Phytochemical Ecology: Allelochemicals, Mycotoxins and Insect Pheromones and Allomones. Institute of Botany, Academia Sinica, Taipei.

Conn, E. E. (ed.) 1981. Secondary Plant Products (The Biochemistry of Plants, Vol. 7). Academic Press, New York.

Harborne, J. B. 1988. Introduction to Ecological Biochemistry, 3rd Ed. Academic Press, London.

Harper, J. L. 1977. Population Biology of Plants. Academic Press, London.

Inderjit, K. M. M. Dakshini, and F. A. Einhellig (eds.). 1995. Allelopathy: Organisms, Processes and Applications. American Chemical Society, Washington, DC.

Inderjit, K. M. M. Dakshini, and C. L. Foy (eds.). 1999. Principles and Practices in Plant Ecology: Allelochemical Interactions. CRC Press, Boca Raton, FL.

Larcher, W. 1995. Physiological Plant Ecology, 3rd Ed. Springer-Verlag, Berlin.

Molisch, H. 1938. The Longevity of Plants. Transl. E. H. Fulling, New York.

Putnam, A. R., and C.-S. Tang. 1986. The Science of Allelopathy. John Wiley & Sons, New York.

Rice, E. L. 1974. Allelopathy. Academic Press, New York.

Rice, E. L. 1984. Allelopathy, 2nd Ed. Academic Press, New York.

Waller, G. R. (ed.). 1987. Allelochemicals: Role in Agriculture and Forestry (ACS Symposium Series, No. 330). American Chemical Society, Washington, DC.

9 Weeds and Other Competitors

Outline

Objectives

1. Understand how plants compete for limiting resources.
2. Describe the ecological importance of competition.
3. Explain the mechanisms by which plants recognize the presence of neighbors.
4. Discuss physiological traits that influence competition above and below ground.
5. Evaluate what traits make plants strong or weak competitiors based on physiological traits and ecological context.
6. Explain some generalized mathematical relationships concerning competitive responses.
7. Review experimental designs utilized to determine competitive relationships

I. INTRODUCTION

All plants require aboveground resources such as sunlight, air, and space and belowground resources such as nutrients, water, and space in order to grow and reproduce. As plants are sedentary, there is a tendency for neighboring plants to share the same pool of resources needed for growth and other life processes. The fitness of a plant (defined by the number of successful offspring) depends on the effectiveness by which it acquires and allocates limited resources to growth, maintenance, and reproduction in the face of similar demands by other plants. Competition (two plants trying the acquire the same resources) for limiting resources among plants is one of the most important mechanisms controlling the establishment and growth of individuals and the size of populations. Competitive interactions between species influence community structure and succession. Therefore, the response of a particular species to competition in a natural system will influence its relative abundance in the community as well as that of other species in the community. Natural selection should favor genotypes that have physiological, morphological, or behavioral traits that maximize the ability of the plant to acquire resources effectively and efficiently.

Competition is defined as an interaction between two plant species that leads to a negative influence on the fitness of both species (see Figure 1.2). In general, competition occurs when resources are limiting and the pool of available resources is shared and depleted. Crowding of plants results in competition for space, light availability, water, and nutrients. The uptake of a resource by one individual can alter the availability of that resource for other plants. Thus, the use of one or more limited resources by an individual can result in reducing the availability of that resource for the growth, survival, or reproductive success of its neighbor. Competition is a density-dependent process, so that the more individuals utilizing a limited pool of resources, the greater the degree of competition.

If competition causes a decrease in resource availability, then there are at least two general ways that plant species can be effective competitiors. As described above, plants can be effective competitiors by drawing down resource availability to low levels quickly, thereby making resource levels too low for other neighboring species. This mechanism requires a large ability to remove resources rapidly. Another way for plants to be effective competitors is by being able to tolerate very low resource availability. In this way, the plants can survive at resource levels lower than their competitiors. This mechanism requires efficient resource acquisition rather than a high capacity for resource acquisition. Theoretically, individual plant species could utilize both mechanisms. For example, a plant species could rapidly grow depleting resources for its neighbors when young, then slow down growth (acclimation) and tolerate the very low resource availability when mature. However, the combination of these two mechanisms in one plant species is not common. When plants are adapted to fast growth and rapid resource depletion, they are generally intolerant of low resource availability. Moreover, plants adapted to efficient resource use at low resource availability usually cannot attain a rapid growth rate at high resource availability. The two syndromes—rapid growth–resource depletion and slow growth–low resource tolerance—may be virtually mutually exclusive.

For a species to be a successful competitor, it must have certain physiological and ecological traits that are optimized in a particular environment. Some plants are better competitors in early successional communities but do not have competitive traits that would make them successful in later successional communities. In early successional communities, plant traits that facilitate rapid growth and rapid utilization of space and resources are likely to be successful. In later successional communities, plant traits that optimize the efficiency of resource capture and retention are more predominant.

Competitive effects on water and nutrient availability and uptake will, in turn, influence rates of photosynthesis and carbon allocation, often resulting in the reduced growth of an individual. Thus, the occurrence and intensity of competition are often inferred from the observed reductions in the growth or productivity of individuals. The growth of an individual is often reduced with increasing density of neighbors. If the competition intensity is high, the ability of an individual to maintain a certain minimum accumulation of carbon required for growth and respiration may be so reduced that mortality occurs.

Competitiveness is related to the individuals' ability to exploit and acquire the resources needed for growth and reproduction when the resources needed are shared and depleted by neighboring plants. When competing for a limiting resource, a plant has four basic options: to increase resource uptake (alter size, morphology, or physiology of canopy or root systems), to increase the efficiency of resource acquisition (alter the behavior or physiology of acquisition systems), to decrease resource loss (maintain long-lived structures), or to increase the efficiency of conversion of internal stores (Chapin et al. 1987). In this chapter, we will discuss some of the morphological, behavioral, and physiological traits associated with competitive ability. Understanding these traits will allow us to investigate the mathematical models that have been developed to study competition theory.

II. RESOURCES

A. Resource Characteristics

A resource is a substance that is depletable or consumable and required for plant growth and maintenance. Resource level is variable, both spatially and temporally, in all ecosystems. Resources available to plants are a subset of the resources in the environment. This is because some of the total amount of resources is outside the ability of plants to reach or accumulate them. Moreover, the availability of resources for utilization by plants is determined by both abiotic environmental attributes (climate, soils) and the nature of the other organisms in the community. The level of each resource needed for optimal plant function varies among species, among individuals in a population, and among developmental stages of each species. The multidimensional space, defined by the sum of all resource requirements of a species, has traditionally been referred to as the *fundamental niche* (see Chapter 1).

Resources vary in quantity and quality, on both a spatial and temporal scale. Soil nutrients may be patchily distributed throughout the soil, or available in pulses at specific times through increased decomposition and mineralization, such as may occur after a disturbance (Bormann and Likens 1979). However, pulses in nutrient availability may occur more rapidly than the plant can respond (Stark 1994). Nutrient heterogeneity in the soil develops from (1) the effect of plants on microtopography and microclimate; (2) the deposition of organic material from plants, and the development of nutrient depletion zones around roots; (3) the amount, location, timing, and chemical composition of mineral and organic inputs; and (4) the effect of microorganisms on decomposition, mineralization, and soil mixing (Stark 1994). How a plant perceives nutrient heterogeneity is a function of its ability to explore its immediate soil volume through root proliferation, mycorrhizal associated uptake (see Chapter 4), or through increased uptake kinetics when it encounters nutrient-rich patches (Jackson and Caldwell 1989).

Water resources for plants are dominated by soil water, although some plants, (such as *Tillansia landbeckii,* can accumulate most of their water from the atmosphere (Rundel et al. 1997). Soil water is derived from a variety of sources such as precipitation (which may occur daily or be seasonally periodic), overland flooding, phreatic zones around and below streams and lakes, and groundwater. The availability of soil water to plants depends on the quantity of water stored in the soil, the location of the water in relation to plant roots, the hydraulic conductivity of the soil, and the soil water potential. Variation in soil water quantity depends on factors such as variation in the supply from the sources, soil texture and chemistry, amount of soil organic matter, soil wettability, and environmental attributes (temperature). Variation in water quality for plant accumulation depends on factors such as salinity, temperature, and water potential. A more thorough description of water availability is presented in Nilsen and Orcutt 1996 (chaps. 7 and 8). Disturbance of a community (reduction of phytomass) where water availability is limited, may result in an increase in soil water availability due to decreased root biomass. Seasonal declines in soil water availability during drought may result in intense competition for this limited resource. Therefore, the availability of water to plants is highly dynamic on both a spatial and temporal scale due to variation in source delivery rates, substrate quality, and plant water accumulation attributes. The competitive ability of plants in relation to water depends on their ability to locate the water resource and effectively accumulate it on a variable temporal scale.

The quantity and quality of light is also variable on the landscape. Seasonal and diurnal changes in light quantity and quality occur in accordance with the changing position of the sun and cloudiness. Light quantity and quality also change as light passes through a canopy and is absorbed and scattered by plants (Ballaré 1994). The structure of the plant communities changes with canopy disturbance and with the infilling of canopy space, creating a very dynamic light environment during the development of a community. At any given time in community development, as light passes through the canopy, the quantity and quality of the light changes, altering its availability for use by the photosynthetic apparatus of other plants. Below plant canopies, the light environment usually consists of low, diffuse radiation with patches of high light called *sunflecks* (Baldocchi and Collineau 1994). Sunflecks are variable in duration, intensity, and frequency in different locations and in different communities. Plant photoreceptors permit the sensing of neighbors, and they are an important mechanism driving architectural plasticity and morphological responses to changing light environments (Ballaré 1994). The physiological traits that cause high competitive ability in relation to light are different at different stages of community development and different locations in a community. Therefore, several different combinations of light absorption and utilization traits provide the best competitive ability for plants in different communities or locations.

Space is an important resource for plants. In any location, the space available for seed germination is a subset of the total space. Some of the total space is already occupied by organisms, some is totally inappropriate (rock surfaces, streams, and lakes), and others locations will not support the proper development of seedlings. Plant requirements for germination vary among species such that different locations (*microsites*) are more favorable for seed germination of different species. For example, in the boreal forest of Hokkaido, Japan, the dominant species of *Picea* can only germinate on rotting logs while codominant species of *Abies* can germinate in the soil around rotten logs. Species that have narrow tolerances of environmental attributes will have a small subset of available microsites for germination. The broader the range of seed germination tolerances for a species, the higher the density of appropriate microsites and the greater the competitive ability. Space is also an important aspect for attaining aboveground resources (light) and belowground resources (nutrient and water).

B. Resource Capture

The quantity of a particular resource needed for the optimal growth or productivity of an individual plant is dependent on its physiology and the balance of resource use within the plant. The degree of competition a plant may experience from its neighbors for a particular resource depends on how much of the resource is needed to optimize its growth, its ability to capture limiting resources, and how it utilizes the resource relative to its neighbors (Goldberg 1990). In competitive environments, many plant species have developed some degree of morphological and physiological plasticity in their ability to obtain a resource. With plasticity, plants vary in their capability to adjust the size, shape, and distribution of their shoot canopies and root systems in order to increase the capture or utilization of resources. Tilman (1988) suggested that a trade-off exists between the ability of a plant to acquire aboveground versus belowground resources. His theory suggests that, in order to acquire one resource, a plant must allocate more biomass to structures involved in the acquisition of that resource at the expense of biomass allocation to acquire the other resource. However, in recent studies of allocation patterns during competition, some species, through plasticity in spatial arrangement of above- and belowground plant parts, have been shown to maximize both above-and belowground responses to resource needs in a manner that increases competitive success in both resource areas (Aerts et al. 1991b).

The intensity of competition for resources may be altered with spatial and temporal fluctuations in resource availability and use. Resource partitioning on a spatial and temporal basis (such as one species obtaining water from a different part of the soil profile than its potential competitor; Weltzin and McPherson 1997) allows plants to coexist without competition. Changes in morphological, physiological, and behavioral characteristics in the spatial and temporal scales are the factors that regulate the competitive ability of species. Many of these factors, along with how the factors determine competitive ability, will be discussed later in the chapter.

C. Intraspecific and Interspecific Competition

Competition that occurs between individuals of the same species is called *intraspecific competition*. Intraspecific competition occurs most often where one species is planted in monoculture, particularly as an arable crop or in forestry plantations. However, many species also occur naturally as swards of individuals of the same species. Intraspecific competition is a prime factor regulating community development in chaparral shrublands (Schlesinger et al. 1982). Invasive weeds often form dense stands in which intraspecific competition occurs. Intraspecific competition can be quite intense, as each individual in the population has the same resource requirements and similar physiology and growth potential as its neighbor.

Intraspecific competition is important in the study of the regulation of population size and changes in the maximum growth potential due to competitive interference. Most of the effects of intraspecific competition are density dependent. That is, the density, or "closeness" of neighbors, will influence the growth rate, reproductive ability, or survival of an individual. Mortality of individuals in a population through self-thinning, explained later in the chapter, is often noted in studies of intraspecific competition.

Competition between individuals of different species is called *interspecific competition.* In temperate, tropical, or other natural forest systems, competition is largely interspecific. In cropping systems, interspecific competition from weeds can result in reduced yields of crop plants.

III. COMPETITION IN NATURAL AND AGRONOMIC SYSTEMS

Agronomic and natural systems differ in their species composition and structure. However, the mechanisms of competition are essentially the same in both systems, though the complexity of competitive interactions is usually smaller in agricultural systems. The relative simplicity of agronomic systems has permitted simple theories and models to be developed concerning the nature of competition. Physiological and morphological responses can be readily measured or observed in individuals of one species under density stress. Experiments performed at different densities are easily accomplished. In agronomic systems, it is often the species with the greatest yield or the species that grows most rapidly under the stress of competition, which is considered indicative of a superior competitor in intercropping. Thus, an understanding of the nature of competitive interactions between individuals must be considered in the design of cropping systems. For example, high planting densities can result in reduced growth or productivity of individuals but maintain a high crop yield, while low planting densities may allow for greater individual plant size but underutilize the site and result in a low crop yield.

In complex natural systems, competition between individuals of different species may have a depressing effect on each species population size and predominance in the community. Competition may lead to the elimination of a species by another species or determine which species can exist together in the same habitat. It has been postulated that similar species growing together in the same habitat and utilizing the same limited resources cannot coexist indefinitely. Competition for the same resources can result in each species ending up with smaller-sized individuals or with lower population numbers, or with one species eventually becoming competitively displaced or excluded from the habitat (Tilman 1988). Competitive stress can slowly affect an evolutionary change in resource use or requirements, altering the distribution of species along environmental gradients, or changing the spatial or temporal patterns of species physiology or phenology. All these changes in species character create a separation of resource use that reduces the intensity of competition (Parrish and Bazzaz 1976). However, in nature, species with similar resource requirements do coexist, often through small ecological differences (Ågren and Fagerström 1984).

The mechanisms of competition are similar in agronomic and natural systems; however, the motives and techniques for studying competition differ between agronomists and ecologists. Many investigators study simple systems to identify the modes and mechanisms of competition between plants and to determine the effects of competition. Many field, greenhouse, and laboratory experiments have been utilized to demonstrate the effect of competition, either with a single species or with limited mixtures. In agricultural settings, one purpose of analyzing the competitive response of species is to assist in the determination of the yields provided by varieties of a species, either in monoculture or in mixture. Another is to determine optimal planting density to obtain the greatest yield. In agronomy, where yield per unit area or mean yield of individual plants is manipulated through management activities, many experiments have been carried out to investigate the relationship between management programs, plant density, and yield. In natural systems, the purpose of studying plant competition is to predict the dynamics (succession, dominance, diversity, invasions, etc.) of natural communities (Damgaard 1998).

There has been some disagreement over the years concerning how to measure the effect of competition in communities. It is sometimes difficult to distinguish between the effect of environmental stress and the effect of competition on plant responses. Generally, the effect of competition on an individual plant may be evaluated in terms of a measure of the effect on its relative productivity, growth or reproduction. Competition can also be viewed in an overall community perspective. The effects of competition can be observed or inferred, and the process of competition can be postulated, from observations of natural communities with or without experimental manipulation. A species may not exist in a certain habitat to which it is adapted because it is unable to compete with other species in that habitat.

In natural systems, the intensity of competition between conspecifics may be measured as the increase in cover of one species following the removal of one or more other neighboring species. Often pot experiments are performed in which plants of two species are sown at different densities with themselves (intraspecific competition) and also in varying densities and proportions with each other (interspecific competition). *Additive* experimental designs usually maintain the study species in the same proportion but vary density, while *substitutive* experiments maintain density but change the proportions of the competing species.

Above- and belowground competition along experimental gradients of nutrient and light availability have been utilized to test contrasting predictions for the relationship between competition intensity and resource availability in natural environments (Wilson and Tilman 1993). In natural systems, it is has been debated whether competition should be stronger in environments of high resource availability or in those of low resource availability. This comes from observations that, in low resource environments, greater tissue longevity, rather than maximum resource uptake, promotes competitive success. In high resource environments, species that exhibit fast growth appear to be better competitors.

IV. MODELS OF COMPETITION

A. Single-Species Population Growth

Single-species population growth in an environment of limited resources is often described under the general assumption that the size of a population tends to increase exponentially until density-dependent processes (intraspecific competition) progressively reduce the rate of increase in population size (Figure 9.1). The relationship between population size and rate of population increase is most often described by the logistic growth equation first formulated by A. J. Lotke and G. F. Gauss. The rate of increase in population size is expressed as:

$$\frac{dN_i}{dt} = rN_i\left(1 - \frac{N_i}{K_i}\right) \tag{9.1}$$

where r is the exponential rate of population increase in the absence of competition, and K is the carrying capacity (the number of individuals that the environment can support. The term N_i/K_i describes the effect of intraspecific competition. As the pop-

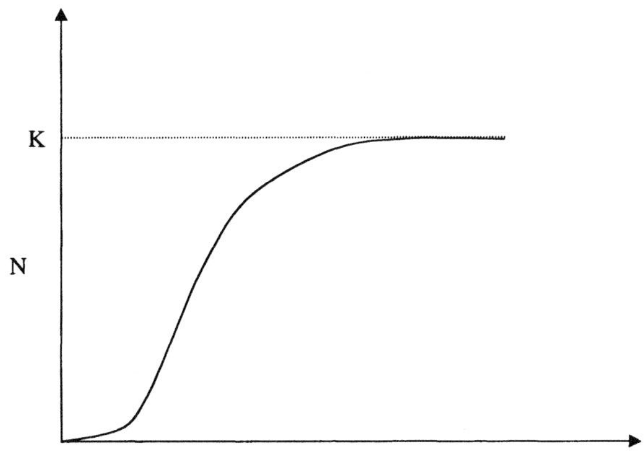

Figure 9.1. Logistic curve of population growth. The increase in number of individuals over time follows a sigmoid curve, where the population growth is slow initially, then grows more rapidly as the number of individuals increases, until it reaches an equilibrium number of individuals determined by the environment.

ulation number N, approaches K ($N/K \rightarrow 1$), the rate of increase of the population approaches zero. Each species will have a different value of K that defines an upper size to which a population can grow due to its unique response to resource limitations (such as space). The logistic growth equation has provided a theoretical foundation for the development of models for competition in multispecies mixtures.

B. Plant Density and Yield Under Competition

Numerous experiments have been carried out to determine the nature of the relationship between plant density and yield per plant or yield per unit area, with and without the effects of competition. To test the effects of density on plant growth and yield, the species of interest is often planted in plots (or pots) of differing densities. Each plot is grown for the same length of time in the same environment, and the final density and yield (often biomass) are measured. These yield-density experiments provide a measure of intraspecific competitive effects and the degree of plastic response to competition within a species. Generally, it has been found that at low planting densities, there is little or no intraspecific competition. When a species is planted at higher densities, individuals begin to interfere with each other and their individual growth rates decline. During intraspecific competition, the mean weight of individual plants at harvest time declines with increasing population density (Figure 9.2).

When density is low and no competition is occurring, plant growth is at a genetically determined maximum (when in an optimal environment). High densities result in reduced individual plant growth rates, and the total biomass on the site does not increase with increasing density. This asymptote on the yield-density diagram (Figure 9.3) is different for each species. A linearly increasing relationship between yield and final density implies that no competition is occurring. When the relationship between yield and final density becomes nonlinear or horizontal, then competition is having an effect on final yield.

As plants grow, increasing in size and utilization of space, competition between crowded individuals may be so severe that mortality occurs. In experiments in which

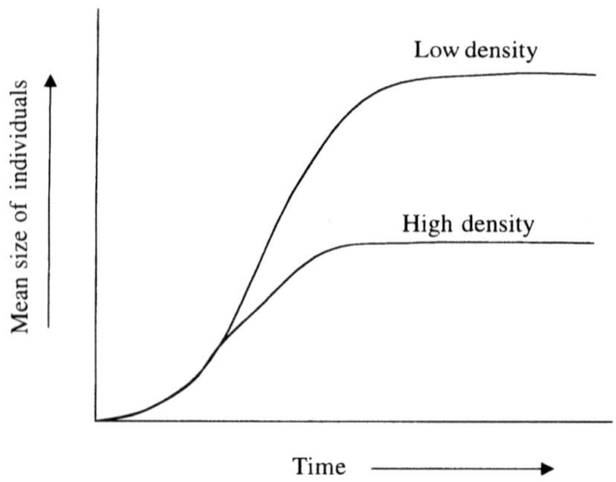

Figure 9.2. Effects of density on mean size of individuals. (Adapted from Weiner and Thomas 1986.)

a gradient of densities of a species is planted, it has been found that there is a limit to the amount of biomass that can be packed into a limited space. This limit has been described by the *self-thinning rule*, also called the $-\frac{3}{2}$ *power rule* (Yoda et al. 1963, White 1981). The self-thinning rule describes the relationship between the mean weight and the number of surviving plants of a population after density-dependent mortality sets in:

$$w = Cp^{-3/2} \tag{9.2}$$

where w is the log mean weight per individual, p is the corresponding log density of surviving plants; and C is a constant that varies with species. At any particular density, mean plant weight will increase until density-dependent mortality occurs. However,

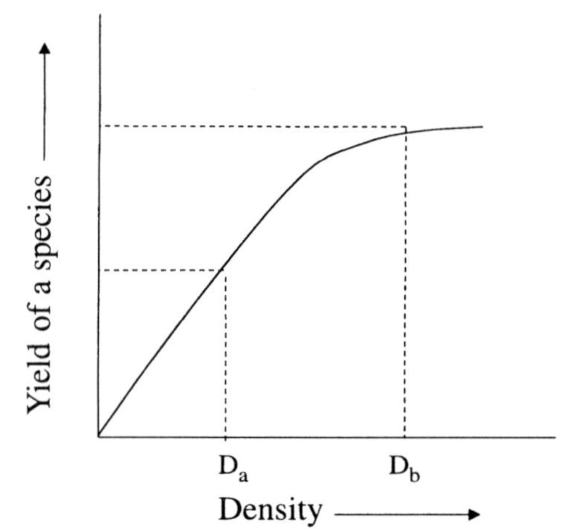

Figure 9.3. Typical yield at final density response curve. At densities below D_a, yield is linearly related to planting density and there is no competition between individuals. Between D_a and D_b, competition is occurring and the growth per individual is declining. After D_b, increasing density has no effect on final biomass yield. (From Goldberg et al. 1995.)

with mortality, density will decline as mean plant weight increases. When log plant weight is plotted against log density, with mortality occurring, the data follow a trajectory over time with a negative $-\frac{3}{2}$ slope (Figure 9.4A). The actual position of this self-thinning line on such a graph varies from species to species (Figure 9.4B). The density at which self-thinning occurs is a measure of a species' tolerance to crowding and intraspecific competition. If individuals in a population continue to grow in size, and the habitat they are in can support no more biomass (i.e., when the carrying capacity is reached), the increasing size of plants must be compensated for by an equal loss of biomass through mortality and the $-\frac{3}{2}$ gradient becomes a -1 gradient. This mathematical relationship between plant size and density is a general rule that has been applied to many species growing in overcrowded populations and has also been used to describe mixed populations of two similar species (White and Harper 1970). For a comprehensive review of the self-thinning rule, see Westoby (1984). These mathematical relationships provide the means for determining interactions between individual plants in a population and between a population and its habitat (Hutchings and Budd 1981).

C. Multispecies Competition Theory

In natural systems, interspecific competition is more likely to occur than intraspecific competition. Moreover, interspecific competition is considered a primary factor associated with succession (community development) and one of the main forces shaping the composition of mature communities. Several models have been developed for describing the processes that determine interspecific competition, and there is evidence in support of all of these models. A detailed analysis of the theory behind these models

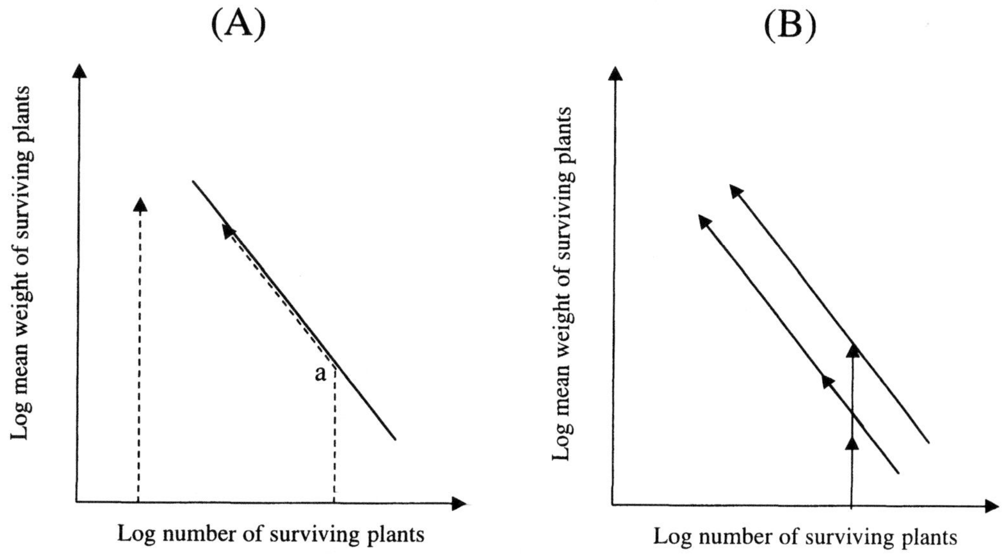

Figure 9.4. Changes in plant density and in mean plant weight over time. (A) As plants increase in size, fewer individuals can be supported per unit area, and density-dependent mortality occurs so that the density of surviving individuals declines along the $-\frac{3}{2}$ relationship of log density to log mean plant size. a = weight when $-\frac{3}{2}$ relationship takes effect (B) Different species reach the $-\frac{3}{2}$ limit at different weights. (After Long and Smith 1984.)

is the subject of any ecology text. Here we introduce these models to establish a framework for our discussion of the physiological basis of competition between species.

One of the first models proposed for interspecific competition was based on the logistic equation of single-species growth. These were the *Lotka-Volterra equations,* which were based on a phenomenological approach. In these models, a competitive coefficient is assigned to each species (see equations 9.3 and 9.4). The effect of one species on the other is dependent upon the population size of the competing species, its competition coefficient, and the population size of the other species. In the Lotka-Volterra models, population growth is dependent solely on population size. Although this is not a mechanistic model, it was able to provide the basis for many studies on species interactions. The following equations describe the population growth model of two species (i and j) that are competing:

$$\text{For species } i: \quad \frac{dN_i}{dt} = r_i N_i \left(1 - \frac{N_i}{K_i} - \frac{a_{ij}N_j}{K_i}\right) \tag{9.3}$$

$$\text{For species } j: \quad \frac{dN_j}{dt} = r_j N_j \left(1 - \frac{N_j}{K_j} - \frac{a_{ji}N_i}{K_j}\right) \tag{9.4}$$

where r is the exponential population growth rate, N is the population size, K is the carrying capacity, a_{ji} is the competition coefficient for the influence of species i on the population size of species j, and a_{ij} is the competition coefficient for the influence of species j on the population size of species i.

The significance of the Lotka-Volterra competition equations can be evaluated in a graphical format (Figure 9.5). The graphical representation comes from the concept that if two species are to coexsit, there must be a time when the populations have a zero growth rate (dN_i/dt and $dN_j/dt = 0$). Under the condition that population growth is zero for both species, the following equations can be derived from equations (9.3) and (9.4):

$$\text{For species } i: \quad \frac{dN_i}{dt} = 0 \quad \text{when} \quad \left(1 - \frac{a_{ij}N_j}{K_i}\right) = 0 \quad \text{or} \quad N_j = \frac{K_i}{a_{ij}} \tag{9.5}$$

$$\text{For species } j: \quad \frac{dN_j}{dt} = 0 \quad \text{when} \quad \left(1 - \frac{a_{ji}N_i}{K_j}\right) = 0 \quad \text{or} \quad N_i = \frac{K_j}{a_{ji}} \tag{9.6}$$

An equilibrium isocline representing the population dynamics of species i can be represented on a two-dimensional graph in which the axes represent the population sizes of species i (Figure 9.5A). If the population size of species i is located in the shaded area below the isocline, then dN_i/dt is positive and the population size will increase (shown as the right-pointing vector in the figure). However, if the population size is in the unshaded area above the isocline, then dN_i/dt is negative and population size will decrease. The same graph can be produced for species j (Figure 9.5B). When the isoclines of both species are graphed together, the change in population size of each species can be determined by the product of the vectors for each species (Figure 9.5C). Several possible results of competition can be represented by these graphs. In one case (Figure 9.5C), the result of competition is the exclusion of species j. (The stable end product of competition is K_i.) In another case (Figure 9.5D), a stable

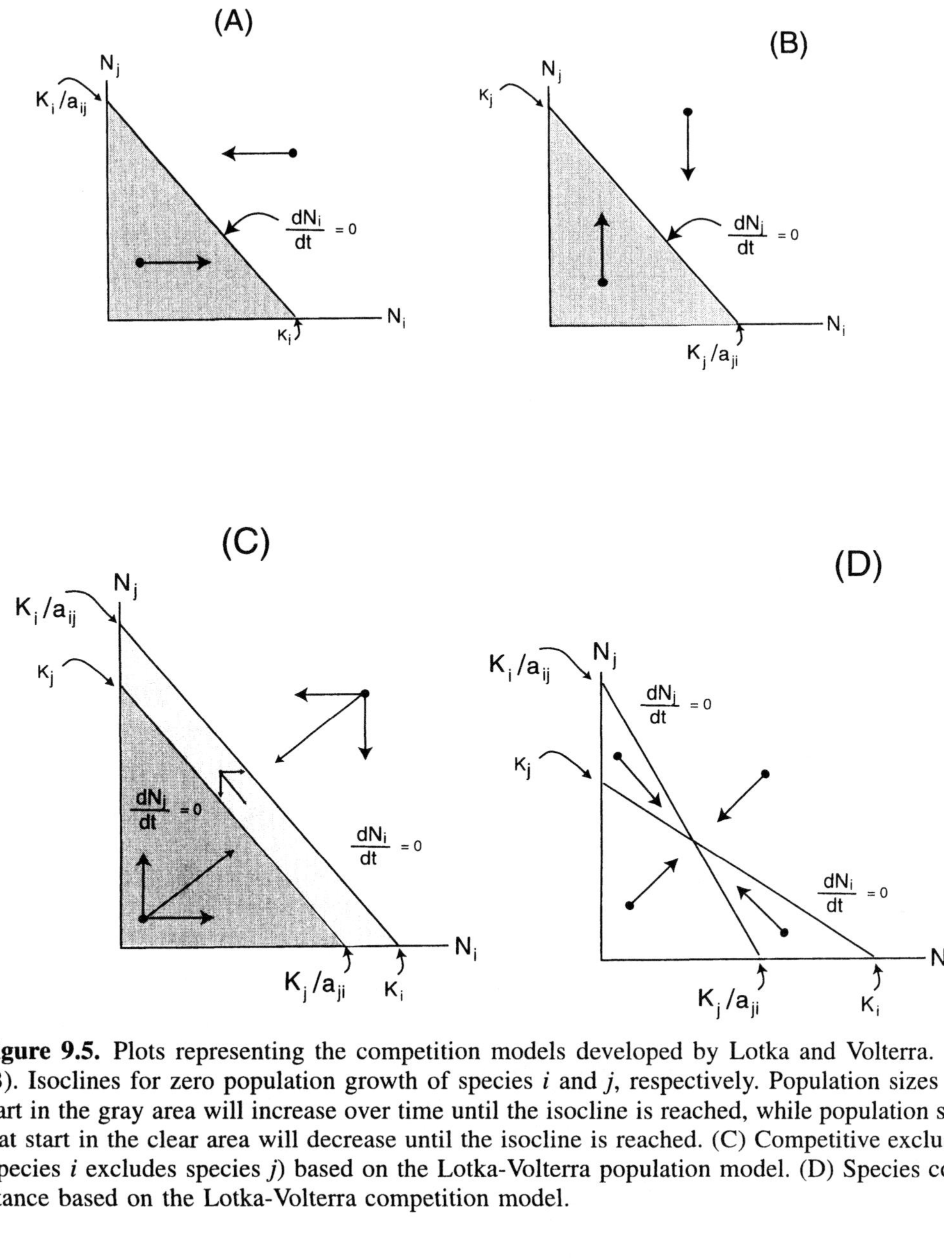

Figure 9.5. Plots representing the competition models developed by Lotka and Volterra. (A), (B). Isoclines for zero population growth of species i and j, respectively. Population sizes that start in the gray area will increase over time until the isocline is reached, while population sizes that start in the clear area will decrease until the isocline is reached. (C) Competitive exclusion (species i excludes species j) based on the Lotka-Volterra population model. (D) Species coexistance based on the Lotka-Volterra competition model.

coexistence point is possible in which both species have a moderate population size. Many other possible interpretations of these isoclines are possible, and they have been the basis of many further theoretical studies. However, these models suffer from lacking a mechanistic explanation.

Several different models have been introduced in an effort to develop a mechanistic theory of interspecific competition. One of the most successful models was developed by Tilman (1982). This model bases the outcome of competition on the availability of limiting resources and the ability of plant species to draw down resource availability in the environment. This model also can be presented in a two-dimensional graph, if one considers the species response to only two resources (Figure 9.6). When resources

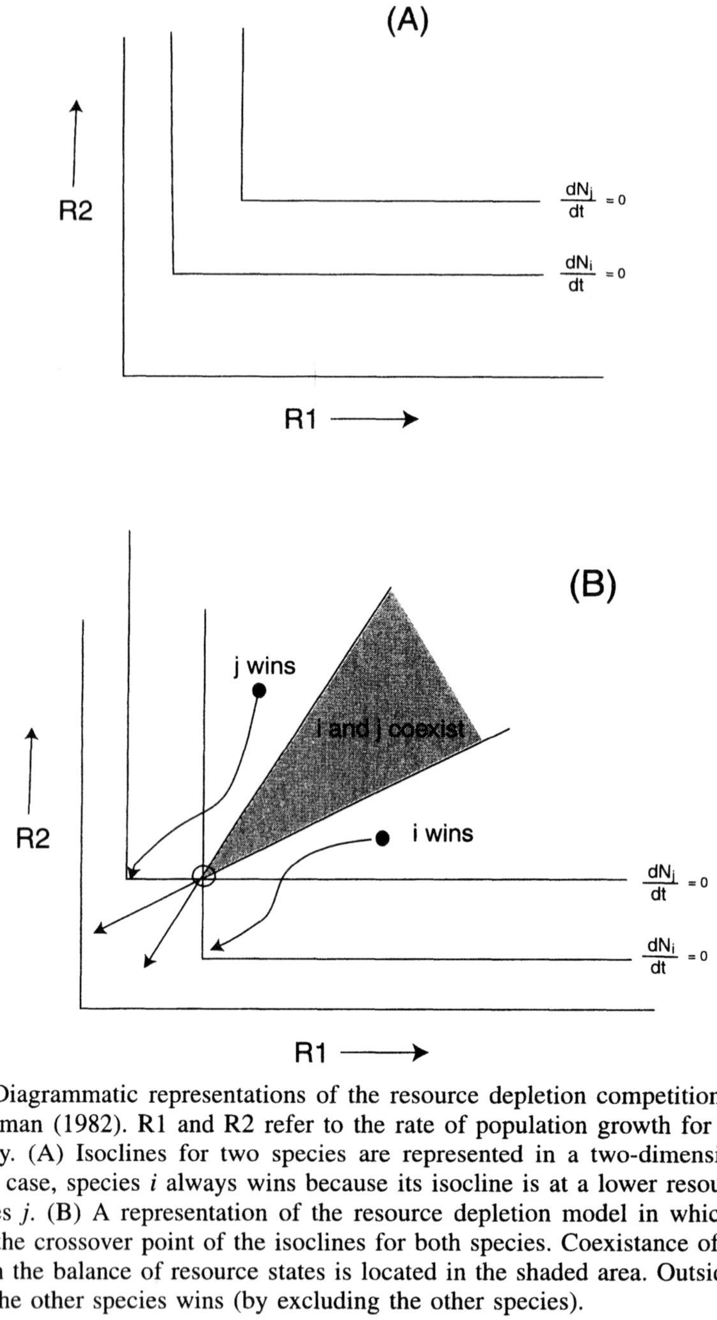

Figure 9.6. Diagrammatic representations of the resource depletion competition model based on that in Tilman (1982). R1 and R2 refer to the rate of population growth for species 1 and 2, respectively. (A) Isoclines for two species are represented in a two-dimensional resource space. In this case, species *i* always wins because its isocline is at a lower resource state than that of species *j*. (B) A representation of the resource depletion model in which coexistance can occur at the crossover point of the isoclines for both species. Coexistance of species *i* and *j* occurs when the balance of resource states is located in the shaded area. Outside this shaded area, one or the other species wins (by excluding the other species).

are relatively high, the population of a species grows, and in doing so places a greater demand on the resources. Population growth continues to rise until resources become limiting; then population growth declines until an equilibrium is reached. This equilibrium point (R^*) occurs when resource supply is exactly matched by resource demand by the population. The population size of the species will increase (i.e., it will deplete resources) if the resource state is above R^*, and decrease (i.e., resources will increase) if the resource state is below R^*. The result of the interaction between population size and resource availability is an isocline of zero population growth ($dN/dt = 0$). A species utilizing two resources will be able to maintain a stable population size defined by one isocline for each resource (Figure 9.6A). Whatever the population size is in the area of the plane above the isoclines, it will drive the resources down to the equilibrium point at the intersection of the two isoclines, given enough time. If the isoclines for one species (species i) are lower than those for the other species, j, then the first species will exclude the latter (Figure 9.6A), because species i will drive resources below the level tolerated by species j. However, if the isoclines cross, there is a possibility of coexistence at a stable resource balance (Figure 9.6B). Of course, there are many resources in the natural environment, and there would be many more planes than the two represented in these graphs. Also, this model is based on a stable and homogeneous resource availability. The simplicity of this model may not apply to the natural environment. However, this model does use a limiting-resource approach to competition interaction, which has led to many studies that examine competition at a mechanistic level. For further investigation of competition modeling at an ecological and population level, refer to the Supplementary Readings (at the end of the chapter).

D. Community Competition Theory

Each plant utilizing a resource tends to locally deplete that resource for itself and for its competitors (Wedin and Tilman 1993). When competition acts on all neighboring individuals equally or in proportion to their size, the mode of competition is *two-sided,* or *size-symmetric.* Two-sided modes of competition are described by two models: (1) the *resource depletion model* described earlier (Weiner and Thomas 1986) and (2) the *resource reduction model* (Wedin and Tilman 1993). Individual plant growth rates vary in a population of individuals and between species in mixed-species communities. In *size-symmetric competition,* competition acts to reduce the growth rates of all individuals by the same proportion so that size variation remains the same among competing individuals. Small plants are as capable of obtaining soil resources as large plants, and the competition for these resources influences each similarly. Often, competition for soil resources such as water and nutrients tends to be size-symmetric (Weiner 1986, Schwinning and Weiner 1998). Wedin and Tilman (1993) provide an example of competition for soil nutrients where the experimental competitive outcome between competing species was a function of species differences in ability to deplete soil nitrogen.

One-sided, or *size-asymmetric,* modes of competition are those in which larger plants obtain more than their share of limiting resources, suppressing the growth of smaller plants while leaving the larger plants relatively unaffected (Weiner 1985). One-sided (asymmetric) modes of competition are characterized by the development of size inequalities among groups or populations of individuals over time. The *resource preemption model* (Weiner and Thomas 1986) describes this mode of competition. Plants that are the first to establish and grow rapidly often gain the largest share of resources. Harper (1977) showed that the emergence time of seedlings in a mixed-species community may have determined which species initially became dominant. Also, differences in the time of emergence of the various species present in a seed bank allow some species to get a "head start" and take up resources before later-emerging species. Differences in the initial densities and timing of establishment of competing species can result in one species capturing a disproportionate share of space and light, eventually suppressing the growth of neighboring plants. Commonly, plants with high allocation of resources to leaves and relatively fast growth rates, or plants that have had a longer period of growth than their neighbors, may be larger than their neighbors and may directly suppress the growth of those neighbors through overtopping and shading.

Resource preemption most often occurs with competition for light. When the canopy closes and plants begin to shade each other, larger plants, with their leaves held high in the canopy, will get more than their share of light (Figure 9.7). A smaller, subordinate individual may not get sufficient light to survive. Resource preemption also occurs under low light conditions and high densities, where the dominant plants usurp what light is available, fully suppressing smaller neighbors (Schmitt et al. 1986).

Difference in the effect of root competition (two-sided) and shoot competition (one-sided) was demonstrated by Weiner (1986). In that study, plants were grown individually, with shoot competition but not root competition, with root competition but not shoot competition, and with both root and shoot competition. Root competition for nutrients had a much larger impact on mean plant weight than did shoot competition for light, but size inequality (measured as the coefficient of variation) among individuals was greater when shoot competition occurred (Figure 9.8). The effect of above- and belowground competition on community succession was demonstrated by Kosola and Gross (1999), who found that belowground competition for resources was the dominant factor in the suppression of early colonizing annuals by later colonizing perennials.

V. PLANT RESPONSES TO COMPETITION

A. Aboveground and Belowground Competition for Resources

The most common response mechanism to a depletion of a resource is for a plant to shift its allocation of biomass to roots or to shoots as it adjusts to resource imbalances (Jackson and Caldwell 1989). The allocation of biomass tends to be divided mostly between the growth of new leaves and roots. New leaves add to the process of pho-

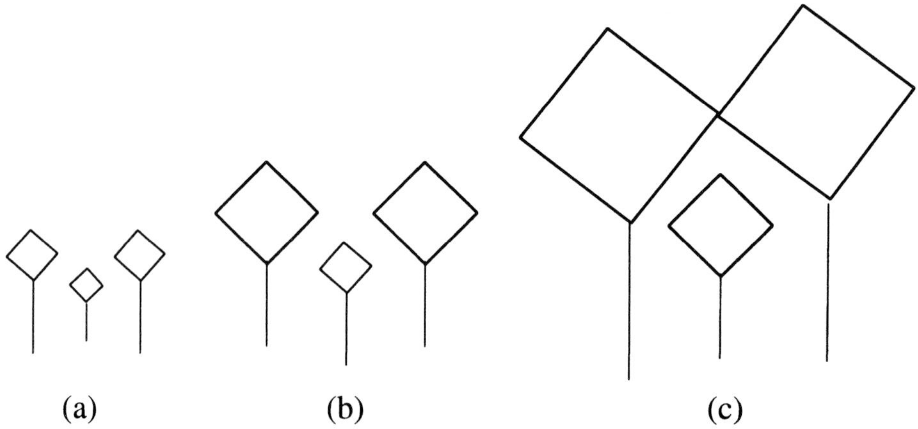

Figure 9.7. Schematic diagram of the decrease in light available to subordinate plants as mean plant size increases. The relative amount of light available to subordinate individuals declines even though the relative size differences between individuals stay the same. (From Weiner 1988, with permission of Blackwell Science Ltd.)

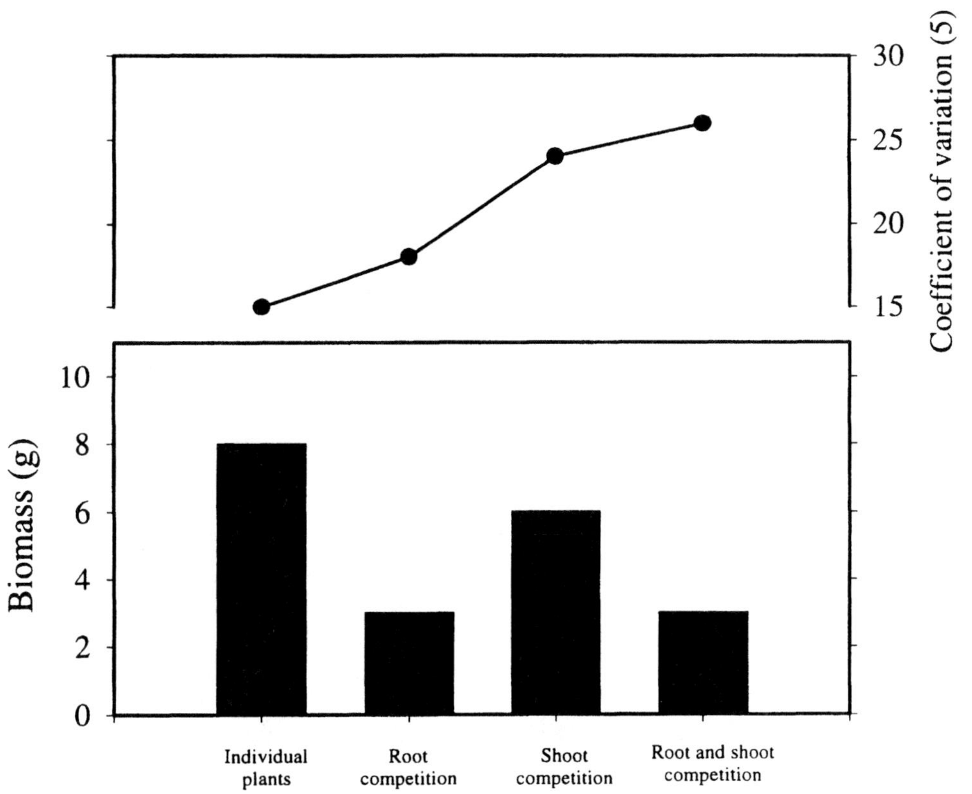

Figure 9.8. The effect of root and shoot competition in *Ipomoea tricolor* on mean biomass (g) and the coefficient of variation (CV%). (Redrawn from Weiner 1986, with permission.)

tosynthesis and biomass production, while new roots contribute to water and nutrient uptake (Schulze et al. 1986). Species difference in capacity to make allocational adjustments may be critical in determining the success of a species in resource capture models (Tilman 1982).

There is an interaction between the effect of above- and belowground resource limitations. Often, reduced light availability leads to a decreased demand for nutrients, an increase in the shoot/root ratio, and lower root function (Jackson and Caldwell 1992). Alternately, if water and nutrient availability are limited but light in not limiting, there is often an increased allocation to root biomass (Schulze et al. 1986). Shoot competition can be significantly more intense and root competition less intense with high nutrient availability (Wilson and Tilman 1995). In low resource environments, where aboveground productivity is low, belowground competition may affect the balance between competing species more than aboveground competition (Wilson 1988).

B. Morphological Responses

Plant competitive ability is partly determined by the ability to acquire resources through development and placement of those plant parts needed to capture the resource. *Plant architecture,* particularly patterns of branching, leaf angles, root angles, and root placement, have been investigated as possible traits important in determining competitive ability. The morphology of a plant species is under genetic control, but meristem activity can be modified by the environment, which influences the architecture of the plant or how plant parts are positioned in space. Increased internode lengths and increased allocation to stem biomass have been noted in studies where physical space was fragmented and resources, such as light availability, were not reduced (McConnaughay and Bazzaz 1992). Flexibility in the morphology and deployment of a foraging organ is advantageous in an environment in which space is limiting.

Plant architecture and physiological processes that influence growth are interrelated. Competition for light favors a crown geometry that maximizes the difference between photosynthetic area and the costs of growing and supporting that leaf area. Branching and leaf exposure patterns that minimize self-shading and maximize shading of a neighbor could provide a competitive advantage to a plant (Givnish 1986). Such patterns are particularly observed in high light or patchy light environments where placement of photosynthetic organs above potentially competing neighbors would provide a competitive advantage. Particular branching displays such as growing branches that overtop neighbors or increased lateral branch growth may assist in the avoidance of other shrubs and may increase light capture. Branch placement has been suggested as a mechanism for increased competitive ability in *Prunus spinosa, Crataegus macrocarpa,* and *Acer campestre* (Schultz et al. 1986). In species with similar architectures, small changes in the canopy structure can affect competitive balance (Küppers 1994). Plastic responses of internode and petiole lengths may be found in competing plants (Ballaré et al. 1988). This was observed in the competition of two conspecific low-lying understory forest herbs (*Polygonatum* and *Smilacina* sp.). Leaves of one species were placed 7 cm higher on sloping ground than the other species, and that difference provided a mechanical advantage that increased competitive ability (Givnish 1986).

These responses to crowding and shading by other plants are called *photomorphogenic* responses (Schmitt 1997). Plastic change in morphology is believed to be controlled, or mediated, by phytochrome sensing of a decrease in the ratio of red to far-red light that occurs with shading. The most common response to low red/far red ratio in subcanopy plants is internode elongation and reduced branching (Ballaré 1994). This plastic response occurs without increasing biomass to the stem. Also, mean petiole length of a plant may be altered depending on the characteristics of the canopy environment (Figure 9.9) (Ballaré et al. 1988). Photomorphogenic responses such as these are advantageous when competition for light develops among individuals of similar size, so that a slight increase in height may result in disproportionately large differences in light capture between neighboring plants (Ballaré et al. 1988, Ballaré 1994). These responses may not benefit plants in a highly mixed size-asymmetric community. For example, in a mixed culture of annual plants with different allocational traits, Tremmel and Bazazz (1995) could not find a consistent architectural or allocation change with neighbor type or density, except those related to changes in biomass. They suggest that with size-asymmetric competition, species differences in morphology or resource uptake rates may be overshadowed by the effects of size differences between individual neighbors.

In a heterogenous soil environment, nutrients are patchily distributed and there may be strong competition among individuals to exploit nutrient rich patches. Nutrients move through the soil via mass flow or diffusion (Stark 1994). Diffusion of nutrients from soil to roots can result in the development of depletion zones in the soil leading to intense competition for limited nutrients. Species differences in root foraging ability for nutrients include plasticity in root growth. Plasticity is reflected in the precision by which growth occurs in nutrient rich patches rather than nutrient poor patches, the mass and extent of root growth, and the sensitivity to and detection of different degrees

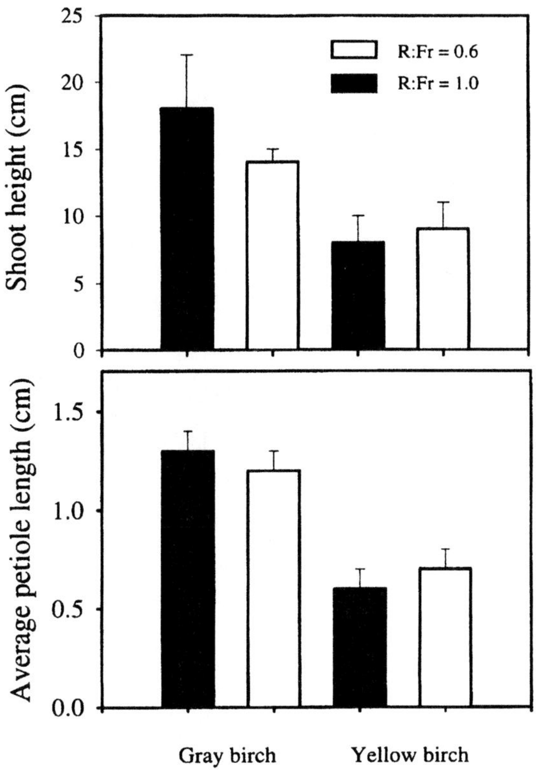

Figure 9.9. Responses of early successional species (gray birch) and later successional species (yellow birch) seedlings to low (0.6) and high (1.0) red/far red light ratios. Error bars represent one standard error. (Redrawn from Bazzaz and Wayne 1994, with permission.)

of nutrient availability (Mou et al. 1997). Generally, morphological mechanisms for resource acquisition include the development of an extensive root system (Aerts et al. 1991b), proliferation of thinner, fine roots in high nutrient patches, narrow root branching angles, and increased fine root elongation (Fitter 1994, Mou et al. 1997).

If soil resources are in limited supply and neighboring plant species are exploring the same soil volume, one species may be more effective in acquiring these resources than another species. Root morphology and architectural characteristics such as high root density, high specific root length and evenly spaced roots may improve ability to uptake soil resources and could accord a strong competitive ability (Robberecht et al. 1983). However, some plants can effectively acquire nutrients from both uniform and patchy soil nutrient distributions (Figure 9.10) with no differential effects on growth (Cui and Caldwell 1998). Caldwell et al. (1991a) found that increased root densities for one species did not result in increased nutrient uptake compared to the competing species. Both species took up a similar amount of soil nutrient (Figure 9.10). Therefore, the importance of root morphological traits to resource accumulation does not always relate to competitive ability.

Recent studies suggest that there is a close relationship between root foraging and light availability and interception. Foraging for soil nutrients may be constrained under conditions of low light availability (Huante et al. 1998). Reduced light availability reduces carbohydrate transfer to roots, thus reducing the total root growth, the proportion of biomass in the roots, and root functions such as respiration (Zangerel and

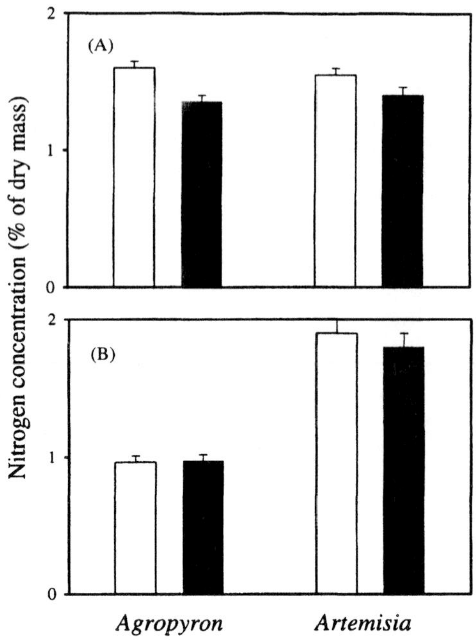

Figure 9.10. Plant shoot tissue nitrogen concentration in balanced (open bars) nutrient and unbalanced (solid bars) nutrient treatments after 30 days (A) and 60 days (B). Both species took up a similar amount of nutrient, although the nutrient patches were differently distributed in each treatment. Error bars represent one standard error above the mean. (Redrawn from Cui and Caldwell, 1998, with permission.)

Bazzaz 1983). Under circumstances of high nutrient demand, particularly in high light environments, there may be portions of the root system that increase nutrient uptake kinetics (Robinson and Rorison 1987). Roots may adjust uptake kinetics corresponding to local soil nutrient levels (Jackson and Caldwell 1992, 1996) (Figure 9.11), or root proliferation may increase (Campbell et al. 1991b) in patches of high nutrient availability.

Interference between the root systems of competing species may occur in high nutrient patches. Fast-growing species may initially allocate more root biomass (greater root lengths and greater densities) into high nutrient patches, through increased growth proliferation, than slower growers. Fast growers can benefit from this high rate of nutrient capture (Campbell et al. 1991a, Fransen et al. 1998). As depletion of the nutrient resource occurs, the predominance of roots of the fast-growing species in a patch may decline (Huante et al. 1998), particularly if fast growers are intolerant of low nutrient conditions. Slow-growing species are often tolerant of low nutrient conditions and can be competitively successful through long tissue life.

Increased root proliferation in nutrient patches is often not related to uptake kinetics. Often, physiological plasticity in nutrient uptake can contribute to higher nutrient contents in both slow and fast growers (Fransen et al. 1998). Attributes such as plasticity in root growth, increased uptake kinetics (Jackson and Caldwell 1992), or mycorrhizal associations will increase belowground competitive ability (Casper and Jackson 1997). Belowground resource demand can also influence aboveground physiology. If water and nutrient availability are limited, there is often an increased allo-

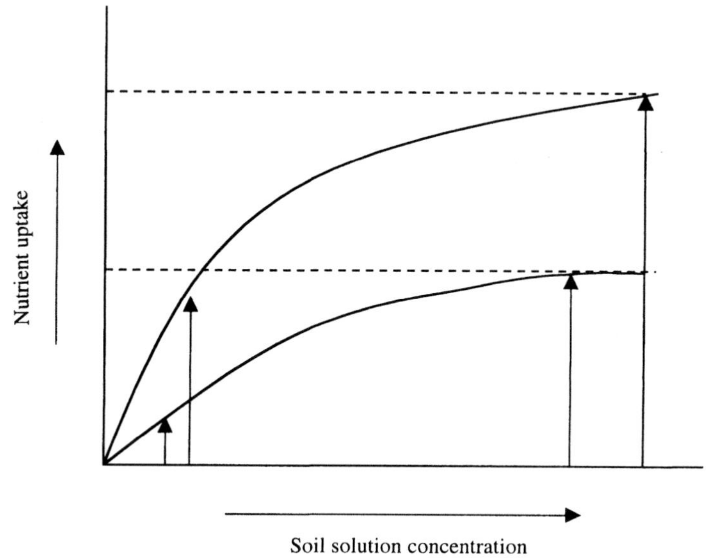

Figure 9.11. Model of a plastic response to nutrient heterogeneity. With increased kinetic uptake ability, a plant gains more nutrients at both low and high nutrient availability than without increased kinetic uptake. (From Jackson and Caldwell 1996, with permission of Blackwell Science, Ltd.)

cation to root biomass. Low water availability also alters the aboveground osmotic potential that changes the rates of plant water use. Reduced transpiration rates from stomatal closure and morphological adjustment in leaves, such as thick cuticles and pubescence, are a consequence of water limitation (Schulze et al. 1987). In replacement series competition experiments for both light and nutrients, neither species increased biomass allocation to leaves or roots in response to competition (Aerts et al. 1991b). However, morphological changes resulting in increased specific leaf area (SLA) and high specific root lengths (SRL) compensated for the small changes in allocation to the biomass of roots and leaves. Therefore, increased growth in a competitive environment is associated more with changes in allocation patterns than with changes in physiological activity.

Generally, the alteration of physiological or morphological characteristics for increased resource uptake does not automatically increase competitive ability. In high nutrient environments, low tissue mass, such as high SLA (leaf area/leaf mass) and high SRL (root length/root mass), may confer a competitive advantage to fast growers. In low nutrient environments, slow growers that invest in more tissue mass (low SLA and low SRL), usually have an increased tissue retention time which lowers the costs of tissue construction or enzyme operation and confers a competitive advantage over faster growers (Poorter et al. 1995). *Scirpus sylvaticus,* a slow-growing plant characteristically found in nutrient-limited conditions, produces an extensive, robust root system, providing a large absorption system that remains functional throughout the year, allowing capture of unpredictable pulses of nutrients (Crick and Grime 1987). A study of morphological traits was not able to link mass traits, such as SLA, leaf area ratio (leaf area/total growth) and, shoot/root ratios (Wardle et al. 1998a) to competitive ability. But those traits related to physiological functioning, particularly traits that are responsive to rapidly changing resource conditions in patchy environments, do appear to have an effect on competitive ability (Hutchings and de Kroon 1994).

C. Physiological Responses

The rates of various physiological activities affect both the use and ability to acquire a resource (Goldberg 1990) and are important in conferring competitive ability. Differences between species in whole-plant growth rates are controlled by physiological differences in the magnitude of photosynthesis, respiration, and patterns of allocation of net photosynthate to different structures within a plant. Studies of plant architecture, net carbon gain, and carbon allocation to those organs that can assist in resource capture may lead to a closer understanding of plant competitive success. For example, a species able to maintain high photosynthetic performance and maximum growth potential in a wide range of favorable and unfavorable environments, particularly where there is large variability in light availability, will be a strong competitor.

The physiological characteristics of plant species common to either initially uncolonized disturbed habitats (early successional) compared with those common to well-occupied habitats (late successional) has been described by Bazzaz (1979). Initial colonizers tend to have high relative growth rates and are adapted to high light so that their light saturation intensity, light compensation point, and photosynthetic rates are high. Colonizing plants also exhibit high plasticity in their photosynthetic capacity to a wide gradient of light availability (Bazzaz and Carlson 1982, Chazdon 1992). These characteristics result in rapid biomass accumulation and allocation to foraging organs and increased resource acquisition rates. These coarse-level foraging characteristics allow colonizing plants to usurp resources quickly prior to the onset of intense competition. However, significant reductions in photosynthetic rates and growth rates can occur when competition becomes intense (Burton and Bazzaz 1995).

Later successional plants exhibit a lower reduction in photosynthesis and growth under competition than do early successional species (Burton and Bazzaz 1995). In later successional habitats, where light availability tends to be patchy or more limited, species tend to predominate, that show progressively lower allocation of biomass to leaves but progressively higher allocation to roots for better nutrient capture.

From an individual plant perspective, competition from neighbors affects a variety of functional processes. The physical presence of plants creates habitats characterized by altered microenvironments. Mechanisms to increase competitive ability, particularly under low light conditions, include adjustment of leaf biomass towards thinner, larger leaves, which increases light interception on a unit leaf area basis and increases the ratio of total leaf area to total plant weight (LAR). Increased LAR enhances growth in low light conditions and effectively compensates for the lower photosynthetic capacity of shade leaves (Pearcy and Sims 1994).

Under low light conditions, physiological changes include alterations in the proportions of light-harvesting and energy-transfer components of the photosynthetic apparatus. For example, increased chlorophyll content, denser grana stacks, lower ribulose-1,5-carboxylase activity, and a lower saturation point for maximum photosynthesis (Boardman 1977, Bazzaz 1979, Goldberg 1990, Nilsen and Orcutt 1996) occur in plants growing in low light environments compared with those in high light environments. The emphasis in low light environments is on increased capacity for light capture rather than increased capacity for photosynthetic reactions. Thus, these changes in physiological traits are designed for tolerating resource limitation, not for competitive advantage. In habitats of high light availability, photosynthetic reactions become limiting to growth capacity. Rubisco activity, the capacity of the electron transport chain and stomatal conductance must be increased to increase photosynthetic capacity and whole-plant performance (Pearcy and Sims 1994). These resources for photosynthesis are relatively costly. Therefore, species that are able to maximize the efficiency of attaining these resources for photosynthesis will have the highest growth rates and have a competitive advantage. A caveat to these physiological processes is that allocation patterns (to leaves and tall stems) are critical to competitive advantage in high light sites. Light foraging through increased shoot elongation (described earlier) is critical to competitive ability in high light environments.

Foraging for nutrients has constraints that are quite different from foraging for light and space above ground. One difference is that there is a larger diversity of resources below ground (all the nutrients, water, space, etc.) than above ground (light, carbon dioxide, oxygen). Another is that different strategies in root systems are required for the maximum accumulation of different resources (particularly nutrients). For example, physiological plasticity is particularly important for maximizing the absorption of mobile ions such as nitrate, while root proliferation (precision and sensitivity) are most important for the maximum absorption of immobile nutrients such as phosphate and calcium. In fact, nutrient transfer to roots depends upon three general processes: (1) root growth into new space (rich patches?), (2) mass flow of nutrients to roots in the water stream that is absorbed for transpiration, (3) diffusion of nutrients along a diffusion gradient from local soil particles to the root surface. The relative importance of these processes varies with the specific nutrient. For example, calcium accumulation is mostly dependent upon root growth into new soil, magnesium and sulfate accumulation is mostly affected by mass flow, and phosphate accumulation is primarily regulated by diffusion to root surfaces. Because different processes determine the ability of plants to acquire different nutrients, different strategies of root allocation and physiological induction also are required for maximizing the accumulation of different nutrients.

Many habitats are nutrient-poor, and nutrients (particularly mobile nutrients like nitrate) may only become available in ephemeral, unpredictable pulses. The ephemeral nature of these soil resources makes induction of physiological activity (uptake kinetics) critical to competitive ability. Nutrient uptake capacity and nutrient uptake kinetics of plants may be increased when roots are exposed to a high nutrient pulse after a period of low nutrient availability (Jackson and Caldwell 1992, 1996). Physiological plasticity in nutrient uptake kinetics is characterized by rapid changes in the capacity of roots to absorb nutrients, and this provides a means to capture nutrient flushes. Grime (1994) showed that slow-growing species common to infertile sites can have a greater capacity to absorb N in short pulses, and exhibit higher relative growth rates under these conditions than do fast-growing species (Figure 9.12). The effective capture of ephemeral pulses of nutrients coupled with persistent roots (Crick and Grime 1987), and internal recycling of nutrients may confer increased competitive ability in low resource environments.

What are the changes in physiological activity that causes the induction of uptake activity during a nutrient pulse? The induction of uptake capacity is related to the balance of nutrient transporters in the root endodermal membrane. Nitrate accumulation is regulated by two inducible transport systems: the high-affinity (HATs) and the low-affinity transporters (LATs). Low-affinity transporters only operate at very high soil nutrition, thus they may only be significant in agricultural systems. In the absence of nitrate in the environment, there is a low abundance of HATs in the root

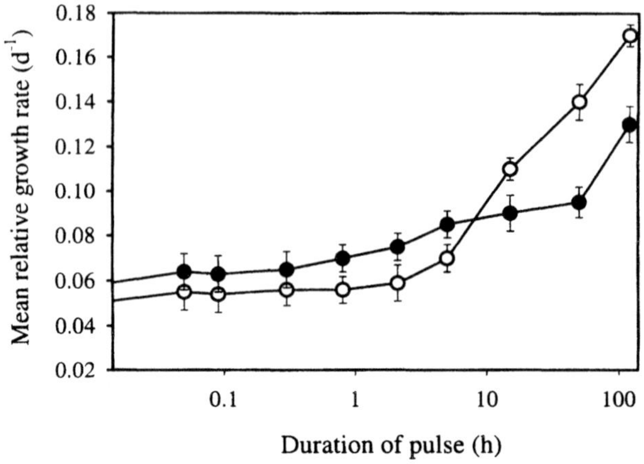

Figure 9.1.2 Mean relative growth rates of a fast-growing (open circles) and slow-growing (closed circles) species exposed to pulses of nutrient enrichment of differing duration. (Redrawn from Grime 1994, with permission.)

membrane (a constitutive population). Upon the arrival of nitrate in the solution, genes that synthesize HATs are upregulated rapidly and the activity of HATs in the root membrane can increase to their maximal levels within a few hours. However, in most situations, there is a low external nitrate concentration in the soil, which inhibits the HATs that are present in the root membrane. Upon a pulse of nitrate in the soil the activity of HATs increases more rapidly than the induction of the genes. Therefore, the constitutive pools of HATs may serve as a sensing system for the signal transduction pathway leading to the induction of HATs genes. Plants with an efficient mechanism of inducing the activity of HATs genes in response to nutrient pulses will have a competitive advantage in regions with low nutrient availability.

In what ways can plants increase their ability to absorb nutrients that are limited by the rates of diffusion to the roots such as phosphates? The net inflow of nutrients to the root (I) is dependent upon the ratio of the maximum accumulation capacity (I_{max}) and the efficiency of the accumulation system (K_m), as shown in equation (9.7):

$$I = \frac{I_{max}(C_s - C_{min})}{K_m(C_s - C_{min})} \tag{9.7}$$

where I is the accumulation of ions, $I_{max}$is the maximum accumulation capacity, C_s is the external ion concentration, C_{min} is the concentration at which net accumulation (influx − effux) equals zero, and K_m is the *Michaelis-Menton coefficient* (the substrate concentration at $\frac{1}{2}$ the maximum velocity of ion accumulation). During nutrient pulses in low nutrient environments, the I_{max} for a specific nutrient can be increased, and this is called a *compensatory response*. This occurs for the accumulation of ammonium, nitrate, phosphate, and sulfate. However, the compensatory response to one nutrient inhibits the ability to accumulate other nutrients (Chapin 1991). Therefore, there is a cost to a compensatory response to a single nutrient. Increased accumulation at low nutrient conditions in the absence of nutrient pulses can be accomplished by decreasing the K_m value of the nutrient transport system. A high-affinity transporter for K^+ (with a low K_m) appears within 1 hour of induction by potassium (Smart et al. 1996). However, this would not be a successful technique for increasing the accumulation of diffusion-limited nutrients such as phosphate because changes in affinity are not important for plant uptake relative to diffusion limitation in the soil. In order to have a competitive advantage in a phosphate-limited environment, root processes other than transporter kinetics would be important.

Competitive advantage in relation to phosphate accumulation involves alternative methods of acquiring phosphate rather than enhanced affinities for phosphate absorption. The most dominant alternative method is the association with mycorrhizae (reviewed in Chapter 4). Other methods include the ability to absorb phosphorus from precipitated sources (rock phosphate). This is accomplished by releasing organic acids and cation chelators into the soil solution from roots. Both of these factors increase the solubility of phosphate (from binding to cations), and the diffusion of phosphate to the root system increases. The ability to release organic acids such as malic, and citric, acids varies among species, and is most prominent in the roots of species that inhabit low nutrient sites, particularly members of the Proteaceae. Some plants also have the ability to utilize soil organic forms of phosphorus such as phytate. The ability of plants to release phosphatase enzymes into the soil increases their ability to absorb inorganic phosphorus in organic soils that are very low in soluble nutrients (Kroeler and Linkins 1991). Many plants, particularly grasses and sedges, can utilize phosphatases (from the plant or microbes) to access a pool of phosphorus unavailable to other plants, thereby attaining a competitive advantage.

D. Summary of Plant Traits and Competitive Ability

The acquisition of resources depends on the size of the foraging organs, the placement of these organs relative to the resource pools, and the physiological ability of the organ to acquire and utilize the resource (Bazzaz 1996). Species with high relative growth rates may be successful competitors as they may be able usurp space and take up most of the available resources. However, where resources are scarce, species that can tolerate low levels of a resource and respond to resource pulses may be the most competitively successful. Recent experimental studies of competition in high resource environments have shown that dominant plants tend to exhibit high rates of nutrient capture from high and low patches of nutrients. In addition, they tend to generate nutrient depletion zones around their roots, and respond to localized nutrient patches by rapid fine-root proliferation (Jackson and Caldwell 1996). In such a way, dominant plants monopolize light and nutrient capture with the development of extensive leaf canopies and root systems (Campbell et al. 1991a).

Later successional communities with lower resource availability (due to depletion by earlier colonizers) are characterized by competitive species with lower photosynthetic capacities than pioneer species (Küppers 1994), slower growth rates, and longer time of tissue retention. With lower light and nutrient availability characteristic of many late successional environments, competitive traits that tend to conserve internal stores of carbon and nutrients are an advantage. In addition, the ability to utilize alternative resource pools may be critical to competitive advantage in low resource sites.

VI. COMPETITION IN STRESSFUL ENVIRONMENTS

When the optimal functioning of a plant is reduced by stressful abiotic conditions, the relative effects of competition can be difficult to detect. There has been conjecture as to whether competition exists between plants growing in stressful environments as in arid ecosystems. These are high-stress habitats characterized by low water availability. Grime (1973) suggests that competition is not a factor in low productivity environments characterized by low standing biomass. Other ecologists suggest that competition for water would be the principle form of competition in an arid environment (Fowler 1986).

Limited available soil resources in unproductive habitats may lead to intense competition Tilman (1988). In dry habitats, water availability is the major factor in plant growth and survival (Fowler 1986). Water availability fluctuates temporally and spatially in these environments. Competition for water is low when this resource is scarcest (Briones et al. 1998). However, competition intensity increases as this resource becomes available, such as after a heavy rain (Allen 1982, Robberecht et al. 1983). Thus, competition intensity fluctuates with fluctuations in resource availability in these environments (Briones et al 1998).

While competition for resources may occur in dry habitats, the presence of a neighbor can also facilitate the growth or fitness of an individual (by reducing air and soil temperature, reducing evapotranspiration, increasing throughfall of nutrients). In fact, in dry environments, increased soil moisture and nutrients under a plant canopy often leads to facilitation (Goldberg 1990).

The role and importance of competition in environments where abiotic factors strongly influence community structure are still being debated. In environments where resources such as water and nutrients are already limiting, competition for such scarce resources may be an added stress limiting survival and growth of a species (Swank and Oechel. 1991). Grace and Tilman (1990) suggest that the effects of abiotic stress on competition should be measured relative to the productivity of the environment. Grime (1973) suggests that competition intensity is greatest in fertile, productive habitats because these habitats support a high standing crop that results in the consumption of space (decreased light) and nutrients. Tilman (1988) suggests that competition occurs in all environments, with belowground competition most intense in unproductive habitats and aboveground competition most intense in fertile habitats. Goldberg (1990) suggests that the type of environments under which competition occurs is an important determinant of community structure that depends on trends in both resource limitation and degree of resource depletion.

In an attempt to resolve this issue, a recent study using gradients in water availability found that that plant competition tended to increase as the standing crop increased with increased water availability (Kadmon 1995). In a short-term experiment utilizing an experimental nutrient gradient, Aerts et al. (1991b) found there was no apparent effect of interspecific competition for nutrients in a low nutrient treatment although all species tested increased their allocation to root biomass. However, over longer periods of time, in nutrient-poor environments, slow-growing species that invest resources into longer-lived tissue (i.e., show low rates of tissue loss) may have a competitive advantage over fast growers with high tissue turnover (Aerts and Van der Peijl 1993).

VII. COMPETITION AND OTHER BIOTIC INFLUENCES

A. Mycorrhizae

Mycorrhizal fungi facilitate nutrient uptake, particularly phosphorus acquisition, in host plants. Mycorrhizal hyphae act similarly to roots in their foraging activities, branching prolifically in high nutrient patches (St. John 1983). Mycorrhizal plants are better able to extract soil resources than nonmycorrhizal plants (Allen and Allen 1990). Often, experiments will show that there is a development of greater biomass production in inoculated plants versus uninoculated plants because of increased nutrition and water uptake (Smith et al. 1986). In nature, plants differ in their degree of mycorrhizal association, ranging from none to being obligate mycotrophs (Janos 1980). Because of these differences in species affinities for mycorrhizal inoculation, it has been suggested that mycorrhizal fungi may mediate plant competition (Allen and Allen 1990).

Hetrick et al. (1994) found that there was a greater intensity of interspecific competition in the presence of mycorrhizal associations than without myccorrhizae in tallgrass prairie species. The competitive ability of *Andropogon* sp. increased in the presence of mycorrhizal fungi when grown in interspecific competition with a nonmycorrhizal species. However, the nonmycorrhizal species dominated when *Andropogon* sp. was not inoculated, indicating that the presence or absence of mycorrhizal fungi could alter the competitive outcome between these co-occurring species (Hartnett et al. 1993). Also, Hartnett et al. found that the degree to which *Andropogon* sp. benefited from the mycorrhizal association was strongly influenced by plant density, with the competitive effect of neighbors (both intra- and interspecific) overriding the mycorrhizal effect at high densities. Similar results were discussed by Watkinson and Freckleton (1997) using data from West (1996), showing that with mycorrhizal association the increase in intensity of competition was greater than without mycorrhizae, and that mean plant size decreased more rapidly as plant density increased when mycorrhizae were present.

Plants benefit from mycorrhizal associations through increased phosphorus (P) uptake (see Chapter 4). The competitive dominance of *Andropogon* with mycorrhizal association in a low P soil was similar to that of *Andropogon* with no mycorrhizal association but with P fertilization (Hartnett et al. 1993). Most of the experiments described in this section were conducted in pots. A field experiment conducted by West et al. (1993) had contrasting results, in which they found no increase in P uptake with mycorrhizal infection, nor any increase in plant fecundity with either increased added P or mycorrhizal infection. Thus, under natural conditions, the benefits of mycorrhizal infection may be abated by local conditions. For mycorrhizal fungi to change the competitive balance between a host and a neighboring species, the mycorrhiza must induce a greater rate of nutrient and water transfer to its host, which changes the competitive ability of the host plant relative to its neighbor. Recent discoveries of nutrient and carbon transfer between neighboring plants though hyphal linkages may modify our thinking on the potential benefits of a mycorrhizal association toward increased competitiveness (Allen and Allen 1990).

B. Herbivory

Herbivory can influence competitive interactions, particularly if one species, adjacent to another, is negatively impacted by herbivory (see discussion in Chapter 7). As more biomass is supported by the community, and with a concomitant increase in plant cover and food availability, herbivory may increase (Bonser and Reader 1995). Therefore, herbivory may have a higher influence on competitive interaction at high plant density than at low plant density.

Herbivory may interact with competition to change the size distribution and degree of asymmetric competition among individuals, particularly if large individuals are grazed selectively. The removal of dominant individuals through herbivory may reduce resource limitation and thus reduce competition. This is important in light-limited environments. Also in species mixtures, one species could benefit from herbivory if another competing species is more frequently grazed (Weiner 1988). Consequently, herbivory can release some plants from competition altogether, and change the competitive outcome in other plant-plant interactions.

VIII. EXPERIMENTAL APPROACHES FOR STUDYING COMPETITION

There have been many studies of the relationship between planting density and yield expectations of crop plants (yield-density experiments). Experiments that have altered the density and proportion of species' grown in mixture have assisted in determining each species response to competition. Experimental designs have been developed to determine the yield of a target species based on the density of neighbors.

The potential for crop loss from weed abundance can be determined from additive experiments. In *additive studies*, crop species are usually grown at a constant density that maximizes economic yield, and invasive weeds are added at different densities relative to the crop (Roush et al. 1989). For example, species A is sown at a preselected density and species B (usually the weed) is sown with species A in a range of densities. Unfortunately, since both proportion and density changes, it is difficult to pinpoint the mechanism of competition. However, if species A is the target species, this approach does provide information on the species A response to competition as the species B density increases.

The competition between two species can also be studied in a *replacement series*. In this method, a constant total plant density is maintained, but the proportion in which the two species are grown together is varied (Roush et al. 1989). This type of experimentation is primarily useful in evaluating the outcome of intercropping systems. In the replacement series design, an arbitrarily chosen density is maintained constant within which the frequency of two species are varied from 0% to 100%. The use of constant density and short experimental time limits extrapolation of results to longer-term predictions of competitive outcome. Also, the results of this type of experimental design vary with the constant density that is selected.

Most commonly, the analysis of competition is based on yield-density responses of plants grown in monoculture, or in two-species mixtures (Watkinson 1980). Connolly (1986) proposed the study of competition among multiple species using a coupling of the additive and replacement techniques. This approach involves varying the densities of species in monocultures and mixtures to create a matrix of total and relative densities. This experimental design is especially useful for studies of crop and weed interactions because the effects of intra- and interspecific competition are separated by systematic variation of total plant density and species proportion.

Plant weight or yield is often used as a measure for determining plant competitive ability or success in competition. For example, the intensity of competition from the joint response of two species to increasing density can be defined as

$$C_i = \frac{w_{aa} + w_{bb}}{w_{ab} + w_{ba}} - 1 \tag{9.8}$$

where w_{aa} is the weight or yield of species a in monoculture, w_{bb} is weight or yield of species *b* in monoculture, w_{ab} is the weight or yield of species a growing in association with species *b*, and w_{ba} is the weight or yield of species *b* growing in association with species *a*. If the yield of each species is unaffected by competition, C_i will be zero. If there is competition, resulting in the reduction in dry weight per plant compared with the control, then the higher the C_i, the greater the amount of competition (Wilson 1988). Relative yields are often determined in a replacement series where the per-plant yield of species *x* in mixture with species *y* is divided by the per-plant yield of species *x* in monoculture. Relative yield can be calculated for each species and is a good measure of competitive ability (Aerts et al. 1990).

Neighborhood experiments locally identify all potential competitors around an individual (Goldberg and Werner 1983). The *neighborhood density approach* assumes the yield of an individual plant is a function of the number or density of neighboring plants. A multiple correlation is performed for each species between individual plant yield and the density of competing plants. Individual plant yield relates to the number of neighbors in concentric neighborhoods using the following equation:

$$W = \sum \frac{1}{d_i^2} \sum C_j N_{ij} \tag{9.9}$$

where W is the competitive effect, d_i is the mean distance to the ith neighborhood, C_j is the mean effects of an individual of the jth species, and N_{ij} is the number of individuals of species j in neighborhood i (Weiner 1982). It has been found that neighborhood density appears to restrict rather than determine plant size where plants from crowded areas are consistently small. The neighborhood density approach ignores variation in the size of neighbors and does not account for asymmetric competition effects.

There are many studies that document the occurrence of competition in planted communities, pots, or microcosms, but few experimental approaches have been able to demonstrate the effects of competition on community structure or how competition is modified by the environment. In some studies, target species are chosen for manipulation. Selective removal of one or more species from a community or from around individuals of a species has been one approach for studying the effect of competition in altering community structure (Fowler 1981). If the target individual increases growth or fitness following the removal of other species, then the target plant may have been released from competition. However, selective removal does alter the environment of the remaining plants, and this could have affected target plant fitness irrespective of resource availability. This approach has also been used in natural communities made up of single species in monoculture to determine the growth and physiological response to a reduction in intraspecific competition (Robberecht et al. 1983).

The *community-density series*, a modification of *yield-density experiments*, has been introduced by Goldberg et al. (1995). It is a direct approach to studying the effect of community-level consequences of competition. In this approach, the density of the entire community and species in it change along a density gradient, but the proportions of each species in the community remains constant.

Natural environmental gradients have also been used for determining the type of competition occurring and the physiological response in terms of biomass allocation patterns. A gradient of soil depth was used to show that when soil resources were limiting, root competition was much more intense than shoot competition (Belcher et al. 1995). Also, resource gradient approaches have been utilized to experimentally study the role of nutrient availability gradient and the effects of multispecies competition on plant community structure. By growing plants in monoculture or in mixture at different nutrient levels and densities, physiological and ecological response curves can be derived (Austin and Austin 1980).

IX. SUMMARY

The importance of competition to the success of a species in a particular habitat depends on a variety of factors, including the species response to the environment, its resource requirements, and its ability to modify its physiological and morphological characteristics in order to obtain a resource (increase resource uptake). Since competition occurs with respect to a variety of resources, overall competitive ability depends on more than one mechanism. Also, the ability to compete for a resource may change according to the availability of that resource. The capacity of species to make plastic adjustments to the size and distribution of their foraging organs within patchily distributed resources may be important in determining success in most competitive environments. Morphological modifications under competitive conditions include the effective placement of fine roots in areas of high nutrient availability or placing leaves in a position to capture more light. Physiological modifications, such as increasing rates of uptake, may be a more prevalent competitive mechanism in low resource environments, where plants must remain viable during periods of resource depletion but also must also react quickly because resource availability occurs unpredictably.

Morphological and physiological characteristics of plants with strong competitive ability include exploitative root systems (Aerts et al 1991b), high rates of photosynthesis, rapid partitioning of photosynthate to new organs, and high morphological plasticity of leaves and roots. Rapid growth assists in obtaining and maintaining space and increases the opportunity to place foraging organs in high resource areas (Grime 1977). Also, a high capacity for physiological acclimation to a changing or unpredictable environment (Patterson 1985, Grime and Hodgon 1987) will increase competitive ability.

Course-grained foraging for light and mineral nutrients, by the development of extensive leaf and root systems, is often characteristic of dominant plants in fertile environments (Grime 1994). Subordinate plants may maintain a place in the community through fine-grained foraging, in which there is precise placement of leaves and roots in small, localized areas of resource enrichment (Campbell et al. 1991b). In this way, diverse community structure and diversity is maintained in high resource environments.

STUDY-REVIEW OUTLINE

Introduction

1. Plants require resources both above and below ground. The two sets of resources are separate in the spatial sense but linked in the ecological or physiological sense.
2. Competition occurs when two species interact and both species experience a decrease in fitness.
3. Resource limitation to the collection of plants in a region is an underlying concept of competition.

4. Competition can regulate the fitness of individual species and its distribution and population size. In addition, competition can have secondary effects on community structure and development.
5. Effective competitiors can be able either to draw down the concentration of resources rapidly or tolerate very low resource availability.
6. In early successional stages, the most competitive species are those that are able to capture resources rapidly, while in late sucessional stages, species that are able to efficiently capture and retain resources are most effective.
7. Competition for resources is manifest in changes of basic physiological processes such as photosynthesis, respiration, and nitrogen metabolism.
8. Competitive effects are usually measured on the basis of comparative growth or fitness in populations of different density or species mixture.
9. Plants have four basic options for being effective competitiors:
   a. Increase resource uptake (alter size, morphology, or physiology of canopy or root systems)
   b. Increase the efficiency of resource acquisition (alter the behavior or physiology of acquisition systems)
   c. Decrease resource loss (maintain long-lived structures)
   d. Increase the efficiency of conversion of internal stores

Resources

Resource Characteristics

1. A *resource* is an aspect of the environment that is required for the growth and fitness of a species and is depletable or consumable.
2. The availability of resources to plants is spatially and temporally variable and is dependent upon the nature of the environment as well as the nature of neighboring organisms.
3. The level of each resource required by species is variable among species and among developmental stages of each species.
4. Resources vary in the environment qualitatively and quantitatively on spatial and temporal scales.
5. The availability of resources to plants is not the same as the resource level, and is determined by biotic and abiotic attributes of the environment.
6. Soil nutrients vary spatially and temporally because of the following:
   a. The effect of plants on microtopography, microclimate, and organic inputs
   b. The development of nutrient depletion zones around roots
   c. The amount, location, timing, and chemical composition of mineral and organic inputs
   d. The effect of microorganisms on decomposition, mineralization, and soil mixing
7. The ability of plants to perceive soil nutrient heterogeneity depends on root proliferation, mycorrhizal association, and root physiological responses to nutrient-rich patches.
8. Soil water availability to plants is variable due to variation in the supply from sources and variation in the physical and chemical attributes of the soil.
9. Water accumulated from the atmosphere occurs only in rare cases and so is not considered in this review.
10. Space is an important resource for seed germination, plant maturation, and fitness.
11. Appropriate *microsites* for germination are restricted in the community and vary among species based on species germination requirements.

Resource Capture

1. The degree of resource competition that a plant experiences from its neighbor depends on its ability to capture limiting resources and on how it utilizes the resources relative to its neighbors.
2. A trade-off may exist between the ability of a plant to acquire aboveground versus belowground resources.

3. Plants that grow in highly competitive environments often have a relatively high phenotypic plasticity, because such plasticity maximizes the plant's ability to acquire resources.
4. Some species are able to use both above- and belowground plasticity to increase competitive ability.

Intraspecific and Interspecific Competition

1. *Intraspecific competition* occurs when two individuals of the same species interfere with the growth and success of each other.
2. Intraspecific competition is most common in agricultural fields and forestry plots because in such areas the species diversity is low and the individual density is high.
3. Natural monocultures do exist in some cases, and these stands would be expected to harbor considerable competition.
4. Intraspecific competition can be intense because all the individuals that interact have the same resource requirements and the same or similar phenology.
5. Most of the effects of intraspecific competition are density dependent.
6. *Interspecific competition* occurs between two different species that each use the same limiting resources.
7. Competitive interactions in natural plant communities are most likely to be interspecific.

Competition in Agronomic and Natural Systems

1. Competition occurs through the same mechanisms in agronomic and natural situations; however, competition is simpler in agronomic systems.
2. Competition is manifest through plant size and reproductive effort in relation to plant density.
3. Agronomic methods are designed to maximize yield even if a significant amount of competition is occurring.
4. In natural systems, competition can lead to competitive exclusion or coexistence in the long term.
5. An evolutionary view of competitive effects is important when studying plant-plant interactions in natural systems.
6. Over time, a separation of resource use can occur among competing species such that they coexist with minimum competition.
7. Species with very similar resource requirements can coexist in natural systems by maintaining small ecological differences during their life cycles.
8. Because the mechanisms of competition are similar in agronomic and natural systems, many researchers have studied simple agronomic systems in order to better understand more complicated natural systems.
9. The purpose of studying competition in natural systems is to understand the mechanisms behind succession, plant distribution, and the dynamics of communities.
10. There has been some disagreement over the methods for measuring competition, much as there is for measuring allelopathy.
11. The effects of competition are often observed (inferred) from plant distributions. There are too many possible interactions that can cause species distribution patterns that confound simple interpretation of observation.
12. Competition experiments are often performed in pots based on two designs:
    a. *Additive,* where two species are mixed and the density increases with the mixture.
    b. *Substitutive,* where two species are mixed in pots at different proportion but the same density is maintained in each pot.
13. The possible factors regulating competitive success vary in environments with high or low resource availability.
14. Plants in high resource sites have a competitive advantage if they grow fast and have maximum resource uptake.
15. Plants in low resource sites have a competitive advantage if they have greater tissue longevity and a greater efficiency of resource acquisition.

Models of Competition

Single-Species Population Growth

1. Single-species population growth is described as exponential until density-dependent factors (competition) limit the rate of increase.
2. The logistic equation is normally used to describe single-species population growth as affected by density-dependent intraspecific competition.
3. The effect of intraspecific competition is described by N/K (population size/carrying capacity). As N/K approaches 1, maximum competition occurs and population growth becomes zero.
4. The value of K varies among species because of individual species' responses to resource availability.
5. The logistic equation has provided the basic concept for the development of multispecies competition theory.

Plant Density and Yield Under Competition

1. Yield-density relationships provide a measure of intraspecific competition and the degree of plastic response to density.
2. As density increases above the levels at which intraspecific competition becomes evident, individual plant size decreases.
3. As density increases above the level of intraspecific competition, the biomass on the site remains the same while individuals become smaller.
4. When planted at very high density, individual plants die as others gain size and maturity. This is the process of *self-thinning*.
5. Yoda's *self-thinning* rule describes the relationship between mean weight and the number of surviving individuals.
6. When the log of mean weight is plotted against the log of mean density, the resulting curve has a slope of $-\frac{3}{2}$, and this relationship is called the $-\frac{3}{2}$ thinning rule.
7. The $-\frac{3}{2}$ thinning law applies to a wide variety of species in many habitats.
8. If individuals continue to grow in size after the point where the habitat cannot support any more biomass, then the $-\frac{3}{2}$ relationship becomes a -1 relationship.

Multispecies Models of Competition

1. There is a long history (dating back to the 1930s) of modeling the interactions between species based on competition.
2. The Lotka-Volterra equations were the first to consider the effect of one species on another through competition.
3. The *Lotka-Volterra equations* can describe situations of coexistence and competitive exclusion.
4. These models are based only on population size and do not relate to mechanisms of competition.
5. Several models have been developed with a mechanistic view of competition.
6. The *resource depletion models* developed by Tilman are the best known and most utilized.
7. The basis for these models is the concept that the species (population) that maintains the fastest growth rate for the most limiting resource will be the superior competitor, given enough time.
8. One species is able to draw down resources to a point that they exclude other species.
9. The isocline of zero population growth in the resource depletion models describes the point at which resource supply exactly matches resource consumption by the population. The lower the isocline a species can maintain, the better it is at competing for those resources.
10. These models can also predict coexistence and competitive exclusion based on the different combinations of responses of populations to limiting resources.
11. The resource depletion models have limited application to natural systems because of the number of different resources and their potential interactions.

12. The resource depletion competition models have been the basis of many more studies in the mechanism of competition in plants.

Community Competition Theory

1. *Two-sided symmetric models* are based on the resource depletion or resource reduction mechanism of competition.
2. In *size symmetric competition,* all individuals are reduced in size proportionately so that size differences remain constant through competition.
3. In size-symmetric competition, small plants are as capable of obtaining resources as large plants.
4. In *asymmetric competition* (*one-sided*), larger plants obtain more than their proportional share of resources, suppressing smaller plants.
5. The asymmetric models of competition are characterized by an increasing difference between plant size over time. The models are called *resource preemption.*
6. In resource preemption models, the plants that first arrive on a site are more likely to obtain a disproportionately large portion of the available resources.
7. Early establishment, fast germination, rapid growth, and large allocation to leaf area all serve to enhance a plant's competitive ability in resource preemption competition.
8. Resource preemption competition often occurs when species compete for light.
9. Root competition tends to be two-sided (resource depletion type) competition, while aboveground competition tends to be resource preemption–type competition.

Plant Responses to Competition

Aboveground and Belowground Competition for Resources

1. The most common response of plants to competition is a shift in the allocation pattern between root and shoot biomass.
2. Investment in leaves increases photosynthesis and growth rates, while investment in roots increases resource capture below ground but reduces growth rate.
3. Changes in allocation patterns are critical to competition in the resource depletion models.
4. An interaction between the effect of above- and belowground resource limitation should be noted: resource limitation above or below ground affects both the organs above and below ground.

Morphological Responses

1. Plant architecture can be dramatically influenced by competition.
2. The architecture of plants is under general genetic control, but flexibility is possible in response to some environmental cues.
3. The environmental cues used to regulate flexibility in plant architecture are also affected by neighboring plants.
4. Flexibility in organ development is critical to performance in plants when space is limiting or resources are patchy.
5. Plant architecture and physiological processes that regulate growth are interrelated.
6. *Branch architecture,* in which leaves are placed above those of competitors, assists in avoiding the canopy of other species and increases light availability.
7. Plasticity in petiole and internode length is a character that may provide competitive advantage.
8. *Photomorphogenic responses* of plants are changes in morphology induced by different ratios of red to far red light. The most dramatic photomorphogenic response is shoot elongation.
9. At decreased red/far red ratios shoots elongate more with no change in biomass.
10. Photomorphogenic responses are important for species that are the same relative height, or for intraspecific competition.

11. Photomorphogenic regulation of internode length is not an important trait for competition among size-asymmetric groups of species.
12. Soil nutrients are patchy, with rich areas and depleted regions.
13. Resource depletion zones develop around roots due to the limitations of nutrient mobility in the soil.
14. The plasticity of root systems to nutrient heterogeneity in the soil is reflected by the precision with which roots can proliferate in nutrient-rich patches (precision), the sensitivity to different degrees of nutrient availability (sensitivity), and the magnitude of the root mass (scale).
15. In a soil volume occupied by the roots of several species, the species best able to locate and accumulate the nutrients in rich patches has a competitive advantage.
16. In some species, there is no association between total root mass and competitive advantage. Physiological traits are also important in determining the relationship between root architecture and plant competitive ability.
17. Reduced light availability may constrain the ability of root systems to forage for nutrients.
18. Increasing the ability to absorb nutrients from a given soil volume depends on the ability to proliferate roots in nutrient-rich patches, and the ability to change nutrient uptake kinetics.
19. Plants that have a high growth rate may locate patches quickly and deplete the patches quickly. These plants usually require high nutrient conditions, so after patch depletion, roots die and the patch is colonized by the root of species with slower growth rates, which can tolerate lower nutrient patches.
20. Root demography can also influence the way fine roots can utilize nutrients in the soil.
21. Belowground resource foraging can influence aboveground physiology.
22. In a competitive environment, changes in morphology and belowground accumulation of nutrients are more important to growth rate than are physiological processes.
23. Generally, the alteration of physiological or morphological characteristics for increased resource uptake does not automatically increase competitive ability.
24. Slow growers have a competitive advantage over fast growers in nutrient-limited environments because increased tissue longevity decreases the cost of constructing the tissues.
25. Characteristics of the root system that enhance the ability to absorb temporal pulses of nutrients confer a competitive advantage to species in low nutrient environments.

Physiological Responses

1. The rates of certain physiological activities affect both the use of resources and the ability to accumulate resources.
2. Plants that are able to maintain positive and relatively stable rates of important physiological processes in a wide variety of environmental conditions are good competitors.
3. Rates of physiological activities need to be integrated with changes in allocation patterns in order to fully understand the mechanisms by which plants are effective competitors.
4. Initial colonizers of disturbed sites tend to be fast growers, with high light saturation rates and high light compensation points for photosynthesis.
5. Initial colonizers also characteristically have a high degree of plasticity in their physiological rates under different environmental conditions.
6. When competition is intense (i.e., plant density is high), these early colonizers have large reductions in growth rate and photosynthetic rates.
7. In later stages of community development, species with lower rates of photosynthesis, lower allocation to leaf area, and greater belowground foraging ability usually predominate.
8. A diversity of traits, both physiological and morphological, are used by plants to grow in low light environments. These traits serve for tolerating of low light conditions, not as mechanisms of enhancing competitive ability.
9. In high light environments, physiological activity must be maximized in order to maintain a competitive advantage. Therefore, physiological rates are important for competition in high resource environments.

10. In symmetric competition, belowground physiological plasticity is critical to competitive advantage.
11. The ephemeral nature of soil nutrients makes the induction of physiological activity critical to competitive advantage below ground when considering ions that are mobile in the soil.
12. Different strategies of roots systems are required to maximize the accumulation of different soil nutrients.
13. Physiological plasticity is most important for absorbing nutrients that are mobile in the soil and particularly ephemeral in time.
14. Physiological plasticity of uptake kinetics involves the balance of activity of different transporters in the root cell membrane.
15. Most nutrients are accumulated into plants in a biphasic manner by a low-affinity transporter (LATs) and a high-affinity transporter (HATs).
16. The relative abundance of HATs in the root membrane determines the capacity of nutrient accumulation.
17. Under very high nutrient conditions, the abundance of LATs increases in some species that can increase total nutrient acquisition.
18. Resources below ground are more diverse that those above ground.
19. The transfer of various elements to roots is regulated by different processes for each element.
20. The transfer of calcium to roots depends upon root growth into new soil.
21. The transfer of magnesium and sulfates to roots is dependent upon the mass flow of water to roots.
22. The transfer of phosphate to roots is limited by diffusion in the soil solution.
23. Nitrate accumulation can be enhanced during nutrient pulses by upregulating the genes that produce high-affinity transporters.
24. During nutrient pulses, the affinity of roots for potassium rapidly increases.
25. The accumulation of mobile ions is dependent upon the maximum capacity of nutrient accumulation and the efficiency of the transporters.
26. Phosphate accumulation cannot be enhanced by increasing the kinetics of membrane transporters.
27. Alternative mechanisms of phosphate accumulation include utilizing mycorrhizae, excreting phosphatases to accumulate phosphate from inorganic sources, and releasing organic acids and chelators to increase the solubility of phosphate from rock phosphate.
28. Plants that have a high competitive ability in low phosphorus soils are those that maximize the alternative methods of phosphate accumulation.

Competition in Stressful Environments

1. There is some controversy over whether competition occurs in stressful environments.
2. Some scientists say that tolerance of resource limitation is critical in stressful environments, so competition is not a factor.
3. Other scientists say that competition for an extremely limiting resource such as water in a desert must be occurring in natural systems.
4. In desert environments, water availability is patchy. When availability is low, no competition is occurring; however, at times when water is abundant, such as immediately after the rare rainfall events, competition is intense.
5. In environments where resources are limiting, competition may be an added stress.
6. Tilman suggests that competition occurs in all environments, but belowground competition is more intense in low resource sites, and above ground competition is more intense in high resource sites.

Competition and Other Biotic Influences

Mycorrhizae

1. Mycorrhizae enhance nutrient and water accumulation by plants in all ecosystems.
2. The relative colonization of species by mycorrhizae varies, and this may be a factor in competitive ability of plants.

3. Mycorrhizae may mediate belowground competition in plants.
4. The presence and absence of mycorrhizae can alter the competitive outcome of a two-species interaction.
5. The benefits of mycorrhizal association can be modified by local environmental conditions.
6. Mycorrhizal connection between members of a species can result in shared resources and a greater competitive ability of the population.

Herbivory

1. Herbivory can influence the outcome of competition.
2. There is a concomitant increase in both competition and herbivory as biomass increases, so these processes interact to a greater degree in high productivity sites.
3. The removal of dominance by one species through herbivory increases resource availability and reduces competition.

Experimental Approaches for Studying Competition

1. Competition in crop systems, between crops and weeds, is usually studied with additive designs.
2. In additive designs, the density of the crop is kept constant and weeds are added in various densities.
3. The additive approach is not effective in determining the mechanism of competition, but it does provide evidence for determining the density of weeds that results in significant reduction in crop yield.
4. In a *replacement series*, the density is maintained constant but the proportion of species contribution to the density is changed.
5. The use of constant density and short experimental times in replacement series experiments limits extrapolation of results to longer-term predictions of competitive outcome.
6. The results of replacement experiments vary with initial density; thus, results vary because the competitive effects are dependent on total density.
7. A common experiment uses monocultures and proportional mixes at various densities to produce a surface of interaction. The results of this combined experiment can be used to evaluate both intra- and interspecific competition.
8. Measuring relative yield in a replacement series can be a good indicator of competitive ability.
9. *Neighborhood analysis* involves a correlation between the yield of plants and the number of neighboring plants.
10. Neighborhood analysis ignores the size of neighbors and does not account for asymmetric competition.
11. In natural communities, competition has been studied by *selective removal.* If the target plant increases growth after removal of other species, this is taken as a release from competition.
12. This technique is difficult because removal of some individuals also causes changes in the abiotic environment that could affect growth of the target species irrespective of the resource availability.
13. *Natural resource gradients* have been probed for information related to competition.
14. Natural resource gradients have been used to evaluate the effect of the physiological response to resources and its relationship to competition.

SELF-STUDY QUESTIONS

1. What is the definition of competition?
2. What are some basic characteristics of resources that are important for competition theory?
3. What factors dominate the availability of water resources?
4. What are the factors that dominate the availability of light resources?
5. Why is space an important resource for plants both above and below ground?

6. Why is plasticity in morphology important for resource capture?
7. Differentiate between the definitions for *intra-* and *interspecific competition.*
8. For what aspects of ecology or agronomy is intraspecific competition important?
9. What processes of the natural community and agronomic field are affected by interspecific competition?
10. Do the processes of competition differ between agronomic fields and natural systems? If so, how does this affect the importance of competition in these two systems?
11. What is the logistic curve, and how does this relate to competition?
12. What is the relationship between plant density, plant size, and plant yield? Why is this important to natural and agronomic systems?
13. What is the $-\frac{3}{2}$ law, and how does this relate to competition?
14. Plot the isoclines for the Lotka-Volterra competition model. Explain the vectors for population change on this surface. Under what conditions can populations coexist?
15. What are the limitations of the Lotka-Volterra competition models?
16. Make a two-dimensional plot of the resource depletion model. Under what conditions can two species coexist? Under what conditions does one species win over the other?
17. Can the resource depletion model be applied directly to natural ecosystems? If not, why not?
18. Describe the difference between the resource preemption and resource depletion models of competition.
19. What is the difference between symmetric and asymmetric competition? To what resources do asymmetric and symmetric competition best apply?
20. In what ways do the morphological aspects of plants vary in response to competition from other species above ground.
21. Describe the meaning of *precision, sensitivity,* and *scale* in relation to root foraging processes.
22. Which is more important for nutrient acquisition—root foraging behavior, or changes in uptake kinetics of nutrient transporters?
23. Which environments favor the importance of physiological factors for competitive interaction above ground?
24. What are the physiological factors that can determine the competitive ability of a species above ground?
25. What are the important constraints on foraging for nutrients below ground that differ from those on foraging for aboveground resources?
26. What three factors determine the transfer of nutrients to roots in the soil?
27. Which transfer processes are important for the accumulation of nitrate, phosphate, calcium, and potassium?
28. What characteristic of nutrient transporters can be changed to increase the amount or efficiency of accumulation?
29. Which nutrients are most affected by changes in nutrient transporter capacity? Which are least affected? Why?
30. What processes is the competitive ability for phosphate accumulation dependent upon?
31. Does competition occur in stressful environments? What are the arguments for and against this concept?
32. How does the presence of mycorrhizae interact with the process of competition?
33. How does herbivory relate to competition?
34. What is the additive experimental design for studying competition? What are its limitations?
35. What is the replacement series experimental design for the study of competition? What are its limitations?
36. What is the advantage of the response surface analysis of competition?
37. What is a neighborhood model, and what are its limitations?
38. How can competition be studied in field situations?

SUPPLEMENTARY READING

Bazzaz, F. A. 1996. Plants in Changing Environments: Linking Physiological, Population, and Community Ecology. Cambridge University Press, Cambridge, UK.

Bergon, M., J. L. Harper, and C. R. Townsend. 1996. Ecology, 3rd Ed., Blackwell Scientific, Oxford, UK.

Bergon, M., M. Mortimer, and D. J. Thompson. 1996. Population Ecology, 3rd Ed. Blackwell Scientific, Oxford, UK.

Connel, J. H. 1961. The influence of interspecific competition and other factors on the distribution of the barnacle *Cthalamus stellatus*. Ecology 42:710–723.

Connolly, J. 1987. The use of response models in mixture experiments. Oecologia 72:95–103.

Grace, J. B., and D. Tilman (Eds.). 1990. Perspectives on Plant Competition. Academic Press, San Diego, CA.

Grime, J. P. 1979. Plant Strategies and Vegetation Processes. John Wiley & Sons, Chichester, UK.

Harper, J. L. 1977. Population Biology of Plants. Academic Press, London.

Hastings, A. 1997. Population Biology. Springer-Verlag, New York.

Silverton, J. W., and L. Lovett Doust. 1993. Introduction to Plant Population Biology. Blackwell Scientific, Oxford, UK.

Smith, T. M., H. H. Shugart, and F. I. Woodward. 1997. Plant Functional Types. Cambridge University Press, Cambridge, UK.

Tilmanm D. 1982. Resource Competition and Community Structure. Princeton University Press, Princeton, NJ.

Tilman, D. 1988. Plant Strategies and the Dynamics and Structure of Plant Communities. Princeton University Press, Princeton, NJ.

Tilman, D., and P. Karieva. 1997. Spatial Ecology: The Role of Space in Population Dynamics and Interspecific Interactions. Princeton University Press, Princeton, NJ.

10 Parasitic Vascular Plants

Outline

Objectives

1. Present the diversity and classification of parasitic vascular plants.
2. Explain the probable evolution of the parasitic nature in vascular plants.
3. Discuss ecological consequences of parasitic plants in natural environments.
4. Introduce the problem of parasitic plants as weeds in agricultural systems, and describe the magnitude of the problem.
5. Describe important aspects of parasitic angiosperm life cycles (such as seed germination and haustoria formation), particularly as they pertain to interaction with their host plants.

6. Explain the general mechanisms by which parasitic plants share resources and defensive compounds with their host plants.
7. Discuss the types and magnitudes of detrimental effects that parasitic vascular plants can have on their hosts.
8. Describe the molecular responses elicited in host plants by parasitic plant attack.
9. Explain the factors that determine host plant resistance to parasitic plant infection.
10. Explain the mechanisms by which parasitic plants influence host plant response to environmental stress.

I. INTRODUCTION

The term *parasite* usually congers an image of a tapeworm or liver fluke in one's mind. Although it is true that there are many animal parasites, this is not an exclusive trait of the animal kingdom. There is an abundance of parasitic angiosperms that constitute a stress to their host plants. Some of these plants are obviously parasitic because they lack any chlorophyll and they are evident in the landscape. Others can be less obvious because their parasitic connections are below ground: they have a normal nonparasitic appearance, they are completely submersed in the tissues of a plant, or they are completely subterranean.

Knowledge of parasitic plants goes back to the writings of Theophrastis, who reported on a dodder covering vegetation in Babylon, and wrote about several mistletoes (Kuijt 1969). Arabian scholars of the 10th century were aware of the parasitic nature of plants, as the following passage translated by Kuijt (1969) indicates: "Another kind, the actions of which correspond to those of the animal soul while its body remains that of a plant, is the parasitic plant, for it attaches itself to trees, seeds, and thorns, and feeds itself as the worm from the juices of the host plant." In the Western world, the first published account of parasitic plants occurred in the early 1700s, when Micheli (1729) realized that *Cynamorium* (Balanoporaceae) was not a fungus but was a parasitic plant. Individual recognition of parasitic species occurred throughout the 18th and 19th centuries, but the botanical field was surprised when Mitten (1847) revealed that *Thesium linophyllum* (a plant well known to botanists) was a root parasite. Western botanists of the early 19th century thought that parasitic plants comprised a unique group, separate from nonparasitic plants. Trattinick (1828), as translated by Kuijt (1969), emphasized that opinion in the following passage: ". . . we therefore have no choice but to cast them [the parasites] together, as oddities, into their own category, much as in an asylum, we bring together the mentally ill, whose mania are extremely varied, but of whom no one is really what he pretends or imagines to be." Modern understanding of parasitism in vascular plants can most likely be traced back to the monograph of the Balanophoraceae by Hooker (1856) and the demonstration of seed germination in *Cynomorium* by Weddell (1858–1861).

There is no doubt that parasites have important effects on individual plants and plant populations in natural systems. This is evident when one observes that the sprawl of dodder completely blots out the other vegetation in some salt marsh communities, or that the top of a *Prosopis* tree in the desert is dripping with large masses of mistletoes. Parasitic plants also have many importances to human populations. They constitute a significant weed to agricultural systems, particularly species of *Striga, Cuscuta,* and *Orobanche*. Parasitic angiosperms have been and are currently used as a source of food. For example, the fleshy stems of *Pholisma sonorae* (sand food) have been used as a food source for native populations in the American West, and in Afghanistan two species of *Orobanche* were once eaten like asparagus. Some parasites, particularly mistletoes, are utilized for animal feed. For example, the poor grade of Christmas mistletoe is used for sheep fodder (Griffis 1956). Woody parasitic plants in the Santalacaceae and Olacaceae are utilized for building materials, furniture, and wooden ornaments. In particular, the fragrance of sandalwood (*Santalacum album*) is world renowned, and carries a very large economic significance. In fact, many populations of this tree were completely eradicated because of their economic value. Of course, there is the cultural association of mistletoe with kissing during the Christmas season in the Western world, which most likely dates back to the Druids (Coghan 1995a).

Other human uses of parasitic plants concern the potential medicinal or economic value of products from parasitic plants. Sandalwood oil is widely sold for anointing the body and for manufacturing soaps and perfumes. Wax from the seeds of several members of the Balanophoraceae is used in the candle and cosmetic industries. Extracts of mistletoes have been used to treat "distempers" in Europe, Africa, and Asia, and in the relief of digestive disorders by the California native Americans (Kuijt 1969). In fact, root parasites, such as *Kramaria* in Central and South America (Simpson 1989), and *Hydnora* in parts of Africa (Musselman 1991), that accumulate tannins are used as antidiarrheal medicines. However, it is also true that extracts of some mistletoes are toxic (Coghan 1995b). Modern medicine is pursuing several extracts of mistletoes for the treatment of various forms of cancer (Rostok and Ungar 1998). The most active component of extracts from *Viscum album* is mistletoe lectin (ML-1), although other components of the extract also may affect activity (Stein et al. 1998). Incubation of peripheral blood lymphocytes in ML-1 stimulates the activity of T-cells (Baxevanis et al. 1998) by a different mechanism than that of interleukin. In fact, the combination of ML-1 lectin and interleukin may be a very effective treatment for adjuvent cancers.

The potential applications of parasitic plants to human culture and the detrimental effects of parasitic plants to agriculture have made this topic an attractive one for scientists throughout the history of botanical sciences. The result is a broad and healthy literature based on the general botany and economic significance of these species. However, the knowledge of the physiological relationships between host and parasite, and the significance of parasitic plants to natural systems, is less well developed. The purpose of this chapter is to review what we know about the relationship of parasites to plant stress physiology and set the context for this knowledge in a broader framework. Both agricultural and natural systems will be included in this discussion. We will cover the diversity of parasitic plants, their potential effects on plants at a molecular, whole-plant, and population level, and their importance as a stress in agricultural and natural systems.

A. Functional Classification of Parasitic Plants

The most important terminology for classifying parasitic plants is that which defines the extent of autotrophism. *Holoparasitic plants* are those species that can not perform photosynthesis and must attain all of their reduced carbon from the host plant. Holoparasites are obligate parasites that depend on their host to complete their life cycle. These parasites often lack the ability to assimilate inorganic nitrogen. In contrast, *hemiparasitic plants* are those species that can perform photosynthesis (usually at a reduced rate) yet may still receive a portion of their reduced carbon from the host.

Some hemiparasites are able to complete their life cycle without a host in a controlled environment, and these species are called *facultative parasites*. All parasitic plants (hemi- or holo-) are connected to their host by a haustorium. The *haustorium* is most likely a modified root that is inserted into the host tissue, eventually forming the connective structures required for transfer of resources from host to parasite. The portion of the haustorium located inside host tissues is called the *endophyte*. The distinction between holoparasites and hemiparasites is not clear-cut. In fact, hemiparasites can range from being almost completely dependent on, to almost completely independent of, photosynthate flow from the host.

One functional classification of parasitic plants is based on the location of their haustarium connection to the host. Stem parasites are connected to hosts above ground usually by attacking through the cortex of the stem or leaves into the phloem and xylem tissues. Examples are the hemiparasite dwarf mistletoe (*Aceuthobium* spp.) and the holoparasite dodder (*Cuscuta* spp.). However, there are some stem parasites that are completely contained in their host (cactus), and only the flowers of the parasite extend out of the host stem. An example of such a parasite is *Tristerix aphyllus* on the cactus *Echinocloa chilensis* in Chile (Maseuth et al. 1984). Root parasites are those that connect to the host root, and the rest of the parasite (above ground) may be an herb, shrub, or tree. Examples of herbaceous root parasites are the hemiparasites witchweed (*Striga* spp.) and yellow rattle (*Rhinanthus sclerotinus*), and the holoparasite broomrape (*Orobanche* spp.). Some examples of shrubby root parasites are castor bean (*Ricinis communis)* and Olax *Olax* sp.). Large tree root parasites are represented by the Christmas tree (*Nuytsia floribunda*) and sandalwood trees (*Santalaca* sp.). Some root parasites are completely subterranean except for the infloresense, such as sand food (*Pholisma sonorae*) and broomrape (*Orobanche* sp.).

The presence of chlorophyll is also used as a classification mechanism. *Achlorophyllous* (without chlorophyll) species are both obligate parasites and holoparasites such as broomrape (*Orobanche* spp.) and Indian pipes (*Monotropa* sp.). Other parasites (chlorophyllous) contain chlorophyll in their leaf tissues; examples are mistletoes (*Loranthus* spp., *Viscum* spp.), although their photosynthetic rates may be very depressed. Some chlorophyllus parasites, such as Indian paintbrush (*Castilleja* spp.), primarily gain nutrients and water from their host. In fact, species such as Indian paintbrush may be facultative parasites because they can survive on their own ability to absorb nutrients and reduce carbon when nutrients are relatively plentiful (Heckard 1962). In such species, parasitism may be engaged only at times of nutrient or water limitation. However, in natural situations, so-called facultative parasites such as *Castilleja* sp. or *Rhinanthus minor* are always found to be parasitic. Thus, the facultative nature of parasitism may be an artifact of growing these species in controlled conditions of high resource availability, which these plants never experience in nature.

Some parasitic plants are connected to their host plants by a secondary organism, mycorrhizae. These associations are termed *epiparasites* because the host plant is not directly supporting the parasitic plant. Epiparasites can also occur when one parasitic plant is connected to another parasitic angiosperm by a haustorium. Members of Monotropaceae are holoparasitic species that are attached to their host by shared ectomycorrhizae (see Chapter 4). The association can be very specific and involve only one mycorrhizal fungus (Cullings et al. 1996). Furthermore, many orchids are parasitic on a host plant through a mycorrhizae during the early stages of their development.

Parasitic plants also are differentiated into generalists or specialists in a similar manner to that for herbivores (see Chapter 7). *Specialist plant parasites* are very selective in their host species (Kuijt 1969). For example, *Epifagus virginiana* (Orobanchaceae) is parasitic only on *Fagus grandifolia, Conopholis americana* (Orobanchaceae) only attacks *Quercus borealis,* and *Striga gesnerioides* (Scrophulareaceae) is parasitic on only dicots (all other *Striga* species are parasitic on a wide diversity of monocots). Also, there are some mistletoes that are parasitic only on other mistletoes, and the dwarf mistletoe *Arceuthobium minutissimum* has only been found on *Pinus griffithii.* Specialist parasites, particularly those that colonize only one host species, are infrequent compared to generalist plant parasites.

An excellent example of a generalist parasite in *Olax phyllanthi* (Olacaceae) a parasitic shrub in Australia that parasitizes herbs (annual or perennial), shrubs, trees, and other parasitic plants. Another generalist is the mistletoe *Dendrophthoe falcata* that is known to have at least 343 host species in nature (Narashima and Rabindranath 1964). *Rhinanthus minor* is a widespread root parasite in Europe. When the host range is carefully examined it is clear that the parasite is not simply attacking the most abundant roots. Yet, this parasite is considered a generalist because different host species are attacked by different individuals in the same population of *Rhinanthus minor* (Gibson and Watkinson 1989). When interpreting the generalist nature of a parasitic plant, it is important to understand the difference between host range and host preference. The root parasite witchweed (*Striga* spp.) will infect a wide array of monocots and dicots in laboratory situations. Thus, witchweed has a wide host range and is classified as a generalist. However, in nature, witchweed is predominantly found on grass roots, and is not found on roots of many other monocots or any dicot species. Therefore, the host preference of witchweed is much smaller than the host range. Similarly, dodder (*Cuscuta*) a generalist parasitic vine, has a very wide host range, and can attack several hosts at the same time, but in the field, the number of hosts this parasite successfully attacks is much smaller than its host range.

Like *Cuscuta* sp., *generalist parasitic plants* can attack several host species simultaneously. It is clear that parasitic plants can derive different substances from each different host species they attack (Grovier et al. 1967). Therefore, it is possible that parasitic plants that attack two different host species simultaneously may have better performance than if they attacked either of the host species alone. Such a result was shown with *Castilleja* (Scrophulariaceae), a generalist hemiparasitic root parasite (Marvier 1998a). *Castilleja wightii* tended to have greater growth on mixed-host species, and a higher flowering frequency than when grown on a single host species. The increased performance of parasites that utilize several hosts may be due to such parasites receiving a more useful balance of nutrients than parasites that attack only one host (Grovier et al. 1967). Moreover, it is possible that utilizing two host species simultaneously results in a lower dose of toxin from either host species or an antagonism between toxins from the two host species, effectively negating the chemical defense mechanisms of the host species (Marvier 1998a). Although these ideas about toxin interaction from multiple hosts are interesting, few data are available to evaluate these relationships.

B. Taxanomic Diversity of Parasitic Plants

Parasitic plants in the broad definition (including hemi- and holoparasites) account for about 1% of all flowering plant species. It is interesting to note that this percentage is very similar to that of halophytic plants (see Chapter 5). Within the angiosperms, 22 different families contain parasitic members (Table 10.1). Of those families, some are dominated by parasitic species such as the Orobanchaceae, Rafflesiaceae, and Cuscutaceae, while others have only a few parasitic members such as the Scrophu-

TABLE 10.1 Most of the Families and Genera of Parasitic Plants

Family	Type of Parasitism	Number of Genera/Species	Representative Genus
Balanophoraceae	Holoparasites	18/100	*Balanophora*
Cuscutaceae	Holoparasites	1/180	*Cuscuta*
Cynomoriaceae	Holoparasites	1/2	*Cynomorium*
Eremolepidaceae	Hemiparasites	3/10	*Eremolepis*
Hydnoraceae	Holoparasites	2/20	*Hydnora*
Krameriaceae	Hemiparasites	1/20	*Krameria*
Lauraceae	Hemiparasites	1/30	*Cassytha*
Lennoaceae	Holoparasites	3/6	*Pholisma*
Loranthaceae	Hemiparasites	50/1000	*Loranthus, Nuytsia*
Myzodendraceae	Hemiparasites	1/10	*Myzodendron*
Olacaceae	Hemiparasites	4/20	*Olax*
Opiliaceae	Hemiparasites	2/2	*Cansjera*
Orobanchaceae	Holoparasites	14/160	*Orobanche, Conopholis*
Podocarpaceae	Hemiparasite	1/1	*Parasitaxis*
Rafflesiaceae	Holoparasites	9/50	*Rafflesia*
Santalacaceae	Hemi and holoparasites	25/450	*Dendrotrophe, Exocarpus*
Scrophulariaceae	Hemi and holoparasites	10/500	*Striga, Rhinanthus*
Viscaceae	Hemi and holoparasites	7/500	*Phoradendron, Viscum*

Source: From Atsatt 1983, and Luther and Becker 1987.

lariaceae, Viscaceae, and Santalacaceae. The total number of parasitic vascular plant species is approximately 4000 (Nickrent et al. 1998).

Although the great majority of parasitic plants are angiosperms, there may be one parasitic gymnosperm. It has been suggested that *Parasitaxus ustus,* a member of Podocarpaceae from New Caledonia, is a hemiparasite (De Laubenfels 1959). The roots of this species are always embedded in its host *Falcatifolium taxoides* (also in the Podocarpaceae). There is some question about the nature of this parasitic relationship. The mechanism by which nutrients and photosynthate are transferred from host to parasite is unknown and unverified. In fact, Kopke et al. (1981) refer to this relationship as an "obligate root graft."

Parasitic angiosperms can be grouped into 11 *clades* (groups of closely related taxa) based upon genetic and molecular similarities (Nickrent et al. 1998). Therefore, the parasitic trait has evolved many times in angiosperms. The Scrophulariaceae is an interesting family to consider because it contains members of all possible nutritional status (autotrophs, facultative parasites, hemiparasites, and holoparasites). One-third of the species in Scrophulariaceae are parasitic; however, based on gene sequences (rps2, matk, and rbcL) this family (and Orobanchaceae) are grouped into a single clade (dePamphalis et al. 1997, Nickrent et al. 1998). Thus, there was a single origin to the evolution of parasitism in the Scrophulariaceae. The systematic relationships among the parasitic members of this family suggest that the first parasites were facultative parasites, and that over time photosynthetic capacity was lost by deletions in the chloroplast genome that were released from constraint by the selective pressure of autotrophism (Yoder 1999). We will consider the chloroplast genome of parasitic plants in a later section.

C. Distribution of Parasitic Plants

Parasitic flowering plants have a worldwide distribution. In fact, it is likely that every terrestrial ecosystem in the world has at least one parasitic flowering plant. The most northerly parasite is probably *Pedicularis dasyantha,* which is found above 80°N latitude in Svalbad, Norway. At the equator, tropical forests contain parasites such as those in the Rafflesiaceae.

The Orobanchaceae and Scropulariaceae are predominantly temperate families. The Orobanchaceae mostly occur in the Northern Hemisphere, where they are concentrated in temperate and Mediterranean habitats. Only a few genera with low species diversity are found in the tropics. Many of the members of the genus *Orobanche* are endemic to small areas of Mediterranean regions. However, there are *Orobanche* species with very broad distributions, such as *Orobanche ramosa,* which is distributed mainly in the Mediterranean, Middle East, and northern Africa, but has also been introduced into North America, South Africa, Australia, Cuba, and Central America (Parker and Riches 1993). Parasitic members of Scrophulariaceae are most frequent in grasslands and temperate semiarid regions such as in Mediterranean zones. These hemiparasites are restricted to regions in which they can receive relatively high radiation and where they can engage their relatively repressed but active photosynthetic apparatus. One genus of the Scrophulariaceae (*Striga*) is most abundant in the grasslands of Africa, but is also found in grasslands of Asia, India, and Australia. *Castilleja* (Indian paintbrush) are hemiparasitic members of the Scrophulaceae that are abundant in regions with a Mediterranean climate.

The traditional Christmas mistletoes of Europe (*Viscum album*) and north America (*Phoradendron serotinum*) are representative of only a few mistletoes (Loranthaceae and Viscaceae) that have adapted to temperate regions. In fact, a large majority of mistletoes are tropical (over 700 species). All genera of mistletoes (except the dwarf mistletoes; *Arcuethobium* spp.) are restricted to either the eastern or western hemisphere. The dwarf mistletoes may have a broader distribution into the temperate and boreal biomes because of their parasitism of conifers. Other families that are almost entirely tropical are Rafflesiaceae and Balanophoraceae. One member of the Rafflesiaceae, *Cytinus hypocistis,* is found in the Mediterranean region and has been recognized since the time of Dioscorides. *Rafflesia arnoldii,* discovered in Sumatra in the early 1800s, is most notorious because of its gigantic flower (the largest known flower). This species represents the epitome of vegetative reduction because the entire plant is composed of a haustorium the size of a fungal hypha, and enormous reproductive organs.

Santalacaceae is often misconstrued to be dominated by woody species like sandalwood. However, most of the members of this family are small perennial herbs or dwarf shrubs. Santalacaceae is composed of approximately 250 species in 30 genera. Most of the species in this family are tropical, but the proportion of species in the temperate zone is higher than that of the Loranthaceae (mistletoes), and some reach the southern Arctic. Olacaceae is typified by *Olax* spp., which are parasitic shrubs of the tropics and subtropics. Most of the members of this group are woody, hemiparasitic shrubs, trees, or vines. The family is loosely aligned taxanomically and may be split or lumped into other taxa in the future. However, this family may be at the edge of parasitism, and many members of the family may not be parasitic. Possibly the most restricted family of parasitic angiosperms is Myzodendraceae. This monotypic family has a single genus Myzodendron with about 10 species only located in the *Nothofagus* forest of South America. The base of the attachment of these parasites to their host (almost always *Nothofagus*) can be three to four times the size of the branch, and the host branch usually dies beyond the parasite attachment. One subgenus (*Gymnophyton*) has an unusual warty stem appearance, and the warts are due to raised stomata above a very large substomatal chamber. There are many other unusual aspects of the vascular and floral development of these plants that makes them unique among parasitic angiosperms. The common names for this group are "feathery mistletoes" and "Barba de Angel" because the seeds are characterized by long extended setae with many long trichomes, giving the pistillate flowers a feathery look upon maturity.

The Cuscutaceae (dodders) are the most recognized parasitic plant group other than mistletoes. These vines vary from course, thick stems to very fine, cobweblike stems. The genus is cosmopolitan, extending from the tropics to southern Greenland. The largest number of species has been recorded in the western hemisphere, where no country from Canada to Chile is without a species. Members of the Cuscutaceae also are abundant in Europe (extending into Sweden) and Africa (extending to South Africa), but they are less abundant in Australia and India. Interestingly, very few species of the Cuscutaceae are located on islands, and most are associated with large continental land masses. For example, there may be only one *Cuscuta* native to New Zealand, and there are none in the Malaysian archipelago or the Philippines, while some continents have many species of *Cuscuta*.

D. Evolution of Parasitism in Vascular Plants

It is clear from the taxonomic diversity of parasitic angiosperms that the parasitic trait has evolved multiple times in many parts of the globe. For example, parasitic plants are found in three orders of flowering plants (Rosidae, Asteridae, and Magnoliidae). Magnoliidae is a relatively ancient order of flowering plants, while Rosidae and Asteridae are relatively modern. In addition, there may have been different pathways of evolution leading to root and stem parasites. In any of the evolutionary trajectories, the development of the haustorium must be central because this is the defining characteristic of all parasitic plants.

One theory of the evolution of the haustorium in root parasites is a gradual emphasis of haustoria over roots (Figure 10.1). This theory could be termed the *gradualism theory*. Weber (1987) suggests that root parasitism originated with very small, weakly parasitic secondary haustoria. Over time, the root system became reduced, and eventually secondary haustoria gave way to the primary haustoria. Following the development of the primary haustorium selective forces favored the complete degeneration of the root system and a large enhancement of the primary haustorium. This gradualism theory is based on the fact that root parasites can be identified with a continuum of root/haustoria ratios, and the continuum of relative dependency on the host is directly associated with the ratio of root system to haustoria. The lower the ratio of root system mass to haustoria mass, the more dependent the parasite is on the host resources.

Another theory is the *root grafting theory*. This concept is based on the fact that roots of tree species can graft together in a swollen structure that looks a little bit like a haustorium-host interface. In this case, selective forces could have favored the evolution of a parasitic relationship (development of an endophyte) from the root graft structure. This theory has little support because many parasitic species are herbaceous, and there is only rare evidence for root grafting in herbaceous species. The abundance of root grafting in woody species in present forests may indicate a possible evolu-

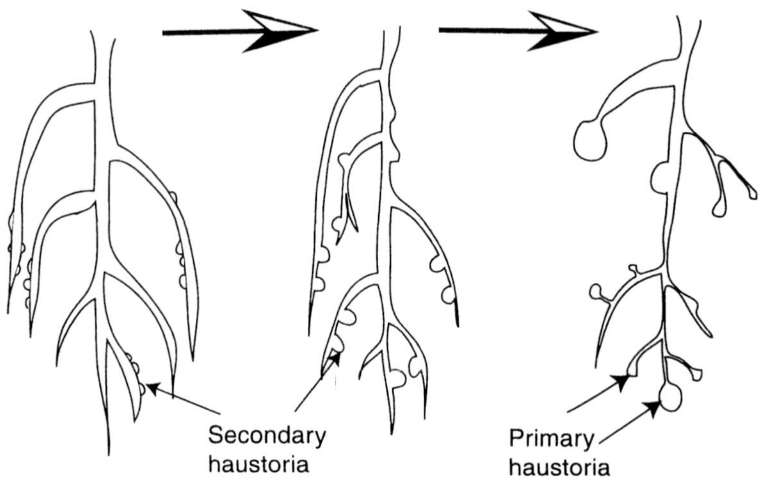

Figure 10.1. A figure representing one view for the evolution of root hemiparasites. Notice the gradual change from secondary haustoria to primary haustoria, and the coordinated decrease in the size of the root system. (Redrawn from Weber 1987, with permission.)

tionary trajectory for the development of parasitic plants in some woody Santalacaceae and the Olacaceae. However, this evolutionary pathway in woody species is not very likely because root grafting usually occurs between closely related species, and woody root parasitic species attack unrelated host species. In fact, the relationship between the two species of Podocarpaceae may be the only case of a root graft that is approaching parasitism.

A different model for the evolution of parasitic plants could be called the *mycorrhizal model.* Some parasitic species such as *Monotropa hypopitys* are connected to their host species by mycorrhizal hyphae (Björkman 1960). These parasites attain their reduced carbon from the host plant species via a shared mycorrhizal mycelium. This means that the mycorrhizal fungus retains some of the carbohydrate from the host, and the parasitic *Monotropa* receives only a fraction of the carbon flowing from the tree to the mycorrhizal fungus. It is possible that selective pressures would favor plants that had aggressive roots that took over the function of the mycorrhizae and thereby increased their supply of carbohydrate from the host.

An alternative theory, the *pathogen model,* was proposed by Atsatt (1973). This model is based on the idea that parasitic plants are the evolutionary products of microbial parasitism. The pathogen model suggests that the haustorium originated as a tumorous growth resulting from infection by bacteria such as *Agrobacterium.* When *Agrobacterium* infects plant tissues, it transfers a DNA plasmid to the plant; the plasmid is incorporated into the host genome and causes the induction of a tumor. This is the basis of one major technique used in bioengineering to transform the genetic structure of plants. It is possible that over evolutionary time, these tumors were transformed into haustoria (Atsatt 1987). In fact, the morphology of some of the most ancient forms of haustoria resemble the morphology of tumors induced by *Agrobacterium* (Figure 10.2).

In the case of *Cuscuta,* the earliest haustoria must have developed from aggressive adventitious roots on a liana (Kuijt 1969). The disappearance of roots from this plant could have occurred from selection to maximize the effectiveness of nutrient and water uptake into the stems from the developing haustoria.

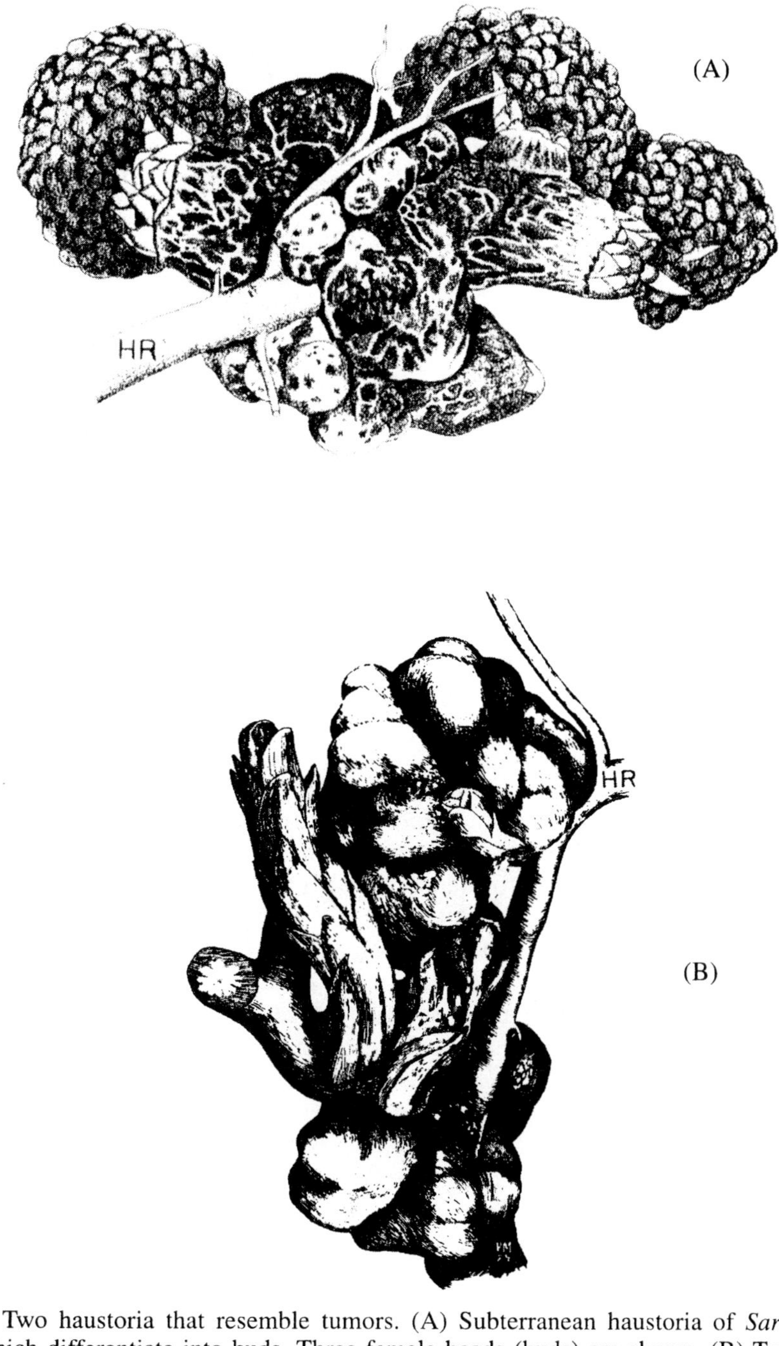

Figure 10.2. Two haustoria that resemble tumors. (A) Subterranean haustoria of *Sarcophyte sanguinea*, which differentiate into buds. Three female heads (buds) are shown. (B) Tumorlike haustoria galls associated with *Mystropetalon thomii*. New shoots originate from these galls. Such haustoria are cited as support for the pathogen-induced evolution of parasitic plants. (Reproduced from several figures in Atsatt 1981.)

E. Effects of Parasitic Plants on Natural Communities and Agriculture

1. Affects on Natural Communities. Parasitic plants can reduce the productivity of natural plant communities. For example, dwarf mistletoe infects conifers (particularly Douglas fir) in the U.S. Pacific Northwest (Tinnin 1984) and severely reduces tree production (up to 50% reduction). In addition, parasitic plants have been used as a biocontrol agent because they can reduce the growth and reproduction of specific trees. For example, *Cassytha filiformis* has been proposed as a biological control agent for the invasive species *Schinus terebinthefolius* in southern Florida (Burch 1992).

Species diversity in natural communities can be changed by parasitic plants. The annual root hemiparasite *Rhinanthus minor* causes a decrease in species diversity of grasslands in Europe (Watkinson and Gibson 1988, 1989; Gibson and Watkinson 1992). In areas of the grassland where *R. minor* is mature, species diversity is less than the surrounding grassland. However, the following spring, at the time that *R. minor* plants are germinating, there is no significant difference between species diversity of these areas and the surrounding grassland. Moreover, removal of *R. minor* from patches of low species diversity results in an increase in diversity to equal that of the surrounding grassland. Another example was shown for a California salt marsh, where the presence of *Cuscuta salina* increases species diversity. In the vegetation zone, where *Cuscuta* is abundant, its preferred host is *Salicornia virginica. Cuscuta salina* reduces the growth and vigor of *Salicornia virginica,* which releases *Limonium californicum* and *Frankenia salina* from competition with *S. virginica* (Pennings and Callaway 1996), causing an increase in species diversity.

The community structure of a coastal prairie in California (Marvier 1998b) is affected by the presence of the hemiparasitic plant *Triphysaria pucillia.* This case demonstrates the influence of a generalist parasitic plant both on the interactions between other nonparasitic species and on community structure. When *T. pucillia* was attached to a host, the total biomass accumulation of host species was reduced in a direct relationship with the suitability of the host for the parasite. Therefore, the hosts that support the highest parasite performance are also the hosts that suffer the most from parasite presence. Given the large influence of the parasite on grass productivity and the low influence of the parasite on the associated legume species (*Lupinus arboreum*), it was logical to predict that the presence of the parasite in mixed stands would allow *Lupinus arboreum* to attain a grander stature than in similar stands where the parasite was absent. This was not the case. In fact, the legume was further inhibited by the combined presence of the parasite and grasses. It was proposed that the combined community of parasite and grass was more wasteful of water resources, because most parasitic plants have a low water-use efficiency and liberally transpire. The reduced water-use efficiency constrained productivity of the legume when the parasite was present in the community.

There are also competitive effects of host plants on the success of the parasitic plants. This relationship was studied for two hemiparasites (*Rhinanthus minor* and *Odontites rubra*) growing in the presence of the host, *Medicago sativa* (Matthies 1995). Both parasitic plants are root parasites that gain some of their total photosynthate from their own leaves. Therefore, the parasites are taking resources below ground (water, nutrients, and some carbohydrates) away from the host, and the host could preempt aboveground resources (light) from the parasites. It has previously been shown that hemiparasite populations perform poorly under high host density (Van Hulst et al. 1987), and competition for light was proposed as the mechanism for this inhibition. The growth of *Medicago sativa* was inhibited when either of the parasites were present. Separating the parasites from light inhibition by the host had no affect on the growth of the host. However, parasites grown with host and separated from host canopy grew 30% more than when they were in contact with the host canopy. Thus, the host competes for light with the parasite.

The relationship between hosts, resources, and parasites has some interesting permutations (Figure 10.3). Once host plants are exposed to parasites, the host plants are under selective pressure to make use of the parasite in a positive manner (Price et al. 1986). For example, plant host competitor A in Figure 10.3 would benefit from actively encouraging the growth and success of parasite C if the detrimental effect of parasite C were greater for competitor B than the competitive effect of competitor B on competitor A. In addition, the influences of resource availability on competition between species A and B could affect the success of parasite C on species A and B. An example of this relationship was shown by Marvier (1998b) and discussed above. Therefore, a complex interaction between parasites, their hosts, and resource availability exists that influences ecosystem productivity, species diversity, and the evolution of competitive interactions among species.

2. Effects on Agriculture. Parasitic plants can have a large economic impact on agriculture. The three most important economic pests are *Orabanche* spp., *Cuscuta* spp., and *Striga* spp. *Cuscuta pentagonia* (North American dodder) is a worldwide weed problem in agriculture. As shown by Wolswinkel (1974), seed set of fava bean can be completely inhibited by the presence of *Cuscuta* because the parasite is such a large sink for the photosynthate produced by the host plant. *Cuscuta* is a significant weed in potato, sugar beet, carrot, tobacco, chickpea, onion, and other important crops of the world. In fact, potato crops planted in rotation after alfalfa crops can be completely devastated by *Cuscuta* parasitism (Dawson et al. 1994). *Cuscuta* has such a wide distribution in agricultural regions because it is very difficult to separate the seeds of the parasite from those of the crop. Seeds of small-seeded legumes such as

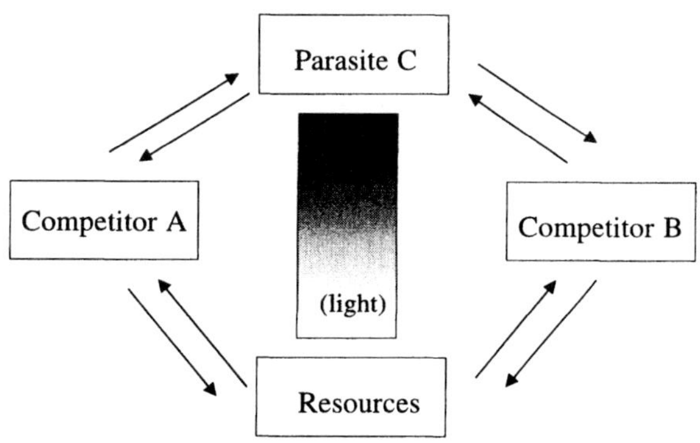

Figure 10.3. A diagrammatic representation of the interactions between two competing plants and a generalist parasite. Most resources travel from soil to host (competitors) to parasite, except for light that is reduced by competitors on its way to the parasite. Encouragement of the parasite population by one species could reduce the competitive ability of the other species. (Redrawn and modified from Price et al. 1986. Adapted with permission, from the Annual Review of Ecology and Systematics, Volume 17, © 1986, by Annual Reviews http://www.annualreviews.org.)

alfalfa, clover, or *Lespedezia* are very similar in size and color to those of *Cuscuta.* One mechanism by which the *Cuscuta* seeds can be removed from alfalfa seeds is to add iron shavings that will stick to the relatively rougher seed coats of *Cuscuta* and not to the smooth seed coats of the crop. A magnet passed through the seed lot will preferentially remove *Cuscuta* seed. This and other removal methods are expensive and labor intensive.

Striga is a particularly important weed in African agriculture. In six West African countries, species of *Striga* reduce agricultural productivity by an average of 24% and the inhibition occurred in 48% of all fields that are infected with these parasites. This translates into 21 million hectares infected and a loss of 4.1 million tons of grain (Sauerborn 1991), valued at approximately U.S. $2.9 billion. Yield reductions up to 100% were reported in sorghum fields infected with *Striga* (Doggett 1965). *Striga asiatica* was introduced into the Carolinas of the United States in the 1950s and remains a threat to corn, sorghum, and sugarcane crops. The *Striga* infection in the United States could cost farmers as estimated $1 billion in reduced yield and cost of controlling the pest (Sand 1987).

Orobanche species (broomrapes) occur predominantly in Mediterranean regions and parts of Northern Africa, where they are parasites on some cash crops, particularly in the Fabaceae and the Solanaceae (e.g., fava bean, potato, eggplant, tobacco, and tomato). Loss of yield for fava beans ranges from 5% to 100%, with an average of 35% in Northern Africa (from Sauerborn 1991). Tomato production is reduced by about 25% due to *Orobanche* infestations in the United States. During the 1950s, the land area under cultivation for sunflower was reduced by 37% in Yugoslavia due to *Orobanche* infestation (Mijatovic and Stojanovic, 1973), and sunflower fields in Spain were taken out of production due to *Orobanche* infestation. The estimated cost of controlling nodding *Orobanche* in sunflower crops is $40 per hectare (Garcia Torres et al. 1996), which translates into billions of dollars worldwide.

When all parasitic weeds are considered together, their effect on agricultural economics is large and worldwide. The billions of dollars spent per year to control these pests and the land areas taken out of production due to parasite infection are strong stimuli for research on the basic biology of parasites and stress relationships between host and parasite.

F. Conservation Issues with Parasitic Plants

Although it is clear that parasitic plants can have a significant detrimental effect on productivity in natural systems and on yield in agricultural systems, there is also some concern that a number of parasitic plants are endangered. When one considers the whole group of parasitic plants, it is easy to marvel at this remarkable set of angiosperms. Simply viewed as an infection system that can enter into another species, avoid the host defense against foreign tissue, and redirect resources into the parasite, its evolution is astounding. The ability to form connections to a host through a third organism (mycorrhizae) is a unique adaptation. The diversity of unique and unusual evolutionary strategies in this group makes parasitic plants a treasure of biological complexity.

Populations of certain parasitic plants have been pushed toward extinction for several reasons. In some cases, plant parasites are systematically exterminated because of their detrimental influences on host plant growth. In other cases, the products of these species are so valuable that parasites such as the sandalwood tree of Hawaii are overharvested. Within the United States alone, 56 species of parasitic plants are currently on the endangered or threatened species lists (Marvier and Smith 1996). Within California, out of the 165 native parasitic plants, 49 are rare or endangered (Skinner and Pavlick 1994). Many of the reasons why parasitic plant populations are endangered relate to the dwindling abundance of host species. For example, the introduction of exotic annual grasses in California reduced the population of perennial grasses, which reduced host availability for a parasitic plant species, *Cordylanthus maritimus* (Coats et al. 1993). Conservation of parasitic plants may depend upon many of the same procedures used for nonparasitic plants, except that host suitability may be of overriding importance.

II. ECOLOGY AND PHYSIOLOGY OF PARASITIC PLANTS

A. Pollination Biology

Floral structures of parasitic plants are highly variable. Some members of the group have very reduced floral structure such that botanists originally thought they were mushrooms (Balanophoraceae), while other parasitic plants have very large and showy flowers (for example: *Rafflesia arnoldii*). In some groups of parasitic plants, there is a remarkable similarity among pollinating vectors; however, among groups there is a great variation in successful pollinators. Hemiparasitic plants generally produce flowers that are pollinated in much the same manner as nonparasitic plants. In contrast, many holoparasitic flowers are produced in unusual manners and have elusive pollination mechanisms.

Members of the mistletoe family (Loranthaceae) are almost always pollinated by birds. One mistletoe species in Equador is pollinated by at least nine species of hummingbirds (Kuijt 1969). In fact, a group of birds in the Discaeidae found in Indonesia and Malaysia are commonly called "mistletoe birds." This family includes 55 species of birds, with different adaptations for using pollen, berries, and insects of mistletoes. Birds visiting African mistletoes, such as *Tapinanthus kraussianus,* in the early morning find no open flowers. The older and larger flowers on the inflorescence are under tension due to a rapid buildup of sugars, which draws water into the flower tissues. The birds squeeze these flowers, which causes them to violently burst open, throwing pollen onto the bird. Almost all bird-pollinated Loranthaceae have complete bisexual flowers, while all insect-pollinated Loranthaceae are dioecious and have inconspicuous floral parts. In contrast, mistletoes in the Viscaceae are almost all pollinated by insects, usually hymopteran bees. Wind pollination is also a possibility in some members of Viscaceae.

Floral biology and pollination of the Rafflesiaceae and Balanophoraceae are the most unusual among parasitic plants, and perhaps among plants in general. Imagine the surprise among botanists when *Rafflesia arnoldii* was found to be the largest flower in higher plants, and a holoparasite with virtually no other tissue except the flower. Many of these flowers develop endogenously, within the parasitic tissues that are imbedded within the host. The only tissues apparent to an observer are the flowers above ground or penetrating from the stem of the host. In one unusual case, the flowers of *Pisolyles haussknechtii* only emerge on each side of the base of the petiole of the host *Astragalus* (Kuijt 1969).

The pollination of *Rafflesia* flowers, particularly *Rafflesia arnoldii,* have been an enigma. Originally, flies were thought to be the main pollinators, because the flowers tend to smell of carrion late in their development. However, some botanists have thought that this carrion smell occurs after flower maturity, is a sign of flower decomposition, and so is unrelated to pollination. There is no doubt that flies visit flowers of the Rafflesiaceae, but some birds and bumblebees also visit the flowers and facilitate pollination.

Members of the Balanophoraceae have a completely different floral structure and pollination biology from the Rafflesiaceae. Balanophoraceae inflorescences are fleshy, club-shaped structures that bear a striking resemblance to fungi. Members of this family have often been identified as fungi. For example, *Cynomorium coccineum* in the Mediterranean was originally named *Fungus melitensis,* or commonly, the Maltese fungus. These inflorescences bear millions of tiny flowers, among the smallest flowers in the plant kingdom. The inflorescence of many members of Balanophoraceae extends from an underground tuber that is unique among the parasitic plants. The tuber is composed of an interlacing of both the parasite and host, in which host xylem cells extend well into the parasitic tissue. Pollination of these extremely reduced flowers has not been well explored but may be accomplished by ants or flies.

B. Seed Dispersal, Germination, and Establishment

Mistletoe seeds are usually distributed by birds in their droppings. This may be the origin of the common name for mistletoes, because the German words for manure (*der mist*) and mistletoe (*der mistel*) are very similar (Kuijt 1969). Many species of mistletoe are consumed by only one species of bird. In fact, birds that feed on only one or a few mistletoe species have evolved in Australia, Africa, and South America. Some of those specialist birds have digestive systems that are tailored to the digestion of the tannin rich mistletoe exocarp. However, when all bird-dispersed mistletoe species are considered, there is a continuum from one bird species to large numbers of bird species dispersing each mistletoe species (Reid 1991). Birds distribute these seeds by three methods: (1) regurgitation of consumed fruits, (2) defecation of consumed fruits, and (3) leaving a seed while chipping at the flesh of the fruit. For example, Chilean mockingbirds feed only on the fruit of *Tristerix aphyllus* (Loranthaceae). The seeds are deposited on the spines of relatively tall members of the host cactus (Martinez Del Rio et al. 1995, 1996). The seeds produce an exceptionally long radicle (10–91 mm long) in comparison with seeds of other parasitic plants, which is adaptive for this host species (very long spines). In another case, a bizarre behavior of a cedar waxwing during defecation actually pastes the mistletoe seed to the branch. Members of Santalacaceae are also dispersed by birds, as is suggested by their fleshy and bright colored fruits.

Not all mistletoe species are dispersed by birds. A specific case is the explosive seed dispersal mechanism of several members of the Viscaceae such as *Arceuthobium* and *Phoradendron.* The pedicel holding the seeds of these mistletoes curves backwards during development and develops significant tension. A transverse tissue at the base of the seed weakens upon maturity, and the seed is shot upward and away from the host tree. The seed is shot at a speed of up to 24 m^{-1} and can travel up to 15 m.

Most of the hemiparasites in Orobanchaceae and Scrophulariaceae have seeds released from capsules. The number of seeds per locule in these capsules is very variable among species, and the seeds are small (0.3–2 mm in diameter). In fact, an average *Orobanche minor* plant may produce 100,000 seeds, and one *Aeginetia* sp. capsule holds about 50,000 seeds. Most of these seeds are dispersed by wind and water, and some are dispersed by ants.

Seed dispersal in tropical *Rafflesia* species is an enigma. The seeds mature inside the massive fruit and are not exposed to the environment. The fruit matures and rots.

Seed dispersal may be carried out by ants and other insects that break into the fruit and take the seeds, or by rats that eat the fruits, or by pigs or elephants that trample the rotting fruits. In the Hydnoraceae, seeds are small and hard, resembling grains of sand. The seeds are embedded in a fleshy fruit that is attractive to mammals as well as humans. In the Balanophoraceae, small collections of tiny seeds are subtended with a white fatty tissue. These small groups are collected by ants and transported to their nests. From the above review, it is clear that many parasitic species have evolved varied and elaborate methods of ensuring seed dispersal.

As stated earlier in this chapter, parasitic plants have preferred hosts. That means the parasite must forage in the environment to find an appropriate host, or that connections are attempted randomly with possible hosts and only those with preferred hosts succeed. There are several stages at which the success or failure of an attempt to form an association could be determined. For example, all the following events need to occur for a successful connection of a root parasite (such as *Orobanche* or *Striga*) with a preferred host:

1. Germination of the parasite seed near the host
2. Location of a host root by the germinating parasite
3. Haustorial initiation when a host root or stem is contacted
4. Host tissue penetration by the the haustoria
5. Physiological compatability of host and parasite tissues
6. Parasite apical meristem growth after the association has been made

Among these possibilities, seed germination is known best, and is the most critical first step to successful parasitism.

Abiotic requirements for hemiparasitic plant seed germination are much the same as for nonparasitic plants (Worsham 1987). Parasite seeds require moisture, oxygen supply, appropriate temperature, and suitable nutrient and light availability. The length of postdehiscence maturation may vary among species, but the patterns for facultative hemiparasites are very similar to those for nonparasitic plants. Following maturation, a preconditioning cycle is required. A *preconditioning cycle* is a period (length varying with the species) of appropriate temperature and moisture conditions that the seeds must experience before they will respond to a germination stimulant. In contrast, seed germination requirements are more specialized in holoparasites and hemiparasites that cannot survive as autotrophs in nature. Seed germination in these obligate parasites is modified by host recognition and signal transduction systems (Press and Graves 1995).

The best-known examples of seed germination regulation are for *Striga* spp. and *Orobanche* spp. Many *Striga* species are restricted to colonizing a limited number of host species. For example, *Striga gesnerioides* seeds collected from indigo, tobacco, and morning glory could only parasitize the species from which the seeds were collected (Musselman and Parker 1981). Although this specificity for host may be related to many aspects of the parasitic cycle, host recognition before germination is a possibility. *Striga asiatica* seeds can remain dormant in the seed bank for many years (up to 20 years), through many preconditioning cycles, in the absence of an appropriate host (Worsham 1987). However, the seeds germinate rapidly after preconditioning in the presence of the host species. Thus, there is something specific about the presence of the host that breaks dormancy of the parasite seed.

Germination of *Striga* seeds is stimulated by a number of compounds released by roots of host and nonhost species. However, before the stimulant can induce germination, the seeds must receive a conditioning period. In fact, Westoby (1984) outlines three requirements for seed germination in *Striga*: (1) a postshedding ripening period, (2) a conditioning period under moist conditions, and (3) exposure to a chemical stimulant that triggers germination. Of course, successful parasitism will also depend on the emerging seedling locating the host tissue, on appropriate haustorium formation, and on the formation of a parasitic interaction in the haustorium.

Both natural exudates of host roots and synthetic hormones can stimulate *Striga* germination. The hormones kinetin, gibberellic acid, and ethylene can stimulate *Striga* seed germination after preconditioning. Saunders (1933) was the first to show that leachates from pots containing sorghum could stimulate germination of preconditioned *Striga* seed. Isolating stimulants from plant root exudates is a difficult task because root exudates are composed of many compounds and stimulants are active at very low concentration. The first stimulant that was isolated and identified was strigol, from a nonhost, cotton (Cook et al. 1966). Strigol (Figure 10.4) has been found to stimulate germination in almost all parasitic plants at a concentration range of 0.1–10 n*M* (Hauck and Schildknecht 1990). Furthermore, strigol stimulates the seed germination of many nonparasitic plants. Although strigol has been generally thought to be a specific agent for inducing germination of parasite seeds when they are close to host plants, many questions remain about its significance to parasite-host interaction under natural conditions.

Why does this compound stimulate the germination of virtually all parasites and many nonparasites? Such a broad stimulation effect suggests that strigol may act as a general growth regulator rather than a specific host-parasite molecule. For if strigol had evolved between host and parasite, why would this compound also be produced by the roots of nonhosts such as cotton? Such production would cause *Striga* seeds to germinate near nonhost plants, which would not result in a successful parasitism. Moreover, strigol is a very active substance and also a relatively stable product in the soil solution. This is unusual in comparison with other germination stimulants. A relatively stable molecule would have a long residence time in the soil, perhaps longer than the root that produced the molecule. Therefore, *Striga* seed germination would not be restricted to the region close to the host root. The stability and activity of strigol makes it likely that *Striga* seed would germinate far from a host root and be unable to reach the root following germination. In addition, terpenoids found in plants have very different structure from that of strigol, casting doubt upon the idea that plants exude strigol. That evidence, plus the fact that strigol has never been isolated from plants in a sterile environment, suggests that strigol may be a microbial product of the rhizosphere. However, in support of strigol, this is a very active molecule, with a pronounced effect on seed germination of parasitic plants. It seems unlikely that parasitic plant seeds would respond so dramatically to a substance that does not confer a relatively high probability of contacting a root. The relative significance of strigol to germination regulation of parasitic plants remains controversial.

The first host-derived stimulant of parasite germination (originally identified as sorgoleone) was 2-hydroxy-5-methoxy-3-[(8′,11′Z)-8′Z′,11′,14′-pentadecatriene]-*p*-hydroxyquinone, and was isolated from sorghum in large quantities (Figure 10.4). Sorgoleone (a quinone) is inactive as a stimulant, but when converted into the hydroxyquinone sorgoleone stimulates *Striga* germination and is more active and less stable than strigol. Several other dihydroxyquinones have been found to stimulate seed germination of parasitic plants. In addition, the activity of sorgoleone (as the hydroxyquinone) is modified by resorcinol, which also is exuded from sorghum roots.

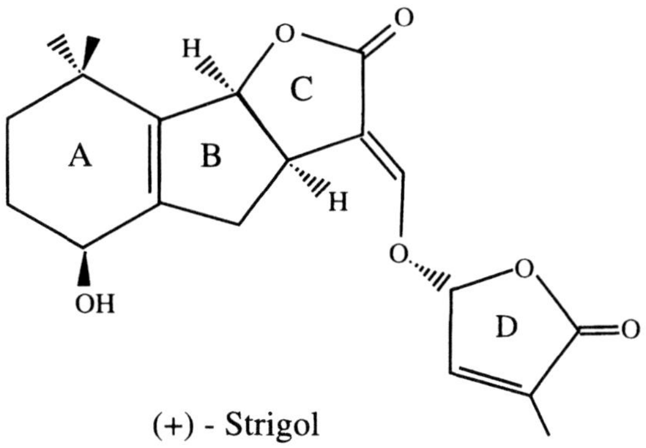

(+) - Strigol

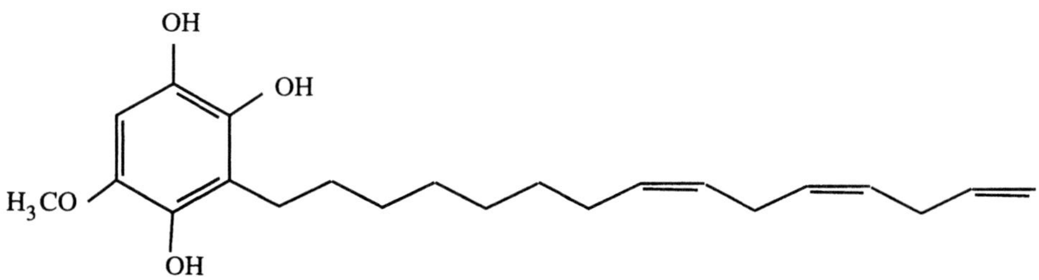

2-hydroxy-5-methoxy-3-[(8',11'Z)-8'Z,11',14'-pentadecatriene]-p-hydroquinone

Figure 10.4. Two molecules that stimulate the germination of parasitic angiosperm seed. (Reproduced from several figures in Boone et al. 1995 with kind permission from Kluwer Academic Publishers.)

Resorcinol is not itself a stimulant of germination but acts to repress autooxidation of hydroxyquinones in the soil solution, and the activity of the hydroxyquinones (and strigol as well) depends upon their oxidation state. The active component of strigol has been isolated to the D-ring. The D-ring of strigol can be synthesized *in vitro.* When the D-ring is used alone as a stimulant of seed germination of *Striga asiatica,* its activity is directly related to its concentration (Figure 10.5). The D-ring of strigol is very similar to the active components of the hydroxyquinones. In fact, sorghum varieties that produce extremely low concentrations of these hydroxyquinones are relatively resistant to *Striga* infection (Butler 1995). Evidence against the effectiveness of these hydroxyquinones is that they are very hydrophobic and have limited movement in the soil. A parasitic plant seed would have to be very close to the host root to receive an active concentration of these compounds. Therefore, these compounds cannot be the only answer to seed germination regulation in parasitic plants. Many compounds may be involved in regulating germination of parasitic plant seeds, and

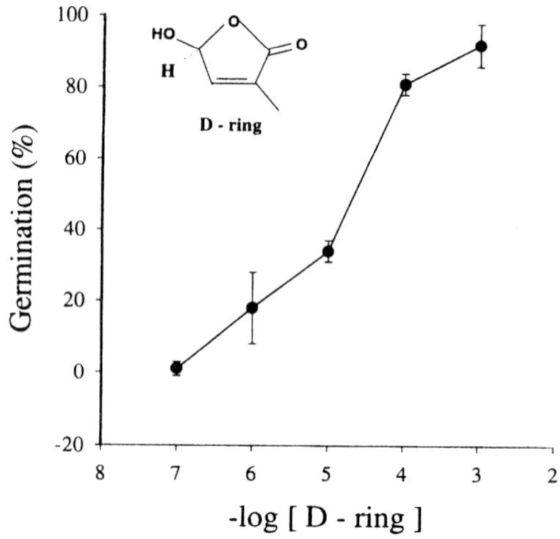

Figure 10.5. Relationship between activity (stimulation of germination) and the concentration of the D-ring of strigol. Error bars represent 2 standard errors. (Redrawn from Boone et al. 1995 with kind permission from Kluwer Academic Publishers.)

the mixture of compounds in association with the soil solution chemistry may be an important factor.

The difference in the stimulation of parasite seed germination among crop species has been used as a "catch crop" agricultural treatment. For example, tomato (*Lycopersicon*) roots induce about 10% germination of *Orobanche aegyptica,* and tomato is highly susceptible to infection. In contrast, pepper (*Capsicum annuum*) roots stimulate more than 50% germination of *O. aegyptica* seed, but only a few parasites were able to form an association with pepper roots. When pepper and tomato are interplanted, tomato plants suffer a far greater infection by *O. aegyptica* than when tomato is planted by itself. Planting pepper before tomato will reduce the *O. aegyptica* infection of tomato (Heshenhorn et al. 1996). These data suggest that differences in resistance are attributable to more than differences in the production of parasite seed germination factors.

Clearly, *Orobanche* seed germination is affected by stimulants and inhibitors found in root exudates (Whitney 1972). Host recognition by *Orobanche* seed before germination is critical because the radicle can only extend about 2 mm before exhausting carbohydrate supplies. Seed germination of *Orobanche* also is sensitive to nutrient concentrations in the soil. In particular, ammonium ion directly inhibits radicle elongation (Westwood and Foy 1999). Therefore, *Orobanche* seed will only germinate after a preconditioning period of moist and warm conditions, in the presence of a stimulant chemical released by the host, and under relatively low ammonium ion concentrations. This interaction has resulted in the concept of fertilization with ammonium nitrate as a procedure for controlling parasite infestation rates in agricultural fields (Jain and Foy 1992). Fertilizing to protect crops from parasitic attack has limited effectiveness, and works unreliably at best against *Orobanche* and *Striga* infection.

C. Structure and Function of the Haustorium

The haustorium is the most defining character of parasitic plants. This unique structure forms the connection between parasite and host, and serves as the conduit for the transfer of water, nutrients, and carbohydrate from the host to the parasite. Three major regions of the parasite haustorium are recognizable:

1. The swollen rounded structure surrounding or connecting to the root or stem of the plant is the *main haustorial tissue*.
2. At the base of this tissue (located close to the host tissue) is the *hymenium*.
3. The *endophyte* is parasitic tissue that penetrates into the host to form a physiological connection with the host vascular tissues.

Haustoria are variable in structure among different parasite-host combinations. In Rafflesiaceae, haustoria are completely embedded in the host tissues. Some haustoria form on roots, others on stems. Some haustroria are derived from the apical meristem (primary haustoria), while others develop from lateral meristems (secondary haustoria). However, all tissues of haustoria develop from primary cell tissues of the parasite. Haustoria have evolved in a number of distantly related plant groups, most likely by different evolutionary pathways, which is a main reason why there is such a large diversity in haustorial structure. The structure of haustoria in a wide variety of parasitic plants is extensively described in other sources (Kuijt 1969, 1977; Musselman 1980, Visser and Dörr 1987).

The development of the haustorium follows three stages: an *adhesive stage,* then an *intrusive stage,* and finally a *conductive stage*. These three stages are true of all haustoria, but the time and effort spent in any one of these stages may vary. For example, haustoria of the Santalacaceae spend a long developmental time in the adhesion stage before penetration occurs. The result can be a swollen haustorium that is much larger than the host organ. In contrast, the haustorium of parasites in the Rafflesiaceae are focused on penetration and proliferation such that the final structure is a fragmented system of strands that permeate through large areas of the host tissues.

The physiological cues that initiate the development of haustoria are critical to the survival of parasites. The main stimulus for the development of haustorial tissue may simply be contact with another surface. In some parasitic plants, such as *Pedicularis,* haustoria can form on dead plant material or pebbles (Piehl 1963). However, the successful haustorium development and appropriate penetration and development of the endophyte require a preferred host. Chemical stimulation of haustorial induction and development is an important factor determining host preference in parasites.

The requirement for chemical stimulation of haustoria formation is variable among parasitic species. For example, *Aurolia pedicularis* forms many haustoria in the absence of a host (no chemical requirement), while *Agalinus purpurea* forms only 25% of its maximum number of haustoria, and *Striga asiatica* forms no haustoria without a host present. Those chemicals that have been found to stimulate seed germination of parasites (strigol, hydroxyquinone, etc.) are not the same as those that stimulate haustorial formation. The first stimulants of haustorial formation identified (*haustorial induction factor,* or HIF) were phenylpropanoids extracted from *Astragalus* (Lynn et al. 1981). In fact, *m*-methoxyphenol compounds are strictly required for haustorial development in *Agalinis purpurea* (Stephens et al. 1982), as well as other parasites. Haustorium formation of *Striga hermonthica* is stimulated by phenolic acids, such as sinapic, vanillic, and syringic acids. Enzymes released by parasite root systems (peroxidases and hydrolases) convert the inactive parahydroxy acids (by oxidative carboxylation) into the active quinones. Another parasite, *Striga asiatica,* will establish a functional haustorium only in the presence of a quinone associated with sorghum roots (DMBQ: 2,6- dimethoxy-*p*-benzoquinone) (Chang and Lynn 1986). DMBQ is most likely not produced by sorghum roots, but it is a product of fungal decomposition of lignin by lacase. Therefore, in order to initiate haustorium formation, the parasites must be able to release lacases that break down lignin, thus releasing DMBQ. Such lacase activity has been demonstrated on the root tip of *Striga* and *Agalinis* (Chang and Lynn 1986). When the HIF is perceived, the parasite root system stops elongating and all energy is directed toward producing the haustorium. The molecular signaling that occurs between host and parasite may be an excellent system for studying how plants communicate in the environment (Yoder 1999).

The chemical stimulation of seed germination and haustorial initiation does provide some degree of host selection. In fact, similarly to herbivorous insects, some parasites may locate preferred hosts based on the defensive compounds of the host, which at low concentrations are attractants and at high concentration are defensive compounds. However, initiation of the haustorium does not assure successful parasitism. A successful haustorium is based on a direct connection between the parasite and the vascular system of the preferred host. At this point, the haustorium enters into the mature conductive phase.

The connection with a preferred host's vascular system is complex, and the xylem and phloem connections differ in structure and function. We will only discuss two examples of mature haustoria structure representing that complexity. The haustorium connection of *Striga* exemplifies the process of parasite connection to the xylem (Figure 10.6). The first successful haustorium of a germinating *Striga* seed is derived from the primary root meristem. Upon contacting a host root, the parasite root meristem is transformed into a globose primary haustorium. The epidermal cells of the parasite elongate much as a root hair elongates penetrating the root tissues. This endophytic tissue (which resembles palisades cells) penetrates through the cortical cells of the host root until a xylem element is reached. At the xylem element wall, the endophyte penetrates the xylem element (most likely through a pore). Once inside the vessel element, the endophyte tissue proliferates radially and longitudinally, and some of the endophyte cell wall ruptures, producing holes through which host xylem constituents can pass into the parasite tissue. Endophytic tissue connected to the host xylem element then transforms into vessel elements themselves, completing the connection between parasite and host vascular system. This entire process occurs before the parasite develops its primary axis, because the vascular connection between host and parasite must be established before the parasite emergence above ground.

Evidence of phloem connections between *Striga* parasites and hosts are not clear; thus, we will use *Cuscuta* as an example of phloem connections. The *Striga* haustorium also contains a large central mass of cells that surrounds the newly differentiated vascular system. This mass of cells called the *hyaline body* is characterized by highly nucleated cells with abundant organelles (mitochondria, vacuoles, endoplasmic reticulum, Golgi apparatus). The abundance of organelles and enzymatic activities of these cells indicates that the hyaline body is characterized by high metabolic activity possibly related to the control of nutrient and carbohydrate exchange.

The haustorium of *Cuscuta* was the first to be described and is therefore best known. In fact, the term *haustorium* was first applied to *Cuscuta* (Kuijt 1969). Haustoria are initiated as small superficial bumps along a coiled part of the stem called the *haustorial coil* (although sometimes haustoria are not located here). Upon contact with the host, a *prehaustorial tissue* seems to sink into the host tissues. The prehaustorial cells release enzymes that soften and decompose host cell walls. At the apex of this tissue in the parasite, a meristem forms that proceeds down the path established

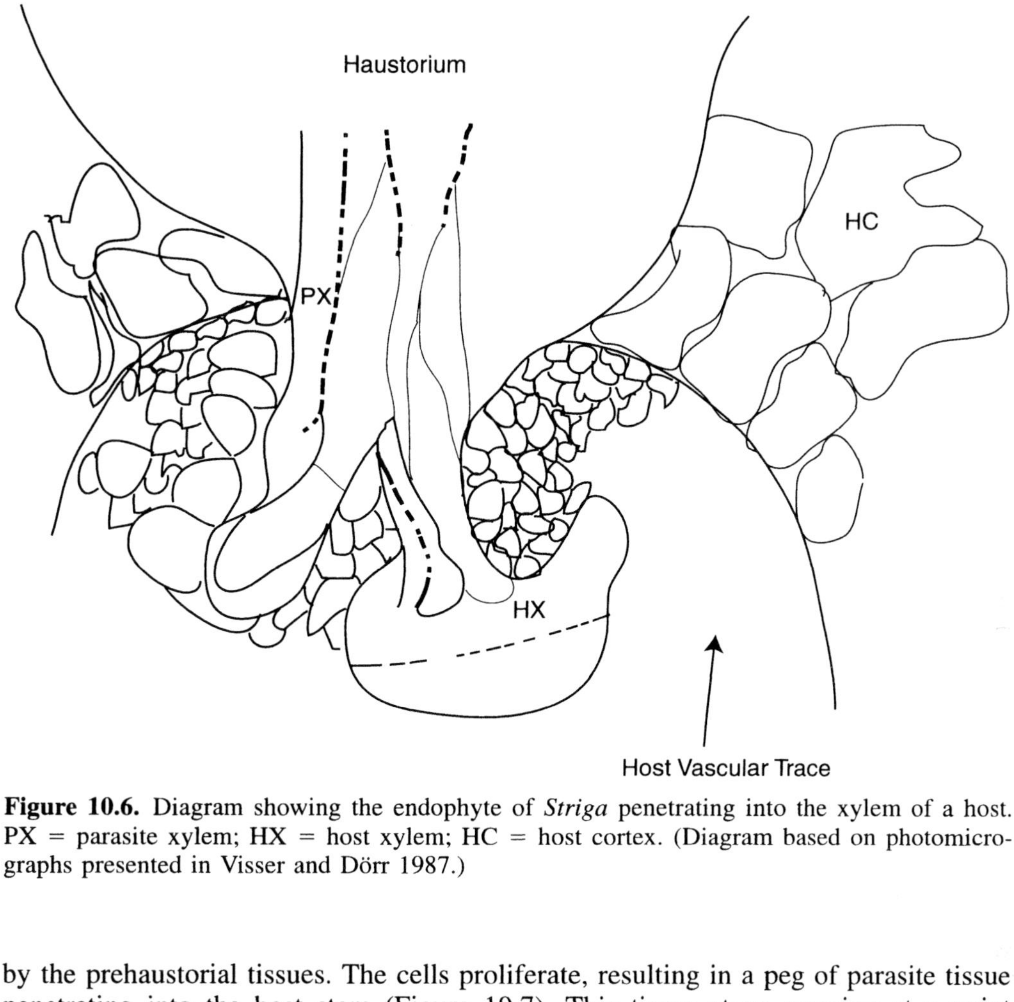

Figure 10.6. Diagram showing the endophyte of *Striga* penetrating into the xylem of a host. PX = parasite xylem; HX = host xylem; HC = host cortex. (Diagram based on photomicrographs presented in Visser and Dörr 1987.)

by the prehaustorial tissues. The cells proliferate, resulting in a peg of parasite tissue penetrating into the host stem (Figure 10.7). This tissue stops growing at a point before connections have been made with host xylem or phloem. Individual cells on the surface of this peg elongate into filaments called *hyphae*. The tip of each hypha penetrates into parenchyma cells and grows directly across to the other cell wall. This penetration system continues across several connected parenchyma cells. Between successive cell penetrations, the hyphal cells divide, resulting in a column of parasitic cells transversing a row of parenchyma cells. Some of these hyphae will penetrate either into a xylem element or between xylem cells. Once thus associated with host xylem, the entire column of cells in the hyphae expands to fill the ruptured parenchyma cells, and differentiates into xylem elements. Some hyphal cells come into contact with phloem cells, where the hyphal cells transform into transfer cells. Once the cells penetrate into a host phloem cell, they differentiate into multiple long lobes (called *rhizopodials*) that enhance transfer of carbohydrates into the parasite cells.

The description of these two haustoria only hints at the diversity of structure and function of these organs. Some haustoria resemble tumors, while others are only small tissues embedded in the host tissue; some are widely distributed in the host tissue, while others are massive structures surrounding host tissues. However, all haustoria serve the same purpose for the parasite. They form a link between the vascular system

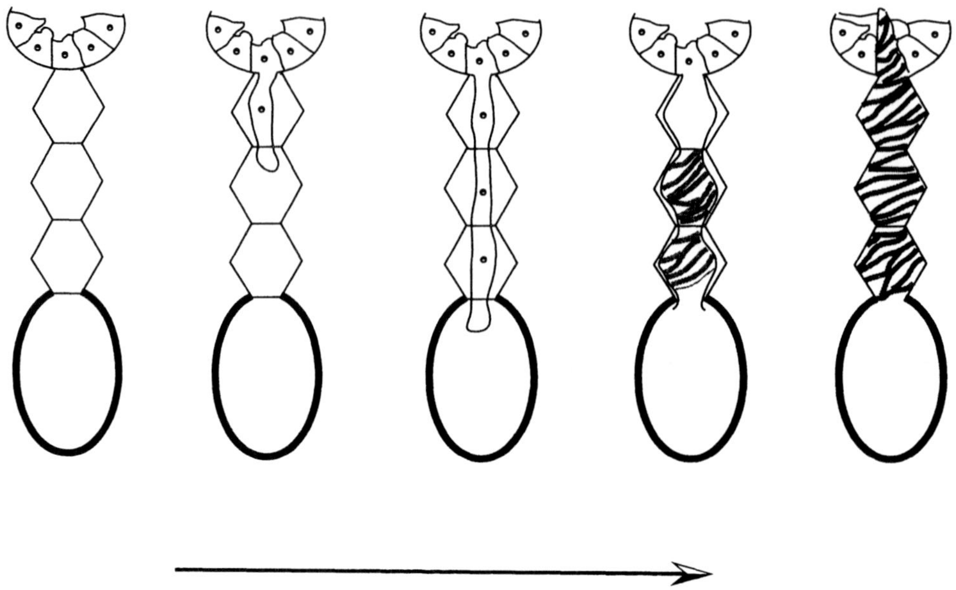

Figure 10.7. Diagrammatic representation of the endophyte penetration of host xylem by *Cuscuta*. Following penetration of the host xylem cell, the parasite cortical cells differentiate into xylem elements. (Redrawn and modeled after a figure in Job Kuijt, *Biology of Parasite Flowering Plants.* Copyright © 1969 The Regents of the University of California.)

of the host and the parasite. In some host-parasite interactions, the links from parasite to host xylem are more prominent than to phloem, and in other cases, the links to xylem and phloem are equally prominent.

D. Sharing Resources (Carbon, Water, and Nitrogen) Between Parasite and Host

Once a successful haustorium has been formed, the possibility of resources flowing between the connected partners is established. Evidence for resource flow from host to parasite abounds. Initially, it was thought that the transfer of water, nutrients, and carbohydrates occurred by a passive process following the stream of water in the xylem from host to parasite (Raven 1983). However, several lines of evidence suggest that the transfer of resources is at least in part an active process. The parenchyma cells of the hyaline body are dense with organelles, and these cells are in contact with the host vascular system. Furthermore, histochemical studies have shown a high enzymatic activity in these hyaline body cells. Both of these observations suggest some active regulation over the transport of resources by the hyaline body from host to parasite. In addition, the quality of carbohydrates, amino acids, and organic acids is different in parasites compared with in hosts. Such a result would not occur if resources were transported by diffusion alone. For example, in *Striga hermonthica* the carbohydrate concentration in the xylem sap is five times that in the host, and the main component is mannitol, which is not found in the host xylem sap. The main organic acid in *Striga* xylem sap is shikimic acid, which is not present in the preferred host, sorghum (Mallaburn 1992). It is clear that there can be a selective barrier between parasite and host in which resources are accumulated against a concentration gradient and are converted into alternative forms before transport within the parasite.

1. Parasite Water Relations. The most prominent aspect of hemiparasite water relations is their high transpiration rate. This contrasts with holoparasites that have a very low stomatal density or no stomata at all (Kuijt and Dong 1990). Stomatal conductance is usually very high in hemiparasites, which may relate to the high concentration of potassium in hemiparasite tissue. (Potassium is associated with stomatal opening.) Water relations of stem parasites are relatively easy to study because they are simply an additional segment along the soil plant–air continuum. In contrast, the water relations of root parasites can be very complex because the parasite may be connected to several hosts with different water potentials, and the parasite can also accumulate soil water. Therefore, the water relations of stem parasites have been studied the most. Stem parasites transpire at higher rates than their hosts. For example, in a study of 18 different host/stem parasite pairs in Australia, the ratio of parasite to host transpiration ranged from 1.5 to 7.9 (Ullmann et al. 1985). In some root parasitic species, stomata remain open at night, resulting in substantial nocturnal transpiration (Shah et al. 1987), as do some stem parasites (Davidson et al. 1989). This does not mean that the stomatal aperture of all parasites is unregulated. Quite the contrary, the diurnal behavior of stomatal aperture of many stem hemiparasites mimics that of the host. These data suggest that short-lived parasites have little regulation of transpiration while long-lived parasites regulate stomatal aperature to mimic their host (Ehleringer and Marshall 1995). Stomatal conductance of hemiparasites is relatively high, is sensitive to atmospheric vapor pressure, and does correlate fairly well with hemiparasite water potential. In contrast, abscisic acid (ABA), which stimulates rapid stomatal closure in nonparasitic plants, has little effect on stomatal conductance in hemiparasites.

The flow of water to the parasite from the host is maintained by a water potential gradient. Hemiparasites have a lower water potential than the host at all times of the day (Glatzel 1983). This is a general characteristic of hemiparasites, but the holoparasite water potential may not differ from the host. The water potential of the hemiparasite will depend upon the water potential of the host, the transpiration rate of the parasite, the parasite leaf area, the magnitude of hydraulic conductance between host and parasite, and the capacitance in tissues of the parasite. Parasite capacitance is relatively unimportant for most stem parasites (Davidson et al. 1989). The water transport across the haustorium is relatively unidirectional because the parasite water potential is usually lower than that of the host. Davidson et al. (1989) also show that about ¾ of the difference in water potential between host and parasite is due to the hydraulic conductance of the haustorium. Therefore, the maintenance of maximum hydraulic conductance across the haustorium may be critical to survival of perennial hemiparasites. The ability to maintain hydraulic flow to the parasite is particularly important. Desert phreatophytes (deep-rooted trees) generally have a higher incidence of stem parasites than shallow-rooted desert shrubs. In Australia, *Olax phyllanthi* during prolonged drought survives only on deep-rooted hosts, not on surface-rooted ones) (Pate et al. 1990).

Dysfunction of hydraulic conductivity is due to the development of embolisms in the vascular tissue. Embolism development (cavitation) of xylem elements results from low water potential, causing air to be sucked into vessels. Although there are no direct measurements of haustorium vulnerability to cavitation, this may be an important aspect of parasite survival. Limited tolerance of parasites to the low water potential of the host has been observed many times. Stem parasites survive on creosote bushes that have relatively high water potentials. Creosote bushes that experience very low water potentials do not have stem parasites.

The lower water potential of the parasite also could result in low turgor potential of the parasite. Turgor is maintained in parasites at lower water potentials than those of their hosts because the osmotic potentials of parasites is also lower than those of their hosts. Many studies have demonstrated that parasites maintain higher concentrations of ions than do their hosts. These high ion concentrations may be an unavoidable result of xylem sap accumulation from the host, or may have evolved as a mechanism for maintaining a lower osmotic potential than the host. The osmotic potentials of parasitic plants are also maintained by relatively high concentrations of organic acids, amino acids, sugar alcohols, and carbohydrates.

*2. **Exchange of Nutrients.*** Obligate parasites and stem hemiparasites have no alternative source of nutrients and water other than the host. In contrast, there is a gradient of dependency on host resources among hemiparasites (Figure 10.8). In general, stem parasites have a greater dependency on host resources than do root parasites. Although parasitic plants should require all the same macro- and micronutrients as nonparasitic plants, their proportional demand for individual nutrients could differ. For example, the major demand for Mg used in the synthesis of chlorophyll and for Cu used in nitrate reductase is absent in achlorophyllous parasites and is severely reduced in hemiparasites.

Many hemiparasitic species contain high concentrations of cations compared with their hosts. For example, mistletoes can contain 20 times the concentration of potassium compared with their hosts. The buildup of these cations in parasite tissue is most likely the result of relatively high transpiration rates and the passive accumulation of ions into the parasite. If the concentration of cations, such as potassium, is higher in the parasite than the host, there must be a filter that retains potassium in the parasite; otherwise, the potassium concentration would come to an equilibrium between host and parasite. One explaination for the high abundance of K in parasitic plants is that K cannot be recycled to the host in the phloem, while K is cycled through the phloem in the host. This means that phloem mobile elements such as K will be concentrated in the parasite, and phloem immobile elements such as Ca will not be concentrated in the parasite (Figure 10.9). Although the effect of phloem cycling is a partial explanation for the difference in ion concentration of parasite and host, other lines of evidence indicate differential ion absorption by the haustorium of parasites. In one study of *Dendrophthoe falcata* (on 48 different hosts) the potassium ion concentration was enriched in the parasite while the concentrations of magnesium, nitrogen, and chloride were not (Glatzel 1983). Another indication that the haustorium may be a selective barrier is the low vascular content of sodium for mistletoe plants growing on mangroves that have high internal salt concentrations (Goldstein, et al. 1989). Therefore, ion concentrations in stem hemiparasites are not simply the result of accumulation through transpiration. Rather, they result from both passive accumulation (coupled to limited return to host through phloem connections) and selective accumulation by the parasite haustorium.

Historically, nitrogen has been a focus of nutritional studies using parasitic angiosperms. Inorganic nitrogen in the form of nitrate and ammonium ion (the major nitrogen source for nonparasitic plants) is not an effective nitrogen source for many parasitic plants. In general, holoparasitic angiosperms have a limited capacity to ac-

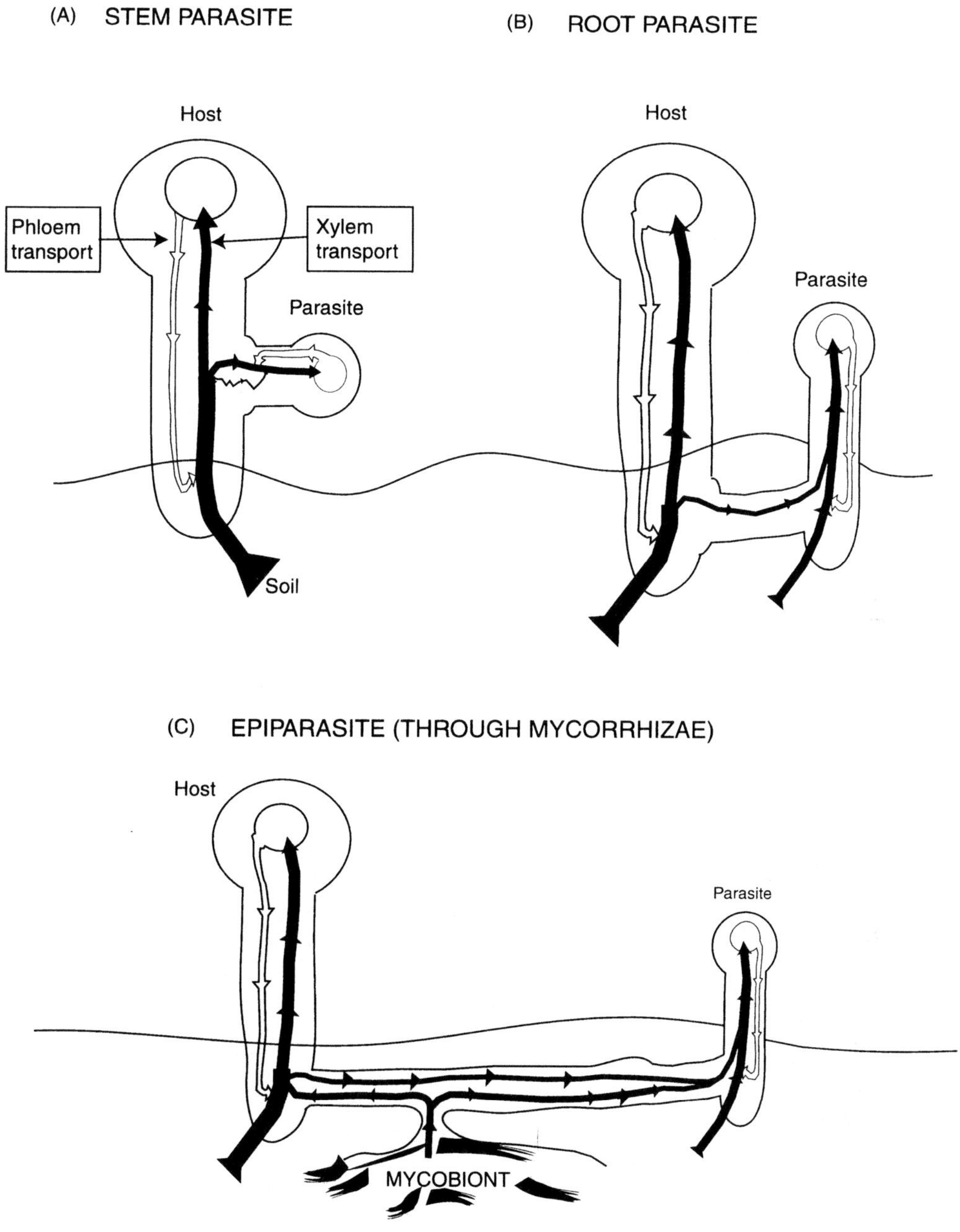

Figure 10.8. Diagrammatic representation of nutrient circulation in several different types of host-parasite association. (A) A stem hemiparasite and its host. (B) A root hemiparasite or holoparasite and its host. (C) An epiparasite through mycorrhizae and its host.

cumulate nitrate and reduce it (using nitrate reductase) to organic nitrogen (Thalouarn et al. 1988). Although stem hemiparasitic plants do have some nitrate reductase activity, their tissue nitrogen concentration does not correlate with the activity of nitrate reductase. Thus, stem parasites depend on both nitrate and organic nitrogen from the host. In addition, when stem parasitic plants are grown on a medium containing only

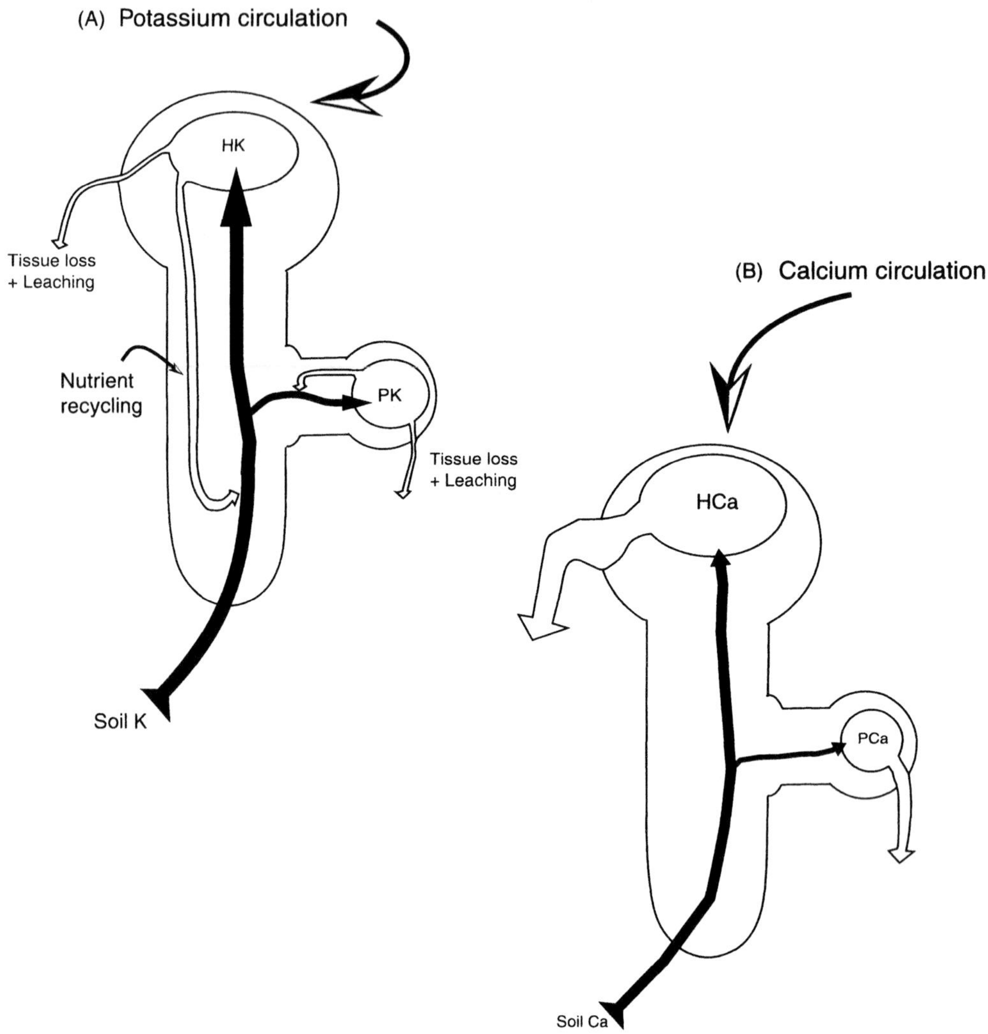

Figure 10.9. Diagrammatic representation of the difference in the circulation patterns of a phloem mobile ion, potassium (A) and a phloem immobile ion calcium (B). The phloem mobile ion concentrates in the parasite. HK = host potassium content; PK = parasite potassium content; HCa = host calcium content; PCa = parasite calcium content (modified from ideas presented in figures in Pate 1995.)

inorganic nitrogen (ammonium nitrate), they have reduced growth. However, when organic nitrogen (amino acids) is added into the growth medium, the growth of stem parasites increases significantly. Accumulation and transport of amino acids and amides from host plants into parasites has been demonstrated many times with labeled molecules (Stewart and Press 1990). In contrast, some hemiparasites grow better on inorganic forms of nitrogen. The root hemiparasite *Striga* can complete its life cycle on media containing only nitrate, and the greater the nitrate availability the greater the parasite success (Igbinnosa et al. 1996). Moreover, adding organic nitrogen or ammonium to the growth media causes a reduction in *Striga* growth. Some root hemiparasites have nitrate reductase activities similar to those of nonparasites, and there can be an inverse relationship between nitrate reductase activity and dependency on the host nitrogen sources in root hemiparasites.

Ammonium accumulation by plants is associated with the glutamate and glutamine synthase systems (see Chapter 4, Figure 4.18). Two different glutamine synthase enzymes are present in cells of nonparasitic plants. One is associated with ammonium transport across the cell membrane (cytasolic isoform), and the other is associated with ammonium ion transport across the chloroplast membrane (chloroplastic isoform). In all three cases, ammonium ion is transported across the membrane and transaminated into glutamine. These ammonium ion transport systems have low activity in parasitic plants when compared to nonparasites (McNally et al. 1983, Thalourn et al. 1988). In fact, only the cytosolic isoform of glutamine synthase has been found in holoparasites. In addition, hemiparasites have a very low activity of chloroplastic glutamine synthase. For example, 5–20% of glutamine synthase activity in *Striga* is due to the chloroplastic isoform, and this proportion is at least twice as large in nonparasitic plants. The chloroplastic isoform of glutamine synthase is thought to serve as a mechanism for reassimilating the ammonium ion produced by photorespiration (Joy 1988) in nonparasitic plants. Since hemiparasites have a relatively low rate of photosynthesis, and a low pool of rubisco, photorespiration is minimal in these plants, and a high activity of chloroplastic glutamine synthase has little importance.

Some studies have indicated that amino acid synthesis may be incomplete in some holoparasites. For example, some holoparasites may not be able to synthesize isoleucine, making them dependent upon the host for this amino acid (Nandakumar et al. 1976). However, the ability of some hemiparasites to complete their life cycle on a growth medium containing only inorganic nitrogen suggests that they are capable of synthesizing all required amino acids.

3. Exchange of Carbohydrates from Host to Parasite. There can be no question that obligate holoparasites gain all their reduced carbon from the host because they have no photosynthetic capacity. In contrast, there is considerable controversy about whether facultative hemiparasites obtain reduced carbon from their host or if they are exclusively mineral and water parasites. Several approaches have been used to verify organic carbon transfer to hemiparasites. Host plants exposed to ^{14}C carbon dioxide will produce ^{14}C-labeled photosynthate. Incorporation of ^{14}C-labeled photosynthate from a host into stem parasites has been demonstrated many times. A novel technique for verifying carbohydrate transfer to a host is to examine the $^{13}C/^{12}C$ ratio of tissue from a C_4 host and a C_3 parasite. The C_4 host produces reduced carbon with a $\delta^{13}C$ of approximately -11, while the host of the C_3 root parasite has a value closer to -30. Therefore, one could use the $\delta^{13}C$ signature of the host and parasite tissues to indicate what proportion of the parasite carbon was derived from the host. Such a study, done with *Striga hermonthica* and *Pennisetum typhoides* (parasite and host, respectively), showed that 49%, 70%, and 87% of carbon in the stem, leaf, and root of the parasite was derived from the host (Graves et al. 1990). Similarly, in a *Striga*-sorghum association, about 30% of the parasite carbon was derived from the host (Press et al. 1987). Thus, root hemiparasites attain a significant proportion of their carbohydrate from the host.

In hemiparasites that have a relatively high rate of photosynthesis, uptake of nitrogen as amino acids from the host could account for a significant accumulation of reduced carbon. For example, if one assumes a 3:1 ratio of carbon to nitrogen in amino acids that are accumulated from the host, and a C/N quotient of 15 in the parasite, then 20% of total reduced carbon in the parasite could be accounted for by amino acid accumulation (Raven 1983). In fact, the high transpiration rates of stem parasites could reflect a mechanism to maximize carbon acquisition in the form of amino acids from the host xylem stream. Moreover, the greater the accumulation of reduced carbon from the host, the lower the demand for nitrogen, because most of the nitrogen demand is destined for use in photosynthesis. Thus, increasing carbon acquisition from the host decreases the need for photosynthesis and nitrogen in the parasite. In contrast, one study on mistletoes indicated that high rates of transpiration primarily supported nitrogen gain, and that acquisition of carbon in the process was an added benefit, not a main mode of carbon assimilation (Marshall et al. 1994).

Carbon is accumulated from the host by the parasite at the interface between the host phloem cells and the parasite endophyte. In some cases, the endophyte tissue enters into the host tissue phloem and elongates into long projections. In other cases, the endophyte tissue does not penetrate the host phloem cells, as with *Cuscuta,* where the endophyte may proliferate around the host phloem cells and differentiate into transferlike cells without host sieve element penetration (Dawson et al. 1994). Plasmodesmata and sieve pores are rarely observed (Dörr 1996) on the interface between the host phloem and endophyte, but the cell wall of the endophyte tissue consistently becomes thin at maturity. Large changes in the ultrastructure of endophyte cells (called hyphae) occasionally occurs after contact with the host phloem; p-protein appears, the nucleus tonoplast and ribosomes disappear, the mitochondria and chloroplasts undergo a transition in form, and an extensive smooth endoplasmic reticulum develops. All of these changes result in endophyte cells that closely resemble the mature structure of the host phloem cells. The smooth endoplasmic reticulum of these hyphal cells is extensively stacked and oriented perpendicular to the cell wall contacting the host phloem. These endophyte cells can be considered to be specialized sieve elements that function without companion cells. Such elaborate transfer cells in association with host phloem have only been found in a few parasitic species. In most cases, parenchyma cells of the endophytic tissues do not go through elaborate ultrastructural changes, but still serve as the probable sites of assimilate transfer from host to parasite.

Assimilate transfer from host to parasite is thought to occur primarily by an apoplastic route. As stated earlier, plasmodesmata have been found only rarely (in *Orobanche* spp.) between endophyte and host cells making a symplastic route of transfer very unlikely. The driving force for assimilate transfer is the sink gradient between the host root and the endophyte tissue. In *Cuscuta,* a high apoplastic activity of acid invertase (which hydrolyzes sucrose into glucose and fructose) has been found, which would increase the rate of sucrose unloading from host phloem cells into the apoplastic space near the endophyte tissue (due to the low sucrose content in the apoplast). Phloem unloading may also be stimulated by relatively high ion concentration in the apoplast. The high ion and hexose contents of the apoplast cause water to enter the apoplast from the phloem cells, reducing the phloem cell turgor pressure. The phloem cells respond by decreasing their osmotic potential further (by increasing sucrose concentration), which enhances sucrose transport out of the host phloem. Endophyte cells near the host phloem cells should have high activity of glucose transporters in order to assimilate the hexoses from the apoplastic space. There is also some evidence that host-derived growth regulators (abscisic acid) may affect assimilate transfer from host to parasite. Very high abscisic acid concentrations are found in the haustoria of *Cuscuta,* and abscisic acid stimulates the accumulation of glucose by the parasite over that of the host (De Bock and Fer 1992).

In hemiparasitic plants such as *Striga,* photosynthesis by the parasite begins to produce a significant amount of carbohydrate for the parasite after only a few days of growth (Pageau et al. 1998). In fact, about 50% of parasite carbon requirements is attained from parasite photosynthesis. Why does the photosynthate produced by the parasite not travel back to the host? Part of the reason that the host does not acquire parasite carbon resources is because the parasite remains a large sink for both carbon and water in comparison with the host. In addition, many parasitic species store carbohydrate as mannitol (Simier et al. 1994), and the host root systems do not have a mannitol accumulation system.

In summary, most photosynthate is transferred from the host to the parasite through an apoplast route to the endophyte, in the form of sucrose. Cells of the endophyte exude invertase into the apoplast, which converts sucrose to hexoses, and the endophyte cells accumulate the hexoses by transfer from parenchyma cells nearby. The sink gradient is maintained by high contents of inorganic ions and abscisic acid in the apoplast. Reverse transfer of assimilates from parasite to host is inhibited by the large sink strength of the parasite and the storage of mannitol by many parasites.

4. Photosynthetic Capacity of Parasites. Holoparasites have no photosynthetic capacity and gain all their reduced carbon from the host. The most interesting question about photosynthesis in holoparasites is whether the genes that synthesize photosynthetic proteins have followed a different evolutionary pathway in these species compared with autotropic plants or hemiparasites. In autotrophic plants, the plastid genome is remarkable similar among species, most likely because of constraints on variation due to the importance of plastid function to carbohydrate assimilation and plant fitness. However, in holoparasites, the constraints on plastid genome structure are relaxed due to the low significance of photosynthesis to plant fitness. In a sense, when photosynthesis is lost in holoparasites, all photosynthesis genes become pseudogenes at the same time, and are subjected to mutation and genetic drift. Thus, one would predict that there would be much more variation in photosynthesis-related chloroplastic genes in holoparasites than in autotrophs or hemiparasitic plants.

The first and perhaps best-studied plastid genome of a holoparasite is that of *Epifagus virginiana,* a parasite of the American beech *(Fagus grandifolia).* The chloroplasts of *E. virginiana* are rudimentary, as is the case for many holoparasites, and they most likely serve as pigment, lipid, starch, or protein storage. A first look at the plastid genome of *E. virginiana* was surprising because the genome was very small and most of the genes directly associated with photosynthesis were missing (de-Pamphilis and Palmer, 1989, 1990). The chloroplastic genome of *E. virginiana* is only half the size of that in tobacco (Figure 10.10). In fact, the *E. virginiana* chloroplast genome contains only 42 known active genes, of which only four are known to be associated with photosynthesis. Moreover, all four genes coding for plastid RNA polymerase subunits are either deleted or truncated (Morden et al. 1991). These results suggest that the genome of *E. virginiana* may be entirely dysfunctional, since RNA polymerase is not present, and the protein synthesis machinery is severely reduced in the chloroplast. However, the remaining 40 or so chloroplast genes are expressed, and their presence and activity can only be explained by invoking strong evolutionary constraints on their structure and function. Since gene expression occurs in these chloroplasts, a functional RNA polymerase, tRNA molecules, and parts of ribosomes must be imported into the chloroplast from the cytoplasm. These results clearly show that the constraints placed on chloroplast evolution by photosynthesis, in autotropic plants, is a powerful process maintaining the uniformity in chloroplast genomes. When these contstraints are lifted, as in holoparasites, plastid genomes with deletions and insertions in genes can survive. The number of deletions and insertions is far greater

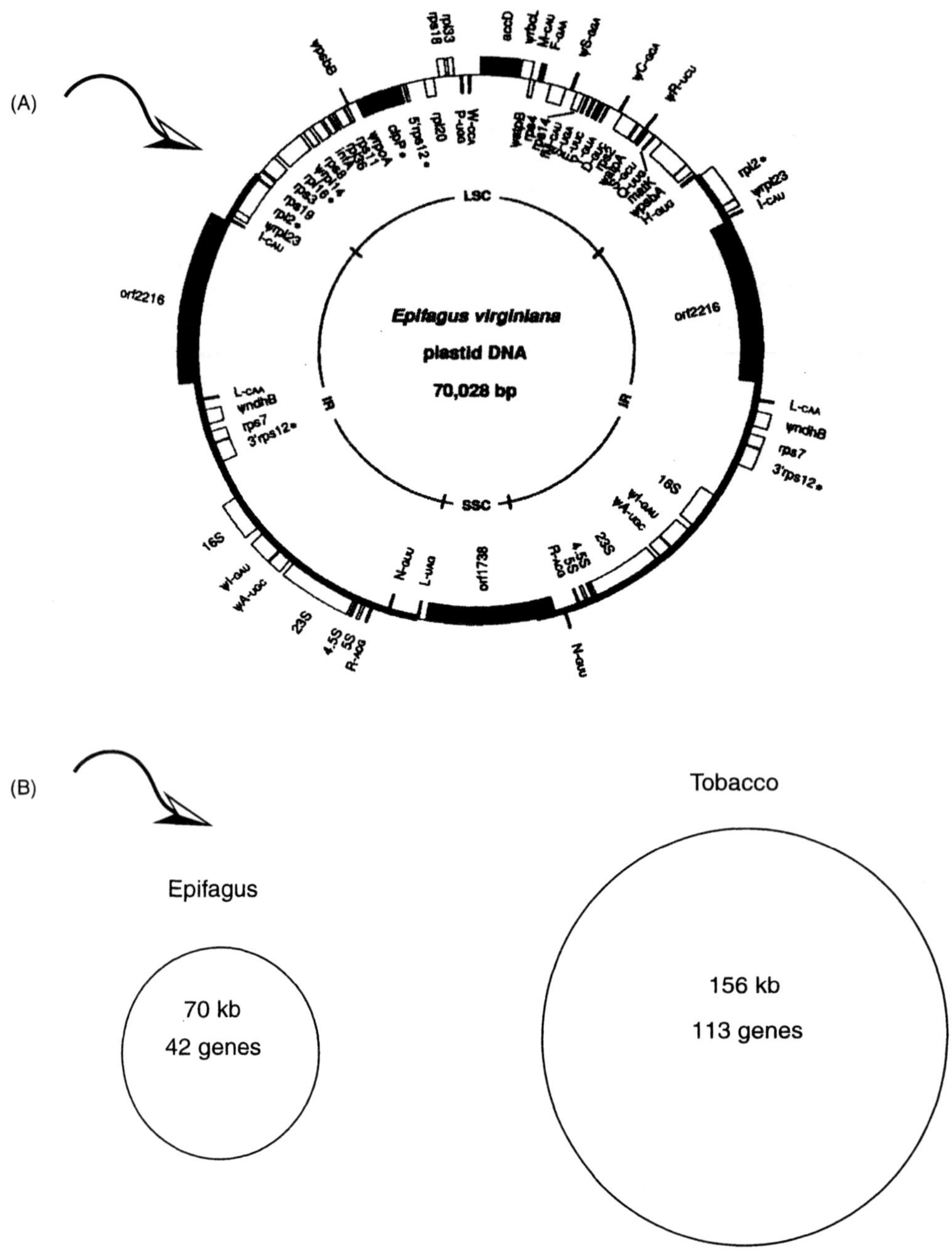

Figure 10.10. (A) Diagrammatic representation of the chloroplast genome from a holoparasite (*Epifagus*). (Reproduced with permission from Wolfe et al. 1992.) (B) A comparison of the genomes of *Epifagus* and tobacco, showing the difference in size and number of actively transcribed genes.

in holoparasites than in nonparasitic plants (Figure 10.11). Moreover, the extent of deletions and truncations in plastid genes varies among different holoparasites. For example, in *Lathraea* both the genes for the small and large subunit of rubisco remain (large in the chloroplast). *Cuscuta europaea,* a true holoparasite, has chloroplasts with no thylakoids but does have a rbcL gene. Another *Cuscuta* species, *C. reflexa,* retains many of the plastid photosynthesis genes found in autotrophic species, although they do not have very high homology to those genes from autotrophic species.

Clearly, the reduction in the significance of photosynthesis in parasitic plants has resulted in major changes in their chloroplast genome. These changes provide a window into the evolution of parasitic plants and the chloroplast genome in general. A recent study comparing the rps2 and rbcL sequences among many parasitic and non-parasitic members of Scrophulariales indicates that a single origin of parasitism occurred in this order, and the loss of photosynthesis genes have been independently derived many times after the onset of parasitism (Wolfe and dePamphilis 1998). Following the origin of parasitism, there have been five different lineages of chloroplast genome evolution in this group. Moreover, the data clearly show that some chloroplast genes in hemiparasitic species have also been lost, which means that the evolutionary loss of some chloroplast genes occurred before complete loss of photosynthesis.

There is an inverse relationship between photosynthetic capacity and carbohydrate flow from host to parasite in hemiparasites. For example, the dwarf mistletoes, *Aecheuthobium* spp., have a low rate of photosynthesis and a high rate of carbohydrate transfer from the host. In contrast, the mistletoes *Phoradendron* spp. have a relatively high rate of photosynthesis (close to that of many C_3 plants) and a very low rate of carbohydrate transfer from the host. Grouping all hemiparasites together, one could

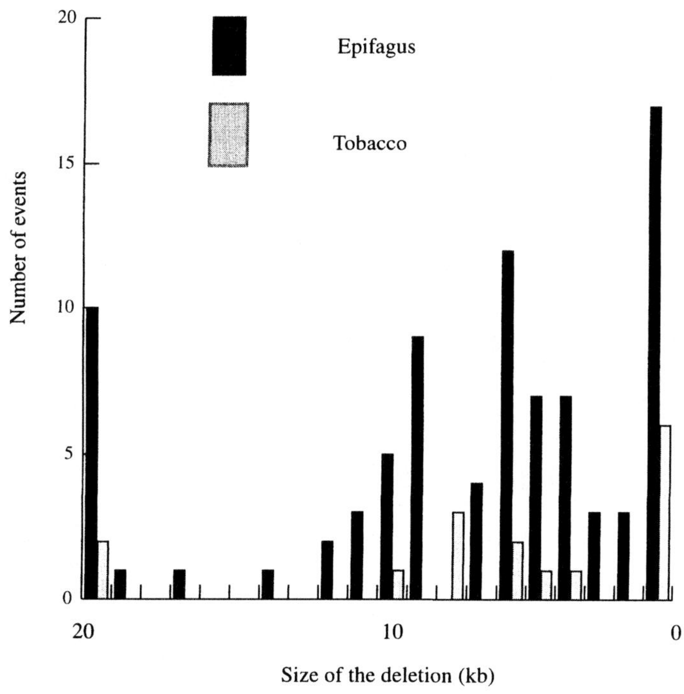

Figure 10.11. Number of deletions in the IR region of *Epifagus* and tobacco chloroplast DNAs. (Redrawn from data presented in Wolfe et al. 1992.)

form a continuum extending from high carbohydrate transfer and low photosynthesis, to low carbohydrate flow and high photosynthesis. However, most parasites have a relatively low rate of photosynthesis, ranging from 0.5 to 5.0 μmol cm^{-2} s^{-1}. These low photosynthetic rates coupled with a relatively high respiration rates produce a net carbon gain, which is normally inadequate to support growth (Press et al. 1987). As a result of the relatively low photosynthetic capacity, most hemiparasites accumulate at least 30%, and possibly up to 90%, of their carbon requirements from the host. Even those parasites that are considered facultative parasites (because of their leafy nature and assimilation rates close to those of their hosts) obtain a significant amount of carbon from the host. For example, *Castilleja linarifolia* has photosynthesis rates indistinguishable from the rate of its host, *Artemesia tridentata,* but attains 40% of its reduced carbon from the host (Ducharme and Ehleringer 1996).

Carbon accumulation by hemiparasites depends in part on the particular host. For example, 19–30% of carbon accumulation by *Olax phyllanthi* is derived from the host when grown on nonnitrogen-fixing C_4 plants, and 51–60% when grown in association with the nitrogen-fixing tree *Acacia littorea* (Tennakoon and Pate 1996b). This result suggests that the higher growth rate of the parasite on N_2-fixing hosts is a result of increased leaf area of the parasite as a result of the increased N availability from the host, causing an increased photosynthesis capacity (increased leaf area) of the parasite. Differential carbon accumulation from different hosts also results in different growth rates of the parasite as seen for *Santalacum album* (Radomiljac et al. 1998,1999a). In this case, a strong relationship was found between *S. album* leaf area and growth rate independently of the host species.

Irrespective of the nitrogen availability, photosynthetic rates per leaf area of parasitic plants are relatively low. In *Striga,* the leaf mesophyll is relatively undifferentiated and there are few chloroplasts. Furthermore, the quantity of chlorophyll and the activity of rubisco is relatively low compared with nonparasitic C_3 plants (Press et al. 1986). Other studies have shown that the activity of photosystem II is also relatively low in the leaves of *Striga* (Salle et al. 1987). In *Cuscuta reflexa,* the concentrations of rubisco and chlorophyll are reduced on a fresh-weight basis because the cells containing active chloroplasts are few and restricted to the region around the vascular bundles (Hibberd et al. 1998a). Those cells are capable of performing normal photochemical and biochemical reactions of photosynthesis. This layer of cells may serve primarily to fix internally generated carbon dioxide, thereby contributing to the overall carbon budget of the parasite. The photosynthetic capacity of parasites is also related to host specificity at the family level. Those families with high host specificity tend to have higher photosynthetic rates (Figure 10.12). In general, all aspects of photosynthetic machinery in hemiparasites are reduced in activity compared with nonparasitic plants, and that of holoparasites is almost nonexistent as indicated by the depauperate chloroplast genome of holoparasites.

E. Uptake of Alkaloids by Parasitic Plants from Their Hosts

In several cases, parasitic plants attack host species that have chemical defenses against herbivores. For example, *Triphysaria pusilla* parasitizes *Lupinus arboreum* in the California coastal dune community, and *L. arboreum* contains alkaloids as chemical defense against herbivores (see Chapter 7). The parasitic plants accumulate the alkaloids from their host. *Triphysaria pusilla* accumulates alkaloids from *L. arboreum* (Marvier 1998b). *Castilleja* accumulates alkaloids from the phloem of composites and legumes (Stermitz and Harris 1987). In addition, *Cuscuta reflexa* accumulates quin-

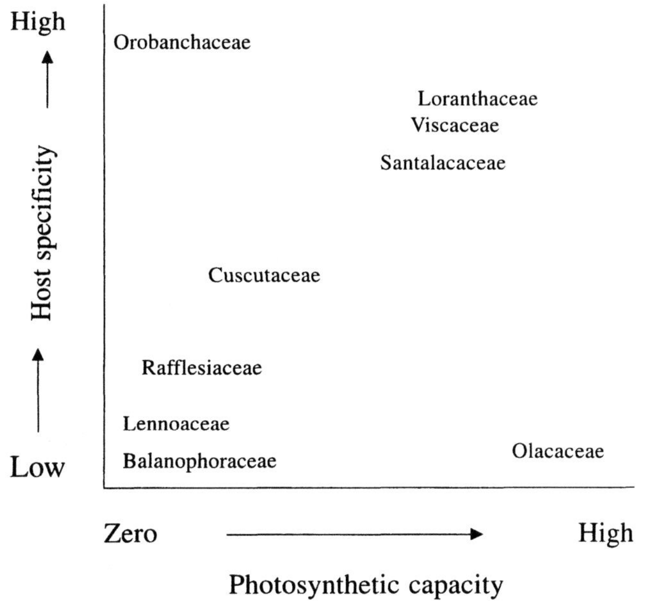

Figure 10.12. Relationship between host specificity and photosynthesis rate for a number of families of parasitic angiosperms. (Redrawn and modified from Gibson and Watkinson 1992 with permission from Springer-Verlag, Inc.)

olizidine alkaloids from *Lupinus album* (Baumel et al. 1995). The parasite may be accumulating the alkaloids preferentially in order to enhance defense against herbivores, or alkaloids could simply be accumulated indiscriminately as a nitrogen source. In the case of *Cuscuta reflexa* and *Lupine spp.,* the parasite accumulates 95% of its alkaloids from the host phloem in a similar proportion to that of other nitrogen-containing compounds. Therefore, the accumulation of alkaloids is most likely a mechanism for accumulating nitrogen, and *C. reflexa* must adapt to the resultant high alkaloid content and catabolize the alkaloids, releasing nitrogen for metabolism (Baumel et al. 1995). Accumulation of alkaloids from the host may confer some protection against herbivores for the parasite.

III. EFFECTS OF PARASITIC PLANTS ON THEIR HOSTS

Parasitic effects on the host plant's form and function are extremely variable. In some cases, the parasite has no noticeable effect and in other cases the host assumes a dramatically changed morphology when attacked by a parasite. Stem parasites often induce a change in the structure (size) of proximal branches to their attachment point because the parasite is sequestering materials required for branch growth. In other cases, the parasite affects the photosynthesis and nitrogen metabolism of the host. Perhaps most important is the effect of parasites on the productivity and reproduction of the host (Silva and Rio 1996). The effect of parasites on hosts involves more than the obvious extraction of carbohydrate and nutrients from the host. For example, shortly after *Striga hermonthica* attacks the host sorghum (when the *S. hermonthica* plants are just a few millimeters long), the predicted parasite-induced weight loss of the host is 30 times the weight gain of the parasites (Parker 1984). The interaction between host and parasite certainly includes competition for host resources, but it also involves a complex physiological interaction.

A. Effects on Plant Architecture

Root parasites often cause an increase in host plant root/shoot ratio. On both sorghum and pearl millet, the parasite *Striga hermonthica* induces radical changes in host root/shoot ratio. The reduced allocation to leaves causes a significant reduction in whole-plant productivity (Graves et al. 1990). The production of the host reduces because there is less leaf area, and the larger mass of roots requires a higher respiratory investment. Infection of tomatos by *Orobanche aegyptica* causes an increase in the root/shoot ratio that is manifest before the reduction in relative aboveground growth rate of the host (Barker et al. 1996). The large changes in root/shoot ratio associated with parasite attacks are most likely the result of root growth stimulation in the host by the parasite infection instead of only a decrease in shoot growth. Increased root growth of the host can be induced by hormones from the parasite, by wounding of host tissue by the haustorium of the parasite, or by changes in soil nutrition near the parasite. In all these cases, root proliferation close to the parasite will favor additional parasite connections to the host root system. In addition, the root/shoot ratio increases in host plants attacked by stem and root parasites, because the demand for water by the host plant is increased by the presence of the parasite, and this causes the root system to proliferate in order to support the increased demand.

Parasites do not always induce a higher root/shoot ratio. For example, some stem parasites may have no impact on the root/shoot ratio unless their leaf area causes a dramatic change in host plant water demand. In addition, some root parasites can have a negative effect on host root/shoot ratio. For example, when *Odontites vulgaris* attacks *Medicago sativa,* the root/shoot ratio of the host reduces in comparison to host plants that are not supporting a parasite (Matthies 1998). Infestation by *Orobanche* species does not induce root production, and the root distal to the attack by *Orobanche* sp. often dies. Parasite infection can change the allocation to leaf area independently from any change in root productivity. In tobacco, parasite-infected plants have a higher allocation to leaf area and a delayed senescence of leaves (Hibberd et al. 1998b) when compared with noninfected plants. As a result, the combined production of host and parasite is the same as the production of uninfected tobacco plants. Therefore, the generality that the root/shoot ratio always increases following parasite attack certainly does not cover all parasite-host interactions and may not be as encompassing a generality as once thought.

Stem parasites can also influence the architecture of the aboveground tissues of the host plant leading to lower production of the host. In fact, except for extremely diffuse endophytes, all parasites cause some sort of host tissue swelling at the site of attack. For example, mistletoes induce swelling of host branches, *Cuscuta* induces gall-like swellings of host leaves and stems, and *Striga* induces swelling of host roots at the site of attachment. This swelling is a result of increased cell division in the host parenchyma near the infection and may be a normal wound response. Wounding of plant tissue caused by bacteria such as *Agrobacterium,* by fungal infections, or by mechanical damage also cause increased cell division by parenchyma near the wound site. Thus, it is not surprising that an attack by a parasite induces tissue swelling of the host. Along with the swelling, *tyloses* (invaginations of parenchyma cells into vascular cells blocking the lumen) often form in vascular cells. Tyloses that develop in response to parasite attack may have been interpreted as endophyte intrusion into phloem tissue in some cases. In stem parasites, the swelling, tyloses, and diversion of resources into the parasite can result in a severe limitation to resources for growth of tissue distal to the parasite. The result is a much reduced branch size and often mortality of the distal branch. In fact, many observations that mistletoes form at the ends of branches are a misinterpretation because the rest of the branch (proximal to the parasite) has died and fallen off. Branches of *Acacia acuminata* attacked by the mistletoe *Amyema preissii* suffer inhibition of growth distal to the parasite. As the severity of parasite attack increases, the ratio of stem diameter proximal and distal to the parasite (*P*/*D*) increases until a ratio of 5–6 occurs, after which the distal branch dies (Tennakoon and Pate 1996a). The distal branch will have a diameter/parasite leaf area ratio that is three times that for branches supporting the host foliage only. This higher diameter/leaf area ratio of branches with parasites occurs because the branch must support the higher transpiration of parasite leaves compared with host leaves.

Extensive infection by stem parasites severely reduces the leaf area of the host, thereby significantly reducing host productivity. Similarly, root parasites also cause a reduction in host leaf area. Cowpea plants attacked by *Striga gesnerioides* have a much reduced leaf area, and the extent of that reduction is proportional to the intensity of the infestation (Aigbokhan and Hussain 1991). The reduction in leaf area occurs rapidly at about 40 days after attack by the parasite, at a time when the host canopy is still expanding (Graves et al. 1992). The result is a severe reduction in growth, and an almost complete eradication of host flowering and seed production.

Undoubtedly, the most dramatic effect of parasites on host plant architecture is the witches'-broom that develops distal to an attack by species of the dwarf mistletoe *Archeuthobium*. The witches'-broom is a large proliferation of branches near the infection that take over most of the growth in the crown of the tree. (See Kuijt 1969 for excellent pictures.) This massive alteration in tree crown structure causes a large reduction in host productivity. Other parasites induce similar broomlike structure in their hosts. For example, members of the genus *Metrastemon,* a root parasite, induces a broom proliferation of host roots, a parasite of Chilean poplar causes irregular witches'-brooms, and several other mistletoe species induce brooms on their hosts (Kuijt 1969).

B. Competition for Resources (Water, Carbon, and Nutrients)

One of the ways parasites can influence the growth and reproduction of the host is by reducing the amount of resources available for host growth. The resources most often considered in this host-parasite competition are water, nutrients, and carbohydrates. Clearly, the branch distal to a stem parasite infection is directly competing with the parasite for resources. Recall that stem parasites have a higher transpiration and lower water potential than the host. Therefore, water is preferentially allocated to the parasite at a cost to the host. Under dry conditions, the host may suffer drought stress more rapidly than does the parasite. This is particularly true in cases where the parasite transpiration is not coordinated with the host water potential, as in the case of *Striga asiatica.* Crops such as sorghum infected with *S. asiatica* often wilt even when soils are quite wet (Parker and Riches 1993). Oak trees infected with the mistletoe *Loranthus europaeus* have reduced stomatal conductance as well as a reduced water potential, compared with trees without parasites. The lower conductance causes a greater limitation to photosynthesis in oak trees with parasites than in those without parasites, and that leads to a lower host productivity (Ehleringer et al. 1986). The potential for host water stress increases with the biomass of parasite supported. For example, in *Striga gesnerioides,* the parasite utilizes a greater proportion of host water as the parasite biomass increases, causing a greater potential of host water stress (Graves et al. 1992). The intensity of parasite effect on plant water relations may depend on whether the parasite is perennial or annual. Annual parasites have evolved to mature rapidly and make the most intense use of host resources for optimal seed production. Such parasites tend to transpire rapidly and accumulate host resources without regulation. In contrast, perennial parasites often regulate resource use in a coordinated manner with the host. For example, the stomata of perennial mistletoes close in response to low humidity shortly after the host stomata close, and perennial shrubby parasites form and detach haustoria in relation to the carbohydrate availability of the host. Stomatal conductance of *Santalacum album* changes in close coordination with its hosts, even though different hosts may have different patterns of stomatal closure (Radomiljac et al. 1999a). Such coordination of resource demand by the parasite with physiological activity of the host is critical to the survival of perennial parasitic plants.

As discussed earlier, inorganic ions tend to become concentrated in the parasite for a number of reasons, including a lack of translocation back to the host and a high transpiration rate of the parasite. The high inorganic ion concentration of parasites does not mean that parasites have an exceptionally high requirement for ions or that parasites cause ion deficiency in the host. In most cases, hosts do not express ion deficiency symptoms when attacked by parasites. Of course, there are exceptions; for example, tobacco expresses potassium ion deficiency when infected with *Orobanche* sp., and nitrogen limitations can occur in some host species when attacked by parasites. For example, sorghum expresses elongated chlorotic blotches, indicating nutrient deficiency, when infected with various *Striga* species (Parker and Riches 1993), as does cowpea when infected with *S. gesnerioides*.

Nitrogen-containing compounds, such as amino acids and alkaloids, are preferentially accumulated by parasitic plants to support the parasite's nitrogen demand. This is critical to root parasites because they often lack the ability to accumulate nitrogen on their own from the soil, and stem parasites because the host is the only source of nitrogen. Even facultative parasites, which can live on their own accumulation of resources for a time, soon expire due to lack of nitrogen in the absence of a host. Thus, parasites are a large drain on host nitrogen resources and can cause a nitrogen deficiency in the host. For example, pearl millet seed produced by plants infected with *Striga hermonthica* is deficient in amino acid content (Wylde 1991), which is likely to reduce the success of seed germination. The high demand for nitrogen by parasites also explains why parasites grow larger when attached to legume hosts. However, the severe reductions in sorghum growth when associated with *Striga* are not associated with N limitation in the host (Cechin and Press 1993). In fact, in many cases, ammonium ion, nitrate ion, and amino acid concentrations in the host are higher when parasites are attached. When *Coleus blumei* plants are parasitized by *Cuscuta reflexa,* nitrate accumulation is stimulated and host nitrogen concentrtation increases compared with uninfected *C. blumei* plants (Jeschke et al. 1997). This seeming contradiction between nitrogen demand by parasites and host nitrogen concentration is due to a reduction in the aboveground growth of the host relative to the belowground nutrient accumulation by the host, which increases the concentration of nutrients relative to host tissue mass when a parasite is attached. In general, the attack of hosts by most parasites causes an increase in nitrogen accumulation and host tissue nitrogen concentration because of the added demand by the parasite on host resources.

C. Effects on Metabolism

Parasites also influence host metabolism other than through direct competition for resources. In general, siphoning resources away from the host by the parasite should stimulate a compensatory increase in host metabolic activity. In particular, the increased sink strength for carbohydrate when parasites are attached to the host should stimulate photoynthesis in the host. Infection of *Coleus blumei* by *Cuscuta reflexa* led to a substantial increase in *C. blumei* photosynthesis per leaf area (Jeschke et al. 1997).

Indirect effects of resource competition also can have a negative effect on metabolism. For example, the water stress induced by parasites can cause a reduction in photosynthesis in the host. Electron transport in photosynthesis was inhibited in sorghum when infected with *Striga hermonthica* (Ramlan and Graves 1996). Consequently, parasite infection made the sorghum plants more susceptible to photoinhibition. In the case of oaks infected with *Loranthus europaeus,* parasite infection increased stomatal limitation of photosynthesis (Ehleringer et al. 1986). Stomatal limitation also reduces photosynthesis in sorghum when infected by *S. hermonthica* (Frost et al. 1997). In this study, it was suggested that an increase in abscisic acid induced by the parasite is a main cause for increased stomatal limitation of photosynthesis in the host plant.

It has also been suggested that reduced stomatal conductance in plants infected with root parasites is due to host plant water stress induced by dry soil, which results from high water use by the parasite. This concept was not supported by a study with maize and *S. hermonthica,* in which maize stomatal conductance and photosynthesis decreased in response to a parasite long before any significant reduction is soil water occurred (Taylor and Seel 1998). Host-parasite interaction can also be balanced so that there is little change in host photosynthesis upon infection by the parasite. The photochemical process of photosynthesis, photosynthetic rates, and intercellular carbon dioxide concentrations of *Mimoso naguirei* (a legume) were identical in unparasitized control plants and plants heavily laden with the stem endophytic parasite *Pilostyles inage* (Fernandes et al. 1998). *Poa alpina* infected with *Rhinanthus minor* for one or two years had a photosynthetic rate that was indistinguishable from *Poa alpina* plants that were never parasitized (Seel and Press 1994, 1996). In general, a negative effect of parasites on photosynthetic metabolism is rare, and is usually associated with increased stress for the host such as water stress. Most commonly, perennial parasites may increase the rate of photosynthesis per leaf area of the host due to the increased host-parasite demand for carbohydrate, and increased host nitrogen concentration. This does not mean that the whole-plant carbon gain is increased in the presence of the parasite. In most cases, the whole-plant carbon gain without the parasite is equal to or greater than in the host plant–parasite association. The reduction in whole-plant carbon gain could be a consequence of decreased host plant leaf area (a common occurrence) and increased host respiration demand (partly due to the increased mass of roots in the host).

Infection by parasites can also influence the normal hormonal balance of plants. Clearly, the witches'-broom response to some parasites is a massive change in hormonal response in the host. The tissues immediately surrounding the haustorium of dwarf mistletoe have an increased level of auxin and cytokinin and a reduced concentration of abscisic acid (Livingston et al. 1984). Auxin and cytokinin control over shoot elongation are the probable cause of the witch's-broom structure. If so, the parasites are inducing an increase in auxin concentration or an increase in tissue sensitivity to auxin. However, the increase in auxin concentration is counter to the release of apical dominance associated with these structures. The concentration of growth promoters such as auxin also increase in the sap and tissues of sorghum infected with *Striga* (Drennan and El Hiweris 1979).

Cytokinin concentrations have been studied in parasite and host. Before attachment to the host, cytokinin concentrations are relatively low in the parasite (*Archeuthobium*) and it rises and falls on a diurnal cycle (Lechowski and Bialczyk 1996). Following attachment to the host, the cytokinin concentration increases greatly in the parasite and is constant over the diurnal cycle. When cytokinins are so high in the parasite, is it possible for them to diffuse into the host and be found in the host sap. Although some scientists feel that cytokinins from parasites enter into the host, this is unlikely because neither ions (K) nor photosynthates diffuse back to the host.

More attention has been paid to the concentration of ABA in host tissues when attacked by parasites. The haustorial tissues of *Cuscuta reflexa* have very high concentrations of ABA, and this may be an important factor in regulating the unloading of sucrose into the parasite tissues (De Bock and Fer 1992). Infected plants can have severely depressed stomatal conductance compared with uninfected hosts. In maize, infection by *Striga hermonthica* causes reduced stomatal conductance that may be attributable to changes in ABA concentration in the host (Taylor et al. 1996). The concentration of ABA in infected maize plants was always higher than in noninfected maize plants. Abscisic acid concentrations are commonly much higher in the parasite than in the host. It is not known whether the parasite is synthesizing and mobilizing ABA to the host, or whether the host is synthesizing ABA as a wound response to the parasite attack and the parasite is accumulating the ABA. The ABA in host and parasite is differentially distributed in leaf tissues. The abaxial (bottom) side of the leaf where most stomata are located had 3.6 times more than the adaxial (top) side of the leaf in both host and parasite (Lechowski 1996). The abaxial concentration of ABA causes a strong reduction in stomatal conductance. Attack of a host plant by a parasite most likely induces a wound response stimulating the ABA signaling pathway (see Chapter 5), inducing changes such as leaf senescence, reduced stomatal conductance, and increased root growth.

IV. PLANT DEFENSES AGAINST PLANT PARASITE ATTACK

Clearly, parasitic plants have evolved elaborate physiological mechanisms to locate, attack, and utilize the host and its resources. Surely, the hosts have evolved mechanisms of defense against parasite attack. Host defense could be achieved in several different ways:

1. Selection could favor genotypes that produce less of the products that initiate parasite seed germination and host localization.
2. Selection could enhance defensive mechanisms against pathogen infection or herbivore attack such that these existing mechanisms could limit parasitic plant attack.
3. Selection could favor the evolution of individuals that produce toxins that specifically inhibit parasitic plants.

The first possibility refers to the hydroxyquinones that stimulate parasite seed germination and the phenolic compounds that stimulate haustoria formation on the parasite at close proximity to the host. More than 8000 different phenolic compounds are produced by plants. The leaching of phenolics into soil serves as a signaling mechanism between plants and other soil organisms. For example, NodD proteins of *Rhizobia* (N_2-fixing bacteria) are stimulated by plant phenolics. The interaction can be specific enough that individual nodD genes are optimally responsive to a particular group of phenolics. Furthermore, phenolics leached into the soil from plants induce the virulence genes of *Agrobacterium tumefaciens.* Virulence gene preference for specific plant phenolic compounds can affect the host range of the bacteria (Heath et al. 1997). Therefore, it is highly possible that phenolics also induce genes of the parasite that enable host location and penetration. The production of phenolics in large and diverse amounts is required for cell wall synthesis, host defense, flower pigmentation, and other important metabolic processes. Therefore, these products are a necessary by-product of root metabolism. In addition, their release is important for root association with beneficial microbes. Thus, it is not likely that plants can be successful in a natural environment with a low level of phenolic production. Consequently, the second and third possibilities for plant defense against parasitic angiosperms are more likely to occur in nature, and these involve the molecular response of host plants to parasite attack.

A. Molecular Responses to Parasitic Plant Attack

Attack of a host by a parasite involves location of the host, adherance to the host, penetration of the host tissue, and formation of the vascular connections between host and parasite. The first molecular responses of the host to parasites could occur when the parasite adheres to the host and/or when the parasitic tissue penetrates the host. The seeds of stem parasitic plants are often sticky, which enhances their ability to adhere to host tissue. There is no known molecular response to the advent of a sticky seed on the bark of trees. Of course, bark that sloughs off frequently may be an effective defense against stem parasite attack. Although "sloughability" was once thought to be important for attachment of bromeliads to cypess in the Okefenokee Swamp (W. H. Schlesinger, personal communication), this branch trait is unlikely to be a particularly important defense mechanism against stem parasites. Root parasites such as *Orobanche* and *Striga* adhere to host roots by exuding a sticky carbohydrate (Joel and Losner-Goshen 1994). Molecular response to the parasite does not occur at this stage of host-parasite association (Westwood et al. 1998). The main host molecular response to parasite attack occurs when the host tissues are penetrated by the parasitic endophyte.

One major molecular defense mechanism in plants is the *wound response*. This is a general response because it occurs following pathogen, parasite, herbivore, and mechanical damage. The wound response is localized and mediated through jasmonic acid (see Chapter 7, Figure 7.10). The localized resistance response to pathogen attack in wounds involves the interaction of plant resistance genes with specific genes of the pathogen or products of the host-pathogen interface. The interaction of resistance gene and pathogen causes a hypersensitive response in cells surrounding the pathogen, ultimately leading to rapid cell death in the immediate region around the wound. The host plant response to parasite attack is also associated with the expression of a class of pathogenesis-related (PR) genes throughout the plant. Some of these genes are related to plant defense against further pathogen attack. The induction of these genes results in a *systemic acquired resistance* (SAR) against a broad diversity of pathogens. An increase in salicylic acid throughout the plant is essential to the SAR response. The parasite induction of SAR also is a very general systemic response that is not focused against any specific organism. Genes responsive to the salicylic signal transduction pathway may be effective against some classes of pathogens (fungi, bacteria, viruses), but are not likely to be effective against angiosperm parasites because such effectiveness would result in autotoxicity. The best defense against a parasitic plant must be a specific molecular response localized at the infection site that is targeted to the endophyte. Therefore, scientists are searching for genes that are specifically induced near the haustorium of parasitic plants (probably related to the hypersensitive response), particularly in the case of incompatible associations between host and parasite (the incompatibility may be due to a parasite-specific host defense gene).

B. Plant Defensive Mechanisms Against Parasitic Plant Attack

The mechanisms of plant defense against parasitic plant attack are not well understood at this time. However, some recent studies may provide the basic stepping stones for understanding defensive mechanisms against parasitic angiosperms at a molecular level. Resistance mechanisms are known to exist because members of some crop species show different levels of resistance that are heritable. For example, the RHA 274 and RHA 801 lines of sunflower crossed with Russian, Romanian, or Turkish lines have strong resistance to *Orobanche* (Dominguez et al. 1996) in all cases when they are part of the breeding lines. Wild species of *Helianthus* have high resistance to *Orobanche* attack relative to sunflower crops (Ruso et al. 1996), and introgressing wild species with crops may introduce more resistance to the crop. In fact, three dominant resistance genes have been identified to control resistance of cowpea to attack by *Striga gesnerioides* (Lane et al. 1997). Moreover, several interesting lines of research with tobacco plants that are either resistant or sensitive to attack by *Orobanche* have resulted in the identification of parasite induction of specific plant defense-related promoters.

The ability of a host to defend itself against an intrusive parasite depends upon the ability of the host to "recognize" the parasitic tissue. If the parasite grows into the host in a coordinated manner (such as that of a pollen tube in a style), then the host cell does not recognize the parasite as an intrusive alien and cannot induce a defense against the parasite. It is thought that some plants that are resistant to *Orobanche* attack can respond to the parasite by induction of the hypersensitive response (Cubero 1991). In addition, sensitive species are also able to induce the PR-1b promoter (serving to induce PR-1 class gene transcription) in localized regions around the parasite haustorium indicating the possibility that parasite attack induces SAR (Joel and Portnoy 1998). Thus, both sensitive and resistant host plants are able to identify a parasitic attack as having occurred. Induction of SAR, including the transciption of PR genes and the hypersensitive response, does not indicate a parasite-specific defensive mechanism, but does indicate the potential for genetic engineering of plant defense against parasitic plants. For example, parasite toxin-producing genes could be fused to the PR-1b promoter and inserted into the genome of a sensitive plant. In such transgenic plants, the intrusion of a parasitic endophyte would induce the PR-1b promoter and cause the transcription of toxin-producing messages for translation.

Another gene that is associated with plant defense against pathogens is hmg2 (from tomato), which is activated in response to various biotic and abiotic (mechanical wounding) stresses. The gene encodes 3-hydroxy-3-methylglutaryl CoA reductase (HMGR). This is considered to be the rate-limiting enzyme in the isoprenoid biosynthesis pathway (Chappel 1995), and is specifically activated during the defensive response associated with the production of sesquiterpene phytoalexins. Expression of hmg2 is induced in cells immediately around pathogen infection sites such as those from the tobacco mosaic virus, the soft rot bacterium, the root knot nematode, and insect herbivory (Cramer et al 1993). In a test of hmg2 induction by parasitic attack of transgenic tobacco, it was clear that hmg2 is induced in cells immediately around the penetration region of *Orobanche* haustoria (Westwood et al. 1998). Induction of the hmg2 promoter increased as the infection became larger, and the hmg2 gene continued to be expressed at all stages of the parasitic attack. Tobacco plants that were transformed with the CaMV 35S promoter fused to the GUS gene (so that tissue stains blue when the gene is activated) showed no induction when attacked by the parasite. Therefore, *Orobanche* attack of host roots does induce the transcription of specific genes in the immediate region of attack.

Host tomato plants (*Lycopersicon esculentum*) that are resistant to attack by *Cuscuta reflexa* utilize a type of programmed cell death (probably associated with the hypersensitive response) to create a barrier to the parasite prehaustoria. Cells of the host elongate under the developing prehaustoria of a *Cuscuta reflexa* plant. These elongating cells then die and form a lignified, suberized plaque. In addition to the physical barrier produced by the plaque, soluble phenolic compounds and peroxidases increased in concentration (Sahm et al. 1995). Such lignified necroses are also associated with resistance after penetration of the host by *Orobanche* (Dörr et al. 1994). In addition, resistant sunflower cells in contact with *O. cumana* developed lignin impregnated layers (Antonova 1994). Similarly, one of the mechanisms that causes resistance against *Striga gesnerioides* in cowpea is the development of lignified necroses around the parasite penetration sites (Moore et al. 1995). Understanding the molecular mechanism behind the development of lignified plaques in response to parasite attack may provide an excellent window into possible general defensive mechanisms against parasitic angiosperms.

V. PARASITIC PLANTS AND OTHER ENVIRONMENTAL STRESSES

A. Interaction with Abiotic Stresses

The most apparent synergistic interaction between abiotic stress and parasite infection is that of water stress. Detrimental effects of *Orobanche* are most often manifest during periods of low water availability. In fava bean crops, infestation by *Orobanche* severely reduces the host's ability to absorb water (Whitney 1972), which drastically reduces growth and reproduction of the host more than water stress alone. Dwarf mistletoe plants transpire rapidly in an unregulated manner and irrespective of host water potential. Thus, the dwarf mistletoes can significantly increase drought severity for the host. Secondarily, dwarf mistletoes can increase conifer susceptibility to pine bark beetle attack because drought-induced resin release can attract pine beetles (Fisher 1983). These and many other cases clearly show that parasitic plants can intensify water stress for the host.

In a general sense, root parasites are more frequent and abundant in soils with low nutrient concentration. In fact, many studies have shown that fertilization of soil reduces the virulence of parasitic plants in agriculture. The preponderance of root parasite successes in low nutrient sites suggests that root parasitism itself is an adaptaton to nutrient stress. The intensity of *Striga hermonthica* infection is positively correlated with an increase in soil stone and gravel content and negatively associated with increasing nutrient content of the soil (Vogt et al. 1991). *Striga hermonthica* infection of rice (*Oryza sativa*) causes a greater reduction in rice growth at lower soil nutrient levels (Cechin and Press 1994). The increase in root parasitism in low nutrient regions suggests that root parasites can exacerbate the detrimental effects of low nutrient soils on host function. However, there are studies that demonstrate the reverse relationship. For example, *Rhinanthus serotinus* has a higher growth rate and greater flower production when growing on host plants in high nutrient environments than on those in low nutrient environments (Salonen and Puustinen 1996). There was no significant relationship between soil nitrogen concentration and the effect of *Striga hermonthica* on maize plants (Aflakpui et al.1998). The differences in the relationships between soil nutrition and parasite virulence may be due to different strains or species of parasites, different responses of the host species to soil nutrition, and differences between patterns observed in natural ecosystems compared with those in fertilized fields (Patterson 1987). Stem parasites are also responsive to variations in nutrition as seen by the greater growth of parasites on legumes than nonlegumes in Australia. Stem parasites are unlikely to enhance nutrient limitation because the growth of stem parasites on nutrient-limited trees is inhibited.

Elevated carbon dioxide may constitute a stress to plants because of the downregulation of photosynthesis at high external carbon dioxide concentrations. Could it be that the extra demand on host carbohydrates when a parasite attacks is large enough to preclude downregulation of photosynthesis at high atmospheric carbon dioxide? The photosynthetic rate of C_4 plants is not responsive to elevated atmospheric CO_2. Therefore, as expected, the influence of a root parasite on the growth of a C_4 plant is not affected by growing plants at an elevated CO_2 concentration (Watling and Press 1998). Similarly, there is no effect of growth in elevated atmospheric CO_2 on the inhibition of the C_4 crop sorghum by species of *Striga* (Watling and Press 1997). In contrast, a C_3 plant, *Trifolium repens,* is differentially affected by *Orobanche minor* when under different atmospheric CO_2 concentrations (Dale and Press 1998). Elevated atmospheric CO_2 partly alleviated the inhibitory effect of the parasite on the host. The photosynthetic rate of *T. repens* was stimulated by elevated CO_2 irrespective of the parasite attachment, and there was no indication of downregulation of photosynthesis in any growth condition. Although the database is small for making generalizations, it does suggest that the effect of parasites on C_3, but not C_4, plants is lessened at elevated CO_2 levels, and that the change is not related to a decrease in downregulation of photosynthesis.

VI. SUMMARY

Parasitic plants are a diverse group of organisms that have evolved in a number of different ways. Within that diverse group there is a continuum of parasite dependency on the host, ranging from complete dependency upon host nutrients and carbohydrate, to an ability to survive and reproduce in the absence of the host. However, in all cases, the parasitic plants have increased growth and reproduction when in association with the host. Effects of parasitic angiosperms on host performance can be slight, or they can be extreme, where 100% of the host population is destroyed. The effects of parasitic angiosperms on plants continue up the scale of ecological processes by influencing competition, diversity, and ecosystem function.

The morphology of parasitic angiosperms is extremely diverse. Some are large trees, others are shrubs, and many are herbaceous species or vines. Many parasites, such as *Orobanche* spp., live underground until the inflorescence is extended above ground; others reside underground, and their inflorescence only barely breaks the ground surface; still others live inside stems, and their inflorescence emerges from the host stem tissue. Some parasitic plant flowers are minute (Balanophoraceae), while others are huge (Rafflesiaceae). The diversity of structure and function of these plants makes them a fascinating group to study.

Interactions between the host and parasite occurs at the haustorium interface. Here, the foreign parasite tissue grows into the host and forms a connection with the vascular system. Following that connection, competition between the host and parasite for resources begins. The parasite creates a low water potential by high transpiration rates in order to keep xylem sap moving from host to parasite. The result is a buildup of specific ions such as potassium in the parasite, while other substances (amino acids and sugars) are selectively accumulated from the host. Intense activity of parenchyma cells in the haustorium, at the host interface, indicates the highly active transport mechanism in this location. Root parasites have become dependent upon nitrogen resources from the host to an extent that they are missing several of their N accumulation enzymes. In general, parasites successfully accumulate some ions by maintaining a flow of xylem sap from the host, and others by selective accumulation.

Parasite seed germination and location of the host can be dependent upon substances that the host releases into the environment. Moreover, when a parasite haustorium breaks into the host cells, a molecular response occurs in the host. In part, this molecular response is a general response to wounding, and in other ways the responses may be specific to parasites. The signaling pathways between parasite and plant before and after attack are a potential resource for plant physiologists to learn about signal transduction in plants.

Parasitic plants are a significant problem for agriculture. Many root parasites have the capacity to reduce crop production to zero. It is difficult to remove the parasite population from a field. In addition, parasite seed is not easy to remove from crop seed. The longevity of parasite seeds, its ease of transport with crop seeds, and the effectiveness of parasitic angiosperms at locating hosts have made these parasites an increasing problem in agriculture. Consequently, interest in how to control parasitic angiosperms in the field is vibrant, as is the interest in understanding the mechanisms by which crops can be engineered for resistance to parasite attack.

Host plant defense against parasite attack is a critical issue to understand, but there is a small base of information on this topic. In general, resistant plants are able to produce plaques at the sites of haustoria attack that will surround the intruding endophyte and block its access to the host vascular system. The molecular regulation of that plaque buildup is not known, but some parasite-specific gene regulation systems, have been isolated as a first step to understanding the molecular regulation of defense against parasites.

STUDY-REVIEW OUTLINE

Introduction

1. Parasitic angiosperms have been recognized as an entity for over 2000 years.
2. Modern understanding of the nature of parasites dates back to the middle of the 19th century.
3. Parasitic angiosperms, particularly mistletoes, have been utilized for medicines and animal feed by many cultures.
4. Woody parasitic angiosperms such as sandalwood have been used for lumber and as a source for fragrances and oils.
5. A mistletoe lectin, ML-1, may have a future as a treatment for certain types of cancer.

Functional Classification of Parasitic Plants

1. *Holoparasites* are those parasitic angiosperms that cannot perform photosynthesis and are entirely heterotrophic.
2. *Hemiparasites* are those species that are able to perform photosynthesis at a reduced rate but also can obtain carbohydrate from their host.
3. All parasitic angiosperms are connected to their host through *haustoria.*
4. There is a continuum of dependence upon the host in parasitic angiosperms, ranging from completely dependent to virtually independent.
5. Some parasites connect by haustoria to the roots of the host, while other parasites connect via haustoria to the stems of the host.
6. Some root parasites are herbs, while others are trees or shrubs.
7. Some stem parasites are connected to their host only by the endophyte of the haustoria, while others are completely embedded in the host tissues.
8. *Epiparasites* are connected to their host through a secondary organism, usually mycorrhizae.
9. *Specialist parasites* attack only one host species, while *generalist parasites* may have a host preference that includes hundreds of species.
10. *Host range* (total diversity of hosts the parasite will attack) is larger than host *preference* (number of hosts the parasite attacks in nature).
11. Parasitic angiosperms can attack several host species simultaneously.
12. Attacking more than one host simultaneously may provide a balanced nutrition for the parasite or may reduce the toxicity of any particular host.

Taxonomic Diversity of Parasitic Plants

1. There are about 4000 parasitic angiosperms located in 22 different families.
2. The parasitic plants can be grouped into 11 clades based upon molecular evidence.
3. Parasitism has evolved multiple times in angiosperms, resulting in the diverse collection of clades.
4. There is only one possible parasitic gymnosperm, and its parasitic nature is in question.
5. The parasites of Scrophulariaceae evolved from one origin, and the parasitic trait followed from facultative parasite to holoparasite based on gene sequence similarities among members of the family.

Distribution of Parasitic Plants

1. Every ecosystem of the world has at least one parasitic angiosperm.
2. Parasites in Orobanchaceae and Scrophulariaceae are characteristic of the temperate and Mediterranean zones.
3. Many members of these groups are found in grassland communities, attacking both monocot and dicot hosts.
4. *Mistletoes* (stem hemiparasites) are most abundant in tropical regions; however, some of the best-known members are from temperate and boreal regions.
5. The parasitic plant families Rafflesiaceae and Balanophoraceae are also primarily tropical in distribution.
6. Most members of the Santalacaceae are herbaceous, but the woody members (sandalwoods) are best known.
7. Many Santalacaceae and Olacaceae are tropical, but some species range up into the Arctic. These species have been extensively studied in the Australian Mediterranean zone.
8. Myzodendraceae is a monotypic family of parasitic plants that is restricted to *Nothofagus* in southern South America.
9. Cuscutaceae is a cosmopolitan family of vine parasites that extends from the tropics to Greenland.
10. Very few species of Cuscutaceae are located on islands.

Evolution of Parasitism in Vascular Plants

1. The *gradualism theory* states that parasitism evolved in root parasites by a gradual emphasis of haustoria over roots.
2. The gradualism theory is based on the continuum of root/haustoria ratios that is negatively associated with parasite dependency on the host.
3. The *root grafting theory* states that haustoria evolved as modifications of root grafts.
4. The root grafting theory has little support because only closely related species form root grafts, and parasites attack unrelated hosts.
5. The root grafting theory may be applicable to large woody parasites such as *Olax.*
6. The *mycorrhizal model* of parasite evolution states that the parasitic trait evolved by a slow reduction in the mycorrhizal importance in epiparasitic relationships.
7. The *pathogen theory* is based on the concept that haustoria evolved from tumors produced by bacterial pathogens such as *Agrobacterium tumefaciens.*
8. The pathogen theory is partly based on the fact that several ancient types of haustoria resemble tumors.
9. The evolution of parasitism in *Cuscuta* may have occurred through the modification of aggressive adventitious roots.

Effects of Parasitic Plants on Natural Communities and Agriculture

Effects on Natural Communities

1. Parasitic plants can reduce ecosystem or population productivity.
2. Dwarf mistletoe reduces productivity of some Pacific Northwest forests by 50%.
3. Parasitic plants can be used as biocontrol agents against the spread of invasive plants.
4. The presence of parasitic plants can cause a change in the species diversity of a community.
5. The diversity of grasslands is decreased in the presence of *Rhinanthus minor,* and the diversity of salt marshes is increased by *Cuscuta salina.*
6. Parasitic plants can also alter the competitive interactions between two other species.
7. Grass and parasite association in a coastal community reduces the growth of a shrubby legume because the grass/parasite association uses more water than a similar density of grass alone.
8. The host plant competes for light with the parasitic plants.
9. Host plants that form associations with parasites can reduce the competitive ability of a competing plant by promoting parasitic attack on the competing species.

Effects on Agricultural Systems

1. Parasitic plants have a large economic effect on agriculture, particularly is Africa.
2. The most important agronomic parasitic plants are *Orobanche, Striga,* and *Cuscuta.*
3. *Cuscuta* has a worldwide distribution because its seeds mimic those of small-seeded legumes.
4. *Cuscuta* is a generalist herbivore, so it is a problem for a number of different crops (potato, sugarcane, carrot, tobacco, onion, and others).
5. *Cuscuta* seed can be removed from the host seed by putting iron shavings into the host seed lot, then passing a magnet through the seed lot.
6. *Striga* infestations of agricultural fields in Africa cause up to 50% loss of crop.
7. *Striga* control and remediation costs billions of dollars of lost revenue worldwide, and remains a problem to U.S. agriculture.
8. *Orobanche* species are parasites of cash crops in the Fabaceae and Solanaceae (fava bean, potato, eggplant, and tomato). *Orobanche* infestation often results in a 35–40% reduction in yield.
9. Tomato production in the United States is reduced by about 25% by *Orobanche* species.

Conservation Issues with Parasitic Plants

1. Some parasitic plants have been driven close to extinction through human intervention of their habitat and host species populations.
2. The unique attributes of the biology of parasitic plants make this group a treasure for plant sciences and worthy of protection.
3. The endangered parasitic plants can be restored by normal conservation procedures except that the restoration of the host is the most important activity.
4. In the United States, there are 56 endangered or threatened parasitic plant species.

Ecology and Physiology of Parasitic Plants

Pollination Biology

1. Flowers of parasitic angiosperms vary in size widely among the different families.
2. Mistletoes in the Loranthaceae are almost always pollinated by birds, and can have relatively large and showy flowers.
3. A variety of different birds can utilize and pollinate the same flowers at different times of the day.
4. One tropical family of birds has 55 species that are adapted to pollinate and utilize mistletoe resources exclusively.
5. In contrast, mistletoes of the Viscaceae are almost all pollinated by bees.
6. *Rafflesia arnoldii* produces the world's largest angiosperm flower, and has virtually no other tissues.
7. Flowers of the Balanophoraceae are minute, and their mechanism of pollination is not well documented.

Seed Dispersal, Germination, and Establishment

1. Most mistletoe seeds are distributed by birds.
2. Distribution of the mistletoe seeds occurs by defecation of eaten undigested seed, regurgitation of seed, and waste from chipping away at seed coats.
3. Seeds of species in the Santalacaceae, such as sandalwood, also are distributed by birds.
4. Some mistletoe species such as *Arcethobium* are shot out of their seed capsules as a dispersal mechanism.
5. Members of Scrophulariaceae and Orobanchaceae have small seeds dispersed from capsules.
6. The number of seeds per capsule can be very high (up to 50,000) for the parasites in Orobanchaceae and Scrophulariaceae.
7. Seeds develop in the flower of *Rafflesia* and remain while the flower rots. Dispersal of the seeds is an enigma but may be associated with ants.
8. Seeds of Balanophoraceae are grouped with a white fleshy material that is collected by ants and transported to their nests.
9. All the following events must occur for a successful colonization of a host by a parasite: (a) germination of the parasite, (b) location of the host root, (c) haustorial initiation, (d) host tissue penetration, (e) compatibility between parasite and host, and (f) growth and reproduction of the parasite.

Chemical Regulation of Seed Germination

1. Seed germination requirements are very specialized for holoparasites, particularly *Orobanche.*
2. *Striga* seeds must locate a suitable host within about 2 mm of radicle growth in order to survive. Thus, finding a host quickly is critical to root hemiparasite survival.
3. *Striga* seed can remain dormant in the soil for many (at least 20) years.
4. Seed germination of *Striga* is stimulated by several compounds including strigol, which was first isolated from the nonhost species cotton.

5. Strigol stimulates seed germination of many parasitic and nonparasitic plants.
6. If strigol has evolved as a host recognition system by parasites, why does strigol stimulate the germination of nonparasites as well?
7. The long half-life of strigol in the soil, even after the host root has died, may confuse the parasite seed into germination without a host root present.
8. Since nonhost species produce strigol, parasite seeds are very likely to germinate near nonhost roots, which would not be adaptive for the parasite population.
9. Strigol may actually be a microbial product of the rhizosphere rather than a product of plant metabolism. If so, then this compound simply identifies the presence of a rhizosphere, not a host root.
10. A more likely candidate for a naturally effective stimulant of parasite seed germination is sorgoleone (a hydroxyquinone), which was originally isolated from sorghum.
11. Hydroxyquinones are more active and less stable than strigol, and are easily isolated from host plants.
12. Resorcinol, also exuded from sorghum roots, protects the hydroxyquinones from oxidation, which would destroy their activity.
13. Chemical recognition of host roots is critical for *Orobanche* seed because they can only grow 2 mm before they run out of carbohydrate for growth.
14. Soil nutrients also affect seed germination of *Orobanche,* such that high ammonium nitrogen availability inhibits germination.
15. The relative significance of strigol and sorgoleone to seed germination cues in parasitic angiosperms is still hotly debated.

Structure and Physiology of the Haustorium

1. The haustorium is the most defining character of parasitic plants.
2. The haustorium is divided generally into three parts: the swollen tissue around the juncture, the hymenium, and the endophyte.
3. Haustoria vary in location of attachment to the host (root or stem), relative size, meristem of origin (primary or secondary), and site (host surface tissue or internal to the host).
4. The development of the haustoria follows from an adhesive stage to an intrusive stage to a conductive stage at maturity.
5. The development of haustoria for all parasites follows the same sequence, but the relative duration of particular stages may be different among species.
6. Chemical signaling between host and parasite is important for proper haustorium development and host preference.
7. Some parasites will form haustoria without a preferred host being present, while others will not.
8. Phenylpropanoids stimulate haustorium formation in parasitic plants.
9. *Striga asiatica* requires DMBQ released from sorghum roots for the initiation of haustoria.
10. DMBQ is most likely a fungal product of lignin digestion by lacase, so parasites must be able to release lacase enzymes to produce DMBQ when a host root is nearby.
11. When the HIF is perceived, the parasite root stops elongating and produces a haustorium.
12. The haustorium is successful when it has formed a direct connection between the host's vascular system and the parasite.
13. The endophyte of a *Striga* haustorium penetrates into the host root until it reaches the vascular tissue of the host.
14. The endophyte cells penetrate the xylem element through a pore and then proliferate. The proliferated endophyte cells lose some cell walls and open up inside the host xylem cell. Endophyte tissue then differentiates into xylem elements and eventually forms a direct connection between host xylem and parasite.
15. The hyaline body, consisting of highly nucleated and vesiculated cells, surrounds the new vascular system of the parasite and may regulate the accumulation of nutrients and carbohydrate from the host.

Sharing Resources (Carbon, Water, and Nitrogen) Between Parasite and Host

1. The transfer of resources from host to parasite is not entirely passive.
2. The parenchyma cells of the hyaline body are densely packed with organelles, indicating high energy output.
3. The hyaline body (in contact with the host xylem) has been shown to have a high metabolic activity.
4. The form of carbohydrate and nitrogen-containing molecules in the host vasculature differs from that in the host. Thus, the resources are changed on their way into the parasite, which indicates active involvement in resource transfer to the host.

Parasite Water Relations

1. Parasites commonly have a very high transpiration rate.
2. Annual parasites have unregulated stomata, which have maximum transpiration all the time.
3. Perennial parasites have higher transpiration than the host, but the stomata will close shortly after the host stomata close. Therefore, there is some regulation of stomata in perennial shoot parasites.
4. Water relations of stem parasites are easy to determine because their water is provided only by the host stem.
5. The water-use patterns of root parasites are complicated because root parasites can have two sources of water: soil and multiple hosts.
6. The diurnal behavior of stem parasite stomata mimics that of the hosts in the American desert.
7. The water potential of the parasite is always lower than that of the host.
8. The flow of water from the host to the parasite is maintained by a water potential gradient.
9. In holoparasitic root parasites, that simplistic model for water flow does not apply.
10. The water potential of a hemiparasite is determined by the water potential of the host cell. The water potential of the host is dependent upon the transpiration of the parasite, the hydraulic conductance of host and parasite, and the capacitance of the tissue.
11. Most of the difference in water potential between host and parasite is due to the difference in hydraulic conductance between the root and the haustoria.
12. The vulnerability of the haustorium to drought-induced xylem embolism may be important to the vegetation in southern California.
13. Parasites maintain a higher osmotic potential than their host due to a high concentration of inorganic ions in the parasite.

Exchange of Nutrients

1. Holoparasites and stem hemiparasites are more dependent upon host nutrient and water than are root hemiparasites and facultative parasites.
2. Some ions can become concentrated in parasites compared with their host.
3. Many nutrients are accumulated by passive means through the haustorium-host xylem connection in the stream of water.
4. One of the reasons why levels of ions such as K are so high in parasites is that phloem mobile ions are concentrated in the parasite and cannot return to the host.
5. The haustorium also has selective permeability capabilities. Some ions (e.g., Na) are selected against and others are selectively accumulated.
6. Holoparasites lack the ability to accumulate inorganic nitrogen forms, and there is a negative relationship between hemiparasite ability to reduce inorganic nitrate and dependency on the host.
7. Hemiparasites that are growing on a medium with only inorganic nitrogen can survive, but they grow better if some organic nitrogen is added.
8. The activity of ammonium ion transport systems are relatively low in parasitic plants.

9. Holoparasites do not have the chloroplastic glutamine synthetase enzyme. In hemiparasites, the activity of this enzyme is very low.
10. Glutamine synthetase in the chloroplast may mostly function to recycle ammonium ion produced by photorespiration. Since photosynthesis (and by default protorespiration) is inhibited in hemiparasitic plants, there is little importance to glutamine synthetase in the chloroplast.
11. Some holoparasites are unable to synthesize isoleucine, but many others can synthesize all their required amino acids.

Exchange of Carbohydrates from Host to Parasite

1. Holoparasites gain all their reduced carbon from the host because they do not perform photosynthesis. Hemiparasites gain some portion (30–75%) from the host and the rest from parasite photosynthesis.
2. The use of stable isotopes to evaluate the accumulation of reduced carbon from a C_4 host by a C_3 parasite revealed that 49%, 70%, and 87% of carbon in the stem, leaf, and root (respectively) of the parasite was derived from the host.
3. Accumulation of large quantities of amino acids from the host is an alternative way of accumulating reduced carbon other than by accumulating photosynthate directly.
4. The accumulation of carbon as amino acids cannot account for a large portion of carbon used to synthesize carbohydrate by the parasite.
5. Accumulation of carbohydrate from the host takes place at the interface between host phloem and endophyte.
6. In some parasites, the endophyte cells penetrate into the host sieve elements and then differentiate into sieve element structures themselves.
7. In other parasitic plants, the endophyte does not penetrate the host but rather encircles the host sieve elements, thins the cell wall, and at times forms plasmodesmata with the host cell.
8. Transfer of photosynthate to the parasite mainly follows an apoplastic route.
9. Sucrose is exported into the space between host phloem and parasite endophyte.
10. Sucrose is broken down to hexoses by invertase excreted by the endophyte. The endophyte cells accumulate the hexoses and transport them to the parasite.
11. The endophyte also releases potassium and abscisic acid into the apoplast near the host sieve elements, which enhances sucrose unloading from the sieve elements.
12. The sucrose gradient is maintained by the size of the parasite sink for carbohydrate.
13. Photosynthate from the parasite is not translocated to the host because many parasites store compounds such as mannitol that are not accumulated by the host cells.

Photosynthetic Capacity of Parasites

1. Holoparasites have no photosynthetic capacity, and little need for the chloroplastic genome.
2. Many genes in the chloroplast genome have been lost in holoparasites so that they are only 50% of the size of a nonparasitic chloroplast genome.
3. Mutations in chloroplast genes occur before and after photosynthesis is completely lost by parasitic plants.
4. Pseudogenes in holoparasitic chloroplast genomes are the result of many additions and deletions.
5. Studies of the similarity between sequences of the rps2 and rbcL genes shows that there was one evolutionary origin to parasitism in the Scrophulariaceae, and five lineages of evolution after this point.
6. The requirement for photosynthesis in autotrophs places powerful evolutionary constraints on chloroplast genome evolution that are removed in parasitic plants.
7. There is an inverse relationship between hemiparasite photosynthesis rate and carbohydrate flow from host to parasite among hemiparasites.

8. Although there is a continuum of photosynthetic capacity in hemiparasites, which extends from that of nonparasitic plants to that of holoparasites, the average hemiparasite photosynthesis rate is depressed below that of nonparasitic plants.
9. Parasites that have high photosynthesis rates still attain significant amounts of carbohydrate from their host.
10. Respiration rates of parasitic plants are relatively high.
11. Carbon accumulation by parasites is different when attached to different hosts.
12. Photosynthesis of parasites attacking dinitrogen-fixing hosts is higher than when attacking nonfixing hosts. Therefore, the growth of hemiparasites is greater on legume hosts than on nonlegume hosts.
13. Hemiparasites such as *Striga* often have a low stomatal frequency, relatively small amounts of spongy mesophyll, and a relatively low activity of rubisco compared with nonparasitic plants.
14. In *Cuscuta,* the chlorophyll and rubisco concentrations are low per stem area or weight because the photosynthetic cells are located in a small region around the vascular tissue.

Uptake of Alkaloids by Parasitic Plants from Their Hosts

1. Parasitic plants attacking host species with alkaloids accumulate the alkaloids in their tissues.
2. Alkaloids could be selectively accumulated or simply accumulated with the rest of the material in the phloem.
3. In the case of *Cuscuta,* alkaloid accumulation is simply another mechanism for accumulating nitrogen.
4. The parasitic plant that atacks host species with alkaloids must adapt to the alkaloids in their own tissues.
5. Accumulating alkaloids from the host may afford parasitic plants some protection from herbivore damage.

Effects of Parasitic Plants on Their Hosts

Effects on Plant Architecture

1. Attack by a parasite may have no effect on host structure (the case of many root parasites), or it can leave a major effect on host structure (the case of several stem parasites).
2. In general, parasites decrease the productivity and reproduction of the host, which lowers the host's fitness.
3. Host reduction in mass is much greater than the parasite increase in mass.
4. Root parasites commonly induce an increase in host root/shoot ratio.
5. The change in root/shoot ratio causes less allocation to leaves and a lower productivity of the host.
6. An increased root/shoot ratio is normally the result of increased root proliferation.
7. Root proliferation increases because host root wounding initiates new root growth, parasites release hormones that induce root growth, and the increased soil nutrients around parasitic plants induced root proliferation.
8. Not all parasite-host relationships result in an increased host root/shoot ratio.
9. All parasites cause some sort of host tissue swelling near the haustorium infection.
10. In many stem parasites, the host stem distal to the parasite withers or dies, and the stem proximal to the parasite swells.
11. The amount of vascular tissue required to support the parasite is greater than that needed to support an equivalent area of host leaves.
12. In some cases, tyloses develop in the host tissue near the haustorium attachment.
13. If the parasite infection rate is large enough, the host leaf area will become too small to support the tree and may all abscise.
14. The extent of leaf area reduction in the cowpea is directly related to the amount of parasite atttack.

Competition for Resources (Water, Carbon, and Nutrients)

1. The host stem distal to a stem parasite is competing with the parasite for resources.
2. Water is preferentially allocated to the parasite over the host.
3. *Striga*-infected crops often wilt even when there is plenty of water.
4. Parasite infection can decrease host plant stomatal conductance.
5. As the parasite biomass increases in a crop, the likelihood of a drought increases.
6. Host plants infested with annual parasites are much more likely to experience water stress than those exposed to perennial stem parasites.
7. The long-term perennial parasite *Santalacum album* closely coordinates its water use to those periods with the most water requirements of the host.
8. Even though parasites often contain high quantities of nutrients, this does not mean that the host is suffering nutrient deficiency.
9. There are only a few cases in which nutrient deficiency is a symptom of parasites.

Effects on Metabolism

Photosynthesis of the Host

1. Removal of resources (carbohydrates) from the host should induce a compensatory response in the host (increased photosynthesis).
2. In some species, the host increases photosynthesis following parasite attack.
3. In some cases, the parasite induces water stress, and this causes a reduction in host photosynthesis.
4. Parasite attack can preferentially increase stomatal limitation to photosynthesis of the host, as is the case for oaks attacked by mistletoe.
5. Since parasite attack is often associated with an increase in ABA, this may be an important factor causing reduced stomatal conductance in the host.
6. In some cases, there is a balance between water relations of host and parasite such that there is no change in host photosynthesis.
7. In general, parasites cause no change in host photosynthesis or an increase in host photosynthesis. The only cases where host plant photosynthesis decreases are those where the parasite induces water stress in the host.

Hormone Relationships of the Host

1. The witches'-broom is caused by a change in auxin and cytokinin concentration in the host, near the parasite infection.
2. The sap concentration of auxin increases in plants parasitized by the root parasite *Striga*.
3. Cytokinins in parasites are relatively low and cyclic before attachment to the host. After attachment, cytokinin concentration in parasites increases greatly and is not cyclic.
4. Parasite attack stimulates a large increase in ABA concentration near the parasite haustorium and in the host plant tissues.
5. The increase in ABA induces the ABA signaling pathway, which causes reduced stomatal conductance, leaf abscission, and increased root growth.

Plant Defenses Against Plant Parasite Attack

Molecular Responses to Parasitic Plant Attack

1. Host defense against parasite attack could be accomplished in three ways:
   a. Decrease the production of compounds that initiate germination or haustorium formation in the parasite.
   b. Enhance current defensive mechanisms against fungi, virus, or bacteria.
   c. Produce toxins that are specific for parasitic plant attack.
2. Phenolics and hydroxyquinones are by-products of cell wall synthesis and are used as signaling molecules for defense against bacteria and fungi.

3. Reducing the abundance of those compounds that stimulate parasite germination and haustorium formation would disturb other plant interactions with beneficial microbes.
4. Molecular response of host plants to parasite attack does not occur until the parasite endophyte breaks the surface of the host stem or root.
5. The SAR wound response may be induced by attack from parasitic angiosperms.
6. Transcription of several PR genes is induced following parasite attack, particularly the PR-1 genes.
7. The induction of SAR is not an effective defense against parasitic plants because it is directed toward nonangiosperm cells and is not specifically targeted to the region where the endophyte is penetrating the host.

Plant Defensive Mechanisms Against Parasitic Plant Attack

1. Resistance to parasitic angiosperm attack is present in certain genetic lines of crops and in some nondomesticated species.
2. The resistance against parasite attack is heritable, often by only a few dominant genes.
3. Parasite attack of tobacco results in the induction of a parasite-specific promoter in the host.
4. Defense against parasite attack is dependent upon the host recognizing the parasitic endophyte as foreign tissue.
5. Sensitive and resistant species are able to induce the PR-1b promoter in localized regions around the endophyte.
6. Induction of SAR or PR-1b does not correlate with resistance against parasites, but it does suggest potential means for bioengineering defense traits in crops.
7. Another gene that is induced by parasite attack in a localized manner is hmg2. This gene codes for an enzyme that is critical for the formation of defense-related lectins in the isoprenoid pathway.
8. The hmg2 promoter is induced at all times of the parasite attack, and the induction intensity increases as the attack becomes more intense.
9. Many resistant host species utilize a programmed-cell-death mechanism to defend against parasitic angiosperms.
10. Cells surrounding the penetrating endophyte die and lignify, creating a barrier that keeps the endophyte from reaching the vascular tissue.

Parasitic Plants and Other Environmental Stresses

Interaction with Abiotic Stresses

1. Infection by parasites can increase the intensity of water stress for host plants.
2. The detrimental effects of *Orobanche* attack are strongly enhanced during periods of low water availability.
3. Dwarf mistletoe plants increase the water stress of their host, and this can result in increased probability of bark beetle attack.
4. Root parasites are more abundant in soils with low nutrient concentration.
5. The demand by parasites for nutrients in low nutrient sites does not result in increased nutrient deficiency for the host.
6. Stem parasites grow best when host trees have ample nutrient availability. These parasites will not increase nutrient limitation for host plants.
7. There is no effect of root parasites on the way C_4 plants respond to elevated levels of atmospheric carbon dioxide.
8. Elevated atmospheric carbon dioxide levels partially alleviate the stress of parasites on C_3 plants.
9. The C_3 plants respond to elevated CO_2 in the same way whether infected with a parasite or not.

SELF-STUDY QUESTIONS

1. What is the history behind the knowledge of parasitic plants?
2. How have parasitic plants been utilized by human cultures around the world?
3. By what criteria are parasitic angiosperms classified? Give two methods.
4. What is the diversity of parasitic angiosperms?
5. How are parasitic angiosperms distributed among ecosystems of the world?
6. Explain the difference between host range and host preference as they apply to parasitic angiosperms.
7. What is the center of distribution for *Orobanche* species?
8. Where are the members of Rafflesiaceae and Santalacaceae most abundant? Are mistletoes mostly temperate or tropical families?
9. What is unique about the Cuscutaceae and Monotropaceae compared with other parasitic angiosperms?
10. Differentiate between the gradualism, root-grafting, mycorrhyzal, and pathogen models of parasitic plant evolution?
11. What are some of the effects that parasitic plants can have on natural communities? Provide some specific examples for these effects.
12. Are parasitic plants important for agriculture? Why, and which parasitic plants are the most important threats?
13. What factors are most important in the conservation of parasitic plant populations?
14. Contrast the floral structure and pollination mechanisms of the mistletoes and species in the Balanophoraceae?
15. How are the seeds of mistletoes distributed? Contrast this with that of the species in the Rafflesiaceae and Balanophoraceae.
16. What events need to occur for the root parasite to make a successful connection with its preferred host?
17. What are the requirements for germination of parasitic angiosperm seeds in general?
18. What are some of the stimulants for *Striga* germination?
19. Compare and contrast the evidence for the hypothesis that strigol and sorgoleone act to stimulate germination.
20. What chemical factors are associated with the initiation of haustorium development on parasitic plant roots?
21. Describe the three stages of haustorium development on parasite roots.
22. Describe the development of the haustorium connection in *Cuscuta*.
23. How does the water potential of stem parasites relate to the host water potential? Why is this so? (Give ultimate and proximate reasons.)
24. Which has a greater hydraulic conductivity—the host or the parasite? Why does this relationship occur?
25. What elements are at relatively high concentration in parasitic plants, and why do they become concentrated?
26. Why do elements such as calcium not become concentrated in parasitic plants?
27. How do the activities of glutamine synthetase compare in parasites and their hosts?
28. What is the major route of carbon accumulation by holoparasites? Why is amino acid absorption from the host xylem an augmentation to carbon acquisition by holoparasites?
29. How do parasitic plants enhance the transport of carbohydrates from host to parasite?
30. What is the relative size of the chloroplast genome in holoparasites compared with autotrophs? What accounts for this difference in size?
31. Why are some of the genes in the chloroplast genome lost by mutation at a higher rate than other genes?
32. How does the water-use efficiency of stem hemiparasites compare with that of the host? Why is this so?
33. What are the changes in root/shoot ratio of plants attacked by stem and root parasites? Why does this change occur?

34. What is the cause of witches'-brooms on plants attacked by parasites?
35. How do parasites influence branch diameter distal and proximal to the infection? What is the response of leaf area to root parasite infection?
36. Does resource competition occur between host and parasite? If so, what is the consequence to the host?
37. How do parasite infections affect the photosynthesis and respiration of host plants?
38. Do parasite infections affect the hormonal balance of plants? If so, which hormones are most affected and why?
39. What are the concentrations of auxin, cytokinin, and ABA relative to the host? Does this have any affect on host morphology and physiology?
40. Differentiate between the localized wound response (hypersensitive) and the systemic response reaction (SAR) in host plants upon attack by parasites.
41. What defenses can plants utilize to prevent or limit parasite attack?
42. What is hmg2, and why is it significant to plant defense against parasite attack?
43. How does the tomato defend itself against attack from *Cuscuta reflexa*?
44. What is the relationship between water stress and the virulence of parasitic angiosperms?
45. What do you expect to happen to host-parasite interactions in an environment that has relatively high carbon dioxide concentrations?

SUPPLEMENTARY READINGS

Hawksworth, F. G., and R. F. Scharpf. 1984. Biology of Dwarf Mistletoes: Proceedings of the Symposium. Fort Collins, CO. USDA Forest Service General Technical Report RM-111.

Kuijt, J. 1969. The Biology of Parasitic Flowering Plants. University of California Press, Berkeley, CA.

Moreno, M. T., J. I. Cubero, D. Berner, D. Joel, L. J. Musselman, and C. Parker (eds.). 1996. Advances in Parasitic Plant Research. 6th International Parasitic Weed Symposium, Cordoba, Spain.

Musselman, L. J. 1993. *Striga* (Parasitic Weeds in Agriculture, Vol. 1). CRC Press, Boca Raton, FL.

Parker, C., and C. R. Riches. 1993. Parasitic Weeds of the World: Biology and Control. CAB International, Oxon, UK.

Parker, C., L. J. Musselman, R. M. Polhill, and A. K. Wilson. 1984. Proceedings of the 3rd International Symposium on Parasitic Weeds, Aleppo, Syria. ICARDA, Aleppo.

Pieterse, A. H., J. A. C. Verkleij, and S. J. ter Borg. 1994. Biology and Management of *Orobanche*. Proceedings of the 3rd International Workshop on *Orobanche* and Related *Striga* Research. Royal Tropical Institute, Amsterdam, Netherlands.

Ransom, J. K., L. J. Musselman, A. D. Worsham, and C. Parker. 1991. Proceedings of the 5th International Symposium of Parasitic Weeds, Nairobi. The International Maize and Wheat Improvement Center (CIMMYT), Nairobi, India.

Sauerborn, J. 1991. Parasitic Flowering Plants: Ecology and Management. Verlag Josef Margraf, Weikersheim, Germany.

Weber, H. C., and W. Forstreuter. 1987. Parasitic Flowering Plants: Proceedings of the 4th International Symposium on Parasitic Flowering Plants. Phillips Universität, Marburg, Germany.

Wegmann, K., and L. J. Musselman. 1991. Progress in *Orobanche* research. Proceedings of the International Workshop on *Orobanche* Research, Obermarchtal, Germany.

Zheng-yi, W., and P. H. Raven. 1995. Gentianaceae through Boraginaceae (Flora of China, Vol. 16). Science Press, Missouri Botanical Garden, St. Louis, MO.

PART III
Anthropogenic Induced Stress

11 Phytotoxicity and Soil Pollution: Heavy Metals and Xenobiotics

Outline

Objectives

1. Discuss the natural and anthropogenic sources of heavy metals as well as types and amounts of metals produced from these sources.

2. Describe the mechanisms and modes of action of heavy-metal toxicity in plants.
3. Establish what is known about avoidance and tolerance mechanisms in plants that allows plants to survive in soils contaminated with high concentrations of metals.
4. Summarize the metals most commonly involved with metal toxicity in plants, indicating mechanisms and modes of action as well as tolerance mechanisms.
5. Discuss organic pollutants relative to sources, types, amounts produced, and means of distribution, with special attention to pesticides.
6. Compare what is known about the effects of pesticides on agricultural plants versus natural plants and communities.
7. Address the many complicating factors (physicochemical, environmental, and soil) that can influence how xenobiotics can affect plants.
8. Using pesticide uptake and metabolism by plants as a model for other xenobiotics, describe the different means by which plants absorb and metabolize pesticides.

I. INTRODUCTION

Elemental toxicity, as it pertains to plants, is complex and depends on plant species, specific element, concentration, form, soil pH, and the composition of the soil. Many elements are considered to be essential for plant growth (see Chapter 2). However, an excess of an essential or nonessential element can have detrimental effects on plant growth and development. Thus, the plant has a dilemma with respect to balancing the need and concentrations of essential elements and yet prevent the accumulation and/or uptake of essential and nonessential elements. This chapter will focus on the sources of toxic elements, sometimes referred to as *metals* or *trace metals,* in the environment and then address the mechanisms by which plants deal physiologically and biochemically with elevated levels of elements that are encountered in the environment.

There are many sources, both natural and anthropogenic (human-made), from which high concentrations of metals can originate. Generally, these can be categorized as natural, agricultural, industrial, municipal, and transportation sources.

II. NATURAL SOURCES OF METALS

The following summarizes some of the major sources of elemental pollution as addressed in reviews by Freedman and Hutchinson (1981), Friedland (1990), Ross (1994), Seaward and Richardson (1990), and Verkleij (1993). One of the most important natural sources of high concentrations of metals is from geologic parent materials or rock outcroppings. The composition and concentrations of metals present will depend on the type of rock and the environmental conditions that act upon the rock during the weathering process. In parent material, igneous rock generally has higher concentrations of the following trace metals than does sedimentary rock: Cr, Mn, Co, Ni, Cu, Zn, Cd, Sn, Hg, and Pb. However, within classes of igneous rock, ultramafic (serpentine) has greater concentrations of these metals compared to mafic rock (basalt), followed by granite. Granite can have higher levels of Pb and Sn than the former igneous rock classes. Within the classes of sedimentary rocks, shale has the highest concentrations of the above elements, followed by limestone and sandstone, which are very similar except that limestone has much higher concentrations of Mn than does sandstone. The relative concentrations (>100 $\mu g\ g^{-1}$) of metals in the three igneous rock classes in decreasing order are as follows. ultramafic: Cr > Ni > Mn > Co; mafic: Mn > Cr > Ni > Zn ≥ Cu; granite: Mn. In the case of sedimentary rock, the relative concentrations (>50 $\mu g\ g^{-1}$) in decreasing order are shale: Mn > Zn > Cr > Ni > Cu; limestone: Mn; sandstone: Mn. The concentrations of metals that comprise various soils depends on the type of parent material from which the soil is derived as well as on the weathering conditions. Igneous and metamorphic rock comprise 95% of the Earth's crust, while sedimentary rock comprises the remaining 5%. The last category contributes the most to soil formation but is only a small source of trace metals since sedimentary rock is not easily weathered. However, igneous rocks such as olivine, augite, and hornblende contribute considerable amounts of Mn, Co, Ni, Cu, and Zn to soils.

Another natural source of metal contamination is volcanic outgasings. Volcanoes have been reported to emit high levels of Al, Zn, Mn, Pb, Ni, Cu, and Hg along with toxic and harmful gases such as SO_2 and H_2S. The effects of such events are usually limited to the local flora and fauna, but certainly windblown contaminants are likely to be spread greater distances. Other natural sources of airborne emissions originate from forest and prairie fires caused by lightning. High levels of carbonaceous particulate matter are produced on which elements such as Ca, Mg, and K, can be carried and from which volatile elements such as Hg, Se, and Cd are emitted. Windblown dust from desert regions, such as the Sahara, carry high levels of Fe and lesser amounts of other metals such as Mn, Cr, Zn, Ni, and Pb. Natural vegetation emits elements into the soil and atmosphere through leachates from leaves and stems, decomposition, and volatilization. Such activity is an important component of the recycling process that occurs in the environment but can become dangerously out of balance if high levels of metal contaminants accumulate as a result of human activity. Elements such as Cu, Mn, Cd, Ni, Pb, and Zn have been detected in inland precipitation near coastal areas and have been attributed to seasprays and aerosols produced in oceanic activities. Table 11.1 outlines some of the contributions of various natural sources of heavy-metal deposition on a global basis.

III. AGRICULTURAL SOURCES OF METALS

Agricultural sources of heavy-metal pollution include inorganic and organic fertilizers, liming applications to raise soil pH, sewage sludge, irrigation waters, and pesticides, particularly fungicides. Inorganic fertilizers, particularly phosphatic fertilizers, have varying levels of Cd, Cr, Ni, Pb, and Zn depending on their source. Cadmium is of particular concern in plants since it can be accumulated to high levels in the leaves of many plants (high soil/plant concentration ratio) that may be consumed as animal or human food. It has been estimated that about 660 metric tons of Cd per year are being applied to soils on a global basis through use of phosphatic fertilizers. Although the levels of trace metals applied to soils are usually small, the fact that repeated applications of fertilizers to agricultural fields and the long residence times for metals such as Cd in the soil (hundreds to >1000 years) should be some cause for concern.

Soil amendments such as domestic animal manure, sewage sludge, phosphate and nitrate fertilizers, composted refuse, and lime are also sources of potential metal contamination of soils. Although the levels of elements in animal manure and sewage sludge vary greatly, both can make significant contributions to soil metal concentra-

TABLE 11.1 Global Emission of Heavy Metals from Natural Sources

Source	Global Production (kg × 10^9 yr^{-1})	Annual Emissions (kg × 10^6)								
		Cd	Co	Cu	Cr	Hg	Mn	Ni	Pb	Zn
Windblown dust	6–1100	0.25	4	12	50	0.03	425	20	10	25
Vulcanism	6.5–150	0.50	1.4	4	3.9	0.03	82	3.8	6.4	10
Forest fires	2–200	0.01	—	0.3	—	0.1	—	0.6	0.5	0.5
Vegetation	75–1000	0.2	—	2.5	—	—	5	1.6	1.6	10
Sea salt	300–2000	0.002	—	0.1	—	0.003	4	0.04	0.1	0.02
Total		0.96	5.4	18.9	8.9	0.16	516	26	18.6	45.5

Source: From Seaward and Richardson 1990. Reprinted with permission from A. Jonathan Shaw (ed.), *Heavy Metal Tolerance in Plants: Evolutionary Aspects*, pp. 75–92, CRC Press, Inc., Boca Raton, FL. Copyright CRC Press, Boca Raton, Florida.

tions, and this is particularly true of sewage sludge. Animal manure is particularly enriched with Mn, Zn, Cu, and Co. However, sewage sludge has very high concentrations of Zn, Cr, Pb, Ni, Cd, and Cu. The application of sewage sludge is regulated in some countries based on concentrations of elements in the sludge as well as the soil characteristic to which it is applied. Increased applications of sewage sludge to agricultural and forested areas will likely increase in the future as restrictions on disposal of these wastes in marine environments increase. However, like the marine environment, considerable care should be taken in how this alternative approach impacts the terrestrial environment. Other sources of metal accumulation in soils as a result of application of soil amendments include composted refuse, nitrate fertilizers, and lime (Table 11.2). Interestingly, lime has a greater probability of increasing metal contamination than do nitrate fertilizers and composted refuse, and, like sewage sludge, is one of the greatest contributors to the heavy-metal pollution of soils (Ross 1994).

Several metal derivatives have been and are, in some instances, still used to control diseases of fruit crops such as apple, cherry, peach and grape. In the orchards in which these compounds have been used, soils may be contaminated with high concentrations of Cu, As, Pb, Zn, Fe, Mn, and Hg. Some chemicals, such as lead arsenate, were used in Canadian orchards for over 65 years (1910–1975) at annual loading rates of 4–8.7, 1–2.7, and 5.5–7.5 kg ha^{-1} of Pb, As, and Zn, respectively. Certainly, such high levels of metals could have dire consequences not only for soil pollution but also for food contamination as well.

Finally, soil contamination by continued irrigation of agricultural lands can lead to the accumulation of Pb and Cd. However, the source of the irrigation water must be considered and may be highly variable in concentrations of toxic metals, depending on whether the water comes from deep wells or from rivers, lakes, or irrigation canals.

IV. INDUSTRIAL SOURCES OF METALS

Industrial sources of pollution represent a myriad of possibilities, including all aspects related to the mining, refinement (spoil heaps and tailings through mining operations, transport of ores, smelting, milling, and metal finishing), and recycling of metals. Mining operations can result in different types of pollutants depending on the type of mining. For example, coal mining can cause an increase in Fe- and S-reducing bacteria, which result in reduced forms of S and iron pyrite. This process generates acid soils, thus making elements such as Al, Fe, and Mn more soluble to the extent they are toxic to plants. Industrial metal contamination occurs in various ways. Smelting of ores probably represents the most important localized forms of metal pollution. High temperature processing of metals (smelting, casting, etc.) can result in metals being emitted as particulates, while more volatile metals can be vaporized (As, Cd, Cu, Pb, Sn, and Zn) and combined with water in the atmosphere to form aerosols. These can then be dispersed by the wind (*dry deposition*) or precipitated in rainfall (*wet deposition*) causing pollution of the soil or bodies of water. Contamination of soil and water can occur as a result of runoff from erosion of mine wastes, dusts produced in mining and the transport of crude ore, corrosion of metals to be recycled, and leaching into soil and groundwater from numerous sources—this last route depending, to a large extent, on the soil type and pH. Other industrial sources of pollution involve various aspects of processing of plastics, textiles, microelectronics, wood preserving, paper processing, and refineries. Each of these industries contributes

TABLE 11.2 Trace Elements in Agricultural Amendments

	Amendments ($\mu g\ g^{-1}$)						
Element	Sewage Sludge	Composted Refuse	Farmyard Manure	Phosphate Fertilizers	Nitrate Fertilizers	Lime	Pesticides
Cr	8–40,600	1.8–410	1.1–55	66–245	3.2–19	10–15	—
Mn	60–3900	—	30–969	40–2000	—	40–1200	—
Co	1–260	—	0.3–24	1–12	5.4–12	0.4–3	—
Ni	6–5300	0.9–279	2.1–30	7–38	7–34	10–20	—
Cu	50–8000	13–3580	2–172	1–300	—	2–125	—
Zn	91–49000	82–5894	15–566	50–1450	1–42	10–450	—
Cd	<1–3410	0.01–100	0.1–0.8	0.1–190	0.05–8.5	0.04–0.1	—
Hg	0.1–55	0.09–21	0.01–0.36	0.01–2.0	0.3–2.9	0.05	0.6–6
Pb	2–7000	1.3–2240	0.4–27	4–1000	2.120	20–1250	11–26

Source: From Ross 1994, and references therein.

metals to the environment that are particular to the respective enterprise (Table 11.3). Industry has contributed significantly to the increased accumulation of metals over the last 100 years. Man-made and natural pollution by metals are frequently expressed as a ratio of anthropogenic to natural sources. Table 11.4 expresses these ratios for several metals and sources in the environment. The very high lead ratio has been attributed to smelting and lead in automobile gasoline, the latter declining, although in some countries lead is still added to petrol.

V. MUNICIPAL SOURCES OF METALS

Municipal sources of metal contamination include refuse incineration, landfills, electric power generation from the burning of fossil fuels (coal and oil), and sources from various forms of transportation (automobiles, diesel-powered vehicles, and aircraft). The two main sources of the metal contamination of soils are (1) fly ash created by the burning of coal, and (2) the corrosion of commercial waste products such as occurs in metal scrap yards, as well as from other sources such as galvanized metals (primarily Zn contamination) used as railings along highways, metal pylons, and electrical conductors (Al-Hiyaly et al. 1988) used to support electric power lines and transmit electricity, respectively. It has been estimated that coal burning contributes about 34%, 33%, 63%, and 51% of the annual inputs of Cd, Hg, Mn, and Ni, respectively, into soils. Other metals produced from the burning of coal include very high levels of Al,

TABLE 11.3 Sources of Heavy Metal and Metalloid Contamination in the Environment

Source	Important Elements
Industry:	
Ore outcrops	Zn, Pb, Cu, Cd, Ni
Metal smelters	Zn, Cu, Cd, Pb, Sn, Ni
Blast furnaces	Fe, Zn, Mn
Electrolysis	Hg
Cement industry	Tl
Traffic:	
Leaded gasoline	Pb
Catalysts	Al
Metal emission from tires	Ni, Zn
Household:	
Waste	Cd, Pb, Cu, Zn, Ni
Sewage sludge	Cd, Pb, Cu, Zn, Ni
Energy supply:	
Coal-burning power stations	Se, B, M, Cd
Petroleum combustion	V, Ni
Nuclear power plants	U, Cs, Pu
High tension lines	Cu, Zn
Agriculture:	
Food additives	Cu, Zn
Phosphate fertilizers	Cd
Pesticides	Cu, Hg, As, Sn

Source: From Verkleij 1993, and references therein.

TABLE 11.4 Ratio of Anthropogenic to Natural Sources of Particulate Atmospheric Emissions

Element	Total Emissions (10^8 g yr^{-1})	Anthropogenic/Natural Ratio
Al	72,000	0.15
Hg	110	0.44
Ni	989	3.46
Cu	2,630	13.63
Cd	55	18.97
Zn	8,400	23.46
Pb	20,300	345.83

Source: From Friedland 1990. Lantzy and MacKenzie 1979. Reprinted with permission from A. Jonathan Shaw (ed.), *Heavy Metal Tolerance in Plants: Evolutionary Aspects,* pp. 7–19, CRC Press, Inc., Boca Raton, Fl. Copyright CRC Press, Boca Raton, Florida.

Fe, and Ti (Table 11.5). Oil burning for heat and power generation contributes V, Fe, Pb, and Ni to the environment as well. Corrosion of commercial waste products contributes 47%, 56%, 42%, and 23% of the Cr, Cu, Pb, and Zn, respectively. Studies conducted on soils around galvanized steel pylons used to support electric power lines (Al-Hiyaly et al. 1988) have shown that levels of Zn ranging from 1250–6500 μg g^{-1} of soil can occur, compared to 170–320 μg g^{-1} in uncontaminated soil. It was noted that, in the vicinity of the pylons, little to no vegetation grew and what vegetation did grow had developed a genetic tolerance for high levels of Zn. It was further noted that, in areas where the towers were constructed on slopes, the Zn tended to

TABLE 11.5 Metal Concentrations of Fly Ash Collected from 23 U.S. Locations

Element	Concentration Range (ppm dry wt.)
Al	59,100–143,400
As	2–312
B	10–600
Cd	0.1–3.8
Co	5–73
Cr	43–259
Cu	45–616
Fe	27,100–290,000
Mn	58–543
Mo	7–41
Ni	2–115
Pb	3–241
Sb	1–13
Se	1–17
Sn	27–334
Sr	59–3,855
Ti	2760–8310
V	68–422
Zn	14–406

Source: From Friedman and Hutchinson 1981, and references therein.

migrate downslope, indicating mobility in the soil. This could be very significant, particularly in acid soils, where metals such as Zn are much more soluble in soil solution, which would enable the contamination to be extended greater distances from the source and possibly even enter the groundwater.

Another study involving Cu contamination (Kraal and Ernst 1976) from the electrical conductors associated with power lines showed that plants growing beneath the lines accumulated levels of Cu to such a high extent that they were considered to be toxic to any sheep that might graze on the plants in the vicinity of the lines. The plants involved in this study appeared not to be affected themselves by the high levels of copper that was accumulated.

Several elements are produced by the incineration of municipal wastes (Table 11.6), but some of the most significant in terms of concentration include Zn, Pb, Al, Sn, Fe, and Cu. Leachates from landfills can include metals such as As, Cd, Fe, and Pb. The burning of leaded gasoline has in the past been an important source of Pb contamination in the environment. Although Pb levels are declining as a result of discontinued use of Pb as a gasoline additive in the United States, Pb is still added to gasoline in other countries. Antiwear protectants (lubricants) containing Cd, Cr, Hg, Ni, Pb, and Zn are also sources of metal pollution, particularly in inefficient engines or due to the improper disposal of engine oils. Additional sources of contamination from transportation vehicles include Ni and Zn from tires, Al from catalysts, Cd, Cu (primarily from diesel engines), Ni, and Zn from aerosol emissions.

VI. PLANTS ADAPTED TO HIGH METAL ENVIRONMENTS

A. Metal Sensitivity, Resistance and Accumulation

In the context of heavy-metal stress, we will define *sensitivity* to high concentrations of metals as that which results in injury or death of the plant. *Resistance* implies that,

TABLE 11.6 Metals in the Combustible Fraction of Five U.S. Municipal Solid Waste Incinerators

Element	Range (ppm)
Ag	<3–7
Al	5400–12,000
Ba	47–450
Cd	2–22
Co	<3–5
Cr	20–100
Cu	80–900
Fe	1000–3500
Hg	0.66–1.9
Mn	50–240
Ni	9–90
Pb	100–1,500
Sb	20–80
Sn	<20–40
Zn	200–2500

Source: From Law and Gordon 1979.

although the plant is affected, it reacts in a way that allows it to survive high concentrations of metals, and to reproduce the next generation of plants. Two types of resistance will be discussed: *avoidance,* which describes mechanisms for external protection of the plant from metal stress, and *tolerance,* in which the plant is able to survive internal stress imposed by high metal concentrations. As indicated in Chapter 3, plants require certain essential elements for normal growth and development. Many of these elements fall into the category of heavy metals and are only required in trace amounts. However, plants can become *hyperaccumulators,* accumulating higher concentrations of these elements than are actually required for normal growth and development. In fact, many elements that are not normally required for growth are accumulated by plants. Accumulation of high levels of elements in plants is species and tissue specific, and is thought to be controlled genetically by specific genes or small groups of genes. Elements can be accumulated in plant tissues at levels severalfold higher than those found in soils. Hence, tolerance and sensitivity to heavy metals are genetically regulated. For example, Zn tolerance is under the control of a single gene and controls the rapid compartmentation of Zn from the cytoplasm to the vacuole. Copper tolerance is also under the control of a single gene but can be enhanced by a second gene. In this case, tolerance is regulated by controlling the integrity of the plasmalemma (Ernst 1996).

The ability of plants to accumulate elements is taken advantage of by those interested in using plants for bioremediation purposes, that is, for cleaning up polluted sites, or as indicator plants to locate deposits of metals used for commercial purposes. Much has been written about the genetics, distribution, and evolution of plants that grow on soils contaminated with high concentrations of metals. However, the primary intent of this section is to address the physiology and biochemistry of resistance and sensitivity of plants to metal toxicity.

B. Dynamics of Metal Recycling

In the previous section, we considered some of the numerous sources of metal pollution in the environment. Prior to dealing with the physiological and biochemical aspects of how plants deal with toxic metals, it is perhaps pertinent to describe how metals are cycled between the physical environment and the biosphere. Figure 11.1 depicts the various inputs, outputs, and fluxes that can occur in the environment. *Inputs* include precipitation, cloud water interception, and dry deposition. *Outputs* include resuspension to the atmosphere, runoff, erosion, and infiltration to stream channels. *Fluxes* are illustrated by the throughfall of vegetation resulting from the leaching of elements from plant tissue; by stemflow resulting from elements washing off stem surfaces; and by litterfall and root tissue decomposition, which results in the liberation of bound elements and soil water infiltration (Friedland 1990). All of these factors interact in complex ways to form a kind of nutrient recycling in which plants play an important role in the extraction and replenishing of elements.

C. Metallophytes (Accumulators, Indicators, and Excluders)

Plants that can grow on soils that are contaminated with high levels of metals are referred to as *metallophytes.* Some are considered *absolute metallophytes,* meaning that they are totally confined to metalliferous soils. Other species may occur on contaminated as well as uncontaminated soils. The survival of plants that can grow on contaminated soils is considered to be the result of tolerance rather than avoidance

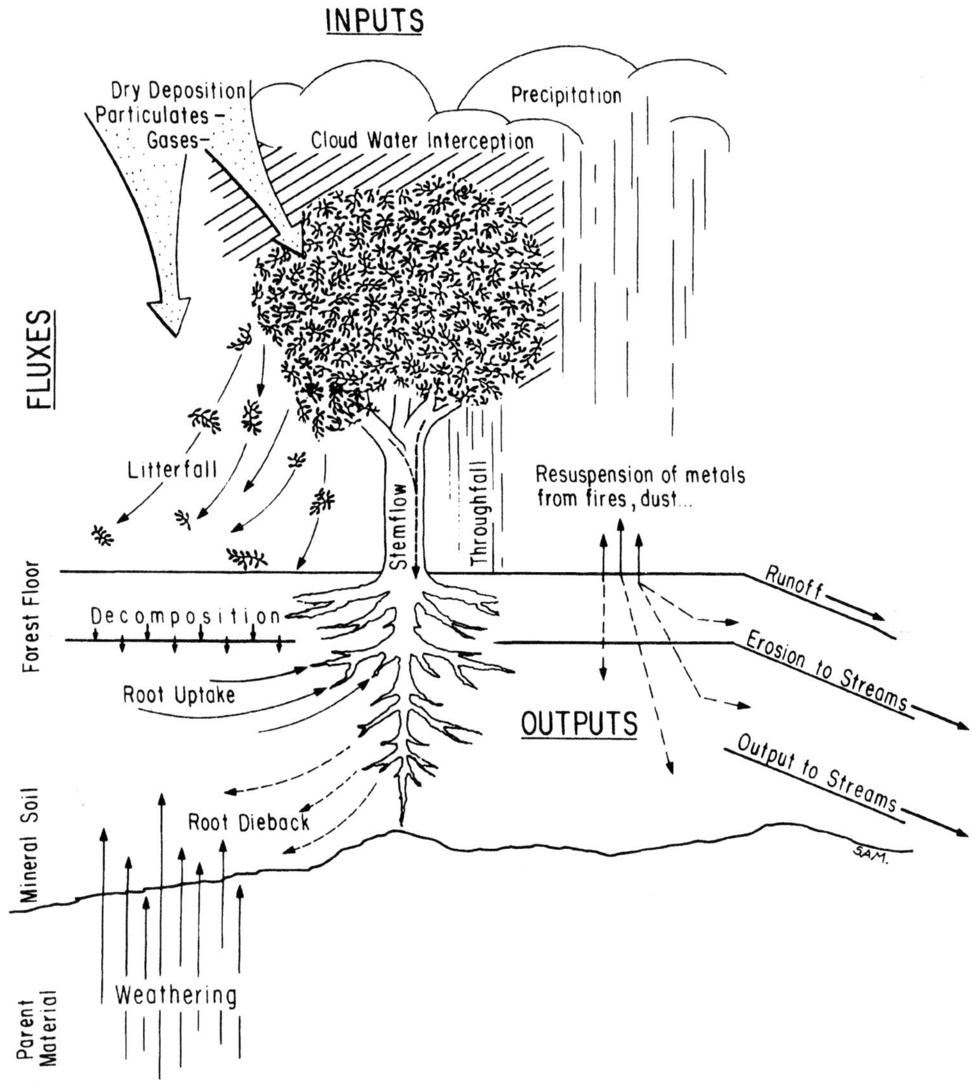

Figure 11.1. Inputs, fluxes, and outputs related to metal recycling between plants, soils, and the atmosphere. (From Friedland 1990. Reprinted with permission from A. Jonathan Shaw (ed.), *Heavy Metal Tolerance in Plants: Evolutionary Aspects,* pp. 7–19, CRC Press Inc., Boca Raton, FL. Copyright CRC Press, Boca Raton, Florida.)

since no plant has the ability to prevent metal uptake but can only restrict it. Plants that are tolerant to metalliferous soils have been categorized into three groups (Baker, 1981). This classification is based on concentrations of metals in aerial parts of the plant versus increasing concentrations of metals in the soil. One type is the *accumulator,* which allows the uptake of very high levels of ions to the extent of exceeding the levels in the soil. Another type is the *indicator,* which takes up metals at a linear rate relative to the concentration of metal in the soil. The third type is the *excluder,* which actually takes up metals but restricts increased concentrations in the shoots until a critical level is reached, above which metals start to increase in the shoots. Figure 11.2 illustrates these three types of tolerant plants. Plants differ among species

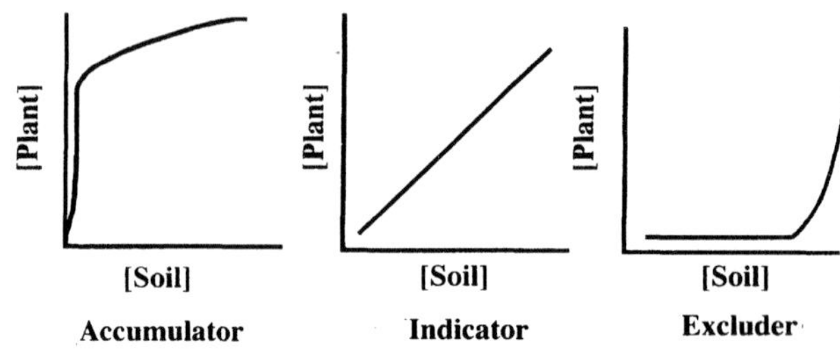

Figure 11.2. Plant strategies for dealing with high metal concentrations in soils. Brackets indicate concentration of elements in the soil and plant. Reprinted from Baker 1981, p. 646, by courtesy of Marcel Dekker, Inc.

with respect to how much of a specific metal they can accumulate in their shoots. Some plants can concentrate high levels of metals in their leaves, while others are highly restrictive. As an example, the 12 plant species depicted in Figure 11.3 illustrate differences in the concentrations of Zn and Pb in leaf tissues. These 12 species were shown to have variations in plant/soil ratios of Zn, Pb, Cd, and Cu of 18, 240, 273, and 9, respectively. A single plant species may also differ with respect to uptake, transport, and accumulation of different metals. For example, in Figure 11.4, the grass *Triodia pungens* could be classified as an accumulator of Zn, while it appears to be an excluder when concentrations of Pb and Cu increase in the soil. Also, the plant appears to have different capacities in its ability to exclude these two metals, based on the slope of the curves and the point at which the exclusion mechanism appears to become saturated. Plants also differ with respect to where in the plant the metals are concentrated. These differences are usually based on leaf/root ratios for specific elements. Table 11.7 illustrates how different plant species may localize different elements in leaves versus roots. This is species specific and metal specific since, in some plants (i.e., *Silene vulgaris*), Pb is accumulated in the leaves, while in others (i.e., *Thlaspi alpestre* and *Armeria maritima*), it is found mainly in the roots.

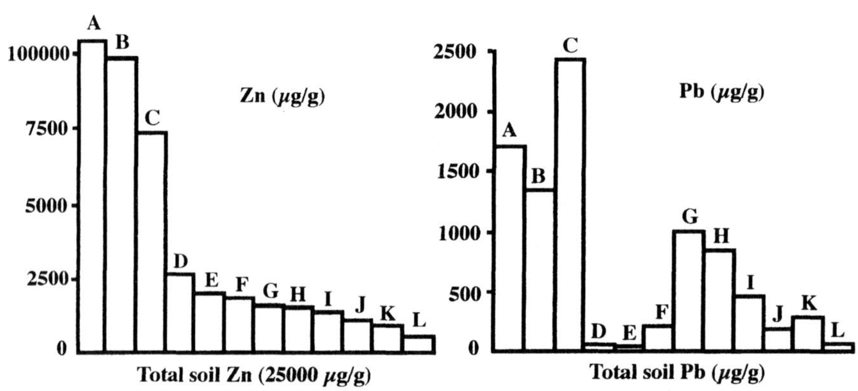

Figure 11.3. Species differences relative to the accumulation of metals in leaf tissues. (A) *Thlaspi alpestre*. (B) *Minuartia vetna*. (C) *Armeria maritima*. (D) *Silene vulgaris*. (E) *Lotus corniculatus*. (F) *Festuca ovina*. (G) *Campanula rotundifolia*. (H) *Thymus serpyllum*. (I) *Rumex acetosa*. (J) *Euphrasia stricta*. (K) *Achillea millefolium*. (L) *Viola calaminaria*. Reprinted from Baker 1981, p. 645, by courtesy of Marcel Dekker, Inc.

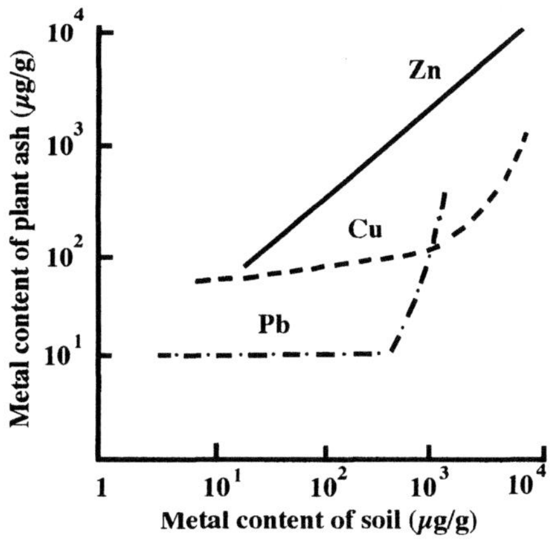

Figure 11.4. Multiple strategies by a single plant grass species for regulating the uptake of different metals. From Baker 1981, p. 646, by courtesy of Marcel Dekker, Inc.

D. Effects of Plant Age, Part, and Species on Metal Accumulation

Table 11.8 illustrates differences in metal accumulation based on tissue type and age in pine trees. Note that Al, Fe, and Mn are accumulated to the greatest extent in all tissues. The first is not essential for plant growth; however, the latter two metals are required. Also, note that older needles accumulate higher concentrations of metals than younger needles except for Al and perhaps Cu and Ni. Also, wood is very low in all elements, while branches have higher levels of ions than knots or bark. Bark, which contains phloem, is generally higher in all elements except Mn when compared with knots. Young roots (1 mm in diameter) accumulate higher concentrations of all elements compared with older roots (5 mm in diameter), and young roots have the highest concentrations of Al and Fe compared with any other tissues.

Specificity of accumulation may also be reflected in different plant families (Baker 1987). Low accumulations of Cd have been observed in Leguminosae, moderate accumulation in Gramineae, Liliaceae, Cucurbitaceae, and Umbelliferaceae, while high accumulations were observed in the Chenopodiaceae, Cruciferae, Solanaceae, and Compositae. In addition, a number of long-term field crop experiments, in which sewage sludge was used as a soil amendment, showed that dicots such as spinach or lettuce absorbed more heavy metals than did monocots such as oats or wheat. Finally,

TABLE 11.7 Ratio of Heavy Metals in Dry Matter of Plants Grown on Contaminated Soils

	Leaf/Roof Ratios		
Species	Zn	Pb	Cu
Thlaspi alpestre	2.54	0.71	1.60
T. cepeaefolium	3.44	—	—
Minuartia verna	1.41	0.44	0.71
Silene vulgaris	0.97	1.23	0.64
Armeria maritima	0.32	0.16	0.83

Source: Reprinted from Baker 1981, p. 647, by courtesy of Marcel Dekker, Inc.

TABLE 11.8 Age and Tissue Variation in Trace Element Content in Pine Trees

	Trace Element (ppm dry wt)										
Plant Part	Al	B	Co	Cr	Cu	Fe	Mn	Ni	Pb	Ti	V
Needles											
1 yr old	400	18.0	0.9	4.8	4.2	150	430	6.0	0.2	15	0.6
Older	200	24.0	0.8	4.0	2.5	370	740	2.1	0.5	30	1.2
Branches	400	6.0	0.6	1.6	3.0	650	430	1.1	0.6	25	1.8
Knots	120	4.5	0.2	0.8	1.2	78	185	0.3	0.1	6	0.8
Bark	230	4.5	0.4	1.0	2.0	100	123	0.4	0.3	15	2.8
Wood	7	0.9	0.1	0.3	0.6	5	61	0.3	0.1	1	0.2
Roots											
1 mm diam.	1430	6.5	0.1	0.9	3.5	7171	134	1.1	0.3	46	0.6
5 mm diam.	82	3.2	0.7	0.6	1.2	46	50	0.4	0.1	6	0.5

Source: From Kabata-Pendias and Pendias 1992. Reprinted with permission from *Trace Elements in Soils and Plants,* 2nd ed., CRC Press, Inc., Boca Raton, FL. Copyright CRC Press, Boca Raton, Florida.

Table 11.9 illustrates how different plant groups, occupying a similar habitat, can accumulate metals differently. Note that almost all of the plants represented in the table accumulate high levels of Mn and Fe, both of which are essential elements. However, mosses and lichens accumulate considerably higher levels of Fe and Pb compared with the other plant groups. Fungi, on the other hand, accumulate higher levels of Cu and Zn than other species. The most toxic metals with respect to plants are Hg, Cu, Ni, Pb, Co, Cd, Ag, Be, and Sn. Table 11.10 illustrates the toxic levels of various metals in plant leaf tissues as well as phytotoxically excessive levels in surface soils.

VII. PHYSIOLOGY AND BIOCHEMISTRY OF METAL TOXICITY

A. Metal Binding to Organic Ligands

The ability of ions to interact with ligands of biological importance and the stability constants of such complexes have led to the classification of metals into three groups: class A, class B, and borderline metals; the last in this list can have representative species with reactivities more like A than B or the converse (Woolhouse 1983). Some class A metals include Li, Na, K, Rb, Cs, Be^{2+}, Mg^{2+}, Ca^{2+}, Sr^{2+}, Ba^{2+}, Cr^{3+}, and Al^{3+}. Class B metals include Ag, Au, Hg^{2+}, and Tl^{3+}. Some borderline metals include Mn^{2+}, Fe^{2+}, Co^{2+}, Ni^{2+}, Cu^{+}, Zn^{2+}, Pb^{2+}, and Bi^{3+}. Table 11.11 outlines some of the ligands with which these metals most likely will react in biological systems. Briefly, class A metals complex with fluoride ions and ligands having oxygen as electron donor atoms. This includes water and reactive groups such as hydroxyl, carbonate, and phosphate. Class B metals bind preferentially with ligands containing S, N, or I as donor atoms. These metals bind strongly with NH_4^+ and CN^- and form insoluble sulfides with S^{2-} and HS^-. Also, class B metals strongly bind with sulfur-containing amino acids. Borderline metals such as Co, Ni, and Pb (which are closer to class A metals) and As, Sn, and Sb (which are closer to class B metals) would have similar reactivities as the respective class to which they are most closely associated. Some of the proposed mechanisms by which metals elicit an effect on biological organisms include changes in membrane permeability (Ag, Au, Br, Cd, Cu, F, Hg, I, Pb, and UO_2), thiol group affinity for cations (Ag, Hg, and Pb), site competition with metabolites (As, Sb, Se, Te, W, and F), reactivity with phosphate groups such as ADP and ATP (Al, Be, Sc, Y, Zr, lanthanides, and possibly all heavy metals), replacement of essential cations (Cs, Li, Rb, Se, and Sr), and occupation of sites for essential groups such as phosphate and nitrate (arsenate, fluorate, borate, bromate, selenate, tellurate, and tungstate) (Kabata-Pendias and Pendias 1992).

B. Metal Antagonisms and Synergisms

Another way metals can impact plant physiology is through interactions with other elements. Both antagonisms and synergisms have been observed in plants, but antagonisms are more common (Kabata-Pendias and Pendias 1992). An *antagonism* is observed when the combined physiological effect of two or more elements is less than the sum of their separate effects, while a *synergism* occurs when the combined effect is greater than the sum of their separate effects. Antagonisms can occur in two ways: when a macronutrient inhibits the uptake of a trace metal, and conversely, when a trace metal inhibits the absorption of a macronutrient. Calcium, P, and Mg are the

TABLE 11.9 Trace Element Comparisons Among Species from One Site in a Forest Ecosystem

	Trace Element (ppm dry wt)								
Plants	B	Cd	Cu	Fe	Mn	Pb	Zn	Cr	Ni
Grass:									
Agrostis alba	3.4	0.6	4.2	80	740	1.2	59	1.0	1.8
Clover:									
Trifolium pratense	9.0	0.7	6.0	115	136	2.8	99	1.0	2.2
Plantain:									
Plantago major	7.0	1.9	9.8	135	100	2.4	97	1.4	3.0
Mosses:									
Polytrichum juniperinum	3.4	0.8	9.2	800	176	22.4	69	2.0	2.0
Entodon schreberi	3.2	0.7	10.3	425	180	13.0	77	2.8	1.6
Lichens:									
Parmelia physodes	2.4	0.4	5.0	1100	62	17	78	3.2	4.8
Lobaria pulmonaria	2.4	0.5	7.5	1450	66	28	74	3.2	2.4
Fungi:									
Cantharellus cibarius	4.0	1.0	24.5	49	19	1.2	150	0.4	2.2
Leccinum scabra	0.8	2.7	18.0	44	6	0.1	125	0.4	1.8
Tylopilus felleus	3.5	1.6	35.0	50	14	0.4	180	0.4	1.2
Russula veternosa	6.4	1.0	32.0	28	18	1.0	175	0.4	1.0

Source: From Kabata-Pendias and Pendias 1992. Reprinted with permission from *Trace Elements in Soils and Plants,* 2nd ed., CRC Press, Inc. Boca Raton, FL. Copyright CRC Press, Boca Raton, Florida.

TABLE 11.10 Tissue and Soil Concentrations of Elements Considered Sufficient and Toxic to Plants

	Concentration (ppm dry wt)		
Element	Sufficient Level in Leaves[a]	Toxic Level in Leaves[a]	Toxic Level in Soil Surface[b]
Ag	0.5	5–10	2.0 (1)
As	1–1.7	5–20	15–50 (5)
B	10–200	50–200	25–100 (4)
Ba	—	500	—
Be	<1–7	10–50	10 (3)
Br	—	—	10–20 (2)
Cd	0.05–0.2	5–30	3–8 (4)
Co	0.02–1	15–50	25–50 (6)
Cr	0.1–0.5	5–30	75–100 (4)
Cu	5–30	20–100	60–125 (6)
F	5–30	50–500	200–1000 (3)
Hg	—	1–3	0.3–5 (4)
Li	3	5–50	—
Mn	20–300	300–500	1500–3000 (2)
Mo	0.2–1	10–50	2–10 (5)
Ni	0.1–5	10–100	100 (5)
Pb	5–10	30–300	100–400 (5)
Se	0.01–2	5–30	5–10 (4)
Sn	—	60	50 (2)
Sb	7–50	150	5–10 (2)
Ti	—	50–200	—
Tl	—	20	1 (1)
V	0.2–1.5	5–10	50–100 (3)
Zn	27–150	100–400	70–400 (6)
Zr	—	15	—

[a] Values not given for sensitive and tolerant species.
[b] Parentheses represent the number of data points for each value or range.
Source: From Kabata-Pendias and Pendias 1992. Reprinted with permission from *Trace Elements in Soils and Plants,* 2nd ed., CRC Press, Inc., Boca Raton, Fl. Copyright CRC Press, Boca Raton, Florida.

TABLE 11.11 Ligands Bound by Heavy Metals

I. Ligands Preferred by Class A Metal Ions[a]	II. Other Important Ligands[a]	III. Ligands Preferred by Class B Metal Ions[a]
F^-, O_2^-, OH^-, H_2O	Cl^-, Br^-, N_3^-, NO^{2-}	H^-, I^-, R^-, CN^-
CO_3^{2-}, SO_4^{2-}, $ROSO_3^-$, NO_3^-	SO_3^{2-}, NH_3, N_2, RNH_2	CO, S^{2-}, RS^-, R_2S, R_3As
HPO_4^{2-}, PO_4^-	R_2, NH, R_3N, =N—	
	—CO-N-R	
ROH, RCOO—, —C=O, ROR	O_2, O_2^-, O_2^{2-}	

[a] R = CH_3, CH_3CH_2, or aromatic moiety; RNH_2 = amine such as CH_3NH_2.
Source: From Woolhouse 1983, with permission of Springer-Verlag, Inc.

main antagonistic elements inhibiting the absorption of a number of trace elements. Calcium and P are particularly effective in inhibiting the uptake of Be, Cd, Pb, and Ni, which can accumulate in plants and be passed on into the food chain. Since Ca and P are important as membrane constituents, the protective effect may in some way be related to changes in membrane permeability. Zinc, Ni, and Cu can suppress the uptake of Fe in plants, thus causing chlorosis, which is indicative of an Fe deficiency. Iron and Zn are strong antagonists of Cd. This may be an important relationship since Zn and Cd occur commonly in sewage sludge used for agricultural purposes. If Zn is high in sewage sludge, it could antagonize the uptake of Cd and prevent the accumulation of the latter in the food chain. Conversely, it has been observed that if low concentrations of Zn are available, a synergism may occur. Other synergisms have been observed in grasses and vegetables with respect to Ni uptake when soils are supplemented with Cd, Cu, Mn, or Zn. Thus, one can see how high concentrations of metals might have serious consequences to the physiology and biochemistry of biological organisms, based on the various types of chemical reactions and interactions that could potentially occur as a result of the presence of these metals.

C. Metal Uptake, Toxicity, and Detoxification

As can be seen from the foregoing, the mechanisms of action of metal toxicity may be numerous and multiple (Ernst 1996). The initial contact a metal has with a plant is usually in the roots and at the membrane level, although heavy atmospheric deposition of elements can cause the effect to be initiated in the shoot. The membrane serves as a barrier of selectivity and sensitivity to high concentration of metals. Depending on the metal to which a root cell may be exposed, the integrity and stability of the membrane may be altered. If the integrity of the membrane is compromised, this might serve as the primary event leading to injury and possibly death of the cell. If the membrane integrity is not compromised, the genetic constitution of the plant will determine how, or if, the plasmalemma of the plant will initiate signals leading to the induction of metabolic events that enable the plant to at least maintain environmental homeostasis with respect to the presence of the metal stressor (Cumming and Tomsett 1992). Once the metal has encountered the plasmalemma, it enters the cell via transport mechanisms similar to those described in Chapter 2 for the transport of essential elements. Antagonisms and synergisms between metals can have considerable impacts on ionic balances, which can cause membrane degradation, affect electrical charge potentials across membranes, and disrupt pH compartmentation, resulting in imbalances in energy and metabolic activity. Hopefully, mechanisms for metal detoxification are in place by the time the metal enters the cytoplasm of the cell; otherwise, the plant is likely to be susceptible to metal toxicity as a result of the direct binding of metals to key metabolic enzymes.

Precipitation, chelation, and isolation of the metal in the cytoplasm are key mechanisms for detoxification. Each organelle has specific requirements for some essential elements, and each may have specific regulatory mechanisms for dealing with excessive levels of ions. The vacuole appears to be an important repository for excess elements, but even there, the vacuolar membrane must have specific means of self-protection. Disturbances in ion balances in membranes, organelles, and cytoplasm and ionic antagonisms contribute in part to complex mechanisms of change in the metabolism and physiology that result in detrimental effects on the growth and development of plants. More specifically, changes in ion balances in cells can alter important processes such as photosynthesis, respiration, and nitrogen metabolism. Probably these changes occur through interference with multiple enzymes required by each of these important metabolic pathways, by competing with or ineffectively replacing essential elements required for their function.

Heavy metals can have negative effects on transpiration and stomatal conductance, probably as a result of their interactions with Ca-mediated signals in guard cells and their well-known antagonist effects on Ca at the membrane level, particularly in roots. Changes in ionic balance may also be important relative to water relations throughout the whole plant system (Barceló and Poschenrieder 1990). These effects translate to decreased growth and yield in plants, with the frequency of symptom development occurring in roots first and much delayed in shoots. The symptomology of metal toxicity is quite complex in view of the antagonistic and synergistic interactions that can occur among ions which can actually result in deficiences in essential elements. Figure 11.5 summarizes the various processes that can be affected by high levels of toxic ions.

D. Metal Avoidance and Tolerance Mechanisms

Plants have evolved multiple mechanisms to deal with high concentrations of metals in the environment (Cumming and Tomsett 1992, Jackson et al. 1990, Kabata-Pendias and Pendias 1992, Prasad 1997, Turner 1994, and Verkleij and Schat 1990). Mycorrhizae may protect plants by preventing the uptake of high concentrations of elements through accumulation of the element in its own cells or cell walls. Differences may also exist in permeability mechanisms of the plasmalemma resulting in the differential

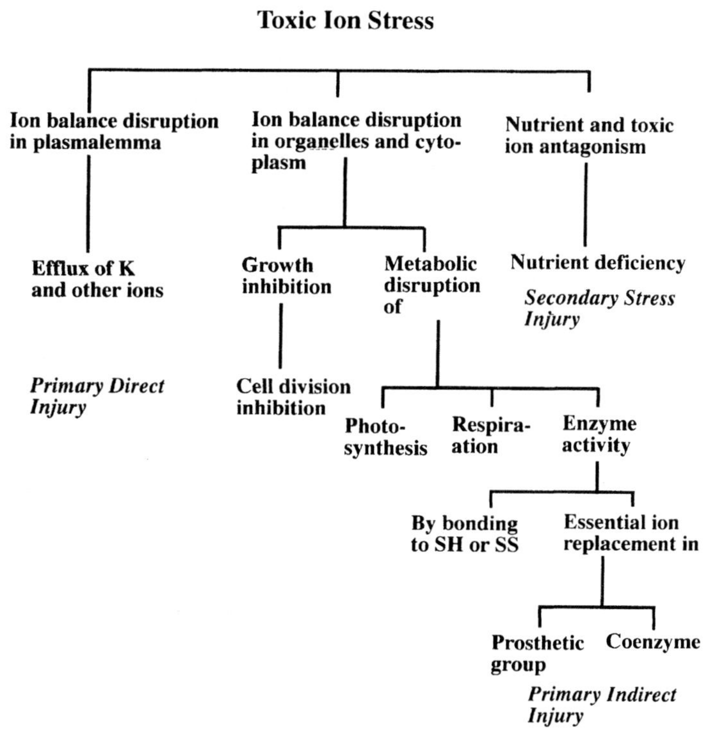

Figure 11.5. Modes of action of toxic ion stress in plants. SH = sulfhydryl group, SS = disulfide group. (From Levitt 1980, with permission.)

uptake of metals. Depending on type of root systems, some plants may avoid toxic levels of metals in surface soils by growing down into deeper horizons. The cell wall has a high cation exchange capacity and provides an important mechanism to bind excess ions, which helps to regulate the uptake of ions across the plasmalemma. Plants with unique cell wall compositions may have differing abilities to exclude metal ions. Some plants may exude organic acids or perhaps small polypeptides or polysaccharides (root mucilages) that act as metal chelators to prevent the uptake of metal ions, while others have the ability to alter the pH and/or redox potential of the rhizosphere resulting in altered uptake of toxic ions. Other plants have the ability to restrict the transport of ions and localize them in areas such as roots, where they may have less of an effect on the biochemistry and physiology of the plant. Other strategies may be leaf shedding or metal excretion into the apoplasts of tissues, where ions can be leached by throughfall from rain. Other ions may be lost through abscission of trichomes on leaf and stem surfaces where high levels of ions may accumulate. Selenium can be lost from some plant leaves and roots through volatilization as dimethylselenide (i.e. *Astragalus* and *Brassica*).

Tolerance mechanisms include the production of metal-binding molecules such as organic acids (e.g., malic acid, citric acid, amino acids) and polypeptides (Robinson 1990, Tomsett and Thurman 1988, and Zenk 1996) such as phytochelatins (PCs) also known by other names such as metallothionein, cadystin, γ-glutamyl metal-binding peptide, poly(γ-glutamylcysteinyl)glycine, and phytoallothionein. These are generally considered to be secondary metabolites and are inducible by a number of different metals. These compounds are produced very rapidly in response to metal stress, suggesting post-translation production; in fact, the enzyme phytochelatin synthase (γ-glutamylcysteine dipeptidyl transpeptidase) is activated by high levels of metals such as Cd. However, some plants have been shown not to produce such chelating compounds but yet are tolerant to specific metals. This suggests that the production of metal-binding compounds may not be the only mechanism of tolerance in such plants. Other tolerance mechanisms include localization of the metals in specific areas of the cell such as the vacuole, production of metal-tolerant enzymes (Van Assche and Clijsters 1990), and alteration in membrane composition of organelles to prevent the entry of metals into sensitive organelles.

VIII. ELEMENTS COMMONLY INVOLVED WITH METAL TOXICITY

The following outlines some elements commonly associated with metal toxicity in plants, along with some of the known mechanisms of toxicity and tolerance (Jackson et al. 1990, Prasad 1997, and Verkleij and Schat 1990).

A. Aluminum

Aluminum toxicity is usually associated with acidic soils. Its toxicity in some plants may be related to its antagonistic effects on the uptake of Mg and Ca, and the reduced ability of plants to utilize phosphate. These would result in nutrient deficiencies of these essential elements. Also, Al may express its toxic effects by altering membrane function, particularly in root cells, and has also been shown to inhibit growth through reduction in DNA synthesis.

Tolerance mechanisms that have been observed in plants with respect to Al include the following: cell wall binding of Al; increasing soil pH with subsequent decreased

Al solubility and availability to the plant; facilitation of greater uptake efficiency and utilization of essential mineral nutrients; adsorption of less NH_4 than NO_3 from the soil, keeping soil pH high and resulting in less available Al; production of chelating agents such as organic acids; and localized accumulation of Al in the roots of tolerant plants.

B. Cadmium

One proposed mechanism of Cd toxicity is the inhibition of respiratory carbohydrate metabolism, possibly by substituting for critical elements in respiratory enzymes. Cadmium also antagonizes the uptake of essential elements such as Cu and Zn, and irreversibly replaces them in enzyme reactions needed in RNA, DNA, and protein metabolism. Cadmium also interferes with chlorophyll synthesis, the chlorophyll a/b ratio in light harvesting complex II (LHC II), water hydrolysis in the early stages of the light reactions, noncyclic photophosphorylation, and carbon fixation in the Calvin's cycle. The last may be due to increased stomatal resistance as a result of high Cd levels.

Tolerance mechanisms that have been observed with respect to high levels of Cd include cell wall accumulation in shoots and roots, synthesis of polypeptide chelating compounds such as phytochelatins, and storage of Cd as complexes in the cytoplasm or as insoluble complexes of phosphate in the epidermal cells of stems. It has been suggested that Cd entering the cell is first bound to phytochelatin molecules and then precipitated as Cd-phosphate or in Ca-oxalate crystals.

C. Copper

Copper toxicity is manifested through its high affinity for sulfhydryl groups, which can alter the activity of enzymes through affecting the tertiary structure of proteins. It has also been shown to stimulate the peroxidation and degradation of membrane lipids, particularly chloroplast membranes. High Cu concentrations can also affect primary charge separation in PSII reactions, causing a loss of energy as heat and thus reducing photosynthetic efficiency (Maksymiec 1997).

Tolerance mechanisms include production of S-containing phytochelatins and organic acids that bind to Cu, storage of Cu in vacuoles bound to S-containing compounds and binding to cell walls, exclusion of Cu from chloroplasts, and reduced uptake rates.

D. Mercury

Mercury is highly toxic to vascular plants. Toxic effects have been observed in young barley plants at levels of 3 ppm dry weight while volatile forms of Hg have been reported to inhibit growth in sensitive plants at 0.5–1 ppm. A critically toxic level in most plants is considered to be 1–8 ppm. The major impact of Hg is that it has a high affinity for sulfhydryl groups and thus can impact many proteins and enzyme systems in plants. It appears to be easily mobilized in plants, hence, it can be accumulated in plant parts that could be consumed by humans or other animals. Consequently, like other heavy metals that can be accumulated in plants, it is a concern relative to animal and human health.

Tolerance to Hg has been reported in plants, although the exact mechanism or mechanisms are not known. Some plants form insoluble S-rich proteins, probably similar to phytochelatins, while other plants may form volatile organic derivatives to remove it from tissues.

E. Lead

Lead is also a highly toxic element for plants. Lead has been reported to inhibit photosynthesis and respiration by affecting electron transport mechanisms. As little as 1 ppm has been reported to affect mitochondrial respiration. Lead tolerance has also been reported for vascular plants and seems to be related to differences in membrane characteristics as well as cell wall binding by pectic acid and the formation of inactive Pb-pyroorthophosphates.

F. Selenium

Selenium replaces and/or mimics the functions of S in plants, leading to the alteration of the tertiary structure of proteins and enzymes. Such alterations are a result of substituting of Se for cysteine in the synthesis of proteins. Selenium alters the transport of S in plants, probably as a result of competition for transport sites in membranes. Selenium interferes with electron transport processes by altering the required Fe binding sites. It also interferes with methylation reactions since it forms Se-adenosylmethionine instead of S-adenosylmethionine, which is used as a methyl donor in many methylation reactions.

Insensitive plants can store large amounts of Se-methylselenocysteine, which is considered to be a nonprotein amino acid. These plants have the ability to exclude these proteins from important proteins critical for metabolism.

G. Zinc

Zinc toxicity is associated with the blockage of xylem elements and the inhibition of photosynthesis through alteration of electron transport and the capacity of rubisco to fix CO_2. Tolerance mechanisms may be related to restricted uptake (which is not common), chelation by organic acids (malic acid and phytic acid) and polypeptides, and isolation in vacuoles, at least in some plants.

IX. ORGANIC POLLUTANTS

Sources of organic pollutants include polycyclic aromatic hydrocarbons (PAHs) that are produced by the incomplete combustion of organic substances, the most significant of which is the burning of fossil fuels. These compounds have the potential of being carcinogenic, and 16 such compounds are considered as being high priority by the EPA. Vehicle emissions account for about 30% of all PAHs produced annually in the U.S. It is estimated that about 1300 tons of the PAH benzo(a)pyrene are produced and emitted to the atmosphere annually, half of which is produced by the burning of fossil fuels produced in the generation of heat and power as well as from refuse burning and coke ovens. Other organic pollutants include chlorinated aromatics with oxygen (polychlorinated biphenyls, or PCBs) and those without oxygen (polychlorophenols, or PCPs), which are used as electrical and mechanical fluids and as wood preservatives, respectively; chlorinated paraffins, used in the chemical and plastics industry; light aromatic solvents, used in chemical and mechanical industries; surfactants, used in the formulation of pesticides and industrial washing products; plastifiers, used in the production of plastics; flame retardants, found in building materials; a myriad of volatile petroleum products; residues associated with the use and manufacture of explosives; and chemicals used in global conflicts such as the defoliant Agent Orange used in the Vietnam War, Sarin gas and V-nerve agents used in the recent Gulf War, and pesticides produced by the agrochemical industry (Tarradellas and Bitton 1997).

Such compounds can enter soils (and hence plants) through various means, including atmospheric deposition from rainwater or dry deposition, leakage from landfills and waste disposal sites, surface leaching and runoff, direct application as with pesticides, and through accidental spillages such as in large oil spills. Very persistent and semivolatile organic substances have been detected in snow and in biological organisms in polar regions of the world. Such organic molecules include PCBs, DDT (still used in parts of the world), PAHs, chlordane, toxaphene, dieldrin, mirex and their metabolites. These compounds apparently can be fractionated naturally; some of them originate at the equator, where the molecules are volatilized, and then they deposit latitudinally based on regional temperature and volatility of the compound (Tarradellas and Bitton 1997). A somewhat sobering statistic is that in the United States it has been estimated that about 1.3 billion kg (2.86 billion lb) of organic pollutants (pesticides and nonpesticides from industry) are released to the environment annually.

X. AGRICULTURAL PESTICIDES

A. Types, Amounts Used, and Economics of Pesticide Usage

Mainly, what will be discussed in this section is the impact of organic pesticides, particularly herbicides, on plant growth and physiology, as well as the way pesticides can impact nontarget plants. The main categories of pesticides used for agricultural purposes include herbicides, fungicides, insecticides, and nematicides. *Pesticides* are used in agriculture, mainly for the purpose of increasing plant yields. It has been estimated that farmers use 320 million kg of pesticides annually, at a cost of $4 billion, and that total pesticide use in the United States, for all purposes, is 434 million kg (Nellessen and Fletcher 1993, Pimentel et al. 1991). *Herbicides* are used for the control of crop-competing weeds, *fungicides* and *bactericides* are used to control fungal and bacterial pathogens, *insecticides* are used to control insect pests, and *nematicides* are used to control nematode infestations in plants and soils. Herbicides constitute the greatest percentage of the pesticides marketed in Canada (81%) the United States (67%), Great Britain (58%), and Germany (54%). Worldwide, the breakdown of pesticide use for 1991 is herbicides 44%, fungicides 20.4%, insecticides 28.9%, and others 5.9%. In the United States, 1986 estimates indicated that 220 million kg of herbicide were applied to about one-half of the nation's croplands. However, 1994 estimates of herbicide sales ($\$13,077 \times 10^6$) are nearly double those reported in 1985 ($\$7075 \times 10^6$) (Tarradellas and Bitton 1997). The main types of herbicides in use include alachlor (chloroacetanilide), atrazine (triazine), chlorsulfuron (sulfonylurea), dinoseb (dinitrophenol), diquat (bipyridyl), glyphosate, isopro proturon (urea), molinate (thiocarbamate), and propanil (anilide). Figure 11.6 gives the chemical structures of these compounds and the general modes of action with respect to plant metabolism.

The focus of the agrochemical industry has been to develop chemicals that are effective against weeds, pathogens, and insects but have little effect on the plants they are designed to protect. The industry has been quite successful in doing so, since it

Common name	Chemical family	Mode of action	Developed formula
Dinoseb	Dinitrophenol	Respiratory electron transport inhibitor	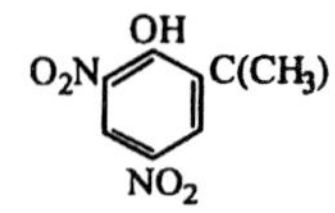
Atrazine	Triazine	Photosynthetic electron transport inhibitor	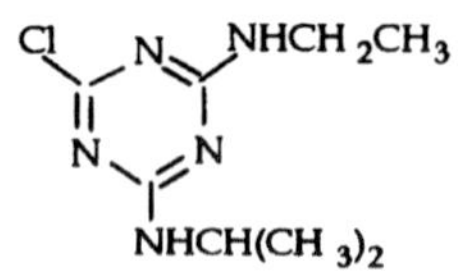
Diquat	Bipyridyl	Electron flow diverter	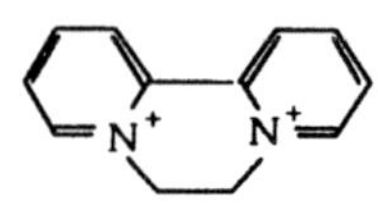
Propanil	Anilide	Photosynthetic electron transport inhibitor	
Alachlor	Chloroacetanilide	Cell division inhibitor	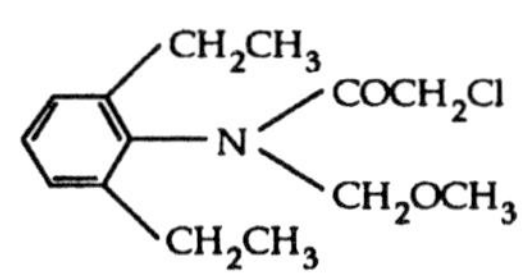
Glyphosate	Glyphosate	Synthesis inhibitor	$HOCOCH_2NHCH_2\overset{O}{\overset{\|}{P}}(OH)_2$
Isopro proturon	Urea	Electron inhibitor	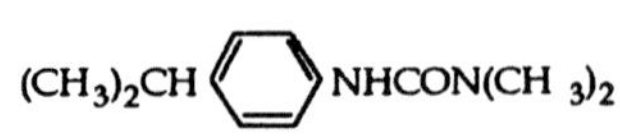
Molinate	Thiocarbamate	Germination inhibitor	
Chlorsulfuron	Sulfonyl-urea	Synthesis inhibitor	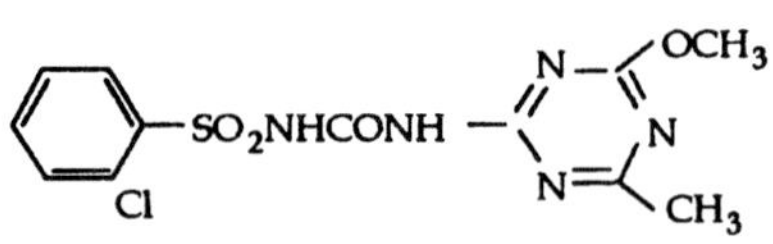

Figure 11.6. Common herbicides used in agricultural production. (From Tarradellas and Bitton 1997. Reprinted with permission from J. Tarradellas, G. Bitton and D. Rossel (eds.), *Soil Ecotoxicology,* pp. 3–32, Lewis Publishers, Boca Raton, FL. Copyright CRC Press, Boca Raton, Florida.)

has been estimated that for every dollar invested in pesticide application, there is a \$3–5 return in crops saved (Pimentel et al. 1991). However, these costs do not reflect the indirect costs of pesticide use when one considers expenses related to the following: human pesticide poisonings; the reduction in fish and wildlife populations; livestock losses; the destruction of susceptible crops, natural vegetation, honeybees, and other beneficial insects; evolved pesticide resistance; and the creation of secondary pest problems. A conservative estimate of expenses incurred as a result of the indirect effects of pesticide use are about \$2.2 billion per year, with a more realistic value of \$4.4 billion. This cost, as well as the annual cost of \$4.1 billion per year for the purchase and application of pesticides, makes pesticide use a very expensive enterprise. Of course, these are all estimates based on data that are, at best, tenuous with respect to true values. How does one assess the long-term impact of pesticides or other organic pollutants on ecological communities and extrapolate the implications to humankind?

B. Effects of Pesticides on Crops and Natural Plants

What is known about the physiological and biochemical effects of herbicides on plants has come largely through studies conducted on perhaps 100 species of important weeds and about five crop species (corn, cotton, rice, small grains, and soybean) of economic importance. Fletcher and Johnson (1988), in surveying the PHYTOTOX database, estimated that phytotoxicity studies have been conducted on 1569 plant species representing 692 genera and 147 families. They point out that this is very small portion of the plant kingdom since there are over 250,000 plant species in 12,000 genera and 300 families. They further point out that most of the phytotoxicological studies have been conducted on economically important crop species with little attention given to tree or other natural plant species. In addition, EPA testing requirements with respect to test plant species used in toxicological evaluations for the purpose of registering new herbicides are restricted to only 11 species of plants representing eight families. Also, testing procedures are conducted only on seed germination and the early vegetative stages of plant development, with no attention given to late vegetative stages, reproduction (flowering, fruiting, seed set), or senescence. This is a very important point since there is considerable variability of response by plant species to phytoxicants at the genus, family, and order levels.

The importance of studying the potential environmental risks associated with sulfonylurea herbicides has been addressed (Fletcher et al. 1993). The importance of studying the reproductive stage of plant development was demonstrated (Fletcher et. al. 1996) when the herbicide chlorosulfuron was studied at levels 1000 times lower than those recommended by the manufacturer for usage during the reproductive stages of canola and soybean. If applied at critical stages of reproduction it was found that seed yield was reduced to 8% and 1% of the control levels, respectively. Concentrations of chlorosulfuron 1000 times lower were considered to be at levels that might be encountered as spray drift when the herbicide is applied in the field and was considered to be potentially harmful to other crop plants and species of plants in the natural environment.

Thus, considering the myriad of organic chemical pollutants that are emitted into the environment, including herbicides, very little information is available regarding how organic pollutants impact natural plants and community structure (Fletcher 1997). However, it is clear that plants have the ability, in some instances, to metabolize and/or detoxify various phytotoxins (*xenobiotics*), as indicated by studies conducted on agricultural plants using various types of pesticides and a few industrially produced organics. Although it is generally assumed that microbial degradation of xenobiotics is the most important process relative to partial or complete mineralization of organic pollutants, it should be noted that 90% of the U.S. landmass and 99% of the landmass in the world is occupied by plants. Comparing the biomass of land plants (827×10^9 tons of carbon) with microorganisms (6.9×10^9 tons of carbon) reveals that, indeed, land plants are potentially very important in the removal of pollutants from the environment and converting them to smaller forms that can be further mineralized by other organisms in the soil environment (Scheunert 1992).

Understandably, the impact of organic xenobiotics on plants in the natural environment is quite complex as the result of many thousands of organics that are emitted into the environment and of the fact that these may be further metabolized in soil by microorganisms and soil fauna into other chemical forms that may or may not be toxic to higher plants. In addition, climatic conditions and soil types will interact to affect degradation and mobility of organic pollutants. Figure 11.7 summarizes the interactions that may ultimately impact the growth of plants as well as community structure.

XI. PESTICIDE MINERALIZATION

Most of the remainder of the discussion will focus on modes/mechanisms of pesticide metabolism in plants based on what is known about the topic relative to weed and crop species. This discussion should be viewed as a model that might be used to better understand how other nonpesticide organic pollutants might be mineralized in the natural environment. Keep in mind, however, that most studies have been conducted on crop and weed species (Nellessen and Fletcher 1993) that have "evolved" in highly selective environments, which in many instances may not reflect the physiology and metabolism of the great diversity of species found in the natural environment.

The first thing that has to occur prior to mineralization of a pesticide is that it has to enter the plant. This may appear to be a simple task, but in reality it is a rather complex problem considering the thousands of chemically diverse organic molecules present in the environment and the differences in complexity of plant organs and tissues (Scheunert 1992). Organic molecules can enter plants through the root system or through the shoot tissues, particularly the leaves, although entry into other plant

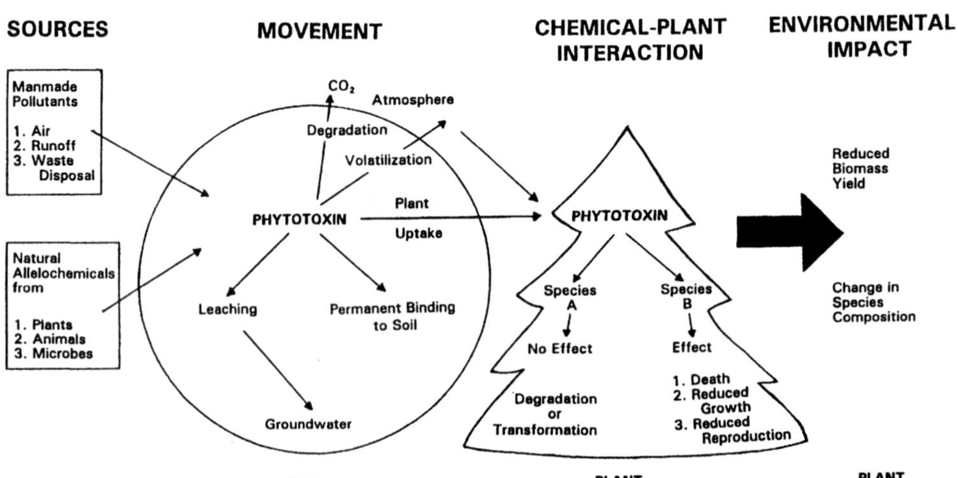

Figure 11.7. Fate of chemical pollutants as they move through the ecosystem. (From Fletcher 1997. Reprinted with permission from p. 273 in J. Tarradellas, G. Bitton and D. Rossel (eds.), *Soil Ecotoxicology*, pp. 271–289. Lewis Publishers, Boca Raton, FL. Copyright CRC Press, Boca Raton, Florida.)

parts, including bark, fruits and seed coats, is also possible. However, depending on the port of entry, the plant possesses different barriers to keep chemicals from actually entering the living tissues (symplast) of the cells, where the molecules can be detoxified or induce detrimental effects. If the chemical enters at the foliage, as may occur with particulate pollution or the intentional application of pesticides, the chemical must pass through several barriers before it enters the cytoplasm of the cell. The first barrier is the cuticle, which is comprised of surface wax (hydrophobic mixtures of straight-chain 21-37-carbon alkanes and fatty acid esters), cutin (a rigid, three-dimensional network of long-chain hydroxy fatty acids attached by ester linkages) embedded in wax, and a mixture of cutin, wax, pectin (pectic acid molecules comprised of 100 or less galactouronic acid molecules), and cellulose (long-chain glucose polymers) (Taiz and Zeiger 1991, McFarlane 1995). The next barrier is the middle lamella, which is comprised of pectin, followed by the primary cell wall of the epidermis, and then the plasmalemma. Figure 11.8 summarizes these various leaf layers. The cuticle is primarily hydrophobic because of its lipid composition. However, mixtures of polar and nonpolar cuticular components result in variable affinities of the cuticle for organic pollutants. The middle lamella and cell wall are more hydrophilic areas. Leaf cuticle structure and composition are highly variable by species (e.g., begonia vs. tomato) and environmental conditions (e.g., increased wax production in high light).

The ability of a pollutant to penetrate the cuticle depends to a large extent on the nature of the chemical. Polar and ionized chemicals will be repelled by the hydrophobic waxy layer of the cuticle, while nonpolar or nonionized pollutants can penetrate the cuticle more readily. In the case of pesticide application, very polar pesticides are formulated with surfactants to enhance penetration into the leaf tissues. However, one can see that, based on the complex composition of the various barriers mentioned, penetration of a chemical into the symplast of the cells of a leaf is a complex problem involving multiple physicochemical interactions. Smaller amounts of pollutants may actually enter the cell because of the various affinities of the cuticle and cell wall, for the pollutant. Once the pollutant reaches the membrane, its chemical nature will determine how it enters the symplast of the cell. In the case of lipophilic substances, entry may occur simply by diffusion through lipophilic pores of the plasmalemma. Other forms of uptake may include the involvement of catalyzed uptake or specific carriers, although the latter may be unlikely since specialized carriers are not thought to be present in phloem cells to facilitate loading of such chemicals in the phloem.

In addition, cell pH plays an important role with respect to ionizable chemicals (Bromilow and Chamberlain 1995). This is particularly important with respect to compartments within the cell or in the apoplast compared with the symplast. Where

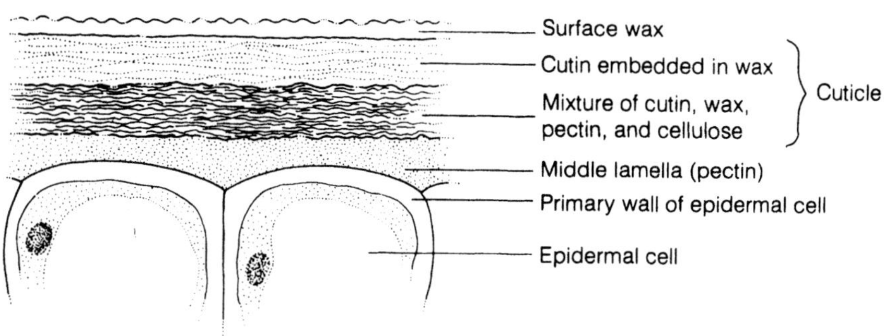

Figure 11.8. Structure of leaf cuticle and epidermal layer. (From Taiz and Zeiger 1991.)

differences in pH exist, such as the apoplast (pH 5) and a phloem cell (pH 8), the undissociated form of the acid will be at highest concentrations in the low pH environment (apoplast), and will diffuse across the membrane more readily and become dissociated in the high pH environment (phloem cell). This process, which has been called *ion trapping,* essentially traps the chemical in the symplast and allows it to be transported into the phloem.

Another way that pesticides may enter the tissues of leaves is through the stomates. Entrance through stomatal pores is considered to be minimal in the case of most pesticides except when the plant is exposed to a vapor phase of a volatile pollutant. In some cases, this may be a very significant port of entry in view of the many volatile pollutants derived from industrial sources. Once the pollutant has entered the symplast of the leaf, it may enter the xylem or the phloem and again, depending on the chemical characteristics of the compound, be transported throughout the plant.

Roots present still another set of barriers to the penetration and uptake of chemicals. As with leaves, the chemistry of the compounds involved and the structure of the root is important. However, another problem with respect to entry into roots is the soil environment. Again, such chemical characteristics as lipophilicity, volatility, ionization, and polarity of the molecules involved are all important factors affecting xenobiotic entry into root tissues. However, the chemical and physical characteristics of the soil will also determine how xenobiotics interact with soil and hence affect their availability to plants; generally, the more organic material there is associated with soils, the more tightly bound are the soil xenobiotics.

The structure and tissues of the root are very different than leaf tissues. Root growth characteristics, such as the depth of the root system, as well as morphological characteristics (fibrous versus taproot) can affect root contact with soil pollutants. These characteristics are important relative to the mobility of the chemical in the soil. In addition, the anatomy and composition of root cells differ significantly from those of leaves. For instance, the root has a specialized tissue called the *endodermis,* which surrounds the vascular cylinder of the root. Each cell of this layer is surrounded by a layer of *suberin* (similar to cutin in its chemical composition), which prevents apoplastic movement of water and nutrients, thus preventing their movement into the xylem and phloem cells of the root via the apoplast. Movement past the endodermis must occur via the symplast, which requires transport or diffusion of substances across cellular membranes and subsequent transport from cell to cell via the plasmodesmata (Chapter 2). The endodermal barrier, thus, forces anything that enters the symplast to be selectively transported across the plasma membrane.

Once the chemical is in the symplast, it is then transported up the xylem with the transpiration stream. In general, nonionized chemicals, regardless of whether they enter via the leaves or the roots, are primarily transported in the xylem transpiration stream. However, the movement of nonionized lipophilic molecules in the xylem is subject to partitioning effects against lipophilic constituents that comprise the cellular components of the xylem and closely associated tissues. This, in effect, is very similar to column chromatography and ultimately leads to a slowing of the movement of substances throughout the plant. Transport in the phloem is primarily restricted to nonionized polar molecules with minimal accumulation of chemicals associated with the phloem sinks (Bromilow and Chamberlain 1995). It is generally thought that transport from the apoplast (xylem) to the symplast (phloem) in the vascular system continues along the entire length of the stem from shoot to root, thus providing additional mechanisms for chemicals to be exchanged between tissues. Some commonly phloem-transported pesticides include benzoic acid, phenyl acetic acid derivative, carbamates, and phenoxy compounds. Compounds transported in the xylem include phenylureas, S-triazines, triazinones, acetanilides, carbaryl, carbofuran, dimethoate, and some systemic fungicides. Chemicals transported both in the xylem and phloem include aliphatic acids, phenoxyphenoxy and heterocyclic phenoxy compounds, cyclohexanedione compounds, and some systemic fungicides (Scheunert 1992).

XII. METABOLISM OF PESTICIDES (XENOBIOTICS)

A. Oxidation, Reduction, and Hydrolysis Reactions

Once the organics enter the symplast, three phases of metabolism have been identified (Komossa et al. 1995, Scheunert 1992). Phase I consists of *oxidation, reduction, and hydrolysis reactions*. Oxidation reactions are the most frequently encountered and include the introduction of OH, NH_2, and SH groups into various organic substrates. These reactions include alkyl and aromatic hydroxylation, sulfur and nitrogen peroxidation and epoxidation, and N- or O-dealkylation processes. Reductive reactions are fewer and can involve electron donations from photosynthetic electron transport and occur with nitroaromatic compounds. Hydrolytic reactions are associated with carboxylic acid esters, organophosphorus compounds, carbamates, anilides, urea groups, nitriles, and amides. Table 11.12 illustrates some of the enzymatic systems operative in plants cells in phase I and II reactions along with representative xenobiotic and natural substrates. It should be noted that these same types of detoxification mechanisms operate in animals and soil flora and fauna as well. In fact, DNA homologies of some phase I and phase II enzymes have suggested the existence of so-called enzyme superfamilies, with similarities in bacteria, plants, and humans. This has led some researchers to refer to plants as not only the "green lungs" of humanity but also the "green liver." Thus, plants not only contribute directly to the mineralization process, but soil organisms provide a complex variety of organic molecules to challenge the plant as well.

B. Conjugation and Isolation Mechanisms

Phase II of metabolizing organics refers to *conjugation reactions*. These can be divided into soluble and insoluble (bound) conjugates. The primary soluble reactions include conjugation with glucose, glutathione, amino acids, and malonyl. Bound conjugates (heteroaromatic and aromatic compounds having hydroxyl, carboxyl, amino, or sulfhydryl groups) are generally incorporated into biopolymers such as lignin and cellulose that comprises the cell wall.

Phase III represents the *storage phase,* where metabolites are stored in the cell either in the vacuole or bound to cell wall components. This further detoxifies the molecules by separating them from critical metabolic sites within the cell.

Figure 11.9 illustrates the various phases that occur in the detoxification of 2,4-D in wheat. Phase I represents hydroxylation followed by conjugation to glucose and malonyl residues (phase II) and compartmentalization as a soluble conjugate in the vacuole (phase III).

Figure 11.10 summarizes the various interactions that occur between a xenobiotic and a plant from the time the former is absorbed and transported, to when it is processed by the plant. It is obvious that plants play a role in the removal and cleansing of organic pollutants from our environment. It is indeed unfortunate that we do

TABLE 11.12 Phase I and II Enzymes Associated with Natural and Xenobiotic Reactions in Plants

Enzyme Class	Xenobiotic Substrates	Natural Substrates
Phase I:		
Cytochrome P-450 Monooxygenases	4-Chloro-N-methylaniline	Cinnamic acid, pterocarpanes
Carboxylesterases	Diethylhexylphthalate	Lipids, acetylcholine
Phase II:		
Glutathion S-transferases	Fluorodifen, alachlor, atrazine	Cinnamic acid
O-Glucosyltransferases	Chlorinated phenols	Flavonoids, coniferyl alcohol
O-Malonyltransferases	β-D-Glucosides of pentachlorophenol and of 4-hydroxy-2,5-dichlorophenoxyacetic acid	β-D-Glucosides of flavonoids and of isoflavonoids
N-Glucosyltransferases	Chlorinated anilines, metribuzin	Nicotinic acid
N-Malonyltransferases	Chlorinated anilines	1-Aminocyclopropylcarboxylic acid, anthranilic acid, D-amino acids

Source: Komossa et al. 1995. Reprinted with permission from S. Trapp and J. McFarlane (eds.), *Plant Contamination: Modeling and Simulation of Organic Chemical Processes,* pp. 69–103, Lewis Publishers, Boca Raton, FL. Copyright CRC Press, Boca Raton, Florida.

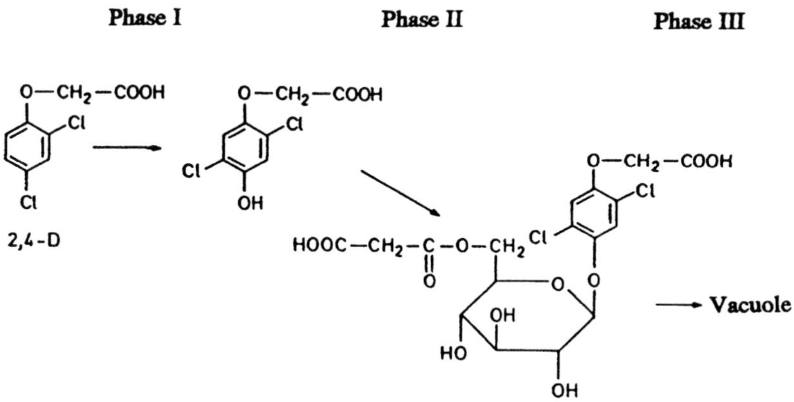

Figure 11.9. Three-phase metabolism of the herbicide 2,4-D. (From Komoßa et al. 1995. Reprinted with permission from S. Trapp and J. Craig McFarlane (eds.), *Plant Contamination: Modeling and Simulation of Organic Chemical Processes,* pp. 69–103, Lewis Publishers, Boca Raton, FL. Copyright CRC Press, Boca Raton, Florida.)

not have a better idea about the impact xenobiotics have on the growth and survival of natural plants and plant communities.

XIII. SUMMARY

There are a variety of sources for potentially phototoxic elements, including natural, agricultural, industrial, and municipal sources. Metals are recycled in the biosphere through complex inputs, outputs and fluxes. Plants growing in high concentrations of elements are called metallophytes and are classified as accumulators, excluders, and indicators depending on how they respond and process the metals in their environments. A single plant may process different elements in different ways and thus may fall under two or more classification designations. Metals have differing degrees of reactivity with biological systems and are classified as class A, class B, and borderline elements. Elements such as Hg, Pb, Cu, Zn, Se, Al, and Cd are considered to be the most toxic to plants. The modes of action of toxic elements include membrane disruption, reaction with thiol groups, site competition with metabolites, reactivity with phosphate groups, replacement of essential cations, and interactions with other elements. Plants have developed avoidance mechanisms to toxic elements by accumulating of ions in cell walls, possessing different membrane ion transport carriers and membrane components, production of organic ion chelators, altering soil redox potentials and pH, restricting ion transport to sensitive tissues, abscission of plant parts containing high levels of elements, and volatilization of complexed metals. Tolerance mechanisms include binding of metals to organic molecules such as organic acids and peptides, isolation of ions in vacuoles, production of metal-tolerant enzymes, and alterations in membrane composition so as to affect membrane transport.

Thousands of organic pollutants are emitted into the environment from sources such as the burning of fossil fuels, coke ovens, refuse burning, wood preservatives, aromatic solvents produced for industrial purposes, flame retardants, pesticides and explosive manufacturing, to name just a few. Little is known about the effects of

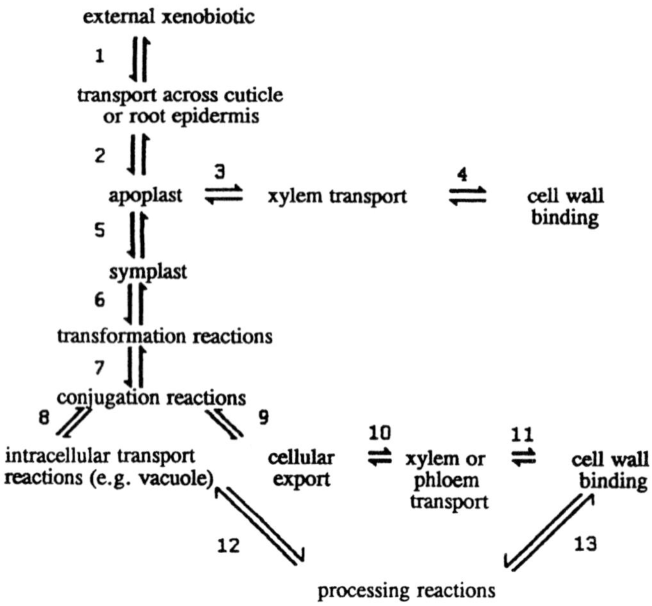

Figure 11.10. Various steps in the uptake and metabolism of xenobiotics in plants. (From Komoßa et al. 1995. Reprinted with permission from S. Trapp and J. Craig McFarlane (eds.), *Plant Contamination: Modeling and Simulation of Organic Chemical Processes,* pp. 69–103, Lewis Publishers, Boca Raton, FL. Copyright CRC Press, Boca Raton, Florida.)

organic pollutants on plants except for what is known about pesticide metabolism in plants. Such information may not be representative of how organics influence plants in natural environments, since agricultural plants have been bred for characteristics relating to increasing yield potentials under well-fertilized and -managed conditions, and only a limited number of plant species have been involved in such studies. Chemicals enter plant tissues in different ways depending on chemical structure and plant barriers that prevent or slow down entry. Once organics enter the plant, they may either be mobile in the phloem or the xylem, or in some cases transported by both. Chemicals may be metabolized by different methods including oxidation, reduction, and hydrolysis; conjugations reactions; and isolation mechanisms.

Considering the diversity of organic pollutants emitted into the environment, little is known as to how they may impact the growth and development of natural and agricultural plant communities. What we do know is primarily based on studies of only a few pesticide chemicals and on restricted numbers of plant species used in food production.

STUDY-REVIEW OUTLINE

Introduction

1. Phytotoxicity of heavy elements is a complex issue involving different sources, concentrations, and interactions of heavy metals as well as the physicochemical aspects of soils and numerous environmental variables.
2. Plants require small amounts of essential elements for normal growth processes and must balance this requirement with excess elements that may be in the environment.

Natural Sources of Metals

1. Geologic parent material represents one natural source of heavy metals. Igneous and sedimentary rocks provide different sources and amounts of elements that comprise soils.
2. Other natural sources of metals include volcanic outgasings, windblown dust from desert regions, forest and prairie fires, decomposition of biological organisms, leachates from plants, volatilization from plants and soils, and seasprays.

Agricultural Sources of Metals

1. Agricultural sources of heavy metals include phosphate fertilizers, organic fertilizers, lime, sewage sludge, irrigation waters, and pesticides.
2. Cadmium is of particular concern with respect to agricultural amendments added to soils since Cd levels can accumulate in the leaves of plants consumed by animals and humans.
3. Application of sewage sludge is of concern in agricultural environments in view of high concentrations of a variety of metals known to accumulate in these waste products. Monitoring of the amounts of these materials is necessary to avoid overloading the soils with high levels of metals.
4. Metal-containing pesticides have been used extensively in the past and are still being used to control diseases of fruit crops. Areas where these types of pesticides have been used extensively are most likely highly contaminated with metals such as Cu, As, Pb, Zn, Fe, Mn, and Hg.

Industrial Sources of Metals

1. Industrial sources of heavy-metal pollution include mining operations, refining processes (spoil heaps, tailings, transport, smelting, milling, and metal finishing), industries associated with processing plastics, textiles, microelectronics, wood preserving, paper processing, and refineries.
2. Several metals have accumulated to exceptionally high levels in the environment over the past 100 years. Among these include Pb, Zn, Cd, and Cu. These increases are primarily attributed to human activities.

Municipal Sources of Metals

1. Municipal sources of metal pollution include refuse incineration, landfills, electric power generation (burning of coal and oil), transportation, metal recycling operations, galvanized metals used along roadways and in the construction of power line towers, as well as metals that comprise the conductors of electrical transmission lines.

Plants Adapted to High Metal Environments

Metal Sensitivity, Resistance, and Accumulation

1. Sensitive plants are injured or die from high levels of metals, while resistant plants can grow and reproduce in the presence of high levels of toxic ions.
2. Two types of resistance are recognized in plants: avoidance and tolerance.
3. Plants can accumulate elements that are not required for growth at levels several times higher than those in soils.
4. Tolerance to metal toxicity is under genetic control and for, some metals, is controlled by a single gene or small groups of gene families.

Dynamics of Metal Recycling

1. Metals are recycled in the biosphere through various complex inputs, outputs, and fluxes.
2. Outputs include runoff, erosion, infiltration of stream channels, and resuspension to the atmosphere.

3. Inputs include precipitation, cloud water, and dry deposition.
4. Fluxes include vegetative leaching, litterfall, root decomposition, and soil water infiltration.

Metallophytes (Accumulators, Indicators, and Excluders)

1. Plants that can grow in high metal concentrations are called *metallophytes,* some of which are *absolute metallophytes.*
2. Plants tolerant of high metal–containing soils have been classified as *accumulators, excluders,* and *indicators.*
3. Single species of plants may have multiple classifications reflecting the different ways they process metals.

Effects of Plant Age, Part, and Species on Metal Accumulation

1. Accumulation of specific metals is species, age, and tissue dependent. Also, different families of plants appear to be more prone than others to accumulation of some metals such as Cd.

Physiology and Biochemistry of Metal Toxicity

Metal Binding to Organic Ligands

1. Metals are classified into class A, class B, and borderline metals based on their reactivity with organic ligands.
2. Class A metals bind with fluoride ions and ligands having oxygen. Class B metals bind with ligands containing S, N, and I as donor atoms.
3. Some borderline metals are closer to class A metals, while others are closer to class B metals and exhibit corresponding reactivities.

Metal Antagonisms and Synergisms

1. The mechanisms and modes of action of metals in plants include membrane disruption, reaction with thiol groups, site competition with metabolites, reactivity with phosphate groups, replacement of essential cations, and synergisms and antagonisms of one metal with another.

Metal Uptake, Toxicity, and Detoxification

1. The composition and integrity of the plasmalemma is important in establishing and maintaining an initial point of defense against the entry of toxic metals.
2. Ionic signaling at the membrane level probably initiates events that lead to the production of important organic compounds (organic acids and polypeptides) that aid in the detoxification of heavy metals.
3. Homeostasis with respect to ionic balances within plant cells is important for maintaining good water relations, stomatal function, and important metabolic processes such as photosynthesis, respiration, and nitrogen metabolism.

Metal Avoidance and Tolerance Mechanisms

1. Avoidance mechanisms developed by plants to deal with toxic levels of metals in the environment include mycorrhizal associations with roots, accumulation of ions in cell walls, differences in membrane composition and permeability, type of root system, exudation of organic chelators into the soil, altering soil pH and redox potentials, limiting ion transport in plant tissues, abscission of plant parts, and organic volatilization.
2. Plant tolerance mechanisms include the production of metal binding organic acids and polypeptides, isolation of metals in vacuoles, production of metal tolerant enzymes, and membrane alteration to change permeability.

Elements Commonly Involved with Metal Toxicity

1. Aluminum, Cd, Cu, Hg, Pb, Se, and Zn are frequently studied toxic elements that can affect plant and animal health. The modes and mechanisms of action of these elements are complex, often suggesting multiple interactive relationships.

Organic Pollutants

1. Thousands of organic pollutants are emitted into the environment from various sources including vehicle exhausts, the burning of oil and coal to generate electricity, refuse burning, coke ovens, materials used as electrical and mechanical fluids, wood preservatives, aromatic solvents, petroleum products, flame retardants, pesticides, and the manufacture and deployment of explosives.
2. Depending on chemical structure, organics can be distributed great distances both in the physical environment and in the biological food chain.

Agricultural Pesticides

Types, Amounts Used, and Economics of Pesticide Usage

1. About one-third (943.8 million lb) of the estimated 2.86 billion lb of organics (pesticide + nonpesticide industrial) emitted annually into the U.S. environment comes from agricultural pesticides.
2. The estimated annual cost of the utilization of pesticides in the United States is, conservatively, about \$6.6 billion and, realistically, about \$8.5 billion. These estimates are based on direct costs related to purchase and application; and to indirect costs related to pesticide poisoning of humans, animals, and beneficial insects, destruction of susceptible crops and natural vegetation, evolved pesticide resistance, and development of secondary pest problems.

Effects of Pesticides on Crops and Natural Plants

1. Most of what is known about the effects of pesticides on the physiology and biochemistry of plants is based on a few agronomically important species representing a limited number of plant families.
2. Studies defining the effects of organic pollutants on natural vegetation and plant communities are urgently needed.
3. Pesticide registration procedures need to be expanded to include pesticide effects on more natural plant species across more plant families and to include more stages of development, especially reproduction and senescence.

Pesticide Mineralization

1. Pesticide uptake and metabolism may be used as a model to study how natural plants may process xenobiotics in the environment, although many of the plants studied have been bred for production purposes and may not be representative of how natural plants and communities respond to organic pollutants.
2. Organic pollutants can enter plants via the shoot tissues, especially leaves, or by the roots.
3. How readily a chemical enters the plant depends on the characteristics of the chemical, the method of entry, temperature, humidity, and the physicochemical characteristics of the soil in which the plant is growing.
4. Chemicals entering plant leaves have to pass through the cuticle, which is comprised of complex mixtures of waxes and other lipids, followed by the middle lamella and cell wall. Entry into the epidermal symplast depends on the ability of the chemical to cross the plasmalemma.
5. Polarized and ionized chemicals will be repelled by the cuticle while nonpolar and nonionized chemicals can penetrate more readily.

6. Surfactants are sometimes used to facilitate the penetration of pesticides into plant tissues.
7. Chemicals may enter the symplast by diffusion, active transport, and as a result of differential pH levels in compartments resulting in ion trapping.
8. Barriers to chemical uptake by roots include suberized cell layers such as the endodermis, mucilaginous materials excreted into the soil, depth of root growth, and affinity of chemicals for cellular solids such as cell walls and other polysaccharide material.
9. The transport of organics can occur in either the xylem or the phloem. Some chemicals are exclusively transported in one or the other or both.
10. Nonionized chemicals are generally transported in the xylem and may be subject to partitioning as they move up the xylem because of affinities of the molecules for cellular components such as cell walls.

Metabolism of Pesticides (Xenobiotics)

Oxidation, Reduction, and Hydrolysis Reactions

1. Oxidation reactions most frequently involve the introduction of OH, NH_2, and SH groups on various substrates.
2. Reductive reactions occur less frequently but involve electron donations from photosynthetic electron transport and occur with nitroaromatic compounds.
3. Hydrolytic reactions are associated with carboxylic acid esters, organophosphorus compounds, carbamates, anilides, urea groups, nitriles, and amides.

Conjugation and Isolation Mechanisms

1. Conjugation reactions involve the binding of the pesticide with soluble or insoluble conjugates such as sugars, amino acids, glutathione, or malonyl. Also, insoluble conjugates can be formed with lignin and cellulose.
2. Isolation involves removal of the molecule from critical metabolic sites and storing it in the vacuole or attaching it to cell walls.

SELF-STUDY QUESTIONS

1. Discuss the dichotomy between the essentiality of nutrient elements and the presence of toxic levels of metals in polluted soils.
2. Differentiate between the elemental composition of igneous and sedimentary rock.
3. What percentage of the Earth's crust is comprised of igneous and sedimentary rock?
4. Describe other natural processes that disseminate elements via atmospheric processes.
5. Outline various agricultural, industrial, and municipal practices that result in pollution of the environment, and indicate the metals involved in each process.
6. Define *sensitivity* and *resistance* in plants as they apply to metal toxicity.
7. Name two types of tolerance.
8. Give two examples of single-gene or gene-family control of metal tolerance.
9. Indicate two practical uses of plants that accumulate high concentrations of metals.
10. Outline the inputs, outputs, and fluxes as they pertain to metal recycling in the environment.
11. Define *metallophyte, accumulator, excluder,* and *indicator*.
12. Describe how plants can differ relative to the accumulation and distribution of metals in plant tissues.
13. What evidence suggests that the accumulation of high concentrations of some elements may be associated with specific plant families?
14. List the class A and B metals, and indicate with what ligands these metals may most likely associate.
15. Why is the association of metals with the ligands referred to in question 14 important to biological systems?
16. Discuss antagonisms and synergisms relative to the metals involved, as well as physiological and biochemical significance of such interactions to plants.

17. Discuss metals relative to how they enter the plant, affect ionic balances, interact with membranes, and impact water relations and general metabolic processes.
18. Outline mechanisms by which plants can avoid and tolerate high metal concentrations.
19. What are the most phytotoxic metals?
20. Select three of the most phytotoxic metals and discuss mechanisms of toxicity and tolerance.
21. Describe several nonpesticide sources of organic pollutants.
22. How are the pollutants in question 21 distributed, and to what extent?
23. What are the different types of pesticides that contribute to organic pollution?
24. Discuss the amounts of pesticides used in the United States and the economics of their use.
25. What are the main types of herbicides used in the United States?
26. Discuss what is known concerning the physiological and biochemical effects of pesticides on agricultural versus natural plants.
27. What are the problems associated with testing requirements relative to the registration of new pesticides?
28. Compare and contrast leaves and roots relative to the various barriers to pesticide uptake.
29. Discuss how pesticides enter the symplast and are transported throughout the plant.
30. There are three phases in the metabolism of a pesticide in plants. Discuss what occurs in each phase as well as its importance to the physiology of the plant.
31. What is the role and significance of vascular plants in the mineralization of organic pollutants in the environment?

SUPPLEMENTARY READING

Clark, R. B. 1982. Plant response to mineral element toxicity and deficiency. Pp. 71–142 in: M. N. Christiansen, and C. F. Lewis (eds.), Breeding Plants for Less Favorable Environments. John Wiley & Sons, New York.

Davis, B. E. 1992. Trace metals in the environment: retrospect and prospect. Pp. 1–18 in: D. C. Adriano (ed.), Biogeochemistry of Trace Metals. Lewis Publishers, Boca Raton, FL.

Foy, C. D. 1978. The physiology of metal toxicity in plants. Annual Review of Plant Physiology 29:511–556.

Haygarth, P. M., and K. C. Jones. 1992. Atmospheric deposition of metals to agricultural surfaces. Pp. 249–276 in: D. C. Adriano (ed.), Biogeochemistry of Trace Metals. Lewis Publishers, Boca Raton, FL.

Henry, C. L., and R. B. Harrison. 1992. Fate of trace metals in sewage sludge compost. Pp. 195–216 in: D. C. Adriano (ed.), Biogeochemistry of Trace Metals. Lewis Publishers, Boca Raton, FL.

Huisingh, D. 1974. Heavy metals: implications for agriculture. Annual Review of Plant Pathology 12:375–388.

Vyas, S. C. 1988. Nontarget effects of agricultural fungicides. CRC Press, Boca Raton, FL.

12 Atmospheric Pollution: SO_2, O_3, NO_x, and "Greenhouse Gases"

Outline

- I. Introduction
  - A. Types, Sources, and Abundance of Gaseous Air Pollutants
  - B. Photochemical Reactions and Formation of Gaseous Air Pollutants
- II. Physiological and Biochemical Effects of Ozone on Plants
  - A. General Considerations and Symptomology of O_3 Injury
  - B. Complexity of Air Pollution Interactions in Plant Communities
  - C. Mechanism of O_3 Entry into Plants
  - D. Chemical Reactions of O_3 with Plant Cellular Components
  - E. Metabolic Reactions Affected by O_3, and Implications for Plant Growth
  - F. Ozone Resistance and Detoxification Mechanisms
  - G. Physiological Functions Affected by O_3
- III. Sulfur Dioxide and Plants
  - A. Sources of SO_2 and Plant Symptomology
  - B. Sulfur Dioxide Uptake, Toxicity, and Metabolism
- IV. Nitrogen Oxides and Plants
  - A. Sources and Generation of Nitrogen Oxides in the Atmosphere
  - B. Plant Symptomology and Relative Toxicity of NO_x and SO_2
  - C. Nutritional, Metabolic, and Ecological Implications of Nitrogen Dioxide
- V. Carbon Dioxide and Other Greenhouse Gases
  - A. Evidence and Sources of Rising Atmospheric CO_2 Levels
  - B. Effects of Elevated CO_2 on C_3, C_4, and CAM Plants
  - C. Plant Water Relations, Carbon Allocation, Nitrogen Metabolism, and High CO_2 Levels
- VI. Carbon Dioxide and the Greenhouse Effect
  - A. Greenhouse Gases and Global Warming
  - B. Global Warming and Plant Growth and Distribution
- VII. Summary
  - Study-Review Outline
  - Self-Study Questions
  - Supplementary Reading

Objectives

1. Discuss the sources and causes of major air pollutants.
2. Describe the physiological and biochemical effects of O_3, SO_2, and NO_x on plants.
3. Review possible effects of air pollutants on natural and agricultural ecosystems.
4. Characterize the greenhouse effect with respect to global warming and consequences for plant growth and community changes.
5. Describe the sources and types of gases that contribute to the greenhouse effect.
6. Examine how and why changes in atmospheric CO_2 levels may contribute to global warming.

7. Describe how elevated CO_2 levels may affect the physiology, biochemistry, and growth of plants.
8. Discuss the implication of increased CO_2 levels on natural and agricultural plant ecosystem structure.

I. INTRODUCTION

A. Types, Sources, and Abundance of Gaseous Air Pollutants

This chapter deals with atmospheric pollutants, exclusive of metals, which were dealt with in Chapter 11. The basic pollutants include ozone, sulfur dioxide, oxides of nitrogen and greenhouse gases (carbon dioxide, methane, nitrous oxide, chlorofluorocarbons, tropospheric ozone, and water vapor).

Air pollution has been variously defined. In one definition, air pollutants are aerial substances that have some adverse effect on plants, animals (including humans), and/or materials such as monuments and historical treasures exposed to the atmosphere (Treshow 1984). Another definition is that air is considered polluted when its natural uses are impaired (Treshow and Anderson 1989). Some have likened air pollution to a social disease—a disease generated primarily from the activities of humans, adversely affecting our health and welfare (Gillette 1984). The social implications emanate from the fact that air pollution is a by-product of industrial progress and affluence usually occurring in countries with high per capita incomes, large gross national products, and large consumers of energy. Ironically, concerns about air pollution appear to be less tolerated by the rich than the poor, who are more concerned about the daily necessities of life. Somewhere along the journey of humankind, we have separated ourselves from being components of the natural ecosystems, from which we derive sustenance, and have become passive, nonnatural entities that voraciously consume without consideration for the sustainability of our environment. However, humankind can no longer afford to ignore our role in protecting and sustaining our environment since its misuse can have detrimental effects on our health and the ability of other animals and plants to survive, the sustainability on which we are so dependent for our food and the very oxygen that we breathe.

Substances emitted into the atmosphere come from both natural and anthropogenic sources. (Use of the latter term in the pollution literature may be a perfect example of how we view ourselves as "passive, nonnatural entities".) Natural sources of substances emitted into the atmosphere that can contribute to contaminated air include volcanic activity, the burning of vegetation, electrical storms, natural degradation and metabolic processes produced in soils by biological organisms, and windblown particulate matter originating from soils or from plant and microbial origins such as pollen and fungal spores. Air pollutants take on various forms and can be emitted as gases, vapors, and liquid or solid particles. Anthropogenic sources of air pollutants are many and varied and reflect industrialization on a local, regional, and with global basis. Table 12.1 outlines many of the common air pollutants of concern today along their sources, most of which are anthropogenic. The pollutants we will be mainly concerned about in this chapter are derived from the burning of fossil fuels (automobile exhausts, power generation from the burning of coal, smelting, and metal refining, and the refinement of petroleum products).

How do we know when air is contaminated? In other words, what is our reference point for deciding when air is polluted or not polluted? Polluted air is apparent in

TABLE 12.1 Air Pollutants and Their Origin

Air Pollutant	Source
Carbon dioxide	Fossil fuel; natural processes; land-use conversion
Ozone	Atmospheric photochemical reactions; storm centers; natural occurrence in the upper atmosphere
Sulphur dioxide	Fossil fuels; petroleum and natural gas industries; metal smelting and refining processes
Hydrogen fluoride	Aluminum industry; steel manufacturing; phosphate fertilizer plants; brick plants
Oxides of nitrogen	Cars and trucks; combustion of natural gas, fuel oil, coal; refining of petroleum; compressors of natural gas; incineration of organic waste; wood burning
Ethylene and other olefins	Motor vehicles; refuse burning; combustion of coal and oil; leaky natural gas heaters; natural occurrence
Ammonia	Leaks or breakdown in industrial operations; spillage of anhydrous ammonia; feedlots and stockyards
Chlorine and hydrogen chloride	Petroleum refineries; glass industries; plastic incineration; scrap burning; accidental spills
Metals	Fossil fuel combustion; metal smelting and refining processes; waste incineration and assorted industries involving some sort of combustion

Source: From Krupa 1996.

large industrial cities. You can see it, taste it, smell it, and, if in high enough concentrations, it can have observable effects on the health and well-being of humans as well as other animals and plants. Even prior to global industrialization, substances were emitted into the atmosphere from natural sources causing changes in its composition. However, such changes were usually local or regional and had limited impact on air quality on a global basis.

It should be emphasized, at this point, that the gaseous primary atmospheric pollutants of concern (CO_2, NO_2, and SO_2) can all be used as nutrients by plants for growth and development. In fact, carbon, nitrogen, and sulfur are all essential elements for the growth of plants, and without them plants will not grow normally. These primary pollutants have always been with us throughout the course of Earth's history. However, the recycling of C, S, and N between the biological and physical realms of the biosphere was kept in balance, and plants and animals had the time and the opportunity to adapt to changing atmospheric conditions. This is because any changes that occurred most likely developed at much reduced rates compared with what has been happening since recent times as a result of industrialization. In fact, the current situation with respect to atmospheric pollutants has taken place within a time frame of perhaps 100–150 years, while similar changes may have required millions of years to occur under natural (nonanthropogenic) influences. The rate at which these changes are taking place should perhaps raise some concern relative to whether plants and animals will be able to adapt or have the plasticity required to tolerate such rapid changes in the environment. The reader is referred to any current biological or botanical text for a review of the nitrogen, carbon, and sulfur cycles in order to better appreciate these important cycling processes.

Since the early to mid-1800s, advances in technology and growing industrialization caused changes in air quality that ultimately resulted in the distribution of air contaminates on a global basis. Thus, it is truly difficult to define what unpolluted air really is since air quality monitoring has been conducted only in relatively recent times. Some evidence, such as the CO_2 trapped in the polar ice caps and recent record keeping, has given us clues as to the levels of CO_2 that existed in the atmosphere in the past (Figure 12.1). Preindustrial levels of CO_2 were thought to be in the range of 260–285 ppm, while levels measured between 1958 and 1990 changed from 315 to 353 ppm, an increase of 0.35% per year. The kind of recent changes observed in CO_2 levels lies at the extreme of all known natural variability for this gas (Gammon et al. 1985). During the past few million years, the level of CO_2 has been lower (200–270 ppm) than levels observed in recent times, and it has fluctuated less during this long period of plant evolution. Carbon dioxide and other greenhouse gases will be discussed in more detail later in this chapter in reference to global warming. In any event, attempts have been made to define the composition of unpolluted air, and Table 12.2 illustrates one such example.

It has been estimated that global emissions of SO_2 are on the order of 294 million mt (metric tons) annually, of which 160 million mt are from anthropogenic activities, and that SO_2 levels are increasing at a rate of 4% annually. In the United States, roughly 23 million mt of SO_2 are emitted into the atmosphere annually, 70% of which is generated by the burning of high-sulfur coal by electrical generating companies. In the case of NO_x pollutants, 150 million mt are emitted annual into the global atmosphere, of which greater than half is from natural sources. However, in industrialized regions, the amount of NO_x pollutants from anthropogenic sources is estimated at 5–10 times higher than from natural sources (UNEP/GEMS 1991). Global emissions of CO_2 from the burning of fossil fuels have been estimated at 5000 million mt yr^{-1}, while anthropogenic changes to Earth biota (forest and grassland destruction and desertification) contribute another 1600 million mt yr^{-1} (Yunus et al. 1996).

B. Photochemical Reactions and Formation of Gaseous Air Pollutants

Primary pollutants are those that come directly from a source such as automobile exhausts, burning coal, refuse, or natural sources such as vegetation fires and volcanic

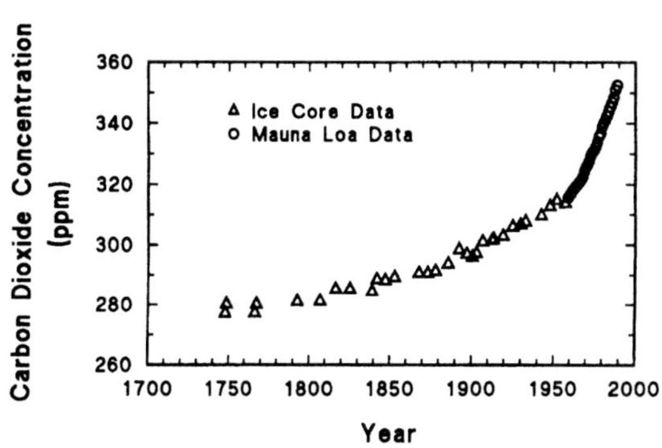

Figure 12.1. Ice core and atmospheric (Mauna Loa, Hawaii) measurements of CO_2 concentration showing increasing levels from 1750 to 1990. (From Neftel et al., 1985 with permission, as reproduced by Keeling et al. 1982 by permission of Oxford University Press.)

TABLE 12.2 Composition of Unpolluted Air

Molecule	Formula	Fraction of the Atmosphere
Nitrogen	N_2	78.09%
Oxygen	O_2	20.94%
Argon	Ar	0.93%
Carbon dioxide	CO_2	0.032%
Neon	Ne	18 ppm
Helium	He	5.2 ppm
Methane	CH_4	1.3 ppm
Nitrous Oxide	N_2O	0.25 ppm
Carbon monoxide	CO	0.1 ppm
Ozone	O_3	0.02 ppm
Sulfur dioxide	SO_2	0.001 ppm
Nitrogen dioxide	NO_2	0.001 ppm

Source: From Hodges 1977.

activity. Oxides of carbon (CO_2 and CO), sulfur (SO_2), and nitrogen (NO and NO_2) are considered to be the primary sources of inorganic gaseous atmospheric pollutants of considerable concern relative to the well-being of all living organisms. Primary organic gaseous pollutants include hydrocarbons, oxygenated hydrocarbons, halocarbons, mercaptans, and sulfides. These are of concern in that either they can affect plants or animals directly, or, through photocatalytic activity and interactions in the atmosphere, they can result in the formation of secondary pollutants with adverse effects on plants and animals.

Secondary inorganic gaseous pollutants derived from primary pollutants include ozone (O_3), hydrogen peroxide (H_2O_2), nitric acid (HNO_3), and nitrous acid (HNO_2). Organic gaseous secondary pollutants include peroxyacetyl nitrate (PAN), oxygenated hydrocarbons, N-nitroso compounds and peracetic acid. *Secondary pollutants* are produced as a result of photocatalytic reactions of primary pollutants and interactions with VOCs. One of the most important—and one that concerns many scientists today, relative to its impact on human health and plant and animal growth and survival—is O_3. Basically, we think of O_3 as being both good and bad. "Good O_3" is associated with the stratosphere where it plays and important role in blocking out harmful UV irradiation. "Bad O_3," on the other hand, is ozone that occurs in the troposphere as a result of photochemical reactions involving NO_2. Basically, NO_2, in the presence of light, dissociates into equal numbers of NO molecules and O atoms; the latter reacts with O_2 to form O_3, and NO can react back with O_3 to reform more NO_2. If this were the only photocatalytic reaction occurring in the atmosphere, then it is easy to see that O_3 levels would be kept in check in the reformation of NO_2. However, VOCs in the atmosphere also can react with NO to form NO_2 which then actually results in competition with O_3 for reactant, causing an accumulation of O_3 in the atmosphere. Highly reactive molecules, called *free-radical initiators*, are activated in the early morning hours by sunlight. These include compounds such as O_3, formaldehyde, peroxyacetyl nitrate, nitrous acid, and O_3 together with ethylene. These then react with VOCs in the presence of O_2 to form *peroxy radicals* (HO_2 and RO_2, where R represents any number of carbon atom arrangements representing different organic molecules), which react with NO to form NO_2, and *alkoxy radicals* (HO, RO), which can react with VOCs or be further oxidized to form more peroxy radicals. In addition, NO_2 can react with peroxy radicals in the light to form hydrogen peroxide (H_2O_2), peroxyacetyl nitrate (PAN), aldehydes (RCHO), acids (RCOOH), and CO_2. Oxidation of NO_x is not well understood, but it is thought that NO reacts with O_3 and $HO_2(RO_2)$ and OH radicals to form HNO_3 and PAN. Sulfur dioxide (SO_2) can also be oxidized by numerous reactions to form H_2SO_4 which is important in acid rain considerations.

The chemicals produced in the above described photooxidative cycle are very reactive with biological systems and form the basis for observed detrimental effects of air pollutants on human health and the well being of other animals and plants. Figure 12.2 illustrates the complexity of the photoreactions that can occur among gaseous pollutants in the atmosphere.

II. PHYSIOLOGICAL AND BIOCHEMICAL EFFECTS OF OZONE ON PLANTS

A. General Considerations and Symptomology of O_3 Injury

Ambient O_3 levels are typically in the range of 20–60 nl l^{-1} with episodes of 250 nl l^{-1} (Sandermann 1996). Current levels are considered to be higher by a factor of 3–4 over 19th-century concentrations (Selldén and Pleijel 1995). Figure 12.3 illustrates how O_3 levels have increased in western Europe at high altitudes from around 1870 into the 1990s. In the United States, an O_3 monitoring station located at Whiteface Mountain in Adirondack State Park (northern New York) provides one of the longest series of O_3 measurements. Based on observations from 1973 to 1993, it appears that in that region of the country, O_3 levels have been increasing at the rate

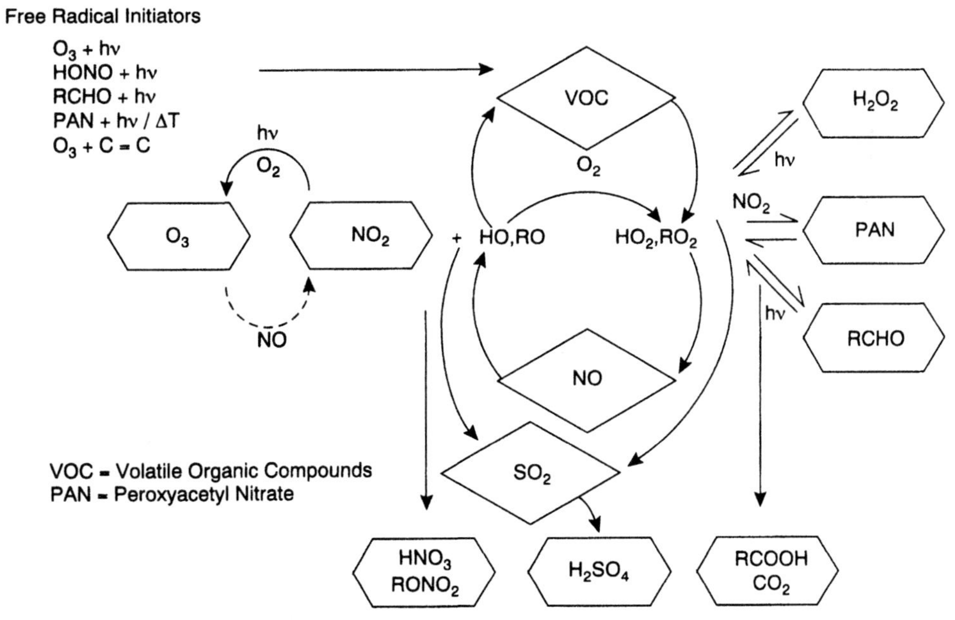

Figure 12.2. Photooxidation cycle showing interactions of primary (*diamond shaped boxes*) and secondary (*hexagonal boxes*) air pollutants. hv = radiant light. (From K. L. Demerjian. 1986. Atmospheric chemistry of ozone and nitrogen oxides. Pp. 105–127 in: A. H. Legge and S. V. Krupa (eds.), *Air Pollutants and Their Effects on the Terrestrial Ecosystems.* Copyright © 1986 by John Wiley & Sons, Inc. Reprinted by permission of John Wiley & Sons, Inc.)

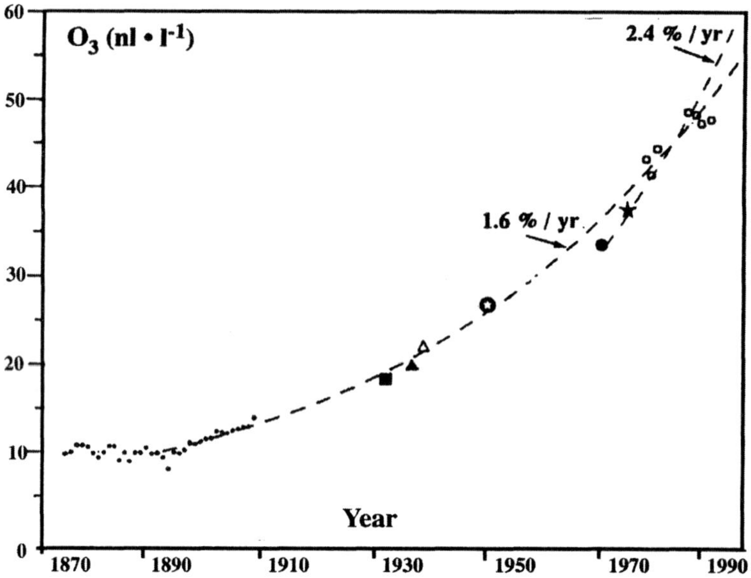

Figure 12.3. Atmospheric O_3 levels over western Europe recorded at high altitude. Different symbols represent various locations. (From A. Marenco, H. Gouget, P. Nedelc, and J. P. Pages, *Journal of Geophysical Research* 99:16617–16632, 1994, copyright at the American Geophysical Union.)

of 0.6% per year (Demerjian et al. 1992). Overall, O_3 concentrations in the United States increased by 1% between 1979 and 1988 (EPA 1991b).

Ozone is considered to be the most phytotoxic pollutant in the atmosphere (Bergmann et al. 1995). *Ozone phytotoxicity* was first discovered in the early 1950s when an unidentified disease of cigar-wrapper tobacco was discovered in the eastern United States; it was initially named "weather fleck" since it was thought to be caused by weather conditions associated with the summer and fall months of the year (Wanta et al. 1961, MacDowall et al. 1963). The symptoms associated with O_3 injury in angiosperms include necrotic spots on the adaxial surface referred to as stipples or flecks. Unlike sulfur and nitrogen oxide, O_3 affects the leaf palisade cells instead of the mesophyll cells. Ultimately, the entire leaf surface takes on a bleached appearance as these flecks coalesce. Necrosis can extend through to the abaxial side of the leaf, and, at high levels of O_3, water soaking can occur. Low levels of O_3 can cause chlorosis in angiosperms.

In gymnosperms, needles will exhibit tip necrosis, tip dieback, and browning on exposure to O_3. Chlorotic flecks and mottling have been observed to be associated with tip necrosis. Other species exhibit dwarf, chlorotic needles followed by needle abscission (Smith 1990). Figure 12.4 illustrates classic examples of weather fleck in tobacco and chlorotic stipple of ponderosa pine, both induced by O_3 exposure. Studies using pinto beans exposed to O_3 (Middleton et al. 1955) resulted in similar symptoms as those observed on tobacco. Soon, similar symptoms were recognized as occurring in grapes, spinach, and numerous other crop species. Not long after these discoveries, declines in eastern white pines were observed in Tennessee, West Virginia, and North Carolina and were eventually found to have spread from Ohio to New York and from southern Appalachia to Maine (Treshow and Anderson 1989, and references therein).

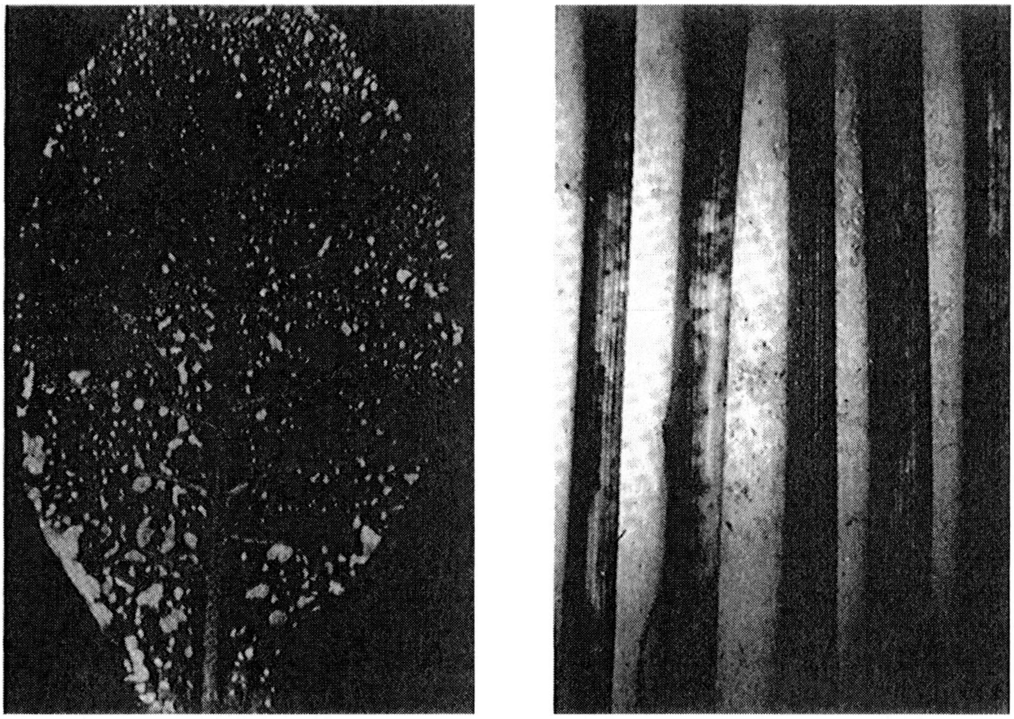

(A) (B)

Figure 12.4. Ozone-induced effects on plants. (A) Weather fleck on tobacco leaves following a 2 h exposure at 100 ppb. (From Treshow and Anderson, 1989.) (B) Chlorotic stipple on ponderosa pine needles. (From Treshow and Anderson 1989.)

Declines in forest ecosystems, as with crop losses, continues to be a concern today. However, forest tree decline most likely involves many factors associated not only with air pollutants but also with other abiotic and biotic stress interactions. It is possible that air pollution could predispose trees to other biotic and abiotic factors and in turn could lead to decline. However, the inverse could be stated as well. Forest tree decline is a very complex issue that apparently has no simple explanation.

Most of the early work regarding the impact of O_3 on plants was based on visual symptoms which, depending on species, developed at much higher O_3 concentrations. However, more recent research indicates that, prior to the onset of any visual symptoms, O_3 at much lower concentrations causes yield and growth reductions. Such effects are sometimes referred to as "invisible injury," "subliminal injury," or "latent injury" since it is not as readily observed as visual symptoms of O_3 injury. Using open-top chambers in order to develop O_3 exposure–yield loss functions, which were used in conjunction with ambient O_3 exposure and crop distribution data, researchers concluded that a 14% yield loss occurred for major U.S. crops for the years 1982–1987 (Tingey et al. 1994). A range of yield loss of 11.4–16% was observed for the same time period, with 15% of the agricultural area considered to have exceeded 20% reduction in yield. Unlike agricultural crops, it is more difficult to determine the impact of air pollutants on economically important forest ecosystems, so accurate estimates relative to losses are not available. Reasons for this lack of information relates to many complex factors, and only a few will be mentioned here. For example, the trees that comprise a forest are perennial woody species that can live many hundreds of years, while crop species are generally herbaceous annuals that complete their life cycles in one year. Crops are grown on relatively uniform soils, are provided with ample nutrients and water, and are kept relatively free of pests. Except for intraspecific competition, plants in agricultural settings have no other plant competitors except for weeds, which in many instances can be controlled. Forest trees, on the other hand, grow on heterogeneous terrain, usually under conditions of low nutrient and water availability, generally are not protected against biotic stresses, and are subjected to interspecific competition from other plant species. Thus, studying large intact ecosystems such as a forest provides many more obstacles and variables than does studying plants in agricultural environments. However, techniques are now being developed using open forest systems to determine how air pollutants, such as high CO_2 levels, may influence developing forest ecosystems over time. However, such systems are quite costly to develop and maintain.

B. Complexity of Air Pollution Interactions in Plant Communities

Figure 12.5 illustrates the many factors that must be considered relative to the effects of any gaseous air pollutant on individual plants as well as the impact on plant communities. One of the most difficult aspects of studying gaseous air pollutants is to determine the dose that the plant is exposed to. This is highly variable and depends on a number of factors including the concentration of the pollutant and the duration and frequency of exposure. It appears that intermittent exposure with high doses may have less of a detrimental effect on plants than exposures for longer, sustained periods of time at low concentrations. Other factors that need to be considered relate to the plant. Consideration needs to be given to stage of development (younger, actively growing plants frequently are more susceptible), genetic constitution (some plants are more tolerant than others), and factors that may predispose the plant to injury from air pollutants (soil, climate, and biotic factors). Plants have different strategies relative to how they deal with air pollutants, some having greater tolerances than others. Although much can be learned from single plants relative to how air pollutants affect their physiology, it is difficult to extrapolate this information to field and natural plant populations. Thus air pollutants can impact plant communities in different ways, affecting their structure and composition and, if severe enough, resulting in simplification and/or destruction of the community. This, then, leads to impairment of ecological function, economic performance, and genetic diversity.

C. Mechanism of O_3 Entry into Plants

Ozone, like other gaseous air pollutants, enters the leaves of plants primarily through the stomates. The cuticle and associated cutins serve as a protective barrier to the penetration of gaseous pollutants as well as to regulate the flux of water vapor, O_2, and CO_2 between the leaf and the atmosphere. It has been estimated that the flux of O_3 through the cuticles of *Ficus elastica* and *Lycopersicon esculentum* was 1/10,000 that through the stomates when treated with O_3 concentrations of 400 nl^{-1} (Kerstiens and Lendzian 1989). However, O_3 can affect cuticular lipids by transforming the lipid from a crystalline array of fine tubules to one that takes on the appearance of a melted amorphous structure. These changes have been observed to occlude the stomatal opening and could result in reduced CO_2 uptake and subsequent photosynthesis and growth. In addition, there are many factors that can affect the synthesis of cuticular lipids

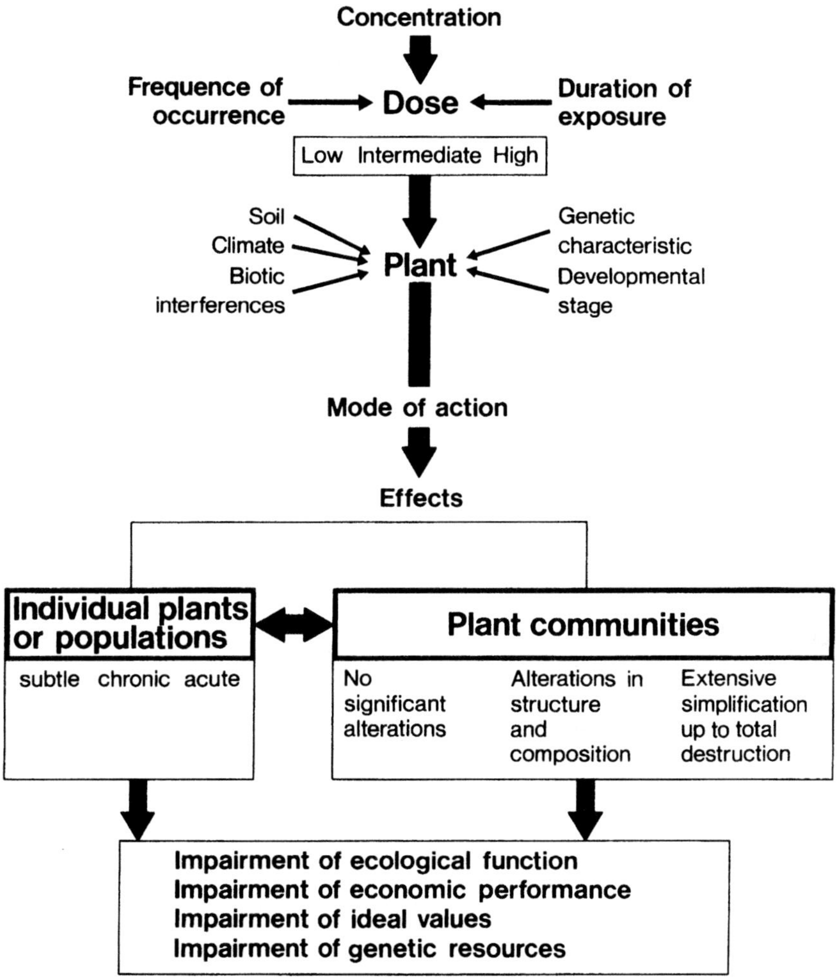

Figure 12.5. Factors interacting to influence individual plant and plant community responses to air pollution. (From R. Guderian, D. T. Tingey, and R. Rabe. 1985. Effects of photochemical oxidants on plants. Pp. 129–295 in: R. Guderian (ed.), *Air Pollution by Photochemical Oxidants*, Springer-Verlag, New York.)

including temperature, relative humidity, irradiance, wind, and acid rain (Huttunen 1994). Differences also exist among species and with stage of plant development. Thus, whether based on genetics, stage of development, or environmental conditions, the ability to produce cuticular lipids could impact the efficiency of the cuticular layer relative to excluding air pollutants.

Several kinds of resistance are encountered with respect to the entry of an air pollutant such as O_3 into plant tissues via what is referred to as the *gaseous phase of conductance*. These are illustrated in Figure 12.6 and include aerodynamic resistance (determined by shape, size, degree of pubescence, leaf surface texture, and size of epidermal cells), boundary layer resistance (determined by water vapor shell around the leaf surface, affected by size, shape, leaf orientation, epidermal characteristics, trichomes, and wind speed), stomatal resistance (determined by stomatal pore size, number of stomates, size, and anatomical characteristics of the guard cells), cuticular

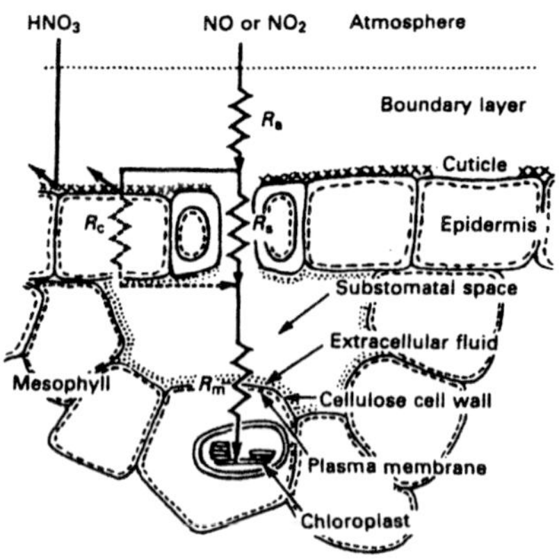

Figure 12.6. Plant resistances to diffusion of air pollutants, such as NO and NO_2, into the inner tissues of a leaf. Ra, Rc and Rs = atmospheric, cytoplasmic and stomatal resistance, respectively. (From Lea et al. 1994, as reproduced from Wellburn 1988, 1990.)

resistance (nature and amount of cuticular lipids), and mesophyll resistance (cellular structure and nature of membranes). Stomatal entry of gaseous air pollutants appears to be the primary mode of entry into plants. Thus, factors that influence stomatal opening and closing (i.e., light, CO_2 levels, drought, ionic balances, etc.), can have an impact on how much of an air pollutant can enter a plant during a given time. In fact, factors such as drought, which causes the stomates to close, have been shown to reduce damage from air pollutants. Obviously, stomatal closure can have detrimental effects on plants, such as restricting the CO_2 uptake needed for photosynthesis and growth.

D. Chemical Reactions of O_3 with Plant Cellular Components

Once O_3 enters the apoplast of the cell, via the gaseous phase of conductance, it is highly reactive and is thought to be converted to other compounds via *liquid phase conductance*. In liquid phase, unsaturated fatty acids, sulfhydryl groups, and amino acids (methionine, tryptophan, tyrosine, histidine, phenylalanine, and cystine) are susceptible to O_3. Reactions with olefinic double bonds cause the formation of aldehydes or ketones and H_2O_2, the latter of which is toxic to cells. The aldehydes can be further oxidized by O_3 to form acids. This process is referred to as *ozonolysis,* and is now thought to be the likely mechanism for the reaction of O_3 with unsaturated fats as opposed to peroxidation, which requires free-radical formation first, followed by the formation of cyclic peroxides and their degradation and release of malonaldehyde (Mudd 1996). Ozone also oxidizes amino acids and proteins. As indicated above, amino acids that have ring structures such as tryptophan, tyrosine, histidine, and phenylalanine are susceptible. Also, amino acids such as methionine and cystine are highly reactive, most likely because these amino acids have reactive S atoms associated with their structures. Reactivity with amino acids is important since many amino acids comprise the structures of important proteins. Some proteins (enzymes) that have been shown to be affected by O_3 include urease, peroxidase, catalase, lysozyme, acid phosphatase, β-glucouronidase and rubisco. Many other enzymes and proteins are likely to be affected as well. It is interesting to note that the amino acid tryptophan is a precursor of the hormone indoleacetic acid (IAA). Ozone has be shown to oxidize tryptophan and several other indole compounds, which could have important implications relative to changes in hormone balances in plants.

Ozone may also increase free radicals in plant tissues indirectly or by decomposition in solution. Some of the proposed reactants of O_3 in water include H_2O_2, hydroxyl radical (HO·) and superoxide (O_2^-). The formation of highly reactive radicals is then involved with various cellular degradation processes. However, the decomposition of O_3 in water is pH dependent and is much more rapid at alkaline pH than at acid pH (Mudd 1996). In addition, the ability of O_3 to oxidize substrates is also pH dependent. At high pH, oxidation is nonselective but at neutral pH, or less, oxidation is selective and temperature dependent (Gorbenkov-Germanov and Kozlova 1973). Such pH-dependent reactivities have important ramifications for tissue and cellular interactions since pH differences exist in the apoplast and symplast compartments.

It is generally believed that O_3 does not enter the cytoplasm, but its high rate of reactivity results in the formation of reactive products in the apoplast and/or at the surface of the plasmalemma. These products can then penetrate the membrane and disrupt metabolic activities leading ultimately to visual injury or subliminal effects affecting growth and productivity of plants. Reactants produced at the membrane may also act as signal transducers that trigger events leading to the induction of detoxification mechanisms in plant tissues. The primary mechanism of injury is highly debatable, although most would agree that O_3 directly or indirectly affects the plasmalemma first, affecting its fluidity and/or permeability with subsequent changes in transport capabilities associated with reduced K^+-stimulated ATPase transport and Ca^{2+} exclusion. It is argued that, since O_3 probably does not penetrate the plasmalemma, its site of action must be at the surface of the membrane, and the most susceptible components on the surface would be proteins involved with ion transport activities. Effects on lipid components are viewed as being more minimal since the aliphatic side chains of the unsaturated lipids of the bilayer are directed inward and are unlikely to be exposed to ozone degradation. However, at higher concentrations of O_3 or at later stages of membrane deterioration, lipid ozonolysis/peroxidation may be accelerated. It is likely that O_3 affects many processes simultaneously and that trying to determine a single primary event attributable to O_3 is difficult or impossible.

E. Metabolic Reactions Affected by O_3, and Implications for Plant Growth

What the candidates are for metabolic toxicants that disrupt normal plant functions is also a matter of debate among scientists; in fact, there may be many different reactants, including O_3 itself. Apart from the formation of aldhydes, H_2O_2, ketones, and various radicals, O_3 induces ethylene production in plants and can react with ethylene to form potentially toxic substances such as formaldehyde, which can react with H_2O_2 (produced by the ozonolysis of ethylene and by other biochemical processes in the cell) to form hydromethyl hydroperoxide, which inhibits peroxidases in plants (Polle and Junkermann 1994). It is further possible that O_3 could interact with other alkenes of biogenic origin to form organic peroxides that could be detrimental to plant growth. The production of ethylene by plants is interesting in that there are general increases in activities of enzymes normally associated with general plant defense mechanisms (controlling phenylpropanoid, flavonoid, and lignin biosynthesis pathways) as well as plant responses that are very similar to those observed when plants are challenged by plant pathogens (Kangasjärvi et al. 1994).

In addition, plants have been observed to show an increase in polyamine synthesis when exposed to O_3. This may be important in view of the possibility that polyamines may act as oxyradical scavengers as well as chelators of metal ions required for the peroxidation of membrane lipids. In addition, polyamine synthesis requires S-adenosylmethionine as a metabolic precursor, as does ethylene. It is possible that polyamines and ethylene exhibit feedback inhibition on one another, controlling key enzymes in their respective synthesis (Figure 12.7). Thus, plants with differential abilities to induce polyamine synthesis, compared with ethylene, when exposed to O_3 appear to be, at least in some cases, more tolerant to O_3. This may be because polyamines inhibit ethylene synthesis, which is felt by some to produce toxicants that lead to O_3 injury in plants.

F. Ozone Resistance and Detoxification Mechanisms

Plants show variations in sensitivity to O_3 inter- and intraspecifically. A variety of mechanisms have been proposed for these differences, but ultimately the ability to detoxify the various reactants produced when O_3 enters the plant is important. Plants have the ability to detoxify reactants by producting of antioxidants (ascorbate and α-tocopherol) and enzymes (peroxidases, catalase, glutathione reductase, and superoxide dismutase). Plants have the innate ability to detoxify reactants by producting a variety of antioxidants. This is a result of being constantly exposed to reduced oxygen species, such as superoxide radicals and H_2O_2, during metabolic processes associated with the electron transport reactions that occur in chloroplasts and mitochondria.

One of the best-studied pathways for the detoxification of hydroxyl radicals in plants is the *Halliwell-Ashada pathway* (Figure 12.8). In this pathway, superoxide is converted by superoxide dismutase to H_2O_2. Ascorbate peroxidase reduces H_2O_2 to water using ascorbic acid as an electron donor, and the resulting dehydroascorbate is recycled back to ascorbate using glutathione (GSH) as the electron donor, and GSSH (oxidized glutathione) is recycled back to GSH using glutathione reductase. However, in the process, NADPH is oxidized costing the plant valuable energy reserves. In some instances, the constitutive levels or the ability of the plant to produce such protectants is correlated with tolerance, while in other instances correlations are weak. However, there comes a point at which the plant is no longer able to deal with high

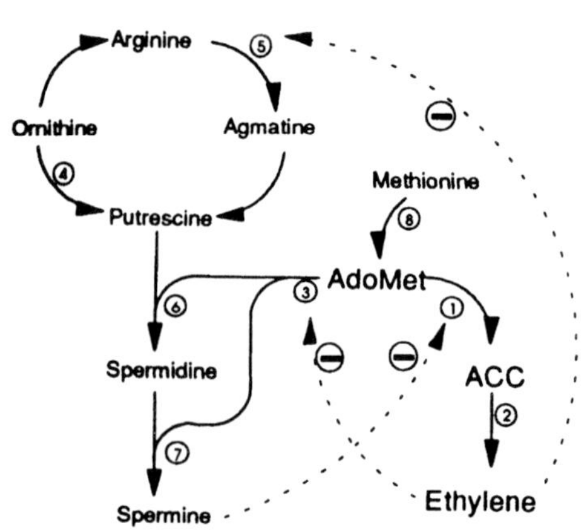

Figure 12.7. Biosynthesis of ethylene and polyamines in plants. Numbers represent different enzymes needed for the various reactions to occur and (−) indicates where various feedback inhibitions can occur. AdoMet = S-adenosyl-L-methionine, ACC = aminocyclopropane-1-carboxylic acid. (From Kangasjärvi et al. 1994, reproduced with permission of Blackwell Science Ltd.)

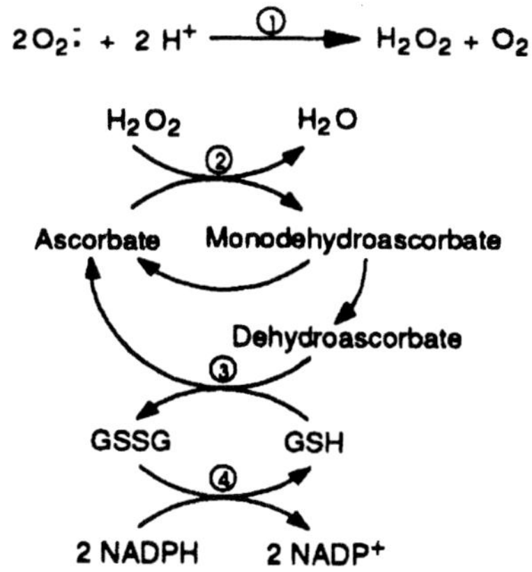

Figure 12.8. The Halliwell-Ashada pathway, used by plants in scavenging hydroxyl radicals when subjected to environmental stress such as high levels of O_3. GSSG = oxidized glutathione, GSH = reduced glutathione. Circled numbers represent specific enzymes required for the various reactions. (From Kangasjärvi et al. 1994, reproduced with permission of Blackwell Science Ltd.)

levels of oxidants and thus responds by producing more of the protectants or succumbing to the toxicant.

G. Physiological Functions Affected by O_3

Physiologically, O_3 can impact plants in numerous ways. The impact of O_3 on photosynthesis is probably the most sensitive. Ozone can affect carbohydrate metabolism directly through its effects on both the light and dark reactions of photosynthesis, and indirectly through its effect on stomatal conductance. It is not completely clear how O_3 affects stomatal regulation, but O_3 has been observed to induce both closing and opening of stomates. The latter usually occurs at low concentrations. These differences in response have been attributed to the cells, that comprise the stomatal apparatus (i.e., guard cells and subsidiary cells), having different sensitivities to O_3. Subsequent water potential adjustments in the subsidiary cells ultimately cause turgor changes in the guard cells leading to opening or closing of the stomata. This mechanism suggests that O_3 affects stomatal behavior directly. In addition, it has been well documented that the plant hormone abscisic acid (ABA) has a function to regulate stomatal aperture under various stress conditions, particularly drought. It is possible that O_3 changes the sensitivity of the guard cells relative to altering the signaling mechanisms thought to be associated with the guard cell membranes, since O_3 or its reactants are thought to have detrimental effects on membrane lipids and/or proteins. Ethylene also increases in plants exposed to O_3, and has been observed to induce stomatal closure and affect membrane permeability characteristics. Another factor that may indirectly affect reduced stomatal conductance is that O_3 entering the stoma first has an impact on the fixation of CO_2 , most likely by inhibiting the synthesis of rubisco or its activity, or by promoting its degradation. This, in turn, would result in the buildup of internal CO_2, resulting in reduced stomatal conductance, which would also limit the further influx of O_3. In some cases, the reduction in stomatal conductance leads to limited carbon fixation, while in other plant species it does not. Both direct and indirect mechanisms may be operative in different plant species. Relative sensitivities of plants to direct or indirect effects of O_3 on stomatal closure may represent different strategies for avoiding the toxic effects of pollutants.

Ozone can inhibit photosynthesis by affecting various reactions in photosynthetic electron transport (light reactions). One of the most obvious effects of O_3 injury is chlorosis, which reflects a reduction in chlorophyll in plant tissues. Chlorophyll b is more susceptible than chlorophyll a, while carotenes and xanthophylls are less susceptible than chlorophylls. In the light reactions, both photosystems (PSI and PSII) are impaired by O_3, as is the Hill reaction involved with the hydrolysis of water. In addition, O_3 has been observed to enhance and degrade the D1 protein associated with the reaction centers of PSII. Reduction in ATP synthesis has also been observed.

In the case of carbon fixation, the very important and abundant protein rubisco is degraded and/or its synthesis is inhibited by O_3. However, the effect may be age dependent, since recovery has been observed to occur in young leaves. Figure 12.9 illustrates how O_3 can affect rubisco levels in two kinds of radish, a hybrid poplar, and the eastern white pine. In this illustration, it is clear that species differences exist relative to the effects of O_3 on plants. In this case, both the wild-type and Cherry Belle radishes exhibited drastic reductions in rubisco, the hybrid poplar exhibited some decline, but the eastern white pine was least affected. The reasons given for these differences in the longevity of the leaves of the respective plants, is that the eastern white pine retained its leaves longer (years) than the others and perhaps had the capability to retain the capacity for synthesis of rubisco for longer periods of time compared with the radish, which is an annual, and the poplar, which is a determinate perennial, and both of which produce leaves with life spans of weeks to months. The efficiency of activity of rubisco has also been shown to be reduced by O_3. Photorespiration, which is considered to be a wasteful process in C_3 plants, is increased by O_3, as indicated by increased pools of intermediate metabolites common to the photorespiration process, namely glycine and serine.

Transport of carbohydrates is affected by O_3 as well. The exact mechanism is not known, but it appears that photosynthates accumulate in the leaves of plants exposed to O_3. Soluble sugars have been observed to increase in leaves, and in some instances starch has been observed to increase in chloroplasts. However, inconsistencies in this general observation have been observed and may be attributed to different stages of

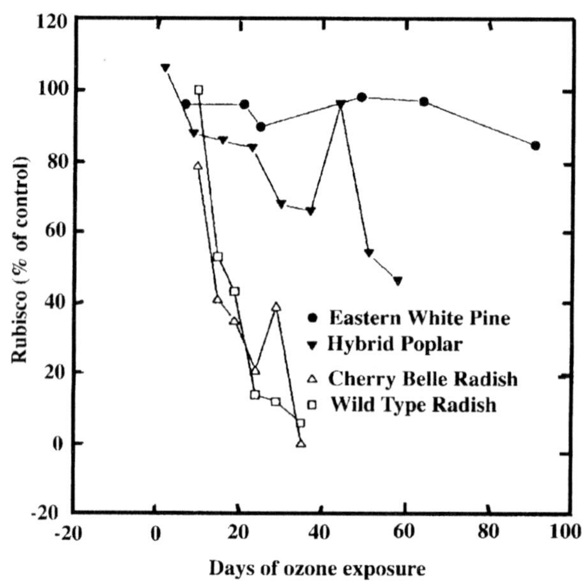

Figure 12.9. Differences in rubisco response to daily chronic levels of O_3 in radish, pine, and poplar plants. (From Pell et al. 1994.)

growth associated with reproduction or the formation of new leaves, which provide a considerable sink for the translocation of carbohydrates. Since sugars are loaded into phloem cells by active transport mechanisms for transport to roots, it is feasible to assume that O_3, or O_3 reactants, may be affecting sieve tube membranes in some way as to prevent loading and/or transport. Thus, nourishment of plant roots can be restricted, which may in turn predispose roots to disease and environmental extremes such as drought and/or freezing.

The effects of O_3 on respiration are variable. Ozone can cause either an increase or a decrease in respiration. In some plants, respiration declines at early stages of O_3 exposure or remains unchanged, and in latter stages respiration increases. In some plant species, increased respiration is associated with visible injury symptoms. Chronic levels of O_3 generally increase respiration and greatly accelerate senescence processes in plants, which are frequently associated with elevated ethylene levels.

Initially, O_3 enters the leaf and initiates a series of reactions with cellular components most likely associated with cell walls and the outer plasmalemma producing radicals, aldhydes, ketones, H_2O_2, etc. The initial site of action, and the region where toxic O_3 reactants are produced, appears to be at the plasmalemma, possibly affecting susceptible amino acids that comprise membrane proteins involved in ion transport. Also affected may be unsaturated lipids in the membrane, but this may occur later if the pollution event is concentrated enough or persists for long periods of time. These effects cause increased permeability of the plasmalemma.

Ionic imbalances in the cell that alter ion transport proteins, cause pH changes to occur that can have negative implications on energy-related metabolic processes in the cell. Ozone-generated reactants that enter the cell cause reductions in photosynthetic carbon fixation resulting in the accumulation of CO_2, which causes the closure of stomates via unknown mechanisms. Perhaps this closure is related to changes in sensitivity of the stomatal apparatus to hormones (ABA or ethylene) and/or to differential direct effects of reactants on the membrane permeability of guard cells or subsidiary cells. Alternatively, O_3 may directly affect the membranes of the stomatal apparatus causing opening or closing of the stomata. Simultaneously, the plant responds by producing antioxidants and enzymes associated mechanisms to detoxify O_3-induced reactants. Secondary plant products, and other proteins similar to those found when plants are attacked by pathogens or subjected to wounding stress, are produced, suggesting common mechanisms for dealing with a variety environmental induced stresses. The hormone ethylene and/or its O_3-generated reactants have been implicated as the cause of the visible injury in plants exposed to O_3 and thus are placed in the same response category as stress ethylene production in the wounding process. Thus, senescence induced by O_3 probably relates to the observed induction of ethylene production. The close association of polyamine and ethylene synthesis also provides possible clues as to potential mechanisms of O_3 resistance in plants where high levels of polyamines are produced that can act as ionic and free-radical scavengers. Reductions in carbohydrate production by interfering with the light reactions of photosynthesis, carbon fixation, increased photorespiration, and reductions in translocation of carbohydrates to the roots lead to reduced growth and increased predisposition of plants to other biotic and abiotic stresses.

III. SULFUR DIOXIDE AND PLANTS

A. Sources of SO_2 and Plant Symptomology

As indicated earlier SO_2 is a major air pollutant generated by the burning of fossil fuels, especially high sulfur coal. It should be noted that other sources of atmospheric sulfur pollution include sulfates such as sulfuric acid, metallic sulfates, and ammonium sulfate which also pose important risks to plants and animals through acid deposition. This discussion will focus only on SO_2 pollution. When emitted into the atmosphere, sulfur dioxide is oxidized in the gaseous phase by hydroxyl radicals and in the aqueous phase by H_2O_2 and O_3 (National Academy of Sciences 1983). In the gaseous phase, the loss of SO_2 to a surface, such as a plant leaf, is considered to be dry deposition. Sulfur dioxide, like other gaseous air pollutants, enters the leaf primarily through stomatal openings. Injury to plants as a result of SO_2, like other air pollutants, has been classified as either chronic or acute. Acute injury occurs when visible symptoms are observed after brief exposures of rather high levels of a pollutant. Chronic injury is associated with lower levels of pollutants over longer periods of time. However, both types of injury can occur at the same time. Symptom initiation is associated with the rapid decomposition of chlorophyll resulting in chlorosis and ultimately necrosis of tissues. Cellular responses to SO_2 are probably very similar, but, as a result of anatomical and morphological differences, symptomology may be variable. Thus, providing a general description of SO_2 symptomolgy in plants is somewhat difficult. However, initial symptoms in angiosperms include the development of interveinal or marginal water-soaked pale green areas followed by the development of necrotic patches that may penetrate completely through the leaf. Necrotic lesions develop in the vicinity of the stomates where tissues are exposed to the highest concentrations of the pollutant and usually results in the collapse of the mesophyll cells in the area. Gymnosperms respond by forming water-soaked areas and eventually turn reddish-brown in color. Usually, developing or newly expanding leaves are most susceptible. Also, in gymnosperms, chlorotic banding can be observed on needles, reflecting episodes of SO_2 pollution during the time course of needle elongation. Some of the most sensitive species of plants to SO_2 are vegetable and broad-leaved species. Alfalfa is very sensitive to SO_2, as are vegetable crops such as tomato and squash. Some sensitive broad-leafed woody species include red and black oaks, white ash, and sumacs. These plants are susceptible to SO_2 at levels of 1–1.5 ppm. Resistant woody species include white oaks, dogwood, and peach, which are resistant at levels in excess of 2 ppm.

Like other air pollutants, SO_2 can be trapped near the ground as a result of temperature inversions. Such inversions are particularly prevalent in valleys that lie between mountain ridges. In such areas, plants located on mountain ridges are particularly prone to SO_2 damage, since higher levels will be concentrated along the ridgetops. Inversions are particularly prevalent during the warmer seasons; however, SO_2 concentrations in the atmosphere are seasonal, with peaks occurring during the winter months and lower levels during the spring and summer (Figure 12.10). For evergreens, this may or may not be an advantage. Since metabolism is low during the winter months, and stomatal conductance is also low, SO_2 influx is probably not as high as would occur in the summer, when levels of SO_2 are lower. Conversely, however, metabolic detoxification is lower in the winter than during warmer periods as a result of differing metabolic rates. Under particularly cold conditions, high SO_2 levels may result in injury to forest trees that is later reflected in reduced growth the following spring. Thus, cold may predispose trees to harmful effects of SO_2 under such conditions.

B. Sulfur Dioxide Uptake, Toxicity, and Metabolism

Ironically, SO_2 can act as a nutrient in plants since sulfur is a required element for plant growth and development. Thus, plants have developed mechanisms, through

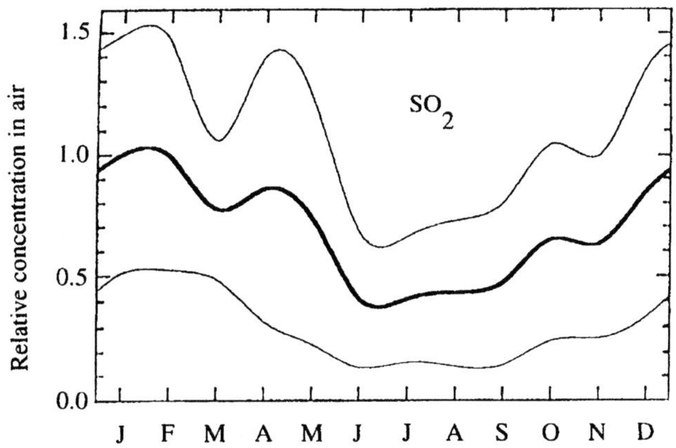

Figure 12.10. Seasonal variations in SO_2 concentrations taken from six locations in Germany. Thin lines represent standard deviations. (From Heber and Hüve 1998 as reproduced from Slovik et al. 1996.)

eons of evolutionary development, that allow them to control the internal levels of sulfur obtained from the soil or as atmospheric pollutants that may occur as gases originating from both natural and anthropogenic sources. Sulfur dioxide enters the stomatal cavity as a result of diffusion gradients maintained as a result of the reaction of SO_2 with water to form bisulfite (HSO_3^-). Bisulfite accumulates in the apoplastic space of the cell wall and is thought to be converted to SO_4^{2-} as a result of sulphite oxidase activity requiring O_2, H_2O_2, and monophenols that are catalyzed by peroxidases. These reactions may also generate radicals such as $HSO_3\cdot$ which could potentially damage cell membranes. Sulfur dioxide can enter the cytoplasm directly by diffusing through the cell membrane, where it is then converted to SO_4^{2-}. Sulfate can likewise be taken up by the cell via active carrier-mediated transport. Figure 12.11 summarizes these steps. Once inside the cell, both SO_4^{2-} and SO_2 can then be incorporated into the normal sulfur metabolic system of the plant. Several scenarios could occur, depending on the sulfur needs of the plant and the levels of SO_4^{2-} available to

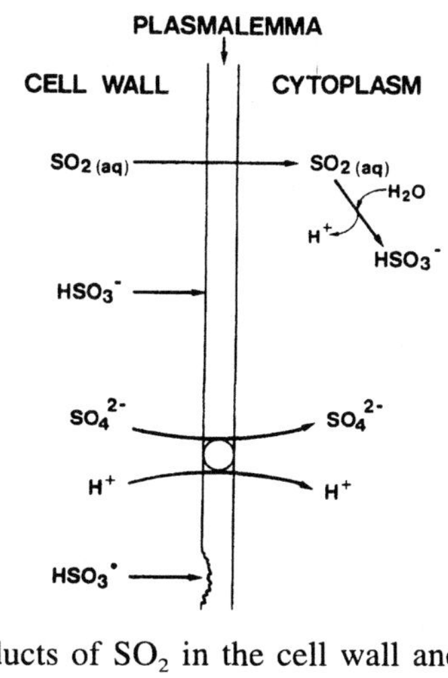

Figure 12.11. Conversion products of SO_2 in the cell wall and cytoplasm. (From Rennenberg and Polle 1994.)

the plant. If SO_4^{2-} levels are high in the plant, SO_4^{2-} can be stored in the vacuole, if there is a need for sulfur-containing amino acids in protein metabolism, the SO_4^{2-} is converted to cysteine, methionine, or reduced glutathione. Glutathione is thought to serve as a transport molecule to the roots, where it can be used in the synthesis of cysteine and proteins in the roots. It also may function to regulate uptake and/or transport of SO_4^{2-} taken up from the soil by roots. Thus, it is hypothesized that, when the plant is exposed to high levels of SO_2 from the atmosphere, more glutathione is produced in the leaves and is transported to the roots and inhibits SO_2 uptake and/or transport. Another checkpoint to prevent excess sulfur accumulation and possibly help regulate cysteine pools in tissues of plants is thought to be the production of H_2S gas derived from the reduction of HSO_3^- . This process is light mediated, occurs in the chloroplast, and may represent a mechanism to prevent acidification and possibly injury to the chloroplast, which would occur if HSO_3^- were oxidized to SO_4^{2-} in the chloroplast instead. The H_2S produced in this process diffuses out of the plant into the atmosphere and accounts for as much as 7–15% of the SO_2 absorbed by the plant depending on species (Rennenberg 1991). Plants showing increased capacities to emit H_2S gas have been correlated with SO_2 tolerance. Figure 12.12 summarizes how SO_2 may interact with the sulfur metabolism pathway in plants, thus affecting various important molecular allocations in the plant.

IV. NITROGEN OXIDES AND PLANTS

A. Sources and Generation of Nitrogen Oxides in the Atmosphere

Nitrogen oxides are formed mainly by the burning of fossil fuels. This results from a chemical reaction between atmospheric N_2 and O_2 in the presence of heat depicted by the following equation: $N_2 + O_2 + \text{heat} \rightarrow 2NO$. The NO then reacts again with O_2 to form NO_2 as follows: $2NO + O_2 \rightarrow 2NO_2$. The rate of reaction is determined by the temperature of combustion. Thus, unlike SO_2, NO_2 is not a component of fossil fuels but results from a catalytic reaction of heat with atmospheric N_2 and O_2 in the combustion process. The likelihood of increased use of coal for electricity generation in the future will undoubtedly cause an increasing level of NO_2 in the future, particularly since coal-fired furnaces require higher temperatures for operation than oil- and gas-fired furnaces. The balance between NO and NO_2 in the atmosphere is variable and unpredictable; thus, the designation of nitrogen oxides as NO_x is often used. Table 12.3 illustrates some of the natural and anthropogenic sources of NO_x in the environment.

B. Plant Symptomology and Relative Toxicity of NO_x and SO_2

Both NO and NO_2 can be toxic to plants in high enough concentrations; however, their relative ability to cause damage relates not so much to the concentration of the gas, to which the plant is exposed, but more to their relative solubility in water. Neither gas is as toxic to plants as SO_2. As an example, the threshold dose for the formation of visible symptoms due to NO_x is 1.6–2.6 ppm for a 48-hour exposure, while plants resistant to SO_2 show visible symptoms after 1-hour exposure at 2 ppm. Plants sensitive to SO_2 show symptoms after exposure for 1 hour at 0.1–1 ppm (Smith 1990).

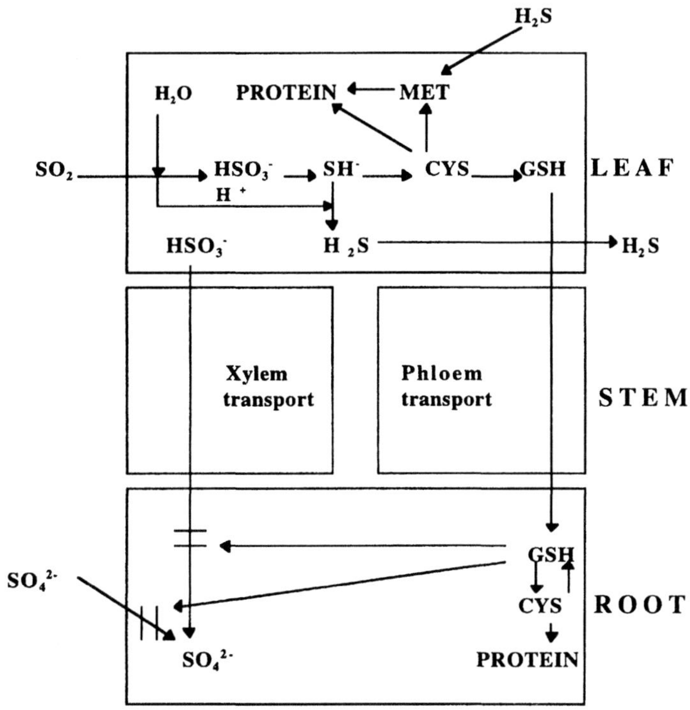

Figure 12.12. Interactions between SO_2 pollution and the interorgan regulation of sulfur metabolism. CYS = cysteine. (From Rennenberg and Herschbach 1996.)

Symptoms of NO_x toxicity in plants are difficult to diagnose in the field. In angiosperms, discolored gray-green or light brown spot form interveinally, at times combining to form stripes, with marginal chlorosis occurring in some species such as oak, maple and locust. High dosages can result in chlorosis of leaves with only the veins retaining their green coloration. In gymnosperms, discoloration occurs in the terminal portion of needles and may extend to the base resulting in red-brown to fuchsia

TABLE 12.3 Estimates of Global NO_x Emissions

Source	Emissions (10^6 mt Na^{-1} yr^{-1})
Fossil fuel burning	
Coal, lignite	5.5
Fuel oil	1.8
Natural gas	1.9
Automobiles	4.3
Soil release	5.5
Biomass burning	11.1
Lightning	5.0
Oxidation of NH_3	3.1
Miscellaneous	0.9
Total	39 ± 20

Source: From Ehhalt and Drummond 1982.

coloration. Long-term exposure may result in no initial visible symptoms but if persistent could lead to general leaf chlorosis (Smith 1990).

C. Nutritional, Metabolic, and Ecological Implications of Nitrogen Dioxide

Nitric oxide is relatively insoluble in water, while NO_2 is readily soluble and forms both nitric and nitrous acids, both of which can create acidic environments within the cell and may be part of the basis for toxicity. The ability to detoxify NO_x involves the activation of the nitrate reductase pathway. In very general terms, the process involves the following steps leading to the production of proteins:

$$NO_x \rightarrow NO_3^- \rightarrow NO_2^- \rightarrow NH_3 \rightarrow \text{amino acids} \rightarrow \text{proteins}$$

This process is the normal NO_3 reduction pathway that occurs in vascular plants. Once NO_3 is taken up by the roots, it is either reduced there or, in some species, is transported to the shoots, where it is reduced. Thus, as was the case for SO_2, NO_x can be incorporated into the normal metabolic systems of the plant and used as a nutrient for growth. However, different plants have varying capacities to assimilate nitrogen dioxide. In 217 plant taxa, including 50 wild herbaceous plants (42 genera, 15 families), 60 cultivated herbaceous plants (55 genera, 30 families), and 107 cultivated wood plants (74 genera, 45 families) fumigated with ^{15}N-labeled NO_2, there was a 657-fold difference in NO_2-N content between the highest and the lowest values in the 217 species. Some of these plants had NO_2-utilization indices greater than 10% and were considered to be "NO_2-philic" (Morikawa et al. 1998). Thus, such differences may be reflected in the abilities of different species to tolerate NO_x and their subsequent competence to compete in natural and agricultural ecosystems.

It is possible that the greatest injury to plants is not the direct effect of gaseous NO_x on plants but the addition of such compounds to the soil in the form of acid rain. It has been stated that NO_x (NO and NO_2) in the ambient atmosphere has rarely been, if ever, associated with *visible* injury in plants (Garner et al. 1989). However, the rate of NO_3^- deposition in soils has been correlated with NO_2 levels in the atmosphere (EPA 1991a). Although most natural ecosystems are thought to be nitrogen limited, relative to requirements for plant growth, it has been suggested that in typically N-limited soils plants have adapted to low N environments and may have limited nitrate reduction capacities, such has been observed in gymnosperms. Since the 1920s, deposition of nitrogen-containing compounds in soils has increased as a result of industrialization and the burning of fossil fuels. Recent estimates of loading of nitrates in U.S. soils indicate that 15 kg ha^{-1} y^{-1} are deposited in midwest soils and 20 kg ha^{-1} y^{-1} in parts of the northeast (EPA 1991a). It has been suggested that there has been a gradual accumulation of N in soils to the extent that in natural ecosystems N may have become saturated for some plants. This is particularly true for old forest stands and species such as gymnosperms with low NO_3 reduction capacities (Garner 1994). Such changes could lead to competition between nitrophiles and plants less capable of using N resulting in shifts in dominant species within ecosystems. Examples where increased soil nitrogen may be affecting plant growth have been identified in Europe, where declines in forest tree species are correlated with increased NO_3 in soil solution compared to nondeclining areas (Waring 1987). In the Netherlands, excess N deposition has been correlated with the replacement of heathland vegetation with nitrophilous grass species. In the Great Smoky Mountains, nitrogen leaching from high-elevation forest soils, the presence of increased NO_3 in leaves and xylem sap of conifers have been considered indicative of N saturation in this area (Garner 1994).

Toxicity due to NO_x most likely is related to the inability of plants to detoxify NO_x via the nitrate reduction pathway. Low reduction capacity could lead to reduced cytoplasmic pH, which could impact numerous ion transport processes and affect general metabolism. Induction of deamination reactions of amino acids and nucleic acids could occur, as well as reactions with unsaturated lipids in membranes resulting in free-radical formation and deterioration of cellular membranes. Competition for reducing power (NADPH) between nitrate reduction and photosynthesis could also lead to reduced photosynthetic efficiency as well as to creating a deficit of carbon components required in the synthesis of amino acids and proteins. Inefficient reduction of NO_3 can result in the accumulation of NO_2 and NH_3, which are toxic. A buildup of NH_3, as a result of not being incorporated into amino acids, results in toxicity, probably through inhibition of the electron transport system of the light reactions of photosynthesis. In some cases, fumigations with NO_2 have been reported to stimulate rubisco activity, while in others, inhibition has been observed. Stimulation of photorespiration has also been observed, which could account, in part, for reduced photosynthetic efficiency.

Although some scientists believe that the current ambient levels of NO_x in the atmosphere play little to no role in causing injury or damage to plants, the deposition of NO_x into soils, as acid rain, could have a tremendous impact on the growth and distribution of plants in complex ecosystems. Some predicted responses to increased soil nitrogen levels include permanent increases in foliar nitrogen and reduced foliar phosphorus and lignin, reduced root biomass, reduced productivity of conifer stands, increased soil nitrification and nitrate leaching, reduced soil fertility as a result of increased cation and nitrate leaching, and increases in toxic levels of ions such as Al (Aber et al. 1989).

V. CARBON DIOXIDE AND OTHER GREENHOUSE GASES

A. Evidence and Sources of Rising Atmospheric CO_2 Levels

Its an indisputable fact that global CO_2 levels are rising. Estimates based on ice core analysis of trapped CO_2, ratios of isotopes of carbon from tree growth rings as well as recent direct measurements of atmospheric CO_2 indicate that, prior to the onset of global industrialization, atmospheric levels were approximately 275 ppm. Currently, CO_2 levels are around 350 ppm. This represents an increase of 25%. Estimates, based on current data, suggest that CO_2 is increasing by 1.5 ppm each year. Thus, by the end of the 21st century, CO_2 levels could be twice the current levels.

What are the sources of CO_2, and why is CO_2 increasing in the atmosphere? Like SO_2 and NO_2, CO_2 is part of a complex cycle (called the carbon cycle) in which carbon is transferred from the atmosphere to different sinks on Earth. Plants are an integral part of this process since they utilize CO_2 in photosynthesis and store CO_2 in the form of organic carbon-containing molecules. Thus, plants are important sinks for the recirculation of carbon from the atmosphere to Earth.

Over the hundreds of millions of years of Earth's geologic history, plants have played a major role in maintaining a balance of carbon between the earth and atmosphere, in that huge amounts of carbon have been "tied up" in geologic strata in the form of oil, coal, and natural gas reserves. Plants most likely played an integral role in reducing CO_2 levels during periods of Earth's history when CO_2 levels were elevated. Now, through the use of fossil fuels for generating electricity and providing fuel for transportation, as well as other needs, these previous reserves of carbon are now contributing to elevated levels of CO_2 compared with preindustrial periods of Earth's history. Thus, the carbon balance has shifted from storing carbon via biological systems on Earth to accumulation of CO_2 in the atmosphere.

The oceans, rivers, lakes, and streams also are very important components of the carbon cycle, in that they also act as sinks for CO_2. Carbon dioxide is soluble in water and is used by photosynthetic algae that live in aquatic systems and are thus involved in the consumption (sinks) of large amounts of CO_2. Plants provide food sources for all animals, and they in turn, like plants, respire to release carbon in the form of CO_2 back to the atmosphere. Thus, every living thing on Earth is part of the carbon cycle and contributes to changes in fluxes of carbon from Earth to the atmosphere. Not only is the burning of fossil fuels contributing to the increase in atmospheric CO_2, but also the removal and depletion of important plant sinks such as the tropical rain forests have become an increasing concern. It has been estimated that 70% of the terrestrial biomass is made up of woody plants that carry out two-thirds of terrestrial photosynthesis. Equally of concern to some is the increased pollution of our oceans, lakes, and streams, which represent the largest biological sinks for the recycling of carbon.

B. Effects of Elevated CO_2 on C_3, C_4, and CAM Plants

So why are rising CO_2 levels of concern to scientists? There are two reasons, both of which can affect plants and consequently all living organisms. One is that CO_2 increases are predicted to cause global warming through a process called the *greenhouse effect*. Secondly, CO_2, like SO_2 and NO_2, can be viewed as a nutrient for plants since it is required for photosynthesis. Thus rising CO_2 levels can affect plant growth directly by affecting photosynthetic rates, and indirectly by causing an increase in temperatures. Let's first consider the direct effect of CO_2 on plants and then consider the aspect of global warming.

Basically there are three strategies that plants use to fix CO_2. These include what are commonly referred to as the C_3, C_4, and CAM (crassulacean acid metabolism) photosynthetic pathways. The designations C_3 and C_4 represent the number of carbon atoms associated with the first stable intermediate formed after CO_2 is initially fixed in the respective metabolic processes. C_3 plants are considered to be the most primitive evolutionarily, and CAM plants perhaps transitional to the more advanced C_4 plants. These different plants are distributed latitudinally on a global basis with ranges overlapping. Very simplisticly, C_4 plants exist predominantly in the tropics and equatorial regions, C_4/C_3 mixes occur in temperate areas, and C_3 plants predominate in higher and polar latitudes. CAM plants are typically associated with dry, desert environments.

These three groups of plants differ in the ways that they acquire and process the CO_2 used in photosynthesis. All three have what is called the *Calvin cycle reactions* (C_3 reactions), where CO_2 is ultimately fixed to form simple sugars. The three groups differ as to how CO_2 is "fed" to the Calvin cycle and when. C_4 plants have essentially two mechanisms that fix CO_2, the Calvin cycle and the so-called C_4 metabolic reactions, or *Hatch and Slack photosynthesis*. The C_4 reactions are associated with the outer bundle sheath layer that surrounds the vascular bundles in the leaf. The C_3 reactions are localized in the mesophyll cells beneath the bundle sheath cells and are directly adjacent to the vascular tissue. This type of vascular arrangement of photosynthetic cells is referred to as *Krantz anatomy* and is unique to C_4 plants. The C_4 pathway, localized in the bundle sheath cells, initially fixes CO_2 using the enzyme phosphoenolpyruvate carboxylase (PEPcase) and forms either malic or aspartic acid, which is then transported from the bundle sheath to the mesophyll layer. There these acids are decarboxylated, releasing previously fixed CO_2 to the Calvin cycle, where it is fixed a second time by rubisco with the subsequent formation of sugars. The anatomical structure of the C_4 plant thus represents a type of pumping system where CO_2 can be transported to the mesophyll via malic and aspartic acid and maintain a relatively high partial pressure of CO_2, which is near saturation levels for rubisco in the mesophyll where the C_3 reactions occur. Why is this important? It is important because rubisco, in C_3 plants (Calvin cycle only) can fix not only CO_2 but also O_2. If O_2 is fixed, less carbon is incorporated into the manufacturing of sugars and, in fact, some carbon is lost since the molecules formed as a result of oxygenase activity enter the photorespiratory pathway, which ultimately leads to the loss of additional carbon as CO_2 during the process. Thus, the Krantz anatomy design, in C_4 plants, provides a mechanism whereby the CO_2/O_2 ratio is kept high and photorespiration is minimized or prevented. Also, the enzyme PEPcase does not exhibit oxygenase activity and is saturated at normal ambient CO_2 concentrations. In C_3 plants, photorespiration is high. The reason for this is that the Krantz anatomy and C_4 reactions do not occur. All photosynthetic cells in C_3 plants fix carbon using the Calvin cycle (C_3) pathway only. Thus, in C_3 plants, during high rates of photosynthesis when O_2 is being produced in the light reactions of photosynthesis, the CO_2/O_2 ratio shifts in favor of O_2, which supports more photorespiration. Loss estimates of net carbon fixed have been as high as 20–50% of previously fixed carbon when plants are grown at near ambient CO_2 levels. Crassulacean acid metabolism (CAM) plants are unique in that they close their stomates during the day and fix CO_2 at night. This mechanism allows desert plants to conserve water during the day while fixing carbon at night. Although CAM plants do not have the typical Krantz anatomy, they do have a mechanism similar to that in C_4 plants for fixing carbon at night, in that PEPcase is involved in the initial fixation process, and formation of malic acid and aspartic acid occurs. These acids are typically stored in the vacuole but are decarboxylated during the day to support the Calvin cycle. Thus, it appears that C_4 plants have developed a spatial strategy (Krantz anatomy) for increasing photosynthetic efficiency, while the CAM plants have developed a temporal strategy (stomatal opening at night). However, what little information is available for CAM plants suggests that they can benefit from increased nocturnal CO_2 levels.

Essentially, what all of the foregoing is trying to establish is that CO_2 is limiting in C_3 plants and that elevated CO_2 levels most likely would stimulate photosynthesis and hence growth in C_3 plants and possibly CAM plants. The stimulation of C_3 plants is a well-known fact that has been established for many years and has been used successfully to increase production in greenhouse operations. It is also known that C_4 plants do not respond to elevated CO_2 levels the way C_3 plants do. This relates, as previously explained, to the fact that the C_3 fixing enzyme, rubisco, is kept at or near CO_2 saturation as a result of the anatomical configuration and associated C_4 reactions surrounding the mesophyll where the C_3 reactions are restricted. So the question remains, how will increasing CO_2 affect plants and plant communities? Some researchers predict that most plants will benefit greatly because increased CO_2 will stimulate growth and plant productivity. In fact, when experiments using C_3 plants are conducted with twice the current levels of CO_2, photosynthesis can be increased by as much as 50%. It has also been observed that some plants maintain these high rates of photosynthesis for only short periods of time before dropping back to lower but still elevated levels of photosynthesis.

C. Plant Water Relations, Carbon Allocation, Nitrogen Metabolism, and High CO_2 Levels

Several factors play a role in how plants will respond to elevated CO_2 levels. One important factor involves stomatal sensitivity to elevated CO_2 concentrations in the atmosphere. It is well known that one of the control mechanisms that plants have developed to control stomatal conductance is a mechanism whereby internal CO_2 levels, in some manner, regulate stomatal guard cell turgor. High levels tend to cause stomates to close, and low levels cause stomates to open. Sensitivity to internal CO_2 levels may differ with different plant species. Such regulation creates a trade-off between water loss and CO_2 uptake by plants. However, in many plant species, partial closure of stomates does not necessarily limit the photosynthetic capacity of the plant. In those instances, partial closure may be beneficial relative to water conservation by the plant and hence higher production rates occur as a result of increased water-use efficiency. This may be of particular advantage to plants growing in drought-prone environments, in that plants subjected to higher CO_2 concentrations may exhibit increased productivity as a result of reduced transpirational water loss. In fact, it has been predicted that plants growing in xeric habitats may stand to gain the most from increased CO_2 levels.

Increased carbon assimilation is not always translated into increased growth or production. Plants have different capabilities, requirements, and mechanisms for transport, storage, and utilization of carbon assimilates. Complex interactions between these processes can translate into increased vegetative growth, increase allocation of resources toward reproduction, or essentially very little change resulting in acclimation to increased levels of CO_2. The assimilation of carbon by plants depends to a large extent on available sinks for the storage and on the ability to accumulate excess carbon, which could be influenced by stage of plant development, storage form and location in the plant, energy requirements for maintenance and growth, as well as environmental limitations to plant growth such as nutrients, light, and water. Nitrogen availability is particularly important in view of the fact that considerable nitrogen is allocated to the production of rubisco in C_3 plants. About 30–50% of the protein produced in C_3 plants is rubisco. Thus, nitrogen availability greatly influences plants' ability to produce this important protein used in the fixation of CO_2. In fact, it has been demonstrated that, in those cases where plants seemingly acclimate to increased CO_2 levels and have reduced capacities to store increased assimilates, the plants actually exhibit a reduction in the amount of rubisco produced. This, in essence, results in an increased C/N ratio in such plants and could influence the nutritive value for humans and other animals. Such changes have many ramifications relative to nutrient recycling since low nitrogen levels could result in lower populations of microorganism required in the degradation of organic plant materials. Herbivory would also require greater amounts of vegetative material to provide for the nutritional needs of insects and other animals, resulting in greater predation on plants and increased stress relative to carbon assimilation.

VI. CARBON DIOXIDE AND THE GREENHOUSE EFFECT

A. Greenhouse Gases and Global Warming

The so-called *greenhouse effect* occurs as a consequence of the accumulation of greenhouse gases in the upper layers of the troposphere surrounding Earth which results in trapping radiant energy within the troposphere (Figure 12.13). Some of the important *greenhouse gases* are CO_2, methane, chlorofluorocarbons (CFCs), nitrous oxide, ozone, and water vapor. These gases are especially effective since many absorb in the infrared region of the electromagnetic spectrum preventing the reradiation of heat out of the troposphere. This process is analogous to light penetrating the glass of a greenhouse and radiant energy in the form of infrared radiation being trapped, causing increased temperature. Figure 12.14 illustrates the relative contribution of each of the major greenhouse gases to the greenhouse effect.

It has been determined that increased CO_2 concentrations have been responsible for a rise in global temperatures of 0.5–0.7°C over the last 140 years. Based on current emissions of CO_2 into the atmosphere, it is anticipated that global temperature could increase as much as 1.5°C in the next 40 years. Greater increases in temperature are predicted to occur in higher latitudes of both hemispheres. Also, temperature differentials will be experienced more in winter months than summer, with subsequent impacts on seasonality. As a consequence, considerable change could occur in the regional distribution of plant and animal species as well as shifts in zones of agricultural production. Figure 12.15 illustrates the predicted change in distribution of two important weed species that occur in the U.S. East and South, if the CO_2 in the atmosphere were to double, causing an increase in temperature. It has been estimated that, for every 3°C increase in temperature, bioclimatic zones will shift latitudinally 600 km toward the poles and rise altitudinally by 150 m. In addition, for every 1°C increase in annual mean temperature, a 15–20-day increase in growing season will

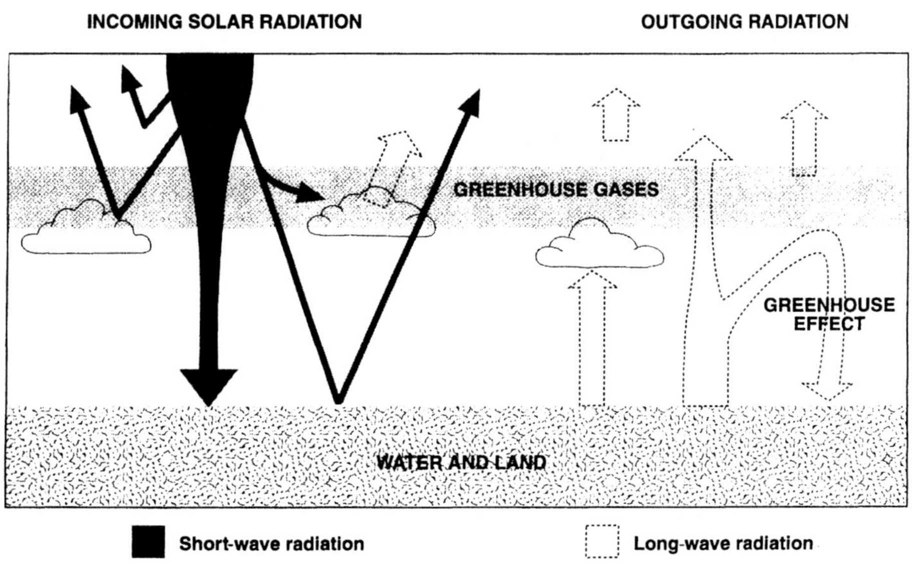

Figure 12.13. Generalized process of atmospheric trapping of radiation energy by greenhouse gases. (From Atkinson 1996.)

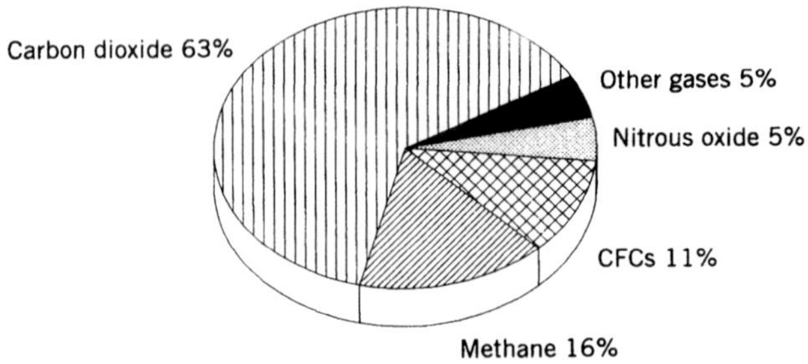

Figure 12.14. Relative percentages of major greenhouse gases. (From K. Strazalka and P. Ketner. Carbon dioxide. Pp. 393–456 in: M. N. V. Prasad (ed.), *Plant Ecophysiology.* Copyright © 1997 by John Wiley & Sons, Inc. Reprinted by permission of John Wiley & Sons, Inc.

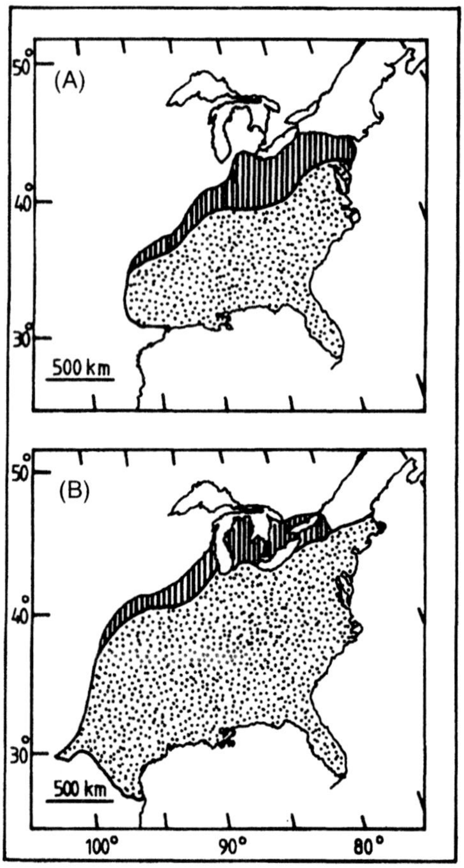

Figure 12.15. Current (*stippled*) and potential (*vertical lines*) ranges of kudzu (A) and Japanese honeysuckle (B) as a result of doubling of atmospheric CO_2 levels and subsequent global warming. (From Atkinson 1996. Reproduced from *Climate Change*, vol. 16 (1990), p. 44, fig. 7, from: Implications of atmospheric CO_2 enrichment and climatic change for the geographical distribution of two introduced vines in the USA, by T. W. Sasek and B. R. Strain. Copyright © 1990 by Kluwer Academic Publishers. Reproduced with kind permission from Kluwer Academic Publishers.)

occur, which will translate into 2–5% increase in global primary productivity (Smith and Lytle 1997).

B. Global Warming and Plant Growth and Distribution

The rate of temperature change is very critical to the survival and redistribution of plant species. Plants have the ability to acclimate to climatic changes on a short-term basis and adapt genetically on a long-term basis. Species plasticity and ability to migrate as climates change will impact plant species distribution and composition. If temperature change occurs faster than the ability of a species to migrate or adapt, it may become extinct. Plants with short life cycles, such as herbaceous annuals, will have an advantage over larger woody perennials relative to temperature increases. Thus, considerable changes in forest ecosystems may be anticipated relative to increased temperatures. It has been estimated that American beech and eastern hemlock can migrate northward at a rate of about 25 km per 100 years. It is estimated that, in order to keep up with the rate of current increase in global temperature, these species would have to migrate at a rate of 100 km per 100 years, or four times faster (Woodward 1992). These kinds of predictions should cause some concern relative to how natural and agricultural ecosystems may be affected by such rapid changes in temperature.

VII. SUMMARY

The result of increasing CO_2 will have both direct and indirect effects on plant growth, survival, and distribution. Some of the predicted direct effects are summarized in Table 12.4. In general, most plants will be affected in a positive manner relative to growth. However, this is a complex issue, and when increases in temperature are superimposed on the effects of elevated CO_2 concentrations, predictions become more complicated. As an example, it has been shown that C_4 grasses, grown under elevated CO_2 levels, were able to tolerate lower temperatures than plants grown under the same temperature and normal CO_2 levels. In the case of wheat, a C_3 plant that grows best in cooler climates, a higher temperature (3°C increase) and a doubling of CO_2 concentration resulted in reduced water usage and increased production compared with reduced yields when temperature was the only parameter increased (Nonhebel 1993).

Indirect effects, as a result of temperature increases, are likely to increase the competitive ability of C_3 plants relative to C_4 plants; when one considers that 95% of species are C_3 plants, this may have considerable impact on agricultural and natural plant ecosystems. It is also predicted that C_4 plants will most likely move out of tropical areas to high latitudes as the climate warms. Certainly geographical ranges of plants will change, with consequent changes in soil qualities and characteristics that will impact the establishment of natural and agricultural ecosystems. In addition, changes in temperature are likely to impact rainfall patterns and distribution, with droughts predicted for some areas and increased rainfall in others.

Plants are faced with an array of environmental pollutants that are increasing at rates unsurpassed in the history of Earth. How well plants are able to adapt, acclimate, or migrate in response to such rapid changes will determine their fate and, ultimately, the fate of all living things on Earth.

TABLE 12.4 Predicted Effects of Elevated CO_2 Levels on the Physiology, Morphology, and Phenology of Plants

Physiology
Increase:
Photosynthesis
Dark CO_2 fixation
CO_2 fixation in CAM plants
Dark respiration
Growth rate (dry weight)
Decrease:
Photorespiration
Transpiration
Stomatal conductance
Change in:
Photosynthate composition
Photosynthate allocation
Carbon partitioning
Photosynthetic pathway ($C_3 \rightarrow C_4$)
Morphology
Increase:
Root/shoot ratio
Lateral branch number
Leaf area
Root length
Fruit size
Petiole length
Decrease:
Flower size
Phenology
Leaf senescence delayed
Flowers produced earlier
Crop maturity accelerated
Fruit numbers increase
Number of seeds per plant increases
Induction of germination

Source: From K. Strazalka and P. Ketner. Carbon dioxide. Pp. 393–456 in: M. N. V. Prasad (ed.), *Plant Ecophysiology.* Copyright © 1997 by John Wiley & Sons, Inc. Reprinted by permission of John Wiley & Sons, Inc.

STUDY-REVIEW OUTLINE

Introduction

Types, Sources, and Abundance of Gaseous Air Pollutants

1. *Air pollution* is defined as aerial substances that have some adverse effects on plants, animals, and/or materials.
2. Substances emitted into the atmosphere come from both natural and anthropogenic sources.
3. Air pollutants can be gases, vapors, liquids, or solids.
4. The gaseous air pollutants of primary concern are CO_2, O_3, SO_2, and NO_x.

5. Major sources of gaseous air pollutants originate directly or indirectly from the burning of fossil fuels.
6. Carbon dioxide, SO_2, and NO_x can have nutritive value for plants.
7. Changes in the levels of atmospheric gases have occurred throughout the history of Earth, but current rates of change may prove difficult for plants relative to acclimation or adaptation.
8. Carbon dioxide levels have increased in the atmosphere from 1958 to 1990 by 0.35% per year, while SO_2 is accumulating at 4 % per year.
9. An estimated 75 million mt yr^{-1} of NO_x are emitted into the atmosphere from anthropogenic sources, and 5000 million mt of CO_2 are emitted from the burning of fossil fuels.

Photochemical Reactions and Gaseous Air Pollutants

1. Oxides of C, N, and S are of primary concern relative to the well-being of plants and animals.
2. *Primary pollutants* are those that come directly from a source, while *secondary pollutants* result from secondary sources such as photochemical reactions in the production of O_3.
3. The primary inorganic pollutants are CO_2, NO_2, and SO_2.
4. Ozone in the stratosphere protects plants and animals from harmful UV irradiation effects, while O_3 generated by photochemical reactions in the troposphere can have detrimental effects on plant and animal health.
5. Ozone, in the troposphere, is generated in a complex series of photocatalytic reactions involving light, NO_2, O_2, and volatile organic molecules (VOCs).

Physiological and Biochemical Effects of Ozone on Plants

General Considerations and Symptomology of O_3 Injury

1. Ozone levels increased by 0.6% per year from 1979 to 1988 in the United States.
2. Symptoms of O_3 injury to plants range from flecking, to stippling of the adaxial surfaces of angiosperm leaves, to chlorosis, mottling, tip dieback, and browning of gymnosperm needles.
3. Ozone, as well as other gaseous pollutants, can produce invisible, subliminal, or latent injury in plants that results in reduced growth and production before signs of visible injury occur.

Complexity of Air Pollution Interactions in Plant Communities

1. Considerable difficulty exists in determining the impact of O_3 on natural ecosystems compared with agricultural systems because of differences in plant composition and growth characteristics and requirements.
2. Ozone dosage, plant age, species differences, and environmental factors all contribute to the way plants respond to O_3 pollution.

Mechanism of O_3 Entry into Plants

1. Ozone enters plants primarily through the leaf stomates.
2. Barriers to diffusion of O_3 into plant leaves include aerodynamic resistance, boundary layer resistance, stomatal resistance, curticular resistance, and mesophyll resistance.
3. Factors affecting stomatal conductance affect how much O_3 can enter a plant leaf and potentially cause damage.

Chemical Reactions of O_3 with Plant Cellular Components

1. Once O_3 enters the apoplast of the cell, it reacts with numerous organic compounds (unsaturated fatty acids, amino acids, proteins, enzymes, etc.), disrupting metabolic and physiological processes.

2. Ozone also increases free-radical production in plants, which are also highly disruptive of biochemical and physiological processes in plants.

Metabolic Reactions Affected by O_3 and Implications for Plant Growth

1. It is likely that no single mechanism can be attributed to O_3 damage in plants, although the integrity of the plasmalemma may be an initial point of perturbation leading to aberrations in membrane function.

Ozone Resistance and Detoxification Mechanisms

1. Plants resistant to O_3 may have mechanisms to control O_3 entry through differences in stomatal physiology or biochemical mechanisms relative to the ability to generate antioxidants.

Physiological Functions Affected by O_3

1. Photosynthesis is very sensitive to O_3, which affects photosynthetic pigments, the light reactions, production and degradation of rubisco, and photorespiration.
2. Carbohydrates may accumulate in leaves of plants exposed to O_3 as a result of interference with phloem transport.
3. Ozone affects respiration variably depending on concentration, plant age, and species.

Sulfur Dioxide and Plants

Sources of SO_2 and Plant Symptomology

1. Sulfur dioxide is a primary pollutant generated by the burning of fossil fuels.
2. Cellular responses by different plant species to SO_2 toxicity are probably similar, but the symptomology varies, probably as a result of differences in anatomy and morphology.
3. General symptoms of SO_2 effects on plants include interveinal chlorosis, water-soaked pale green areas, necrosis near stomatal areas, and chlorotic banding in needles of gymnosperms.
4. Plants sensitive to SO_2 respond at concentrations of 1–1.5 ppm, while those plants that are tolerant are not affected at 2 ppm or higher.
5. Levels of SO_2 are higher in the winter months and concentrated near ridgetops.
6. Since plant metabolism is lower in winter months, the effects of SO_2 on plants are difficult to assess since lower metabolism could be either beneficial or detrimental to plants.

Sulfur Dioxide Uptake, Toxicity, and Metabolism

1. Since sulfur is normally required for plant growth, plants have developed metabolic pathways to incorporate sulfur into tissues and to control its uptake and distribution.
2. Rapid conversion of SO_2 into amino acids and transport compounds, such as glutathione, is essential; otherwise, the accumulation of toxic sulfur ions can lead to acidic conditions in the cell and the formation of free radicals that can affect enzyme activity and degrade membranes.
3. Sulfur metabolism can be regulated by transport of glutathione to the roots, which can be used to produce amino acids or possibly regulate the uptake of sulfate from the soil.
4. Sulfur can also be regulated in plants by storing SO_4^{2-} ions in the vacuole of the plant, while HSO_3^- can be reduced to form H_2S, which can escape from the plant into the atmosphere.

Nitrogen Oxides and Plants

Sources and Generation of Nitrogen Oxides in the Atmosphere

1. Nitrogen oxides are formed by a heat-mediated reaction between N_2 and O_2 that occurs normally in the atmosphere when fossil fuels are burned.

Plant Symptomology and Relative Toxicity of NO_x and SO_2

1. Nitrogen oxides are less toxic to plants than is SO_2.
2. Symptoms to NO_x vary in angiosperms and gymnosperms and are difficult to diagnose in the field unless plants are exposed to high levels near point sources.
3. Toxicity from NO_x can occur as a result of increased cell acidity in plant tissues.

Nutritional, Metabolic, and Ecological Implications of Nitrogen Dioxide

1. Nitrogen oxides are detoxified in the nitrate reduction pathway that is normally involved in the reduction of nitrates taken up in the roots of plants.
2. Varying abilities of plants to reduce nitrates may be a factor in their resistance to nitrogen oxides.
3. The inability of plants to reduce nitrates could lead to the accumulation of NH_3 or NO_2, which are toxic to plants tissues.
4. Although little direct evidence for NO_x injury to plants is available in natural or agricultural communities, the effects of NO_x may come through acidification and addition of excess nitrogen to soils, which could affect plant communities typically adapted to low nitrogen inputs.

Carbon Dioxide and Other Greenhouse Gases

Evidence and Sources of Rising Atmospheric CO_2 Levels

1. Atmospheric carbon dioxide levels are rising at a rate of 1.5 ppm yr^{-1} as a result of burning of fossil fuels.
2. The accumulation of CO_2 in the atmosphere results from a shift in carbon balance involving the burning of stored carbon resources in the form of fossil fuels and from the disturbance of important CO_2 sinks such as the destruction of rain forests and the pollution of aquatic environments that sustain important algal populations.

Effects of Elevated CO_2 on C_3, C_4 and CAM Plants

1. Increased CO_2 levels can affect plants directly by influencing photosynthesis, respiration, water relations, carbon allocation, and nitrogen metabolism which frequently impact growth in a positive manner.
2. C_3, C_4, and CAM plants respond in different ways to increased CO_2 levels. This reflects the different ways in which these plants acquire CO_2 and incorporate it into the Calvin cycle.
3. Elevated CO_2 levels will undoubtedly change the distribution and community structure of C_3, C_4, and CAM plants.

Carbon Dioxide and the Greenhouse Effect

Greenhouse Gases and Global Warming

1. Increased CO_2 will affect plant growth and distribution indirectly through global warming, commonly referred to as the *greenhouse effect.*
2. The greenhouse effect occurs as a result of CO_2 and other *greenhouse gases* accumulating in the upper layers of the troposphere, where they absorb far-red irradiation reflected back from Earth's surface and, in effect, trap heat energy in the troposphere.

Global Warming and Plant Growth and Distribution

1. As a result of the greenhouse effect, temperature will likely increase at higher latitudes, resulting in the migration of normally cool climate plants farther north and the invasion of warmer climate plants from more southern latitudes.

2. The rate of temperature increase is of some concern since plants may not be able to migrate, acclimate, or adapt fast enough to survive.
3. It is predicted that considerable reduction in genetic diversity may occur as a result of increasing temperatures and that herbaceous annuals may have a competitive advantage over perennials.

SELF-STUDY QUESTIONS

1. Define *air pollution,* and indicate various sources of air pollutants.
2. What are the air pollutants of major concern relative to plant and animal health?
3. Indicate how the major air pollutants have changed in recent years as well as through the course of Earth's history.
4. What is the difference between primary and secondary pollutants, and what are the primary inorganic pollutants?
5. Outline the photochemical reactions that lead to the production of O_3 in the troposphere.
6. Outline the symptoms of O_3 injury to plants, and discuss the various factors that could influence how plants respond to O_3.
7. Discuss plant factors that can influence O_3 entry into plants.
8. Describe the mechanisms by which O_3 damages plants.
9. Outline important physiological functions that are sensitive to O_3, and indicate why some plants may be able to resist higher O_3 levels than others.
10. Describe the symptoms of SO_2 toxicity in plants, and give the toxicity ranges for sensitive and tolerant plants.
11. Discuss the mechanism(s) of SO_2 metabolism and detoxification in plants.
12. How does SO_2 damage plants?
13. In what way might increased atmospheric SO_2 influence the uptake of SO_4^{2-} from the soil by the roots of a plant?
14. How are nitrogen oxides formed in the atmosphere, and what is their relative toxicity to plants compared with O_3 and SO_2?
15. How can NO_x act as a nutrient to plants?
16. How are NO_x gases detoxified by plants, and how can they damage plants if not detoxified?
17. What likely characterizes the ability or lack of ability of a plant to detoxify NO_x?
18. How might NO_x adversely affect plants indirectly through affecting their nutritional status?
19. Discuss how and why CO_2 levels are increasing in the atmosphere.
20. Describe how elevated CO_2 levels can impact the physiology of plants.
21. Describe how differences in photosynthetic strategies might influence plant community composition and distribution as a result of increasing CO_2 levels.
22. Describe how greenhouse gases may cause increased temperatures on a global basis.
23. Discuss how global warming will likely affect the distribution of plant species and the survival of plant communities.

SUPPLEMENTARY READING

Allen, Jr., L. H., M. B. Kirkham, D. M. Olszyk, and C. E. Whitman (eds.). 1997. Advances in Carbon Dioxide Effects Research. American Society of Argonomy, Madison, WI.

Alscher, R. G., and J. R. Cumming (eds.). 1990. Stress Responses in Plants: Adaptation and Acclimation Mechanisms. Wiley-Liss, New York.

Desjardins, R. L., R. M. Gifford, T. Nilson, and E. A. N. Greenwood. 1992. Advances in Bioclimatology. Springer-Verlag, New York.

Landis, W. G., and M.-H. Yu. 1995. Introduction to Environmental Toxicology: Impacts of Chemicals upon Ecological Systems. Lewis Publishers, Boca Raton, FL.

Schulte-Hostede, S., N. M. Darrall, L. W. Blank, and A. R. Wellburn (eds.). 1988. Air Pollution and Plant Metabolism. Elsevier Applied Science, New York.

Trapp, S., and J. C. McFarlane (eds.). 1995. Plant Contamination: Modeling and Simulation of Organic Chemical Processes. Lewis Publishers, Boca Raton, FL.

Unsworth, M. H., and D. P. Ormrod. 1982. Effects of Gaseous Air Pollution in Agriculture and Horticulture. Butterworth Scientific, Boston, MA.

PART IV
Conclusion

13 Generalities, Trends, and Future Directions

Outline

Objectives

1. Develop a case for the importance and reasons for studying plant stress physiology.
2. Establish why biodiversity is important to the health of ecosystems.
3. Discuss the importance of the integration of natural and agroecosystems.
4. Provide evidence that ecosystems, once degraded, are difficult to reclaim and are self-reinforcing.
5. Discuss how global temperatures may change quicker than previously thought and their impact on natural and agroecosystems.
6. Compare hierarchical levels of plant-ecosystem interactions with those of human society.
7. Discuss how new technologies can advance or hinder future efforts to conserve biodiversity and maintain healthy ecosystems.
8. Provide suggestions for further needs and directions of research in the area of plant stress physiology.

I. INTRODUCTION

In our first book, *Physiology of Plants Under Stress: Abiotic Factors* (Nilsen and Orcutt, 1996) we addressed the abiotic environment (light, temperature, drought, flooding, how other factors, and how they impact the physiology, growth, and development of a variety of plants associated with both "natural" and agricultural ecosystems. Although each stressful factor was treated separately, as we have done in this volume as well, we tried to give the reader, whenever possible, a sense of the importance of studying the interactions of multiple environmental factors and their effects on the physiology of plants. The current volume expands the emphasis to include soil processes (soil-plant interactions, nutrient and salinity stress, mycorrhizal effects), biotic factors (plant pathogens, herbivory, plant competition, parasitic angiosperms), and anthropogenic factors (soil and atmospheric pollution). To the reader who reads both volumes, it should become apparent that, after considering the abiotic factors that influence plant growth and then superimposing on these the biotic and soil factors, the complexity of such interactions, how individual plants and plant communities respond to such a multitude of insults, become staggering. Indeed, it becomes a wonderment as to how plants are able to acclimate and adapt to such a diversity of environments.

II. COMMUNICATING THE SCIENCE OF PLANT STRESS PHYSIOLOGY

The layperson, with only a passing interest in plants, may wonder why it is important to study plant stress physiology, or better yet, spend five to six years preparing two books on the subject. Certainly, there are many reasons to write books, but hopefully the most important is to provide an educational resource in a subject or discipline area. Indeed, we have experienced in recent years two U.S. Presidents who did not understand the meaning or implications of the importance of plant stress physiology relative to our agricultural economy and the sustainability of our natural ecosytems. One of us (D. M. Orcutt) is also reminded of an occasion on one of the morning talk shows when an economist was being interviewed. The discussion centered around how global climate change, in particular global warming, might impact the agricultural economy of the United States. The economist rightly stated that there was considerable controversy regarding whether or not global temperatures were increasing or decreasing, but astonishingly the economist went on to remark that it did not matter one way or another since agriculture was such a small part of the U.S. economy. As if this were not enough, the individual went on to suggest that we would be better off, economically, to spend our resources on developing and building townhouses on the beaches of the United States and that such endeavors would result in greater economic returns.

Another instance occurred during a seminar given by a visiting professor some years ago. During his discussion, he told about a course he taught to undergraduate students in which he liked to take students out to one of his field plots, let them plant some seed of their choice, and have them watch and observe the plants as they grew. One student was observed planting some radish seeds in a peculiar fashion. Rather than placing the seeds in linear rows, the student was sowing them densely in small, patchlike circles. When asked the reason for sowing the seeds thusly, the student replied that radishes come in the supermarket in bunches about the same size as the student was sowing.

None of these examples is intended to reflect negatively on politicians, economists, or students in any way. However, it should suggest to us, as plant scientists, the need to communicate the importance of our subject to all kinds of people—from highest political levels of our society to the grade school student just beginning to explore and learn about this extremely complex planet on which we live.

III. RECENT REVELATIONS AND PREDICTIONS

The importance of communicating with the public at large is further emphasized by the current alarming rate at which environmentally sensitive areas of the Earth are being affected by anthropogenic activities. This is best reflected in the initial statement by F. S. Chapin III and colleagues (Chapin, et al. 1998), in which the first three sentences state:

> The earth is currently in the midst of the sixth major extinction event in the history of life. The causes of earlier extinction events (e.g., the extinction of dinosaurs) are uncertain but probably reflected changes in the physical environment caused by such factors as meteor impacts or volcanism. By contrast, the current extinction event is biotically driven—specifically by human impact on land use, species invasions, and atmospheric and climatic change.

In this paper, Chapin and his colleagues point out the need and importance of research related to the retention and promotion of biodiversity with respect to the health of ecosystems. They suggest, hypothetically, that the biodiversity of ecosystems has important implications relative to

1. Maximizing and retaining resources at each trophic level
2. Reducing the risk of large environmental changes as a result of directional or random events in the environment
3. Reducing the probability of large changes in ecosystem processes in response to invasions of pathogens and other species
4. Understanding specific habitat requirements in heterogenic landscapes that promote the existence of unique biota in isolated areas that may be affected by habitat fragmentation

Currently, there are an estimated 10 million species of plants and animals in the world, of which about 380,000 are plant species (including fungi, bacteria, lichens, diatoms, vascular plants, liverworts, and mosses). It is estimated that species are being eliminated at the rate of 150 per day, and that in 20 years, one-tenth of all plant and animal species will be extinct (Pimentel et al. 1992). Interestingly, the largest proportion of biodiversity is represented by arthropods, which comprise about 9 million species. When one considers global biodiversity, one cannot neglect the biodiversity that occurs in managed ecosystems that are used for human food and fiber sources. Such sources represent 70% of the temperate and tropical ecosystems and provide 98% of human food and fiber. Table 13.1 shows how biodiversity compares in agricultural and forested ecosystems and the variation that occurs in the respective ecosystems relative to arthropod populations. Unfortunately, it is from agroecosystems that many problems arise for other ecosystems. These problems include pollution from agrochemicals, disease dissemination, soil erosion, water quality and amount, land utilization, fragmentation of habitat, among others. Hopefully, the implementation of "sustainable" practices in agriculture, the recognition of the importance of biodiversity in agroecosystems, and a better integration and increased compatibility of agriculture within existing biomes will make for a more balanced and stable environment that can sustain humankind.

Rapport and Whitford (1999) classify anthropogenic stresses on ecosystems into four categories: (1) physical restructuring (changes resulting from land use); (2) introduction of exotic species; (3) discharge of toxic substances to air, land, and water; and (4) overharvesting. The authors suggest these stresses can result in degradation of the ecosystem, resulting in a "distress syndrome" characterized by less biodiversity,

TABLE 13.1 Arthropod Biological Diversity in Various Ecosytems

Ecosystem	Location	Number of Arthropod Species
Collards	New York	262 ha^{-1}
Corn and woodland	Italy	200–500 ha^{-1}
Alfalfa	New York	600 ha^{-1}
Corn	Hungary	600 $field^{-1}$
Pasture	Britain	1000 ha^{-1}
Forest, tropical	Bornea	2800 $(10\ trees)^{-1}$
Forest, beech	West Germany	1500–1800 $forest^{-1}$
Forest, park	Hungary	4433–8847 $forest^{-1}$
Forest, tropical	Costa Rica	13,000 $(10,800\ ha)^{-1}$

Source: p. 356 in D. Pimentel, U. Stachow, D. A. Takacs, H. W. Brubaker, A. R. Dumas, J. J. Meaney, J. A. S. O'Neal, D. E. Onsi, and D. B. Corzilius. 1992. *BioScience* 42(5):354–362. © 1992 American Institute of Biological Sciences.

reduced primary and secondary production, and a low ability of the ecosystem to recover from natural perturbations. Based on studies they conducted on three very different regional ecosystems having histories of exposure to multiple anthropogenic and natural stresses, they made the following conclusions:

1. Mechanisms that promote degradation of ecosystems, after a certain point, set in motion a self-reinforcing system that creates further degradation, even after the stress that set the process into motion is removed.
2. Ecosystems in alternate transformed states may be highly resistant to further alteration, such as attempts at reclamation.
3. The most effective way to ensure healthy ecosystems is to limit stress pressures before self-reinforcing degradative processes are set in motion.

Potentially compounding the future scenario of these degradative processes, is the recent realization that climatic change in the past may have occurred more quickly than scientists originally estimated (Taylor 1999). Based on ice cores from Greenland, the Antarctic, and ocean sediment analysis, scientists have discovered what appears to be evidence that, in the past 110,000 years of Earth's history, there have been episodes of increases in temperature of as much as 5–10°C in as little as a 20-year period. In one instance, a 4°C increase in surface temperature was recorded in a 10-year period. These types of changes in temperature were likened to shifting agricultural activities 500 miles south of their current locations. The reasons for such shifts in temperature were attributed to changes in the ocean circulation patterns involved in the exchange of heat energy between the oceans and the atmosphere. It is not exactly clear what can trigger changes in ocean circulation patterns, but such changes may be due to changing surface temperature related to atmospheric trapping of solar energy or to a change in Earth's orbital patterns. If increases in atmospheric temperature occur, it is anticipated that the hydrologic cycle will be stimulated causing dilution of Earth's oceans, which may trigger changes in ocean circulation patterns. These would then affect atmospheric and oceanic heat-exchange patterns in different areas of the world, making some areas in northern latitudes dryer and cooler and shifting growing regions farther south.

This hypothesis predicts that global warming, ironically, is setting the stage for another ice age. Further, it is pointed out, that human activity in the last 100 years has increased the concentration of greenhouse gases to their highest levels in 260,000 years, setting the stage for increased global warming. If, indeed, significant climatic changes of 5–10°C were able to occur in a time span of 10–20 years, the authors question how human society will be able to deal with the obvious social and economic disruption that most likely would occur in such a short time span, and how natural ecosystems will be able to adjust. The ability to adjust to such rapid changes certainly will depend, in part, on the degree to which various ecosystems are degraded. In addition, paleocological records indicate that plant migration was able to occur throughout the Quaternary period (Allen 1996), which was characterized by 10 complete glacial-interglacial cycles over the last 2 million years, each cycle lasting approximately 10,000 years. During these cycles it has been estimated, based on pollen identification and carbon dating, that trees had the capacity to migrate anywhere from 100 to 1000 m per year. Assuming the best-case scenario (1000 m yr^{-1}) means that the fastest-migrating tree species would be able to migrate about 12 miles in 20 years if one assumes that Earth's climate can change as quickly as suggested by Taylor (1999). It is indeed questionable whether plant migrations could keep pace with such rapid changes in temperature considering that temperature and growth zones would be shifted as much as 500 miles to the south.

IV. PLANT-HUMAN HIERARCHICAL RELATIONSHIPS AND "SILVER BULLETS"

The ability to protect our ecosystems from irreversible degradation will depend much on the priorities humankind places on individual standards of living as well as decisions made by groups in various hierarchical levels of society. Figure 13.1 illustrates such a relationship on a spatiotemporal scale. In this diagram, Giampietro (1994) illustrates the increasing complexity of interacting relationships that occur in our society, all of which directly or indirectly ultimately impact the biosphere. Giampietro states in this paper that

> Within a hierarchical perspective, human-generated ecological problems are generally the result of a lack of historical perspective (the adoption of too small a scale) when assessing technological innovations. By adopting a narrow anthropocentric point of view and ignoring the rest of the natural system to which they belong, humans analyze only what bothers them at the moment and look for solutions that fix the problem as quickly as possible.

These quick fixes to problems Giampietro refers to as "silverbullets." Examples given of this syndrome included references to higher costs of fuel resulting in more fuel-efficient cars, which promoted more leisure driving, which encouraged the building of more roads, which further encouraged more driving of personal vehicles, which ultimately led to traffic problems and greater pollution. The development of higher-yielding varieties of wheat, rice, and other foods associated with the "green revolution" for the purpose of feeding the world's growing population only stimulated

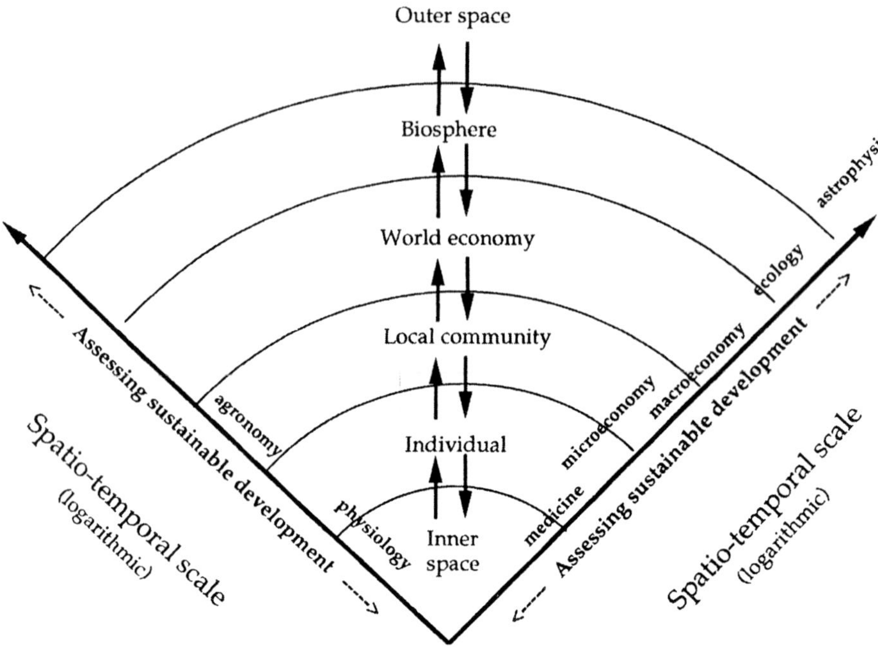

Figure 13.1. Hierarchical relationships among socioeconomic factors that affect biosphere integrity as expressed on a spaciotemporal scale. (From fig. 2 in M. Giampietro. 1994. *BioScience* 44(1):667–689. © 1994 American Institute of Biological Sciences.)

further increases in population growth. Giampietro sees genetic engineering as potentially another "silver bullet" in that "Genetic engineering aimed only at increasing economic return and technological efficiency is likely to further lower the compatibility of human activity and natural ecosystem processes." He suggests that the ability to rapidly insert genes into plants and animals and then releasing them into the environment changes the dynamics of how such organisms interact with other components within that ecosystem. These components have coevolved on a much longer time scale, which allowed for the stabilization and integration of each organism within its ecological niche. Introduction of a genetically altered organism that has not coevolved with other organisms in the ecosystem, during a similar time scale, could potentially destabilize the ecosystem. This could then select for populations of deleterious organisms that, in an environment of reduced biodiversity, be difficult to control.

It is interesting to note that the hierarchical structure depicted in Figure 13.1 has similarities to the one illustrated in Figure 13.2. However, in the latter scenario, the hierarchy suggests relationships at varying levels of complexity relative to stomatal function and its impact on higher levels of plant organization and subsequently on ecosystem function. This progression of suggested relationships illustrates very clearly why it is important to study plants at all levels of organization, and to further carry that understanding to a broader scale of organization in an attempt to explain the importance of stomatal function at the ecosystem level. Extrapolation of function from the biochemical to the ecosystem level is a formidable task, but one that is necessary if we are to understand the complex relationships that exist between individual plants and ecosystem function. Figure 13.2 is similar to Figure 13.1 in that each portrays a different lifeform and how different hierarchical levels affect the biosphere.

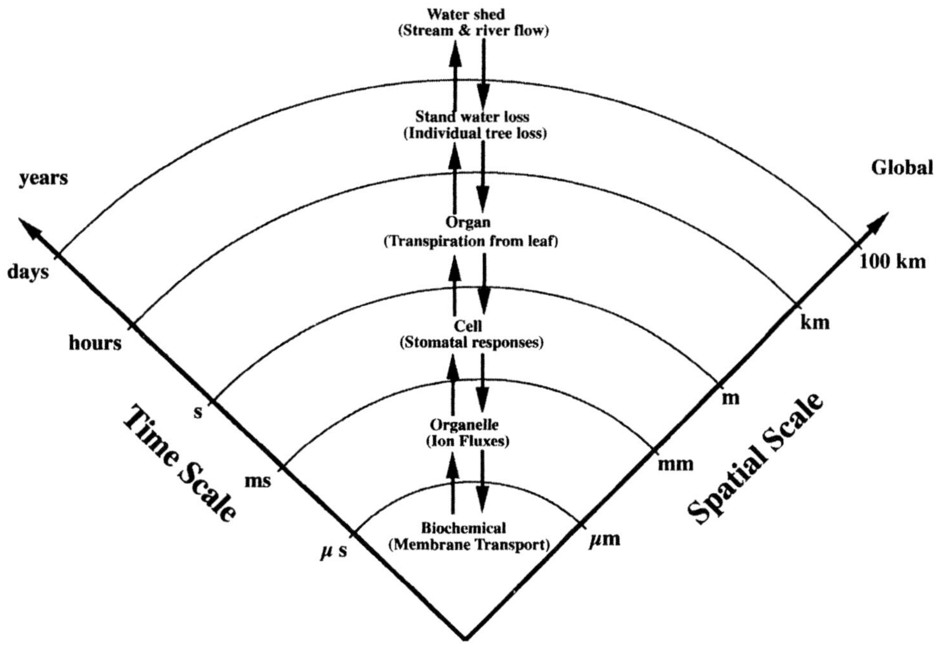

Figure 13.2. Relationship of stomatal function to higher levels of organization at the whole plant, community and ecosystem levels.

Why can't we humans see the forest for the trees? Perhaps part of the problem is that we may be a bit short sighted and lack perspective on such issues, or perhaps we elevate ourselves above other lifeforms on Earth, which results in an attitude of superiority and a false sense that we can understand, fix, and adjust to any problem. On a more scientific level, we tend to study plants in isolation since it is easier to control variables that way. In reality, plants are exposed to numerous interactions in natural ecosystems (combined effects of all the biotic and abiotic factors together). Such ecosystems are in constant flux, some changing faster than others depending, particularly, on anthropogenic factors. Thus, the challenge of predicting and modeling how ecosystems interact is obvious. Only a few years ago, scientists were in a kind of gridlock with respect to studying how hormones control developmental processes. We appeared to have reached a point at which not much progress could be made in understanding how plant hormones (plant growth regulators) are regulated and, specifically, in getting definitive answers to how they interact and control plant processes. Development of new technologies related to isolation, chemical detection, and, particularly, biotechnology have opened new doors in studying the effects of plant processes and understanding growth and development in plants. How this new technology will advance our understanding at the ecosystem level has yet to be determined. Hopefully, studying multiple stresses on plants and achieving a better understanding of how plants control development at the molecular, biochemical, and physiological levels will provide new and better information, which can be used for developing mathematical models to more accurately predict responses to changes in ecosystems. Hopefully, these new technologies, as pointed out by Giampietro (1994), will be used beyond the level of quick fixes, individual self-indulgence, and corporate profits—but instead at the level of preserving the natural capital stored in the complex structure and functions of the biosphere, which all humankind depends on for its existence and which cannot be substituted for by human-made capital.

Its interesting to note the results of a recent survey, conducted annually since 1966, by the Cooperative Institutional Research Program of the Higher Education Research Institute at the University of California, Los Angeles. This survey indicates that, at least at our institution (and this most likely reflects attitudes at many other colleges and universities), the number one motivation among students for getting a college education was to make more money and to get a better job. This priority was followed by becoming an authority in my field, obtaining recognition from colleagues, developing a meaningful philosophy of life, promoting racial understanding, making a theoretical contribution to science, and, finally, being involved in environmental cleanup. All categories have declined in importance to students since 1994 except the number one priority—making more money—which has *increased* in importance. (Is it old fashioned or to idealistic to consider that one goes to college to further one's education and to acquire new knowledge? Interestingly, this was not even one of the questions asked on the survey.) Such attitudes further support the notion, presented by Giampietro (1994), that "Because the aspirations of individuals for a better standard of living generally have a stronger effect on decision makers than the requirement of ecological systems for lower human pressure, technological fixes tend to never solve the problems at the higher level (i.e. ecological)."

V. FUTURE TRENDS AND DIRECTIONS

It should be stated that all biotic stresses covered in this book certainly could benefit from additional research efforts, and we leave it to researchers working in those fields to identify more specific needs. However, as we prepared this book, several general concepts or areas struck us as important areas of concern, and a few of these will be discussed below.

First, it appears not as much is known about the interactions of biotic stresses on plants as is known about how abiotic stresses impact the physiology of plants. This perhaps reflects the way plant physiologists, plant pathologists, entomologists, and others with interests in how plants are affected by biotic factors are trained. In most instances, scientists are trained as specialists in a particular field and, as a consequence, plant physiologists have an expertise in how plants function but little knowledge concerning pathogens, insects, and so on. The converse can be said for pathologists and entomologists. As a result, the literature reflects a less in-depth understanding of how biotic stresses impact the physiology of plants compared with abiotic stress. Most information centers around genetics, plant breeding for resistance, and mechanisms of resistance to pathogens and insects, with little definitive evaluation of physiological responses. Consequently, there is a great need both to educate students in a broader sense and also to promote a greater cooperative effort in teaching and research in these areas.

Second, we know a great deal about how individual plant processes respond to single components of the environment but very little about how multiple interactions can affect individual plants or plant communities. Is ecophysiology at a similar stage in its evolution as plant hormone research was several years ago—waiting for technological breakthroughs so that the science could again advance? Can biotechnology have similar influences on the understanding of how plants interact in complex ecosystems? To do so will mean fostering a cooperative—and not a competitive—spirit among molecular biologists, ecologists, and ecophysiologists, in much the same way as plant physiologists, pathologists, and entomologists need to cooperate if these new technologies are to have a significant impact on higher-order ecosystem interactions. Again, Giampietro (1994) summarizes it well: "Will agricultural crops engineered to fix nitrogen, resist pests, and/or tolerate dry conditions be used to reduce the pressure on the environment (the higher hierarchical level) or rather to alleviate the pressure on the poorer part of humankind—the lower hierarchical level?"

Third, our understanding of soil-plant interactions is very much lacking. This area represents one of the least-studied topics, in particular the interactions of vascular plants and soil biota. This is better expressed by a quote from Wall (1999): "The exploration and discovery of species in soils and in fresh water and marine sediments is just beginning. These interconnected subsurface habitats remain a frontier when compared to the above ground world." Wall goes on to document that there may be more subsurface species of organisms than aboveground species and that 98% of those are unknown. Current increases in soil and water pollution make this area an extremely important research emphasis. Degradation of soils through heavy-metal pollution from pesticides, fertilizers, and airborne pollutants could potentially cause problems for hundreds, if not thousands, of years into the future. Heavy metals can be accumulated in plants and passed on to higher trophic levels eventually reaching humans. Increased CO_2 levels can potentially affect belowground organisms and their associations by increasing carbon exudation by roots or simply by changing biomass accumulation in soils. How are soil pollutants going to affect mycorrhizal fungi and nitrogen-fixing bacteria, which are so important for supplying nutrients to both natural and agroecosystems? How gaseous air pollutants will affect soil biota and their interaction with vascular plants are virtually unknown. Soil degradative processes such as salination, nutrient leaching, and wind and water erosion further compound the problems of conserving our important soil resources.

Much is known about how pesticides affect the physiology of a limited number of crop plants and weed species associated with their cultivation. However, much less is known about the influence of such chemicals on natural plant ecosystems. In addition, other organics are being emitted into our environment from industrial sources for which we have no indication what their impact might be on plants in natural or agroecosystems.

The concept of allelochemical interactions between plants has been of longstanding importance and discussion in plant research. However, isolating and identifying specific agents involved in allelochemical interactions has been complicated and elusive. This is an extremely important area of research, with implications for plant succession as well as potential for applied applications in agroecosystems. Perhaps genetic engineering will provide the necessary technology to identify genes that may be responsible for the production of plant exudates involved with allelochemical interactions. Such findings could be important in developing "naturally produced" herbicides in agricultural systems as well as providing insights into how such chemicals operate in natural ecosystems.

Much greater insight is needed with respect to how agricultural and natural ecosystems interact and how the two can be integrated so that, ideally, each compliments the other. This is not out of the realm of possibility, for it is being done on some sustainable farms in the tropical rain forests of Australia and the temperate rain forests of New Zealand. There is now an even greater need for the scientists of the agricultural community and scientists studying natural plant systems to cooperate and exchange ideas as to how this can be best accomplished. Making this to work will necessitate cooperation among scientific disciplines, better education of the general public, and a readjustment of personal attitudes and goals related to individual standards of living, particularly in more affluent societies.

VII. SUMMARY

There is an urgent need to communicate to the public at large the importance of scientific research in the area of plant stress physiology. This research is fundamental to the survival and health of complex ecosystems on which every living organism, including humans, depends for survival.

Human attitudes regarding the use and consumption of resource capital without providing mechanisms to sustain or replenish the resource base will undoubtedly result in rapid degradation of ecosystems. Whether the use of new technologies, such as genetic engineering, will help solve problems associated with conservation and management of resources will depend in a large part on shifting attitudes away from anthropocentric concerns to a more holistic emphasis on the conservation and sustainability of living organisms. The potential for rapid changes in climatic conditions and the current rates of ecosystem degradation should be cause for considerable concern for future generations of humans as well as other living creatures.

In the future, a more concerted effort should be placed on educating all sectors of society, globally, regarding issues relating to the importance of maintaining healthy ecosystems. Unfortunately, many of the decisions that affect the environment are economically and politically motivated. Individuals, politicians, and corporations should recognize that being environmentally prudent is economically benefical now and in the long term. For example, how many billions of dollars are spent globally every year for treatment of air pollution–related health problems? How much air pollution–related productivity is being lost in argoecosystems that would aid in feeding and providing shelter for the world's population and generate revenues for those industries. What are the costs of using fertilizers and pesticides relative to human health, degradative processes associated with groundwater contamination, pollution of the world's fisheries, to name only a few? In many cases, we may not even know what the implications might be.

Finally, study of plant-soil interactions, multiple plant stress–ecosystem interactions, biotic stress effects on the general physiology of plants, the impact of pesticides and other organics on natural and agroecosystems, new approaches for studying allelochemcial interactions in plants, and a better understanding of the interactions and integration of natural and agroecosystems are suggested as future research needs.

STUDY-REVIEW OUTLINE

Introduction

1. The study of plant stress physiology entails multiple facets of the biotic and abiotic environment.
2. The study of abiotic and biotic factors and their effects on plants is extremely complex and poses many problems relative to understanding plant responses.

Communicating the Science of Plant Stress Physiology

1. Considerable misunderstanding of the meaning and importance of plant stress physiology exists in the public at large.

2. A concerted effort by scientists needs to be made to educate the public about the relationship of plant stress physiology to ecosystem functioning and health.

Recent Revelations and Predictions

1. The Earth, according to some, is experiencing the sixth major extinction event in its history.
2. Anthropogenic activities are the major cause of the current depletion of biodiversity.
3. Biodiversity is important relative to the retention of resources, prevention of environmental degradation, prevention of invasion by exotic species, and maintenance of landscape niches.
4. It is estimated that 150 species become extinct daily and that $\frac{1}{10}$ of current species will be extinct in 20 years.
5. Agroecosystems contribute significantly to biodiversity but also are a cause of considerable degradation of natural ecosystems.
6. Hopefully, sustainable agriculture and a better integration of natural and agricultural systems will help retain and replenish natural resource bases and increase biodiversity.
7. Degrading natural ecosystems reach a point where they become self-reinforcing and difficult to reclaim.
8. The best approach to preventing environmental degradation is to remove the source of the problem as soon as possible.
9. Global warming could potentially increase as much as 5–10°C in a 10- to 20-year period, based on ice core and ocean sediment analysis.
10. Changes in global temperature patterns are related to exchange of heat between the surface and ocean waters, which in turn controls ocean current patterns.
11. Ocean current patterns impact rainfall and temperature patterns on landmasses.
12. Changes in surface temperatures can be affected by changes in insolation related to atmospheric gases, as well as shifts in the Earth's orbit around the Sun.
13. Based on paleoecological studies, some tree species are able to migrate 100–1000 m yr^{-1}.
14. Based on a worse-case scenario, if global temperatures increased 10°C in 20 years, growth zones will shift 500 miles south, while maximum tree migration would be only about 12 miles.

Plant-Human Hierarchical Relationships and "Silver Bullets"

1. According to some scientists, quick fixes to immediate problems have historically been the focus of technological developments, rather than dealing with long-term holistic consequences.
2. Some scientists believe that the ability to rapidly change the genetics within and across species using genetic engineering technologies has the potential for selecting for populations of deleterious organisms that, in a population of declining biodiversity, could be difficult to control.
3. An examination hierarchical relationships of human interactions with the biosphere, and individual plant relationships with higher, more complex ecosystems, suggest a lack of understanding at higher scales of complexity.
4. Understanding ecosystem function will require a concerted effort toward elucidating the effects of multiple environmental stresses on the physiology of plants.
5. The wise use of genetic engineering could provide the necessary technological breakthroughs to expand our knowledge of complex ecosystem function.
6. Conversely, the unwise use of biotechnology could enhance the degradation of already fragile ecosystems.

Future Trends and Directions

1. A greater cooperative effort is required between specialists in the fields of plant stress physiology, pathology, entomology, biotechnology, herbivory, and other fields in order to expand our knowledge of how biotic stresses impact the general physiology of plants.
2. More carefully designed studies need to be initiated that deal with multiple stresses on the physiology of single plants and plant communities.

3. Greater emphasis should be placed on soil-plant interactions, particularly in the areas of heavy metal pollution from pesticides, fertilizers, and air pollution.
4. The impact of gaseous air pollutants on soil microflora and -fauna and the relationship of this pollution to vascular plant growth needs further clarification.
5. Little is known regarding the impact of pesticides and other organic pollutants on natural plant communities.
6. New approaches need to be developed for detecting, isolating, and determining the functions of allelochemicals in plants.

SELF-STUDY QUESTIONS

1. What are the major problems associated with studying plant stress physiology in a natural field setting?
2. Cite some examples suggesting that more needs to be done to educate the general public on the importance of research dealing with plant stress physiology and ecophysiology.
3. What evidence is there that we may be experiencing the sixth great extinction event in the history of the world?
4. How many plant and animal species exist on Earth, and what is the current estimated rate of species extinction?
5. How do natural and agroecosystems interact, and what might be the advantages and disadvantages of each on the other?
6. What four anthropogenic stresses may influence ecosystem degradation?
7. Once ecosystem degradation is set in motion, what problem can occur, and what is the best way to prevent degradation from starting?
8. What does recent ice core and ocean sediment data tell us about global temperature changes in the past history of Earth?
9. How might changes mentioned in the above questions impact growing regions and migration of plant species?
10. According to some scientists, how have members of our society used technological advances in solving problems? Give some examples.
11. Indicate how genetic engineering might accentuate the degradation of already fragile ecosystems.
12. How might biotechnology further advance our understanding of complex interactions within ecosystems?
13. Indicate several directions, potentially important for the advancement of the science, that plant physiological research might take in the future.

SUPPLEMENTARY READING

Baldwin, I. T., and C. A. Preston. 1999. The eco-physiological complexity of plant responses to insect herbivores. Planta 208:137–145.

Chapin F. S. III, A. J. Bloom, C. B. Field, and R. H. Waring. 1987. Plant responses to multiple environmental factors. BioScience 37:49–57.

Chen, J., S. C. Saunders, T. R. Crow, R. J. Naiman, K. D. Brosofske, G. D. Mroz, B. L. Brookshire, and J. F. Franklin. 1999. Microclimate in forest ecosystem and landscape ecology. BioScience 49:288–297.

Dahlberg, K. A. 1991. Sustainable agriculture—fad or harbinger? BioScience 41:337–340.

Groffman, P. M., and P. J. Bohlen. 1999. Soil and sediment biodiversity. BioScience 49:139–148.

Osmond, C. B., M. P. Austin, J. A. Berry, W. D. Billings, J. S. Boyer, J. W. H. Dacey, P. S. Nobel, S. D. Smith, and W. E. Winner. 1987. Stress physiology and the distribution of plants. BioScience 37:38–48.

Pimentel, D., M. S. Hunter, J. A. LaGro, R. A. Efroymson, J. C. Landers, F. T. Mervis, C. A. McCarthy, and A. E. Boyd. 1989. Benefits and risks of genetic engineering in agriculture. BioScience 39:606–613.

Pimentel, D., J. Houser, E. Preiss, O. White, H. Fang, L. Mesnick, T. Barsky, S. Tariche, J. Schreck, and S. Alpert. 1997. Water resources: agriculture, the environment, and society. BioScience 47:97–106.

Poster, S. L. 1998. Water for food production: Will there be enough in 2025? BioScience 48: 629–637.

PART V
Appendices

APPENDIX A
Symbols and Abbreviations

ABA	abscisic acid
ACC	aminocyclopropane-1-carboxylase
A_p	plow layer
ARSK1	*Arabidopsis* root specific kinase gene 1
ATP	adenosine triphosphate
BA	benzylisoquinoline alkaloids
BADH	betaine aldehyde dehydrogenase
c	species-specific constant used to calculate density-dependent competition
^{14}C	carbon isotope 14
C_i	intensity of competition from the joint response of two species to increase density
^{45}Ca	calcium isotope 45
CAD	cinnamyl alcohol dehydrogenase
CAM	crassulacean acid metabolism
CaN	calcineurin
CDPK	calcium-dependent protein kinase
cm	centimeter
CMO	choline monooxygenase
d	day
2,4D	2,4-dichlorophenoxyacetic acid
DDT	dichlorodiphenyltrichloroethane
D_e	diffusion coefficient
DHAP	dihydroxyacetone phosphate
DIBOA	2,4-dihydroxy-(2H)-1,4-benzoxazolin-one
dN/dt	instantaneous population size at time t
DNA	deoxyribonucleic acid
dS	decisieman
E_c	electrical conductivity
E_h	redox potential
ENA1 (PMR2)	gene that clones for a Na^+/H^+ antiporter in the plasma membrane of yeast
EPA	Environmental Protection Agency
FIR	fluoroindole resistance
g	gram
GDH	glutamate dehydrogenase
GS	glutamine synthetase
GOGAT	glutamate synthetase
GS-GOGAT	glutamine synthetase–glutamate synthetase
ha	hectare

HKT1	H^+/K^+ transporter 1 gene
IAA	indoleacetic acid
Imt1	gene for inositol-O-methyltransferase
JA	jasmonic acid
JH	juvenile hormone
K_i	carrying capacity of the ith population, number of individuals an environment can support
kD	kilo Dalton
kg	kilogram
K_m	substrate concentration at 0.5 maximum velocity
l	liter
LA	linolenic acid
LAR	leaf area ratio (leaf area/total plant weight)
LH II	light harvesting complex II
m	meter
M	mycorrhizal
MA	monoterpenoid alkaloids
mEq	milliequivalent
mg	milligram
MIP	membrane intrinsic protein
MLO	mycoplasma-like organism
m*M*	millimolar
mmho	millimhos
mmol	millimoles
mol	mole
molal	molality
MPa	megapascal
^{15}N	nitrogen isotope 15
N_i	number of individuals in the ith population
NADPH	(reduced) nicotinamide adenine dinucleotide phosphate
NDGA	nordihydroguiairetic acid
nl	nanoliter
NM	nonmycorrhizal
NO_x	nitrous oxides
ρ	log density of plants surviving after density-dependent mortality
^{32}P	phosphorus isotope 32
P.t.	*Pisolithus tinctorius*
P5CDH	Δ-pyrroline-5-carboxylase dehydrogenase
P5CR	Δ-pyrroline-5-carboxylase reductase
P5CS	Δ-pyrroline-5-carboxylase synthetase
PA	pyrrolizidine alkaloids
PAHs	polycyclic aromatic hydrocarbons
PAL	phenylalanine ammonia lyase
PAN	peroxyacetyl nitrate
PAR	photosynthetically active radiation
PATAG	periodic acid-thiocarboxydrezide-silver proteinate
PCBs	polychlorinated biphenyls
PCPs	polychlorophenols
PDA	phytodienoic acid
PDH	proline dehydrogenase

PEPcase	phosphopenolpyruvate carboxylase
PGAL	phosphogylceraldehyde
P_I	inorganic phosphorus
PIN	proteinase inhibitor gene
PIP	plasma membrane intrinsic protein
PKABA1	ABA-dependent phosphokinase gene 1
POD	peroxidase
PPO	polyphenol oxidase
PPP	pentose phosphate pathway
PR	pathogen-related gene
PSI	photosystem I
PSII	photosystem II
QA	quinilizidine alkaloids
QAM	quaternary amine
r	exponential rate of population growth
RNA	ribonucleic acid
RLO	Rickettsia-like organism
RP	rock phosphate
rubisco	ribulose bisphosphate carboxylase-oxygenase
s	second
S	sieman (a unit); sulfur
^{35}S	sulfur isotope 35
SAM	S-adenosylmethionine
SEF	substitution error frequency
SLA	specific leaf area (cm^2 g^{-1} dry weight)
SLW	specific leaf weight (g of dry weight cm^{-2})
SOD	superoxide dismutase
SOD2	gene that codes for an Na^+/H^+ antiporter in yeast
sp., spp.	species
SR	symbiosis-related gene
SRL	specific root length (cm g^{-1} dry weight)
TIP	tonoplast intrinsic protein
α-TIP	α-tonoplast intrinsic protein
VAM	vesicular-arbuscular mycorrhizae
V_{max}	maximum velocity of an enzymatic reaction or a transport process
VOCs	volatile organic compounds
w	log mean weight of plants surviving after density-dependent mortality
w_{aa}	weight or yield of species a in monoculture
w_{bb}	weight or yield of species b in monoculture
w_{ab}	weight or yield of species a and b in mixed culture
ZSOD2	gene that codes for a Na^+/H^+ antiporter in a salt-tolerant yeast variety
μm	micrometer

APPENDIX B
Literature Cited

Aber, J. D., K. J. Nadelhoffer, P. Steudler, and J. M. Melillo. 1989. Nitrogen saturation in northern forest ecosystems: excess nitrogen from fossil fuel combustion may stress the biosphere. Bioscience 39:378–386.

Abrahamson, W. G., and G. Melika. 1998. Gall inducing insects (Cynipinae) provide insights into plant systematics relationships. American Journal of Botany 85(6, suppl.): 111.

Abuzinadah, R. A., and D. J. Read. 1989a. Carbon transfer associated with assimilation of organic nitrogen sources by silver birch (*Betula pendula* Roth.). Trees 3:17–23.

Abuzinadah, R. A., and D. J. Read. 1989b. The role of proteins in the nitrogen nutrition of ectomycorrhizal plants. IV. The utilization of peptides by birch (*Betula pendula* L.) infected with different mycorrhizal fungi. New Phytologist 112:55–60.

Adams, P., D. E. Nelson, S. Yamada, W. Chmara, R. G. Jensen, H. J. Bohnert, and H. Griffiths. 1998. Growth and development of *Mesembreanthemum crystallinum* (Aizoaceae). Tansley Review No. 97. New Phytologist 138:171–190.

Aerts, R., and M. J. Van der Peijl. 1993. A simple model to explain the dominance of low-productive perennials in nutrient poor habitats. Oikos 66:144–147.

Aerts, R., F. Berendse, H. de Caluwe, and M. Schmitz. 1990. Competition in heathland along an experimental gradient of nutrient availability. Oikos 57:310–318.

Aerts, R., J. W. Snoeijer, E. V. D. Meijden, and R. Verpoorte. 1991a. Allelopathic inhibition of seed germination by *Cinchona* alkaloids? Phytochemistry 30:2947–2951.

Aerts, R., R. G. A. Boot, and P. J. M. var der Aart. 1991b. The relationship between above and below ground biomass allocation patterns and competitive ability. Oecologia 87:551–559.

Airy Shaw, H. K. 1973. A Dictionary of Flowering Plants and Ferns. Cambridge University Press, Cambridge, UK.

Aflakpui, G. K. S, P. J. Gregory, and R. J. Froud-Williams. 1998. Uptake and partitioning of nitrogen by maize infected with *Striga hermonthica.* Annals of Botany 81:287–294.

Agerer, R. 1987. Colour atlas of the ectomycorrhizae. Einhorn-Verlag Eduard Dietenberger, Schwäbash Gmünd, Germany.

Agrawal, A. A. 1998. Induced responses to herbivory and increased plant performance. Science 279:1201–1202.

Ågren, G. I., and T. Fagerström. 1984. Limiting dissimilarity in plants: randomness prevents exclusion of species with similar competitive abilities. Oikos 43:369–375.

Agrosystems. Technical Corp. of Germany, Eschborn, Germany.

Aigbokhan, E. I., and H. S. N. Hussain. 1991. Reduction in leaf area in cowpeas due to attack by *Striga gesnerioides*. Pp. 287–292. in: J. K. Ransom and L. J. Musselman (eds.), 5th International Symposium on Parasitic Weeds, Nairobi, CIMMYT, Nairobi, India.

Alberico, G. J., and G. R. Cramer. 1993. Is the salt tolerance of maize related to sodium exclusion? I. Preliminary screening of seven cultivars. Journal of Plant Nutrition 16:2289–2303.

Alborn, T., T. C. J. Turlings, T. H. Jones, G. Stenhagen, J. H. Loughrin, and J. H. Tumlinson. 1997. An elicitor of plant volatiles from beet armyworm oral secretion. Science 276:945–949.

Albrecht, C., T. Burgess, B. Dell, and F. Lapeyrie. 1994. Chitinase and peroxidase activities are induced in *Eucalyptus* roots according to aggressiveness of Australian ectomycorrhizal strains of *Pisolithus* sp. New Phytologist 127:217–222.

Alder, J. H., and R. J. Grebenok. 1995. Biosynthesis and distribution of insect-molting hormones in plants—a review. Lipids 30:257–262.

Alexander, I. J., and K. Hardie. 1981. Surface phosphatase activity of Sitka spruce mycorrhizas from serpentine soils. Soil Biology and Biochemistry 13:301–305.

Alexandre, J., and J.-P. Lassalles. 1992. Hydrostatic and osmotic pressure activated channel in the plant vacuole. Biophysics Journal 60:1326–1336.

Alexopoulos, C. J., C. W. Mims, and M. Blackwell. 1996. Introductory Mycology, 4th Ed. John Wiley & Sons, New York.

Al-Hiyaly, S. A., T. McNeilly, and A. D. Bradshaw 1988. The effects of zinc concentration from electricity pylons—evolution in a replicated situation. New Phytologist 110:571–580.

Allaway, W. H. 1957. pH, soil acidity, and plant growth. Pp. 67–79 in: Soil: The Year Book of Agriculture. U.S. Department of Agriculture, Washington, DC.

Allen, E. B. 1982. Water and nutrient competition between *Salsola ali* and two native grass species (*Agropyron smithii* and *Bouteloua gracilis*). Ecology 63:732–41.

Allen, E. B., and M. F. Allen. 1990. The mediation of competition by mycorrhizae in successional and patchy environments. Pp. 367–385 in: J. B. Grace and D. Tilman (eds.), Perspectives in Plant Competition. Academic Press, New York, NY.

Allen, G. J., and D. Sanders. 1995. Calcineurin, a type 2b protein phosphatase, modulates the Ca^{2+} -permeable slow vacuolar ion-channel of stomatal guard-cells. Plant Cell 7:1473–1483.

Allen, J. R. M. 1996. Climate change—How do plants cope? Biologist 43:65–68.

Allen, M. F. 1991. The Ecology of Mycorrhizae. Cambridge University Press, Cambridge, UK.

Alsaadawi, I. S., S. M. Al-Hadithy, and M. B. Arif. 1986. Effects of three phenolic acids on chlorophyll content and ion uptake in cowpea seedlings. Journal of Chemical Ecology 12: 221–227.

Alscher, R. G., J. L. Donahue, and C. L. Cramer. 1997. Reactive oxygen species: Relationships in green cells. Physiologia Plantarum 100:224–233.

Alway, F. J., T. E. Maki, and W. J. Methley. 1934. Composition of the leaves of some forest trees. Soil Survey Association Bulletin 15:81–84.

Amerzllag, G. N., H. R. Lerner, and A. Poljakoffmayber. 1992. Interaction between mineral nutrients, cytokinin and gibberellic acid during growth of sorghum at high NaCl salinity. Journal of Experimental Botany 43:81–87.

Ames, R. N., C. P. P. Reid, L. Porter, and C. Cambardella. 1983. Hyphal uptake and transport of nitrogen from two ^{15}N-labeled sources by *Glomus mossea*, a vesicular-arbuscular mycorrhizal fungus. New Phytologist 95:381–396.

An, M., I. R. Johnson, and J. V. Lovett. 1993. Mathematical modeling of allelopathy: biological response to allelochemicals and its interpretation. Journal of Chemical Ecology 19:2379–2388.

Antonova, T. S. 1994. Biochemical aspects of the development of new virulent forms in the Moldovian population (race) of *Orobanche cumana* Wallr. against the background of resistant sunflower cultivars. Pp. 290–292 in: A. H. Pieterse, J. A. C. Verkleij, and S. J. ter Borg (eds.), Biology and Management of *Orobanche*. Proceedings of the Third International Workshop on *Orobanche* and Related *Striga* Research. Royal Tropical Institute, Amsterdam, Netherlands.

Appel, H. M. 1993. Phenolics in ecological interactions: the importance of oxidation. Journal of Chemical Ecology 19:1521–1552.

Appel, H. M., and M. M. Martin. 1990. The significance of metabolic load in the evolution of host specificity of *Manduca sexta*. Ecology 73:216–228.

Appel, H. M., and J. C. Schultz. 1992. Activity of phenolics in insects: the role of oxidation. Pp. 609–620 in: R. W. Hemingway and P. E. Laks (eds.), Plant Phenolics: Synthesis, Properties, and Significance. Plenum Press, New York.

Arbol, I. P., J. S. P. Yadov, and M. I. Massoud (eds.). 1988. Salt-Affected Soils and Their Management. Food and Agriculture Organization of the United Nations, Rome.

Archer, S., and L. L. Tieszen. 1980. Growth and physiological response of tundra plants to defoliation. Arctic and Alpine Research 12:531–532.

Armillas, P. 1961. Land use in pre-Columbian America. Pp. 255–276 in: L. D. Stamp (ed.), A History of Land Use in Arid Regions (Arid Zone Research, vol. 17) UNESCO, Paris.

Aronson, J. A. 1989. Haloph. A data base of salt tolerant plants of the world. Office of Arid Lands Studies, University of Arizona, Tempe, AZ.

Ashford, A. E., S. Ryde, and K. D. Barrow. 1994. Demonstration of short chain polyphosphate in *Pisolithus tinctorius* and the implications for phosphate transport. New Phytologist 126: 239–247.

Ashraf, M. 1994. Breeding for salinity tolerance in plants. CRC Critical Reviews in Plant Sciences 13:17–42.

Atkinson, C. J. 1996. Global changes in atmospheric carbon dioxide: The influence on terrestrial vegetation. Pp. 99–133 in: M. Yunus and M. Iqbal (eds.), Plant Response to Air Pollution. John Wiley & Sons, New York.

Atsatt, P. R. 1973. Parasitic flowering plants: how did they evolve? American Naturalist 107: 502–510.

Atsatt, P. R. 1983. Host-parasite interaction in higher plants. Pp. 519–535 in: O. L. Lange, P. S. Nobel, C. B. Osmond, and H. Ziegler (eds.), Encyclopedia of Plant Physiology (12C New Series) Springer Verlag, Berlin.

Atsatt, P. R. 1987. Evolution of parasitic plants: are haustoria modified tumors? Pp. 127–141 in: L. J. Musselman (ed.), *Striga* (Parasitic Weeds in Agriculture. Vol. I). CRC Press, Boca Raton, FL.

Augé, R. M., K. A. Schekel, and R. L. Wample. 1987a. Rose leaf elasticity changes in response to mycorrhizal colonization and drought acclimation. Physiologia Plantarum 70:175–182.

Augé, R. M., K. A. Schekel, and R. L. Wample. 1987b. Leaf water and carbohydrate status of VA mycorrhizal rose exposed to drought stress. Plant and Soil 99:291–302.

Augustyniak, J., M. Dabert, and K. Wypijewski. 1997. Transgenes in plants: protection against viruses and insects. Acta Physiologia Plantarum 19:561–569.

Austin, M. P., and B. O. Austin. 1980. Behaviour of experimental plant communities along a nutrient gradient. Journal of Ecology 68:891–918.

Awada, S., W. F. Campbell, L. M. Dudley, and J. J. Jurinak. 1995. Interactive effects of sodium chloride, sodium sulfate, calcium sulfate, and calcium chloride on snapbean growth, photosynthesis, and ion uptake. Journal of Plant Nutrition 18:889–900.

Ayala, F., J. W. O'Leary, and K. S. Schumaker. 1996. Increased vacuolar and plasma membrane H^+-ATPase activities in *Salicornia biglovii* Tott. in response to NaCl. Journal of Experimental Botany 47:25–32.

Ayers, M. P., T. P. Clousen, S. F. MacLean Jr., A. M. Redman, and P. B. Reichardt. 1997. Diversity of structure and antiherbivore activity of condensed tannins. Ecology 78:1696–1712.

Bach, C. E. 1994. Effects of a specialist herbivore (*Attica sublicata*) on *Salix cordata* and sand dune succession. Ecological Monographs 64:423–445.

Baker, A. J. M. 1981. Accumulators and excluders—strategies in the response of plants to heavy metals. Journal of Plant Nutrition 3:643–654.

Baker, A. J. M. 1987. Metal tolerance. New Phytologist 106:93–111.

Baldocchi, D., and S. Collineau. 1994. The physical nature of solar radiation in heterogeneous canopies: spatial and temporal attributes. Pp. 21–71 in: M. M. Caldwell and R. W. Pearcy (eds.), Exploitation of Environmental Heterogeneity by Plants. Academic Press, San Diego, CA.

Baldwin, I. T. 1989. The mechanism of damage-induced alkaloids in wild tobacco. Journal of Chemical Ecology 151:1661–1680.

Baldwin, I. T. 1994. Chemical changes rapidly induced by folivory. Pp. 1–24 in: E. A. Bernays (ed.) Insect-Plant Interactions Vol V. CRC Press, Boca Raton, FL, USA.

Baldwin, I. T., and C. A. Preston. 1999. The eco-physiological complexity of plant response to insect herbivores. Planta 208:137–145.

Baldwin, I. T., C. L. Simms, and S. E. Kean. 1990. The reproductive consequences associated with inducible alkaloidal responses in wild tobacco. Ecology 71:252–262.

Balke, N. E. 1985. Effects of allelochemicals on mineral uptake and associated physiological processes. Pp. 163–178 in: A. C. Thompson (ed.), The Chemistry of Allelopathy (American Chemical Society Symposium 268). American Chemical Society, Washington DC.

Balke, N. E., M. Davis, and C. C. Lee. 1987. Conjugation of allelochemicals by plants. Enzymatic glucosylation of salicylic acid by *Avena sativa*. Pp. 214–227 in: G. R. Waller (ed.), Allelochemicals: Role in Agriculture and Forestry (ACS Symposium Series, Vol. 330). American Chemical Society, Washington, DC.

Ball, M. C. 1988. Salinity tolerance in the mangroves, *Aegceras corniculatum* and *Avicennia marina*. I. Water use in relation to growth, carbon partitioning and salt tolerance. Australian Journal of Plant Physiology 15:447–464.

Ball, M. C. 1996. Comparative ecophysiology of mangrove forest and tropical lowland moist rainforest. Pp. 461–496 in: S. S. Mulkey, R. L. Chazdon, and A. P. Smith (eds.), Tropical Forest Plant Ecophysiology. Chapman & Hall, New York.

Ball, M. C., and J. M. Anderson. 1986. Sensitivity of photosystem II to NaCl in relation to salinity tolerance. Comparative studies with thylacoids of the salt tolerant mangrove, *Avicennia marina* and the salt sensitive pea, *Psium sativum*. Australian Journal of Plant Physiology 13:689–698.

Ball, M. C., and G. D. Farquhar. 1984. Photosynthetic and stomatal responses of two mangrove species, *Aegceras corniculatum* and *Avicennia marina*, to long term salinity and humidity conditions. Plant Physiology 74:1–6.

Ball, M. C., W. S. Chow, and J. M. Anderson. 1987. Salinity-induced potassium deficiency causes loss of functional photosystem II in leaves of the grey mangrove, *Avicennia marina*, through depletion of the atrazine-binding polypeptide. Australian Journal of Plant Physiology 14:351–361.

Ballaré, C. L. 1994. Light gaps: sensing the light opportunities in highly dynamic canopy environments. Pp. 73–110 in: M. M. Caldwell and R. W. Pearcy (eds.), Exploitation of Environmental Heterogeneity by Plants. Academic Press, San Diego, CA.

Ballaré, C. L., R. A. Sánchez, A. L. Scopel, J. J. Casal, and C. M. Chersa. 1987. Early detection of neighbor plants by phytochrome perception of spectral changes in reflected sunlight. Plant, Cell and Environment 10:551–557.

Ballaré, C. L., R. A. Sánchez, A. L. Scopel, and C. M. Ghersa. 1988. Morphological responses of *Datura ferox* L. seedlings to the presence of neighbors: their relationships with canopy microclimate. Oecologia 76:288–293.

Ballaré, C. L., A. L. Scopel, and R. A. Sánchez. 1997. Foraging for light: photosensory ecology and agricultural implications. Plant, Cell and Environment 20:820–825.

Barber, S. A. 1974. Influence of the plant root on ion movement in soil. Pp. 523–563 in: E. W. Carson (ed.), The Plant Root and Its Environment. University Press of Virginia, Charlottesville, VA.

Barber, S. A. 1984. Soil Nutrient Bioavailability. John Wiley & Sons, New York.

Barber, S. A. 1995. Soil Nutrient Bioavailability: A Mechanistic Approach. John Wiley & Sons, New York.

Barceló, J., and C. Poschenrieder. 1990. Plant water relations as affected by heavy metal stress: a review. Journal of Plant Nutrition 13:1–37.

Barea, J. M., and C. Azcón-Aguilar. 1983. Mycorrhizas and their significance in nodulating nitrogen-fixing plants. Advances in Agronomy 36:1–54.

Barker, E. R., J. D. Scholes, M. C. Press, and W. Quick. 1995. Effects of the holoparasite *Orobanche aegyptiaca* on the growth and photosynthesis of its tomato host. Pp. 141–148 in: Physiological Responses of Plants to Pathogens, Aspects of Applied Biology Symposium 42, 11–13 September 1995, University of Dundee, Wellesbourne, U.K.

Barker, E. R., M. C. Press, J. D. Scholes, and W. P. Quick. 1996. Interactions between the parasitic angiosperm *Orobanche aegyptica* and its tomato host: growth and biomass allocation. New Phytologist 133:637–642.

Barrett, A. J. 1987. The cystatins: a new class of peptidase inhibitors. Trends in Biochemical Sciences 12:193–196.

Barrieu, F., F. Chaumont, and M. J. Chrispeels. 1998. High expression of the tonoplast aquaporin ZmTIP1 in epidermal and conducting tissues of maize. Plant Physiology 117:1153–1163.

Barry, T. N., and T. R. Manley. 1986. Interrelationships between the concentrations of total condensed tannin and lignin in *Lotus* sp. and their possible consequences in ruminant nutrition. Journal of Science Food and Agriculture 37:248–254.

Baruah, N. C., J. C. Sarma, N. C. Barua, S. Sarma, and R. P. Sharma. 1994. Germination and growth inhibitory sesquiterpene lactones and a flavone from *Tithonia diversiflora*. Phytochemistry 36:29–36.

Bauer, R., P. Feeny, and E. Staadler. 1993. Oviposition stimulants for the black swallowtail butterfly: identification of electrophysiologically active compounds in carrot volatiles. Journal of Chemical Ecology 19:919–937.

Baumel, P., W. D. Jeschke, N. Rath, F. C. Czygan, and P. Proksch. 1995. Modeling of quinolizidine alkaloid net flows in *Lupinus album* and between *L. album* and the parasite *Cuscuta reflexa*: new insights into the site of quinolizidine alkaloid synthesis. Journal of Experimental Botany 46:1721–1730.

Baxevanis, C. N., L. F. Voutsas, M. H. Soler, A. D. Gritzapis, O. E. Tsitsilonis, S. Stoeva, W. Voelter, P. Arsenis, and M. Papamichail. 1998. Mistletoe lectin I–induced effects on human cytotoxic lymphocytes. I. Synergism with IL-2 in the induction of enhanced LAK cytotoxicity. Immunopharmacology and Immunotoxicology 20(3):355–372.

Bazzaz, F. A. 1979. Physiological ecology of plant succession. Annual Review of Ecology and Systematics 10:351–371.

Bazzaz, F. A. 1996. Plants in Changing Environments: Linking Physiological, Population, and Community Ecology. Cambridge University Press, Cambridge, UK.

Bazzaz, F. A., and R. W. Carlson. 1982. Photosynthetic acclimation to variability in the light environment of early and late successional plants. Oecologia 54:313–316.

Bazzaz, F. A., and P. M. Wayne. 1994. Coping with environmental heterogeneity: the physiological ecology of tree seedling regeneration across the gap-understory continuum. Pp. 349–390 in: M. M. Caldwell and R. W. Pearcy (eds.), Exploitation of Environmental Heterogeneity by Plants. Academic Press, San Diego, CA.

Bedbrook, J. R., and R. E. F. Matthews. 1973. Changes in the flow of early products of photosynthetic carbon fixation associated with the replicaton of TYMV. Virology 53:84–91.

Beggs, C. J., and E. Wellmann. 1994. Photocontrol of flavenoid biosynthesis. Pp. 733–750. in: R. E. Kendrick, and G. H. M. Kronenberg (eds.), Photomorphogenesis in plants, Vol. 2. Klewer Academic, Dordrecht, Germany.

Belcher, J. W., P. A. Keddy, and L. Twolan-Strutt. 1995. Root and shoot competition intensity along a soil depth gradient. Journal of Ecology 83:673–682.

Bell, D. T. 1974. The influence of osmotic pressure in tests for allelopathy. Transactions of the Illinois State Academy of Science 67:312–317.

Bell, D. T., and C. H. Muller. 1973. Dominance of California annual grasslands by *Brassica nigra*. American Midland Naturalist 90:277–299.

Benedict, J. H., and J. L. Hatfield. 1988. Influence of temperature-induced stress on host plant suitability to insects. Pp. 139–165 in: E. A. Heinrichs (ed.), Plant Stress-Insect Interactions. John Wiley & Sons, New York.

Benepal, P. S., and C. V. Hall. 1967. The influence of mineral nutrition of varieties of *Cucurbita pepo* L. on the feeding response of squash bug *Anasa tristis* DeGeer. Proceedings of the American Society of Horticultural Science 90:304–312.

Bennett, W. F. 1993. Plant nutrient utilization and diagnostic plant symptoms. Pp. 1–7 in: W. F. Bennett (ed.), Nutrient Deficiencies and Toxicities in Crop Plants. APS Press, St. Paul, MN.

Ben-Zioni, A., Y. Vaadia, and S. H. Lips. 1971. Nitrate uptake by roots as regulated by nitrate reduction products of the shoot. Physiologia Plantarum 24:288–290.

Berenbaum, M. R., and A. R. Zangerl. 1988. Stalemates in the coevolutionary arms race: synthesis, synergisms, and sundry other sins. Pp. 113–132 in: K. C. Spencer (ed.), Chemical Mediation of Coevolution. American Institute of Biological Sciences, Washington DC.

Bergey, D. R., G. A. Howe, and C. A. Ryan. 1996. Polypeptide signaling for plant defensive genes exhibits analogies to defense signaling in animals. Proceedings of the National Academy of Sciences (USA). 93:12052–12058.

Bergmann, E., J. Bender, and H. J. Weigel. 1995. Growth responses and foliar sensitivities of native herbaceous species to ozone exposures. Water, Air, and Soil Pollution. 85:1437–1442.

Berhow, M. A., and S. F. Vaughn. 1999. Higher plant flavonoids: biosynthesis and chemical ecology. Pp. 423–438. in: K. Inderjit, M. M. Dakshini, and C. L. Foy (eds.), Principles and Practices in Plant Ecology: Allelochemical Interactions. CRC Press, Boca Raton, FL.

Bernays, E. A., and R. F. Chapman. 1994. Host-Plant Selection by Phytophagous Insects. Chapman & Hall, New York.

Bernstein L. 1975. Effects of salinity and sodicity on plant growth. Annual Review of Phytopathology. 13:295–312.

Bhowmik, P. C., and J. D. Doll. 1983. Growth analysis of corn and soybean response to allelopathic effects of weed residues at various temperatures and photosynthetic photon flux densities. Journal of Chemical Ecology 9:1263–1280.

Bidwell, R. G. S. 1979. Plant Physiology, 2nd edition. Macmillan Publishing Company, New York.

Bieleski, R. L., and I. B. Ferguson. 1983. Physiology and metabolism of phosphate and its compounds. Pp. 422–449 in: A. Läuchli and R. L. Bieleski (eds.), Inorganic Plant Nutrition (A. Pirson and M. H. Zimmerman [series eds.], Encyclopedia of Plant Physiology, Vol. 15A). Springer-Verlag, New York.

Bildusas, I. J., P. L. Phleger, E. L. Stewart, and R. K. Dixon. 1986. Response of *Bromus intermis* innoculated with *Glomus fasciculatum* to potassium fertilization and drought stress. Plant and Soil 95:441–444.

Binzel, M. L. 1995. NaCl-induced accumulation of tonoplast and plasma membrane H^+-ATPase message in tomato. Physiologia Plantarum 94:722–728.

Bishop, P. D., D. J. Makus, G. Pearce, and C. A. Ryan. 1981. Proteinase inhibitor-inducing factor activity in tomato leaves resides in oligosaccharides enzymatically released from cell walls. Proceedings of the National Academy of Sciences (USA) 78:3536–3640.

Björkman, E. 1949. The ecological significance of the ectotrophic mycorrhizal association in forest trees. Svensk Botanisk Tidskrift 43:223–262.

Björkman, E. 1960. *Monotropa hypopitys* L., an epiparasite on tree roots. Physiologia Plantarum 13:308–327.

Björkman, O., B. Demming, and T. J. Andrews. 1988. Mangrove photosynthesis: response to high irradiance stress. Australian Journal of Plant Physiology 15:43–61.

Blackford, M., B. Clarke, and L. Dinan. 1996. Tolerance of the Egyptian cotton leafworm *Spodoptera littoralis* (Lepidoptera: Noctuidae) to ingested phytoecdysteroids. Journal of Insect Physiology 42:931–936.

Blau, W. S. 1981. Life history variation in the black swallowtail butterfly. Oecologia 48:116–122.

Bleiler, J. A., G. A. Rosenthal, and D. H. Janzen. 1988. Biochemical ecology of canavanine-eating seed predators. Ecology 69:427–433.

Blits, K. C., and J. L. Gallagher. 1990. Effect of NaCl on lipid content of plasma membranes of the dicot halophyte *Kosteletzks virginica* (L.) Pprestl. Plant Cell Reports 9:156–159.

Blomstrom, D. C., E. Knight, W. D. Phillips, and J. F. Weiher 1964. The nature of iron in ferredoxin. Proceedings of the National Academy of Sciences (USA) 51:1085–1092.

Bloom, A. J., and J. H. Troughton. 1979. High productivity and photosynthetic flexibility in a CAM plant. Oecologia 38:35–43.

Blum, A. 1988. Salinity resistance. Pp.163–178 in: A. Blum (ed.), Plant Breeding for Stress Environments. CRC Press, Boca Raton, FL.

Blum, U. 1998. Effects of microbial utilization of phenolic acids and their phenolic acid breakdown products on allelopathic interactions. Journal of Chemical Ecology 24:685–708.

Blum, U., B. R. Dalton, and J. R. Shann. 1985. Effects of various mixtures of ferulic acid and some of its microbial metabolic products on cucumber leaf expansion and dry matter in nutrient. Journal of Chemical Ecology 11:619–641.

Boardman, N. K. 1977. Comparative photosynthesis of sun and shade plants. Annual Review of Plant Physiology 28:355–377.

Bohnert, H. J., and R. G. Jensen. 1996. Strategies for engineering water-stress tolerance in plants. Trends in Biotechnology 14:89–97.

Bohnert, H. J., D. E. Nelson, and R. G. Jensen. 1995. Adaptations to environmental stresses. Plant Cell 7:1099–1111.

Bolan, N. S. 1991. A critical review on the role of mycorrhizal fungi on the uptake of phosphorus by plants. Plant and Soil 134:189–207.

Boulter, C. J., and M. Latoszek-Green. 1997. Effect of chronic ingestion of the cytstein proteinase inhibitor, E-64, on Colorado potato beetle gut proteinases. Entomologia Experimentalis et Applicata 83:295–303.

Bonfante-Fasolo, P., and V. Gianinazzi-Pearson. 1982. Ultrastructural aspects of endomycorrhiza in the Ericaceae. III. Morphology of the dissociated symbionts and modifications occurring during their reassociation in axenic culture. New Phytologist 91:691–704.

Bonser, S. P., and R. J. Reader. 1995. Plant competition and herbivory in relation to vegetation biomass. Ecology 76:2176–2183.

Boone, L. S., G. Fate, M. Change, and D. G. Lynn. 1995. Seed germination. Pp. 14–38 in: M. C. Press and J. D. Graves (eds.), Parasitic plants. Chapman and Hall, London.

Bordas, M., C. Montesinos, M. Dabauza, A. Salvador, L. A. Roig, R. Serrano, and V. Moreno. 1997. Transfer of the yeast salt tolerance gene HAL1 to *Cucumis melo* L. cultivars and *in-vitro* evaluation of salt tolerance. Transgenic Research 6:41–50.

Bormann, F. H., and G. E. Likens. 1979. Pattern and Process in a Forested Ecosystem. Springer-Verlag, Berlin.

Botella, M. A., V. Martinez, M. Nieves, and A. Cerd . 1997. Effects of salinity on the growth and nitrogen uptake by wheat seedlings. Journal of Plant Nutrition 20:793–804.

Boto, K. G., and J. T. Wellington. 1984. Soil characteristics and nutrient status in a north Australian mangrove forest. Estuaries 7:61–69.

Botton, B., and M. Chalot. 1995. Nitrogen assimilation: enzymology in ectomycorrhizas. Pp. 325–363 in: A. Varma, and B. Hack (eds.), Mycorrhizas: Structure, Function, Molecular Biology, and Biotechnology. Springer-Verlag, Berlin.

Bougher, N. L., T. S. Grove, and N. Malajczuk. 1990. Growth and phosphorus acquisition of karri (*Eucalyptus diversicolor* F. Muell.) seedlings inoculated with ectomycorrhizal fungi in relation to phosphorus supply. New Phytologist 114:77–85.

Bould, C. E., J. Hewitt, and P. Needham. 1983. Principles (Diagnosis of Mineral Disorders in Plants, Vol 1). Her Majesty's Stationary Office, London.

Bowen, G. D., and A. D. Rovira. 1991. The rhizosphere: the hidden half of the hidden half. Pp. 641–670 in: Y. Waisel, A. Eshel, and U. Kafkafi (eds.), Plant Roots: The Hidden Half, Marcel Dekker, New York.

Bowers, W. S. 1976. Discovery of insect antiallatotropins. Pp. 394–408 in: L. I. Gilbert (ed.), The Juvenile Hormones, Plenum Press, New York.

Bowers, W. S. 1991. Insect hormones and antihormones in plants. Pp. 431–456 in: G. A. Rosenthal and M. R. Berenbaum (eds.), The Chemical Participants (Herbivores: Their Interactions with Secondary Plant Metabolites, 2nd Ed., Vol. 1). Academic Press, San Diego, CA.

Boyer, J. S. 1982. Plant productivity and environment. Science 218:443–448.

Bradow, J. M. 1991. Relationships between chemical structure and inhibitory activity of C6 through C9 volatiles emitted by plant residues. Journal of Chemical Ecology 17:2193–2212.

Brady, N. C. 1974. The Nature and Properties of Soils. Macmillan Publishing Company, New York.

Brattsen, L. B., J. H. Samuelian, K. Y. Long, S. A. Kincaid, and C. K. Evans. 1983. Cyanide as a feeding stimulant for the southern army worm *Spodoptera eridania*. Ecological Entomology 8:125–132.

Breedlove, D. E., and P. R. Ehrlich. 1972. Coevolution: patterns of legume predation by a lycenid butterfly. Oecologia 10:98–104.

Briones, O., C. Montaña, and E. Excurra. 1998. Competition intensity as a function of resource availability in a semiarid ecosystem. Oecologia 116:365–372.

Broadway, R. M. 1996. Dietary proteinase inhibitors alter complement of midgut proteases. Archives of Insect Biochemistry 32:39–53.

Broadway, R. M., S. S. Duffey, G. Pearce, and C. A. Ryan. 1986. Plant proteinase inhibitors: a defense against insects? Entomologia Experimentalis et Applicata 41:33–38.

Bromilow, R. H., and K. Chamberlin 1995. Principles governing uptake and transport of chemicals. Pp. 37–68 in: S. Trapp and J. Craig McFarlane (eds.), Plant Contamination: Modeling and Simulation of Organic Chemical Processes. Lewis Publishers, Boca Raton, FL.

Brown, D. G. 1988. The cost of plant defense: an experimental analysis with inducible proteinase inhibitors in tomato. Oecologia 76:467–470.

Brown, D. J., and F. M. DuPont. 1989. Lipid composition of plasma membranes and endomembranes prepared from roots of barley (*Hordeum vulgare* L.). Effects of salt. Plant Physiology 90:955–961.

Brown, M. S., and G. J. Bethlenfalvay. 1988. The *Glycine*, *Glomus*, *Rhizobium* symbiosis. VII. Photosynthetic nutrient use efficiency in nodulated, mycorrhizal soybeans. Plant Physiology 86:1292–1297.

Brown, S. A. 1981. Coumarins. Pp. 269–300 in: E. E. Conn (ed.), The Biochemistry of Plants. Academic Press, New York.

Brown, R. T. 1967. Influence of naturally occurring compounds on germination and growth of jack pine. Ecology 48:542–546.

Bryant, J. P., F. S. Chapin III, and D. R. Kline. 1983. Carbon/nutrient balance of boreal plants in the role of plant chemistry. Annual Review of Ecology and Systematics. 11:261–285.

Bryant, J. P., J. Toumi, and P. Niemalä. 1988. Environmental constraints of constitutive and long-term inducible defenses in woody plants. Pp. 367–389 in: K. C. Spencer (ed.), Chemical Mediation of Coevolution. American Institute of Biological Sciences, Washington, DC.

Burch, J. N. 1992. *Cassytha filiformis* and limits to growth and reproduction of *Schinus terebinthefolius* in southern Florida. Florida Scientist 55:28–34.

Burgess, T. I., N. Malajczuk, and T. S. Grove. 1993. The ability of 16 ectomycorrhizal fungi to increase growth and phosphorus uptake of *Eucalyptus globulus* Labill. and *E. diversicolor* F. Muell. Plant and Soil 153:155–164.

Burgess, T. I., B. Dell, and N. Malajczuk. 1994. Variation in the mycorrhizal development and growth stimulation of 20 isolates of *Pisolithus* inoculated onto *Eucalyptus grandis* W. Hill ex Maiden. New Phytologist 127:731–739.

Burgess, T. I., P. Laurent, B. Dell, N. Malajczuk, and F. Martin. 1995. Effect of fungal-isolate aggressivity on the biosynthesis of symbiosis-related polypeptides in differentiating eucalypt ectomycorrhizas. Planta 195:408–417.

Burkholder, P. R. 1952. Cooperation and conflict among primitive organisms. American Scientist 40:601–631.

Burton, P. J., and F. A. Bazzaz. 1995. Ecophysiological responses of tree seedlings invading different patches of old-field vegetation. Journal of Ecology 83:99–112.

Bush, D. S. 1995. Calcium regulation in plant cells and its role in signaling. Annual Review of Plant Physiology and Plant Molecular Biology 46:95–122.

Butler, L. G. 1989. Effects of condensed tannin on animal nutrition. Pp. 391–402 in: R. W. Hemmingway and J. J. Karchesy (eds.), Chemistry and Significance of Condensed Tannins. Plenum Press, New York.

Butler, L. G. 1995. Chemical communication between the parasitic weed *Striga* and its crop host—a new dimension in allelochemistry. American Chemical Society Symposium Series 582:158–168.

Cairney, J. W. G., and R. M. Burke. 1994. Fungal enzymes degrading plant cell walls: their possible significance in the ectomycorrhizal symbiosis. Mycological Research 98:1345–1356.

Caldwell, M. M., and J. H. Richards. 1986. Competing root systems: morphology and models of absorption. Pp. 251–273. in: T. Givnish (ed.), On the Economy of Plant Form and Function (Proceedings of the 6th Maria Moors Cabot Symposium: Evolutionary Constraints on Primary Productivity, Adaptive Patterns of Energy Capture in Plants, Harvard Forest, August 1983). Cambridge University Press, Cambridge, UK.

Caldwell, M. M., J. H. Richards, D. A. Johnson, R. S. Nowak, and R. S. Dzuree. 1981. Coping with herbivory: photosynthetic capacity and resource allocation in two semiarid *Agropyron* bunchgrasses. Oecologia 50:14–24.

Caldwell, M. M., T. J. Dean, R. S. Nowak, R. S. Dzurec, and J. H. Richards. 1983. Bunchgrass architecture, light interception and water-use efficiency: assessment by fiber optic point quadrats and gas exchange. Oecologia 59:178–84.

Caldwell, M. M., J. H. Manwaring, and S. L. Durham. 1991a. The microscale distribution of neighbouring plant roots in fertile soil microsites. Functional Ecology 5:765–772.

Caldwell, M. M., J. H. Manwaring, and R. B. Jackson. 1991b. Exploitation of phosphate from fertile soil microsites by the Great Basin perennials when in competition. Functional Ecology 5:757–764.

Callow, J. A. 1983. Biochemical Plant Pathology. John Wiley & Sons, New York.

Campbell, B. D., J. P. Grime, and J. M. L. Mackey. 1991a. A trade-off between scale and precision in resource foraging. Oecologia 87:532–538.

Campbell B. D., J. P. Grime, J. M. L. MacKay, and A. Jalili. 1991b. The quest for a mechanistic understanding of resource competition in plant communities: the role of experiments. Functional Ecology 5:241–253.

Cao, W., and D. L. Crawford. 1993a. Carbon nutrition and hydrolytic and cellulolytic activities in the ectomycorrhizal fungus *Pisolithus tinctorius*. Canadian Journal of Microbiology 39: 529–535.

Cao, W., and D. L. Crawford. 1993b. Purification and some properties of β-glucosidase from the ectomycorrhizal fungus *Pisolithus tinctorius* strain SMF. Canadian Journal of Microbiology 39:125–129.

Casper, B. B., and R. B. Jackson. 1997. Plant competition underground. Annual Review of Ecology and Systematics 28:545–570.

Cates, R. G., and D. F. Rhodes. 1977. *Prosopis* leaves as a resource for insects. Pp. 61–83 in: B. B. Simpson (ed.), Mesquite: Its Biology in Two Desert Ecosystems. Dowden, Hutchinson, & Ross, Stroudsburg, PA.

Cavalieri, A. J. 1983. Proline and glycinebetaine accumulation in *Spartina alterniflora* Loisel. in response to NaCl and nitrogen in a controlled environment. Oecologia 57:20–24.

Cechin, I., and M. C. Press. 1993. Nitrogen relations of the sorghum–*Striga hermonthica* host-parasite association: growth and photosynthesis. Plant Cell and Environment. 16:237–247.

Cechin, I., and M. C. Press. 1994. The effect of nitrogen on growth and photosynthesis of a C-3 cereal, *Oryza sativa,* infected with root hemiparasite *Striga hemonthica.* Journal of Experimental Botany 45:925–930.

Cen, Y. P., and J. F. Bornman. 1993. The effect of exposure to enhanced UV-B radiation on the penetration of monochromatic and polychromatic UV-B radiation in leaves of *Brassica napus*. Physiologia Plantarum 87:249–255.

Cendrin, F., J. Chroboczek, G. Zaccai, H. Eisenberg, and M. Mevarech. 1993. Cloning, sequencing, and expression in *Escherichia coli* of the gene coding for malate dehydrogenase of the extremely halophilic archaebacterium *Haloarcula marismortui*. Biochemistry 32:4308–4313.

Chalot, M., A. Brun, and B. Botton. 1990. Occurrence and distribution of the fungal NADP-dependent glutamate dehydrogenase in spruce and beech ectomycorrhizas. Pp. 324–327 in: D. Werner and P. Müller (eds.), Fast Growing Trees and Nitrogen Fixing Trees. Gustav Fischer, Stuttgart.

Chang, M., and D. G. Lynn. 1986. Haustoria and the chemistry of host recognition in parasitic angiosperms. Journal of Chemical Ecology 12:561–579.

Chapin, F. S. III 1989. The cost of tundra plant structures—evaluation of concepts and currencies. American Naturalist 133:1–19.

Chapin, F. S. III, A. J. Bloom, C. B. Field, and R. H. Waring. 1987. Plant responses to multiple environmental factors. Bioscience 37:49–57.

Chapin, F. S. III. 1991. Effects of multiple environmental stresses on nutrient availability and use. Pp. 67–88 in: H. A. Mooney, W. E. Winner, E. J. Pell (eds.), Response of Plants to Multiple Stresses. Academic Pres, San Diego, CA.

Chapin, F. S. III, O. E. Sala, I. C. Burke, J. P. Grime, D. U. Hooper, W. K. Lauenroth, A. Lombard, H. A. Mooney, A. R. Mosier, S. Naeem, S. W. Pacala, J. Roy, W. L. Steffen, and D. Tilman. 1998. Ecosystem consequences of changing biodiversity. BioScience 48:45–52.

Chapman, R. F., E. A. Bernays, and T. Wyatt. 1988. Chemical aspects of host-plant specificity in three *Larrea*-feeding grasshoppers. Journal of Chemical Ecology 14:561–579.

Chapman, V. J. 1974. Salt marshes and salt deserts of the world. Pp. 3–19 in: R. J. Reimold and W. H. Queen (eds.), Ecology of Halophytes. Academic Press, New York.

Chappell, J. 1995. Biochemistry and molecular biology of the isoprenoid biosynthetic pathway in plants. Annual Review of Plant Physiology and Plant Molecular Biology 46:1399–1402.

Chaturvedi, O. P., and A. N. Jha. 1992. Studies on allelopathic potential of an important agroforestry species. Forest Ecology and Management 53:91–98.

Chaves, N., and J. C. Escudero. 1999. Variation in flavonoid synthesis induced by ecological factors. Pp. 267–285 in: K. M. Inderjit, M. Dakshini, and C. L. Foy (eds.), Principles and Practices in Plant Ecology: Allelochemical Interactions. CRC Press, Boca Raton, FL.

Chazdon, R. L. 1992. Photosynthetic plasticity of two rain forest shrubs across natural gap transects. Oecologia 92:586–595.

Chazen, O., and P. M. Neumann. 1994. Hydraulic signals from the roots and rapid cell wall hardening in growing maize leaves are primary responses to PEG induced water deficits. Plant Physiology 87:547–550.

Cheeseman, J. M. 1988. Mechanisms of salinity tolerance in plants. Plant Physiology 87:547–550.

Chesson, A. 1981. Effects of sodium hydroxide on cereal straws in relation to the enhanced degredation of structural polysaccharides in rumen microorganisms. Journal Science of Food and Agriculture 32:745–758.

Chiang, H. S., and D. M. Norris. 1983. Morphological and physiological parameters of soybean resistance to agromyzid beanflies. Environmental Entomology 12:260–265.

Chiang, H. S., D. M. Norris, A. Ciepiela, A. Shapiro, and A. Osterwyk. 1986. Inducible vs. constitutive PI 227687 soybean resistance to Mexican bean beetle, *Epilachna varivestis*. Journal of Chemical Ecology 13:741–749.

Chiapusio, G., A. M. Sanchez, M. J. Reigosa, L. Gonzalez, and F. Pellissier. 1997. Do germination indices adequately reflect allelochemical effects on the germination process? Journal of Chemical Ecology 23:2445–2453.

Chiarello, N., J. C. Hickman, and H. A. Mooney. 1982. Endomycorrhizal role for interspecific transfer of phosphorus in a community of annual plants. Science 217:941–943.

Choesin, D. N., and R. E. J. Boerner. 1991. Allyl isothiocyanate release and the allelopathic potential of *Brassica napus* (Brassicaceae). Americal Journal of Botany 78:1083–1090.

Chou, C.-H. 1987. Allelopathy in subtropical vegetation and soils in Taiwan. Pp. 102–117 in: G. R. Waller (ed.), Allelochemicals: Role in Agriculture and Forestry. American Chemical Society, Washington, DC.

Chou, C.-H., and K. J. Lee. 1988. Effects of nitrogen fertilizer levels on the allelopathic potential of pangola grasses and weeds. Botanical Bulletin of Academia Sinica 29:39–47.

Chou, C. H., and G. R. Waller (eds.). 1989. Phytochemical Ecology: Allelochemicals, Mycotoxins and Insect Pheromones and Allomones. Institute of Botany, Academia Sinica, Taipei.

Chou, C.-H., C.-Y. Fu, S.-Y. Li, and Y.-F. Wang. 1998. Allelopathic potential of *Acacia confusa* and related species in Taiwan. Journal of Chemical Ecology 24:2131–2150.

Chrispeels, M. J., and C. Maurel. 1994. Aquaporins: the molecular basis of facilitated water movement through living plant cells. Plant Physiology 105:9–13.

Christopherson, C. 1985. Marine alkaloids. Alkaloids 24:25–111.

Chung, I.-M., and D. A. Miller. 1995a. Allelopathic influence of nine forage grass extracts on germination and seedling growth of alfalfa. Agronomy Journal 87:767–772.

Chung, I.-M., and D. A. Miller. 1995b. Effect of alfalfa plant and soil extracts on germination and growth of alfalfa. Agronomy Journal 87:762–767.

Cipollini, D. F. 1997. Wind-induced mechanical stimulation increases pest resistance in common bean. Oecologia 111:84–90.

Cipollini, D. F. Jr., and A. H. Redman. 1999. Age-dependent effects of jasmonic acid treatment and wind exposure on foliar oxidase activity and insect resistance in tomato. Journal of Chemical Ecology 25(2):271–281.

Clarholm, M. 1985. Possible roles for roots, bacteria, protozoa and fungi in supplying nitrogen to plants. Pp. 355–356 in: A. H. Fitter, D. Atkinson, D. J. Read, and M. B. Usher (eds.), Ecological Interactions in Soil: Plants, Microbes, and Animals. Blackwell Scientific Publications, Boston, MA.

Clark, C. W., and C. D. Harvell. 1992. Inducible defenses and the allocation of resources: a minimal model. American Naturalist 139:521–539.

Clark, D. B., and D. A. Clark. 1985. Seedling dynamics of a tropical tree: impacts of herbivory and meristem damage. Ecology 66:1884–1892.

Clark, R. B. 1982. Plant response to mineral element toxicity and deficiency. Pp. 71–142 in: M. N. Christiansen and C. F. Lews (eds.), Breeding Plants for Less Favorable Environments. John Wiley & Sons, New York.

Clarkson, D. T. 1991. Root structure and sites of ion uptake. Pp. 417–453 in: Y. Waisel, A. Eshel, and U. Kafkafi (eds.), Plant Roots: The Hidden Half. Marcel Dekker, New York.

Clausen, T. P., F. D. Provenza, E. A. Burritt, P. B. Reichardt, and J. P. Bryant. 1990. Ecological implications of condensed tannin structure: a case study. Journal of Chemical Ecology 16: 2381–2392.

Close, T. J. 1996. Dehydrins: Emergence of a biochemical role for a family of plant dehydration proteins. Physiologia Plantarum 97:795–803.

Coats, R., M. A. Showers, and B. Pavlick. 1993. Management plan for an alkali sink and its endangered plant *Cordylanthus palmatus*. Environmental Management 17:115–127.

Coghan, A. 1995a. Keep kissing to save the British mistletoe. New Scientist (2009):5.

Coghan, A. 1995b. Mistletoe's kiss of death revealed. New Scientist (2002):21.

Cole, J. D. 1994. Detoxification and activation of agrochemicals in plants. Pesticide Science 42:209–222.

Coleman, D. C. 1985. Through a ped darkly: an ecological assessment of root-soil-microbial-faunal interactions. Pp. 1–22 in: A. H. Fitter, D. Atkinson, D. J. Read, and M. B. Usher (eds.), Ecological Interactions in Soil: Plants, Microbes, and Animals. Blackwell Scientific Publications, Boston, MA.

Coleman, J. E. 1992. Zinc proteins: enzymes, storage protein, transcription factors, and replication proteins. Annual Reviews of Biochemistry 61:897–946.

Coleman, J. S., and C. G. Jones. 1991. A phytocentric perspective of phytochemical induction by herbivores. Pp. 3–45 in: D. W. Tallamy, and M. J. Raupp. (eds.), Phytochemical induction by Herbivores. John Wiley & Sons. New York.

Coleman, P. M., H. C. Freeman, J. M. Guss, M. Murata, V. A. Norris, J. A. M. Ranshaw, and M. P. Venkatappa. 1978. X-ray crystal structure analysis of plastocyanin at 2.7 Å resolution. Nature 272:319–324.

Coley, P. D. 1983. Herbivory and defense characteristics of tree species in a lowland tropical forest. Ecological Monographs 53:209–233.

Coley, P. D. 1986. Costs and benefits of defense by tannins in a neotropical tree. Oecologia 70: 238–241.

Coley, P. D., J. P. Bryant, and F. S. Chapin III. 1985. Resource availability and plant antiherbivore defense. Science 230:895–899.

Conjeaud, C., P. Scheromm, and D. Mousain. 1996. Effects of phosphorus and ectomycorrhiza on maritime pine seedlings (*Pinus pinaster*). New Phytologist 133:345–351.

Conn, E. E. (ed.). 1981. Secondary Plant Products (The Biochemistry of Plants, Vol. 7). Academic Press, New York.

Conn, E. E. 1981. Cyanide and cyanogenic glycosides. Pp. 387–411 in: Rosenthal, G. A., and M. R. Berenbaum (eds.), Herbivores: Their Interactions with Secondary Plant Metabolites, 2nd. Ed., Vol. 1. The Chemical Participants. Academic Press, San Diego, CA.

Connell, J. H. 1983. On the prevalence and relative importance of interspecific competition: evidence from field experiments. American Naturalist 122:661–696.

Connolly, J. 1986. On difficulties with replacement-series methodology in mixture experiments. Journal of Applied Ecology 23:125–137.

Cook, C. E., L. P. Wichard, B. Turner, M. E. Wall, and G. H. Egley. 1966. Germination of witchweed (*Striga Lutea Lour.*): Isolation and properties of a potent stimulant. Science 154: 1189–1190.

Cooper, K. M., and P. B. Tinker. 1978. Translocation and transfer of nutrients in vesicular-arbuscular mycorrhizas. II. Uptake and translocation of phosphorus, zinc, and sulfur. New Phytologist 91:43–52.

Cooper, S. M., and N. Owen-Smith. 1986. Effects of plant spinescence on large mammalian herbivores. Oecologia 68:446–455.

Cosgrove, D. J. 1993. Wall extensibility: its nature, measurement, and relationships to plant cell growth. Tansley review 46. New Phytologist 124:1–23.

Cote, J. F., and J. R. Thibault. 1988. Allelopathic potential of raspberry foliar leachates on growth of ectomycorrhizal fungi associated with black spruce. American Journal of Botany 75:966–970.

Coulson, R. N. 1979. Population dynamics of bark beetles. Annual Review of Entomology 24: 217–246.

Cowan, I. R. 1982. Regulation of water use in relation to carbon gain in higher plants. Pp. 589–613 in: O. L. Lange, P. S. Nobel, C. B. Osmond, and H. Ziegler (eds.), Water Relations and Carbon Assimilation (A. Pirson and M. H. zimmerman [series ed.], Encyclopedia of Plant Physiology, Vol. 12B), Springer, Berlin.

Cowie, R. H., and T. G. Wood. 1989. Damage to crops, forestry and rangeland by fungus-growing termites (Termitidae, Mactotermitinae) in Ethiopia. Sociobiology 15:168–173.

Craig, T. P., P. W. Price, and J. K. Itami. 1986. Resource regulation by a stem galling sawfly on the arroyo willow. Ecology 67:419–425.

Cramer, C. L., D. Weissenborn, C. K. Cottingham, C. J. Denbow, J. D. Eisenbeck, D. N. Radin, and X. Yu. 1993. Regulation of defense related gene expression during plant-pathogen interactions. Journal of Nemotology 25:507–518.

Cramer, G. R., G. J. Alberico, and C. Schmidt. 1994a. Salt tolerance is not associated with the sodium accumulation of two maize hybrids. Australian Journal of Plant Physiology 21:675–682.

Cramer, G. R., G. J. Alberico, and C. Schmidt. 1994b. Leaf expansion limits dry matter accumulation of salt stressed maize. Australian Journal of Plant Physiology 21:663–674.

Crawley, M. J. 1987. Benevolent herbivores? Trends in Ecology and Evolution 2:167–168.

Cress, W. A., G. O. Throneberry, and D. L. Lindsey. 1979. Kinetics of phosphorus absorption by mycorrhizal and non-mycorrhizal tomato roots. Plant Physiology 64:484–487.

Crick, J. C., and J. P. Grime. 1987. Morphological plasticity and mineral nutrient capture in two herbaceous species of contrasted ecology. New Phytologist 107:403–414.

Crominski, A., R. B. Bhat, D. J. Weber, and B. N. Smith. 1988. Osmotic stress-dependent conversion of 1-aminocyclopropane-1-carboxylic acid (ACC) to ethylene in the halophyte *Allenrolfea occidentalis.* Environmental and Experimental Botany 28:171–174.

Cronin, G. 1998. Between-species and temporal variation in *Acacia*-ant-herbivore interactions. Biotropica 30:135–139.

Csonka, L. N., S. B. Gelvin, B. W. Goodner, C. S. Orser, D. Siemieniak, and J. Slightom. 1988. Nucleotide-sequence of a mutation in the probe gene of *Escherichia coli* that confers proline overproduction and enhanced tolerance to osmotic stress. Gene 64:166–205.

Cubero, J. I. 1991. Breeding for resistance to *Orobanche* species: a review. Pp. 257–277 in: K. Wegmann, and L. J. Musselman (eds.), Progress in *Orobanche* Research. Eberhand-Karls-Universität, Tübingen, Germany.

Cui, M., and M. M. Caldwell. 1998. Nitrate and phosphate uptake by *Agropyron desertorum* and *Artemisia tridentata* from soil patches with balanced and unbalanced nitrate and phosphate supply. New Phytologist 139:267–272.

Cullings, K., T. Szaro, and T. Burnes. 1996. Evolution of extreme specialization within a lineage of ectomycorrhizal epiparasites. Nature 379:63–66.

Cumming, J. R., and A. B. Tomsett. 1992. Metal tolerance in plants: signal transduction and acclimation mechanisms. Pp. 329–364 in: D. C. Adriano (ed.), Biogeochemistry of Trace Metals, Lewis Publishers, Baco Raton, FL.

Curran, M. 1985. Gas movement in the roots of *Avicennia marina* (Forsk.) Vierh. Australian Journal of Plant Physiology 12:97–108.

Cushman, J. C., and H. J. Bohnert. 1992. Salt stress alters A/T-rich DNA-binding factor interactions within the phosphoenolpyruvate carboxylase promoter from *Mesembryanthemum crystallinum.* Plant Molecular Biology 20:411–424.

Cushman, J. C., C. B. Michalowski, and H. J. Bohnert. 1990. Developmental control of Crassulacean acid metabolism inducibility by salt stress in the common ice plant. Plant Physiology 94:1137–1142.

Dakora, F. D. 1995. Plant flavonoids: biological molecules for useful exploitation. Australian Journal of Plant Physiology 22:87–99.

Dale, D. 1988. Plant-mediated effects of soil mineral stresses on insects. Pp. 35–110 in: E. A. Heinrichs (ed.), Plant Stress–Insect Interactions. John Wiley & Sons, New York.

Dale, H., and M. C. Press. 1998. Elevated CO_2 influences the interaction between the parasitic angiosperm *Orobanche minor* and its host *Trifolium repens*. New Phytologist 140:65–73.

Dale, J. E., and R. G. Wilson. 1979. The effects of photoperiod and mineral nutrient supply on growth and primordia production at the stem apex of barley seedlings. Annals of Botany 44: 537–546.

Daly J. W., and T. F. Spande. 1986. Amphibian alkaloids: chemistry, pharmacology, and biology. Alkaloids, Chemical and Biological Perspectives 4:1–274.

Dalton, B. R. 1999. The occurrence and behavior of plant phenolic acids in soil environments and their potential involvement in allelochemical interference interactions: methodological limitations in establishing conclusive proof of allelopathy. Pp. 57–74 in: Inderjit, K. M. M. Dakshini, and C. L. Foy (eds.), Principles and Practices in Plant Ecology: Allelochemical Interactions. CRC Press, Boca Raton, Fl.

Damgaard C. 1998. Plant competition experiments: testing hypotheses and estimating the probability of coexistence. Ecology 79:1760–1767.

Datta, K. S., S. K. Varma, R. Angrish, P. Kumar, and P. Kumari. 1998. Alleviation of salt stress by plant growth regulators in *Triticum aestivum* L. Biologia Plantarum 40:269–275.

Davidson, N. J., K. C. True, and J. S. Pate. 1989. Water relations of the parasite-host relationship between the mistletoe *Amyema linophyllum* (Fenzl) Tieghem and *Casurina obesa Miq.* Oecologia 80:321–330.

Davis, R. M. 1980. Influence of *Glomus fasciculatus* on *Thielaviopsis basicala* root of citrus. Plant Disease 64:839–841.

Davis, R. M., and J. A. Menge. 1980. The influence of *Glomus fasciculatus* and soil phosphorus on *Phytophthora* root rot of citrus. Phytopathology 70:447–452.

Davis, S. D., and H. A. Mooney. 1985. Comparative water relations of adjacent California shrub and grassland communities. Oecologia 66:522–529.

Dawson, J. H., L. J. Musselman, P. Wolswinkel, and I. Dörr. 1994. Biology and control of *Cuscuta*. Review of Weed Science 6:265–317.

Day, T. A., T. C. Vogelmann, and E. H. DeLucia. 1992. Are some plant forms more effective than others in screening out ultraviolet-β radiation? Oecologia 92:513–519.

DeAngelis, J. D., R. E. Berry, and G. W. Krantz. 1983. Evidence for spider mite (Acari: Tetranychidae) injury-induced leaf water deficits and osmotic ajustment in peppermint. Environmental Entomology 12:336–339.

DeBell, D. S. 1971. Phytotoxic effects of cherrybark oak. Forest Science 17:180–185.

DeBock, F., and A. Fer. 1992. Effects of absiscic acid on the transfer of sucrose from host, *Pelargonium zonale* (L.) Aiton, to a phanerogamic parasite, *Cuscuta reflexa* Roxb. Australian Journal of Plant Physiology 19:679–691.

DeJong, T. M. 1978. Comparative gas exchange and growth responses of C_3 and C_4 beach species grown at different salinities. Oecologia 36:59–68.

Dekker, J. H., Meggitt, W. F., and A. R. Putnam. 1983. Experimental methodologies to evaluate allelopathic plant interactions: The *Abutilon theophrasti-Glycine max* model. Journal of Chemical Ecology 9:945–981.

De Laubenfels, D. J. 1959. Parasitic conifer found in new Caledonia. Science 130:97.

Delauney, A. J., and D. P. S. Verma. 1993. Proline biosynthesis and osmoregulation in plants. Plant Journal 4:215–223.

Delauney, A. J., C. A.-A. Hu, P. B. Kavi Kishor, and D. P. S. Verma. 1993. Cloning of ornithine delta-aminotransferase cDNA from *Vigna aconitifolia* by transcomplementation in *Escherichia coli* and regulation of proline biosynthesis. Journal of Biological Chemistry 268:18673–18678.

Dell, B., B. Botton, F. Martin, and F. le Tacon. 1989. Glutamate dehydrogenases in ectomycorrhizas of spruce (*Picea excelsa* L.) and beech (*Fagus sylvatica* L.). New Phytologist 111: 683–692.

Del Moral, R. 1972. On the variability of chlorogenic acid concentration. Oecologia 9:289–300.

Del Moral, R., and R. G. Cates. 1971. Allelopathic potential of the dominant vegetation of western Washington. Ecology 52:1031–1037.

Del Moral, R., R. J. Willis, and D. H. Ashton. 1978. Suppression of coastal heath vegetation by *Eucalyptus baxteri*. Australian Journal of Botany 26:203–219.

del Rio, C. M., M. Hourdequin, A. Silva, and R. Medel. 1995. The influence of cactus size and previous infection on bird deposition of mistletoe seeds. Australian Journal of Ecology 20: 571–576.

del Rio, C. M., A. Silva, R. Medel, and M. Hourdequin. 1996. Seed dispersers as disease vectors: bird transmission of mistletoe seeds to plant hosts. Ecology 77:912–921.

Demerjian, K. L. 1986. Atmospheric chemistry of ozone and nitrogen oxides. Pp. 105–127 in: A. H. Legge and S. V. Krupa (eds.), Air Pollutants and Their Effects on the Terrestrial Ecosystem. John Wiley & Sons, New York.

Demerjian, K. L., U. K. Roychowdhury, S. Jin, and T. R. Lanni. 1992. Measurements of trace chemical constituents at Whiteface Mountain. Pp. 25–34 in: M. Trout (ed.), ASRC Report 1988–1991. Atmospheric Science Research Center, Albany, NY.

DeMoraes, C. M., and M. C. Mescher. 1999. Interactions in entomology: plant-parasitoid interaction in tritrophic systems. Journal of Entomological Science 34:31–39.

DeMoraes, C. M., W. J. Lewis, P. W. Pare, H. T. Alborn, and J. H. Tumlinson. 1998. Herbivore-infested plants selectively attract parasitoids. Nature 393:570–573.

Denno, R. F., and M. S. McClure (eds.). 1983. Variable Plants and Herbivores in Natural and Managed Ecosystems. Academic Press, San Diego, CA.

dePamphilis, C. W., and J. D. Palmer. 1989. Evolution and function of plastid DNA: a review with special reference to nonphotosynthetic plants. Pp. 182–202 in: C. D. Boyer, J. C. Shannon, and R. C. Hardison (eds.), The Physiology, Biochemistry, and Genetics of Nongreen Plastids. American Society of Plant Physiologists, Rockville, MD.

dePamphilis, C. W., and J. D. Palmer. 1990. Loss of photosynthetic and chlororespiratory genes from the plastid genome of a parasitic flowering plant. Nature 348:337–339.

dePamphilis, C. W., N. D. Young, and A. D. Wolfe. 1997. Evolution of the plastid gene rps2 in a lineage of hemiparasitic and holoparasitic plants: many losses of photosynthesis and complex patterns of rate variation. Proceedings of the National Academy of Sciences (USA) 94:7367–7372.

De Scisciolo, B., D. J. Leopold, and D. C. Walton. 1990. Seasonal patterns of juglone in soil beneath *Juglans nigra* (black walnut) and influence of *J. nigra* on understory vegetation. Journal of Chemical Ecology 16:1111–1130.

Dethier, V. G. 1941. The chemical factors determining the choice of food plants by Papilio larvae. American Naturalist 75:61–73.

Detling, J. K., and M. I. Dyer. 1981. Evidence for potential growth regulators in grasshoppers. Ecology 52:485–488.

Deus-Neumann, B., and M. H. Zenk. 1986. Accumulation of alkaloids in plant vacuoles does not involve an ion-trap mechanism. Planta 167:44–53.

Devlin, R. M., and F. H. Witham. 1983. Plant Physiology, 4th ed. Wadsworth Publishing Company, Belmont, CA.

Diamond, J. M., and J. M. Tormey. 1966. Role of long extracellular channels in fluid transport across epithelia. Nature 210:817–820.

Diaz, S. 1995. Elevated CO_2 responsiveness, interactions at the community level and plant functional types. J. Biogeography 22:289–295.

Dickson, J. G. 1923. Influence of soil temperature and moisture on the development of the seedling-blight of wheat and corn caused by *Gibberella saubinetii*. Journal of Agricultural Ressearch 23:837–870.

Dinan, L. 1995. Distribution and levels of phytoecdysteroids within individual plants of species of the Chenopodiaceae. European Journal of Entomology 92:295–300.

Dirzo, R. 1984. Herbivory: a phytocentric overview. Pp. 141–165 in: R. Dirzo and J. Sarukhán (eds.), Perspectives in Plant Population Biology. Sinauer, Sunderland, MA.

Dodd, A. J. 1940. The Biological Campaign Against Prickly Pear. Commonwealth Prickly Pear Board, Brisbane, Australia.

Doggett, H. 1965. *Striga hermonthica* on sorghum in East Africa. Journal of Agricultural Science 65:83–194.

Dolobovskaya, A. S. 1975. The character of biogenic accumulation of minor elements in forest litter. Pochvovendeniye 7:189–198

Dominguez, J., J. M. Melero Vara, J. Ruso, J. Miller, and J. M. Fernandez Martinez. 1996. Screening for resistance to broomrape (*Orobanche cernua*) in cultivated sunflower. Plant Breeding 115:201–202.

Dopico, B., A. L. Lowe, I. D. Wilson, C. Merodio, and D. Grierson. 1993. Cloning and characterization of avocado fruit messenger-RNAs and their expression during ripening and low-temperature storage. Plant Molecular Biology 21:437–449.

Dörr, I. 1996. New results on interspecific bridges between parasites and their hosts. Pp. 257–277 in: M. T. Moreno, J. I. Cubero, D. Berner, D. Joel, and L. J. Musselman (eds.), Advances in Parasitic Plant Research. Junta de Andalusia, Cordoba, Spain.

Dörr, I., A. Staack, and R. Kollmann. 1994. Resistance of *Helianthus* to *Orobanche*. Histological and cytological studies. Pp. 276–289 in: A. H. Pieterse, J. A. C. Verkleij, and S. J. ter Borg (eds.), Biology and Management of *Orobanche*: Proceedings of the 3rd International Workshop on *Orobanche* and Related *Striga* Research. Royal Tropical Institute, Amsterdam, Netherlands.

Dorschner, K. W., R. C. Johnson, R. D. Eikenbary, and J. D. Ryan. 1986. Insect-plant interactions: greenbugs (Homoptera: Aphididae) disrupt acclimation of winter wheat to drought stress. Environmental Entomology 15:118–121.

Dosskey, M. G., R. G. Linderman, and L. Boersma. 1990. Carbon sink stimulation of photosynthesis in Douglas-fir seedlings by some ectomycorrhizas. New Phytologist 115:269–274.

Douglas, T. J., and R. R. Walker. 1984. Phospholipids, free sterols, and adenosine triphosphate of plasma membrane enriched preparation from roots of citrus genotypes differing in chloride exclusion ability. Physiologia Plantarum 62:51–58.

Dowd, P. F. 1994. Enhanced maize (*Zea mays* L.) pericarp browning: associations with insect resistance and involvement of oxidizing enzymes. Journal of Chemical Ecology 20:2777–2803.

Drennan, D. S. H., and S. O. El Hiweris. 1979. Changes in growth regulating substances in *Sorghum vulgare* infected with *Striga hermonthica*. Pp. 144–155 in: L. J. Musselman, A. D. Worsham, and R. E. Eplee (eds.), 2nd International Symposium on Parasitic Weeds. North Carolina State University, Raleigh, NC.

Ducharme, L. A., and J. R. Ehleringer. 1996. Gas exchange, delta C-13, and heterotrophy for *Castilleja linarifolia* and *Orthocarpus tolmiei*, facultative root hemiparasites on *Artemesia tridentata*. Great Basin Naturalist 56:333–340.

Duchesne, L. C., R. L. Peterson, and B. E. Ellis. 1988a. Interactions between the ectomycorrhizal fungus *Paxillus involutus* and *Pinus resinosa* induces resistance to *Fusarium oxysporum*. Canadian Journal of Botany 66:558–562.

Duchesne, L. C., R. L. Peterson, B. E. Ellis. 1988b. Pine root exudate stimulates antibiotic synthesis by the ectomycorrhizal fungus *Paxillus involutus*. New Phytologist 108:470–476.

Duddridge, J. A., and D. J. Read. 1982. Ultrastructural analysis of the development of mycorrhizas in *Rhododendron ponticum*. Canadian Journal of Botany 60:2345–2356.

Duddridge, J. A., A. Malibari, and D. J. Read. 1980. Structure and function of mycorrhizal rhizomorphs with special reference to their role in water transport. Nature 287:834–836.

Dugger, W. M. 1983. Boron in plant metabolism. Pp. 626–650 in: A. Läuchli and R. L. Bieleski (eds.), Inorganic Plant Nutrition (A. Pirson and M. H. Zimmerman [series eds.], Encyclopedia of Plant Physiology, Vol 15A). Springer-Verlag, New York

Duniway, J. M., and R. D. Durbin. 1971. Some effects of *Uromyces phaseoli* on the transpiration rate and stomatal response of bean leaves. Phytopathology 61:114–119.

Dupont, F. M. 1992. Salt induced changes in ion transport: regulation of primary pumps and secondary transporters. Pp. 91–100 in: D. T. Cooke and D. T. Clarkson (eds.), Transport and Receptor Proteins of Plant Membranes. Plenum Press, New York.

Durand, N., J. C. Debaud, L. A. Casselton, and G. Gay. 1992. Isolation and preliminary characteristics of 5-fluoroindole resistant and IAA overproducer mutants of the ectomycorrhizal fungus *Hebeloma cylindrosporum* Romagnesi. New Phytologist 121:545–553.

Dvorak, J., E. Epstein, A. Galvez, P. J. Gulick, and J. A. Omielan. 1992. Genetics of the physiology and molecular mechanisms associated with salt tolerance in wheat and *Lophopyrum elongatum*. Pp. 125–136 in: Proceedings of the International Conference of Agricultural Management in Salt-affected areas, Agdir, Morocco.

Dvorak, J., M. M. Noaman, S. Goyal, and J. Gorham. 1994. Enhancement of salt-tolerance of *Triticum turgidum* L. by the Knal locus transferred from the *Triticum aestivum* L. chromosome 4D by homologous recombination. Theoretical and Applied Genetics 87:872–877.

Dyer, M. I., T. R. Seastedt, and C. L. Turner. 1993. Herbivory and its consequences. Ecological Applications 3:654–665.

Dyer, M. I., A. M. Moon, M. R. Brown, and D. A. Crossley Jr. 1995. Grasshopper crop and midgut extract effects on plants: an example of reward feedback. Proceedings of the National Academy of Sciences (USA) 92:5475–5478.

Eastmond, P. J., and J. D. Ross. 1997. Evidence that the induction of crassulacean acid metabolism by water stress in *Mesembreanthemum crystallinum* (L.) involves root signaling. Plant Cell and Environment 20:1559–1565.

Ebel, R. C., X. Duan, D. W. Still, and R. M. Auge. 1997. Xylem sap abscisic acid concentration and stomatal conductance of mycorrhizal *Vigna unguiculata* in drying soil. New Phytologist 135:755–762.

Egan, T. P., and I. A. Ungar. 1998. Effect of different salts of sodium and potassium on the growth of *Atriples prostrata* (Chenopodiaceae). Journal of Plant Nutrition 21:2193–2205.

Egger, K. N., R. M. Danielson, and J. A. Fortin. 1991. Taxonomy and population structure of E-strain mycorrhizal fungi inferred from ribosomal and mitochondrial DNA polymorphisms. Mycological Research 95:866–872.

Ehhalt, D. H., and J. W. Drummond 1982. The tropospheric cycle of NO_x. Pp. 219–251 in: H. W. Georgii and W. Jaeschke (eds.), Chemistry of the Unpolluted and the Polluted Troposphere. Reidel, Dordrecht.

Ehleringer, J. R. 1984. Intraspecific competitive effects on water relations, growth, and reproduction in *Encelia farinosa*. Oecologia 63:153–58.

Ehleringer J. R., and O. Bjorkman. 1978. Pubescence and leaf spectral characteristics of a desert shrub *Encelia farinosa.* Oecologia 36:151–162.

Ehleringer, J. R., and J. D. Marshall. 1995. Water relations. Pp. 125–140 in: M. C. Press and J. D. Graves (eds.), Parasitic Plants. Chapman & Hall, London.

Ehleringer, J. R., C. S. Cook, and L. L. Tieszen. 1986. Comparative water use and nitrogen relations in a mistletoe and its host. Oecologia 68:279–284.

Ehrlich, P. R., and P. H. Raven. 1964. Butterflies and plants: a study in coevolution. Evolution 18:586–608.

Ehrlich, P. R., and P. H. Raven. 1967. Butterflies and plants. Scientific American 216:106–113.

Einhellig, F. A. 1986. Mechanisms and modes of action of allelochemicals. Pp. 171–188 in: A. R. Putnam, and C.-S. Tang (eds.), The Science of Allelopathy. John Wiley & Sons, New York.

Einhellig, F. A. 1987. Interactions among allelochemicals and other stress factors of the plant environment. Pp. 343–357 in: G. R. Waller, (ed.), Allelochemicals: Role in Agriculture and Forestry. American Chemical Society, Washington, DC.

Einhellig, F. A. 1989. Interactive effects of allelochemicals and environmental stress. Pp. 101–116 in: C. H. Chou and G. R. Waller (eds.), Phytochemical Ecology: Allelochemicals, Mycotoxins, and Insect Phermones and Allomones. Institute of Botany, Academia Sinica, Taipei.

Einhellig, F. A. 1995. Mechanisms of action of allelochemicals in allelopathy. Pp. 96–116 in: K. M. Inderjit, M. Dakshini, and F. A. Einhellig (eds.), Allelopathy: Organisms, Processes, and Application. American Chemical Society, Washington, DC.

Einhellig, F. A. 1996. Interactions involving allelopathy in cropping systems. Agronomy Journal 88:886–893.

Einhellig, F. A., M. K. Schon, and J. A. Rasmussen. 1982. Synergistic effects of four cinnamic acid compounds on grain sorghum. Journal of Plant Growth Regulation 1:251–258.

Einhellig, F. A., J. A. Rasmussen, A. M. Hejl, and I. F. Souza. 1993. Effects of root exudates sorgoleone on photosynthesis. Journal of Chemical Ecology 19:369–375.

Eissenstadt, D. M., J. H. Graham, J. P. Syvertsen, and D. L. Drouillard. 1993. Carbon economy of sour orange in relation to mycorrhizal colonization and phosphorus status. Annals of Botany 71:1–10.

Elzam, O. E., and E. Epstein. 1969. Salt relations of two grass species differing in salt tolerance. I. Growth and salt content at different salt concentrations. Agrochemica 13:187–195.

EPA (U.S. Environmental Protection Agency). 1991a. Air Quality Criteria for Oxides of Nitrogen. Review draft EPA-600/8-91-049aA. Office of Health and Environmental Assessment, Environmental Criteria and Assessment Office, Research Triangle Park, NC.

EPA (US Environmental Protection Agency). 1991b. National Air Quality and Emissions Trends Report, 1989. EPA/450/4-91-003. Office of Air Quality Planning and Standards, Research Triangle Park, NC.

Epstein, E. 1972. Mineral Nutrition of Plants: Principles and Perspective. John Wiley & Sons, New York.

Ericson, M. C., and S. H. Alfinito. 1984. Proteins produced during salt stress in tobacco cell culture. Plant Physiology 74:506–509.

Ernst, W. H. O. 1996. Phytotoxicity of heavy metals. Pp. 423–430 in: C. Rodrigues-Barrueco (ed.), Fertilizers and Environment, Kluwer Academic Publishers, Boston, MA.

Espartero, J., J. A. Pintor-Toro, and J. M. Pardo. 1994. Differential accumulation of S-adenosylmethionine synthase transcripts in response to salt stress. Plant Molecular Biology 25:217–227.

Eüsackers, H., and C. F. VandeBund. 1980. Effects on soil fauna. Pp. 255–305 in: R. J. Hance (ed.), Interactions Between Herbicides and the Soil. Academic Press, NY.

Fahn, A. 1990. Plant Anatomy, 4th Ed. Pergamon Press. New York.

Farmer, E. E., and C. A. Ryan. 1990. Interplant communication: airborne methyl jasmonate induces synthesis of proteinase inhibitors in plant leaves. Proceedings of the National Academy of Science (USA) 87:7713–7716.

Farmer, E. E., R. R. Johnson, and C. A. Ryan. 1992. Regulation of expression of proteinase inhibitor genes by methyl jasmonate and jasmonic acid. Plant Physiology 98:995–1002.

Feeney, P. 1970. Seasonal changes in oak leaf tannins and nutrients as a cause of spring feeding by winter moth caterpillars. Ecology 51:561–581.

Feeney, P. 1976. Plant apparancy and chemical defense. Recent Advances in Phytochemistry 10:1–40.

Feeney, P., L. Rosenberry, and M. Carter. 1983. Chemical aspects of oviposition behavior in butterflies. Pp. 27–76. in: S. Ahmed (ed.), Host Seeking Behavior and Mechanisms. Academic Press, New York.

Feeney, P., K. Sachdev, L. Rosenbury, and M. Carter. 1988. Luteolin 7-O-(6′-O-malonyl)-β-D-glucoside and *trans*-chlorogenic acid: oviposition stimulants for the black swallowtail butterfly. Phytochemistry 27:3439–3448.

Felton, G. W., and S. S. Duffey. 1991. Reassesment of the role of gut alkalinity and detergency in insect herbivory. Journal of Chemical Ecology 17:1821–1836.

Felton, G. W., K. Donato, R. J. del Vecchio, and S. S. Duffey. 1989. Activation of plant foliar oxidases by insect feeding reduces the nutritive quality of foliage for herbivores. Journal of Chemical Ecology 15:2667–2694.

Feng, J. N. and A. V. Barker. 1992. Ethylene evolution and ammonium accumulation by tomato plants under water and salinity stress. Journal of Plant Nutrition 15:2471–2490.

Ferguson, D. E. 1991. Allelopathic potential of western coneflower (*Rudbeckia occidentalis*). Canadian Journal of Botany 69:2806–2808.

Fernandes, G. W., E. A. deMattos, A. C. Franco, U. Luttge, and H. Zeigler. 1998. Influence of the parasite *Pilostyles inge* (Rafflesiaceae) on some physiological parameters of the host plant, *Mimosa naguirei* (Mimosaceae). Botanica Acta 111:51–54.

Field, C. B. 1983. Allocating leaf nitrogen for the maximizing of carbon gain: leaf age as a control on the allocation program. Oecologia 56:341–347.

Firbank, L. G., and A. R. Watkinson. 1987. On the analysis of competition at the level of the individual plant. Oecologia 71:308–317.

Fischer, N. H. 1986. The function of mono and sesquiterpenes as plant germination and growth regulators. in: A. R. Putnam, and C.-S. Tang (eds.), 1986. The Science of Allelopathy. John Wiley & Sons, New York.

Fischer, N. H., N. Tanrisever, and G. B. Williamson. 1988. Allelopathy in the Florida scrub community as a model for natural herbicide actions. Pp. 233–249 in: H. G. Cutler (ed.), Biologically Active Natural Products: Potential Use in Agriculture (ACS Symposium Vol. 380), American Chemical Society, Washington, DC.

Fischer, N. H., J. D. Weidenhamer, J. L. Riopel, L. Quijano, and M. A. Menelaou. 1994. In search of allelopathy in the Florida scrub: the role of terpenoids. Journal of Chemical Ecology 20:1355–1380.

Fisher, J. T. 1983. Water relations of mistletoes and their hosts. Pp. 161–184 in: M. Calder and P. Bernhardt (eds.), The Biology of Mistletoes. Academic Press, Sydney, Australia.

Fisher, R. F. 1987. Allelopathy: a potential cause of forest regeneration failure. Pp. 176–184 in: G. R. Waller (ed.), Allelochemicals: Role in Agriculture and Forestry. American Chemical Society, Washington, DC.

Fitter, A. H. 1988. Water relations of red clover, *Trifolium pratense* L., as affected by VA mycorrhizal infection and phophorus supply before and during drought. Journal of Experimental Botany 39:595–604.

Fitter, A. H. 1994. Architecture and biomass allocation as components of the plastic response of root systems to soil heterogeneity. Pp. 305–324 in: M. M. Caldwell and R. W. Pearcy (eds.), Exploitation of Environmental Heterogeneity by Plants: Ecophysiological Processes Above and Belowground, Academic Press, San Diego, CA.

Fitter, A. H., and K. M. Hay. 1987. Environmental Physiology of Plants. Academic Press, New York.

Fletcher, J. S. 1997. Testing and knowledge of organic phytotoxins. Pp. 271–289 in: J. Tarradellas, G. Bitton, and D. Rossel (eds.), Soil Ecotoxicology, Lewis Publishers, Boca Raton, FL.

Fletcher, J. S., and F. L. Johnson. 1988. Database assessment of phytotoxicity data published on terrestrial vascular plants. Environmental Toxicology and Chemistry 7:615–622.

Fletcher, J. S., T. G. Pfleeger, and H. C. Ratsch. 1993. Potential environmental risks associated with the new sulfonylurea herbicides. Environmental Science and Technology 27:2250–2252.

Fletcher, J. S., T. G. Pfleeger, H. C. Ratsch, and R. Hayes 1996. Potential impact of low levels of chlorsulfuron and other herbicides on growth and yield of nontarget plants. Environmental Toxicology and Chemistry 15:1189–1196.

Flowers, T. J., and J. L. Hall. 1972. Biochemical and cytological studies of *Sueda maritima*. Pp. 357–377 in: W. P. Anderson (ed.), Ion Transport in Plants. Academic Press, New York.

Flowers, T. J., and A. Läuchli. 1983. Sodium versus potassium: substitution and compartmentation. Pp. 651–681 in: A. Läuchli and R. L. Bieleski (eds.), Inorganic Plant Nutrition (A. Pirson and M. H. Zimmerman [series eds.], Encyclopedia of Plant Physiology, Vol. 15A). Springer-Verlag, New York.

Fogel, R. 1985. Roots as primary produces in below-ground ecosystems. Pp. 23–42 in: A. H. Fitter, D. Atkinson, D. J. Read, and M. B. Usher (eds.), Ecological Interactions in Soil: Plants, Microbes, and Animals. Blackwell Scientific Publications, Boston, MA.

Foley, W. J., and I. D. Hume. 1987. Digestion and metabolism of high-tannin *Eucalyptus* foliage by the brushtail possum (*Trichosurus-Vulpecula*) (Marsupialia, Phalangeridae). Journal of Comparative Physiology B 157:67–76.

Fonteyn, P. J., and B. E. Mahill. 1981. An experimental analysis of structure in a desert plant community. Journal of Ecology 69:883–96.

Fortmeier, R., and S. Schubert. 1995. Salt tolerance of maize (*Zea mays* L.): the role of sodium exclusion. Plant Cell and Environment 18:1041–1047.

Fosket, D. E. 1994. Plant Growth and Development: A Molecular Approach. Academic Press, New York.

Fowden, L., P. J. Lea, and E. A. Bell. 1979. The nonprotein amino acids in plants. Advances in Enzymology 50:117–175.

Fowler, N. 1981. Competition and coexistence in a North Carolina grassland. II. The effects of the experimental removal of species. Journal of Ecology 69:843–854.

Fowler, N. 1986. The role of competition in plant communities in arid and semiarid regions. Annual Review of Ecological Systematics 17:89–110.

Fox, L. R., and P. A. Morrow. 1986. On comparing herbivory damage in Australia and north temperate systems. Australian Journal of Ecology 11:387–393.

Foy, C. L. 1999. How to make bioassays for allelopathy more relevant to field conditions. Pp. 25–33 in: K. M. Inderjit, M. Dakshini, and C. L. Foy (eds.), Principles and Practices in Plant Ecology: Allelochemical Interactions. CRC Press, Boca Raton, FL.

Fraenkel, G. S. 1959. The raison dêtre of secondary plant substances. Science 129:1466–1470.

Francis, R., and D. J. Read. 1984. Direct transfer of carbon between plants connected by vesicular-arbuscular mycorrhizal mycelium. Nature (London) 307:53–56.

Fransen, B., H. de Kroon, and F. Berendse. 1998. Root morphological plasticity and nutrient acquisition of perennial grass species form habitats of different nutrient availability. Oecologia 115:351–358.

Freckman, D. W., and R. A. Virginia. 1989. Plant feeding nematodes in deep rooting desert systems. Ecology 70:1665–1678.

Fredeen, A. L., and N. Terry. 1988. Influence of vesicular-arbuscular mycorrhizal infection and soil phosphorus level on the growth and carbon metabolism of soybean. Canadian Journal of Botany 66:2311–2316.

Freedman, B., and T. C. Hutchinson. 1981. Sources of metal and elemental contamination of terrestrial environments. Pp. 35–94 in: N. W. Lepp (ed.), Effect of Heavy Metal Pollution on Plants, Vol. 2. Applied Sciences Publishers, Englewood, NJ.

Fried, M., and H. Broeshart. 1967. The Soil-Plant System. Academic Press, New York.

Friedel, M. H., G. N. Bastin, and G. F. Griffen. 1988. Range assesment & monitoring in arid lands: the derivation of functional groups simplify data. Journal of Environmental Management 27:85–97.

Friedland, A. J. 1990. The movement of metals through soils and ecosystems. Pp. 7–19 in: A. Jonathan Shaw (ed.), Heavy Metal Tolerance in Plants: Evolutionary Aspects. CRC Press, Boca Raton, FL.

Friedman, J. 1987. Allelopathy in desert ecosystems. Pp. 53–68 in: G. R. Waller (ed.), Allelochemicals: Role in Agriculture and Forestry. American Chemical Society, Washington, DC.

Friedman, J., and G. R. Waller. 1983. Caffeine hazards and their prevention in germinating seeds of coffee (*Coffea arabica* L.). Journal of Chemical Ecology 9:1099–1106.

Friedman, J., and G. R. Waller. 1985. Allelopathy and autotoxicity. Trends in Biochemical Sciences 10:47–50.

Friend, J. 1981. Alterations in secondary metabolism. Pp. 179–200 in: P. G. Ayers (ed.), Effects of Disease on the Physiology of the Growing Plant. Cambridge University Press, New York.

Fritig, B., L. Hirth, and G. Ourisson. 1970. Biosynthesis of the coumarins: scopoletin formation in tobacco tissue cultures. Phytochemistry 9:1963–1975.

Frost, D. L., A. L. Gurney, M. C. Press, and J. D. Scholes. 1997. *Striga hermonthica* reduces photosynthesis in sorghum: the importance of stomatal limitations and a potential role of ABA? Plant Cell and Environment 20:483–492.

Fuerst, E. P., and A. R. Putnam. 1983. Separating the competitive and allelopathic components of interference: theoretical principles. Journal of Chemical Ecology 9:937–944.

Fukuda, A., Y. T. Yazaki, S. Ishikawa, T. Koike, and Y. Tanaka. 1998. Na^+/H^+ antiporter in tonoplast vessicles from rice roots. Plant Cell Physiology 39:196–201.

Gallet, C. 1994. Allelopathic potential in bilberry-spruce forests: influence of phenolic compounds on spruce seedlings. Journal of Chemical Ecology 20:1009–1024.

Gaffney, T., L. Friedrich, B. Vernooij, D. Negrotto, G. Nye, S. Uknes, E. Ward, H. Kessman, and J. Ryals. 1993. Requirement of salicylic acid for the induction of systemic acquired resistance. Science 261:754–756.

Gammon, R. H., E. T. Sundquist, and P. J. Fraser. 1985. Pp. 25–62 in: J. R. Trabalka (ed.), Atmospheric Carbon Dioxide and the Global Carbon Cycle. DOE/ER-0239, U.S. Department of Energy, Washington, DC.

Garbaye, J. 1994. Helper bacteria: a new dimension to the mycorrhizal symbiosis. New Phytologist 128:197–210.

Garcia-Torres, L., M. C. Munoz, M. J. Exposito, and F. L. Granados. 1996. Modelling the economics of controlling nodding broomrape (*Orobanche cernua*) in sunflower (*Helianthus annuus*). Weed Science 44:591–595.

Gardes, M., and T. Bruns. 1993. ITS primers with enhanced specificity for basidiomycetes—application to the identification of mycorrhizae and rusts. Molecular Ecology 2:113–118.

Gardes, M., and T. Bruns. 1996. Community structure of ectomycorrhizal fungi in a *Pinus muricata* forest: above- and below-ground views. Canadian Journal of Botany 74(10):1572–1583.

Gardes, M., G. Mueller, J. A. Fortin, and B. A. Kropp. 1991a. Mitochondrial DNA polymorphism in *Laccaria bicolor*, *L. Laccata*, *L. proxima*, *L. amethystina*. Mycological Research 95:206–216.

Gardes, M., G. Mueller, J. A. Fortin, and B. A. Kropp. 1991b. Restriction fragment length polymorphisms in ribosomal DNA from four species of *Laccaria*. Phytopathology 80:1312–1317.

Garfield, E. 1987. Citation classics in plant sciences and their impact on research. Current Contents 40:3.

Garner, J. H. B. 1994. Nitrogen oxides, plant metabolism and forest ecosystem response. Pp. 301–338 in: R. G. Alscher and A. R. Wellburn (eds.), Plant Responses to the Gaseous Environment: Molecular, Metabolic and Physiological Aspects. Chapman & Hall, New York.

Garner, J. H. B., T. Pagano, and E. B. Cowling. 1989. An Evaluation of the Role of Ozone, Acid Deposition, and Other Airborne Pollutants in the Forests of Eastern North America. General technical report SE-59. U.S. Department of Agriculture, Forest Service, Southeastern Forest Experiment Station, Asheville, NC.

Garwood, V. A., and J. C. Rogler. 1987. Responses of growth selected Japaneese quail lines to tannin levels in grain sorghum diets with supplemental protein. Poultry Science 66:1095–1100.

Gassmann, W., F. Rubio, and J. I. Schroeder. 1996. Alkali cation selectivity of the wheat root high-affinity potassium transporter HKT1. Plant Journal 10:869–882.

Gay, G., and J. C. Debaud. 1987. Genetic study on indole-3-acetic acid production by ecto-mycorrhizal *Hebeloma* species: inter- and intraspecific variability in homo- and di-karyotic mycelia. Applied Microbiology and Biotechnology 26:141–146.

Gay, G., L. Normand, R. Marmeisse, B. Sotta, and J. C. Debaud. 1994. Auxin overproducer mutants of *Hebeloma cylindrosporum* Romagnesi have increased mycorrhizal activity. New Phytologist 128:645–657.

Gelburd, D. E. 1985. Managing salinity lessons from the past. Journal of Soil and Water Conservation 40(4):329–331.

Gentle, C., and J. Duggin. 1997. Allelopathy as a competitive strategy in persistent thickets of *Lantana camara* L. in three Australian forest communities. Plant Ecology 132:85–95.

Georgiadis, N. J., and S. J. McNaughton. 1988. Interactions between grazers and a cyanogenic grass, *Cynadon plectostachyus*. Oikos 51:343–350.

Gerloff, G. C. 1987. Intact-plant screening for tolerance of nutrient-deficiency stress. Pp. 55–68 in: W. H. Gabelman and B. C. Loughman (eds.), Genetic Aspects of Plant Mineral Nutrition. Martinus Nijhoff Publishers, Boston, MA.

Gershenzon, J. 1984. Changes in levels of plant secondary metabolites under water and nutrient stress. Pp. 273–320 in: B. N. Timmermann, C. Steelink, and F. A. Loewus (eds.), Phytochemical Adaptations to Stress (Recent Advances in Phytochemistry, Vol. 18). Plenum Press, New York.

Giampietro, M. 1994. Sustainability and technological development in agriculture. BioScience 44:677–689.

Gianinazzi-Pearson, V., and P. Bonfante-Fasolo. 1986. Variability in wall structure and mycorrhizal behaviour of ericoid fungal isolates. Pp. 563–567 in: V. Gianinazzi-Pearson and S. Gianinazzi (eds.), Physiological and Genetical Aspects of Mycorrhizae. Institut National de la Recherche Agronomique, Paris.

Gianinazzi-Pearson, V., and S. Gianinazzi. 1983. The physiology of vesicular-arbuscular mycorrhizal roots. Plant and Soil 71:197–209.

Gianinazzi-Pearson, V., S. E. Smith, S. Gianinazzi, and F. A. Smith. 1991. Enzymatic studies on the metabolism of vesicular-arbuscular mycorrhizas. V. Is H^+-ATPase a component of ATP-hydrolysing enzyme activities in plant-fungus interfaces? New Phytologist 117:61–74.

Gibson, C. C., and A. R. Watkinson. 1989. The host range and selectivity of a parasitic plant: *Rhinanthus minor* L. Oecologia 78:401–406.

Gibson, C. C., and A. R. Watkinson. 1992. The role of the hemiparasitic annual *Rhinanthus minor* in determining grassland community structure. Oecologia 89:62–68.

Gillette, D. G. 1984. Concern about atmospheric pollution. Pp. 7–13 in: M. Treshow (ed.), Air Pollution and Plant Life. John Wiley & Sons, New York.

Gilmore, A. E. 1971. The influence of endotrophic mycorrhizae on the growth of peach seedlings. Journal of the American Society of Horticulture Science 96:35–38.

Gilmore, A. R. 1977. Effects of soil moisture stress on monoterpenes in loblolly pine. Journal of Chemical Ecology 3:667–676.

Gitay, H., and I. R. Nobel. 1997. What are functional types and how should we seek them? In: T. M. Smith, H. H. Shugart, and F. I. Woodward (eds.), Plant Functional Types. Cambridge Univesity Press, Cambridge, UK.

Givnish, T. J. 1986. Biomechanical constraints on crown geometry in forest herbs. Pp. 525–583 in: T. J. Givnish (ed.), On the Economy of Plant Form and Function: Proceedings of the 6th Maria Moors Cabot Symposium (Evolutionary Constraints on Primary Productivity, Adaptive Patterns of Energy Capture in Plants), Harvard Forest, August 1983. Cambridge University Press, Cambridge, UK.

Glass, A. D. M. 1976. The allelopathic potential of phenolic acids associated with the rhizosphere of *Pteridium aquilinum*. Canadian Journal of Botany 54:855–858.

Glass, A. D. M. 1989. Plant Nutrition: An Introduction to Current Concepts. Jones and Bartlett Publishers, Sudbury, MA.

Glass, A. D. M., and M. Y. Siddiqi. 1984. The control of nutrient uptake rates in relation to the inorganic composition of plants. Pp. 103–147 in: P. B. Tinker, and A. Läuchli (eds.), Advances in Plant Nutrition, Vol. 1. Praeger Scientific, New York.

Glass, D. J. 1993. Commercialization of soil microbial technologies. Pp. 595–618 in: F. B. Metting, Jr. (ed.), Soil Microbial Ecology. Marcel Dekker, New York.

Glass, D. M., and J. Dunlop. 1974. Influence of phenolic acids on ion uptake IV. Depolorization of membrane potentials. Plant Physiology 54:855–858.

Glatzel, G. 1983. Mineral nutrition and water relations of hemiparasitic mistletoes: a question of partitioning. Experiments with *Loranthus europaeus,* on *Quercus petraea* and *Quercus robur*. Oecologia 56:193–201.

Glenn, E. P. 1987. Relationship between cation accumulation and water content of salt tolerant grasses and a sedge. Plant Cell and Environment 10:205–212.

Glenn, E. P. 1997. Mechanisms of salt tolerance in higher plants. Pp. 83–110 in: A. S. Basra and R. K. Basra (eds.), Mechanisms of Environmental Stress Resistance in Plants. Harwood Academic Publishers, Amsterdam.

Glenn, E. P., and J. W. O'Leary. 1984. Relationship between salt accumulation and water content of dicotyledonous halophytes. Plant, Cell and Environment 7:253–261.

Glenn E. P., M. W. Olsen, R. J. Frye, and D. W. Moore. 1994. How much sodium accumulation is necessary for salt tolerance of subspecies of the halophyte, *Atriplex canascens*. Plant Cell and Environment 17:711–719.

Gliessman, S. R. 1976. Allelopathy in a broad spectrum of environments as illustrated by bracken. Botanical Journal of the Linnaean Society 73:95–104.

Gliessman, S. R., and C. H. Muller. 1978. The allelopathic mechanisms of dominance in bracken (*Pteridium aquilinum*) in southern California. Journal of Chemical Ecology 4:337–362.

Gogala, N. 1991. Regulation of mycorrhizae by hormonal factors produced by hosts and fungi. Experientia 47:331–340.

Godoy, J. A., J. M. Pardo, and J. A. Pintor-Toro. 1990. A tomato cDNA inducible by salt stress and abscisic acid: Nucleotide sequence and expression pattern. Plant Molecular Biology 15: 695–705.

Goldberg, D. E. 1990. Components of resource competition in plant communities. Pp. 27–49 in: J. B. Grace, and D. Tilman (eds.), Perspectives on Plant Competition. Academic Press, San Diego, CA.

Goldberg, D. E., and P. A. Werner. 1983. Equivalence of competitors in plant communities: a null hypothesis and a field experimental approach. American Journal of Botany 70:1098–1104.

Goldberg, D. E., R. Turkington, and L. Olsvig-Whittaker. 1995. Quantifying the community-level consequences of competition. Folia Geobotanica Phytotaxica 30:231–242.

Goldner, W. R., and R. J. Medve. 1986. Allelopathic effects of *Cladonia cristalla* on ectomycorrhizal fungi common to bituminous strip-mine spoils. Canadian Journal of Botany 64: 1586–1590.

Goldstein, G., F. Rada, L. Sternberg, J. L. Burguera, M. Burguera, A. Orozco, M. Montilla, O. Zabala, A. Azocar, M. J. Canales, and A. Celis. 1989. Gas-exchange and water-balance of a mistletoe species and its mangrove hosts. Oecologia 78:176–183.

Gollan, T., J. B. Passioura, and R. Munns. 1986. Soil water status affects the stomatal conductance of fully turgid wheat and sunflower leaves. Australian Journal of Plant Physiology 13: 459–464.

Gomez-Cadenas, A., F. R. Tadeo, E. Primo-Millo, and M. Tallon. 1998. Involvement of abscisic acid in the response of citrus seedlings to salt shock. Physiologia Plantarum 103:475–484.

Gonzales-Coloma, A., C. S. Wisdom, and P. W. Rundel. 1988. Ozone impact on the antioxidant nordihydroguaiaretic acid content in the external leaf resin of *Larrea tridentata*. Biochemical Systematics and Ecology 16:59–64.

Gonzalez, V. M., J. Kazimir, C. I. Nimbal, L. A. Weston, and G. M. Cheniae. 1997. Inhibition of a photosystem II electron transfer reaction by the natural product sorgoleone. Journal of Agricultural and Food Chemistry 45:1415–1421.

Goodman, R. N., Z. Kiraly, and K. R. Wood. 1986. The Biochemistry and Physiology of Plant Disease. University of Missouri Press, Columbia, MO.

Gopal, B., and U. Goel. 1993. Competition and allelopathy in aquatic plant communities. Botanical Review 59:155–210.

Gorbenkov-Germanov, D. S., and I. V. Kozlova. 1973. Mechanism of the decomposition of ozone in basic aqueous media. Doklady Akademii Nauk USSR 210:851–854.

Gorham J. 1992. Salt tolerance of plants. Science Progress 76:273–285.

Gorham J. 1993. Genetics and physiology of enhanced K/Na discrimination. Pp. 151–159 in: P. Randall (ed.), Genetic Aspects of Plant Mineral Nutrition. Kluwer Academic Publishers, Dordrecht, Netherlands.

Gorham, J. 1994. Salt tolerance in the Triticae: K/Na discrimination in some perennial wheat grasses and their amphiploids with wheat. Journal of Experimental Botany 45:441–447.

Gorham, J., R. G. Wyn-Jones, and E. McDonnell. 1985. Some mechanisms of salt tolerance in crop plants. Plant and Soil 89:15–40.

Gorin M. B., S. B. Yancy, J. Cline, J. R. Revel, and J. Horwitz. 1984. The major intrinsic protein (MIP) of the bovine lens fiber membrane: characterization of structure based on cDNA cloning. Cell 39:49–59.

Gould, F. 1988. Genetics of pairwise and multispecies, plant-herbivore coevolution. Pp. 13–55 in: K. C. Spencer (ed.), Chemical Mediation of Coevolution. American Institute of Biological Sciences, Washington, DC.

Grace, J. B. 1990. On the relationship between plant traits and competitive ability. Pp. 51–65 in: J. Grace and D. Tilman (eds.), Perspectives on Plant Competition. Academic Press, New York.

Grace, J. B., and D. Tilman. 1990. Perspectives on plant competition: some introductory remarks. Pp. 3–7 in: J. B. Grace, and D. Tilman (eds.), Perspectives on Plant Competition. Academic Press, San Diego, CA.

Graham, J. H., J. P. Syvertson, and M. L. Smith. 1987. Water relations of mycorrhizal and phosphorus-fertilized non-mycorrhizal citrus under drought stress. New Phytologist 105:411.

Graham, R. D. 1975. Male sterility in wheat plants deficient in copper. Nature 254:514–515.

Graves, J. D., A. Wylde, M. C. Press. 1990. Growth and carbon allocation in *Pennisetum typhoides* infected with the parasitic angiosperm *Striga hermonthica*. Plant Cell and Environment 13:367–373.

Graves, J. D., M. C. Press, S. Smith, and G. R. Stewart. 1992. The canopy carbon economy of the association between cowpea and the parasitic angiosperm *Striga gesnerioides*. Plant Cell and Environment 15:283–288.

Greenway, H., and R. Munns. 1980. Mechanisms of salt tolerance in nonhalophytes. Annual Review of Plant Physiology 31:149–190.

Greenway, H., and C. B. Osmond. 1972. Salt responses of enzymes from species differing in salt tolerance. Plant Physiology 49:256–259.

Gregory, R. A., and P. M. Wargo. 1986. Timing of defoliation and its effect on bud development, starch reserves, and sap sugar concentration in sugar maple. Canadian Journal of Forest Research 16:10–17.

Griffis, F. C. 1956. Mistletoe myth develops industry. Journal of Geography 55:251–254.

Grime, J. P. 1973. Competitive exclusion in herbaceous vegetation. Nature 242:344–347.

Grime, J. P. 1977. Evidence for the existence of three primary strategies in plants and its relevance to ecological and evolutionary theory. American Naturalist 111:1169–1194.

Grime, J. P. 1994. The role of plasticity in exploiting environmental heterogeneity. Pp. 1–20 in: M. M. Caldwell and R. W. Pearcy (eds.), Exploitation of Environmental Heterogeneity by Plants. Academic Press, San Diego, CA.

Grime, J. P., and J. G. Hodgson. 1987. Botanical contributions to contemporary ecological theory. Pp. 283–296 in: I. H. Rorison, J. P. Grome, R. Hunt, G. A. F. Hendry, and D. H. Lewis (eds.), Frontiers of Comparative Plant Physiology 106:283–296.

Grote, E. M., G. Ejeta, and D. Rhodes. 1994. Inheritance of glycine-betaine deficiency in sorghum. Crop Science 34:1217–1220.

Grovier, R. N., M. D. Neldon, and J. S. Pate. 1967. Hemiparasitic nutrition in angiosperms. I. The transfer of organic compounds from host to *Odontites verna* (Bell.) Dum. (Scrophulariaceae). New Phytologist 66:285–297.

Grzywacz, A., and J. Wazny. 1973. The impact of industrial air pollutants on the occurrence of several important pathogenic fungi of forest trees in Poland. European Journal of Forest Pathology 3:129–141.

Guderian, R., D. T. Tingey, and R. Rabe. 1985. Effects of photochemical oxidants on plants. Pp. 129–295 in: R. Guderian (ed.), Air Pollution by Photochemical Oxidants. Springer-Verlag, New York.

Guerrero, M. G., J. M. Vega, and M. Losda. 1981. The assimilatory nitrate-reducing system and its regulation. Annual Reviews of Plant Physiology 32:169–204.

Hadley, G. 1970. Non-specificity of symbiotic infection in orchid mycorrhiza. New Phytologist 69:1015–1023.

Hagerman, A. E., and L. G. Butler. 1991. Tannins and Lignins. Pp. 355–385 in: G. A. Rosenthal and M. R. Berenbaum (eds.), The Chemical Participants (Herbivores: Their Interactions with Secondary Plant Metabolites, 2nd Ed., Vol 1). Academic Press, San Diego, CA.

Hagerman, A. E., and C. T. Robbins. 1993. Specificity of tannin-binding salivary proteins relative to diet selection by mammals. Canadian Journal of Zoology 71:628–633.

Hahlbrook, K., and D. Scheel. 1989. Physiology and molecular biology of phenolpropanoid metabolism. Annual Reviews of Plant Physiology and Plant Molecular Biology 40:347–369.

Hajibagheri, M. A., D. M. R. Harvey, and T. J. Flowers. 1984. Photosynthetic oxygen evolution in relation to ion contents in the chloroplasts of *Sueda maritima*. Plant Science Letters 34: 353–362.

Hajibagheri, M. A., D. M. R. Harvey, and T. J. Flowers. 1987. Quantitative ion distribution within root cells of salt-sensitive and salt-tolerant maize varieties. New Phytologist 105:367–379.

Hall, A. B., U. Blum, and R. V. Fites. 1983. Stress modification of allelopathy of *Helioanthus annuus* L. debris on seedling biomass production of *Amaranthus retroflexus* L. Journal of Chemical Ecology 9:1213–1222.

Hamberlandt, G. 1914. Physiological Plant Anatomy. MacMillan and Co., London.

Handley, L. L., M. J. Daft, J. Wilson, C. M. Scrimgeour, K. Ingleby, and M. A. Sattarma. 1993. Effects of the ectomycorrhizal and VA mycorrhizal fungi *Hydnagium carneum* and *Glomus clarum* on the delta N^{15} and delta C^{13} values of *Eucalyptus globulus* and *Ricinus communis.* Plant Cell and Environment 16:375–382.

Hanson, A. D., K. M. Ditz, G. W. Singletary, and T. J. Leland. 1983. Gramine accumulation in leaves of barley grown under high temperature stress. Plant Physiology 71:896–904.

Hanson, J. B. 1984. The functions of Ca in plant nutrition. Pp. 149–208 in: P. B. Tinker and A. Läuchli (eds.), Advances in Plant Nutrition, Vol 1. Praeger Scientific, New York.

Hanson, P. J., and R. K. Dixon. 1987. Allelopathic effects of interrupted fern on northern red oak seedlings: amelioration by *Suillus luteus* L.: Fr. Plant and Soil 98:43–51.

Hanson, A. D., B. Rathinasabapathi, J. Rivoal, M. Burnet, M. O. Dillon, and D. A. Gage. 1994. Osmoprotective compounds in the Plumbaginaceae—A natural experiment in metabolic engineering of stress tolerance. Proceedings of the National Academy of Science of the United States of America 91:306–310.

Harborne, J. B. 1988a. The coevolutionary arms race: plant defense and animal response. Pp. 235–236 in: J. B. Harborne (ed.), The Introduction to Ecological Biochemistry, 3rd Ed. Academic Press, London.

Harborne, J. B. 1988b. Introduction to Ecological Biochemistry, 3rd Ed. Academic Press, London.

Hardin, J. W. 1979. Patterns of variation in foliar trichomes of eastern North American *Quercus*. American Journal of Botany 66:576–585.

Hare, P. D., and W. A. Cress. 1997. Metabolic implications of stress-induced proline accumulation in plants. Plant Growth Regulators 21:79–102.

Harley, J. L. 1965. Mycorrhiza. Pp. 218–230 in: K. F. Baker and W. C. Snyder (eds.), Ecology of Soil-Borne Plant Pathogens: Prelude to Biological Control. University of California Press, Berkley, CA.

Harley, J. L., and S. E. Smith. 1983. Mycorrhizal Symbiosis. Academic Press, London.

Harper, J. L. 1977. Population Biology of Plants. Academic Press, London.

Harper, J. F., L. Manney, N. D. DeWitt, M. H. Yoo, M. Sussman. 1990. The *Arabidopsis thaliana* plasma membrane H-ATPase multi-gene family. Journal of Biological Chemistry 256:13601–13608.

Harris, D., and E. A. Paul. 1987. Carbon requirements of vesicular-arbuscular mycorrhizae. Pp. 93–105 in: G. R. Safir (ed.), Ecophysiology of VA Mycorrhizae. CRC Press, Boca Raton, FL.

Hartmann, T. 1999. Chemical ecology of pyrrolizidine alkaloids. Planta 207:483–495.

Hartmann, T. 1991. Alkaloids. Pp. 79–121 in: G. A. Rosenthal and M. R. Berenbaum (eds.), The Chemical Participants (Herbivores: Their Interactions with Secondary Plant Metabolites, 2nd Ed., Vol 1). Academic Press, San Diego, CA.

Hartnett, D. C., B. A. D. Hetrick, G. W. T. Wilson, and D. J. Gibson. 1993. Mycorrhizal influence on intra- and interspecific neighbour interactions among co-occurring prairie grasses. Journal of Ecology 81:787–795.

Hartung, A. C., and C. T. Stephens. 1983. Effects of allelopathic substances produced by asparagus on incidence and severity of asparagus decline due to *Fusarium* crown rot. Journal of Chemical Ecology 8:1163–1174.

Hasenstein, K. H., and M. L. Evans. 1988. Effects of cations on hormone transport in primary roots of *Zea mays*. Plant Physiology 86:890–894.

Haslam, E. 1988. Plant polyphenols (syn. vegetable tannins) and chemical defense—a reappraisal. Journal of Chemical Ecology 14:1789–1805.

Hass, G. M., and C. A. Ryan. 1981. Carboxypeptidase inhibitors from potatoes. Methods in Enzymology 80:778–791.

Hassidim, M., Y. Braun, H. R. Lerner, and L. Reinhold. 1990. Na^+/H^+ and K^+/H^+ antiport in root membrane vessicles isolated from the halophyte *Atriplex* and the glycophyte cotton. Plant Physiology 94:1795–1801.

Hassall, C. H., and K. Reyle. 1955. Hypoglycine A and B, two biologically active polypeptides from *Blighta sapida*. Biochemical Journal 60:334–339.

Hassan, M. S., I. S. Alsaadawi, and A. El-Behadli. 1989. Citrus replant problem in Iraq. II. Possible role of allelopathy. Plant and Soil 116:157–160.

Haukioja, E., and S. Neuvonen. 1987. Insect population dynamics and induction of plant resistance: the testing of hypotheses. Pp. 411–432 in: P. Barbosa and L. Schultz (eds.), Insect Outbreaks. Academic Press, New York.

Hauck, C., and H. Schildknecht. 1990. Separation of enantiomers of the germination stimulant strigol on cellulose triacetate and determination of their biological acitvity. Journal of Plant Physiology 136:126–128.

Hauser, H. 1991. Effect of inorganic cations on pulse transitions. Chemistry and Physics of Lipids 57:309–325.

He, T., and G. R. Cramer. 1996. Abscisic acid concentrations are correlated with leaf area reductions in two salt-stressed rapid-cycling *Brassica* species. Plant and Soil 179:25–33.

Heath, J. D., M. I. Boulton, M. I. Raineri, S. L. Doty, A. R. Mushegian, T. C. Charles, J. W. Davies, and E. W. Nester. 1997. Discrete regions of the sensor protein VirA determine the strain-specific ability of *Agrobacterium* to agroinfect maize. Molecular Biology—Microbiology Interactions`10:221–227.

Heber, U., and K. Huve. 1998. Action of SO_2 on plants and metabolic detoxification of SO_2. International Review of Cytology 177:255–286.

Heckard, L. R. 1962. Root parasitism in *Castilleja*. Botanical Gazette 124:21–29.

Hegarty, M. P., E. E. Hegarty, and A. H. Gentry. 1991. Secondary compounds in vines with an emphasis on those with defensive functions. Pp. 287–310 in: F. E. Putz and H. A. Mooney (eds.), The Biology of Vines. Cambridge University Press, Cambridge, MA.

Hegde, R., and D. Miller. 1992. Concentration dependency and stage of crop growth in alfalfa autoxicity. Agronomy Journal 84:940–946.

Heinrich, E. A. (ed.). 1988. Plant Stress-Insect Interactions. John Wiley & Sons, New York.

Heisey, R. M. 1990. Evidence for allelopathy by tree-of-heaven. Journal of Chemical Ecology 16:2039–2055.

Heisey, R. M. 1996. Identification of an allelopathic compound from *Ailanthus altissima* (Simaroubaceae) and characterization of its herbicidal activity. American Journal of Botany 83: 192–200.

Heisey, R. M., and C. C. Delwiche. 1983. A survey of California plants for water-extractable and volatile inhibitors. Botanical Gazette 144:382–390.

Hejl, A. M., F. A. Einhellig, and J. A. Rasmussen. 1993. Effects of juglone on growth, photosynthesis, and respiration. Journal of Chemical Ecology 19:559–568.

Herde, O., H. Fuss, H. Penacortes, J. Fisahn. 1995. Proteinase-inhibitor-II gene expression induced by electrical-stimulation and control of photosynthetic activity in tomato plants. Plant and Cell Physiology 36:737–742.

Heshenhorn J., Y. Goldwasser, D. Plakhine, G. Herzlinger, S. Golan, R. Russo, and Y. Kleifeld. 1996. Role of pepper (*Capsicum annuum*) as a trap and cash crop for control of *Orobanche aegyptica* and *O. cernua*. Weed Science 44:948–951.

Hetrick, B. A. D., D. C. Harnett, G. W. T. Wilson, and D. J. Gibson. 1994. Effects of mycorrhizae, phosphorus availability, and plant density on yield relationships among competing tallgrass prairie grasses. Canadian Journal of Botany 72:168–176.

Hibberd, J. M., R. Whitbread, and J. F. Farrar. 1996. Effect of elevated concentrations of CO_2 on infection of barley by *Erysiphe graminis*. Physiological and Molecular Plant Pathology 48:37–53.

Hibberd, J. M., R. A. Bungard, M. C. Press, W. D. Jeschke, L. D. Scholes, and W. P. Quick. 1998a. Localization of photosynthetic metabolism in the parasitic angiosperm *Cuscuta reflexa*. Planta. 205:506–513.

Hibberd, J. M., W. P. Quick, M. C. Press. 1998b. Can source-sink relations explain responses of tobacco to infection by the root hemiparasitic angiosperm *Orobanche ceruna*? Plant Cell and Environment 21:333–340.

Hilbert, J. L., and F. Martin. 1988. Regulation of gene expression in ectomycorrhizas. I. Protein changes and the presence of ectomycorrhiza specific polypeptides in the *Pisolithus-Eucalyptus* symbiosis. New Phytologist 110:339–346.

Hilbert, J. L., G. Costa, and F. Martin. 1991. Ectomycorrhizin synthesis and polypeptide changes during the early stage of eucalypt mycorrhiza development. Plant Physiology 97:977–984.

Hill, A. E., and B. S. Hill. 1976. Mineral Ions. Pp. 225–243 in: U. Lüttge and M. G. Pittman (eds.), Transport in Plants. II. Pt. B. Tissues and Organs (Encyclopedia of Plant Physiology, Vol. 2B). Springer Verlag, New York.

Hillel, D. 1980. Fundamentals of soil physics, Academic Press, New York.

Hobbs, R. J. 1984. Possible chemical interactions among heathland plants. Oikos 43:23–29.

Hodges, L. 1977. Environmental pollution, 2nd Ed. Holt, Rinehart & Winston, New York.

Hooker, J. D. 1856. On the structure and affinities of Balophoreae. Transactions of the Linnaean Society of London 22:1–68.

Hopkins, W. G. 1995. Introduction to Plant Physiology. John Wiley & Sons, New York.

Horsley, S. B. 1987. Allelopathic interference with regeneration of the Allegheny hardwood forest. Pp. 205–212 in: G. R. Waller (ed.), Allelochemicals: Role in Agriculture and Forestry. American Chemical Society, Washington, DC.

Howeler, R. H., D. G. Edwards, C. J. Asher. 1979. The effect of soil sterilisation and mycorrhizal inoculation on the growth, nutrient uptake and critical phosphorus concentration in Cassava. In: Proceedings of 5th International Tropical Root and Tuber Crops Symposium. Manilla, Philippines, 17–21 September.

Hruska, A. J. 1988. Cyanogenic glycosides as defense compounds: a review of the evidence. Journal of Chemical Ecology 14:2213–2217.

Huang, R. S., W. K. Smith, and R. J. Yost. 1985. Influence of vesicular arbuscular mycorrhizae on growth, water relations, and leaf orientation in *Leucena leucocephala* (Lam) DeWit. New Phytologist 99:229–243.

Huante, P., E. Rincon, and F. S. Chapin III. 1998. Foraging for nutrients, responses to changes in light, and competition in tropical deciduous tree seedlings. Oecologia 117:209–216.

Humble, G. D., and K. Raschke. 1971. Stomatal opening quantitatively related to potassium transport: evidence from electron probe analysis. Plant Physiology 48:447–453.

Hunter, M. D. 1992. A variable insect-plant interaction: the relationship between tree budburst phenology and population levels of insect herbivores among trees. Ecological Entomology 17:91–95.

Hunter, M. D., T. Ohgushi, and P. W. Price. 1992. The Effects of Resource Distribution on Plant-Animal Interaction. Academic Press, San Diego, CA.

Hutchings, M. J., and C. S. J. Budd. 1981. Plant competition and its course through time. Bioscience 31:640–645.

Hutchings, M. J., and H. De Kroon. 1994. Foraging in plants: the role of morphological plasticity in resource acquisition. Pp. 159–238 in: M. Begon and A. H. Fitter (eds.), Advances in Ecological Research, Vol. 25. Academic Press, San Diego, CA.

Hutchings, P., and P. Saenger. 1987. Ecology of Mangroves. University of Queensland Press, St. Lucie, Australia.

Huttunen, S. 1994. Effects of air pollutants on epicuticular wax structure. Pp. 81–96 in: K. E. Percy, J. N. Cape, R. Jagels, and C. J. Simpson (eds.), Air Pollutants and the Leaf Cuticle. Springer-Verlag, New York.

Hwang, I. W., and H. M. Goodman. 1995. An *Arabidopsis thaliana* root-specific kinase homolog is induced by dehydration, ABA, and NaCl. Plant Journal 8:37–43.

Igbinnosa, I., K. F. Cardwell, and S. N. C. Okonkwo. 1996. The effect of nitrogen on the growth and development of giant witchweed, *Striga hermonthica* Benth: Effect on cultured germinated seedlings in host absence. European Journal of Plant Pathology 102:77–86.

Inderjit. 1998. Influence of *Pluchea lanceolata* (Asteraceae) on selected soil properties. American Journal of Botany 85:64–69.

Inderjit, and K. M. M. Dakshini. 1992. Formononetin 7-*O*-glycoside (ononin), an additional growth inhibitor in soils associated with the weed *Pluchea lanceolata* (DC) C. B. Clarke (Asteraceae). Journal of Chemical Ecology 18:713–718.

Inderjit, and K. M. M. Dakshini. 1994a. Algal allelopathy. Botanical Review 60:182–196.

Inderjit, and K. M. M. Dakshini. 1994b. Allelopathic effect of *Pluchea lanceolata* (Asteraceae) on characteristics of four soils and tomato and mustard growth. American Journal of Botany 81:799–804.

Inderjit, and R. Del Moral. 1997. Is separating resource competition from allelopathy realistic? Botanical Review 63:221–230.

Inderjit, and A. Mallik. 1997. Effects of *Ledum groenlandicum* amendments on soil characteristics and black spruce seedling growth. Plant Ecology 133:29–36.

Inderjit, K. M. M. Dakshini, and F. A. Einhellig (eds.). 1995. Allelopathy: Organisms, Processes and Applications. American Chemical Society, Washington, DC.

Inderjit, M. Muramatsu, and H. Nishimura. 1997. On the allelopathic potential of certain terpenoids, phenolics, and their mixtures, and their recovery from soil. Canadian Journal of Botany 75:888–891.

Ingersoll, R. B., R. E. Inman, and W. R. Fisher. 1974. Soils potential as a sink for atmospheric carbon monoxide. Tellus 26:151–158.

Ishitani, M., T. Nakamura, S. Y. Han, and T. Takabe. 1995. Expression of betaine aldehyde dehydrogenase gene in barley in response to osmotic-stress and abscisic acid. Plant Molecular Biology 27:307–315.

Ismail, B. S., and L. S. Mah. 1993. Effects of *Mikania micrantha* H. B. K. on germination and growth of weed species. Plant and Soil 157:107–113.

Itoh, L. S., and S. Uwano. 1986. Characteristics of the Cl^- action site in the O_2 evolving reaction in PS II particles: electrostatic interaction with ions. Plant and Cell Physiology 27:25–36.

Jackson, J. R., and R. W. Willemsen. 1976. Allelopathy in the first stages of secondary succession on the piedmont of New Jersey. American Journal of Botany 63:1015–1023.

Jackson, P. J., P. J. Unkefer, E. Delhaize, and N. J. Robinson. 1990. Mechanisms of trace metal toleance in plants. Pp. 231–255 in: F. Katterman (ed.), Environmental Injury to Plants. Academic Press, New York.

Jackson, R. B., and M. M. Caldwell. 1989. The timing and degree of root proliferation in fertile-soil microsites for three cold-desert perennials. Oecologia 81:149–153.

Jackson, R. B., and M. M. Caldwell. 1992. Shading and the capture of localized soil nutrients: nutrient contents, carbohydrates, and root uptake kinetics of a perennial tussock grass. Oecologia 91:457–462.

Jackson, R. B. and M. M. Caldwell. 1996. Integration resource heterogeneity and plant plasticity: modelling nitrate and phosphate uptake in a patchy soil environment. Journal of Ecology 84:891–903.

Jackson, R. B., J. H. Manwaring, and M. M. Caldwell. 1990. Rapid physiological adjustment of roots to localized soil enrichment. Nature 344:58–59.

Jacobs, P. E., R. L. Peterson, H. B. Massicotte. 1989. Altered fungal morphogenesis during the early stages of ectomycorrhizae formation in *Eucalyptus pilularis*. Scanning Microscopy 3: 249–255.

Jain, R., and C. L. Foy. 1992. Nutrient effects of parasitism and germination of Egyptian broomrape (*Orobanche aegyptica*). Weed Technology 6:269–275.

Jakobsen, I., and L. Rosendahl. 1990. Carbon flow into soil and external hyphae from roots of mycorrhizal cucumber plants. New Phytologist 115:77–83.

James, R. L., F. W. Cobb Jr., W. W. Wilcox, and D. L. Rowney. 1980. Effects of photochemical oxidant injury of ponderosa and Jeffery pines on susceptibility of sapwood and freshly-cut stumps to *Fomes annosus*. Phytopathology 70:704–708.

Janos, D. P. 1980. Mycorrhizae influence tropical succession. Biotropica. 12:56–64.

Janssen, A. 1999. Plants with spider-mite prey attract more predatory mites than clean plants under greenhouse conditions. Entomologia Experimentalis et Applicata 90:191–198.

Jansson, R. K., and Z. Smilowitz. 1985. Influence of nitrogen on population parameters of potato insects: abundance, development, and damage of the Colorado potato beetle, *Leptinotarsa decemlineata* (Coleoptera: Chrysomelidae). Environmental Entomology 14:500–506.

Jeffrey, D. W. 1987. Soil-Plant Relationships: An Ecological Approach. Timber Press, Portand, OR.

Jeffries, R. L., T. Rudmik, and E. M. Dillon. 1979. Responses of halophytes to high salinities and low water potentials. Plant Physiology 64:989–994.

Jenkins, J., and J. Aber. 1999. Hemlock wooly adelgid impacts on community structure and N cycling rates in eastern Hemlock forest. Canadian Journal of Forest Resarch 29:630–645.

Jermy, T. 1976. Insect-host-plant-relationship-coevolution or sequential coevolution? Pp. 109–113 in: T. Jeremy (ed.), The Host Plant in Relation to Insect Behaviour and Reproduction. Plenum Press, New York.

Jermy, T. 1984. Evolution of insect-host plant relationships. American Naturalist 124:609–630.

Jeschke, W. D., E. A. Kirkby, A. D. Peuke, J. S. Pate, and W. Hartung. 1997. Effects of P deficiency on assimilation and transport of nitrate and phosphate in the intact plants of castor bean (*Ricinis communis* L.). Journal of Experimental Botany 48:75–92.

Jobidon, R. 1986. Allelopathic potential of coniferous species to old-field weeds in eastern Quebec. Forest Science 32:112–118.

Joel, D. M., and D. Losner-Goshen. 1994. The attachment organ of the parasitic angiosperm *Orobanche crumana* and *O. aegyptica* and its development. Canadian Journal of Botany 72: 564–574.

Joel, D. M., and V. H. Portenoy. 1998. The angiospermous root parasite *Orobanche crumana* L. (Orobanchaceae) induces expression of a pathogenesis related (PR) gene in susceptible tobacco roots. Annals of Botany 81:779–781.

Johnson, C. M., M. N. Hanson, and M. S. Rohrbach. 1986. Toxicity to endothelial cells mediated by cotton bract tannin: potential contribution to the pathogenesis of byssinosis. American Journal of Pathology 122:399–409.

Johnson, K. D., and M. J. Crispeels. 1992. Tonoplast-bound protein kinase phosphorylates tonoplast intrinsic protein. Plant Physiology 100:1785–1795.

Johnson, R., J. Navarez, G. An, and C. A. Ryan. 1989. Expression of proteinase inhibitors I and II in transgenic tobacco plants: effects on natural defense against *Manduca sexta* larvae. Proceedings of the National Academy of Sciences (USA) 86:9871–9875.

Jones, C. G., and R. D. Firn. 1991. On the evolution of plant secondary chemical diversity. Philosophical Transactions of the Royal Society of London, Series B 333:273–280.

Jones, M. D., D. M. Durall, and P. B. Tinker. 1991. Fluxes of carbon and phosphorus between symbionts in willow ectomycorrhizas and their change with time. New Phytologist 119:99–106.

Jordan, B. R., P. E. James, A. Strid, and R. G. Anthony. 1994. The effect of ultraviolet-β radiation on gene espression and pigment composition in etiolated and green pea leaf tissue: UV-B induced changes are gene-specific and dependent upon the developmental stage. Plant Cell and Environment 17:45–54.

José, S., and A. Gillespie. 1998a. Allelopathy in black walnut (*Juglans nigra* L.) alley cropping. I. Spatio-temporal variation in soil juglone in a black walnut-corn (*Zea mays* L.) alley cropping system in the midwestern USA. Plant and Soil 203:191–197.

José, S., and A. Gillespie. 1998b. Allelopathy in black walnut (*Juglans nigra* L.) alley cropping. II. Effects of juglone on hydroponically grown corn (*Zea mays* L.) and soybean (*Glycine max* L. Merr.) growth and physiology. Plant and Soil 203:199–205.

Joslyn, M. A., and Z. Glick 1969. Comparative effects of gallotannic acid and related phenolics on the growth of rats. Journal of Nutrition 98:119–126.

Jouanin, L., M. Bonade-Bottino, C. Girard, G. Morrot, and M. Giband. 1998. Transgenic plants for insect resistance. Plant Science 131:1–11.

Joy, K. W. 1988. Ammonium, glutamine, and asparagine: a carbon nitrogen interface. Canadian Journal of Botany 66:2103–2109.

Jungk, A. O. 1991. Dynamics of nutrient movement at the soil-root interface. Pp. 455–481 in: Y. Waisel and A. Eshel (eds.), Plant Roots: The Hidden Half. Marcel Dekker, New York.

Kabata-Pendias, A, and K. Pendias. 1992. Trace Elements in Soils and Plants, 2nd Ed. CRC Press, Boca Raton, FL.

Kadmon, R. 1995. Plant competition along soil moisture gradients: a field experiment with the desert annual *Stipa capensis*. Journal of Ecology 83:253–262.

Kaminsky, R., and W. H. Muller. 1977. The extraction of soil phytotoxins using a neutral EDTA solution. Soil Science 124:205–210.

Kaminsky, R. 1981. The microbial origin of the allelopathic potential of *Adenostoma fasciculatum* H&A. Ecological Monograph 51:365–382.

Kanchan, S., and Jayachandra. 1980. Pollen allelopathy—a new phenomenon. New Phytologist 84:739–746.

Kangasjärvi, J., J. Talvinen, M. Utriainen, and R. Karjalainen. 1994. Plant defence systems induced by ozone. Plant Cell Environment 17:783–794.

Karban, R. 1991. Inducible resistance in agricultural systems. Pp. 403–419 in: D. W. Tallamy, and M. J. Raupp. (eds.), Phytochemical Induction by Herbivores. John Wiley & Sons, New York.

Karban, R., A. A. Agrawal, and M. Mangel. 1997. The benefits of induced defenses against herbivores. Ecology 78:1351–1355.

Kasamo, K. 1990. Mechanism for the activation of plasma membrane H^+-ATPase from rice (*Oryza sativa* L.) culture cells by molecular species of a phospholipid. Plant Physiology 93: 1049–1052.

Kato-Noguchi, H., J. Mizutani, and K. Hasagawa. 1994. Allelopathy of oats: II. Allelochemical effect of L-tryptophan and its concentration in oat root exudates. Journal of Chemical Ecology 20:315.

Keeley, J. E. 1987. The role of fire in the germination of woody taxa in California chaparral. Ecology 68:434–443.

Keeling, C. D., R. B. Bacastow, and T. P. Whorf. 1982. Measurements of the concentrations of carbon dioxide at Mauna Loa Observatory, Hawaii. Pp. 377–385 in: W. C. Clark (ed.), Carbon Dioxide Review. Oxford University Press, New York.

Keever, C. 1950. Causes of succession on old-fields of the Piedmont, North Carolina. Ecological Monographs 20:229–250.

Keller, P., and H. Deuel. 1957. Kationenaustauschk-apazität und Pektengehalt von Pflanzenwurz-eln. Zeitschrift fur Pflanzenernahrung, Duengung, Boden Kunde. 79:119–131.

Kelsey, R. G., G. W. Reynolds, and E. Rodriguez. 1984. The chemistry of biologically active constituents secreted and stored in plant glandular trichomes. Pp. 187–241 in: E. Rodriguez, P. L. Healy, and I. Metha (eds.), Biology and Chemistry of Plant Trichomes. Plenum Press, New York.

Kerstiens, G., and K. J. Lendzian. 1989. Interactions between ozone and plant cuticles. I. Ozone deposition and permeability. New Phytologist 112:13–19.

Killham, K. 1994. Soil Ecology, Cambridge University Press, New York.

Kingsbury, R. W., E. Epstein, and R. W. Pearcy. 1984. Physiological responses to salinity in selected lines of wheat. Plant Physiology 74:417–423.

Kinney, K. K., R. L. Lindroth, S. M. Jung, and E. V. Nordheim. 1997. Effects of CO_2 and NO_3-availability on deciduous trees: phytochemistry and insect performance. Ecology 78:215–230.

Kirkby, E. A., and A. H. Knight. 1977. Influence of the level of nitrate nutrition on ion uptake and assimilation, organic acid accumulation, and cation-anion balance in whole tomato plants. Plant Physiology 60:349–353.

Kishor, K. P. B., Z. Hong, G.-H. Miao, C. A.-A. Hu, and D. P. S. Verma. 1995. Overexpression of delta-pyrroline-5-carboxylate synthetase increases proline production and confers osmotolerance intransgenic plants. Plant Physiology 108:1387–1394.

Klepzig, K. D., K. F. Raffa, and E. B. Smalley. 1991. Association of an insect fungal complex with red pine decline in Wisconsin. Forest Science 37(4):1119–1139.

Klepzig, K. D., E. L. Kruger, E. B. Smalley, and K. F. Raffa. 1995. Effects of biotic and abiotic stress on induced accumulation of terpenes and phenolics in red pines inoculated with bark beetle-vectored fungus. Journal of Chemical Ecology 21:601–626.

Knapp, A. K., J. M. Blair, J. M. Briggs, S. L. Collins, D. C. Hartnett, L. C. Johnson, and E. G. Towne. 1998. The keystone role of bison in north American tallgrass prairie—Bison increase habitat heterogeneity and alter a broad array of plant, community, and ecosystem processes. Bioscience 49:39–50.

Kobza, J., and F. A. Einhellig. 1987. The effects of ferulic acid on the mineral nutrition of grain sorghum. Plant and Soil 98:99–109.

Koeppe, D. E., L. M. Rohrbaugh, and S. H. Wender. 1970. The effect of age and chilling temperatures on the concentration of scopolin and caffeoylquinic acids in tobacco. Physiologia Plantarum 23:258–266.

Koeppe, D. E., I. M. Southwick, and J. E. Bittell. 1976. The relationship of tissue chlorogenic acid concentrations and leaching of phenolics from sunflowers grown under varying phosphate nutrient conditions. Canadian Journal of Botany 54:593–599.

Koide, R. 1991. Nutrient supply, nutrient demand, and plant responses to mycorrhizal infection. New Phytologist 117:365–386.

Koide, R. 1993. Physiology of the mycorrhizal plant. Advances in Plant Pathology 9:33–54.

Koide, R., and G. Elliot. 1989. Cost, benefit and efficiency of the vesicular-arbuscular mycorrhizal symbiosis. Functional Ecology 3:4–7.

Koide, R. T., and M. Li. 1991. Mycorrhizal fungi and the nutrient ecology of three old field annual plant species. Oecologia 85:403–412.

Kolb, T. E. 1988. Allelopathic effects of Kentucky bluegrass on northern red oak and yellow-poplar. Journal of Arboriculture 14:281–283.

Kolodziej, H. 1990. Oligomeric flavan-3-ols from medicinal willow bark. Phytochemistry 29: 955–960.

Komoßa, D., C. Langebartels, and H. Sandermann, Jr. 1995. Metabolic processes for organic chemicals in plants. Pp. 69–103 in: S. Trapp and J. Craig McFarlane (eds.), Plant Contamination: Modeling and Simulation of Organic Chemical Processes, Lewis Publishers, Boca Raton, FL.

Kopke, E., L. J. Musselman, and D. J. de Laubenfels. 1981. Studies on the anatomy of *Parasitaxus ustus* and its root connections. Phytomorphology 31:85–92.

Koske, R. E., and J. N. Gemma. 1992. Fungal reactions to plants prior to mycorrhizal formation. Pp. 3–36 in: M. F. Allen (ed.), Mycorrhizal Functioning. Chapman & Hall, New York.

Kosola, K. R., and K. L. Gross. 1999. Resource competition and suppression of plants colonizing early successional old fields. Oecologia 118:69–75.

Kothari, S. K., H. Marschnerand, and E. George. 1990. Effect of VA mycorrhizal fungi and rhizosphere microorganisms on root and shoot morphology, growth and water relations in maize. New Phytologist 116:303–311.

Kovda, V. A. 1976. Arid land irrigation and soil fertility: problems of salinity, alkalinity, and compaction. Pp. 211–235 in: E. B. Worthington (ed.), Arid Land Irrigation in Developing Countries. Environmental Problems and Effects. Pergamon Press, Oxford, UK.

Kovda, V. A., B. G. Rozonov, and S. K. Onishenko. 1976. On the probability of droughts and secondary salinization of world soils. Pp. 237–238 in: E. B. Worthington (ed.), Arid Land Irrigation in Developing Countries: Environmental Problems and Effects. Pergamon Press, Oxford, UK.

Kraal, H., and W. Ernst 1976. Influence of copper high tension lines on plants and soils. Environmental Pollution 11:131–135.

Krishnan, H. B., and S. G. Pueppke. 1993. Flavonoid inducers of nodulation genes stimulate *Rhizobium fredii* USDA257 to export proteins into the environment. Molecular Plant-Microbe Interactions 6:107–113.

Kroeler, C. J., and A. E. Linkins. 1991. The absorption of inorganic phosphate from ^{32}P-labeled inositol hexaphosphate by *Eriophorum vaginatum*. Oecologia 85:424–428.

Krupa, S. V. 1996. The role of atmospheric chemistry in the assessment of crop growth and productivity. Pp. 35–73 in: M. Yunus and M. Iqbal (eds.), Plant Response to Air Pollution. John Wiley & Sons, New York.

Krupnick, G. A., and W. E. Weis. 1999. The effect of floral herbivory on male and female reproductive success in *Isomeris arborea*. Ecology 80(1):135–149.

Krupnick, G. A., W. E. Weis, and D. R. Campbell. 1999. The consequences of floral herbivory for pollinator service to *Isomeris arborea*. Ecology 80(1):125–134.

Kuijt, J. 1969. The Biology of Parasitic Flowering Plants. University of California Press, Berkeley, CA.

Kuijt, J. 1977. Haustoria of phanerogamic parasites. Annual Reviews of Phytopathology 11:91–118.

Kuijt, J., and W. X. Dong. 1990. Surface features of the leaves of Balanophoraceae—A family without stomata. Plant Systematics and Evolution 170:29–35.

Kuiper, P. C. 1984. Funtioning of plant cell membranes under saline conditions: membrane lipid composition and ATPases. Pp. 77–91 in: R. C. Staples and G. H. Toenniessen (eds.), Salinity Tolerance in Plants: Strategies for Crop Improvement. John Wiley & Sons, New York.

Kumar, H. D. 1985. Introductory Phycology. Affiliated East-West Press, New Delhi.

Kunitz, M. 1945. Crystallization of a trypsin inhibitor from soybeans. Science 101:668–669.

Küppers, M. 1994. Canopy gaps: competitive light interception and economic space filling—a matter of whole plant allocation. Pp. 111–144 in: M. M. Caldwell and R. W. Pearcy (eds.), Exploitation of Environmental Heterogeneity by Plants: Ecophysiological Processes Above and Belowground. Academic Press, San Diego, CA.

Lambers, H., F. S. Chapin III, and T. Pons. 1998. Plant Physiological Ecology. Springer-Verlag, New York.

Landsberg, J., and C. P. Ohmart. 1989. Levels of insect defoliation in forests: patterns and concepts. Trends in Ecology and Evolution 4:96–100.

Lane, J. A., D. V. Child, G. C. Reiss, V. Entcheva, and J. A. Baily. 1997. Crop resistance to parasitic plants. Pp. 81–97 in: I. R. Crute, E. B. Holub, and J. J. Burdon (eds.), The Gene-for-Gene Relationship in Plant-Parasite Interactions. CAB International, New York.

Langenheim, J. H. 1994. Higher plant terpenoids: a phytocentric overview of their ecological roles. Journal of Chemical Ecology 20:1223–1280.

Lantzy, R. J. and F. T. MacKenzie. 1979. Atmospheric trace metals: Global cycles and assessments of man's impact. Geochimic aet Cosmochimica Acta 43:511–525.

Lapeyrie, F., and K. Mengden. 1993. Quantitative estimation of the surface carbohydrates of ectomycorrhizal fungi in pure culture and during *Eucalyptus* root infection. Mycological Research 97:603–609.

Larcher, W. 1980. Physiological Plant Ecology, 1st Ed., Springer-Verlag, New York.

Larcher, W. 1995. Physiological Plant Ecology, 3rd Ed. Springer-Verlag, New York.

Laskowski, M., and I. Kato. 1980. Protein inhibitors of proteinases. Annual Review of Biochemistry 49:593–626.

Lauda, S. M. 1982. Distribution ecology: variation of plant recruitment over a gradient in relation to insect seed predation. Ecological Monographs 52:25–41.

Lawlor, D. W. 1991. Concepts of nutrition in relation to cellular processes and environment. Pp. 1–54 in: J. R. Porter and D. W. Lawlor (eds.), Plant Growth: Interaction with Nutrition and Environment. Cambridge University Press, New York.

Lawrence , J. G., A. Colwell, and O. J. Sexton. 1991. The ecological impact of allelopathy in *Ailanthus altissima* (Simaroubaceae). American Journal of Botany 78:948–958.

LaRosa, P. C., N. K. Sing, P. M. Hasegawa, and R. A. Bressan. 1989. Stable NaCl tolerance of tobacco cells is associated with enhanced accumulation of osmotin. Plant Physiology 91: 855–861.

Lea, P. J., J. Wolfenden, and A. R. Wellburn. 1994. Influence of air pollutants upon nitrogen metabolism. Pp. 279–299 in: R. G. Alscher and A. R. Wellburn (eds.), Plant Responses to the Gaseous Environment: Molecular, Metabolic and Physiological Aspects. Chapman & Hall, New York.

Leather, G. B., and F. A. Einhellig. 1988. Bioassay of naturally occurring allelochemicals for phytotoxicity. Journal of Chemical Ecology 14:1821–1828.

Lechowski, Z. 1996. Abscisic acid content in the root hemiparasite *Melampyrum arvense* L. before and after attachment to the host plant. Biologia Plantarum 38:489–494.

Lechowski, Z., and Bialczyk. 1996. Cytokinins of the hemiparasite *Melampyrum arvense* L. before and after attachment to the host. Biologia Plantarum 38:481–488.

Lee, D. W., and J. B. Lowry. 1980. Young-leaf anthocyanin and solar ultraviolet. Biotropica 12:75–76.

Lee, T. D., and F. A. Bazzaz. 1980. Effects of defoliation and competition on growth and reproduction in the annual plant *Abutilon theophrasti*. Journal of Ecology 68:813–821.

Lehtilä, K., and S. Y. Strauss. 1997. Leaf damage by herbivores affects attractiveness to pollinators in wild radish, *Raphanus raphanistrum*. Oecologia 111:396–403.

Lehtilä, K., and S. Y. Strauss. 1999. Effects of foliar herbivory on male and female reproductive traits of wild radish, *Raphanus raphanistrum*. Ecology 80(1) 116–124.

Lei, J. G., F. Lapeyrie, N. Mlajczuk, and J. Dexheinmer. 1990a. Infectivity of pine and eucalypt isolates of *Pisolithus tinctorius* (Peers.) Coker and Couch on roots of *Eucalyptus urophylla* S. T. Blake *in vitro*. II. Ultrastructural and biochemical changes in the early stage of mycorrhiza formation. New Phytologist 116:115–122.

Lei, J. G., Y. Piché, H. Ding, F. Lapeyrie, N. Malajczuk, and J. Dexenheimer. 1990b. The events in ectomycorrhizal formation on primary roots of *Eucalyptus globulus* with two isolates of *Pisolithus tinctorius*: structural and cytochemical observations. Pp. 19–22 in: P. Nardon, V.

Gianinazzi-Pearson, A. M. Greiner, L. Margulis, and D. C. Smith (eds.), Endocytobiology, Vol. 4, Institut National de la Recherche Agronomique, Paris.

Lei, J., G. Bécard, J. G. Catford, and Y. Piché. 1991. Root factors stimulate ^{32}P uptake and plasmalemma ATPase activity in vesicular-arbuscular mycorrhizal fungus, *Gigaspora margarita*. New Phytologist 118:289–294.

Leigh, R. A., and R. Storey. 1991. Nutrient compartmentation in cells and its relevance to the nutrition of the whole plant. Pp. 33–53 in: J. R. Porter, and D. W. Lawlor (eds.), Plant Growth: Interactions with Nutrition and Environment. Cambridge University Press, New York.

Leopold, A. C., and R. P. Willing. 1984. Evidence for toxicity effects of salt on membranes. Pp. 67–75 in: R. C. Staples and G. H. Toenniessen (eds.), Salinity Tolerance in Plants: Strategies for Crop Improvement. John Wiley & Sons, New York.

Le Roux, X., T. Bariac, and A. Mariotti. 1995. Spatial partitioning of the soil water resource between grass and shrub components in a West African humid savanna. Oecologia 104:147–155.

Levering, C. A., and W. W. Thomson. 1971. The Ultrastructure of the salt gland of *Spartina foliosa*. Planta 97:183–196.

Levering, C. A., and W. W. Thomson. 1972. Studies on the ultrastructure and mechanism of secretion of the salt gland of the grass *Spartina*. Pp. 222–223 in: Proceedings of the 30th Electron Microscopy Society of America.

Levin, D. A. 1973. The role of trichomes in plant defense. Quarterly Review of Biology 48:3–15.

Levitt, J. 1980. Responses of Plants to Environmental Stresses: Water, Radiation, Salt and Other Stresses, Vol. 2. Academic Press, New York.

Levitt, J. 1980. Responses of Plants to Environmental Stresses: Chilling, Freezing and High Temperature Stress, Vol. 1. Academic Press, New York.

Levitt, J. 1982. Stress Terminology. Pp. 437–439 in: N. C. Turner and P. J. Kramer (eds.), Adaptations of Plants to Water and High Temperature Stress. Wiley Interscience, New York.

Lewis, A. C. 1984. Plant quality and grasshopper feeding: effects of sunflower condition on preference and performance in *Melanoplous differentialis*. Ecology 65:836–843.

Lewis, D. H., and J. L. Harley. 1965a. Carbohydrate physiology of mycorrhizal roots of beech. I. Identity of endogenous sugars and utilisation of exogenous sugars. New Phytologist 64: 224–237.

Lewis, D. H., and J. L. Harley. 1965b. Carbohydrate physiology of mycorrhizal roots of beech. II. Utilisation of exogenous sugars by uninfected and mycorrhizal roots. New Phytologist 64:338–356.

Lewis, D. H., and J. L. Harley. 1965c. Carbohydrate physiology of mycorrhizal roots of beech. III. Movement of sugars between host and fungus. New Phytologist 64:256–269.

Lewis, W. M. J. 1986. Evolutionary interpretations of allelochemical interactions in phytoplankton algae. American Naturalist 127:184–194.

Li, X.-L., H. Marschner, and E. George. 1991. Acquisition of phosphorus and copper by VA-mycorrhizal hyphae and root-to-shoot transport in white clover. Plant and Soil 136:49–57.

Lieberei, R., D. Selmar, and B. Biehl. 1985. Metabolization of cyanogenic glucosides in *Hevea-brasiliensis*. Plant Systematics and Evolution 150:49–63.

Lieberei, R., B. Biehl, A. Giesemann, and N. T. V. Junqueria. 1989. Cyanogenesis inhibits active defense reactions in plants. Plant Physiology 90:33–36.

Lieberman, D., and M. Lieberman. 1984. The causes and consequences of synchronous flushing in a dry tropical forest. Biotropica 16:193–201.

Linderman, R. G. 1992. Vesicular-arbuscular mycorrhizae and soil microbial interactions. Pp. 45–70 in: G. L. Bethlenfalvay and R. G. Linderman (eds.), Mycorrhizae in Sustainable Agriculture. American Society of Agronomy, Madison, WI.

Lindroth, R. L., G. A. Arteel, and K. K. Kinney. 1995. Response of three saturniid species to paper birch grown under enriched CO_2 atmospheres. Functional Ecology 9:306–311.

Liu, D., K. G. Raghothama, P. M. Hasegawa, and R. A. Bressan. 1993. Disease responses to fungal infection in the transgenic tobacco and potato plants that overexpress the osmotin gene. Plant Physiology (Suppl.) 102:165.

Livingston, W. H., M. L. Brenner, and R. A. Blanchette. 1984. Altered concentrations of abscisic acid, indole-3-acetic acid, and zeatin riboside associated with eastern dwarf mistletoe infections on black spruce. Pp. 53–61 in: F. G. Hawksworth and R. F. Scharpf (eds.), Biology of Dwarf Mistletoes: Proceedings of the Symposium. Fort Collins, CO. USDA, Forest Service General Technical Report RM-111.

Lloyd, J., J. P. Syvertsen, and P. E. Kriedemann. 1987. Salinity effects on leaf water relations and gas exchange of Valencia orange, *Citrus sinensis* (L.) Osbeck, on rootstocks with different salt exclusion characteristics. Australian Journal of Plant Physiology 15:717–726.

Lodhi, M. A. K. 1976. Role of allelopathy as expressed by dominating trees in a lowland forest in controlling the productivity and pattern of herbaceous growth. American Journal of Botany 63:1–8.

Lodhi, M. A. K. 1978. Allelopathic effects of decaying litter of dominant trees and their associated soil in a lowland forest community. American Journal of Botany 65:340–344.

Lodish, H., Darnell, J., and D. Baltimore. 1990. Molecular Cell Biology, 2nd Ed. Scientific American Books, New York.

Long, J., and F. W. Smith. 1984. Relation between size and density in developing stands: a description and possible mechanisms. Forest Ecology and Management 7:191–206.

Long, S. P., and N. R. Baker. 1986. Saline terrestrial environments. Pp. 63–102 in: N. R. Baker and S. P. Long (eds.), Photosynthesis in Contrasting Environments. Elsevier Science Publishers, Amsterdam.

Longnecker, N. 1994. Nutrient deficiencies and vegetative growth. Pp. 137–172 in: A. S. Basra (ed.), Mechanisms of Plant Growth and Improved Productivity: Modern Approaches. Marcel Dekker, New York.

Longstreth, D. J., and B. R. Strain. 1977. Effects of salinity and illumination on photosynthesis and water balance of *Spartina alterniflora* Loisel. Oecologia (Berlin) 31:191–199.

Longstreth, D. J., and P. S. Nobel. 1979. Salinity effects on leaf anatomy, consequences for photosynthesis. Plant Physiology 63:700–703.

Lorber, P., and W. H. Muller. 1980. Volatile growth inhibitors produced by *Salvia leucophylla*: Effects on cytological activity in *Allium cepa*. Comparative Physiology and Ecology 5:60–67.

Lorillou, S., B. Botton, and F. Martin. 1996. Nitrogen source regulates the biosynthesis of NADP-glutamate dehydrogenase in the ectomycorrhizal basidiomycete *Laccaria bicolor*. New Phytologist 132:289–296.

Lovett, J. V. 1989. Allelopathy research in Australia: an update. Pp. 49–66 in: C. H. Chou, and G. R. Waller (eds.), Phytochemical Ecology: Allelochemicals, Mycotoxins and Insect Pheromones and Allomones. Institute of Botany, Academia Sinica, Taipei.

Lovett, J. V., and H. F. Jackson. 1980. Allelopathic activity of *Camelia sativa* (L.) Crantz in relation to its phyllosphere bacteria. New Phytologist 86:273–277.

Luther, P., and H. Becker. 1987. Die Mistel. Springer-Verlag, Berlin.

Lüttge, U., and M. G. Pitman. 1976. Transport in plants II (Encyclopedia of Plant Physiology, new series, Vol. 2A). Springer-Verlag, New York.

Lynch, J. M. 1985. Microbial saprophytic activity on straw and other residues: consequences for the plant. Pp. 181–191 in: A. H. Fitter, D. Atkinson, D. J. Read, and M. B. Usher (eds.), Ecological Interactions in Soil: Plants, Microbes, and Animals, Blackwell Scientific Publications, Boston, MA.

Lynn, D. G., J. C. Steffens, V. S. Kamut, D. W. Graden, J. Shabanowitz, and J. L. Riopel. 1981. Isolation and characterization of the first host recognition substances for parasitic angiosperms. Journal of the American Chemical Society 103:1868–1870.

Maas, E. V. 1987. Salt tolerance of plants. Pp. 57–75 in: B. Christie (ed.), Handbook of Plant Science in Agriculture, Vol. 2. CRC Press, Boca Raton, FL.

Maas, E. V. 1990. Crop Salt Tolerance. Pp. 262–304 in: K. Tanji (ed.), Agricultural Salinity Assesment and Management. American Society of Civil Engineers, New York.

MacDowall, F. D. H., L. S. Vickery, V. C. Runecles, and Z. A. Patrick. 1963. Ozone damage to tobacco in Canada. Canadian Plant Disease Survey 43:131–151.

Mack, R. N. 1996. Predicting the identity and fate of plant invaders: emergent and emerging approaches. Biological Conservation 78:107–121.

MacNae, W. 1968. A general account of the fauna and flora of mangrove swamps and forests in the Indo-West Pacific region. Advances in Marine Biology 6:73–270.

Maddox, D. M., and M. Rhyne. 1975. Effects of induced host-plant mineral deficiencies on attraction, feeding and fecundity of the alligator flea beetle. Environmental Entomology 4: 682–686.

Magyarosy, A. C., P. Schurmann, and B. B. Buchanan. 1976. Effect of powdery mildew infection on photosynthesis by leaves and chloroplasts of sugar beets. Plant Physiology 57:486–489.

Mahall, B. C., and R. M. Callaway. 1991. Root communication among desert shrubs. Proceedings of the National Academy of Sciences 88:874–876.

Mahler, H. R., and E. H. Cordes. 1966. Biological Chemistry. Harper & Row, New York.

Makela, P., R. Munns, T. D. Colmer, A. G. Condon, and P. Peltonen-Sainio. 1998. Effect of foliar applications of glycinebetaine on stomatal conductance, abscisic acid and solute concentrations in leaves of salt- or drought-stressed tomato. Australian Journal of Plant Physiology 25:655–663.

Maksymiec, W. 1997. Effect of copper on cellular processes in higher plants. Photosynthetica 34:321–342.

Malajczuk, N., F. Lapeyrie, and J. Garbaye. 1990. Infectivity of pine and eucalypt isolates of *Pisolithus tinctorius* on the roots of *Eucalyptus urophylla in vitro*. New Phytologist 114: 627–631.

Mallaburn, P. S. 1992. The physiological anatomy of the haustorium of *Striga hermonthica* (Del.) Benth. (Scrophulariaceae). Ph.D. thesis, University of London.

Mallik, M. A. B., and K. Tesfai. 1987. Stimulation of *Bradyrhizobium japonicum* by allelochemicals from green plants. Plant and Soil 103:227–231.

Mallik, M. A. B., and K. Tesfai. 1988. Allelopathic effect of common weeds on soybean growth and soybean-radyrhizobium symbiosis. Plant and Soil 112:177–182.

Mallik, M. A. B., R. Puchala, and F. A. Grosz. 1994. A growth inhibitory factor from lambsquarters (*Chenopodium album*). Journal of Chemical Ecology 20:957–967.

Manners, J. G. 1993. Principles of Plant Pathology. Cambridge University Press, New York.

Mansour, M. M. F., P. R. Hassalt, and P. J. C. Kuiper. 1994. Plasma membrane lipid alterations induced by NaCl in winter wheat roots. Physiologia Plantarum 92:473–478.

Marenco, A., H. Gouget, P. Nedelec, and J. P. Pages. 1994. Evidence of a long-term increase in tropospheric ozone from Pic Du Midi data series: consequences: positive radiative forcing. Journal of Geophysical Resarch 99:16617–16632.

Mares M., B. Meloun, M. Pavlik, V. Kostka, and M. Baudys. 1989. Primary structure of cathepsin-D inhibitor from potatoes and its structure relationship to soybean trypsin-inhibitor family. FBES Letters 251:94–98.

Marschner, H. 1995. Mineral Nutrition of Plants, 2nd Ed. Academic Press, New York.

Marshall, D. R., and S. K. Jain. 1969. Interference in pure and mixed populations of *Avena fatua* and *A. barbata*. Journal of Ecology 57:251–270.

Marshall, J. D., T. E. Dawson, and J. R. Ehleringer. 1994. Integrated nitrogen, carbon, and water relations of a xylem tapping mistletoe following nitrogen fertilization of the host. Oecologia 100:430–438.

Martin, F., and B. Botton. 1993. Nitrogen metabolism of ectomycorrhizal fungi and ectomycorrhiza. Advances in Plant Pathology 9:83–102.

Martin, F., and D. Tagu. 1995. Ectomycorrhiza development: a molecular perspective. Pp. 29–58 in: A. Varma and B. Hock (eds.), Mycorrhiza: Structure, Function, Molecular Biology, 1st Ed. and Biotechnology, Springer-Verlag, Berlin.

Martin, F., and D. Tagu. 1999. Developmental biology of a plant-fungus symbiosis: the ectomycorrhiza. Pp. 51–73 in: A. Varma and B. Hock (eds.), Mycorrhiza: Structure, Function, Molecular Biology, and Biotechnology, 2nd Ed. Springer-Verlag, Berlin.

Martin, J. S., M. M. Martin, and E. A. Bernays. 1987. Failure of tannic acid to inhibit digestion or reduce digestibility of plant protein in gut fluids of insect herbivores: implications for theories of plant defense. Journal of Chemical Ecology 13:605–621.

Martinoia, E. 1992. Transport processes in vacuoles of higher plants. Botanica Acta 105:232–245.

Marty-Mazars, D., M.-C. Clemencet, P. Dozolme, and F. Marty. 1995. Antibodies to the tonoplast from the storage parenchyma cells of beet root recognize a major intrinsic protein related to the TIPs. European Journal of Cell Biology 66:106–118.

Marvier, M. 1998a. A mixed diet improves performance and herbivore resistance of a parasitic plant. Ecology 79:1272–1280.

Marvier, M. 1998b. Parasite impacts on host communities: plant parasitism in a California coastal prairie. Ecology 79:2616–2623.

Marvier, M. A., and David Smith. 1996. Conservation implications of host use for rare parasitic plants. Conservation Biology 11:839–848.

Marx, D. H. 1979. Synthesis of ectomycorrhizae by different fungi in Northern Red Oak seedlings. USDA, Forest Service Research Note, SE282.

Marx, D. H. 1991. Practical significance of the ectomycorrhizae in forest establishment. Marcus Wallenberg Foundation Symposium Proceedings 7:54–90.

Maseuth, J. D., G. Montenegro, and A. M. Walckowiak. 1984. Studies of the holoparasite *Tristerix aphyllus* (Loranthaceae) infecting *Trichocerus chilensis* (Cactaceae). Canadian Journal of Botany 62:847–857.

Massey, A. B. 1925. Antagonism of walnuts. Phytopathology 15:773–925.

Massicotte, H. B., R. L. Peterson, C. A. Ackerley, and Y. Piché. 1986. Structure and ontogeny of *Alnus crispa—Alpova diplophloeus* ectomycorrhizae. Canadian Journal of Botany 64:177–192.

Massicotte, H. B., L. H. Melville, R. Molina, and R. L. Peterson. 1993. Structure and histochemistry of mycorrhizae synthesized between *Arbutus menziesii* (Ericaceae) and two basidiomycetes, *Pisolithus tinctorius* (Pisolithaceae) and *Piloderma bicolor* (Corticiaceae). Mycorrhiza 3:1–11.

Matsushita, N., and T. Matoh. 1991. Characterization of Na^+ exclusion mechanisms for salt-tolerant reed plants in comparison to salt-sensitive rice plants. Physiologia Plantarum 83: 170–176.

Matsushita, N., and T. Matoh. 1992. Function of the shoot base of salt tolerant reed (*Phragmites communis* Trinius) plants from Na^+ exclusion from the shoots. Soil Science and Plant Nutrition 38:565–571.

Matthies, D. 1995. Parasitic and competitive interactions between the hemiparasites *Rhinanthus serotinus* and *Odontites rubra* and their host *Medicago sativa*. Journal of Ecology 83:245–251.

Matthies, D. 1998. Influence of the host on growth and biomass allocation in the two facultative root hemiparasites *Odontites vulgaris* and *Euphrasia minima*. Flora 193:187–193.

Mattocks, A. R. 1986. Chemistry and Toxicity of Pyrolizidine Alkaloids. Academic Press, London.

Maurel, C. 1997. Aquaporins and water permeability of plant membranes. Annual Review of Plant Physiology and Plant Molecular Biology 48:399–429.

Maurel, C., J. Reizer, J. L. Schroeder, and M. J. Chrispeels. 1993. The vacuolar membrane protein γ-TIP creates water specific channels in *Xenopus* oocytes. EMBO Journal 12:2241–2247.

Maurel, C., J. Reizer, J. L. Schroeder, M. J. Chrispeels, and M. H. J. Saier. 1994. Functional characterization of the *Escherichia coli* glycerol facilitator, GlpF, in *Xenopus* oocytes. Journal of Biological Chemistry 269:11869–11872.

May, F. E., and J. E. Ash. 1990. An assessment of the allelopathic potential of *Eucalyptus*. Australian Journal of Botany 38:245–254.

McConnaughay, K. D. M., and F. A. Bazzaz. 1992. The occupation and fragmentation of space: consequences of neighbouring roots. Functional Ecology 6:704–710.

McCue, K. F., and A. D. Hanson. 1990. Drought and salt tolerance: towards understanding and application. Trends in Biotechnology 8:358–362.

McCue, K. F., and A. D. Hanson. 1992. Salt-inducible betaine aldehyde dehydrogenase from sugar beet-cDNA cloning and expression. Plant Molecular Biology 18:1–11.

McDermitt, D. K., and R. S. Loomis. 1981. Elemental composition of biomass and its relation to energy content, growth efficiency, and growth yield. Annals of Botany 48:275–290.

McFarlane, J. C. 1995. Anatomy and physiology of plant conductive systems. Pp. 13–34 in: S. Trapp and J. Craig McFarlane (eds.), Plant Contamination: Modeling and Simulation of Organic Chemical Processes. Lewis Publishers, Boca Raton, FL.

McGurl, B., and C. A. Ryan. 1992. The organization of the prosystemin gene. Plant Molecular Biology 20:405–409.

McKee, K. L. 1993. Soil physiochemical patterns and mangrove species distribution—reciprocal effects? Journal of Ecology 81:477–487.

McKerns, K. W., H. H. Bird, E. Kaleita, B. S. Coulomb, E. C. deRenzo. 1960. Effects of hypoglycine on certain aspects of glucose and fatty acid metabolism in the rat. Biochemical Pharmacology 3:305–315.

McLaughlin, S. B., and R. J. Norby. 1991. Atmospheric pollution and terrestrial vegetation: Evidence of changes, linkages, and significance to selection processes. Pp. 61–101 in: G. E. Taylor Jr., L. F. Pitelka, and M. T. Clegg (eds.), Ecological Genetics and Air Pollution. Springer-Verlag, New York.

McNally, S. F., T. O. Orebamjo, B. Hirel, and G. R. Stewart. 1983. Glutamine synthase isoenzymes of *Striga hermonthica* and other angiosperm root parasites. Journal of Experimental Botany 34:610–619.

McNaughton, S. J. 1979. Grazing as an optimization process: Grass-ungulate relationships in the Serengeti. American Naturalist 113:691–703.

McNaughton, S. J. 1985a. Ecology of a grazing ecosystems: the Serengeti. Ecological Monographs 55:259–294.

McNaughton, S. J. 1985b. Interactive regulation of grass yield and chemical properties by defoliation, a salivary chemical, and inorganic nutrition. Oecologia 65:478–486.

McNaughton, S. J. 1986. On plants and herbivores. American Naturalist 128:765–770.

McPherson, J. K., and C. H. Muller. 1969. Allelopathic effects of *Adenostoma fasciculatum*, "chamise," in the California chaparral. Ecological Monographs 39:177–198.

McPherson, J. K., and G. L. Thompson. 1972. Competitive and allelopathic suppression of understory by Oklahoma oak forests. Bulletin of the Torrey Botanical Club 99:293–300.

Mehansho, H., D. K. Ann, L. G. Butler, J. C. Rogler, and D. M. Carlson. 1987. Induction of proline-rich proteins in hamster salivary glands by isoproterol treatment and an unusual growth inhibition by tannins. Journal of Biological Chemistry 262:12344–12350.

Meharg, A. A. 1994. Integrated tolerance mechanisms: constitutive and adaptive plant responses to elevated metal concentrations in the environment. Plant Cell and Environment 17:989–993.

Melin, E., and H. Nilsson. 1957. Transport of C^{14} labeled photosynthate to the fungal associate of pine mycorrhiza. Svensk Botanisk Tidskrift 49:119–121.

Mendelssohn, I. A., K. L. McKee, and W. H. Patrick. 1981. Oxygen deficiency in *Spartina alterniflora* roots: Metabolic adaptation to anoxia. Science 214:439–441.

Mengel, K., and E. A. Kirkby. 1982. Principles of Plant Nutrition, 3rd Ed. International Potash Institute, Bern, Switzerland.

Mersie, W., and M. Singh. 1988. Effects of phenolic acids and ragweed parthenium (*Parthenium hysterophorus*) extracts on tomato (*Lycopersicon esculentum*) growth and nutrient and chlorophyll content. Weed Science 36:278–281.

Metting, F. B. 1993. Structure and physiological ecology of soil microbial communities. Pp. 3–25 in: F. B. Metting Jr. (ed.), Soil Microbial Ecology. Marcel Dekker, Inc., New York.

Meyer, F. H. 1973. Distribution of ectomycorrhizae in native and man-made forests. Pp. 79–105 in: G. C. Marks and T. T. Kozlowski (eds.), Ectomycorrhizae. Academic Press, New York.

Micheli, P. A. 1729. Nova plantarum genera.

Michelsen, A., I. K. Schmidt, S. Jonasson, J. Dighton, H. E. Jones, and T. V. Callaghan. 1995. Inhibition of growth and effects on nutrient uptake of arctic graminoids by leaf extracts—Allelopathy or resource competition between plants and microbes. Oecologia 103:407–418.

Middleton, J. T., J. B. Kendrick Jr., and E. F. Darley. 1955. Air-borne oxidants as plant damaging agents. Pp. 191–198 in: Proceedings of the 3rd National Air Pollution Symposium. Stanford Research Institute, Menlo Park, California.

Mijatovic, K., and D. Stojanovic. 1973. Distribution of *Orabanche* spp. on the agricultural crops in Yugoslavia. Pp. 28–34 in: The Proceedings of the International Symposium of Parasitic Weeds. European Weed Research Council, Malta.

Mikola, P. 1965. Studies on ectendotrophic mycorrhiza of pine. Acta Forestalia Fennica 79:1–56.

Miles, J. 1985. Soil in the ecosystem. Pp. 407–427 in: A. H. Fitter (ed.), Ecological Interactions in Soil: Plants, Microbes and Animals (British Ecological Society, No. 4). Blackwell Scientific Publications, Boston, MA.

Miller, D. A. 1996. Allelopathy in forage crop systems. Agronomy Journal 88:854–859.

Miller, P. R. 1977. Photochemical Oxidant Air Pollutant Effects on a Mixed Conifer Ecosystem, A Progress Report. EPA-600/3-77-104. U.S. Environmental Protection Agency, Corvallis, OR.

Mitchell, R. J., H. E. Garrett, G. S. Cox, and A. Ataly. 1986. Boron and ectomycorrhizal development influences on indole-3-acetic acid levels, and indole-3-acetic acid oxidase and peroxidase activities of *Pinus echinata* roots. Tree Physiology 1:1–8.

Mitten, W. 1847. On the economy of the roots of *Thesium linophyllum* Hook. London Journal of Botany 6:146–148.

Mizusaki, S., Y. Tenabe, M. Roguchi, and E. Tamaki. 1973. Changes in the activities of ornithine decarboxylase, putrescine *N*-methyltransferase and *N*-methylputrescine oxidase in tobacco roots in relation to nicotine biosynthesis. Plant Cell Physiology 14:103–110.

Moerschbacher, B. M., U. Noll, L. Gorrichon, and H.-J. Reisener. 1990. Specific inhibition of lignification breaks hypersensitive resistance of wheat to stem rust. Plant Physiology 93:465–470.

Mohr, H., and P. Schopfer 1995. Plant Physiology. G. Lawlor and D. W. Lawlor (trans.). Springer-Verlag, New York.

Mole, S., and P. G. Waterman. 1987. A critical analysis of techniques for measuring tannins in ecological studies: I. Techniques for chemically defining tannins. Oecologia 72:137–147.

Molisch, H. 1937. Der Einfluss einer Pflanzen auf die andere-Allelopathie. Fischer, Jena, Germany.

Molisch, H. 1938. The Longevity of Plants. Transl. E. H. Fulling. New York.

Mooney, H. A., and S. L. Gulmon. 1982. Constraints on leaf structure and function in reference to herbivory. BioScience 32:198–206.

Moorby, J., and R. T. Besford. 1983. Mineral nutrition and growth. Pp. 481–527 in: A. Läuchli and R. L. Bieleski (eds.), Inorganic Plant Nutrition (A. Pirson and M. H. Zimmerman [series eds.], Encyclopedia of Plant Physiology, Vol. 15A). Springer-Verlag, New York.

Moore, R., and W. D. Clark. 1995. Botany: Plant Form and Function, William C. Brown Publishers, Dubuque, IA.

Moore, T. H. M., J. A. Lane, D. V. Child, G. M. Arnold, J. A. Bailey, and G. Hofmann. 1995. New sources of resistance of cowpea (*Vinga unguiculata*) to *Striga gesnerioides*, a parasitic angiosperm. Euphytica 84:165–174.

Morden, C. W., K. H. Wolfe, C. W. dePamphilis, and J. D. Palmer. 1991. Plastid translation and transcription genes in a nonphotosynthetic plant: intact, missing and pseudogenes. EMBO Journal 10:3281–3288.

Moreland, D. E., and W. P. Novitzky. 1987. Effects of phenolic acids, coumarins, and flavonoids on isolated chloroplasts and mitochondria. Pp. 247–261 in: G. R. Waller (ed.), Allelochemicals: Role in Agriculture and Forestry. American Chemical Society, Washington, DC.

Morgan, D. C., and H. Smith. 1981. Control of development in *Chenopodium album* L. by shadelight: the effect of light quantity (total fluence rate, and light quality (red:far red ratio). New Phytologist 88:239–248.

Morikawa, H., A. Higaki, M. Nohno, M. Takahashi, M. Kamada, M. Nakata, G. Toyohara, Y. Okamura, K. Matsui, S. Kitani, K. Fujita, K. Irifune, and N. Goshima. 1998. More than 600 fold variation in nitrogen dioxide assimilation among 217 plant taxa. Plant and Cell Environment 21:180–190.

Mosse, B., and D. J. Hayman. 1971. Plant growth responses to vesicular arbuscular mycorrhizae. II. In unsterilised field soils. New Phytologist 70:29–34.

Mosse, B., D. S. Hayman, and D. J. Arnold. 1973. Plant growth responses to vesicular-arbuscular mycorrhiza. V. Phosphate uptake by three plant species from P-deficient soils labeled with ^{32}P. New Phytologist 72:809–815.

Mou, P., R. J. Mitchell, and R. H. Jones. 1997. Root distribution of two tree species under a heterogeneous nutrient environment. Journal of Applied Ecology 34:645–656.

Mozafar, A., and J. R. Goodin. 1970. Vesiculated hairs: A mechanism for salt tolerance in *Atriplex hamilis* L. Plant Physiology 45:62–65.

Mudd, J. B. 1996. Biochemical basis for the toxicity of ozone. Pp. 267–283 in: M. Yunus and M. Iqbal (eds.), Plant Response to Air Pollution. John Wiley & Sons, New York.

Muller, C. H. 1966. The role of chemical inhibition (allelopathy) in vegetational composition. Bulletin of the Torrey Botanical Club 93:332–351.

Muller, C. H. 1969. Allelopathy as a factor in ecological process. Vegetatio 18:48–357.

Muller, C. H., and R. Del Moral. 1966. Soil toxicity induced by volatiles from *Salvia leucophylla*. Bulletin of the Torrey Botanical Club 93:130–136.

Muller, W. H., and C. H. Muller. 1956. Association patterns involving desert plants that contain toxic products. American Journal of Botany 43:354–361.

Mundy, J., and N.-H. Chua. 1988. Abscisic acid and water-stress induce the expression of a novel rice gene. The European Molecular Biology Organization Journal 7:2279–2286.

Munna, R., and A. Termaat. 1986. Whole plant response to salinity. Australian Journal of Plant Physiology 13:143–160.

Munns, R. 1993. Physiological processes limiting plant growth in saline soils: some dogmas and hypotheses. Plant Cell and Environment 16:15–24.

Munns, R., D. P. Schachtman, and A. G. Condon. 1995. The significance of two-phase growth response to salinity in wheat and barley. Australian Journal of Plant Physiology 22:561–569.

Munther, W. E., and D. E. Fairbrothers. 1980. Allelopathy and autotoxicity in three eastern North American ferns. American Fern Journal 70:124–135.

Murphy, S. D., and L. W. Aarssen. 1995a. In vitro allelopathic effects of pollen from three *Hieracium* species (Asteraceae) and pollen transfer to sympatric Fabaceae. American Journal of Botany 82:37–45.

Murphy, S. D., and L. W. Aarssen. 1995b. Reduced seed set in *Elytrigia repens* caused by allelopathic pollen from *Phleum pratense*. Canadian Journal of Botany 73:1417–1422.

Musselman, L. J. 1980. The biology of *Striga*, *Orobanche*, and other root-parasitic weeds. Annual Reviw of Phytopathology 18:463–489.

Musselman, L. J. 1991. The genus *Hydonora* (Hydnoraceae). Pp. 247–250 in: J. K. Ransom, L. J. Musselman, A. D. Worsham, and C. Parker (eds.), Proceedings of the 5th International Symposium of Parasitic Weeds, Nairobi, June 1991. International Maize and Wheat Improvement Center (CIMMYT), Nairobi, India.

Musselman, L. J., and C. Parker. 1981. Studies in indigo witchweed, the American strain of *Striga gesnerioides* (Scrophulariaceae). Weed Science 29:594.

Nakanishi K., M. Koreeda, S. Sasaki, M. L. Chang, and H. Y. Hsu. 1966. Insect hormones. The structure of ponasterone, an insect-moulting hormone from the leaves of *Podocarpus nakaii* Hay. Journal of the Chemical Society Chemical Communications 24:915–917.

Nambiar, E. K. S. 1976. Genetic differences in the copper nutrition of cereals. I. Differential responses of genotypes to copper. Australian Journal of Research 27:453–463.

Nandakumar, S., D. N. Kachru, and P. S. Krishnan. 1976. Threonine-serine dehydratase activity in angiospermous parasites. New Phytologist 18:613–618.

Narashima, V. L., and V. Rabindranath. 1964. A further contribution to the host range on *Dendrophthoe falcata* (L.f.) Ettingsh. Bulletin of the Botanical Survey of India 6:103.

Narváez-Vásquez, J., G. Pearce, M. L. Orozco-Cárdenas, V. R. Francheschi, and C. A. Ryan. 1995. Autoradiographic and biochemical evidence for the systemic translocation of systemin in tomato plants. Planta 195:593–600.

Nash, D., L. G. Paleg, and J. T. Wiskich. 1982. Effect of proline, betaine, and some other solutes on the heat stability of mitochondrial enzymes. Australian Journal of Plant Physiology 9:47–57.

National Academy of Sciences. 1983. Acid Deposition Atmospheric Processes in Eastern North America. National Academy Press, Washington, DC.

Neftel, A., E. Moor, H. Oeschger, and B. Stauffer. 1985. Evidence from polar ice cores for the increase in atmospheric CO_2 in the last two centuries. Nature 315:45–47.

Nellessen, J. E., and J. S. Fletcher 1993. Assessment of published literature pertaining to the uptake/accumulation, translocation, adhesion and biotransformation of organic chemicals by vascular plants. Environmental Toxicology and Chemistry 12:2045–2052.

Neumann, P. M. 1995. Inhibition of root growth by salinity stress: toxicity or adaptive biophysical response? Pp. 299–304 in: F. Baluska, M. Ciamporova, O. Gasparikova, and P. W. Barlow (eds.), Structure and Function of Roots: Developments in Plant and Soil Sciences. Kluwer Academic Publishers, Dordrecht, Netherlands.

Neumann, P. M. 1997. Salinity resistance and plant gowth revisited. Plant Cell and Environment 20:1193–1198.

Newman, E. I. 1978. Allelopathy; adaptation or accident? in: J. B. Harborne (ed.), Biochemical Aspects of Plant and Animal Coevolution. Academic Press, London.

Newman, E. I. 1983. Interactions between plants. Pp. 679–710 in: Encyclopedia of Plant Physiology, New Series, Vol. 12C. Springer-Verlag, Berlin.

Newsham, K. K., A. H. Fitter, and A. R. Watkinson. 1994. Root pathogenic and arbuscular mycorrhizal fungi determine fecundity of asymptomatic plants in the field. Journal of Ecology 82:805–814.

Newsham, K. K., A. H. Fitter, J. W. Merryweather. 1995. Multifunctionality and biodiversity in arbuscular mycorrhizas. Tree 10:407–411.

Nichols-Orians, C. 1991. Condensed tannins, attine ants, and the performance of a symbiotic fungus. Journal of Chemical Ecology 17:1177–1196.

Nickrent, D. L., R. J. Duff, A. E. Colewell, A. D. Wolfe, N. D. Young, K. E. Steiner, and C. W. dePamphilis. 1998. Molecular phylogenetic and evolutionary studies of parasitic plants. Pp. 211–241 in: P. S. Soltis, D. E. Soltis, J. J. Doyle (eds.), Molecular Systematics of Plants II. DNA Sequencing. Klewer Academic Publishers, Boston, MA.

Nicolson, T. H. 1975. Evolution of vesicular-arbuscular mycorrhizas. Pp. 25–34 in: F. E. Sanders, B. Mosse, and P. B. Tinker (eds.), Endomycorrhizas. Academic Press, London.

Niklas, K. J. 1994. Plant Allometry. The Scaling of Form and Process. University of Chicago Press, Chicago.

Nilsen, E. T., and W. H. Muller. 1980a. The relative naturalization ability of two *Schinus* species in southern California. I: Seed germination. Bulletin of the Torrey Botanical Club 107:51–56.

Nilsen, E. T., and W. H. Muller. 1980b. The relative naturalization ability of two *Schinus* species in southern California. II: Seedling establishment. Bulletin of the Torrey Botanical Club 107: 232–237.

Nilsen, E. T., and D. M. Orcutt. 1996. The Physiology of Plants Under Stress: Abiotic Factors. John Wiley & Sons, New York.

Nilsen, E. T., M. R. Sharifi, R. A. Virginia, and P. W. Rundel. 1984. Phenology of warm desert phreatophytes: seasonal growth and herbivory in *Prosopis glandulosa* var *torreyana* (honey mesquite) Journal of Arid Environments 13:217–229.

Nilsen, E. T., P. W. Rundel, M. R. Sharifi, I. N. Forseth, and J. R. Ehleringer. 1990. Water relations of stem succulent trees of north central Baja, California. Oecologia 82:299–303.

Nilsson, M.-C. 1994. Separation of allelopathy and resource competition by the boreal dwarf shrub *Empetrum hermaphroditum* Hagerup. Oecologia 98:1–7.

Nimbal, C. I., J. Pederson, C. N. Yerkes, L. A. Weston, and S. C. Weller. 1996. Phytotoxicity and distribution of sorgoleone in grain sorghum germplasm. Journal of Agricultural Food Chemistry 44:1343–1347.

Niu, X., M. Narasimhan, R. Salzman, R. Bressan, and P. M. Hawegawa. 1993. NaCl regulation of plasma membrane H^+-ATPase gene expression in a glycophyte and a halophyte. Plant Physiology 103:713–718.

Niu, X., B. Damsz, A. K. Kononowicz, R. Bressan, and P. M. Hawegawa. 1996. NaCl-induced alternations in both cell structure and tissue-specific plasma membrane H-ATPase gene expression. Plant Physiology 111:679–686.

Niu, X., R. Bressan, P. M. Hawegawa, and J. M. Pardo. 1995. Ion homeostasis in NaCl stress environments. Plant Physiology 109:735–742.

Noggle, G. R., and G. J. Fritz. 1976. Introductory Plant Physiology. Prentice-Hall Inc., Edgewood Cliffs, NJ.

Nomura, M., T. Hibino, T. Takabe, T. Sugiyama, H. Miyaki, and T. Takabe. 1998. Transgenically produced glycinebetaine protects ribulose 1,5-bisphosphate cabroxylase/oxygenase from inactivation in *Synechococcus* sp. PCC7942 under salt stress. Plant Cell Physiology 39:425–432.

Nonhebel, S. 1993. The importance of weather data in crop growth simulation models and assessment of climate change effects. Ph.D. dissertation, Wageningen Agricultural University, Netherlands.

Numata, A., and T. Ibuka. 1987. Alkaloids from ants and other insects. Alkaloids 31:193–315.

Nylin, S., N. Janz, and N. Wedell. 1996. Oviposition plant preference and offspring performance in the common butterfly: correlations and conflicts. Entomology Experimental Applications 80:141–144.

Oba, M., and M. S. Allen. 1999. Effects of brown midrib 3 mutation in corn silage on dry matter intake and productivity of high yielding dairy cows. Journal of Dairy Science 82: 135–142.

Odum, E. P. 1971. Fundamentals of Ecology, 3rd Ed. W. B. Saunders Company, Philadelphia.

Okeefe, D. M., and D. M. Sylvia. 1991. Mechanisms of the vesicular-arbuscular mycorrhizal plant growth response. Pp. 35–53 in: D. K. Arora, B. Rai, K. G. Mukerji, and G. R. Knudsen (eds.), Soil and Plants (Handbook of Applied Mycology, Vol. 1). Marcel Dekker, New York.

Okonkwo, S. N. C. 1987. Developmental studies of witchweeds. Pp. 63–74 in: L. J. Musselman (ed.), *Striga* (Parasitic Weeds in Agriculture, Vol. 1). CRC Press, FL.

O'Leary, J. W. 1984. The role of halophytes in irrigated agriculture. Pp. 285–300 in: R. C. Staples, and G. H. Toenniessen (eds.), Salinity Tolerance in Plants: Strategies for Crop Improvement. John Wiley & Sons, New York.

O'Leary J. W., and E. P. Glenn. 1994. Global distribution and potential for halophytes. Pp. 7–17 in: V. R. Squires and A. Ayoub (eds.), Halophytes as a Resource for Livestock and for Rehabilitation of Degraded Land (Tasks in Vegetation Science series). Kluwer Academic Publishers, Dordrecht, Netherlands.

Oleszek, W., M. Junkuszew, and A. Stochmal. 1999. Determination and toxicity of saponins from *Amaranthus cruentus* seeds. Journal of Agriculture and Food Chemistry 47:3685–3687.

Olmos, E., and E. Hellin. 1996. Mechanisms of salt tolerance in a cell line of *Psium sativum*: biochemical and physiological aspects. Plant Science 120:37–45.

Olmos, E., J. A. Hernandez, F. Sevilla, and E. Hellin. 1994. Induction of several antioxidant enzymes in the selection of a salt-tolerant line of *Psium sativum*. Journal of Plant Physiology 144:594–598.

Olsen, R. A., G. Odham, and G. Lindeberg. 1971. Aromatic substances in leaves of *Populus tremula* as inhibitors of mycorrhizal fungi. Physiologia Plantarum 25:122–129.

O'Reagan, B. P., W. A. Cress, and J. van Staden. 1993. Root growth water relations, abscisic acid and proline levels of drought resistant and drought sensitive maize cultivars in response to water stress. South African Journal of Botany 59:98–104.

Osmond, C. B., M. P. Austin, J. A. Berry, W. D. Billings, J. S. Boyer, J. W. H. Dacey, P. S. Nobel, S. D. Smith, and W. E. Winner. 1987. Stress physiology and the distribution of plants. Bioscience 37:38–48.

Overall, R. L., J. Wolfe, and B. E. S. Gunning. 1982. Intercellular communication in *Azolla* roots: I. Ultrastructure of plasmodesmata. Protoplasma 111:134–150.

Pageau, K. P. Simier, N. Naulet, R. Robins, and A. Fer. 1998. Carbon dependency of the hemiparasite *Striga hermonthica* on *Sorghum bicolor* determined by carbon isotopic and gas exchange analyses. Australian Journal of Plant Physiology 25:685–700.

Palmgren, M. G., and G. Christensen. 1994. Functional comparisons between plant plasma membrane H-ATPase isoforms expressed in yeast. Journal of Biological Chemistry 269: 3027–3033.

Panda, N., and G. S. Khush. 1995. Host Plant Resistance to Insects. CAB International, Oxon, UK.

Pardo, J. M., M. P. Reddy, S. L. Yang, A. Maggio, G. H. Huh, T. Matsumoto, M. A. Coca, M. Paino-D'Urzo, H. Koiwa, D. J. Yun, A. A. Watad, R. A. Bressan, and P. M. Hasegawa. 1998. Stress signaling through Ca^{2+}/calmodulin-dependent protein phosphatase calcineurin mediates salt adaptation in plants. Proceedings of the National Academy of Sciences (USA) 95:9681–9686.

Pare, P. W., H. T. Alborn, and J. H. Tumlinson. 1998. Concerted biosynthesis of an insect elicitor of plant volatiles. Proceedings of the National Academy of Sciences (USA) 95: 13971–13975.

Parker, C. 1984. The influence of *Striga* species on sorghum under varying nitrogen fertilization. Pp. 90–98 in: C. Parker, L. J. Musselman, R. M. Polhill, and A. K. Wilson (eds.), Proceedings of the 3rd International Symposium of Parasitic Weeds, Aleppo, Syria. ICARDA, Aleppo.

Parker, C., and C. R. Riches. 1993. Parasitic Weeds of the World: Biology and Control. CAB International, Oxon, UK.

Parrish, J. A. D., and F. A. Bazzaz. 1976. Underground niche separation in successional plants. Ecology 57:1281–1288.

Parrish, J. A. D., and F. A. Bazzaz. 1982. Responses of plants from three successional communities to a nutrient gradient. Journal of Ecology 70:233–248.

Passioura, J. B., M. C. Ball, and J. H. Knight. 1992. Mangroves may salinize the soil and in so doing limit their transpiration rate. Functional Ecology 6:476–481.

Pasternak, D., M. Twersky, and Y. De Malach. 1979. Salt resistance in agricultural crops. Pp. 127–142 in: H. Mussell and R. C. Staples (eds.), Stress Physiology in Crop Plants. John Wiley, New York.

Pate, J. S. 1975. Exchange of solute between phloem and xylem and circulation in the whole plant. Pp. 451–473 in: M. H. Zimmerman and J. A. Milburn (eds.), Transport in Plants. I. Phloem Transport (A. Pirson and M. H. Zimmermann [series eds.], Encyclopedia of Plant Physiology, 2A). Springer-Verlag, Berlin.

Pate, J. S. 1995. Mineral relationships of parasites and their hosts. Pp. 80–102 in: M. C. Press and J. D. Graves (eds.), Parasitic plants. Chapman & Hall, London.

Pate, J. S., N. J. Davidson, J. Kuo, and J. A. Milburn. 1990. Water relations of the root hemiparasite *Olax phyllanthi* (Labill) R. Br. (Olacaceae) and its multiple hosts. Oecologia 84: 186–193.

Patterson, D. T. 1985. Comparative ecophysiology of weeds and crops. Pp. 101–129 in: Stephen O. Duke (ed.), Weed Physiology. CRC Press, Boca Raton, FL.

Patterson, D. T. 1987. Environmental factors affecting witchweed growth and development. Pp. 45–62 in: L. J. Musselman (ed.), *Striga* (Parasitic Weeds in Agriculture, Vol. I). CRC Press, Boca Raton, FL.

Pearce, G. D., C. A. Ryan, and D. Liljegren. 1988. Proteinase inhibitors I and II in fruit of wild tomato species: transient components of a mechanism for defense and seed dispersal. Planta 175:527–531.

Pearce, G. D., D. Strydom, S. Johnson, and C. A. Ryan. 1991. A polypeptide from tomato leaves induced wound-inducible inhibitor proteins. Science 253:895–898.

Pearce, G. D., S. Johnson, and C. A. Ryan. 1993. Structure-activity of deleted and substituted systemin, an eighteen amino acid polypeptide inducer of plant defensive genes. Journal of Biological Chemistry 268:212–216.

Pearcy, R. W., and S. L. Ustin. 1984. Effects of salinity on growth and photosynthesis of three California tidal marsh species. Oecologia 62:68–73.

Pearcy, R. W. 1990. Sunflecks and photosynthesis in plant canopies. Annual Review of Plant Physiology 41:421–53.

Pearcy, R. W., and D. A. Sims. 1994. Photosynthetic acclimation to changing light environments: scaling from the leaf to the whole plant. Pp. 145–174 in: M. M. Caldwell and R. W. Pearcy (eds.), Exploitation of Environmental Heterogeneity by Plants: Ecophysiological Processes Above and Belowground. Academic Press, San Diego, CA.

Pearcy, R. W., N. Tumosa, and K. Williams. 1981. Relationships between growth, photosynthesis and competitive interactions for a C_3 and a C_4 plant. Oecologia 48:371–376.

Pearson, J. N., and I. Jakobsen. 1993. Symbiotic exchange of carbon and phosphorus between cucumber and three arbuscular mycorrhizal fungi. New Phytologist 124:481–488.

Pearson, V., and D. J. Read. 1975. The physiology of the mycorrhizal endophyte of *Calluna vulgaris*. Transactions of British Mycological Society 64:1–7.

Pell, E. J., L. G. Landry, N. A. Eckardt, and R. E. Glick. 1994. Air pollution and rubisco: effects and implications. Pp. 239–253 in: R. G. Alscher, and A. R. Wellburn (eds.), Plant Responses to the Gaseous Environment: Molecular, Metabolic and Physiological Aspects. Chapman & Hall, New York.

Peng, Z., Q. Lu, and D. P. S. Verma. 1996. Reciprocal regulation of Delta(1)-pyrroline-5-carboxylate synthetase and proline dehydrogenase genes controls proline levels during and after osmotic stress in plants. Molecular and General Genetics 253:334–341.

Pennings, S. C., and R. M. Callaway. 1996. Impact of a plant parasite on the structure and dynamics of salt marsh vegetation. Ecology 77:1410–1419.

Pennings, S. C., and V. J. Paul. 1992. Effect of plant toughness, calcification, and chemistry on herbivory by *Dolabella auricularia*. Ecology 73:1606–1619.

Pennisi, E. 1997. Research news: plants decode a universial signal. Science 278:2054–2055.

Perera, L. K. R., T. A. Mansfield, and A. J. C. Malloch. 1994. Stomatal responses to sodium ions in *Aster trifolium*: a new hypothesis to explain salinity regulation in aboveground tissues. Plant Cell and Environment 17:335–340.

Perez-Prat, E., M. Narasimhan, X. Niu, M. A. Botella, R. Bressan, and V. Valpuesta. 1994. Growth cycle stage-dependent NaCl induction of plasma membrane H^+-ATPase mRNA accumulation in de-adapted tobacco cells. Plant Cell and Environment 17:327–333.

Perotto, S., R. Perotto, A. Facio, A. Schubert, A. Varma, and P. Bonfante. 1995. Ericoid mycorrhizal fungi: cellular and molecular bases of their interactions with the host plant. Canadian Journal of Botany 73(suppl. 1):557–568.

Peterson, R. L., and M. L. Farquar. 1994. Mycorrhizas: integrated development between roots and fungi. Mycologia 86:311–326.

Petranka, J. W., and J. K. McPherson. 1979. The role of *Rhus copallina* in the dynamics of the forest-prairie ecotone in north-central Oklahoma. Ecology 60:956–965.

Piehl, M. A. 1963. Mode of attachment, haustorium structure, and hosts of *Pedicularis canadensis*. American Journal of Botany 50:978–985.

Pimentel, D. 1976. World food crisis: energy and pests. Bulletin of the Entomological Society of America 22:20–26.

Pimentel, D., L. McLaughlin, A. Zepp, B. Lakitan, T. Kraus, P. Kleinman, F. Vancini, W. J. Roach, E. Grapp, W. S. Keeton and G. Selig. 1991. Environmental and economic effects of reducing pesticide use. BioScience 41:402–409.

Pimentel, D., U. Stachow, D. A. Takacs, H. W. Brubaker, A. R. Dumas, J. J. Meaney, J. A. S. O'Neal, D. E. Onsi, and D. B. Corzilius. 1992. Conserving biological diversity in agricultural/forestry systems. BioScience 42:354–362.

Pitman, M. G., A. Läuchli, and R. Stelzer. 1981. Ion distribution in roots of barley seedlings measured by electron probe X-ray microanalysis. Plant Physiology 68:673–679.

Plaut, Z., and C. M. Grieve. 1988. Photosynthesis of salt-stressed maize as influenced by Ca/Na ratios in the nutrient solution. Plant and Soil 105:283–286.

Plaut, Z., C. M. Grieve, and E. Federman. 1989. Salinity effects on photosynthesis in isolated mesophyll cells of cowpea leaves. Plant Physiology 91:493–499.

Pollak, G, and Y. Waisel. 1970. Salt secretion in *Aeluropus litoralis* (Willd.) Parl. Annals of Botany 34:879–888.

Polle, A., and W. Junkermann. 1994. Inhibition of apoplastic and symplastic peroxidase activity from Norway Spruce by the photooxidant hydroxymethyl hydroperoxide. Plant Physiology 104:617–621.

Poorter, H., C. A. D. M. Van der Vijver, R. G. A. Boot, and H. Lambers. 1995. Growth and carbon economy of a fast-growing and a slow-growing grass species as dependent on nitrate supply. Plant and Soil 171:217–227.

Popp, M., J. Polina, and M. Weiper. 1993. Physiological adaptations to different salinity levels in mangroves. Pp. 217–224 in: H. Leith and A. A. A. Masoom (eds.), Towards the Rational Use of High Salinity Tolerant Plants. Kluwer Academic Publishers, Dordrecht, Netherlands.

Prakash, L., and G. Prathapasenan. 1990. Interactive effect of NaCl salinity and gibberellic acid on shoot growth, content of abscisic acid and gibberellin-like substances and yield of rice

(*Oryza sativa* L. var. GR-3). Proceedings of the Indian Academy of Sciences—Plant Sciences 100:173–181.

Prasad, M. N. V. 1997. Trace metals. Pp. 207–249 in: M. N. V. Prasad (ed.), Plant Ecophysiology, John Wiley & Sons, New York.

Press, M. C., and J. D.. Graves (eds.). 1995. Parasitic Plants. Chapman & Hall, London.

Press, M. C., and G. Stewart. 1987. Growth and photosynthesis in *Sorghum bicolor* infected with *Striga hermonthica*. Annals of Botany 60:657–662.

Press, M. C., N. Shah, and G. R. Stewart. 1986. The parasitic habit: trends in metabolic reductionism. Pp. 96–107 in: S. J. ter Borg (ed.), Biology and Control of *Orobanche*. Proceedings of a workshop in Wageningen, 13–17 January. Dept. of Vegetation Science, Plant Ecology & Weed Sicence, Agricultural University, Wageningen, Netherlands..

Press, M. C., J. M. Tuohy, and G. R. Stewart. 1987. Gas exchange characteristics of the *Sorghum-Striga* host parasite association. Plant Physiology 84:814–819.

Press, M. C., N. Shah, J. M. Touhy, and G. R. Stewart. 1987. Carbon isotope ratios demonstrate carbon flux from C-4 host to C-3 parasite. Plant Physiology 85:1143–1145.

Prestidge, R. A. 1982. Instar duration, adult consumption, oviposition, and nitrogen utilization efficiencies of leaf hoppers feeding of different quality food. (*Auchenorryncha:* Homoptera). Ecological Entomology 7:91–101.

Preszler, R. W., and P. W. Price. 1988. Host quality and sawfly populations: a new approach of life table analysis. Ecology 69:2012–2020.

Price, P. W., M. Westoby, B. Rice, P. R. Atsatt, R. S. Fritz, J. N. Thompson, and K. Mobley. 1986. Parasite mediation in ecological interactions. Annual Review of Ecology and Systematics 17:487–505.

Putnam, A. R., and C.-S. Tang. 1986. The Science of Allelopathy. John Wiley & Sons, New York.

Qasem, J. R., and T. A. Hill. 1989. Possible role of allelopathy in the competition between tomato, *Senecio vulgaris* L. and *Chenopodium album* L. Weed Science 29:349–356.

Rabinowitch, H. D., and I. Fridovich. 1983. Superoxide radicals, superoxide dismutase and oxygen toxicity in plants. Photochemisty and Photobiology 37:679–690.

Radomiljac, A. M., J. A. McComb, and J. S. Pate. 1998. Xylem transfer of organic solutes in *Santalacum album* L. (Indian sandalwood) in association with legume and non-legume hosts. Annals of Botany 82:675–682.

Radomiljac, A. M., J. A. McComb, and J. S. Pate. 1999a. Gas exchange and water relations of the root hemiparasite *Santalacum album* L. in association with legume and non-legume hosts. Annals of Botany 83:215–224.

Radomiljac, A. M., J. A. McComb, and J. F. McGrath. 1999b. Intermediate host influences on the root hemiparasite *Santalacum album* L. biomass partitioning. Forest Ecology and Management 113:143–153.

Rains, D. W., and E. Epstein. 1967. Preferential absorption of potassium by leaf tissues of the mangrove, *Avicennia marina*: an aspect of halophytic competance in coping with salt. Australian Journal of Biological Science 20:847–857.

Ramlan, M. F., and J. D. Graves. 1996. Estimation of the sensitivity to photoinhibition in *Striga hermonthica*-infected sorghum. Journal of Experimental Botany 47:71–78.

Rapport, D. J., and W. G. Whitford. 1999. How ecosystems respond to stress. BioScience 49: 193–203.

Raven, P. H., R. F. Evert, and S. E. Eichhorn. 1992. Biology of Plants, Worth Publishers, New York.

Raven, J. A. 1983. Phytophages of xylem and phloem: a comparison of animal and plant sapfeeders. Advances in Ecological Research 13:135–234.

Rayner, A. D. M., G. S. Griffith, and A. M. Ainsworth. 1994. Mycelial interconnectedness. Pp. 21–40 in: N. A. R. Gow and G. M. Gadd (eds.), The Growing Fungus. Chapman & Hall, London.

Rayner A. D. M., M. Ramsdale, and Z. R. Watkins. 1995. Origins and significance of genetic and epigenetic instability in mycelial systems. Canadian Journal of Botany 73(suppl.1):1241–1248.

Read, D. J. 1984. The structure and function of the vegetative mycelium of mycorrhizal roots. Pp. 215–240 in: D. H. Jennings and A. D. M. Rayner (eds.), The Ecology and Physiology of the Fungal Mycelium. Cambridge University Press, Cambridge, UK.

Read, D. J., and D. P. Stribley. 1975. Effect of mycorrhizal infection on nitrogen and phosphorus nutrition of ericaceous plants. Nature 244:81.

Read, D. J., R. Francis, and R. D. Finlay. 1985. Mycorrhizal mycelia and nutrient cycling in plant communities. Pp. 193–217 in: A. H. Fitter, D. Atkinson, D. J. Read, and M. B. Usher (eds.), Ecological Interactions in Soil: Plants, Microbes and Animals. Blackwell Scientific Publications, Boston, MA.

Reid, N. 1991. Coevolution of mistletoes and frugivorous birds. Australian Journal of Ecology. 16:457–469.

Renegal, Z. 1992. The role of calcium in salt toxicity. Plant Cell and Environmenmt 15:625–632.

Rennenberg, H. 1991. The significance of higher plants in the emission of sulphur compounds from terrestrial ecosystems. Pp. 217–260 in: T. D. Sharkey, E. A. Holland, and H. A. Mooney (eds.), Trace Gas Emissions by Plants. Academic Press, San Diego, CA.

Rennenberg, H., and C. Herschbach. 1996. Responses of plants to atmospheric sulphur. Pp. 285–291 in: M. Yunus, and M. Iqbal (eds.), Plant Response to Air Pollution. John Wiley & Sons, New York.

Rennenberg, H., and A. N. Polle. 1994. Metabolic consequences of atmospheric sulphur influx into plants. Pp. 165–180 in: R. G. Alscher, and A. R. Wellburn (eds.), Plant Responses to the Gaseous Environment: Molecular, Metabolic and Physiological Aspects. Chapman & Hall, New York.

Renwick, J. A. A., and F. S. Chew. 1994. Oviposition behavior in Lepidoptera. Annual Review of Entomology 39:377–400.

Reports from the Academic Assessment Program. 1999. Quality Improvement in Action. Entering New Students: Changing Values with High Expectations. Virginia Polytechnic Institute and State University, Blacksburg, VA.

Reuveni, M., A. B. Bennett, R. A Bressen, and P. M. Hasegawa. 1990. Enhanced H^+ transport capacity and ATP hydrolysis activity of the tonoplast H^+-ATPase after NaCl adaptation. Plant Physiology 94:524–530.

Reynolds, H. T., P. L. Adkisson, R. F. Smith, and R. E. Frisbie. 1982. Cotton insect pest management. Pp. 375–441 in: R. L. Metcalf and W. H. Luckmann (eds.), Introduction to Insect Pest Management, 2nd Ed. John Wiley & Sons, New York.

Rhodes, D. F. 1977. The antiherbivore chemistry of *Larrea.* Pp. 135–175 in: T. J. Mabry, J. H. Hunziker, and D. R. DiFeo (eds.), Creosote Bush: Biology and Chemistry of *Larrea* in New World Deserts. Dowdon, Hutchinson and Ross, Stroudsburg, PA.

Rhodes, L. H., and J. W. Gerdemann. 1975. Phosphate uptake zones of mycorrhizal and non-mycorrhizal onions. New Phytologist 75:555–561.

Rhodes, L. H., and J. W. Gerdemann. 1978a. Translocation of calcium and phosphate by external hyphae of vesicular-arbuscular mycorrhizae. Soil Science 126:125–126.

Rhodes, L. H., and J. W. Gerdemann. 1978b. Hyphal translocation and uptake of sulphur by vesicular-arbuscular mycorrhizae of onions. Soil Biology and Biochemistry 10:355–360.

Rice, E. L. 1974. Allelopathy. Academic Press, New York.

Rice, E. L. 1984. Allelopathy, 2nd Ed. Academic Press, New York.

Richards, J. H., and M. M. Caldwell. 1985. Soluble carbohydrates, concurrent photosynthesis, and efficiency in regrowth following defoliation: A field study with *Agropyron* species. Journal of Applied Ecology 22:907–920.

Richardson, M. 1977. The proteinase inhibitors of plants and microorganisms. Phytochemistry 16:159–169.

Richardson, D. R., and G. B. Williamson. 1988. Allelopathic effects of shrubs of the sand pine scrub on pines and grasses of the sandhills. Forest Science 34:592–605.

Richharia, A., K. Shah, and R. S. Dubey. 1997. Nitrate reductase from rice seedlings: Partial purification, characterization and the effects of in-situ and in-vitro NaCl salinity. Journal of Plant Physiology 151:316–322.

Ricklefs, R. 1993. The Economy of Nature. W. H. Freeman, New York.

Rink, G., and J. W. V. Sambeek. 1987. Variation among four white ash families in response to competition and allelopathy. Forest Ecology and Management 18:127–134.

Robberecht, R., B. E. Mahall, and P. S. Nobel. 1983. Experimental removal of intraspecific competitors—effects on water relations and productivity of a desert bunchgrass, *Hilaria rigida*. Oecologia 60:21–24.

Robbins C. T., T. A. Hanley, A. E. Hagerman, O. Heljord, D. L. Baker, C. C. Schwartz, and W. W. Mautz. 1987. Role of tannins in defending plants against ruminants: reduction in protein availability. Ecology 68:98–107.

Robbins, C. T., A. E. Hagerman, P. J. Austin, C. McArthur, and T. A. Hanley. 1991. Variation in mammalian physiological responses to a condensed tannin and its ecological implications. Journal of Mammology 72:480–486.

Robinson, N. J. 1990. Metal-binding polypeptides in plants. Pp. 195–214 in: A. Jonathan Shaw (ed.), Heavy Metal Tolerance in Plants: Evolutionary Aspects. CRC Press, Boca Raton, FL.

Robinson, D., and I. H. Rorison. 1987. Root hairs and plant growth at low nitrogen availabilities. New Phytologist 107:681–693.

Robinson, R. K. 1972. The production by roots of *Calluna vulgaris* of a factor inhibitory to growth of some mycorrhizal fungi. Journal of Ecology 60:219–224.

Robinson, S. P., W. J. S. Downton, and J. A. Millhouse. 1983. Photosynthesis and ion content of leaves and isolated chloroplasts of salt-stressed spinach. Plant Physiology 73:238–242.

Roessingh, P., E. Stadler, R. Schoni, and P. Feeney. 1991. Tarsal contact chemoreceptors of the black swallowtail butterfly, *Papilio polyxenes*: reponses to phytochemicals from host- and non-host plants. Physiological Entomology 16:485–495.

Root, R. B. 1967. The niche exploration pattern of the blue gray gnatcatcher. Ecological Monographs 37:317–350.

Rose, S. L., D. A. Perry, D. Pilz, and M. M. Schoeneberger. 1983. Allelopathic effects of litter on the growth and colonization of mycorrhizal fungi. Journal of Chemical Ecology 9:1153–1162.

Rosenthal, G. A. 1982. Plant Nonprotein Amino Acids: Biological, Biochemical, and Toxicological Properties. Academic Press, New York.

Rosenthal, G. A., M. A. Berge, J. A. Bleiler, and T. P. Rudd. 1987. Aberrant canavalyl protein formation and the ability to tolerate or utilize L-canavanine. Experimentia 43:558–561.

Rosenthal, G. A. 1991. Nonprotein amino acids as protective allelochemicals. Pp. 1–34 in: G. A. Rosenthal, and M. R. Berenbaum (eds.), The Chemical Participants (Herbivores: Their Interactions with Secondary Plant Metabolites, 2nd. Ed., Vol. 1). Academic Press, San Diego, CA.

Rosenthal, G. A., and M. R. Berenbaum (eds.). 1991. The Chemical Participants (Herbivores: Their interactions with Secondary Plant Metabolites, 2nd Ed., Vol. 1). Academic Press, San Diego, CA.

Rosenthal, G. A., and M. R. Berenbaum (eds.). 1992. Ecological and Evolutionary Processes (Herbivores: their interactions with secondary plant metabolites, 2nd Ed., Vol. 2). Academic Press, San Diego, CA.

Rosenzweig, M. L. 1973. Exploitation in three tropic levels. American Naturalist 107:275–294.

Ross, S. M. 1994. Sources and forms of potentially toxic metals in soil-plant systems. Pp. 3–25 in: S. M. Ross (ed.), Toxic Metals in Soil-Plant Systems. John Wiley & Sons, New York.

Rostock, M., and C. Unger. 1998. Utility of mistletoe therapy in cancer patients. Internist 39(12): 1294–1295.

Roush, M. L., S. R. Radosevich, R. G. Wagner, B. D. Maxwell, and T. D. Petersen. 1989. A comparison of methods for measuring effects of density and proportion in plant competition experiments. Weed Science 37:268–275.

Rousseau, J. V. D., and C. P. P. Reid. 1990. Effects of phosphorus and ectomycorrhizas on the carbon balance of loblolly pine seedlings. Forest Science 36:101–112.

Ruiz-Lozana, J. M., and R. Azcón. 1995. Hyphal contribution to water uptake in mycorrhizal plants as affected by the fungal species and water status. Physiologia Plantarum 95:472–478.

Rundel, P. W., M. R. Sharifi, and A. Gonzales-Coloma. 1994. Resource availability and herbivory in *Larrea tridentata*. Pp. 105–114 in: M. Arianoutsou and R. H. Graves (eds.), Plant-Animal Interactions in Mediterranean-Type Ecosystems. Kluwer Academic Publishers, Amsterdam.

Rundel, P. W., B. Palma, M. O. Dillon, M. R. Sharifi, E. T. Nilsen, and K. Boonpragob. 1997. *Tillandsia ludbeckii* in the coastal Atacamba desert of northern Chile. Revista Chilena de Historia Natural 70:341–349.

Ruso, J., S. Sukino, J. Dominguez-Gimenez, J. M. Melero-Vara, and J. Fernandes-Martinez. 1996. Screening of Wild *Helianthus* species and derived lines for resistance to several populations of *Orobanche cernua*. Plant Disease 80:1165–1169.

Russell, G. B., and P. G. Fenmore. 1970. Insect moulting hormone activity of some New Zealand gymnosperms. New Zealand Journal of Science 13:61–63.

Russness, D. G., and G. G. Still. 1977. Partial purification and properties of S-cysteinyl-hydrochlororopham transferase from *Avena sativa* L. Pesticide Biochemistry and Physiology 7:220–231.

Rutherford, M. C., and L. W. Powrie. 1993. Allelochemic control of biomass allocation in interacting shrub species. Journal of Chemical Ecology 19:893–906.

Ryan, C. A. 1974. Assay and biochemical properties of the proteinase inhibitor inducing factor, a wound hormone. Plant Physiology 54:328–332.

Ryan, C. A. 1990. Proteinase inhibitors in plants: genes for improving defenses against insects and pathogens. Annual Review of Phytopathology 28:425–449.

Ryan, C. A., and G. Pearce. 1998. Systemin: A polypeptide signal for plant defense genes. Annual Review of Cell and Developmental Biology 14:1–17.

Saarikoski, P., D. Clapham, and S. von Arnold. 1996. A wound-inducible gene from *Salix viminalis* coding for a trypsin inhibitor. Plant Molecular Biology 31:465–478.

Sacher, R. F., and R. C. Staples. 1984. Chemical microscopy for the study of plants in saline environments. Pp. 17–36. in: R. C. Staples and G. H. Toenniessen (eds.), Salinity Tolerance in Plants. Strategies for Crop Improvement. John Wiley & Sons, New York.

Sachs, J. 1873. Lehrbuch der Botanik. W. Engelmann, Leipzig, Germany.

Safir, G. R., J. S. Boyer, and J. W. Gerdemann. 1971. Mycorrhizal enhancement of water transport in soybean. Science 172:581–583.

Safir, G. R., J. S. Boyer, and J. W. Gerdemann. 1972. Nutrient status and mycorrhizal enhancement of water transport in soybean. Plant Physiology 49:700–703.

Sahm, A., H. Pfanz, M. Grunsfelder, F. C. Czygan, and P. Proksch. 1995. Anatomy and phenylpropanoid metabolism in the incompatible interaction of *Lycopersicon esculentum* and *Cuscuta reflexa*. Botanica Acta 108:358–364.

Salin, M. L. 1988. Toxic oxygen species and protective systems of the chloroplast. Physiologia Plantarum 72:681–689.

Salisbury, F. B., and C. W. Ross. 1992. Plant Physiology, 4th Ed. Wadsworth Publishing Company, Belmont, CA.

Salle, G., B. Dembele, A. Raynal-Roques, and M. F. Hallis. 1987. Biological aspects of *Striga* species, pest of food crops (Scrophulariaceae). Pp. 719–731 in: H. C. Weber, W. Forstreuter (eds.), Parasitic Flowering Plants: Proceedings of the 4th International Symposium on Parasitic Flowering Plants. Phillips Universität, Marburg, Germany.

Salmanowicz, B., and J.-E. Nylund. 1988. High performance liquid chromatography determination of ergosterol as a measure of ectomycorrhizal infection in Scots pine. European Journal of Forest Pathology 18:291–298.

Salonen, V., and S. Puustinen. 1996. Success of a root hemiparasitic plant is influenced by soil quality and by defoliation of its host. Ecology 77:1290–1293.

Salunkhe, D. K., J. K. Chavan, and S. S. Kadam. 1990. Nutritional consequences of dietary tannins. Pp. 113–146 in: D. K. Salunkhe, J. K. Chavan, and S. S. Kadam (eds.), Dietary Tannins: Consequences and Remedies. CRC Press, Boca Raton, FL.

Salzer, P., and A. Hager. 1993. Characterization of wall bound invertase isoforms of *Picea abies* cells and regulation by ectomycorrhizal fungi. Physiologia Plantarum 88:52–59.

Sanchez-Diaz, M., and M. Honrubia. 1994. Water relations and alleviation of drought stress in mycorrhizal plants. Pp. 167–178 in: S. Gianinazzi and H. Schupp (Eds.), Impact of Arbuscular Mycorrhizas on Sustainable Agriculture and Natural Ecosystems. Birkhäuser Verlag, Basel, Germany.

Sand, P. F. 1987. The American witchweed quarantine and eradication program. Pp. 207–223 in: H. C. Weber, and W. Forstreuter (eds.), Parasitic Flowering Plants: Proceedings of the 4th International Symposium on Parasitic Flowering Plants. Phillips Universität, Marburg, Germany.

Sanderman, E. Jr. 1996. Ozone and plant health. Annual Review of Phytopathology 34:347–366

Sanders, F. E., P. B. Tinker, R. L. B. Black, and S. M Palmerley. 1977. The development of endomycorrhizal root systems. I. Spread of infection and growth promoting effects with four species of vesicular-arbuscular mycorrhizas. New Phytologist 78:257–268.

San Pietro, A. 1982. Biosaline Research: A Look to the Future. Plenum Press, New York.

Sarjus, A., and M. Durand. 1996. Lipid changes in soybean root membranes in response to salt treatment. Journal of Experimental Botany 47:17–23.

Sasek, T. W. and B. R. Strain. 1990. Implications of atmospheric CO_2 enrichment and climatic change for the geographical distribution of two introduced vines in the USA. Climatic Change 16:31–51.

Sauerborn, J. 1991. Parasitic Flowering Plants: Ecology and Management. Verlag Josef Margraf, Weikersheim, Germany.

Saunders, A. R. 1933. Studies in Phanerogamic Parasitism with Particular Reference to *Striga lutea* Lour. South African Department of Agricultural Science Bulletin #128.

Schaeffer, C., T. Wallenda, M. Guttenberger, and R. Hampp. 1995. Acid invertase activity in mycorrhizal and non-mycorrhizal roots of Norway spruce (*Picea abies* [L.] Karst.) seedlings. New Phytologist 129:417–424.

Schafer, W. E., and J. Schönherr. 1985. Accumulation and transport of phenol, 2-nitrophenol, and 4-nitrophenol in plant cuticles. Ecological toxicology and Environmental Safety 10:239.

Schenck, N. C., and Y. Perez. 1990. Manual for the Identification of VA Mycorrhizal Fungi, 3rd Ed. Synergistic Publications, Gainesville, FL.

Scheumann, W. 1997. Managing Salinization. Institutional Analysis of Public Irrigation Systems. Springer-Verlag, Berlin.

Scheunert, I. 1992. Fate of pesticides in plants and in soil fauna. Pp. 77–103 in: W. Ebing (ed.), Chemistry of Plant Protection, Springer-Verlag, New York.

Schimper, A. F. W. 1898. Pflanzengeographie auf Physiologischer Grundlage. Gustav Fischer Verlag, Jena, Germany.

Schlesinger, W. H. 1991. Biogeochemistry: An Analysis of Global Change. Academic Press, San Diego, CA.

Schlesinger W. H., J. T. Gray, D. S. Gill, and B. E. Mahall. 1982. *Ceanothus megacarpus* chaparral: a synthesis of ecosystem processes during development and annual growth. Botanical Review 48:71–117.

Schlichting, C. D., and M. Pigliucci. 1998. Phenotypic Evolution: A Reaction Norm Perspective. Sinaurr Associates, Sunderland, MA.

Schmialek, V. P. 1961. Die identifizierung zweier im tenebriokot und in hefe vorkommender substanzen mit juvenilhormonwirkung. Zeitscrift fur Naturforschung Tiel B. 16:461–464.

Schmidt, I., A. Michelsen, and S. Jonasson. 1997a. Effects of labile soil carbon on nutrient partitioning between an arctic graminoid and microbes. Oecologia 112:557–565.

Schmidt, I., A. Michelsen, and S. Jonasson. 1997b. Effects on plant production after addition of labile carbon to arctic/alpine soils. Oecologia 112:305–313.

Schmidt, S. K. 1988. Degradation of juglone by soil bacteria. Journal of Chemical Ecology 14: 1561–1571.

Schmidt, S. K. 1990. Ecological implications of the destruction of juglone (5-hydroxy-1,4-naphthoquinone) by soil bacteria. Journal of Chemical Ecology 16:3547–3548.

Schmitt, J. 1997. Is photomorphogenic shade avoidance adaptive? Perspectives from population biology. Plant, Cell and Environment 20:826–830.

Schmitt, J., D. W. Ehrhardt, and M. Cheo. 1986. Light-dependent dominance and suppression in experimental radish populations. Ecology 67:1502–1507.

Schmidt, S. K., and F. B. Reeves. 1989. Interference between *Salsola kali* L. seedlings: implications for plant succession. Plant and Soil 116:107–110.

Schmitt, J., A. C. McCormac, and H. Smith. 1996. A test of the adaptive plasticity hypothesis using transgenic mutant plants disabled in phytochrome-mediated elongation responses to neighbors. American Naturalist 146:937–953.

Schnapp, S. R., W. R. Curtis, R. A. Bressan, and P. M. Hasagawa. 1991. Growth yields and maintenance coefficients of unadapted and NaCl-adapted tobacco cells grown in semicontinuous culture. Plant Physiology 96:1289–1293.

Schneider, M. J. 1990. Uptake of host plant alkaloids by root parasitic *Pedicularis* species. Phytochemistry 29:1811–1814.

Schneider, M. J. S., and F. R. Stermitz. 1990. Uptake of host plant alkaloids by root parasitic *Pedicularis* species. Phytochemistry 29:1811–1814.

Schuler, I., A. Milon, Y. Nakatani, G. Ourisson, A. M. Albrecht, P. Benveniste, and M. A. Hartmann. 1991. Differential effects of plant sterols on water permeability and on acyl chain ordering of soybean phosphatidylcholine bilayers. Proceedings of the National Academy of Sciences (USA) 88:6926–6930.

Schulz, M., and A. Friebe. 1999. Detoxification of allelochemicals in higher plants and enzymes involved. Pp. 383–400 in: K. M. Inderjit, M. Dakshini, and C. L. Foy (eds.), Principles and Practices in Plant Ecology: Allelochemical Interactions. CRC Press, Boca Raton, FL.

Schulz, M., A. Friebe, P Kück, M Seipel, and H. Schnabl. 1994. Allelopathic effects of living quackgrass (L.). Identification of inhibitory allelochemicals exuded from rhizome borne roots. Angewandte Botanik 68:195.

Schulze, E.-D., M. Küppers, and R. Matyssek. 1986. The roles of carbon balance and branching pattern in the growth of woody species. Pp. 585–601 in: T. J. Givnish (ed.), On the Economy of Plant Form and Function: Proceedings of the 6th Maria Moors Cabot Symposium, (Evolutionary Constraints on Primary Productivity, Adaptive Patterns of Energy Capture in Plants), Harvard Forest, August 1983. Cambridge University Press, Cambridge, UK.

Schulze, E. D., R. H. Robichaux, J. Grace, P. W. Rundel, and J. R. Ehleringer. 1987. Plant water balance: In diverse habitats, where water often is scarce, plants display a variety of mechanisms for managing this essential resource. Bioscience 37:30–37.

Schwind, P., V. Wray, and A. Nahrstedt. 1990. Structure elucidation of an acylated cyanogenic triglycoside, and further cyanogenic constituents from *Xeranthemum cylindraceum*. Phytochemistry 29:1903–1911.

Schwinning, S., and J. Weiner. 1998. Mechanisms determining the degree of size asymmetry in competition among plants. Oecologia 113:447–455.

Seaward, M. R. D., and D. H. S. Richardson. 1990. Atmospheric sources of metal pollution and effects on vegetation. Pp. 75–92 in: A. Jonathan Shaw (ed.), Heavy Metal Tolerance in Plants: Evolutionary Aspects. CRC Press, Boca Raton, FL.

Seel, W. E., and M. C. Press. 1994. Influence of the host on 3 sub-arctic annual facultative root hemiparasites. 2. Gas exchange characteristics and resource use efficiency. New Phytologist 127:37–44.

Seel, W. E., and M. C. Press. 1996. Effects of repeated parasitism by *Rhinanthus minor* on the growth and photosynthesis of a perennial grass, *Poa alpina*. New Phytologist 134:495–502.

Seemann, J. R. and C. Critchley. 1985. Effects of salt stress on the growth ion content, stomatal behavior and photosynthetic capacity of a salt sensitive species, *Phaseolus vulgaris* (L.). Planta 164:151–162.

Seemann, J. R., and T. D. Sharkey. 1986. Salinity and nitrogen effects on photosynthesis, ribulose-1,5-bisphosphate carboxylase and metabolite pool size in *Phaseolus vulgaris* L. Plant Physiology 82:555–560.

Seigler, D. S. 1981. Secondary metabolites and plant systematics. Pp. 139–176 in: E. E. Conn (ed.), The Biochemistry of Plants. Academic Press, New York.

Seigler, D. S. 1991. Cyanide and cyanogenic glycosides. Pp. 35–77 in: G. A. Rosenthal, and M. R. Berenbaum (eds.), The Chemical Participants (Herbivores: Their Interactions with Secondary Plant Metabolites, 2nd Ed., Vol. 1). Academic Press, San Diego, CA.

Seigler, D. S. 1996. Chemistry and mechanisms of allelopathic interactions. Agronomy Journal 88:876–885.

Seiler, W. 1974. The cycle of atmospheric CO. Tellus 26:116–135.

Sekler, I., M. Weiss, and U. Pick. 1994. Activation of *Dunaliella acidophila* plasma membrane H^+ -ATPase by trypsin cleavage of a fragment that contains a phosphorylation site. Plant Physiology 105:1125–1132.

Selldén, G., and H. Pleijel. 1995. Photochemical oxidant effects on vegetation: Response in relation to plant strategy. Water, Air and Soil Pollution 85:111–122.

Semones, S. W., and D. R. Young. 1995. VAM association in the shrub *Myrica cerifera* on a Virginia, USA barrier island. Mycorrhiza 5:423–429.

Shachar-Hill, B., and A. E. Hill. 1970. Ion and water transport in *Limonium*. Biochemical and Biophysical Acta 211:313–317.

Shah, N., N. Smirnoff, and G. R. Stewart. 1987. Photosynthesis and stomatal characteristics of *Striga hermonthica* in relation to its parasitic habit. Physiologia Plantarum 69:699–703.

Shann, J. R., and U. Blum. 1987. The utilization of exogenously supplied ferulic acid in lignin biosynthesis. Phytochemistry 26:2977–2982.

Sharifi, M. R., E. T. Nilsen, and P. W. Rundel. 1987. Biomass and net primary production of *Prosopis glandulosa* (Fabaceae) in the Sonoran desert of California. American Journal of Botany 69:760–767.

Shaw, P. J. A. 1985. Grazing preferences of *Onychiurus armatus* (Insecta: Collembola) for mycorrhizal and saprophytic fungi of pine plantations. Pp. 333–337 in: A. H. Fitter, D.

Atkinson, D. J. Read, and M. B. Usher (eds.), Ecological Interactions in Soil: Plants, Microbes and Animals, Blackwell Scientific Publications, Boston, MA.

Sheen, J. 1996. Ca^{2+}-dependent protein kinases and stress signal transduction in plants. Science 274:1900–1902.

Shepherd, V. A., D. A. Orlovich, and A. E. Ashford. 1993a. Cell to cell transport via motile tubules in growing hyphae of a fungus. Journal of Cell Science 105:1173–1178.

Shepherd, V. A., D. A. Orlovich, and A. E. Ashford. 1993b. A dynamic continuum of pleiomorphic tubules and vacuoles in growing hyphae of a fungus. Journal of Cell Science 104: 495–507.

Shettel, N. L., and N. E. Balke. 1983. Plant growth response to several allelopathic chemicals. Weed Science 31:293–298.

Shrift, A. 1967. Microbial research with selenium. Pp. 241–271 in: O. Muth, J. Oldfield, and P. Weswig (eds.), Symposium: Selenium in Biomedicine. Avi Publishers, Westport, CT.

Shulaev, V., P. Silverman, and I. Raskin. 1997. Airborne signalling by methyl salicylate in plant pathogen resistance. Nature 385:718–721.

Shuman, L. M. 1994. Mineral nutrition, Pp. 149–182 in: R. E. Wilkinson (ed.), Plant-Environment Ineractions. Marcel Dekker, New York.

Sieverding, E. 1991. Vesicular-Arbuscular Mycorrhiza Management in Tropical Agroecosystems. Technical Corporation of Germany, Eschborn, Germany.

Silva, A., and C. Rio. 1996. Effects of the mistletoe *Tristerix aphyllus* (Loranthaceae) on the reproduction of its cactus host *Echinopsis chilensis*. Oikos 75:437–442.

Simard, S. W., D. A. Perry, M. D. Jones, D. D. Myrold, D. M. Durall, and R. Molina. 1997. Net transfer of carbon between ectomycorrhizal tree species in the field. Nature 388:579–582.

Simier, P., S. Renaudin, and A. Fer. 1994. Characteristics of the mannitol pathway in a root hemiparasitic species, *Thesium humile* Vahl (Santalacaceae). Journal of Plant Physiology 143: 33–38.

Simms, E. L. 1992. Costs of plant resistance to herbivory. Pp. 392–425 in: R. S. Fritz and E. L. Simms (eds.), Plant Resistance to Herbivores and Pathogens. University of Chicago Press, Chicago.

Simms, E. L., and M. D. Rausher. 1987. Costs and benefits of plant defense to herbivory. American Naturalist 130:570–581.

Simpson, B. B. 1977. Mesquite: Its Biology in Two Desert Ecosystems (US/IBP Synthesis Series). Dowden, Hutchinson and Ross, Stroudsburg, PA.

Simpson, B. B. 1989. Kramariaceae. Flora Neotropica Monograph 49. New York Botanical Garden, New York.

Simpson, D., and M. J. Daft. 1990. Interactions between water-stress and different mycorrhizal inocula on plant growth and mycorrhizal development in maize and sorghum. Plants and Soil 121:179–186.

Sinclair, J. B., and P. A. Backman 1989. (eds.). Compendium of soybean diseases, 3rd Ed. APS Press, St. Paul, MN.

Sivak, M. N., and D. A. Walker. 1986. Photosynthesis *in vivo* can be limited by phosphate supply. New Phytologist 102:499–512.

Skinner, M. W., and B. M. Pavlik. 1994. California native plant society's inventory of rare and endangered plants of California, 5th Ed. California Native Plant Society, Sacramento, CA.

Skogley, E. O., S. J. Georgitis, J. E. Yang, and B. E. Schaff. 1990. The phytoavailability soil test—PST. Communications Soil Science Plant Analysis 21:1229–1243.

Skogsmyr, I., and T. Fagerstrom. 1992. The cost of antiherbivory defense: an evaluation of some ecological and physiological factors. Oikos 64:451–457.

Sláma, K., and C. M. Williams. 1965. Juvenile hormone activity for the bug *Pyrrhocoris apterus*. Proceedings of the National Academy of Sciences (USA) 54:411–414.

Slankis, V. 1973. Hormonal relationships in mycorrhizal development. Pp. 232–298 in: G. C. Marks, and T. T. Kozlowski (eds.), Ectomycorrhizae. Academic Press, New York.

Slansky, F. Jr., and P. Feeney. 1977. Stabilization of the rate of nitrogen accumulation by larvae of the cabbage butterfly on wild and cultivated food plants. Ecological Monographs 47:209–228.

Slatyer, R. O. 1977. Altitudinal variation in the photosynthetic characteristics of snow gum, *Eucalyptus pauciflora* Sieb. ex. Spreng. III. Temperature response of material grown in contrasting thermal environments. Australian Journal of Plant Physiology 4:301–312.

Slovik, S., A. Siegmund, H. W. Führer and U. Heber. 1996. Stomatal uptake of SO_2, NO_x and O_3 by spruce crowns (*Picea abies*) and canopy damage in Central Europe. New Phytologist 132:661–676.

Smart, C. J., D. F. Garvin, J. P. Prince, W. J. Lucas, and L. V. Kochian. 1996. The molecular basis of potassium nutrition in plants. Plant and Soil 187:81–89.

Smedegaard-Petersen, V., and O. Støllen 1981. Effect of energy-requiring defense reactions on yield and grain quality in a powdery mildew-resistant barley cultivar. Phytopathology 71: 396–399.

Smirnoff, N. 1993. The role of active oxygen in the response of plants to water deficit and dessication. New Phytologist 125:27–58.

Smirnoff, N., and Q. J. Cumbes. 1989. Hydroxyl radical scavenging activity of compatible solutes. Phytochemistry 28:1057–1060.

Smith, A. E., and L. D. Martin. 1994. Allelopathic characteristics of three cool-season grass species in the forage ecosystem. Agronomy Journal 86:243–246.

Smith, B. N., and C. M. Lytle. 1997. Air pollutants. Pp. 375–392 in: M. N. V. Prasad (ed.), Plant Ecophysiology. John Wiley & Sons, New York.

Smith, C. R. Jr., D. Weisleder, R. W. Miller, I. S. Palmer, and O. E. Olson. 1980. Linustatin + neolinustatin: cyanogenic glycosides of linseed meal that protect animals against selenium toxicity. Journal of Organic Chemistry 45:507–510.

Smith, C. W., J. C. McCarty, T. P. Altamarino, K. E. Lege, M. F. Schuster, J. R. Phillips, and D. J. Lopez. 1992. Condensd tannins in cotton and boleworm-budworm (Lepidoptera, Noctuidae) resistance. Journal of Economic Entomology 85:2211–2217.

Smith, D. C., and A. E. Douglas. 1987. The Biology of Symbiosis. Edward Arnold, London, UK.

Smith, F. A., S. Dickson, C. Morris, R. J. Reid, M. Tester, and S. E. Smith. 1994a. Phosphate transfer in VA mycorrhizas—special mechanisms or not. Pp. 156–161 in: F. Baluska, M. Ciamporova, O. Gasparikova, and P. W. Barlow (eds.), Structure and Function of Roots. Kluwer Academic Publishers, Dordrecht, Netherlands.

Smith, I. K., T. L. Vierheller, and C. A. Thorne. 1989. Properties and functions of glutathione reductase in plants. Physiologia Plantarum 77:449–456.

Smith, J. L., R. I. Papendick, D. F. Bezdicek, and J. M. Lynch. 1993. Soil organic matter dynamics and crop residue management. Pp. 65–94 in: F. B. Metting Jr. (ed.), Soil Microbial Ecology. Marcel Dekker, New York.

Smith, M. T., and J. Van Staden. 1995. Infochemicals: the seed-fungus root continuum. A review. Environmental and Experimental Botany 35:113–123.

Smith, S. E., and D. J. Read. 1997. Mycorrhizal Symbiosis, 2nd Ed. Academic Press, London.

Smith, S. E., B. J. St. John, F. A. Smith, and J.-L. Bromley. 1986. Effects of mycorrhizal infection on plant growth, nitrogen and phosphorus nutrition in glasshouse-grown *Allium cepa* L. New Phytologist 103:359–373.

Smith, S. E., S. Dickson, C. Morris, and F. A. Smith. 1994b. Transport of phosphate from fungus to plant in VA mycorrhizas: calculations of the area of symbiotic interface and of fluxes from two different fungi to *Allium porrum* L. New Phytologist 127:93–97.

Smith, W. H. 1990. Air Pollution and Forests: Interaction Between Air Contaminants and Forest Ecosystems. Springer-Verlag, New York.

Snellgrove, R. C., W. E. Splittstoesser, D. B. Stribley, and P. B. Tinker. 1982. The distribution of carbon and the demand of the fungal symbiont in leek plants with vesicular-arbuscular mycorrhizas. New Phytologist 92:75–87.

Sonnenwald, U., J. Lerchel, R. Zrenner, W. Frommer. 1994. Manipulation of sink-source relationships in transgenic plants. Plant, Cell and Environment 17:649–658.

Sopory, S. K., and M. Munshi. 1998. Protein kinases and phosphatases and their role in cellular signaling in plants. Critical Reviews in Plant Sciences 17:245–318.

Southon, I. W., and J. Buckingham (eds.). 1989. Dictionary of Alkaloids. Chapman & Hall, London.

Souto, X. C., L. Gonzalez, and M. J. Reigosa. 1994. Comparative analysis of allelopathic effects produced by four forestry species during decomposition process in their soils in Galicia (NW Spain). Journal of Chemical Ecology 20:3005–3015.

Spencer, K. C. 1988. Chemical mediation of coevolution in the *Passiflora-Heliconius* interaction. Pp. 167–240 in: K. C. Spencer (ed.), Chemical Mediation of Coevolution. Academic Press, San Diego, CA.

Spencer-Phillips, P. T. N., and J. L. Gay. 1981. Plasma membrane ATPase domains and transport through infected cells. New Phytologist 89:393–400.

Spiller, S. C., and N. Terry 1980. Limiting factors in photosynthesis. II. Iron stress diminishes photochemical capacity by reducing the number of photosynthetic units. Plant Physiology 65:121–125.

Stafford, H. A. 1990. Flavonoid Metabolism. CRC Press, Boca Raton, FL.

Stahl, P. D., and S. E. Williams. 1986. Oil shale process water affects activity of vesicular-arbuscular fungi and *Rhizobium* 4 years after application to soil. Soil Biology and Biochemistry 18:451–455.

Stahl, P. D., and M. Cristensen. 1991. Population variation in the mycorrhizal fungus *Glomus mosseae*: breadth of environmental tolerance. Mycological Research 95:300–307.

Stamp, N. 1996. Developing a theory of plant-insect herbivory interactions: are we there yet? Bulletin of the Ecological Society of America 77:51–61.

Stankovic, B., and E. Davies. 1998. The wound response in tomato involves rapid growth and electrical responses, systemically up-regulated transcription of proteinase inhibitor and calmodulin and down-regulated translation. Plant and Cell Physiology 39:268–274.

Stark, J. 1994. Causes of soil nutrient heterogeneity at different scales. Pp. 255–284 in: M. M. Caldwell, and R. W. Pearcy (eds.), Exploitation of Environmental Heterogeneity by Plants. Academic Press, San Diego, CA.

Stein, G. M., M. Schietzel, and A. Bussing. 1998. Mistletoe in immunology and the clinic (short review). Anticancer Research 18:3247–3249.

Steinsiek, J. W., L. R. Oliver, and F. C. Collins. 1982. Allelopathic potential of wheat (*Triticum aestivum*) straw on selected weed species. Weed Science 30:495–497.

Stenlid, G. 1976. Effects of substituents in the A-ring on the physiological activity of flavones. Phytochemistry 15:911–912.

Stephens, J. C., D. G. Lynn, V. Kamat, and J. L. Riopel. 1982. Molecular specificity of haustorial induction in *Agalinis purpurea* (L.) Raf. (Scrophulareaceae). Annals of Botany 50:1–7.

Sterling, T. M., R. L. Houtz, and A. R. Putnam. 1987. Phytotoxic exudates from velvetleaf (*Abutilon theophrasti*) glandular trichomes. American Journal of Botany 74:543–550.

Stermitz, F. R., and G. H. Harris. 1987. Transfer of pyrrolizidine and quinolizidine alkaloids to *Castilleja* (Scrophulareaceae) hemiparasites from composite and legume host plants. Journal of Chemical Ecology 13:1917–1925.

Sternberg, L. D. L., and P. K. Swart. 1987. Utilization of fresh water and ocean water by coastal plants of southern Florida. Ecology 68:1898–1905.

Stewart, G. R., and M. C. Press. 1990. The physiology and biochemistry of parasitic angiosperms. Annual Review of Plant Physiology 41:127–151.

Stewart, N. C. Jr., and E. T. Nilsen. 1995a. Phenotypic plasticity and genetic heterogeneity of *Vaccinium macrocarpon* (Ericaceae), the American cranberry. I. Reaction norms of central and marginal clones in a common garden. International Journal of Plant Sciences 156:687–697.

Stewart, N. C. Jr., and E. T. Nilsen. 1995b. Phenotypic plasticity and genetic heterogeneity of *Vaccinium macrocarpon* (Ericaceae), the American cranberry. II. Reaction norms and clonal patterns of two marginal populations. International Journal of Plant Sciences 156:698–708.

Stirmitz, F. R., and G. H. Harris. 1987. Chemistry of the Scrophulariaceae. 10. Transfer of pyrrolizidine and quinolizidine alkaloids to *Castilleja* (Scrophulariaceae) hemiparasites from composite and legume host plants. Journal of Chemical Ecology 13:1917–1925.

St. John, T. V., D. C. Coleman, and C. P. P Reid. 1983. Growth and spatial distribution of nutrient-absorbing organs: selective exploitation of soil heterogeneity. Plant and Soil 71:487–493.

Stockwell, W. R., G. Kramm, H.-E. Scheel, V. A. Mohnen, and W. Seiler. 1997. Ozone formation, destruction and exposure in Europe and the United States. Pp. 1–38 in: H. Sandermann, A. R. Wellburn, and R. L. Heath (eds.), Forest Decline and Ozone: A Comparison of Controlled Chamber and Field Experiments. Springer-Verlag, New York, USA.

Stowe, K. A., V. Sork, and A. W. Farrell. 1994. Effects of water availability on the phenotypic expression of herbivore resistance in northern red oak seedlings (*Quercus rubra* L.) Oecologia 100:309–315.

Stowe, L. G. 1979. Allelopathy and its influence on the distribution of plants in an Illinois old-field. Journal of Ecology 67:1065–1085.

Strazalka, K., and P. Ketner 1997. Carbon dioxide. Pp. 393–456 in: M. N. V. Prasad (ed.), Plant Ecophysiology. John Wiley & Sons, New York.

Stribley, D. P., P. B. Tinker, and J. H. Rayner. 1980. Relation of internal phosphorus concentration and plant weight in plants infected by vesicular-arbuscular mycorrhizas. New Phytologist 86:261–266.

Strong, D. R., J. H. Lawton, and R. Southwood. 1984. Insects on Plants. Blackwell Scientific Publications, Oxford, UK.

Sugiyama, T., C. Matsumoto, T. Akazawa, and S. Miyachi 1969. Structure and function of chloroplast proteins. VII. Ribulose-1,5-diphosphate carboxylase of *Chlorella ellipsoidea*. Archives of Biochemisty and Biophysics 129:597–602.

Surjus, A., and M. Durand. 1996. Lipid changes in soybean root membranes in response to salt treatment. Journal of Experimental Botany 47:17–23.

Swank, S. E., and W. C. Oechel. 1991. Interactions among the effects of herbivory, competition, and resource limitation on chaparral herbs. Ecology 72:104–115.

Taiz, L., and E. Zeiger. 1991. Plant Physiology. Benjamin/Cummings Publishing Company. Redwood City, CA.

Tang, C., K. Komai, and R. S. Huang. 1989. Allelopathy and the chemistry of the rhizosphere. Pp. 217–226 in: C. H. Chou and G. R. Waller (eds.), Phytochemical Ecology: Allelochemicals, Mycotoxins and Insect Pheromones and Allomones. Institute of Botany, Academia Sinica, Taipei.

Tang, C.-S., and C.-C. Young. 1982. Collection and identification of allelopathic compounds from the undisturbed root system of bigalta limpograss (*Hemarthria altissima*). Plant Physiology 69:155–160.

Tanji, K. K. (ed.). 1990. Agricultural Salinity Assesment and Management. Irrigation and Drainage Division, American Society of Civil Engineers, New York.

Tanka, M., T. Nakashima, A. Benson, A. F. Mower, and K. T. Yasunobu. 1964. The amino acid sequence of *Clostridium pasteurianum* ferredoxin. Biochemical and Biophysical Research Communications, 16:422–427.

Tanrisever, N., F. R. Fronczek, N. H. Fischer, and G. B. Williamson. 1987. Ceratiolin and other flavonoids from *Ceratiola ericoides*. Phytochemistry 26:175–179.

Tarafdar, J. C., and A. Jungk. 1987. Phosphatase activity in the rhizosphere and its relation to the depletion of soil organic phosphorus. Biological Fertility of Soils 3:199–204.

Tarayre, M., J. D. Thompson, J. Escarre, and Y. B. Linhart. 1995. Intra-specific variation in the inhibitory effects of *Thymus vulgaris* (Labiatae) monoterpenes on seed germination. Oecologia 101:110–118.

Tarradellas, J., and G. Bitton. 1997. Chemical pollutants in soils. Pp. 3–32 in: J. Tarradellas, G. Bitton, and D. Rossel (eds.), Soil Ecotoxicology, Lewis Publishers, Boca Raton, FL.

Taylor, A., J. Martin, and W. E. Seel. 1996. Physiology of the parasitic association between maize and witchweed (*Striga hermonthica*): is ABA involved? Journal of Experimental Botany 47:1057–1065.

Taylor, A., and Seel, W. E. 1998. Do *Striga hermonthica*-induced changes in soil matric potential cause the reduction in stomatal conductance and growth of infected maize plants? New Phytologist 138:67–73.

Taylor, K. 1999. Rapid climate change. American Scientist 87:320–327.

Taylor, R. J., and D. C. Shaw. 1983. Allelopathic effects of Engelmann spruce bark stilbenes and tannin-stilbene combinations on seed germination and seedling growth of selected conifers. Canadian Journal of Botany 61:279–289.

Tennakoon, K. U., and J. S. Pate. 1996a. Effects of parasitism by a mistletoe on the structure and function of branches of its host. Plant Cell and Environment 19:517–528.

Tennakoon, K. U., and J. S. Pate. 1996b. Heterotrophic gain of carbon from the hosts of the xylem-tapping root hemiparasite *Olax phyllanthi* (Olacaceae). Oecologia 105:369–376.

Terborgh, J. 1986. Community aspects of frugivory in tropical forests. Pp. 371–384 in: A. Estrada, and T. H. Flemming (eds.), Frugivores and Seed Dispersal. W. Junk, Dordrecht, Netherlands.

Termaat, A., and R. Munns. 1986. The use of concentrated macronutrient solutions to separate osmotic from NaCl-specific effects on plant growth. Australian Journal of Plant Physiology. 13:509–522.

Termaat, A., J. B. Passioura, and R. Munns. 1985. Shoot turgor does not limit shoot growth of NaCl-affected wheat and barley. Plant Physiology 77:869–872.

Terry, N., and I. R. Rao. 1991. Nutrients and photosynthesis: iron and phosphorus as case studies. Pp. 55–79 in: J. R. Porter, and D. W. Lawlor (eds.), Plant Growth: Interactions with Nutrition and Environment. Cambridge University Press, New York.

Thaler, J. S. 1999. Jasmonate-inducible plant defences cause increased parasitism of herbivores. Nature 399:686–688.

Thalouarn, P., V. Philouze, and S. Renaudin. 1988. Nitrogen metabolism key enzymes in the Scrophulariaceae holoparasite *Lathrea clandestina* L. Journal of Plant Physiology 132:63–66.

Thijs, H., J. R. Shann, and J. D. Weidenhamer. 1994. The effect of phytotoxins on competitive outcome in a model system. Ecology 75:1959–1964.

Thomas, J. C., and H. J. Bohnert. 1993. Salt stress perception and plant growth regulators in the halophyte *Mesembryanthemum crystallinum*. Plant Physiology 103:1299–1304.

Thomas, J. C., E. F. McElwain, and H. J. Bohnert. 1992. Convergent induction of osmotic-stress responses: ABA and cytokinin and the effects of NaCl. Plant Physiology 100:416–423.

Thomas, J. C., A. M. Smigocki, and H. J. Bohnert. 1995. Light-induced expression of *Ipt* from *Agrobacterium tumefaciens* results in cytokinin accumulation and osmotic stress symptoms in transgenic tobacco. Plant Molecular Biology 27:225–235.

Thompson, B. D., D. T. Clarkson, and P. Brain. 1990. Kinetics of P uptake by the germ-tubes of the vesicular-arbuscular mycorrhizal fungus *Gigaspora margarita* New Phytologist 116: 647–653.

Thompson, I. D., and A. U. Mallik. 1989. Moose browsing and allelopathic effects of *Kalmia angustifolia* on balsam fir regeneration in central Newfoundland. Canadian Journal of Forest Research 19:524–526.

Thompson, J. N. 1988. Variation in preference and specificity in monophagous and oligophagous swallowtail butterflies. Evolution 42:118–128.

Thomson, W. W. 1973. The structure and function of salt glands. Pp. 118–146 in: A. Poljakoff-Mayber and J. Gale (eds.), Plants in Saline Environments (Ecological Studies, Vol. 15). Springer-Verlag, New York.

Tilman, D. 1982. Resource Competition and Community Structure. Princeton University Press, Princeton, NJ.

Tilman, D. 1988. Plant Strategies and the Dynamics and Structure of Plant Communities. Princeton University Press, Princeton, N.J.

Ting, I. P., A. Patel, S. Kaur, J. Hann, and L. Walling. 1996. Ontogenetic development of crassulacean acid metabolism as modified by water stress in *Peperomia*. Pp. 135–158 in: K. Winter and J. A. C. Smith (eds.), Crassulacean Acid Metabolism: Biochemistry, Ecophysiology, and Evolution. Springer-Verlag, New York.

Tingey, D. T., D. M. Olszyk, A. A. Herstrom, and E. H. Lee. 1994. Effects of ozone on crops. Pp. 175–206 in: D. J. McKee (ed.), Tropospheric Ozone. Lewis Publishers, Boca Raton, FL.

Tinnin, R. O. 1984. The effect of dwarf mistletoe on forest community ecology. U.S. Forest Service General Technical Report RM-111.

Titarenko, E., E. Rojo, J. Leon, and J. J. Sanchez-Serrano. 1997. Jasmonic acid-dependent and -independent signaling pathways control wound-induced gene activation in *Arabidopsis thaliana*. Plant Physiology 115:817–826.

Tomlinson, P. B. 1986. The Botany of Mangroves. Cambridge University Press, Cambridge, UK.

Tomsett, A. B., and D. A. Thurman 1988. Molecular biology of metal tolerances of plants. Plant, Cell and Environment. 11:383–394.

Toong, Y. C., D. A. Schooley, and F. C. Baker. 1988. Isolation of insect juvenile hormone-III from a plant. Nature 333:170-171.

Toth, R., D. D. Toth, D. Starke, and D. R. Smith. 1990. Vesicular-arbuscular colonization in *Zea mays* affected by breeding for resistance to fungal pathogens. Canadian Journal of Botany 68:1039–1044.

Trappe, J. M. 1977. Selection of fungi for mycorrhizal inoculation in nurseries. Annual Review of Phytopathology 15:203.

Trattinick, 1928. Nachtrag zu der *Ichtysoma wehdemanni*. Linnaea 3:194–198.

Tremmel, D. C., and F. A. Bazzaz. 1995. Plant architecture and allocation in different neighborhoods: implications for competitive success. Ecology 76:262–271.

Treshow, M. 1984. Introduction. Pp. 1–6 in: M. Treshow (ed.), Air Pollution and Plant Life. John Wiley & Sons, New York.

Treshow, M., and F. K. Anderson. 1989. Plant Stress from Air Pollution. John Wiley & Sons, New York.

Tugwell, S., and G. M. Branch. 1992. Effects of gut surfactants on kelp polyphenol defenses. Ecology 73:205–215.

Turner, A. P. 1994. The responses of plants to heavy metals. Pp. 153–187 in: S. M. Ross (ed.), Toxic Metals in Soil-Plant Systems. John Wiley & Sons, New York.

Turner, R. E. 1976. Geographic variation in salt marsh macrophyte production: a review. Contributions in Marine Sciences 20:47–68.

Tyerman, S. D., M. Skerrett, A. Garrill, G. P. Findlay, and R. A. Leigh. 1997. Pathways for the permeation of Na^+ and Cl^- into protoplasts derived from the cortex of wheat roots. Journal of Experimental Botany 48:459–480.

Ucko, D. A. 1986. Living Chemistry: An Introduction to General, Organic, and Biological Chemistry. Harcourt Brace and Jovanovich, New York.

Ullmann, I., O. L. Lange, H. Zeigler, J. Ehleringer, E.-D. Schulze, and I. R. Cowan. 1985. Diurnal courses of leaf conductance and transpiration of mistletoes and their hosts in central Australia. Oecologia 67:577–587.

Ullrich, W. R. 1983. Uptake and reduction of nitrate: Algae and fungi. Pp. 376–397 in: A. Läuchli and R. L. Bieleski (eds.), Inorganic Plant Nutrition (A. Pirson and M. H. Zimmerman [series eds.], Encyclopedia of Plant Physiology, Vol. 15A). Springer-Verlag, New York.

UNEP/GEMS (UN Environment Programme/Global Environmental Monitoring System). 1991. Urban Air Pollution. UNEP, Oxford, UK.

Ungar, I. A. 1991. Ecophysiology of Vascular Halophytes, CRC Press, Boca Raton, FL.

Uno, Y., T. Urao, K. Yamaguchi-Shinozaki, M. Kanechi, N. Inagaki, S. Maekawa, and K. Shinozaki. 1998. Early salt effects on expression of genes for aquaporin homologues in the halophyte sea aster (*Aster trifolum* L.). Journal of Plant Research 111:411–419.

Urwin, P. E., M. J. McPhearson, and H. J. Atkinson. 1998. Enhanced transgenic plant resistance to nematodes by dual proteinase inhibitor constructs. Planta 204:472–479.

Ustin, S. L., R. W. Pearcy, and D. E. Bayer. 1982. Plant water relations in a San Francisco Bay salt marsh. Botanical Gazette 143:368–373.

Valladares, G., and L. H. Lawton. 1991. Host plant selection in the holly leaf-miner: Does mother know best? Journal of Animal Ecology 60:227–240.

Valle, B. L., and K. H. Falchuk. 1993. The biochemical basis of zinc physiology. Physiological Reviews 73:79–118.

Van Assche, F., and H. Clijsters 1990. Effects of metals on enzyme activity in plants. Plant Cell Environment 13:195–206.

Van Hulst, R., B. Shipley, and A. Theriault. 1987. Why is *Rhinanthus minor* such a good invader? Canadian Journal of Botany 65:2373–2379.

Van de Koppel, J., J. Huisman, R. van der Wal, and H. Olff. 1996. Patterns of herbivory along a productivity gradient: an empirical and theoretical investigation. Ecology 77:736–745.

van Loon, J. J. A. 1996. Chemosensory basis of feeding and oviposition behavior in herbivorous insects: a glance at the periphery. Entomologia Experimental et Applicata 47:3–14.

Varma , A. K., and P. Bonfante-Fasolo. 1994. Utilisation of cell wall related carbohydrates by ericoid mycorrhizal endophytes. Symbiosis 16:301–313.

Verkleij, J. A. C. 1993. The effects of heavy metal stress on higher plants and their use as biomonitors. Pp. 415–424 in: B. Markert (ed.), Plants as Biomonitors: Indicators for Heavy Metals in the Terrestrial Environment, VCH, New York.

Verkleij, J. A. C., and H. Schat. 1990. Mechanisms of metal tolerance in higher plants.

Vernon, D. M., J. A. Ostrem, and H. J. Bohnert. 1993. Stress perception and response in a facultative halophyte; the regulation of salinity induced genes in *Mesembryanthemum crystallinum.* Plant Cell and Environment 16:437–444.

Virupaksha, T. K., and A. Shrift. 1965. Biochemical differences between selenium accumulator and non-accumulator *Astragalus* species. Biochemica Biophysica Acta 107:69–80.

Visser, J., and I. Dörr. 1987. The haustorium. Pp. 91–106 in: L. J. Musselman (ed.), *Striga* (Parasitic Weeds in Agriculture, Vol. I). CRC Press, FLA.

Vitousek, P. M., C. M. D'Antonio, L. L. Loope, M. Rejmanek, and R. Westbrooks. 1997. Introduced species: a significant component of human-caused global change. New Zealand Journal of Ecology 21:1–16.

Vogt, W., J. Sauerborn, and M. Honisch. 1991. *Striga hermonthica* distribution and infestation in Ghana and Togo on grain crops. Pp. 372–377 in: J. K. Ransom, L. J. Musselman, A. D. Worsham, and C. Parker (eds.), Proceedings of the 5th International Symposium on Parasitic Weeds, Nairobi, India, CIMMYT, Nairobi.

Voltura, M. B., and B. A. Wunder. 1994. Physiological responses of the Mexican woodrat (*Neotoma mexicana*) to condensed tannins. American Midland Naturalist 132:405–409.

Waisel, Y. 1972. Biology of Halophytes. Academic Press, New York.

Walker, D. A. 1980. Regulation of starch synthesis in leaves—the role of orthophosphate. Pp. 195–207 in: Proceedings of the 15th Colloquium of the International Potash Institute, Bern, Switzerland.

Wall, D. H. 1999. Biodiversity and ecosystem functioning. BioScience 49:107–108.

Wall, D. H., and J. C. Moore. 1999. Interactions underground. BioScience 49:109–117.

Wallace, A., R. T. Mueller, and G. V. Alexander. 1978. Influence of phosphorus on zinc, iron, manganese and copper uptake in plants. Soil Science 126:336–341.

Wallander, H., J.-E. Nylund, and B. Sundberg. 1994. The influence of IAA, carbohydrate, and mineral concentration in host tissue on ectomycorrhizal development on *Pinus sylvestris* L. in relation to nutrient supply. New Phytologist 127:521–528.

Waller, G. R. (ed.). 1987. Allelochemicals: Role in Agriculture and Forestry (ACS Symposium Series, No. 330). American Chemical Society, Washington, DC.

Wallwork, J. A. 1970. Ecology of Soil Animals, McGraw Hill, New York,USA.

Walter, M. H., J. Grima-Pettenati, C. Grand, A. M. Boudet, and C. J. Lamb. 1988. Cinnamyl-alcohol dehydrogenase, a molecular marker specific for lignin synthesis: cDNA cloning and mRNA induction by fungal elicitor. Proceedings of the National Academy of Sciences (USA) 85:5546–5550.

Wanta, R. C., W. B. Moreland, and H. E. Heggestad. 1961. Tropospheric ozone: an air pollution problem arising in the Washington D.C. metropolitan area. Monthly Weather Review 89: 289–296.

Warcup, J. H. 1985. *Rhizanthella gardneri* (Orchidaceae)—Its *Rhizoctonia* endophyte and close association with *Melaluca uncinata* (Myrtaceae) in Western Australia. New Phytologist 99: 273–280.

Warcup, J. H. 1991. The *Rhizoctonia* endophytes of *Rhizanthella* (Orchidaceae) Mycological Research 95:656–659.

Ward, H. A., and L. H. McCormick. 1982. Eastern hemlock allelopathy. Forest Science 28: 681–686.

Wardle, D. A., K. S. Nicholson, and M. Ahmed. 1992. Comparison of osmotic and allelopathic effects of grass leaf extracts on grass seed germination and radicle elongation. Plant and Soil 140:315–319.

Wardle, D. A., and M. C. Nilsson. 1997. Microbe-plant competition, allelopathy and arctic plants. Oecologia 109:291–293.

Wardle, D. A., G. M. Barker, K. I. Bonner, and K. S. Nicholson. 1998a. Can comparative approaches based on plant ecophysiological traits predict the nature of biotic interactions and individual plant species effects in ecosystems? Journal of Ecology 86:405–420.

Wardle, D. A., M. Nilsson, C. Gallet, and O. Zackrisson. 1998b. An ecosystem-level perspective of allelopathy. Biological Reviews 73:305–319.

Waring, R. H. 1987. Nitrate pollution: a particular danger to boreal and subalpine coniferous forests. Pp. 93–105 in: T. Fujimori and M. Kimura (eds.), Human Impacts and Management of Mountain Forests. Forestry and Forest Products Research Institute, Ibaraki, Japan.

Watanabe, Y., A. Miwa, and Y. Tamai. 1995. Characterisation of a Na^+ / H^+ antiporter gene closely-related to the salt tolerance of yeast *Zygosaccharomyces rouxii*. Yeast 11:829–838.

Watkinson, A. R. 1980. Density-dependence in single-species populations of plants. Journal of Theoretical Biology 83:345–357.

Watkinson, A. R., and C. C. Gibson. 1988. Plant parasitism: the population dynamics of parasitic plants and their affect upon plant community structure. Pp. 393–343 in: A. J. Davy, M. J. Hutchings, and A. R. Watkinson (eds.), Plant Population Ecology. Blackwell, Oxford, UK.

Watkinson, A. R., and C. C. Gibson. 1989. The host range and selectivity of a parasitic plant: *Rhinanthus minor* L. Oecologia 78:401–406.

Watkinson, A. R., and R. P. Freckleton. 1997. Quantifying the impact of arbuscular mycorrhiza on plant competition. Journal of Ecology 85:541–545.

Watling, J. R., and M. C. Press. 1997. How is the relationship between the C-4 cereal *Sorghum bicolor* and the C-3 root hemiparasites *Striga hermonthica* and *Striga asiatica* affected by elevated CO_2? Plant Cell and Environment 20:1292–1300.

Watling, J. R., and M. C. Press. 1998. How does the C-4 grass *Eragrostis pilosa* respond to elevated carbon dioxide and infection with the parasitic angiosperm *Striga hermonthica*? New Phytologist 140:667–675.

Wearing, C. H. 1972. Responses of *Myzus persicae* and *Brevicoryne brassicae* to leaf age and water stress in brussel sprouts grown in pots. Entomologia Expimentalis Applicata 15:61–80.

Weber, H. C. 1987. Evolution of the secondary haustoria to a primary haustorium in the parasitic Scrophulariaceae/Orobanchaceae. Plant Systematics and Evolution 156:127–131.

Weddell, H. A. 1858–1861. Mémoire sur le *Cynomorium coccineum*, parasite de l'ordre des Balanophorées. Archives Museum d'Histoire Naturelle 10:269–308.

Wedin, D., and D. Tilman. 1993. Competition among grasses along a nitrogen gradient: initial conditions and mechanisms of competition. Ecological Monographs 63:199–229.

Weidenhamer, J. D. 1996. Distinguishing resource competition and chemical interference: overcoming the methodological impasse. Agronomy Journal 88:866–875.

Weidenhamer, J. D., D. C. Hartnett, and J. T. Romeo. 1989. Density-dependent phytotoxicity: distinguishing resource competition and allelopathic interference in plants. Journal of Applied Ecology 26:613–624.

Weidenhamer, J. D., F. A. Macias, N. H. Fisher, and G. B. Williamson. 1993. Just how insoluble are monoterpenes. Journal of Chemical Ecology. 19:1799–1807.

Weiler, E. W. 1997. Octadecanoid-mediated signal transduction in higher plants. Naturwissenschaften 84:340–349.

Weimberg, R., H. R. Lerner, and A. Poljakoff-Mayber. 1982. A relationship between potassium and proline accumulation in salt-stressed *Sorghum bicolor*. Physiologia Plantarum 55:5–10.

Weiner, J. 1982. A neighborhood model of annual-plant interference. Ecology 63:1237–1241.

Weiner, J. 1985. Size hierarchies in experimental populations of annual plants. Ecology 66: 743–752.

Weiner, J. 1986. How competition for light and nutrients affects size variability in *Ipomoea tricolor* populations. Ecology 76:1425–1427.

Weiner, J. 1988. Variation in the performance of individuals in plant populations. Pp. 59–82 in: A. J. Davy, M. J. Hutchings, A. R. Watinson (eds.), Plant Population Ecology. Blackwell Scientific, London, UK.

Weiner, J., and S. C. Thomas. 1986. Size variability and competition in plant monocultures. Oikos 47:211–222.

Weissenhom, I., A. Glashoff, C. Leyval, and J. Berthelia. 1994. Differential tolerance to Cd and Zn of arbuscular mycorrhizal (AM) fungal spores isolated from heavy metal polluted and unpolluted soils. Plant and Soil 167:189–196.

Wellburn, A. R. 1988. Air Pollution and Acid Rain: The Biological Impact. Longman Scientific and Technical, London, UK.

Wellburn, A. R. 1990. Why are atmospheric oxides of nitrogen usually phytotoxic and not alternative fertilizers. New Phystologist 115:395–429.

Weltzin, J. F., and G. R. McPherson. 1997. Spatial and temporal soil moisture resource partitioning by trees and grasses in a temperate savanna, Arizona, USA. Oecologia 112:156–164.

Wender, S. H. 1970. Effects of some environmental stress factors on certain phenolic compounds in tobacco. Recent Advances in Phytochemistry 3:1–29.

Werner, A., and R. Stelzer. 1990. Physiological response of the mangrove *Rhizophora mangle* grown in the absence and presence of NaCl. Plant, Cell & Environment 13:243–255.

Werner, C., and P. Maitile. 1985. Accumulation of coumaryl glycosides in vacuoles of barley mesophyll protoplasts. Journal of Plant Physiology 118:237–249.

West, H. M., A. H. Fitter, and A. R. Watkinson. 1993. Response of *Vulpia ciliata* spp. to removal of mycorrhizal infection and to phosphate application under natural conditions. Journal of Ecology 81:351–358.

West, H. M. 1996. Influence of arbuscular mycorrhizal infection on competition between *Holcus lanatus* and *Dactylis glomerata*. Journal of Ecology 84:429–438.

Westoby, M. 1984. The self-thinning rule. Advances in ecological research 14:167–225.

Weston, L. A., and A. R. Putnam. 1985. Inhibition of growth, nodulation, and nitrogen fixation of legumes by quackgrass. Crop Science 25:561–565.

Weston, L. A. 1996. Utilization of allelopathy for weed management in agroecosystems. Agronomy Journal 88:860–866.

Westwood, J. H., X. S. Yu, and C. L. Foy. 1998. Expression of a defense-related 3-hydroxy-3-methylglutaryl CoA reductase gene in response to parasitism by *Orobanche* spp. Molecular Plant Microbe 11:530–536.

Westwood, J. H., and C. L. Foy. 1999. Influence of nitrogen on germination and early development of broomrape (*Orobanche* spp.). Weed Science 47:2–7.

Whitaker, R. H. 1975. Communities and Ecosystems. Macmillan Publishing Company, New York.

White, J. 1981. The allometric interpretation of the self-thinning rule. Journal of Theoretical Biology 89:475–500.

White, J., and J. L. Harper. 1970. Correlated changes in plant size and number in plant populations. Journal of Ecology 58:467–485.

Whitham, T. G., and S. Mopper. 1985. Chronic herbivory: impacts on architecture and sex expression of pinyon pine. Science 228:1089–1091.

Whitham, T. G., J. Maschinski, K. C. Larson, and K. N. Paige. 1991. Plant response to herbivory: the continuum from negative to positive and underlying physiological mechanisms. Pp. 227–256 in: P. W. Price, T. M. Lewinshon, G. W. Fernandes, and W. W. Benson (eds.), Plant-Animal Interactions: Evolutionary Ecology in Tropical and Temperate Regions. John Wiley & Sons, New York.

Whitney, P. J. 1972. The carbohydrate and water balance of beans (*Vicia faba*) attacked by broomrape (*Orobanche crenata*). Annals of Applied Biology 70:59–66.

Whittaker, R. H. 1970. The biochemical ecology of higher plants. Pp. 43–70 in: E. Sonheimer and J. B. Simeone (eds.), Chemical Ecology. Academic Press, New York.

Whittaker, R. H., and P. P. Feeney. 1971. Allelochemicals: chemical interactons between plant species. Science 171:757.

Wibawa, A., J. B. Baon, and Nurkholis. 1995. Growth of shade trees for coffee and cacao as affected by mycorrhizal and rhizobial inoculation. Pp. 209–214 in: Supriyanto and J. T. Kartana (eds.), Proceedings of the 2nd Symposium on Biology and Biotechnology of Mycorrhizae and 3rd Asian Conference on Mycorrhizae (Biotrop Special Publication, No. 56). Seameo Biotrop, Bogor, Indonesia.

Williams, C. M. 1956. The juvenile hormones of insects. Nature 178:212–213.

Williams, W. P. 1988. Lipid topology and the role of lipids in plant membrane structure. Pp. 47–64 in: J. L. Harwood and T. J. Walton (eds.), The Biochemical Society of London, London.

Williamson, G. B., and D. Richardson. 1988. Bioassays for allelopathy: measuring treatment responses with independent controls. Journal of Chemical Ecology 14:181–187.

Williamson, G. B., Obee, E. M., and Weidenhamer, J. D. 1992. Inhibition of *Schizachyrium scoparium* (Poaceae) by the allelochemical hydrocinnamic acid. Journal of Chemical Ecology 18:2095–2105.

Willis, R. J. 1985. The historical bases of the concept of allelopathy. Journal of the History of Biology 18:71–102.

Wilson, J. B. 1988. Shoot competition and root competition. Journal of Applied Ecology 25: 279–296.

Wilson, M. F., and A. E. Bell. 1978. The determination of the changes in the free amino acid content of the elute from germinating seeds of *Glycine wightii* (L.) and its effects on the growth of lettuce fruits. Journal of Experimental Botany 29:1243–1247.

Wilson, S. D., and D. Tilman. 1993. Plant competition and resource availability in response to disturbance and fertilization. Ecology 74:599–611.

Wilson, S. D., and D. Tilman. 1995. Competitive responses of eight old-field plant species in four environments. Ecology 76:1169–1180.

Wink, M., and L. Witte. 1984. Turnover and transport of quinolizidine alkaloids. Diurnal fluctuations of lupinine in the phloem sap, leaves and fruits of *Lupinus albus* L. Planta 161: 519–524.

Wink, M. 1987. Quinolizidine alkaloids—biochemistry, metabolism, and function in plants and cell-suspension cultures. Plant Medica 53:509–514.

Wink, M. 1988. Plant-breeding—importance of plant secondary metabolites for protection against pathogens and herbivores. Theoretical Applied Genetics. 75:225–233.

Winter, A. G. 1961. New physiological and biological aspects in the interrelationships between higher plants. Symposium for the Society of Experimental Biology 15:229–244 .

Winter, K., and R. Gadermann. 1991. Daily changes in carbon dioxide and water vapour exchange, chlorophyll fluorescence, and leaf water relations in the halophyte *Mesembreanthemum crystallinum* during the induction of crassulacean acid metabolism in response to high sodium chloride salinity. Plant Physiology 95:768–776.

Winter, K., and D. J. von Willert. 1972. NaCl-induzierter Crassulaceensaurestoffwechsel bei *Mesembreanthemum crystallinum.* Zeitschrift für Pflanzenphysiologie and Bodenernährung 67:166–170.

Winter, K., and H. Ziegler. 1992. Indiction of crassulacean acid metabolism in *Mesembreanthemum crystallinum* increases reproductive success under conditions of drought and salinity stress. Oecologia 92:474–479.

Wolfe, A. D., and C. W. dePamphilis. 1998. The effect of relaxed functional constraints on the photosynthetic gene rbcL in photosynthetic and nonphotosynthetic parasitic plants. Molecular Biology and Evolution 15:1243–1258.

Wolfe, K. H., C. W. Morden, and J. D. Palmer. 1992. Function and evolution of a minimal plastid genome from a nonphotosynthetic parasitic plant. Proceedings of the National Academy of Science (USA) 89:10648–10652.

Wolfson, J. L. 1980. Oviposition response of *Pieris rapae* to environmentally induced variation in *Brassica nigra.* Entomological Experimental Applications 27:223–232.

Wolswinkel, P. 1974. Complete inhibition of setting and growth of fruits of *Vicia faba* L. resulting from the drainage of the phloem system by *Cuscuta* species. Acta Botanica Nederlandica 23:177–188.

Woodhead, S. 1981. Environmental and biotic factors affecting the phenolic content of different cultivars of *Sorghum bicolor.* Journal of Chemical Ecology 6:1035–1047.

Woodward, F. I. 1992. Predicting plant responses to global environmental change. New Phytologist 122:239–251.

Woolhouse, H. W. 1983. Toxicity and tolerance in the responses of plants to metals. Pp. 245–300 in: O. L. Lang, P. S. Nobel, C. B. Osmondt, and H. Ziegler (eds.), Physiological Plant

Ecology III. Responses to the Chemical and Biological Environment, Springer-Verlag, New York.

Worsham, A. D. 1987. Germination of Witchweed seeds. Pp. 45–61 in: L. J. Musselman (ed.), *Striga* (Parasitic Weeds in Agriculture, Vol. 1). CRC Press, FLA.

Worsham, A. D. 1989. Current and potential techniques using allelopathy as an aid in weed management. Pp. 275–289 in: C. H. Chou, and G. R. Waller (eds.), Phytochemical Ecology: Allelochemicals, Mycotoxins and Insect Pheromones and Allomones. Institute of Botany, Academia Sinica, Taipei, Taiwan.

Wright, D. P., J. D. Scholes, and D. J. Read. 1997. Effects of VA mycorrhizal colonisation on photosynthesis and biomass production of *Trifolium repens* L. Plant Cell Environment 21: 209–216.

Wu, C., and J. R. Whittaker. 1991. Homology among trypsin/chymotrypsin inhibitors from red kidney bean, brazilian pink bean, lima bean, and soybean. Journal of Agricultural and Food Chemistry 39:1583–1589.

Wu, J., D. M. Seliskar, and J. L. Gallagher. 1998. Stress tolerance in the marsh plant *Spartina patens*: impact of NaCl on growth and root plasma membrane lipid composition. Physiologia Plantarum 102:307–317.

Wu, S., L. Ding, and J. Zhu. 1996. SOSI, a genetic locus essential for salt tolerance and potassium acquisition. Plant Cell 8:617–627.

Wu, Y., J. Kuzma, E . Marechal, R. Graeff, H. C. Lee, R. Foster, and N. H. Chua. 1997. Abscisic acid signaling through cyclic ADP-ribose in plants. Science 278:2126–2130.

Wylde, A. 1991. Quantitative and qualitative studies on the effect of *Striga hermonthica* upon the quality and quantity of grain yield from millet (*Pennisetum typhoides*). Pp. 322–328 in: K. Wegmann and L. J. Musselman (eds.), Progress in Orobanche research, Proceedings of the International Workshop on *Orobanche* Research, Wageningen, Germany.

Wyn-Jones, R. G., J. Gorham, and E. McDonnell. 1984. Organic and inorganic solute content as selective criteria for salt tolerance in the Triticeae. Pp. 1890–212 in: R. C. Staples and G. H. Toenniessen (eds.), Salinity Tolerance in Plants: Strategies for Crop Improvement. John Wiley & Sons, New York.

Yamada, S., M. Katsuhara, W. Kelley, C. Michalowski, and H. Bohnert. 1995. A family of transcripts encoding the water channel proteins: tissue specific expression in the common ice plant. Plant Cell 7:1129–1142.

Yamaguchi-Shinozaki, K., M. Koizumi, S. Urao, and K. Shinozaki. 1992. Molecular cloning and characterization of 9 cDNAs for genes that are responsive to dessication in *Arabidopsis thaliana*: sequence analysis of one cDNA clone that codes for a putative transmembrane channel protein. Plant Cell Physiology 33:217–224.

Yamasaki, S., J. Fyles, K. Egger, and B. Titus. 1998. The effect of *Kalmia angustifolia* on the growth, nutrition, and ectomycorrhizal symbiont community of black spruce. Forest Ecology and Management 105:197–207.

Yang, C. S., and H. E. Wilcox. 1984. An E-strain ectomycorrhiza formed by a new species of *Tricharina mikolae*. Mycologia 76:675–684.

Yeo, A. 1998. Molecular biology of salt tolerance in the context of whole-plant physiology. Journal of Experimental Botany 49:915–929.

Yeo, A., and T. J. Flowers. 1986. Ion transport in *Sueda maritima*: its relation to growth and implications for the pathway of radial transport of ions across the root. Journal of Experimental Botany 37:143–159.

Yoda, K., T. Kira, H. Ogawa, and H. Hozumi. 1963. Self-thinning in overcrowded pure stands under cultivated and natural conditions. Journal of Biology 14:107–129.

Yoder, J. L. 1999. Parasitic plant responses to host plant signals: a model for subterranean plant-plant interactions. Current Opinion in Plant Biology 2:65–70.

Yoshiba, Y., T. Kiyosue, T. Katagiri, H. Ueda, K. Mizoguchi, K. Yamaguchi-Shinozaki, K. Wada, Y. Harada, and K. Shinozaki. 1995. Correlation between the induction of a gene for delta(1)-pyrroline-5-carboxylate synthetase and the accumulation of proline in *Arabidopsis thaliana* under osmotic-stress. Plant Journal 7:751–760.

Yoshiba, Y., T. Kiyosue, K. Nakashima, K. Yamaguchi-Shinozaki, and K. Shinozaki. 1997. Regulation of levels of proline as an osmolyte in plants under water stress. Plant Cell Physiology 38:1095–1102.

Young, D., D. Erickson, and S. Semones. 1994. Salinity, water relations and zonation of three common barrier island shrubs. Canadian Journal of Botany 72:1365–1372.

Younger, P. D., and L. A. Kapustka. 1983. $N_2(C_2H_2)$ase activitiy by *Alnus incana* spp. on *Rugosa* (Betulaceae) in the northern hardwood forest. American Journal of Botany 70:30–39.

Yun, K. W., and B.-S. Kil. 1992. Assessment of allelopathic potential in *Artemisia princeps* var. *orientalis* residues. Journal of Chemical Ecology 18:1933–1940.

Yunus, M., N. Singh, and M. Iqbal. 1996. Global status of air pollution: an overview. Pp. 1–34 in: M. Yunus and M. Iqbal (eds.), Plant Response to Air Pollution. John Wiley & Sons, New York.

Zackrisson, O., and M.-C. Nilsson. 1992. Allelopathic effects by *Empetrum hermaphroditum* on seed germination of two boreal tree species. Canadian Journal of Forest Research 22: 1310–1319.

Zangerl, A. R., and F. A. Bazzaz. 1983. Responses of an early and a late successional species of *Polygonum* to variations in resource availability. Oecologia 56:397–404.

Zangerl, A. R., and F. A. Bazzaz. 1992. Theory and pattern of plant defense allocation. Pp. 363–391 in: R. S. Fritz and E. L. Simms (eds.), Plant Resistance to Herbivores and Pathogens. University of Chicago Press, Chicago.

Zangerl, A. R., and C. E. Rutledge. 1996. The probability of attack and patterns of constitutive and induced defense: a test of optimal defense theory. American Naturalist 147:599–608.

Zenk, M. H. 1996. Heavy metal detoxification in higher plants—a review. Gene 179:21–30.

Zhang, Q. 1993. Potential role of allelopathy in the soil and the decomposing root of Chinese-fir replant woodland. Plant and Soil 151:205–210.

Zhang, W. H., F. Q. Diao, B. J. Yu, and Y. L. Lin. 1998. H^+-ATPase and H^+-transport on tonoplast vacuoles from barley roots under salt stress and the influence of calcium and abscisic acid. Journal of Plant Nutrition 21:447–458.

Zhu, H., and A. U. Mallik. 1994. Interactions between *Kalmia* and black spruce: isolation and identification of allelopathic compounds. Journal of Chemical Ecology 20:407–421.

Zhu, J. K., J. P. Liu, and L. M. Xiong. 1998. Genetic analysis of salt tolerance in *Arabidopsis*: evidence for a critical role of potassium nutrition. Plant Cell 10:1181–1191.

Ziska, L. H., J. R. Seeman, and T. M. DeJong. 1990. Salinity induced limitations on photosynthesis in *Prunus salicina*, a deciduous tree species. Plant Physiology 93:864–870.

Zobel, A. M., and S. A. Brown. 1991. Psoralens on the surface of seeds of Rutaceae and fruits of Umbelliferae and Leguminosae. Canadian Journal of Botany 69:485–488.

Zucker, W. V. 1983. Tannins: does structure determine function? An ecological perspective. American Naturalist 121:335–365.

APPENDIX C
Glossary

Abaxial. Referring to the upper surface of an organ or organism (often used in relation to leaves).

Abiotic. Not involving or produced by organisms. Used loosely for factors of the environment and their effects on plants producing stress and stress symptoms; physiogenic.

Abscisic acid. A phytohormone that generally acts to inhibit metabolic function and promote dormancy.

Absorbance. The fraction of the radiation that impinges on a leaf and is absorbed.

Acclimation. Nonheritable modification caused by exposure to new environmental conditions; organisms undergoing changes in physiological processes, rendering them tolerant or resistant to stress; nonheritable alterations of phenotype.

Accumulators. Plants that have the ability to accumulate ion concentrations severalfold above the levels observed in soils.

Acidifuge. The avoidance of acidic soils.

Acropetal. Describes movement or translocation of solutes and hormones from the base to the tip of the plant.

Actinomycetes. Colony-forming, branch-chained bacteria resembling fungi. Most are free-living soil forms important in mineralization, with some species important in the production of antibiotics.

Activation energy. The initial investment of energy in a reaction that is required before the reaction will occur on a thermodynamic basis. Only those molecules with energy above the activation energy can react at a given time.

Active site. The specific aspect of an enzyme that is involved in the chemical reaction that the enzyme facilitates.

Acute. Referring to an injurious event that usually involves necrosis, such as fleck, scorch, or bifacial necrosis; injury caused by an air pollutant over a short time period; the opposite of chronic.

Adaptation. Processes of evolution causing a plant to be fit or adapted for the environment in which it thrives; if not successful, the species perishes.

Adaxial. Referring to the lower side of an organism or organ (often used in discussion of leaf morphology and anatomy).

Adenolates. High-energy-containing compounds in cells that serve to promote energy-requiring reactions. Included in this group are ATP, GTP, and NADH.

Aerenchyma. Large cells that die upon maturity and fill with gas. These cells serve as a conduit for gas diffusion, particularly oxygen, to the roots.

Age structure. The proportional distribution of individuals in a population among age classes. An age structure may refer to a population of individuals, or to a population of organs such as leaves on a plant.

Agroecosystems. Ecosystems dominated by agricultural enterprises involved in the production of food and fiber.

Air pollution. Gaseous or particulate substances released into the atmosphere in sufficient quantities or concentrations to cause injury to plant life.

Aldehyde. An organic molecule with a carbonyl group located at the end of the molecule.

Alkaloid. Referring to a group of nitrogen-containing organic compounds based on a phenol structure.

Alkanes. Straight-chained saturated hydrocarbons consisting of carbon and hydrogen.

Alleles. Alternative forms of genes. There are normally two alleles for each gene in each individual.

Allelochemicals. Chemical compounds released from a living organism that influence the structure or function of another organism.

Allelopathy. Biochemical interactions between living organisms; any direct or indirect effect by one plant on another through production of chemical compounds that escape into the environment.

Allocation. The proportional distribution of a particular chemical or resource among the various different functional organs in plants.

Amino acids. Organic molecules possessing both amino and carboxylic acid groups. Between these two groups is a small variable organic constituent.

Amino group. A combination of atoms that contains a nitrogen atom bonded to two hydrogen atoms. This functional group can act as a base in a solution by accepting a hydrogen to form a positively charged moiety.

Ammonification. The process by which proteins are degraded to ammonia by bacteria.

Amphipathic. Referring to phospholipids exhibiting polar charges on one end and nonpolar characteristics on the other end.

Anatomy. The internal structure of organs and organisms.

Anion. A negatively charged ion.

Anoxia. An environmental condition in which oxygen concentration is zero.

Anthropogenic. Refers to a process that is ultimately caused by human population or culture as in sulfur dioxide in the atmosphere.

Antiport. A permease that transports two different substances in opposite directions simultaneously across membranes and is often associated with H^+ transport.

Apical meristem. An embryonic region at the apex of plant shoots in which cells are undergoing rapid cell division. This is the region that regulates shoot growth in length.

Apoplast. The portion of space between cells, and in cell walls, all space outside of the cytoplasm of cells.

Appressorium A flattened, hyphal, pressing organ from which a minute infection peg usually grows and enters the host.

Aquaporin. A membrane-bound protein that forms into complexes that constitute channels for the movement of water across membranes.

ATPase. An enzyme complex that catalyzes the hydrolysis of ATP to ADP and P_i; the resulting energy is used to transport protons across a membrane forming an electrochemical gradient.

ATPase/ATPsynthase. A complex of proteins in the mitochondrial and chloroplast membranes that serve to synthesize ATP from ADP and P_i by chemiosmotic or photophosphorylation.

ATPases. Enzymes involved in the production or degradation of adenosine triphosphate.

Autotoxicity. Secondary metabolites released by a plant producing toxicity to other individuals of the same species.

Auxins. A general category of plant growth regulators that have similar effects on plants as the phytohormone indoleacetic acid.

Avirulent. Not pathogenic.

Avoidance. Stress avoidance caused by the plant not coming to thermodynamic equilibrium with the stress; prevention or decrease in the penetration of a stress into tissues.

Bacillariophyceae. The algal classes to which diatoms belong.

Bacteriods. Enlarged nonmotile nitrogen-fixing bacteria formed in nodules of legumes.

Basipetal. Describing the movement or translocation of solutes and hormones from the tip to the base of the plant.

Beneficial element. An element not considered essential for plant growth and reproduction but which enhances some aspects of development.

Bioassay. Laboratory testing of relative toxicity of substances using common native or crop seed.

Biocontrol. The control of an invasive plant species by releasing a nonindigenous herbivore that is specific for the invasive plant species.

Biodiversity. Refers to the numerous species of living creatures.

Bioengineering. The use of molecular manipulation techniques to change the genetic makeup or regulation of genes in a plant.

Biomass. The total mass of plants.

Bioremediation. The utilization of biological organisms to remove pollutants such as heavy metals, pesticides, hydrocarbons, etc.

Biosphere. The entire portion of the Earth inhabited by living organisms.

Biotechnology. The industrial use of living organisms to improve human health, lessen environmental contamination, and increase food production.

Biotic. Induced or caused by the action of living organisms; a combining form used to complete words (e.g., antibiotic, endobiotic, exobiotic), and meaning "of or relating to life."

Boundary layer. A layer of undisturbed air or solvent next to a surface; commonly, the layer of air at the surface of a leaf.

Broom species. A group of invasive legume species (particularly in Mediterranean habitats) that have many thin branches with photosynthetic stems.

Buffer. A substance that consists of acid and base forms that are interchangeable. In solution this substance is able to minimize the change in pH when an acid or base is added.

Bulk density. The mass of oven-dry soil per unit volume.

Bundle sheath cells. The cells surrounding the vascular traces in the leaf.

C_3 photosynthesis. A type of biochemical pathway for photosynthesis where the first stable product is a three-carbon molecule produced by a reaction controlled by rubisco. Plants having the cycle are called C_3 plants.

C_4 photosynthesis. Carbon assimilation that results in the formation of four-carbon compounds such as malate or aspartate due to carbon dioxide fixation by PEPcarboxylase. Plants having the cycle are called C_4 plants.

Calcicole. Plants able to grow and develop on calcareous soils.

Calcifuge. Plants injured, killed, or inhibited on calcareous soils.

Callose. A glucose polymer produced in developing sieve elements; also produced in response to wounding of the phloem, presumably to prevent leakage of photoassimilate.

Calmodulin. A Ca^{2+} binding protein whose binding to other proteins depends upon calcium concentration; a common component of signal transduction pathways.

CAM. *See* crassulacean acid metabolism.

Capillary water. The quantity of water that is held in small spaces between soil particles or in the apoplast of plant tissues. This water is held in place against gravity because of the adhesive and cohesive attractive forces of water.

Carbonyl group. A functional group of atoms consisting of a carbon atom double bonded to an oxygen atom; these groups are found in many molecules, including ketones and aldehydes.

Carboxydobacteria. A group of bacteria that oxidize CO as an energy source and are thought to be related to the genus *Pseudomonas.*

Carboxylic group. A functional group of atoms consisting of a carbon double bonded to an oxygen atom and also single bonded to a hydroxyl (OH) group. This grouping of atoms is found in organic acids, amino acids, and a number of other molecules.

Carotenes. A group of plant pigments, typically yellow-orange in color, that are considered accessory pigments that may protect plants against photooxidation.

Carotenoids. A group of plant pigments consisting of both xanthophylls and carotenes. Both are considered hydrocarbons, but xanthophylls also have oxygen in their structures.

Carrier. A molecule involved with the transport of another molecule across a membrane; also called a **permease.**

Casparian strip. The heavily suberized transverse and radial walls of the endodermis. Water and solutes are unable to leak past the endodermis through the apoplast and must pass through the cytoplasm.

Cation exchange capacity (CEC). The amount of cations released from the soil when a soil solution is flooded with ammonium sulfate.

Cavitation. The disruption of a water column by an air bubble or other barrier to water movement in a xylem element.

Chaparones. A class of proteins that bind to and stabilize polypeptide chains allowing for appropriate folding of the enzyme.

Chelate. To bind tightly and reversibly to a metal ion.

Chelators. Molecules capable of binding to metal ions such as copper and iron. Some important metal chelators are citric acid, malic acid, chlorophyll, the heme group, and phytometallophores. Chelators play an important role in metabolism and in the detoxification of heavy metals in the soil. *See also* ligand.

Chemoheterotrophic. The ability of an organism to utilize chemical elements in reduction processes to produce energy for growth.

Chemotaxis. The attraction of a biological organism to a chemical stimulus.

Chemotaxonomy. The classification of plant groups using their similarity in chemical compounds.

Chemotrophic. Autotrophic organisms that produce their own organic energy sources from a variety of inorganic substances.

Chitinase. An enzyme found mostly in bacteria that decompose chitin; also found in plants and may serve to break down fungal cell walls.

Chitin. A component of the exoskeleton of arthropods composed of many *N*-acetylglucosamine units.

Chlorinity. The total amount of chloride ions in a solution.

Chlorophyceae. A class of algae comprised of the "green algae," from which land plants are thought to evolved.

Chlorophyll. The green pigment, located inside chloroplasts, that absorbs light energy needed for photosynthesis.

Chloroplast. The organelle where C_3 photosynthesis and the light reactions occur.

Chlorosis. The disappearance of chlorophyll as a result of the altered metabolism brought on by disease or stress.

Chlorotic. Resulting in a loss of chlorophyll resulting in a pale yellow appearance of leaves.

Chronic. Describing an injurious event that occurs over time; exposure to an unfavorable environmental factor for a prolonged time, resulting in injury; the opposite of **acute** in an air pollution injury.

Coevolution. The evolution of two or more organisms such that each depends on the other for some aspect of its growth and development.

Common garden. A site in which species that naturally grow in different regions or climatic conditions are grown together. The site may be an agricultural field, a greenhouse, or a growth chamber.

Compaction. Usually refers to soils and their lack of porosity.

Companion cell. A type of cell in the phloem that is involved in phloem loading and unloading.

Compartmentalization. Isolation of specific molecules or groups of molecules in a specific location of the cell or tissue.

Compatible solute. Molecules in solution that in relatively high concentration have no detrimental effect on metabolism.

Competitive ability. The capacity to obtain resources needed for successful establishment and growth under competitive conditions.

Competitive inhibition. In the context of enzymes, the inhibition of an enzyme's activity by a molecule that binds to the active site of the enzyme, thereby blocking the substrate(s) from binding.

Conspecifics. Individuals of the same species.

Constitutive. Produced constantly during all developmental stages of a tissue. Proteins can have a constitutive transcription pattern.

Construction cost. The amount of material (carbon or nutrients) or energy (ATP or glucose) needed to build the molecule, tissue, or organ.

Convective. The ability of a warmed substance (air or fluid) to move from the surface of a body or an object.

Coordination number. The number of positions within each central atom's (metal) coordination sphere that can be occupied by ligands, irrespective of whether the ligands are neutral or charged; the maximal number of functional groups with which a metal will complex

Copepods. A type of crustacean consisting of about 7500 species, usually planktonic about 2 mm in length with a single eye in the middle of the head.

Crassulacean acid metabolism (CAM). Metabolism that occurs in species of plants that have open stomates at night and that fix CO_2 into organic acids such as malic acid, which is stored until daylight occurs. Decarboxylation occurs in daylight, releasing CO_2. CAM plants are able to effectively photosynthesize without open stomates during periods favoring high transpiration rates.

Critical concentration. The concentration of a nutrient in plant tissue that is just at the level conferring optimal growth.

Cryoprotectant. A chemical compound whose presence prevents injury during freezing; may occur naturally or be applied.

Cultivar. A specific variety or type of a crop population that has been selected from previous populations of the crop.

Cuticle. The waxy deposit on the surface of the plant outside the epidermis that is strongly hydrophobic.

Cuticular resistance. A measure of the reduction in diffusion through the plant tissues as a result of the presence of the cuticle; the reciprocal of conductance. Cuticle may vary in thickness, porosity, and composition, rendering more or less resistance to diffusion of gases such as water vapor and carbon dioxide.

Cuticular transpiration. The outward diffusion of water vapor through the cuticle.

Cutin. The waxy component of the cuticle comprised of complex polymer of hydroxylated fatty acids.

Cyanobacteria. Photosynthetic eubacteria belonging to the Cyanophyceae, which some researchers consider to be blue-green algae.

Cytokinins. A class of phytohormones involved with cell growth and division, of which the phytohormone zeatin is an example.

Cytoskeleton. Cellular structure comprised of protein microfilaments, intermediate filaments, and microtubules that are involved with some aspects of cell division, cytoplasmic streaming, and cell wall production.

Cytosol. The material in cells that organelles are suspended within.

Darcey's laws. The first law describes the rate of isothermal, steady-state liquid water flow in a porous medium. The second law states that water will flow across an air–water interface only when the pressure potential is sufficiently greater than zero to overcome the surface tension of the fluid.

Decarboxylation. The removal of a carboxyl group from an acid; a particularly important function in C_4 plants for acquiring CO_2 from malic or aspartic acids for photosynthesis.

Decomposers. Organisms that are frequently associated with soils responsible for degrading organic material into small chemical subunits that can be used by other organisms.

Decomposition. The breakdown of biomass by fungi and other microbes.

Dehydration avoidance. The prevention of water loss by the closing of stomates, resulting in the maintenance of turgor during periods favoring high rates of transpiration.

Dehydration postponement. The reduction of transpiration or an increase in water absorption because of morphological characteristics of drought-stressed plants.

Dehydration tolerance. The ability of plants to withstand injury at the protoplasmic level when plants are under drought stress.

Dematiaceous. Used to describe hyphae with dark pigmentation.

Denitrification. The conversion of nitrate to dinitrogen or ammonia carried out by microbes. The reaction is used as an energy source for these bacteria.

Desiccation resistance. Protoplasmic resistance to injury under extreme water-stress conditions. Upon rehydration, respiration and photosynthesis begin within a relatively short time.

Developmental stage. The particular anatomical, morphological, and physiological status of the organism. Some developmental stages are juvenile, mature, senescent.

Dicot. A plant characterized by two cotyledonary leaves, net veination in the leaves, and cambial activity in woody species and floral parts arranged in fours, fives, or multiples of these.

Dieback. A condition, particularly prevalent in woody trees, in which the tips of branches die, and can be caused by a combination of abiotic and biotic stresses.

Differential permeability. The differential propensity of different molecules to move across a membrane. Those with the best capacity to move are small hydrophobic molecules.

Differentiation. The specialization of cell form and function through a developmental sequence.

Diffusion. Net movement of a substance from a location of high concentration to one of low concentration. The molecules move based on random kinetic energy and entropy.

Disease. A condition of the living animal or plant body or one of its parts that impairs the performance of a vital function; impairment or modification of the performance of a vital function or functions; response to environmental factors or inherent genetic defects.

Diurnal. A cycle occurring on a daily basis, such as the opening and closing of flowers.

Diversity. The number of different species in a given area; indices of diversity include a richness aspect and an evenness component.

Dormancy. A state of inactive growth; typically used in reference to buds and seeds; important in resistance to stresses.

Dose. Exposure to a measured concentration of a toxicant for a known duration of time.

Downregulation. A decrease in the activity of a reaction or metabolic pathway often associated with the suppression or inactivation of gene transcription.

Drought. A meteorological term for prolonged periods of reduced precipitation.

Drought resistance. Innate mechanisms by which injury is prevented or minimized during drought stress; some scientists prefer the term *drought tolerance.*

Drought tolerance. The result of postponing of dehydration (*see* dehydration postponement), or of innate dehydration tolerance characteristics.

Ecophysiology. The study of the physiological mechanisms involved in ecological phenomena; often associated with the mechanisms by which plants cope with their environment.

Ecosystem. The sum of the physical and biotic factors in a region. The region is usually delimited by a watershed. The fluxes and regulation of fluxes among components of the ecosystem is an important topic in ecosystem ecology.

Ecotype. A population, found in a specific location, that is genetically distinct from another population; usually associated with variation in a functional response to the environment among populations.

Ectomycorrhiza. An association between a fungus and the roots of a plant in which the fungus can enhance the acquisition of soil resources by plant roots, and the plant provides carbohydrate for the fungus. The association is primarily external to the root tissues.

Ectoparasitic. Describing a parasite that lives on the external cells or tissues of a host.

Edaphic. Relating to factors associated with soils.

Efficiency. The rate of a physiological process per unit of plant resource. Resources such as carbohydrate, nitrogen, water, and phosphorus are often considered.

Electrical conductivity. The relative ability of a solution to conduct electrical charge; used as a secondary measure of salinity.

Electrochemical potential. An electrical potential that results from the relative concentration of anions and cations on each side of a membrane.

Electropotential. In biological systems, generally refers to the gradient of current across biological membranes that develop as a result of differences in concentrations of cations and anions.

Electrotaxis. A biological response to electrical current.

Eluvial soil layer. The uppermost mineral layer of the A horizon of a soil profile, characterized by maximum leaching.

Embolism. The development of an air bubble in a xylem element that causes cavitation of the vessel.

Endodermis. A single layer of cells surrounding the vascular cylinder in the center of the root. This cell layer has associated with it suberin layers called Casparian strips, which aid in the selectivity and formation of root pressure in plants.

Endomycorrhiza. A fungus that forms part of its mass inside plant roots and can enhance root acquisition of nutrient resources.

Endoplasmic reticulum. A cellular organelle responsible for protein synthesis and the synthesis of lipid and carbohydrate material required in the production of membranes and cell walls.

Entropy. A measure of the degree to which a system is removed from equilibrium. As the degree of randomness increases in a system, entropy increases.

Ephemeral. Short-lived; often used when discussing protein, leaf, or root survival.

Epidemiology. The study of the distribution and spread of disease.

Epinasty. The lowering of leaves in response to differential growth of the upper surface of leaf petioles, often in response to ethylene or indoleacetic acid.

Epiphyte. An organism that resides upon another organism for support but does not attain any other nourishment from the supporting organism.

Escape. The avoidance of stress by some plants because of the timing sequence in development and growth. The life cycle may be completed during a period or season when a particular stress does not occur.

Ethylene. One of six classes of growth regulators in plants. This gaseous molecule influences aerenchyma formation, tissue abscission, and other processes in plants.

Euhalophytes. Halophytes that grow in conditions of extreme salinity.

Euxerophytes. Plants that grow under extremely dry conditions, such as in deserts.

Evolution. The change in plant form and function over time due to a change in genetic properties regulating that form and function. The products of evolution are races, ecotypes, and species.

Excluders. Plants that have the ability to regulate the uptake of elements even when highly concentrated in the soil.

Exclusion. Prevention of the net entry of a molecule into a space (e.g., an apoplast or symplast).

Excretion. Active secretion of molecules (such as enzymes) out of the symplast into the soil or apoplast (e.g., salt from leaves, oxalic acid from roots).

Extensin. A cell wall glycoprotein that increases the rigidity of the primary wall in plants.

Extracellular. Maintained outside to the cell membrane.

Exudates. Usually, organic substances such as sugars or amino acids lost from roots to the soil.

Facultative halophytes. Species that switches to a saline-tolerant phenotype when the environment become salty. Otherwise the species is not salt tolerant.

Fauna. The animal life, as in the arthropods, that might occupy a soil ecosystem.

Feedback. The regulation mechanism for metabolic pathways in which one of the final products of the pathway affects the activity of an enzyme at the beginning of the pathway. The mechanism that inhibits enzyme activity is called *negative feedback,* while one that increases enzyme activity is called *positive feedback.*

Feedforward. Change in a process before any deleterious changes in the regulator have occurred.

Fermentation. A process that oxidizes NADH in the absence of oxygen and produces NAD^+ which allows glycolysis to continue phosphorylating ADP to ATP.

Ferredoxin. An iron-containing protein associated with thylakoid membranes of the chloroplasts and involved in electron transport in the light reactions of photosynthesis.

Field capacity. The total amount of water that the soil can hold against the forces of gravity.

Fitness. A measure of the success of a population or individual, usually defined on the basis of reproductive success.

Flavonoids. Nonnitrogenous carbon compounds distributed throughout the plant kingdom comprised of three groups; anthocyanins, flavanols, and flavones. Anthocyanins are colored pigments and occur in colors of red, purple, and blue. Some may protect plants against harmful UV-B irradiation.

Flora. The plant life associated with a specific ecosystem.

Fluorescence. Light that is produced after a molecule that has been excited into a triplet state returns to the ground state; often used in reference to the light absorptance by chlorophyll.

Flux. The volume flow across a unit area per unit time; usually given in moles per meter per second (mol m^{-2} s^{-1}).

Freezing avoidance. Exclusion of the effects of freezing from the protoplasm.

Freezing point depression. When the temperature at which a solution freezes is below that of pure water; directly correlated to the amount of solutes in solution.

Frost hardiness. A measure of frost tolerance; may be induced by gradual exposure to lowered temperatures.

Frost resistance. The ability of a plant or tissue to avoid cellular damage during frost conditions.

Frost tolerance. When injury to cells and cell function does not occur at temperatures of 1°–3°C.

Gene. A sequence of nucleotides in the DNA of an organisms that codes for a particular polypeptide and contains other sequences that regulate the transcription of the sequence coding for the polypeptide.

Gene regulation. The control of gene transcription.

Genetic trait. A morphological, anatomical, physiological, or behavioral characteristic of an organism that is determined by the transcription of specific genes.

Genotypic. Constituting or comprised of a specific genetic composition.

Gibberellins. Growth regulators that cause or influence responses in plants such as cell enlargement, flowering, and seed germination.

Gibbs free energy. The isothermal (at constant temperature) energy available for conversion to work; expressed as $G = E + PV - TS$, where G is given in energy per mole, E = internal energy, PV = pressure-volume constant, T = absolute temperature, and S = entropy.

Glaucus. Describing the light-color, shiny nature of leaf surfaces due to a thick layer of wax.

Glucoside or glycoside. Compounds that are attached to a glucose molecule by a side chain.

Glycolipid. A lipid associated with membranes containing a carbohydrate but no phosphorus.

Glycolysis. The initial steps in aerobic and anaerobic respiration in which glucose is broken down to two molecules of pyruvate with a net production of two ATP molecules. The reactions occur in the cytosol.

Glycophytes. Plants that are sensitive to relatively low soil salt concentrations.

Glycoproteins. Proteins that have covalently linked oligosaccharide molecules; often located in the plasma membrane.

Glyoxysomes. Organelles involved with lipid metabolism.

Golgi body. The organelle responsible for the production and/or deposition of carbohydrate and lipid in the formation of cell walls and membranes.

Graviperception. The ability of a plant part to perceive gravity.

Green revolution. Two eras in history: The first occurred between 1950 and 1970, with dramatic increases in crop production in industrialized nations. The second came in 1967, when new varieties of wheat and rice held promise for feeding the world's populations in underdeveloped countries in the tropics and subtropical climates.

Growth inhibitors. Organic chemicals that slow growth; may be naturally occurring or anthropogenic.

Growth regulators. Organic chemicals that promote some form of growth when present.

Guard cells. Specialized epidermal cells that surround the stomata. These cells may be directly involved in regulating stomatal aperture.

Guild. A group of species that utilize similar resources or have similar physiological or behavioral traits.

Guttation. The forcing of water from the tips of leaves (as in grasses), through structures called hydathodes, as a result of root pressure.

Half sib. The degree of relatedness between breeding lines; usually refers to a population of offspring that are derived from one fruiting individual.

Halomorphic. Referring to soil characteristics generated as a consequence of high salinity.

Halophytes. Plants that are able to grow in the presence of high-salt soil concentration. These plants often require some salt to grow their best. *See also* euhalophytes, facultative halophytes, obligate halophytes.

Hardening. Acclimation; increased tolerance of stress. Hardening may be caused by gradual exposure to stress; changes may be phenotypic or occur at the protoplasmic level.

Hardiness. The ability to withstand or tolerate stress conditions.

Hartig net. The portion of the hyphae of ectomycorrhizal fungi that have penetrated (intercellularly) into the cortex of a root between root cells.

Haustorium. The parasitic organ that serves to locate, attach to, and penetrate the root and to establish a mechanism for water, nutrient, and carbohydrate exchange with the root; or a structure of fungal hyphae that attaches to and penetrates host plant tissues.

Heat tolerance. The ability to tolerate high-temperature either by dissipating heat or by physiological changes that prevent injury.

Heavy metal. A metal element with a density greater than 5 g ml^{-1}.

Hemicellulose. A structural polysaccharide that holds cellulose fibrils together in the cell wall.

Hemoproteins. Proteins containing iron.

Herbivores. Organisms that consume plant material for internal digestion to support their nutrient and energy needs.

Herbivory. The ecological process by which animal populations consume plants or parts of plants as a source of nutrients and energy.

Heterotrophic. An organism that depends on preformed organic material derived from plants, fungi, protistans, or animals as an energy source.

Homeostasis. A relatively stable state of equilibrium or tendency toward such a state between the different but interdependent elements or groups of elements of an organism or group.

Homogeneous nucleation. Refers to the temperature that pure water or a solution freezes without the help of heterogeneous nucleation (dust, bacteria, fungi, agitation) and involves supercooling of the liquid or solution well below its freezing point.

Hormone. A specific organic product produced in one part of a plant or animal and transported to another part where it is effective in small amounts at controlling or stimulating processes.

Humus. An amorphous material comprised of highly oxidized aromatic and polymeric organic material derived from the degradation of lignin by microbial activity.

Hydathodes. Specialized structures in leaf tips through which guttation water passes.

Hydrogen bonding. The elements N, O, or F, are the three most electronegative elements; because of this, they apparently attract the electrons of the covalent bond to such an extent as to leave the H nucleus almost completely exposed. H is the smallest of all atoms; therefore, it is possible for it to come very close to an electronegative atom of another molecule. The resulting electrostatic attraction, while weak relative to a covalent bond, is much stronger than the attraction between dipoles or between molecules held together by ordinary van der Waals forces, and for this reason is called a hydrogen bond.

Hydrolysis. The decomposition of a complex molecule by adding a water molecule while the original molecule is parted.

Hydrophobic. Describing substances that do not dissolve in water, are soluble in lipids, and are often a hydrocarbon.

Hydrophytes. Plants that live in water or in virtually water-saturated soils.

Hyperaccumulators. Plants that accumulate ions to much higher levels than occur in the soil in which they grow.

Hypersaline. A concentration of salinity higher than what the plants under consideration can tolerate without detrimental effects.

Hyphae. The vegetative structures of fungi.

Hypoxia. A condition in which there is relatively low oxygen but some oxygen is present.

Ice nucleators. Materials that cause water to freeze when it is supercooled. Ice crystals may serve as nucleators when introduced into supercooled water.

Illuvial soil layer. The B horizon characterized by maximum deposition of leachates from above.

In situ. Used to describe a process studied in a living organism or cell that is located in the region where it normally resides.

In vitro. Used to describe a process studied outside of a living cell, usually in a test tube.

In vivo. Used to describe a process studied in living cells.

Indicator plants. Plants that reflect the ion concentration of the soil in which they grow.

Indigenous. Having to do with the local region, such as indigenous people.

Inducible defense. A mechanism of defense against a biotic or an abiotic stress that occurs only after the plant experiences some of the stress.

Injury. A change in metabolism that results in altered functions and/or appearance with deleterious effects.

Intercellular. Within the cell as opposed to outside the cell membrane.

Intercropping. Growing two or more crops in the same field, usually in separate rows.

Interspecific competition. Competition between plants of different species.

Intraspecific competition. Competition between plants of the same species.

Invaginated. Enfolded, as would occur in the inner membrane of a mitochondrion.

Invasive species. A species that establishes a naturalized population that increases in numbers very rapidly, often replacing native species and changing the nature of the ecosystem.

Ion. A charged atom or molecule formed by the dissociation of ionically bonded atoms or molecules.

Ionic bond. The strong electrostatic attraction between two oppositely charged ions.

Ionic concentration. The summation of all the solutes in solution irrespective of their charge or elemental composition.

Ionization constant. An indication of the constant strength of an acid or base. The greater the degree to which the acid or base ionizes at a given concentration, the stronger it is, and the larger is the value of its ionization constant, expressed as K or K_b.

Ion species. The particular ionic molecule under consideration.

Irradiance. The radiant flux density on a given surface usually expressed in watts m^{-2}.

Isotope fractionation. A process by which metabolic reactions differentially utilize different isotopes of the same atom.

Isotope. An atom with a different number of neutrons compared to the most common form of the atom.

Jasmonic acid. A molecule involved in stress signaling pathways in plants particularly related to pathogen, herbivore, and parasite attack; 3-oxo-2-(2′-*cis*-pentenyl)-cyclopropane-1-acetic acid.

Krebs cycle. The part of aerobic respiration where reducing power is generated and used to produce ATP.

***K* strategy.** A suite of characteristics that allows a plant to live late into the succession of the community.

Lateral phase separation. The separation of phospholipids in a heterogeneous mixture into more-homogeneous groups in membranes when the temperature is reduced.

Leachate. The solution derived from passing water over plant tissue; the resultant solution from the act of leaching or anything that can be lost through the soil profile such as nutrients.

Leaf area index. The total amount of leaf area per given amount of ground surface.

Leaf area ratio. The ratio between leaf area and mass.

Lehemoglobin. A pink pigment, associated with the nodules of legumes, that is involved with providing oxygen for respiration of the bacterial symbiont.

Lenticels. Cracks in the epidermis and cortex of stems and roots. They may serve as entry points for oxygen.

Lewis acid. Any substance capable of acting as an electron pair acceptor: yields a hydrogen ion in solution.

Lewis base. Any substance possessing a pair of electrons available for sharing with a proton to form a covalent bond; yields a hydroxyl ion in solution.

Ligand. Can be either neutral molecules or negative ions, both of which possess unshared electron pairs and are capable of functioning as electron donors, that is, as Lewis bases. Each molecule or ion possesses, therefore, at least one donor center capable of occupying a position in the coordination sphere of the metal atom. A ligand with one donor center is described as unidentate, while a ligand with two or more donor centers is said to be multidentate. Multidentate ligands, such as EDTA, are commonly called **chelating agents** because they enclose the central atom in a pincerlike manner reminiscent of the action of a crab's-claw (Gk. *chele* = claw). The complexes are called **chelates**.

Light-harvesting complex. The complex of pigments that intercept light impinging on leaves; consists of chlorophyll plus other pigments.

Light saturation. The light intensity beyond which no further increase in intensity can cause an increase in photosynthesis.

Lignification. The process in which cell walls are thickened, particularly in woody plants, by the deposition of lignin.

Lipophilicity. The degree of solubility in a nonpolar solvent.

Liquid crystalline. Describing the fluid state of pure phospholipids; related to phase transitions in membranes; caused by temperature.

Luxury consumption. Absorption of mineral nutrients above and beyond those that cause increases in growth or yield.

Lysate. Material lost from cells as a result of cell disruption.

Macronutrients. Elements required by plants for normal growth, development, and reproduction in concentrations of 1000 mg kg^{-1} dry weight.

Mafic. A group of dark-colored minerals rich in magnesium and iron.

Mangrove swamp. A wetland community at the margin of the land near the ocean in tropical and subtropical environments.

Marginal land. Land not well suited for agriculture based upon a number of possible criteria such as being drought-prone, saline, or mountainous.

Mass flow. Large-scale (bulk) transport of nutrients along with the flow of water; a passive process.

Matric potential. The term is used to account for all the forces of capillarity, adsorption, and hydration that cause imbibition, or the holding of water in a matrix of any sort.

Mega. A prefix, when applied to measurement units, meaning $\times 10^6$.

Membrane carriers. Molecules or compounds that assist materials to cross membranes; associated with the transport of charged particles such as ions, many are proteins.

Membrane fluidity. Changes with temperature and amount of cross-linkage.

Membrane transition. A change in the physical state of membrane components as temperature is lowered; the change from a liquid-crystalline state to gel structure.

Meristematic. A group of cells that give rise to new cells that ultimately differentiate into specialized tissues.

Mesofauna. Include organisms ranging in size from 200 μm to 1 cm and include such organisms as mites, springtails, large nematodes, rotifers, spiders, pseudoscorpions, harvestmen, pot-worms, insect larvae, and small millipedes and isopods.

Mesophyll. The majority of cells in a leaf including palisades and the spongy parenchyma cells.

Mesophytes. Plants that thrive best with a moderate water supply.

Metalloenzymes. Enzymes that contain a transition metal as the prosthetic group.

Metallophyte. A plant that grows in soils containing high levels of heavy metals and may have a requirement for high concentrations of a specific metal.

Metallothionein. A low-molecular weight metalobinding protein.

Metaxylem. A type of primary xylem formed from procambium; forms recognizable secondary wall thickenings as hoops, bands, or helices.

Micro. A prefix, when applied to measurement units, meaning $\times 10^{-6}$.

Microbiota. Microscopic plants and animals frequently associated with soils or aquatic systems.

Microclimate. Climate typically considered at a lower scale level, as would occur in soils, lower-canopy areas of field crops, or other small ecological niches.

Microfauna. Organisms that range in size from 20 μm to 200 μm and consisting primarily of protozoans, but also the upper size range of nematodes, rotifers, microscopic arthropods, and copepods.

Microfibrils. A complex of cellulose molecules twisted and integrated to form plant cell walls.

Microhabitat. A small-scale habitat such as would occur between soil particles including all aspects of the physicochemcial, environmental, and biotic elements that comprise it.

Micronutrients. Elements required by plants for normal growth, development, and reproduction in concentrations equal to or less than 100 mg kg^{-1} dry weight.

Microphytic feeders. Microscopic organisms that feed on plant tissues.

Microtubules. The largest of three protein components that comprise the cytoskeleton. Microtubles also function to move the cell nucleus during cell division and make up the internal structure of flagella.

Milliequivalent. In neutralization reactions, one milliequivalent is the weight of the molecule that contributes or reacts with 1 gram formula weight of hydrogen ions. In oxidation-reduction reactions, 1 milliequivalent is the weight that directly or indirectly consumes or produces 1 mole of electrons. Abbreviated mEq.

Mineralization. The decomposition (breakdown) of organic matter that releases nutrients into the soil solution.

Mistletoe. A group of xylem-tapping stem parasitic angiosperms in the families Loranthaceae and Viscaceae.

Mites. Small arachnids that infest plants and animals and can be vectors of disease.

Miticides. Chemicals effective in killing mites.

Molarity. The number of gram weights (moles) of a particular molecule in 1 liter of solution; abbreviated *M*.

Molecular biology or molecular research. Research on the structure, function, and regulation of DNA.

Mollusca. The phylum to which invertebrate animals such as snails and shellfish belong; characterized by soft unsegmented bodies usually enclosed in a calcareous shell.

Monocot. A type of angiosperm characterized by a single cotyledon, parallel veination, and no cambium.

Morphology. The form and structure of a plant or an animal.

Murograph. A diagrammatic representation of the spore wall groups and types; used in the identification of glomalean spores.

Mycorrhizae. A symbiotic association between a fungus and usually the root of a higher plant; the fungus normally makes available forms of elements (especially N and P) that would otherwise be unavailable to the root; mycorrhizal associations occur throughout the plant kingdom.

Nano. A measurement prefix meaning $\times 10^{-9}$.

Natural habitats. Regions in which the collection of available resources have not been altered by the action of human population.

Necrosis. The death of tissues within organs.

Nectaries. Structures in plant flowers that secret nectar.

Nematodes. Cylindrical worms with tapered ends called roundworms; found in aquatic habitats, moist plant tissues, wet soils, and animal body fluids and tissues.

Nematophagous. Nematode-eating organisms, such as some fungi.

Niche. A theoretic "space" defined by the physiological tolerances of a species. The physiological tolerances usually refer to resource availability, but they may also refer to biotic or abiotic stresses of the environment,

Nitrification. The microbial process that transforms ammonium ion into nitrate; this transformation is used as an energy source for the microbes.

Nitrogen fixation. Tthe reduction of dinitrogen gas into ammonium and eventually amino acids or amides.

Nitrophilous plants. Plants having a high requirement for nitrogen, or nitrogen-loving plants.

Nodulation. The formation of nodules on leguminous plant roots by nitrogen-fixing bacteria.

Noncompetitive inhibition (enzymes). Inhibition of enzyme function by a molecule that binds to a location on the enzyme that is not an active site. By binding, the inhibitor changes the shape and/or chemistry of the active site, rendering the enzyme dysfunctional.

Noninfectious disease. Disease caused by an environmental factor, not by a pathogen; caused by a physiogen; a physiogenic disease.

Nonionized. Without charge characteristics.

Nonobligate parasite. A parasite that is not as restricted with respect to growth requirements as are obligate parasites such as viruses, which require living hosts.

Nonpolar molecule. A molecule that has no positive or negative ends; or the center of the negative charge of the molecule corresponds to the center of the positive charge. Molecules such as H_2, O_2, N_2, Cl_2 are nonpolar since atoms of these molecules have the same electronegativity and share electrons equally.

Nonprotein amino acids. Amino acids not found in protein molecules; often harmful due to their ability to replace analogous amino acids in proteins. Many other nonprotein amino acids also are nontoxic.

Normality. A solution measure expressed as the number of milliequivalents of dissolved solute contained in 1 ml of solution, or the number of equivalents in 1 liter of the solution; abbreviated *N*. Thus, a 0.20 *N* solution contains 0.20 milliequivalents of the substance in each milliliter of solution

Nutrient. An ionic moiety in the soil that when accumulated by plants is utilized in metabolism and growth processes.

Nutrient balance. The total number, on an equivalent basis, of cations and anions absorbed by a plant; plants under uniform environmental conditions tend to take in a constant number of cations and anions. Since the total number of cations and anions absorbed remains constant, if K^+ in the plant were increased, Ca^{2+} and Mg^{2+} would tend to decrease and vice-versa.

Nutrient use efficiency. A measure of the growth of a plant per unit nutrient absorbed.

Nyctinastic. Refers to sleep movements in plants, such as the downward curling of bean leaves at night.

Obligate halophytes. Plants (halophytes) that require salt in the growth medium to reach their maximum growth potential and have their highest fitness.

Obligate shade plants. Plants that grow well in reduced light intensities but are injured when placed in relatively high light intensity.

Oligochaetes. Hermaphroditic terrestrial or aquatic annelids, such as earthworms, that lack locomotive organs and a specialized head.

Osmolality. A measure of the total number of osmotically active substances in solution; often used in reference to soil and tissue water solutions (osm or moles L^{-1}).

Osmometer. A device for measuring osmotic pressure.

Osmoprotectants. Organic solutes produced as a result of salt, water, or cold stress and that serve to protect tissues from loss of turgor, freezing damage, or salinity damage.

Osmoregulation. The control of plant functions by the regulation of water potentials across differentially permeable barriers or membranes. Water diffusing down a potential gradient is losing energy and can be made to do work. An example would be stomatal control by the regulation of the turgidity of guard cells.

Osmotic adjustment. An increase in organic solute content as a result of stress; lowers the osmotic potential of cell sap.

Osmotic potential. The component of water potential that becomes increasingly negative with the addition of solute.

Oxidation. The process of chemical transformation in which a pair of electrons are removed from the original molecule; is often associated with an oxygen atom.

Oxygen-evolving complex. A protein complex in the light reactions of photosynthesis responsible for the photolysis of water.

Oxyradical scavengers. Compounds such as ascorbic acid, carotenes, and tocopherol that aid in the removal of reactive species of oxygen from biological tissues.

Paleoecological. The study of past ecological systems of the Earth through observations of the paleontological record.

Parent material. The rock material and other substances that are weathered to generate the basic particles of soil.

Pathogen. A disease-initiating biotic entity; usually restricted to organismic causes (e.g., bacteria, fungi, viruses).

Pathogenesis. The origination and development of a pathogenic disease.

Pectin. A material of the cell wall that is composed of a carbohydrate polymer rich in galactouronic acid.

Pectolytic enzymes. Enzymes frequently produced by plant pathogens that hydrolyze pectin associated with plant cell walls.

Pedology. The study of soil structure and composition.

Pentose phosphate pathway. A biosynthetic pathway important in the production of nucleic acids, hormones, and defensive compounds in plants.

Perennial. Species composed of individuals that live for more than two years.

Pericycle. A specialized tissue in the roots of plants that is periodically meristematic giving rise to lateral roots.

Permanent wilting percentage (PWP). The moisture content in soils at which plants can no longer accumulate water. The result is a bulk tissue turgor potential of zero. The PWP is taken to be −1.5 MPa for many herbaceous crops.

Permeability. A property of membranes that defines the ability of materials to move across the membrane.

Peroxisomes. A microbody involved with photorespiration processes and the removal of hydrogen peroxide in plant tissues.

pH. The negative log of the hydrogen ions in solution.

Phase separation. As water in tissues freezes, the separation of the solid phase (ice) from the liquid (unfrozen water) phase.

Phase transition. The change from liquid to solid, as from liquid water to solid ice. *See also* lateral phase transition.

Phenol. A molecule that has a hydroxyl group attached to an aromatic ring structure.

Phenolics. Hydrocarbons that contain phenol groups (aromatic hydrocarbons).

Phenology. The study of the time course of developmental events in plants, particularly in association with the time course of climatic conditions.

Phenotypic plasticity. An alteration in a trait expression for a given genotype. The change in trait expression is not due to a change in genetic composition, rather to a change in gene expression.

Phloem. The cellular structure that serves for long-distance transport of carbohydrates and other substances in plants.

Phosphatase. An enzyme that hydrolyzes organic molecules containing phosphate groups.

Phospholipid. A lipid consisting of a glycerol backbone, two fatty acyl groups, and a phosphorus head group containing a variety of amino groups, and found associated with biological membranes.

Phosphorylation. A reaction in which a phosphate group is covalently bonded to a molecule. Often the phosphate group is a high-energy phosphorus from ATP. Such phosphorylation of an enzyme can often change its catalytic capabilities.

Photoassimilates. Organic compounds produced through photosynthesis that are used by various metabolic processes in plants.

Photoautotrophic. An organism capable of producing its own food source through the process of photosynthesis.

Photocatalytic. A chemical reaction catalyzed by light.

Photoinhibition. Light-reduced activity such as growth or an enzymatic reaction; often used in reference to a light-induced decrease in photosynthesis.

Photolysis. The process by which water is hydrolyzed in the light reactions of photosynthesis.

Photomorphogenetic. Referring to the ability to change morphology as a result of exposure to light.

Photon. A unit of light energy that has particle-like attributes. The energy level of the particle is defined as the wavelength of the particle; the shorter the wavelength, the higher the photon energy.

Photoperiod. The length of the light period during a diurnal cycle; the photoperiod varies with latitude and season.

Photoperiodic. A physiological or phenological process that varies according to the photoperiod.

Photophosphorylation. The synthesis of ATP in illuminated chloroplasts; the use of photon energy to phosphorylate ADP to ATP.

Photorespiration. A pathway that releases CO_2 and consumes energy-producing amino acids; the process originates with a fixation of O_2 with rubisco and involves the chloroplast, peroxisome, and mitochondria.

Photosynthesis. A metabolic process that utilizes the energy from light and the electrons from water to reduce CO_2 to trios phosphate.

Photosynthetic quotient. The ratio between the amount of CO_2 reduced in photosynthesis to the amount of O_2 released by photosynthesis.

Photosynthetically active radiation (PAR). The wavelengths of light (particle energy levels) that are utilized for photosynthesis. This is taken to be the part of the visible spectrum between 400 and 700 nm.

Photosystem. An organization of proteins and pigments in the inner chloroplast membranes that serve to absorb photon energy and create a hydrogen gradient across the chloroplast membrane that will be used for ATP synthesis.

Phreatophytes. Plants that have very deep root systems and are able to access the phreatic zone of saturated soil just above the ground water table.

Phyla. A taxonomic group of organisms below the kingdom and division levels and above the family level.

Phylogenetic. Having to do with the evolutionary lineage of a particular process or organism.

Physiogen. A disease-initiating abiotic entity or event; caused by physical or chemical factors in the environment.

Physiogenesis. The origination and development of a physiogenic disease.

Phytate. Inositol hexaphosphate; an organic form of phosphorus found in soil solutions; also the form of P storage in seeds and the endodermal tissues of plants.

Phytoalexins. Secondary compounds produced in plants in response to pathogens and herbivory.

Phytochelatins. Sulfur-rich, small polypeptides and organic acids produced in roots that aid in the uptake of essential ions.

Phytohormone. A substance, produced in one part of a plant, that travels to other locations in the plant and regulates developmental process of these other regions. Examples are auxin, cytokinin, gibberellin, brassinosteroid, abscisic acid, systemin, ethylene.

Phytometallophores. Substances exuded from the roots of grasses that chelate heavy metals in the soil.

Phytopathogens. Biotic organisms that parasitize plants.

Phytophages. Organisms that feed on plants.

Phytophagous. Utilizing plant tissue as the sole source of nutrition for herbivores.

Phytosiderophores. Substances produced by grasses that are able to chelate iron in the soil solution.

Phytotoxicity. A substance toxic to plants that could be a pollutant or produced by another organism.

Pico. A measurement prefix meaning $\times 10^{-12}$.

Plasmalemma. The outer cell membrane; also called the **plasma membrane.**

Plasmodesmata. Membrane channels that traverse the plasma membrane and the cell wall, forming a continuous cytoplasmic link between to plants; often found clumped into fields. These passageways are large enough to transport viral particles.

Plasmolysis. The condition of a cell in which the turgor has been reduced by removal of water by osmosis. The cytoplasm may separate from the cell wall.

Plastocyanin. A copper-containing electron transport protein in the thylakoid membranes important in the light reactions of photosynthesis.

Plow layer horizon. A soil horizon characterized by mixing of the A and B soil horizons as a result of mixing during the plowing process.

Pneumatophores. Vertical structures that rise from roots upward above the water level in swamps.

Poikilohydric. Referring to plants and plant organs that can become completely dry and still function soon after rehydration.

Poikilotherms. Organisms that attain the temperature of the environment while still alive. Plants and some animals are poikilotherms.

Polyamines. Polycationic nitrogen-containing compounds found in plants, particularly under drought stress. Some scientists consider them to be hormonal in nature.

Polycation. A molecule possessing multiple cationic sites such as in polyamines.

Polymerase chain reaction. A reaction that utilizes a thermally stable DNA polymerase to multiply particular fragments of a DNA molecule. The reaction involves thermal cycling between high and low temperatures to induce DNA to replicate, and then melt for the next replication cycle.

Population. All the individuals of one species that are located in a specific region; the delineation of the population is often based on environmental attributes, but the actual criterion defining the individuals in a population is the likelihood for interbreeding.

Porphyrins. Derivatives of tetrapyrrole ring structures that are typical of chlorophyll. There are at least five different classes of porphyrins differing in their side-chain substituents.

Pressure potential. The influence of internal pressure on the water potential of cells; sometimes referred to as the turgor potential for plant cells.

Pressure probe. A mechanical device that can insert a glass needle manometer into plant cells to measure their internal pressure.

Productivity. The net increase in plant mass over a specified time, usually one year.

Prop root. Roots that serve to elevate the rest of the plant above the water line in swamps and mangrove forests.

Propagules. Methods of dissemination and reproduction of plant or pathogen (e.g., seeds, spores, etc.).

Proteinase or **protease.** An enzyme that functions to hydrolyze (break down) proteins.

Proteoid root. A morphological proliferation of roots in regions of high nutrition in the soil; also called a **cluster root.**

Proton-motive force. The force across a membrane due to the differential concentration of ions on each side of the membrane; often associated with hydrogen ions.

Pseudoscorpions. An arachnid of the order Pseudoscorpiones that feeds on small insects, mites, and other minute animals.

Q_{10}. The proportionate change in an enzyme-mediated reaction rate in response to a 10°C change in temperature.

Qualitative defense compound. A substance that can have some defensive capabilities at relatively low concentration in plant tissues.

Quantitative defense substance. A molecule that confers some defense against herbivores when the substance is at relatively high concentration in plant tissues.

Quantitative traits. Traits that are regulated by many genes and are represented by a continuum of expression intensities in the population.

Quantum yield. The moles of CO_2 reduced by photosynthesis per mole of quanta (photons) absorbed. The same can be determined for amount of O_2 evolution or number of electrons generated.

Radiative. The propensity to emit, absorb, and transmit radiation.

Receptor protein. A protein embedded in the plasma membrane that has a high affinity for a signaling molecule in the apoplast. Upon binding to the signaling molecule, the receptor initiates a signal transduction pathway inside the cell.

Reciprocal transplant experiment. An experiment in which two species that naturally grow in different habitats are each transplanted into the other species' habitat. This experiment also requires that both species be transplanted into their own habitat as controls.

Recombinant DNA. Genetically engineered DNA prepared by inserting or replacing a fragment of the DNA molecule with DNA fragments from other organisms.

Redox potential. The tendency of a reducing agent to lose electrons (or of an oxidizing agent to gain them).

Reducing power. Said of molecules that have the ability to reduce (transfer electrons) other substances.

Reductant. A reducing agent.

Relative humidity (RH). The percentage of water vapor present in the atmosphere relative to saturation (100%) at the existing temperature.

Relative water content. In a tissue, the relative amount of water compared with the maximum amount of water at full hydration.

Resistance. Tolerance or avoidance; the ability of an organism to withstand or overcome partially or completely the injury caused by a physiogenic or a pathogenic factor; the means of resisting, fighting against, or counteracting the injurious environmental effects.

Resource. A characteristic of the environment that is obtained by plants and used in metabolic processes.

Resource competition. The situation when two or more species vie for the same pool of growth-regulating resources.

Respiratory quotient. The ratio between the CO_2 released by respiration and the O_2 consumed during the respiratory electron transport chain.

Rhizomorph. A thick strand of somatic hyphae that have lost their individuality, with the whole mass behaving as an organized unit. The structure of the growing tip of the rhizomorph somewhat resembles a root tip, hence the name; also known as a **mycelial cord.**

Rhizoplane. The immediate surface of the root.

Rhizosphere. The volume of soil near the root surface that is affected by the root's presence and by the chemical and physical interactions that occur between the root and its immediate environment.

Rickettsia-like-organism (RLO). Walled microorganisms detected in vascular tissues that resemble small bacteria but are not classified in the Rickettsiales.

Root knees. Outgrowths from the roots of plants growing in swampy locations.

Rosetting. A condition in which the nodes are in close proximity, resulting in leaves being arranged in tight aggregations.

Rotation crop. Sequentially alternating different crop species in the same field.

Rotifers. Microscopic many-celled aquatic invertebrate animals whose anterior end is modified into a retractile disc bearing circles of strong cilia that often give the appearance of rapidly revolving wheels.

Salination. The process by which a soil becomes saline, as in over irrigation.

Saline. Referring to soil with high concentrations of salts.

Salt flat. A region of arid or semiarid locations that are relatively flat in topography, are often surrounded by mountains, and have extremely high soil salinity.

Salt gland. A group of epidermal cells that serve to extrude salts out of the cytoplasm onto the leaf or stem surface.

Salt marsh. Temperate zone communities dominated by monocots and located at the margin of the land and ocean.

Salt pruning. The alteration of plant architecture (usually dwarfing and stunting of trees) due to the toxic effects of salt spray from the ocean.

Salt spray. Salt transferred from the ocean by wind and fog onto land.

Saltwater intrusion. The disruption of a freshwater aquifer by salt water from the ocean.

Saprophages. Organisms that use dead or decaying organic matter as a food source.

Saprophyte. Organisms responsible for breaking down dead organic tissue and using the energy released to support their own metabolism.

Saturated flow. Water flow in a soil in which all pores are filled with water. The hydraulic conductance K in a specific saturated soil is at its maximum and is constant.

Scale. The magnitude in volume, area, or time of the particular process under consideration. The study of whole-leaf photosynthesis occurs on a greater scale than the study of rubisco activation.

Sclerenchyma. A plant tissue characterized by thick cell walls, such as would occur in seeds or around vascular tissue to give strength to petioles or leaves.

Secondary metabolites. Substances that are produced by plants and are not directly involved with the main physiological pathways required for metabolism.

Signal transduction. The transfer of information from a signaling molecule or occurrence to the enzymes that mediate the response pathway.

Solar constant. The intensity of unfiltered sunlight entering the ionosphere; 1.39 kW m^{-2}.

Spatiotemporal scale. Reference to a space and time scale.

Sporulation. The process of spore release, as occurs with fungi.

Springtails. Small, primitive wingless arthropods (collembolan) related to or classed among the insects.

Stability. As applied to chelates, thermodynamically evaluates the feasibility and predictability of resistance or destruction of complexes under various conditions.

Static. Describing a condition when a resource, compound, or process is stable. This is not a dynamic condition because positive gains and negative losses are both equal to zero.

Steady state. A condition under which a resource, compound, or physiological process has reached stability. This is a dynamic condition, where positive gains equal negative losses resulting in a stable process or quantity.

Stele. The central cylinder inside the cortex of roots and stems of vascular plants; composed of tissues inside the endodermis.

Steroid. Organic compounds characterized by four rings consisting of 17 carbon atoms.

Stomata or stomates. Structures in the epidermis of leaves and stems that are formed by specialized cells called guard cells; often used to identify the pores surrounded by the guard cells.

Stomatal conductance. The rate per unit area at which gases diffuse through the stomates.

Stomatal resistance. The reciprocal of stomatal conductance.

Storage. The accumulation of a substance in a metabolically inactive pool. At other times, the substances in the pool can be mobilized for metabolic processes.

Strain. An injury of a body part or organ resulting from excessive tension, force, influence, or factors causing excessive physical tension.

Stress. A state or condition caused by factors that tend to alter a balanced equilibrium in an organism's function.

Stress proteins. Proteins induced to form as a result of an environmental stress (e.g., high-temperature-induced proteins, pathogen-induced proteins).

Stressor. A factor (biotic or abiotic) of the environment capable of causing bodily injury, disease, or aberrant physiology.

Suberin. A complex polymer of alcohols, phenolics, carboxylic acids, and other molecules that can come to high concentration in the primary cell walls of certain tissues. For example, suberin is highly concentrated in the cell walls of the endodermis in roots.

Suberization. The process by which suberin (a complex fatty material) is produced in cells, the basis for cork production.

Succession. The change in species composition in one locale over time often ending in the same stable species composition.

Succulence. The tissues of a plant that are thick and contain a high quantity of water. Succulence could occur in all tissues or be restricted to leaves, stems, or roots only.

Sulfolipid. A molecule containing a glycerol backbone, two fatty acyl tails, and a sulfo group attached to a sugar molecule. The sulfolipid sulfoquinovosyldiacylglycerol is found in the lamellar membranes of the chloroplast stroma.

Sunfleck. A short period of high radiation that interrupts a general background of low, diffuse radiation. This is a common characteristic of the subcanopy environments of deciduous, conifer, and tropical forests.

Surfactant. A substance that increases the solubility of hydrophobic compounds in water.

Sustainability. In the context of agriculture and natural ecosystems, the ability of two entities to coexist, with one supporting the other in a type of symbiosis.

Symbiosis. An intimate association between two different organisms; may relate to the relationship between fungi and algae in a lichen, or between bacteria and a root in a dinitrogen-fixing nodule. Each species derives a long-term advantage from being associated with the other.

Symbiotic. A relationship between two organisms that is mutually advantageous.

Symplast. The living component of the cell including everything inside the plasmalemma. The symplast is continuous in living cells as a result of plasmodesmatal connections.

Symplastic. Referring to the pathway that water and solutes may follow upon entering the root if they cross the plasmalemma and move through the interconnecting plasmodesmata from cell to cell; occurring in the symplasm.

Symport. A transport protein that carries ions simultaneously in one direction across a membrane.

Symptom. Evidence of disease or physical disturbance; something that indicates the presence of a bodily disorder; an evident reaction of a plant to a pathogen or physiogen.

Syndrome. A group of factors, signs, and symptoms that collectively indicate a disease or disorder.

Synergistic. Describing two substances that when used together, cause a greater effect than the sum of the two acting separately.

Systemically. To move throughout, as in a disease or a pesticide that moves systemically in a plant.

Tannins. A class of secondary plant chemical molecules that can interfere with the structure and function of proteins.

Tardigrades. Microscopic arthropods with four pairs of legs and living in water or damp moss.

Terpenoids. A general term for a number of compounds produced in the mevalonic acid pathway and includes such chemicals such as sterols, gibberellins, carotenoids, abscisic acid, rubber, etc.

Tetrapyrrole. A four-member ring structure in which each ring contains a nitrogen atom, which in chlorophyll is connected to magnesium to form a closed ring of pyrrole structures.

Thigmonastic. Referring to the response by plants to touch, such as with the Venus flytrap or the sensitive plant.

Thylakoids. Disclike structures comprising the grana of chloroplasts and where the light reactions of photosynthesis occur.

Tidal inundation. The covering of a site with oceanic water due to the action of tides.

Tile drainage system. A system of tiles placed under the agricultural field that serves to drain away excess precipitation and can be used to flush salts out of the soil.

Tolerance. The ability of a plant to survive a stress with little or no injury; the relative capacity of an organism to grow, thrive, or survive when subjected to an unfavorable environmental factor; the capacity to resist or sustain the effects of a disease or stress without dying or suffering irreparable injury.

Tonoplast. The membrane surrounding the vacuole.

Topography. The shape, aspect, and slope of the land surfaces; also used to describe the surfaces of proteins or tissues.

Tracheid. A cell type in the xylem vasculature of plants.

Trade-off. A balancing of processes or factors that cannot be accomplished at the same time; usually refers to the allocation of resources among physiological processes.

Trait. A definable physiological, anatomical, morphological, or behavioral characteristic of a species commonly regulated by genetic factors.

Transduction. The passing of information down a pathway of molecules called a signal transduction pathway.

Translocation. The movement of carbohydrates and solutes in the phloem or xylem.

Transpiration. The loss of water through the stomata of leaves and stems.

Trichome. Hair that develops from an epidermal cell; often found on the lower surface of leaves; can be modified into a storage bladder in some species.

Trophic level. One of several energy or feeding levels in the food chain.

Troposphere. That part of the Earth's atmosphere below the stratosphere, generally extending 7–10 miles above the surface of the Earth.

Turgor pressure. The pressure that develops within the cell wall of plant cells.

Tyloses. Outgrowths in parenchyma cell walls in response to pathogen invasion; thought to restrict the growth of the pathogen to localized areas.

Uncoupling agents. Chemicals that allow electron transport to continue but prevent phosphorylation of ADP to ATP as occurs in oxidative phosphorylation.

Undissociated. Referring to acids that are fully protonated.

Uniport. A permease that transports only one substance in one direction across a membrane.

Unsatuation. The condition in which a fatty acid has one or more double bonds; the number of double bonds of a molecule.

Upregulation. An increase in the rate of a reaction or process above the norm; for example, the transcription of particular genes can be upregulated in response to a stress.

V_{max}. The maximum reaction velocity for a given quantity of an enzyme; the rate of substrate saturation of the enzyme.

Vacuole. A storage sac inside cells composed of a single outer phospholipid membrane; constitutes the central vacuole in plants as well as smaller vacuoles in animal cells; serves in the storage of many types of molecules.

Van der Waals forces. Forces that arise from attractive and repulsive forces between molecules. The repulsive forces diminish with distance.

Vapor pressure deficit. The difference between the actual vapor pressure and that of air saturated with water at the same temperature and pressure.

Vapor pressure gradient. The difference in the vapor pressure of the intercellular spaces of a leaf and the turbulent air outside of the leaf.

Variety. A genetically defined strain or race of a crop.

Vascular plant. Plants that have vascular tissue (xylem and phloem).

Vessel. A conducting pathway (group of vessel elements) in the xylem. Vessels may be as long as 1 meter.

Vessel element. A cell in the vascular system of angiosperms that serves to transport water in the plant.

Volatile organic compounds. Represented by a large range of low-molecular-weight compounds emitted into the environment both from natural and anthropogenic sources.

Water channel. A pore in a membrane lined with protein through which water can traverse the membrane easily.

Water potential. The difference between the activity of water molecules in a solution and that of pure distilled water at atmospheric pressure and 30°C.

Water stress. Usually reserved for short-term situations in which the transpiration rate exceeds the absorption rate of water; may be caused by a number of different combinations of soil and atmospheric factors.

Water use efficiency. Instantaneous: the ratio between the accumulation of CO_2 and the loss of water through the stomata. Integrated: the gain of plant mass divided by the total amount of water lost by the plant.

Wilting. A reduction in cell turgor, which reduces the rigidity of a plant; results from water stress.

Xanthophyceae. A class of algae representing the yellow-green algal species.

Xenobiotics. Chemicals introduced into the environment.

Xeric. Low or deficient in available moisture to support life.

Xerophytes. Plants that are able to resist drought and live in very dry habitats; often with thick cuticles, water storage tissue, and other morphological features that conserve water.

Xylem water. The solution contained within the vessels or tracheids of the xylem.

Yield. The amount of merchantable material (fruit, seeds, timber) produced by a crop in one rotation.

Yield threshold. The minimum amount of turgor potential required to induce cell expansion.

Zeatin. A cytokinin (the first identified in plants) found in *Zea mays* seeds and other species.

Zero stress. In concept, that level of exposure to an environmental factor that leads neither to injury nor to reduction in growth, yield, or value.

Zonation. The regular organization of plant species and communities such that definable regions of specific communities can be located along a distance; usually refers to the gradual change in community composition with distance away from the ocean shore.

INDEX

ABOUT THE AUTHORS

Dr. David M. Orcutt is a Professor of Plant Physiology in the Department of Plant Pathology, Physiology, and Weed Science at Virginia Polytechnic Institute and State University, Blacksburg, Virginia.

Dr. Orcutt's teaching and research career spans 28 years of experience. He has taught undergraduate courses in general botany, plant physiology, and plant growth regulation as well as graduate courses in plant stress physiology and plant growth and development.

Dr. Orcutt's research experience has focused on the impact of environmental stress on developmental responses in plants, particularly as stress vectors influence cellular membrane lipids and sterols. Also, he has conducted research relative to the potential uses of natural and synthetic plant growth regulators to alter plant response to environmentally imposed stresses.

Dr. Erik T. Nilsen is a Professor of Ecology in the Biology Department at Virginia Polytechnic Institute and State University, Blacksburg, Virginia.

Dr. Nilsen has been involved with teaching for over 18 years. During that time he has taught general biology to freshmen, plant biology to sophomores, principles of ecology to juniors, plant ecophysiology and plant ecology to seniors, and plant stress physiology to graduate students. He also has extended his teaching into high schools, elementary schools, and adult education courses.

Dr. Nilsen's main interest in research over the years has concerned the various mechanisms found in plants that cope with naturally stressful habitats. Some of his research has covered studies on water relations in desert systems, growth regulation of plants in chaparral systems, nitrogen cycling in desert and chaparral systems, the significance of leaf movements to cold tolerance, and various aspects of stem photosynthesis in legumes.